OXFORD REFERENCE

Concise
Science
Dictionary

D0594861

Concise Science Dictionary

SECOND EDITION

Oxford   New York

OXFORD UNIVERSITY PRESS

Oxford University Press, Walton Street, Oxford OX2 6DP

Oxford New York Toronto
Delhi Bombay Calcutta Madras Karachi
Kuala Lumpur Singapore Hong Kong Tokyo
Nairobi Dar es Salaam Cape Town
Melbourne Auckland Madrid

and associated companies in
Berlin Ibadan

Oxford is a trade mark of Oxford University Press

British Library Cataloguing in Publication Data
Data available

Library of Congress Cataloging in Publication Data
Concise science dictionary.—2nd ed.
1. Science
500
ISBN 0-19-866167-3
ISBN 0-19-286102-6 (Pbk.)

5 7 9 10 8 6 4

Printed in Great Britain by
Clays Ltd
Bungay, Suffolk

Preface

The second edition of the *Concise Science Dictionary* aims to provide school and first-year university students with explanations of unfamiliar words they might come across in the course of their studies, in their own or adjacent disciplines. For example, chemistry students will find all they are likely to need to know about the life sciences, and vice versa. The dictionary will also provide a handy reference source for non-scientists who require accurate definitions of the scientific terms they are likely to encounter in their work or general reading.

At this level the dictionary provides full coverage of terms and concepts relating to physics, chemistry, biology, biochemistry, palaeontology, and the earth sciences. There is also coverage of key terms in astronomy, mathematics, and computer technology. For this new edition, many terms have been added, particularly in the fields of nuclear and particle physics, quantum theory, chemical techniques and materials, environmental science, genetics (including genetic engineering), molecular biology, and immunology.

In compiling the dictionary, the contributors and editors have made every effort to make the entries as concise and comprehensible as possible, always bearing in mind the needs of the readers. Particular features of the book are its lack of unnecessary scientific jargon and its extensive network of cross references. An asterisk placed before a word used in an entry indicates that this word can be looked up in the dictionary and will provide further explanation or clarification. However, not every word that appears in the dictionary has an asterisk placed before it. Some entries simply refer the reader to another entry, indicating either that they are synonyms or abbreviations or that they are most conveniently discussed in one of the dictionary's longer articles. Synonyms and abbreviations are usually placed within brackets immediately after the headword.

Where appropriate, the entries have been supplemented by fully labelled line drawings or tables *in situ*. In addition, the Appendices at the end of the book include the periodic table, tables of SI units and conversion tables to and from other systems of units, and summary classifications of the plant and animal kingdoms.

A.I.

J.D.

E.A.M.

Editors

Alan Isaacs BSc, PhD, DIC   John Daintith BSc, PhD
Elizabeth Martin MA

Advisers

B. S. Beckett BSc, BPhil, MA (Ed)
R. A. Hands BSc   Michael Lewis MA

Contributors

Richard Batley PhD, MInstP

Tim Beardsley BA

Lionel Bender BSc

H. M. Clarke MA, MSc

W. M. Clarke BSc

Derek Cooper PhD, FRIC

R. Cutler BSc

E. K. Daintith BSc

D. E. Edwards BSc, MSc

A. J. H. Goddard PhD, MIMechE, FInstNucE, FInstP, MSRP

Malcolm Hart BSc, MIBiol

Robert S. Hine BSc, MSc

Ann Lockwood BSc

J. Valerie Neal BSc, PhD

R. A. Prince MA

Richard Rennie PhD, MSc, BSc

Jackie Smith BA

Brian Stratton BSc, MSc

Elizabeth Tootill BSc, MSc

David Eric Ward BSc, MSc, PhD

A

AAS *See* atomic absorption spectroscopy.

ab- A prefix attached to the name of a practical electrical unit to provide a name for a unit in the *electromagnetic system of units, e.g. abampere, abcoulomb, abvolt. The prefix is an abbreviation of the word 'absolute' as this system is also known as the *absolute system*. Compare stat-. In modern practice both absolute and electrostatic units have been replaced by *SI units.

abdomen The posterior region of the body trunk of animals. In vertebrates it contains the stomach and intestines and the organs of excretion and reproduction. It is particularly well defined in mammals, being separated from the *thorax by the *diaphragm. In many arthropods, such as insects and spiders, it may be segmented.

Abelian group *See* group.

aberration 1. (in optics) A defect in the image formed by a lens or curved mirror. In *chromatic aberration* the image formed by a lens (but not a mirror) has coloured fringes as a result of the different extent to which light of different colours is refracted by glass. It is corrected by using an *achromatic lens. In *spherical aberration*, the rays from the object come to a focus in slightly different positions as a result of the curvature of the lens or mirror. For a mirror receiving light strictly parallel with its axis, this can be corrected by using a parabolic surface rather than a spherical surface. Spherical aberration in lenses is minimized by making both surfaces contribute equally to the ray deviations, and can (though with reduced image brightness) be reduced by the use of diaphragms to let light pass only through the centre part of the lens. *See also* astigmatism; coma. **2.** (in astronomy) The apparent displacement in the position of a star as a result of the earth's motion round the sun. Light appears to come from a point that is slightly displaced in the direction of the earth's motion. The angular displacement $\alpha = v/c$, where v is the earth's orbital velocity and c is the speed of light.

abiogenesis The origin of living from nonliving matter, as by *biopoiesis. *See also* spontaneous generation.

abiotic factor Any of the nonliving factors that make up the *abiotic environment* in which living organisms occur. They include all the aspects of climate, geology, and atmosphere that may affect the *biotic environment.

ABO system One of the most important human *blood group systems. The system is based on the presence or absence of *antigens A and B on the surface of red blood cells and *antibodies against these in

Group	Antigens on red cell surface	Antibodies in serum	Blood group of people donor can receive blood from	Blood group of people donor can give blood to
A	A	anti-B	A,O	A, AB
B	B	anti-A	B,O	B, AB
AB	A and B	none	A,B, AB, O	AB
O	neither A nor B	anti-A and anti-B	O	A, B, AB, O

The ABO blood group system

1

blood serum. A person whose blood contains either or both these antibodies cannot receive a transfusion of blood containing the corresponding antigens as this would cause the red cells to clump (*see* agglutination). The table illustrates the basis of the system: people of blood group O are described as 'universal donors' as they can give blood to those of any of the other groups. *See also* immune response.

abscisic acid A naturally occurring plant *growth substance that promotes leaf ageing, leaf fall (*see* abscission), and *apical dominance and induces *dormancy in seeds and buds.

abscissa *See* Cartesian coordinates.

abscission The separation of a leaf, fruit, or other part from the body of a plant. The process is controlled by growth substances, notably *abscisic acid; it involves the formation of an *abscission zone*, at the base of the part, within which a layer of cells (*abscission layer*) breaks down.

absolute 1. Not dependent on or relative to anything else, e.g. *absolute zero. 2. Denoting a temperature measured on an *absolute scale*, a scale of temperature based on absolute zero. The usual absolute scale now is that of thermodynamic *temperature; its unit, the kelvin, was formerly called the degree absolute (°A) and is the same size as the degree Celsius. In British engineering practice an absolute scale with Fahrenheit-size degrees has been used: this is the Rankine scale.

absolute alcohol *See* ethanol.

absolute configuration A way of denoting the absolute structure of an optical isomer (*see* optical activity). Two conventions are in use: The D−L convention relates the structure of the molecule to some reference molecule. In the case of sugars and similar compounds, the dextrorotatory form

of glyceraldehyde ($HOCH_2CH(OH)CHO$, 2,3-dihydroxypropanal) was used. The rule is as follows. Write the structure of this molecule down with the asymmetric carbon in the centre, the −CHO group at the top, the −OH on the right, the −CH_2OH at the bottom, and the −H on the left. Now imagine that the central carbon atom is at the centre of a tetrahedron with the four groups at the corners and that the −H and −OH come out of the paper and the −CHO and −CH_2OH groups go into the paper. The resulting three-dimensional structure was taken to be that of *d*-glyceraldehyde and called D-glyceraldehyde. Any compound that contains an asymmetric carbon atom having this configuration belongs to the D-series. One having the opposite configuration belongs to the L-series. It is important to note that the prefixes D- and L- do not stand for dextrorotatory and laevorotatory (they are not the same as *d*- and *l*-). In fact the arbitrary configuration assigned to D-glyceraldehyde is now known to be the correct one for the dextrorotatory form, although this was not known at the time. However, all D-compounds are not dextrorotatory. For instance, the acid obtained by oxidizing the −CHO group of glyceraldehyde is glyceric acid (1,2-dihydroxypropanoic acid). By convention, this belongs to the D-series, but it is in fact laevorotatory; i.e. its name can be written as D-glyceric acid or *l*-glyceric acid. To avoid confusion it is better to use + (for dextrorotatory) and − (for laevorotatory), as in D-(+)-glyceraldehyde and D-(−)-glyceric acid.

The D−L convention can also be used with alpha amino acids (compounds with the −NH_2 group on the same carbon as the −COOH group). In this case the molecule is imagined as being viewed along the H−C bond between the hydrogen and the asymmetric carbon atom. If the clockwise planar formula   structure in 3 dimensions   Fischer projection

D—(+)— glyceraldehyde (2,3–dihydroxypropanal)

D—alanine (R is CH_2 in the CORN rule). The molecule is viewed with H on top

R–configuration   S–configuration

R–S system. The lowest priority group is behind the chiral carbon atom

order of the other three groups is —**COOH**, −**R**, −**NH₂**, the amino acid belongs to the D-series; otherwise it belongs to the L-series. This is known as the *CORN rule*.

The R–S convention is a convention based on priority of groups attached to the chiral carbon atom. The order of priority is I, Br, Cl, SO_3H, $OCOCH_3$, OCH_3, OH, NO_2, NH_2, $COOCH_3$, $CONH_2$, $COCH_3$, CHO, CH_2OH, C_6H_5, C_2H_5, CH_3, H, with hydrogen lowest. The molecule is viewed with the group of lowest priority behind the chiral atom. If the clockwise arrangement of the other three groups is in descending priority,

the compound belongs to the R-series; if the descending order is anticlockwise it is in the S-series. D-(+)-glyceraldehyde is R-(+)-glyceraldehyde. See illustration.

absolute expansivity *See* expansivity.

absolute humidity *See* humidity.

absolute permittivity *See* permittivity.

absolute pitch (perfect pitch) The ability of a person to identify and reproduce a note without reference to a tuned musical instrument.

absolute temperature *See* absolute; temperature.

absolute value (modulus) The square root of the sum of the squares of the real numbers in a *complex number, i.e. the absolute value of the complex number $z = x + iy$ is $|z| = \sqrt{(x^2 + y^2)}$.

absolute zero Zero of thermodynamic *temperature (0 kelvin) and the lowest temperature theoretically attainable. It is the temperature at which the kinetic energy of atoms and molecules is minimal. It is equivalent to $-273.15°C$ or $-459.67°F$. *See also* zero-point energy; cryogenics.

absorptance Symbol α. The ratio of the radiant or luminous flux absorbed by a body to the flux falling on it. Formerly called *absorptivity*, the absorptance of a *black body is by definition 1.

absorption 1. (in chemistry) The take up of a gas by a solid or liquid, or the take up of a liquid by a solid. Absorption differs from *adsorption in that the absorbed substance permeates the bulk of the absorbing substance. **2.** (in physics) The conversion of the energy of electromagnetic radiation, sound, streams of particles, etc., into other forms of energy on passing through a medium. A beam of light, for instance, passing through a medium, may lose intensity because of two effects: *scattering of light out of the beam, and absorption of photons by atoms or molecules in the medium. When a photon is absorbed, there is a transition to an excited state. **3.** (in biology) The movement of fluid or a dissolved substance across a cell membrane. In animals, for example, soluble food material is absorbed into the circulatory system through cells lining the alimentary canal. In plants, water and mineral salts are absorbed from the soil by the *roots. *See* osmosis.

absorption coefficient *See* Lambert's laws.

absorption indicator A type of indicator used in reactions that involve precipitation. The yellow dye fluorescein is a common example, used for the reaction

$$NaCl(aq) + AgNO_3(aq) \rightarrow$$
$$AgCl(s) + NaNO_3(aq)$$

As silver nitrate solution is added to the sodium chloride, silver chloride precipitates. As long as Cl^- ions are in excess, they adsorb on the precipitate particles. At the end point, no Cl^- ions are left in solution and negative fluorescein ions are then adsorbed, giving a pink colour to the precipitate.

absorption spectrum *See* spectrum.

absorptivity *See* absorptance.

ABS plastic Any of a class of plastics based on acrylonitrile–butadiene –styrene copolymers.

abundance 1. The ratio of the total mass of a specified element in the earth's crust to the total mass of the earth's crust, often expressed as a percentage. For example, the abundance of aluminium in the earth's crust is about 8%. **2.** The ratio of the number of atoms of a particular isotope of an element to the total number of atoms of all the isotopes present, often expressed as a percentage. For example, the abundance of uranium–235 in natural uranium is 0.71%. This is the *natural abundance*, i.e. the abundance as found in nature before any enrichment has taken place.

abyssal zone The lower depths of the ocean (below approximately 2000 metres), where there is effectively no light penetration. Abyssal organisms are adapted for living under high pressures in cold dark conditions.

a.c. *See* alternating current.

acceleration Symbol a. The rate of increase of speed or velocity. It is measured in $m\,s^{-2}$. For a body moving linearly with constant acceleration a from a speed u to a speed v,

$$a = (v - u)/t = (v^2 - u^2)/2s$$

where t is the time taken and s the distance covered.

If the acceleration is not constant it is given by $dv/dt = d^2s/dt^2$. If the motion is not linear the vector character of displacement, velocity, and acceleration must be considered.

acceleration of free fall Symbol g. The acceleration experienced by any massive object falling freely in the earth's gravitational field. Experimentally this is almost constant for all positions near the earth's surface, independent of the nature of the falling body (provided air resistance is eliminated). This is taken to indicate the strict proportionality of *weight (the force causing the acceleration) and *inertial mass, on the basis of *Newton's second law of motion. There is some variation of g with latitude, because of the earth's rotation and because the earth is not completely spherical. The standard value is taken as 9.806 65 m s^{-2}. The acceleration of free fall is also called the *acceleration due to gravity*.

accelerator 1. (in physics) An apparatus for increasing the kinetic energies of charged particles, used for research in nuclear and particle physics. *See* cyclotron; linear accelerator; synchrocyclotron; synchrotron. **2.** (in chemistry) A substance that increases the rate of a chemical reaction, i.e. a catalyst.

acceptor 1. (in chemistry) A compound, molecule, ion, etc., to which electrons are donated in the formation of a coordinate bond. **2.** (in physics) A substance that is added as an impurity to a *semiconductor because of its ability to accept electrons from the valence bands, causing p-type conduction by the mobile positive holes left.

accommodation 1. (in animal physiology) The process by which the focal length of the *lens of the eye is changed so that clear images of

objects at a range of distances are displayed on the retina. In man and some other mammals accommodation is achieved by reflex adjustments in the shape of the lens brought about by relaxation and contraction of muscles within the *ciliary body. **2.** (in botany) The ability of a plant to gradually changing environmental conditions. **3.** (in animal behaviour) The psychological adjustment made by an animal in response to continuously changing environmental conditions.

accretion disc A disc-shaped rotating mass formed by gravitational attraction. *See* black hole; neutron star; white dwarf.

accumulator (secondary cell; storage battery) A type of *voltaic cell or battery that can be recharged by passing a current through it from an external d.c. supply. The charging current, which is passed in the opposite direction to that in which the cell supplies current, reverses the chemical reactions in the cell. The common types are the *lead–acid accumulator and the *nickel–iron accumulator.

acellular Describing tissues or organisms that are not made up of separate cells but often have more than one nucleus. Examples of acellular structures are muscle fibres. *Compare* unicellular.

acentric Describing an aberrant chromosome fragment that lacks a centromere. Such fragments are normally lost because they are unable to orientate properly during cell division.

acetaldehyde *See* ethanal.

acetaldol *See* aldol reaction.

acetals Organic compounds formed by addition of alcohol molecules to aldehyde molecules. If one molecule of aldehyde (RCHO) reacts with one molecule of alcohol (R'OH) a *hemiacetal* is formed (RCH(OH)OR'). The rings of aldose sugars are hemiacetals. Further reaction with a

acetamide

$$R-C{\overset{O}{\underset{H}{\lessgtr}}} + R'OH \longrightarrow {\overset{R}{\underset{H}{\gtrless}}}C{\overset{OR'}{\underset{OH}{\lessgtr}}}$$

aldehyde   alcohol   hemiacetal

$${\overset{R}{\underset{H}{\gtrless}}}C{\overset{OR'}{\underset{OH}{\lessgtr}}} + R'OH \longrightarrow {\overset{R}{\underset{H}{\gtrless}}}C{\overset{OR'}{\underset{OR'}{\lessgtr}}} + H_2O$$

acetal

Formation of acetals

second alcohol molecule produces a full acetal ($RCH(OR')_2$). The formation of acetals is reversible; acetals can be hydrolysed back to aldehydes in acidic solutions. In synthetic organic chemistry aldehyde groups are often converted into acetal groups to protect them before performing other reactions on different groups in the molecule. *See also* ketals.

acetamide *See* ethanamide.

acetate *See* ethanoate.

acetic acid *See* ethanoic acid.

acetone *See* propanone; ketone body.

acetylation *See* acylation.

acetyl chloride *See* ethanoyl chloride.

acetylcholine A substance that is released at some (*cholinergic*) nerve endings (*see* neurotransmitter). Its function is to pass on a nerve *impulse to the next nerve (i.e. at a *synapse) or to initiate muscular contraction. Once acetylcholine has been released it has only a transitory effect because it is rapidly broken down by the enzyme *acetylcholinesterase*.

acetylene *See* ethyne.

acetylenes *See* alkynes.

acetyl group *See* ethanoyl group.

acetylide *See* carbide.

achene A dry indehiscent fruit formed from a single carpel and containing a single seed. An example is the feathery achene of clematis. Variants of the achene include the *caryopsis, *cypsela, *nut, and *samara. *See also* etaerio.

Acheson process An industrial process for the manufacture of graphite by heating coke mixed with clay. The reaction involves the production of silicon carbide, which loses silicon at 4150°C to leave graphite. The process was patented in 1896 by the US inventor Edward Goodrich Acheson (1856–1931).

achromatic lens A lens that corrects for chromatic *aberration by using a combination of two lenses, made of different kinds of glass, such that their *dispersions neutralize each other although their *refractions do not. The aberration can be reduced further by using an *apochromatic lens*, which consists of three or more different kinds of glass.

acid 1. A type of compound that contains hydrogen and dissociates in water to produce positive hydrogen ions. The reaction, for an acid HX, is commonly written:

$$HX \rightleftharpoons H^+ + X^-$$

In fact, the hydrogen ion (the proton) is solvated, and the complete reaction is:

$$HX + H_2O \rightleftharpoons H_3O^+ + X^-$$

The ion H_3O^+ is the *oxonium ion* (or *hydroxonium ion* or *hydronium ion*). This definition of acids comes from the *Arrhenius theory*. Such acids tend to be corrosive substances with a sharp taste, which turn litmus red and give colour changes with other *indicators. They are referred to as *protonic acids* and are classified into *strong acids*, which are almost completely dissociated in water (e.g. sulphuric acid and hydrochloric acid), and *weak acids*, which are only partially dissociated (e.g. ethanoic acid and hydrogen sulphide). The strength of an acid depends on the extent to which it dissociates, and is measured by its *dissociation constant. *See also* base.

2. In the *Lowry–Brønsted theory* of acids and bases (1923), the definition was extended to one in which an acid is a proton donor, and a base is a proton acceptor. For example, in

$$HCN + H_2O \rightleftharpoons H_3O^+ + CN^-$$

the HCN is an acid, in that it donates a proton to H_2O. The H_2O is acting as a base in accepting a proton. Similarly, in the reverse reaction H_3O^+ is an acid and CN^- a base. In such reactions, two species related by loss or gain of a proton are said to be *conjugate*. Thus, in the reaction above HCN is the *conjugate acid* of the base CN^-, and CN^- is the *conjugate base* of the acid HCN. Similarly, H_3O^+ is the conjugate acid of the base H_2O. An equilibrium, such as that above, is a competition for protons between an acid and its conjugate base. A strong acid has a weak conjugate base, and vice versa. Under this definition water can act as both acid and base. Thus in

$$NH_3 + H_2O \rightleftharpoons NH_4^+ + OH^-$$

the H_2O is the conjugate acid of OH^-. The definition also extends the idea of acid–base reaction to solvents other than water. For instance, liquid ammonia, like water, has a high dielectric constant and is a good ionizing solvent. Equilibria of the type

$$NH_3 + Na^+Cl^- \rightleftharpoons Na^+NH_2^- + HCl$$

can be studied, in which NH_3 and HCl are acids and NH_2^- and Cl^- are their conjugate bases.

3. A further extension of the idea of acids and bases was made in the *Lewis theory* (G. N. Lewis, 1923). In this, a *Lewis acid* is a compound or atom that can accept a pair of electrons and a *Lewis base* is one that can donate an electron pair. This definition encompasses 'traditional' acid–base reactions. In

$$HCl + NaOH \rightarrow NaCl + H_2O$$

the reaction is essentially

$$H^+ + :OH^- \rightarrow H:OH$$

i.e. donation of an electron pair by OH^-. But it also includes reactions that do not involve ions, e.g.

$$H_3N: + BCl_3 \rightarrow H_3NBCl_3$$

in which NH_3 is the base (donor) and BCl_3 the acid (acceptor). The Lewis theory establishes a relationship between acid–base reactions and *oxidation–reduction reactions.

carboxylic acids   acid anhydride

Formation of a carboxylic acid anhydride

acid anhydrides (acyl anhydrides) Compounds that react with water to form an acid. For example, carbon dioxide reacts with water to give carbonic acid:

$$CO_2(g) + H_2O(aq) \rightleftharpoons H_2CO_3(aq)$$

A particular group of acid anhydrides are anhydrides of carboxylic acids. They have a general formula of type R.CO.O.CO.R', where R and R' are alkyl or aryl groups. For example, the compound ethanoic anhydride $(CH_3.CO.O.CO.CH_3)$ is the acid anhydride of ethanoic (acetic) acid. Organic acid anhydrides can be produced by dehydrating acids (or mixtures of acids). They are usually made by reacting an acyl halide with the sodium salt of the acid. They react readily with water, alcohols, phenols, and amines and are used in *acylation reactions.

acid–base balance The regulation of the concentrations of acids and bases in blood and other body fluids so that the *pH remains within a physiologically acceptable range. This is achieved by the presence of natural *buffer systems, such as the haemoglobin, bicarbonate ions, and carbonic acid in mammalian blood. By acting in conjunction, these effectively mop up excess acids and bases and therefore prevent any large shifts in blood pH. The acid–base balance is also influenced by the selective removal of certain ions by the kidneys and the rate of removal of carbon dioxide from the lungs.

acid–base indicator *See* indicator.

acid dissociation constant *See* dissociation.

acid dye *See* dye.

acid halides *See* acyl halides.

acidic 1. Describing a compound that is an acid. **2.** Describing a solution that has an excess of hydrogen ions. **3.** Describing a compound that forms an acid when dissolved in water. Carbon dioxide, for example, is an acidic oxide.

acidic hydrogen (acid hydrogen) A hydrogen atom in an *acid that forms a positive ion when the acid dissociates. For instance, in methanoic acid

$$HCOOH \leftrightharpoons H^+ + HCOO^-$$

the hydrogen atom on the carboxylate group is the acidic hydrogen (the one bound directly to the carbon atom does not dissociate).

acidic stains *See* staining.

acidimetry Volumetric analysis using standard solutions of acids to determine the amount of base present.

acidity constant *See* dissociation.

acid rain *See* pollution.

acid salt A salt of a polybasic acid (i.e. an acid having two or more acidic hydrogens) in which not all the hydrogen atoms have been replaced by positive ions. For example, the dibasic acid carbonic acid (H_2CO_3) forms acid salts (hydrogencarbonates) containing the ion HCO_3^-. Some salts of monobasic acids are also known as acid salts. For instance, the compound potassium hydrogendifluoride, KHF_2, contains the ion $[F...H-F]^-$, in which there is hydrogen bonding between the fluoride ion F^- and a hydrogen fluoride molecule.

acid value A measure of the amount of free acid present in a fat, equal to the number of milligrams of potassium hydroxide needed to neutralize this acid. Fresh fats contain glycerides of fatty acids and very little free acid, but the glycerides decompose slowly with time and the acid value increases.

acoustics 1. The study of sound and sound waves. **2.** The characteristics of a building, especially an auditorium, with regard to its ability to enable speech and music to be heard clearly within it. For this purpose there should be no obtrusive echoes or resonances and the reverberation time should be near the optimum for the hall. Echoes are reduced by avoiding sweeping curved surfaces that could focus the sound and by breaking up large plane surfaces or covering them with sound-absorbing materials. Resonance is avoided by avoiding simple ratios for the main dimensions of the room, so that no one wavelength of sound is a factor of more than one of them. If the reverberation time is too long, speech will sound indistinct and music will be badly articulated, with one note persisting during the next. However, if it is too short, music sounds dead. It is long in a bare room with hard walls, and can be deliberately reduced by carpets, soft furnishings and sound-absorbent ('acoustic') felt. Reverberation times tend to be reduced by the presence of an audience and this must be taken into account in the design of the building.

acoustoelectronic devices (electroacoustic devices) Devices in which electronic signals are converted into acoustic waves. Acoustoelectronic devices are used in constructing *delay lines and also in converting digital data from computers for transmission by telephone lines.

acquired characteristics Features that are developed during the lifetime of an individual, e.g. the enlarged arm muscles of a tennis player. It is a basic tenet of current evolutionary thought that such characteristics are not genetically controlled and cannot be passed on to the next generation. *See also* Lamarckism; neo-Lamarckism.

acquired immune deficiency syndrome *See* AIDS.

Acrilan A trade name for a synthetic fibre. *See* acrylic resins.

acrosome *See* spermatozoon.

acrylate *See* propenoate.

acrylic acid *See* propenoic acid.

acrylic resins Synthetic resins made by polymerizing esters or other derivatives of acrylic acid (propenoic acid). Examples are poly(propenonitrile) (e.g. *Acrilan*), and poly(methyl 2-methylpropenoate) (polymethyl methacrylate, e.g. *Perspex*).

acrylonitrile *See* propenonitrile.

ACTH (adrenocorticotrophic hormone, corticotrophin) A hormone produced by the anterior *pituitary gland in response to stress that controls secretion of certain hormones (the *corticosteroids) by the adrenal glands. It can be administered by injection to treat such disorders as rheumatic diseases and asthma, but it only relieves symptoms and is not a cure.

actinic radiation Electromagnetic radiation that is capable of initiating a chemical reaction. The term is used especially of ultraviolet radiation and also to denote radiation that will affect a photographic emulsion.

actinides *See* actinoids.

actinium Symbol Ac. A silvery radioactive metallic element belonging to group IIIB of the periodic table; a.n. 89; mass number of most stable isotope 227 (half-life 21.7 years); m.p. 1050 ± 50°C; b.p. 3300°C (estimated). Actinium–227 occurs in natural uranium to an extent of about 0.715%. Actinium–228 (half-life 6.13 hours) also occurs in nature. There are 22 other artificial isotopes, all radioactive and all with very short

half-lives. Its chemistry is similar to that of lanthanum. Its main use is as a source of alpha particles. The element was discovered by A. Debierne in 1899.

actinium series *See* radioactive series.

actinoid contraction A smooth decrease in atomic or ionic radius with increasing proton number found in the *actinoids.

actinoids (actinides) A series of elements in the *periodic table, generally considered to range in atomic number from thorium (90) to lawrencium (103) inclusive. The actinoids all have two outer *s*-electrons (a $7s^2$ configuration), follow actinium, and are classified together by the fact that increasing proton number corresponds to filling of the $5f$ level. In fact, because the $5f$ and $6d$ levels are close in energy the filling of the $5f$ orbitals is not smooth. The outer electron configurations are as follows:

89 actinium (Ac) $6d^17s^2$
90 thorium (Th) $6d^27s^2$
91 protactinium (Pa) $5f^26d^17s^2$
92 uranium (Ur) $5f^36d7s^2$
93 neptunium (Np) $5f^57s^2$ (or $5f^46d^17s^2$)
94 plutonium (Pu) $5f^67s^2$
95 americium (Am) $5f^77s^2$
96 curium (Cm) $5f^76d^1s^2$
97 berkelium (Bk) $5f^86d7s^2$ (or $5f^97s^2$)
98 californium (Cf) $5f^{10}7s^2$
99 einsteinium (Es) $5f^{11}7s^2$
100 fermium (Fm) $5f^{12}7s^2$
101 mendelevium (Md) $5f^{13}7s^2$
102 nobelium (Nb) $5f^{14}7s^2$
103 lawrencium (Lw) $5f^{14}6d^1s^2$

The first four members (Ac to Ur) occur naturally. All are radioactive and this makes investigation difficult because of self-heating, short lifetimes, safety precautions, etc. Like the *lanthanoids, the actinoids show a smooth decrease in atomic and ionic radius with increasing proton number. The lighter members of the series (up to americium) have *f*-electrons that

actinometer

can participate in bonding, unlike the lanthanoids. Consequently, these elements resemble the transition metals in forming coordination complexes and displaying variable valency. As a result of increased nuclear charge, the heavier members (curium to lawrencium) tend not to use their inner f-electrons in forming bonds and resemble the lanthanoids in forming compounds containing the M^{3+} ion. The reason for this is pulling of these inner electrons towards the centre of the atom by the increased nuclear charge. Note that actinium itself does not have a $5f$ electron, but it is usually classified with the actinoids because of its chemical similarities. *See also* transition elements.

actinometer Any of various instruments for measuring the intensity of electromagnetic radiation. Recent actinometers use the *photoelectric effect but earlier instruments depended either on the fluorescence produced by the radiation on a screen or on the amount of chemical change induced in some suitable substance.

actinomorphy *See* radial symmetry.

actinomycetes A group of Gram-positive mostly anaerobic nonmotile bacteria. All species are fungus-like, with filamentous cells producing reproductive spores on aerial branches similar to the spores of certain moulds. The group includes bacteria of the genera *Actinomyces*, some species of which cause disease in animals and man; and *Streptomyces*, which are a source of many important antibiotics (including streptomycin).

action potential The change in electrical potential that occurs across a cell membrane during the passage of a nerve *impulse. As an impulse travels in a wavelike manner along the *axon of a nerve, it causes a localized and transient switch in electrical potential across the cell membrane from -60 mV (millivolts) to $+45$ mV. Nervous stimulation of a muscle fibre has a similar effect.

action spectrum A graphical plot of the efficiency of electromagnetic radiation in producing a photochemical reaction against the wavelength of the radiation used. For example, the action spectrum for photosynthesis using light shows a peak in the region 670–700 nm. This corresponds to a maximum absorption in the absorption *spectrum of chlorophylls in this region.

activated alumina *See* aluminium hydroxide.

activated charcoal *See* charcoal.

activated complex The association of atoms of highest energy formed in the *transition state of a chemical reaction.

activation analysis An analytical technique that can be used to detect most elements when present in a sample in milligram quantities (or less). In *neutron activation analysis* the sample is exposed to a flux of thermal neutrons in a nuclear reactor. Some of these neutrons are captured by nuclides in the sample to form nuclides of the same atomic number but a higher mass number. These newly formed nuclides emit gamma radiation, which can be used to identify the element present by means of a gamma-ray spectrometer. Activation analysis has also been employed using charged particles, such as protons or alpha particles.

activation energy Symbol E_a. The minimum energy required for a

Reaction profile
(for an endothermic reaction)

chemical reaction to take place. In a reaction, the reactant molecules come together and chemical bonds are stretched, broken, and formed in producing the products. During this process the energy of the system increases to a maximum, then decreases to the energy of the products (see illustration). The activation energy is the difference between the maximum energy and the energy of the reactants; i.e. it is the energy barrier that has to be overcome for the reaction to proceed. The activation energy determines the way in which the rate of the reaction varies with temperature (*see* Arrhenius equation). It is usual to express activation energies in joules per mole of reactants.

active device 1. An electronic component, such as a transistor, that is capable of amplification. **2.** An artificial *satellite that receives information and retransmits it after amplification. **3.** A radar device that emits microwave radiation and provides information about a distant body by receiving a reflection of this radiation. *Compare* passive device.

active mass *See* mass action.

active site 1. A site on the surface of a catalyst at which activity occurs. **2.** The site on the surface of an *enzyme molecule that binds the substrate molecule. The properties of an active site are determined by the three-dimensional arrangement of the polypeptide chains of the enzyme and their constituent amino acids. These govern the nature of the interaction that takes place and hence the degree of substrate specificity and susceptibility to *inhibition.

active transport The movement of substances through membranes in living cells, often against a concentration gradient: a process requiring metabolic energy. Organic molecules and inorganic ions are transported into and out of both cells and their organelles. The substance binds to a carrier protein embedded in the membrane, which carries it through the membrane and releases it on the opposite side. Active transport serves chiefly to maintain the normal balance of ions in cells, especially the concentration gradients of sodium and potassium ions crucial to the activity of nerve and muscle cells. *Compare* facilitated diffusion.

activity 1. Symbol *a*. A thermodynamic function used in place of concentration in equilibrium constants for reactions involving nonideal gases and solutions. For example, in a reaction

$$A \rightleftharpoons B + C$$

the true equilibrium constant is given by

$$K = a_B a_C / a_A$$

where a_A, a_B, and a_C are the activities of the components, which function as concentrations (or pressures) corrected for nonideal behaviour. *Activity coefficients* (symbol γ) are defined for gases by $\gamma = a/p$ (where p is pressure) and for solutions by $\gamma = aX$ (where X is the mole fraction). Thus, the equilibrium constant of a gas reaction has the form

$$K_p = \gamma_B p_B \gamma_C p_C / \gamma_A p_A$$

The equilibrium constant of a reaction in solution is

$$K_c = \gamma_B X_B \gamma_C X_C / \gamma_A X_A$$

The activity coefficients thus act as correction factors for the pressures or concentrations. *See also* fugacity. **2.** Symbol *A*. The number of atoms of a radioactive substance that disintegrate per unit time. The *specific activity* (*a*) is the activity per unit mass of a pure radioisotope. *See* radiation units.

acyclic Describing a compound that does not have a ring in its molecules.

acyl anhydrides *See* acid anhydrides.

acylation The process of introducing an acyl group (RCO–) into a compound. The usual method is to react acylglycerol

an alcohol with an acyl halide or a carboxylic acid anhydride; e.g.

RCOCl + R'OH → RCOOR' + HCl

The introduction of an acetyl group (CH_3CO-) is *acetylation*, a process used for protecting $-OH$ groups in organic synthesis.

acylglycerol *See* glyceride.

acyl group A group of the type RCO−, where R is an organic group. An example is the acetyl group CH_3CO-.

$$R \over X \rangle C = O$$

Acyl halide: X is a halogen atom

acyl halides (acid halides) Organic compounds containing the group −CO.X, where X is a halogen atom (see formula). Acyl chlorides, for instance, have the general formula RCOCl. The group RCO− is the *acyl group*. In systematic chemical nomenclature acyl-halide names end in the suffix *-oyl*; for example, ethanoyl chloride, CH_3COCl. Acyl halides react readily with water, alcohols, phenols, and amines and are used in *acylation reactions. They are made by replacing the −OH group in a carboxylic acid by a halogen using a halogenating agent such as PCl_5.

adaptation 1. (in evolution) Any change in the structure or functioning of an organism that makes it better suited to its environment. *Natural selection of inheritable adaptations ultimately leads to the development of new species. Increasing adaptation of a species to a particular environment tends to diminish its ability to adapt to any sudden change in that environment. 2. (in physiology) The alteration in the degree of sensitivity (either an increase or a decrease) of a sense organ to suit conditions more extreme than normally encountered. An example is the adjustment of the eye to vision in very bright or very dim light.

adaptive radiation (divergent evolution) The evolution from one species of animals or plants of a number of different forms. As the original population increases in size it spreads out from its centre of origin to exploit new habitats and food sources. In time this results in a number of populations each adapted to its particular habitat: eventually these populations will differ from each other sufficiently to become new species. A good example of this process is the evolution of the Australian marsupials into species adapted as carnivores, herbivores, burrowers, fliers, etc. On a smaller scale, the adaptive radiation of the Galapagos finches provided Darwin with crucial evidence for his theory of evolution (*see* Darwin's finches).

addition polymerization *See* polymerization.

addition reaction A chemical reaction in which one molecule adds to another. Addition reactions occur with unsaturated compounds containing double or triple bonds, and may be *electrophilic or *nucleophilic. An example of electrophilic addition is the reaction of hydrogen chloride with an alkene, e.g.

HCl + CH_2:CH_2 → CH_3CH_2Cl

An example of nucleophilic addition is the addition of hydrogen cyanide across the carbonyl bond in aldehydes to form *cyanohydrins. *Addition−elimination* reactions are ones in which the addition is followed by elimination of another molecule (*see* condensation reaction).

additive process *See* colour.

adduct A compound formed by an addition reaction. The term is used particularly for compounds formed by coordination between a Lewis acid (acceptor) and a Lewis base (donor). *See* acid.

12

adenine A *purine derivative. It is one of the major component bases of *nucleotides and the nucleic acids *DNA and *RNA.

adenosine A nucleoside comprising one adenine molecule linked to a D-ribose sugar molecule. The phosphate-ester derivatives of adenosine, AMP, ADP, and *ATP, are of fundamental biological importance as carriers of chemical energy.

adenosine diphosphate (ADP) See ATP.

adenosine monophosphate (AMP) See ATP.

adenosine triphosphate See ATP.

adenovirus One of a group of DNA-containing viruses found in rodents, fowl, cattle, monkeys, and man. In man they produce acute respiratory-tract infections with symptoms resembling those of the common cold. They are also implicated in the formation of tumours (see oncogenic).

ADH (antidiuretic hormone) See vasopressin.

adhesive A substance used for joining surfaces together. Adhesives are generally colloidal solutions, which set to gels. There are many types including animal glues (based on collagen), vegetable mucilages, and synthetic resins (e.g. *epoxy resins).

adiabatic demagnetization A technique for cooling a paramagnetic salt, such as potassium chrome alum, to a temperature near *absolute zero. The salt is placed between the poles of an electromagnet and the heat produced during magnetization is removed by liquid helium. The salt is then isolated thermally from the surroundings and the field is switched off; the salt is demagnetized adiabatically and its temperature falls. This is because the demagnetized state, being less ordered, involves more energy than the magnetized state. The extra energy can come only from the inter-

nal, or thermal, energy of the substance.

adiabatic process Any process that occurs without heat entering or leaving a system. In general, an adiabatic change involves a fall or rise in temperature of the system. For example, if a gas expands under adiabatic conditions, its temperature falls (work is done against the retreating walls of the container). The *adiabatic equation* describes the relationship between the pressure (p) of an ideal gas and its volume (V), i.e. $pV^\gamma = K$, where γ is the ratio of the principal specific *heat capacities of the gas and K is a constant.

adipic acid See hexanedioic acid.

adipose tissue A body tissue comprising cells containing *fat and oil. It is found chiefly below the skin (see subcutaneous tissue) and around major organs (such as the kidneys and heart), acting as an energy reserve and also providing insulation and protection.

admittance Symbol Y. The reciprocal of *impedance. It is measured in siemens.

ADP See ATP.

adrenal glands A pair of endocrine glands situated immediately above the kidneys (hence they are also known as the *suprarenal glands*). The inner portion of the adrenals, the *medulla*, secretes the hormones *adrenaline and *noradrenaline; the outer *cortex* secretes small amounts of sex hormones (*androgens and *oestrogens) and various *corticosteroids, which have a wide range of effects on the body. See also ACTH.

adrenaline (epinephrine) A hormone, produced by the medulla of the *adrenal glands, that increases heart activity, improves the power and prolongs the action of muscles, and increases the rate and depth of breathing to prepare the body for 'fright, flight, or fight'. At the same time it inhibits digestion and excretion. Similar effects are produced by stimulation of the *sympathetic nervous system. Adrenaline can be administered by injection to relieve bronchial asthma and reduce blood loss during surgery by constricting blood vessels.

adrenergic Describing a nerve fibre that either releases *adrenaline or *noradrenaline when stimulated or is itself stimulated by these substances. *Compare* cholinergic.

adrenocorticotrophic hormone *See* ACTH.

adsorbate A substance that is adsorbed on a surface.

adsorbent A substance on the surface of which a substance is adsorbed.

adsorption The formation of a layer of gas, liquid, or solid on the surface of a solid or, less frequently, of a liquid. There are two types depending on the nature of the forces involved. In *chemisorption* a single layer of molecules, atoms, or ions is attached to the adsorbent surface by chemical bonds. In *physisorption* adsorbed molecules are held by the weaker *van der Waals' forces. The property is utilized in adsorption *chromatography.

advanced gas-cooled reactor (AGR) *See* nuclear reactor.

adventitious Describing organs or other structures that arise in unusual positions. For example, ivy has adventitious roots growing from its stems.

aerial (antenna) The part of a radio or television system from which radio waves are transmitted into the atmosphere or space (*transmitting aerial*) or by which they are received (*receiving aerial*). A *directional* or *directive aerial* is one in which energy is transmitted or received more effectively from some directions than others, whereas an *omnidirectional aerial* transmits

and receives equally well in all directions.

aerobe *See* aerobic respiration.

aerobic respiration A type of *respiration in which foodstuffs (usually carbohydrates) are completely oxidized to carbon dioxide and water, with the release of chemical energy, in a process requiring atmospheric oxygen. The reaction can be summarized by the equation:

$$C_6H_{12}O_6 + 6O_2 \rightarrow 6CO_2 + 6H_2O + energy$$

Most organisms have aerobic respiration (i.e. they are *aerobes*); exceptions include certain bacteria and yeasts. *Compare* anaerobic respiration.

aerogenerator *See* wind power.

aerosol A colloidal dispersion of a solid or liquid in a gas. The commonly used aerosol sprays contain an inert propellant liquefied under pressure. Halogenated alkanes containing chlorine and fluorine (*chlorofluorocarbons, or CFCs) have been used in aerosol cans. This use has been criticized on the grounds that these compounds persist in the atmosphere and lead to depletion of the *ozone layer.

aerospace The earth's atmosphere and the space beyond it.

aerotaxis *See* taxis.

aestivation 1. (in zoology) A state of inactivity occurring in some animals, notably lungfish, during prolonged periods of drought or heat. Feeding, respiration, movement, and other bodily activities are considerably slowed down. *See also* dormancy. *Compare* hibernation. **2.** (in botany) The arrangement of the parts of a flower bud, especially of the sepals and petals.

afferent Carrying (nerve impulses, blood, etc.) from the outer regions of a body or organ towards its centre. The term is usually applied to types of nerve fibres or blood vessels. *Compare* efferent.

aflatoxin Any of four related toxic compounds produced by the mould *Aspergillus flavus*. Aflatoxins bind to DNA and prevent replication and transcription. They can cause acute liver damage and cancers: man may be poisoned by eating stored peanuts and cereals contaminated with the mould.

AFM *See* atomic force microscope.

afterbirth The *placenta, *umbilical cord, and *extraembryonic membranes, which are expelled from the womb after a mammalian fetus is born. In most nonhuman mammals the afterbirth, which contains nutrients and might otherwise attract predators, is eaten by the female.

after-heat Heat produced by a nuclear reactor after it has been shut down. The after-heat is generated by radioactive substances formed in the fuel elements.

agamospermy *See* apomixis.

agar An extract of certain species of red seaweeds that is used as a gelling agent in microbiological *culture media, foodstuffs, medicines, and cosmetic creams and jellies. *Nutrient agar* consists of a broth made from beef extract or blood that is gelled with agar and used for the cultivation of bacteria, fungi, and some algae.

agate A variety of *chalcedony that forms in rock cavities and has a pattern of concentrically arranged bands or layers that lie parallel to the cavity walls. These layers are frequently alternating tones of brownish-red. *Moss agate* does not show the same banding and is a milky chalcedony containing mosslike or dendritic patterns formed by inclusions of manganese and iron oxides. Agates are used in jewellery and for ornamental purposes.

agglutination The clumping together by antibodies of microscopic foreign particles (*antigens), such as red blood cells or bacteria, so that they form a visible pellet-like precipitate. Agglutination is a specific reaction, i.e. a particular antigen will only clump in the presence of its specific antibody; it therefore provides a means of identifying unknown bacteria and determining *blood group. When red cells of incompatible blood groups (e.g. group A and group B – *see* ABO system) are mixed together agglutination of the cells occurs (*haemagglutination*).

aggression Behaviour aimed at intimidating or injuring another animal of the same or a competing species. Aggression between individuals of the same species often starts with a series of ritualized displays or contests that can end at any stage if one of the combatants withdraws, leaving the victor with access to a disputed resource (e.g. food, a mate, or *territory) or with increased social dominance (*see* dominant). It is also often seen in *courtship. Aggression or threat displays usually appear to exaggerate the performer's size or strength; for example, many fish erect their fins and mammals and birds may erect hairs or feathers. Special markings may be prominently exhibited, and *intention movements* may be made: dogs bare their teeth, for example. Some animals have evolved special structures for use in aggressive interactions (e.g. antlers in deer) but these are seldom used to cause actual injury; the opponent usually flees first or adopts *appeasement postures. Fights 'to the death' are comparatively rare. *See* display behaviour; ritualization.

Agnatha A class of marine and freshwater vertebrates that lack jaws. They are fishlike animals with cartilaginous skeletons and well-developed sucking mouthparts with horny teeth. The only living agnathans are lampreys and hagfishes (order Cyclostomata), which are parasites or scavengers. Fossil agnathans, covered in an armour of bony plates, are the oldest known fossil vertebrates. They have been dated from the Silurian and Devonian periods, 440–345 million years ago.

AGR Advanced gas-cooled reactor. *See* nuclear reactor.

AIDS (acquired immune deficiency syndrome) A disease of humans characterized by defective cell-mediated *immunity and increased susceptibility to infections. It is caused by the retrovirus *HIV (human immunodeficiency virus). This attacks and destroys helper *T cells, which are essential for combating infections. HIV is transmitted in blood, saliva, and semen; the major routes of infection are unprotected vaginal and anal intercourse, intravenous drug abuse, and the administration of contaminated blood and blood products.

air *See* earth's atmosphere.

air bladder *See* swim bladder.

air mass (in meteorology) An area of the atmosphere that in the horizontal field possesses more or less uniform properties, especially temperature and humidity, and extends for hundreds of kilometres. The transition zone at which one air mass meets another is known as a *front. Air masses develop over extensive areas of the earth's surface, known as source regions, where conditions are sufficiently uniform to impart similar characteristics to the overlying air. These areas are chiefly areas of high pressure. As an air mass moves away from its source region it undergoes modification.

alabaster *See* gypsum.

alanine *See* amino acid.

albedo **1.** The ratio of the radiant flux reflected by a surface to that falling on it. **2.** The probability that a neutron entering a body of material will be reflected back through the same surface as it entered.

albinism Hereditary lack of pigmentation (*see* melanin) in an organism. Albino animals and human beings have no colour in their skin, hair, or eyes (the irises appear pink from underlying blood vessels). The *allele responsible is *recessive to the allele for normal pigmentation.

albumen *See* albumin.

albumin One of a group of globular proteins that are soluble in water but form insoluble coagulates when heated. Albumins occur in egg white (the protein component of which is known as *albumen*), blood, milk, and plants. Serum albumins, which constitute about 55% of blood plasma protein, help regulate the osmotic pressure and hence plasma volume. They also bind and transport fatty acids. α-lactalbumin is one of the proteins in milk.

albuminous cell *See* companion cell.

alburnum *See* sapwood.

alcohols Organic compounds that contain the –OH group. In systematic chemical nomenclature alcohol names end in the suffix *-ol*. Examples are methanol, CH_3OH, and ethanol, C_2H_5OH. *Primary alcohols* have two hydrogen atoms on the carbon joined to the –OH group (i.e. they contain the group $-CH_2-OH$); *secondary alcohols* have one hydrogen on this carbon (the other two bonds being to carbon atoms, as in $(CH_3)_2CHOH$); *tertiary alcohols* have

primary alcohol (methanol)

secondary alcohol (propan-2-ol)

tertiary alcohol (2-methylpropan-2-ol)

Examples of alcohols

no hydrogen on this carbon (as in $(CH_3)_3COH$): see formulae. The different types of alcohols may differ in the way they react chemically. For example, with potassium dichromate(VI) in sulphuric acid the following reactions occur

primary alcohol → aldehyde → carboxylic acid

secondary alcohol → ketone

tertiary alcohol – no reaction

Other characteristics of alcohols are reaction with acids to give *esters and dehydration to give *alkenes or *ethers. Alcohols that have two $-OH$ groups in their molecules are *diols* (or *dihydric alcohols*), those with three are *triols* (or *trihydric alcohols*), etc.

Aldehyde structure

aldehydes Organic compounds that contain the group $-CHO$ (the *aldehyde group*; i.e. a carbonyl group $(C=O)$ with a hydrogen atom bound to the carbon atom). In systematic chemical nomenclature, aldehyde names end with the suffix *-al*. Examples of aldehydes are methanal (formaldehyde), HCOH, and ethanal (acetaldehyde), CH_3CHO. Aldehydes are formed by oxidation of primary *alcohols; further oxidation yields carboxylic acids. They are reducing agents and tests for aldehydes include *Fehling's test and *Tollen's reagent. Aldehydes have certain characteristic addition and condensation reactions. With sodium hydrogensulphate(IV) they form addition compounds of the type $[RCOH(SO_3)H]^- \, Na^+$. Formerly these were known as *bisulphite addition compounds*. They also form addition compounds with hydrogen cyanide to give *cyanohydrins and with alcohols to give *acetals and undergo condensation reactions to yield *oximes, *hydrazones, and *semicarb-

azones. Aldehydes readily polymerize. *See also* ketones.

aldohexose *See* monosaccharide.

aldol *See* aldol reaction.

aldol reaction A reaction of aldehydes of the type
$$2RCH_2CHO \leftrightharpoons RCH_2CH(OH)CHRCHO$$
where R is a hydrocarbon group. The resulting compound is a hydroxyaldehyde, i.e. an aldehyde–alcohol or *aldol*, containing alcohol ($-OH$) and aldehyde ($-CHO$) groups on adjacent carbon atoms. The reaction is base-catalysed, the first step being the formation of a carbanion of the type RHC^-CHO, which adds to the carbonyl group of the other aldehyde molecule. For the carbanion to form, the aldehyde must have a hydrogen atom on the carbon next to the carbonyl group.

Aldols can be further converted to other products; in particular, they are a source of unsaturated aldehydes. For example, the reaction of ethanal gives 3-hydroxybutenal (*acetaldol*):
$$2CH_3CHO \leftrightharpoons CH_3CH(OH)CH_2CHO$$
This can be further dehydrated to 2-butenal (*crotonaldehyde*):
$$CH_3CH(OH)CH_2CHO \rightarrow H_2O + CH_3CH:CHCHO$$

aldose *See* monosaccharide.

aldosterone A hormone produced by the adrenal glands (*see* corticosteroid) that controls excretion of sodium by the kidneys and thereby maintains the balance of salt and water in the body fluids. *See also* angiotensin.

algae A large and diverse group of simple plants that contain chlorophyll (and can therefore carry out photosynthesis) and live in aquatic habitats and in moist situations on land. The plant body may be unicellular or multicellular (filamentous, ribbon-like, or platelike). The algae are not now regarded as a single taxonomic group and are usually assigned to separate taxonomic divisions based primarily

algebra

The human alimentary canal

on the composition of the cell wall, the nature of the stored food reserves, and the other photosynthetic pigments present. *See* Bacillariophyta; Chlorophyta; Euglenophyta; Phaeophyta; Rhodophyta.

The organisms commonly known as *blue-green algae are now classified as bacteria.

algebra The branch of mathematics in which variable quantities and numbers are represented by symbols. Statements are usually made in the form of equations, which are manipulated into convenient forms and solved according to a set of strictly logical rules.

algebraic sum The total of a set of quantities paying due regard to sign,

e.g. the algebraic sum of 3 and −4 is −1.

algorithm A method of solving a problem, involving a finite series of steps. In computing practice the algorithm denotes the expression on paper of the proposed computing process (often by means of a flowchart) prior to the preparation of the program. If no algorithm is possible a *heuristic solution has to be sought.

alicyclic compound A compound that contains a ring of atoms and is aliphatic. Cyclohexane, C_6H_{12}, is an example.

alimentary canal (digestive tract; gut) A tubular organ in animals that is divided into a series of zones specialized for the ingestion, *digestion, and *absorption of food and for the elimination of indigestible material (see illustration). In most animals the canal has two openings, the mouth (for the intake of food) and the *anus (for the elimination of waste). Simple animals, such as coelenterates (e.g. *Hydra* and jellyfish) and flatworms, have only one opening to their alimentary canal, which must serve both functions.

aliphatic compounds Organic compounds that are *alkanes, *alkenes, or *alkynes or their derivatives. The term is used to denote compounds that do not have the special stability of *aromatic compounds. All noncyclic organic compounds are aliphatic. Cyclic aliphatic compounds are said to be *alicyclic*.

alkali A *base that dissolves in water to give hydroxide ions.

alkali metals (group 1 elements) The elements of the first group of the *periodic table (group IA): lithium (Li), sodium (Na), potassium (K), rubidium (Rb), caesium (Cs), and francium (Fr). All have a characteristic electron configuration that is a noble gas structure with one outer *s*-electron. They are typical metals (in the chemical sense) and readily lose their outer electron to form stable M^+ ions with noble-gas configurations. All are highly reactive, with the reactivity (i.e. metallic character) increasing down the group. There is a decrease in ionization energy from lithium (520 kJ mol^{-1}) to caesium (380 kJ mol^{-1}). The second ionization energies are much higher and divalent ions are not formed. Other properties also change down the group. Thus, there is an increase in atomic and ionic radius, an increase in density, and a decrease in melting and boiling point. The standard electrode potentials are low and negative, although they do not show a regular trend because they depend both on ionization energy (which decreases down

the group) and the hydration energy of the ions (which increases).

All the elements react with water (lithium slowly; the others violently) and tarnish rapidly in air. They can all be made to react with chlorine, bromine, sulphur, and hydrogen. The hydroxides of the alkali metals are strongly alkaline (hence the name) and do not decompose on heating. The salts are generally soluble. The carbonates do not decompose on heating, except at very high temperatures. The nitrates (except for lithium) decompose to give the nitrite and oxygen:

$$2MNO_3(s) \rightarrow 2MNO_2(s) + O_2(g)$$

Lithium nitrate decomposes to the oxide. In fact lithium shows a number of dissimilarities to the other members of group I and in many ways resembles magnesium (*see* diagonal relationship) In general, the stability of salts of oxo acids increases down the group (i.e. with increasing size of the M^+ ion). This trend occurs because the smaller cations (at the top of the group) tend to polarize the oxo anion more effectively than the larger cations at the bottom of the group.

alkalimetry Volumetric analysis using standard solutions of alkali to determine the amount of acid present.

alkaline 1. Describing an alkali. **2.** Describing a solution that has an excess of hydroxide ions (i.e. a pH greater than 7).

alkaline-earth metals (group II elements) The elements of the second group of the *periodic table (group IIA): beryllium (Be), magnesium (Mg), calcium (Ca), strontium (Sr), and barium (Ba). The elements are sometimes referred to as the 'alkaline earths', although strictly the 'earths' are the oxides of the elements. All have a characteristic electron configuration that is a noble-gas structure with two outer *s*-electrons. They are typical metals (in the chemical sense) and readily lose both outer electrons to form stable M^{2+} ions; i.e. they are strong reducing agents. All are reactive, with the reactivity increasing down the group. There is a decrease in both first and second ionization energies down the group. Although there is a significant difference between the first and second ionization energies of each element, compounds containing univalent ions are not known. This is because the divalent ions have a smaller size and larger charge, leading to higher hydration energies (in solution) or lattice energies (in solids). Consequently, the overall energy charge favours the formation of divalent compounds. The third ionization energies are much higher than the second ionization energies, and trivalent compounds (containing M^{3+}) are unknown.

Beryllium, the first member of the group, has anomalous properties because of the small size of the ion; its atomic radius (0.112 nm) is much less than that of magnesium (0.16 nm). From magnesium to radium there is a fairly regular increase in atomic and ionic radius. Other regular changes take place in moving down the group from magnesium. Thus, the density and melting and boiling points all increase. Beryllium, on the other hand, has higher boiling and melting points than calcium and its density lies between those of calcium and strontium. The standard electrode potentials are negative and show a regular small decrease from magnesium to barium. In some ways beryllium resembles aluminium (*see* diagonal relationship).

All the metals are rather less reactive than the alkali metals. They react with water and oxygen (beryllium and magnesium form a protective surface film) and can be made to react with chlorine, bromine, sulphur, and hydrogen. The oxides and hydroxides of the metals show the increasing ionic character in moving down the group: beryllium hydroxide is amphoteric, magnesium hydroxide is only very slightly soluble in water and is weakly basic, calcium hydroxide is sparingly soluble and distinctly basic, strontium and barium hydroxides are quite soluble and basic. The hydroxides decompose on heating to give the oxide and water:

$$M(OH)_2(s) \rightarrow MO(s) + H_2O(g)$$

The carbonates also decompose on heating to the oxide and carbon dioxide:

$$MCO_3(s) \rightarrow MO(s) + CO_2(g)$$

The nitrates decompose to give the oxide:

$$2M(NO_3)_2(s) \rightarrow 2MO(s) + 4NO_2(g) + O_2(g)$$

As with the *alkali metals, the stability of salts of oxo acids increases down the group. In general, salts of the alkaline-earth elements are soluble if the anion has a single charge (e.g. nitrates, chlorides). Most salts with a doubly charged anion (e.g. carbonates, sulphates) are insoluble. The solubilities of salts of a particular acid tend to decrease down the group. (Solubilities of hydroxides increase for larger cations.)

alkaloid One of a group of nitrogenous organic compounds derived from plants and having diverse pharmacological properties. Alkaloids include morphine, cocaine, atropine, quinine, and caffeine, most of which are used in medicine as analgesics (pain relievers) or anaesthetics. Some alkaloids are poisonous, e.g. strychnine and coniine, and *colchicine inhibits cell division.

alkanal An aliphatic aldehyde.

alkanes (paraffins) Saturated hydrocarbons with the general formula C_nH_{2n+2}. In systematic chemical nomenclature alkane names end in the suffix -*ane*. They form a *homologous series (the *alkane series*) methane (CH_4), ethane (C_2H_6), propane (C_3H_8), butane (C_4H_{10}), pentane

(C_5H_{12}), etc. The lower members of the series are gases; the high-molecular weight alkanes are waxy solids. Alkanes are present in natural gas and petroleum. They can be made by heating the sodium salt of a carboxylic acid with soda lime:

$$RCOO^-Na^+ + Na^+OH^- \rightarrow$$
$$Na_2CO_3 + RH$$

Other methods include the *Wurtz reaction and *Kolbe's method. Generally the alkanes are fairly unreactive. They form haloalkanes with halogens when irradiated with ultraviolet radiation.

alkanol An aliphatic alcohol.

alkenes (olefines; olefins) Unsaturated hydrocarbons that contain one or more double carbon–carbon bonds in their molecules. In systematic chemical nomenclature alkene names end in the suffix *-ene*. Alkenes that have only one double bond form a homologous series (the *alkene series*) starting ethene (ethylene), $CH_2:CH_2$, propene, $CH_3CH:CH_2$, etc. The general formula is C_nH_{2n}. Higher members of the series show isomerism depending on position of the double bond; for example, butene (C_4H_8) has two isomers, which are (1) but-1-ene $(C_2H_5CH:CH_2)$ and (2) but-2-ene $(CH_3CH:CHCH_3)$; *see formulae*.

Butene isomers

Alkenes can be made by dehydration of alcohols (passing the vapour over hot pumice):

$$RCH_2CH_2OH - H_2O \rightarrow RCH:CH_2$$

An alternative method is the removal of a hydrogen atom and halogen atom from a haloalkane by potassium hydroxide in hot alcoholic solution:

$$RCH_2CH_2Cl + KOH \rightarrow KCl +$$
$$H_2O + RCH:CH_2$$

Alkenes typically undergo *addition reactions to the double bond. *See also* hydrogenation; oxo reaction; ozonolysis; Ziegler process.

alkoxides Compounds formed by reaction of alcohols with sodium or potassium metal. Alkoxides are salt-like compounds containing the ion $R-O^-$.

alkylbenzenes Organic compounds that have an alkyl group bound to a benzene ring. The simplest example is methylbenzene (toluene), $CH_3C_6H_5$. Alkyl benzenes can be made by the *Friedel–Crafts reaction.

alkyl group A group obtained by removing a hydrogen atom from an alkane, e.g. methyl group, CH_3-, derived from methane.

alkyl halides *See* haloalkanes.

alkynes (acetylenes) Unsaturated hydrocarbons that contain one or more triple carbon–carbon bonds in their molecules. In systematic chemical nomenclature alkyne names end in the suffix *-yne*. Alkynes that have only one triple bond form a *homologous series: ethyne (acetylene), $CH\equiv CH$, propyne, $CH_3CH\equiv CH$, etc. They can be made by the action of potassium hydroxide in alcohol solution on haloalkanes containing halogen atoms on adjacent carbon atoms; for example:

$$RCHClCH_2Cl + 2KOH \rightarrow 2KCl +$$
$$2H_2O + RCH\equiv CH$$

Like *alkenes, alkynes undergo addition reactions.

allantois One of the membranes that develops in embryonic reptiles, birds, and mammals as a growth from the hindgut. It acts as a urinary bladder for the storage of waste excretory products in the egg (in reptiles and birds) and as a means of providing the embryo with oxygen (in reptiles, birds, and mammals) and food (in mammals; *see* placenta). *See also* extraembryonic membranes.

allele (allelomorph) One of the alternative forms of a gene. In most organisms there are two alleles of any one gene (one from each parent), which occupy the same relative position on *homologous chromosomes. One allele is often *dominant to the other (known as the *recessive), i.e. it determines which aspects of a particular characteristic the organism will display.

allelomorph *See* allele.

allergy A condition in which the body produces an abnormal *immune response to certain *antigens (called *allergens*), which include dust, pollen, certain foods and drugs, or fur. In allergic individuals these substances, which in a normal person would be destroyed by antibodies, stimulate the release of *histamine, leading to inflammation and other characteristic symptoms of the allergy (e.g. asthma or hay fever). This response is a type of *anaphylaxis.

allogamy Cross-fertilization in plants. *See* fertilization.

allometric growth The regular and systematic pattern of growth such that the mass or size of any organ or part of a body can be expressed in relation to the total mass or size of the entire organism according to the allometric equation: $Y = bx^{\alpha}$, where Y = mass of the organ, x = mass of the organism, α = growth coefficient of the organ, and b = a constant.

allopatric Describing or relating to groups of similar organisms that could interbreed but do not because they are geographically separated. *Compare* sympatric. *See* speciation.

all-or-none response A type of response that may be either complete and of full intensity or totally absent, depending on the strength of the stimulus; there is no partial response.

For example, a nerve cell is either stimulated to transmit a complete nervous impulse or else it remains in its resting state; a stinging *thread cell of a coelenterate is either completely discharged or it is not.

allotropy The existence of elements in two or more different forms (*allotropes*). In the case of oxygen, there are two forms: 'normal' dioxygen (O_2) and ozone, or trioxygen (O_3). These two allotropes have different molecular configurations. More commonly, allotropy occurs because of different crystal structures in the solid, and is particularly prevalent in groups IV, V, and VI of the periodic table. In some cases, the allotropes are stable over a temperature range, with a definite transition point at which one changes into the other. For instance, tin has two allotropes: white (metallic) tin stable above 13.2°C and grey (nonmetallic) tin stable below 13.2°C. This form of allotropy is called *enantiotropy*. Carbon also has two allotropes – diamond and graphite – although graphite is the stable form at all temperatures. This form of allotropy, in which there is no transition temperature at which the two are in equilibrium, is called *monotropy*. *See also* polymorphism.

allowed bands *See* energy bands.

alloy A material consisting of two or more metals (e.g. brass is an alloy of copper and zinc) or a metal and a nonmetal (e.g. steel is an alloy of iron and carbon, sometimes with other metals included). Alloys may be compounds, *solid solutions, or mixtures of the components.

alloy steels *See* steel.

alluvial deposits Sediments deposited in a river, which range in particle size from fine silts to coarse gravels.

allyl group Formerly, the organic group $CH_2{:}CHCH_2{-}$. It was used in names such as *allyl alcohol* ($CH_2{:}CHCH_2OH$, prop-2-en-1-ol).

Alnico A trade name for a series of alloys, containing iron, aluminium, nickel, cobalt, and copper, used to make permanent magnets.

alpha-iron *See* iron.

alpha-naphthol test A biochemical test to detect the presence of carbohydrates in solution, also known as *Molisch's test* (after the Austrian chemist H. Molisch (1856–1937), who devised it). A small amount of alcoholic alpha-naphthol is mixed with the test solution and concentrated sulphuric acid is poured slowly down the side of the test tube. A positive reaction is indicated by the formation of a violet ring at the junction of the liquids.

alpha particle A helium–4 nucleus emitted by a larger nucleus during the course of the type of radioactive decay known as *alpha decay*. As a helium–4 nucleus consists of two protons and two neutrons bound together as a stable entity the loss of an alpha particle involves a decrease in *nucleon number of 4 and decrease of 2 in the *atomic number, e.g. the decay of a uranium–238 nucleus into a thorium–234 nucleus. A stream of alpha particles is known as an *alpha-ray* or *alpha-radiation*.

alternating current (a.c.) An electric current that reverses its direction with a constant *frequency (f). If a graph of the current against time has the form of a *sine wave, the current is said to be *sinusoidal*. Alternating current, unlike direct current, is therefore continuously varying and its magnitude is either given as its peak value (I_0) or its *root-mean-square value ($I_0/\sqrt{2}$ for a sinusoidal current). This r.m.s. value is more useful as it is comparable to a d.c. value in being a measure of the ability of the current to transmit power. The instantaneous value of a sinusoidal current (I) is given by $I = I_0\sin 2\pi ft$. If a direct current is supplied to a circuit the only opposition it encounters is the circuit's *resistance. However, an alternating current is opposed not only by the resistance of the circuit but also by its *reactance. This reactance is caused by *capacitance and *inductance in the circuit. In a circuit consisting of a resistance (R), an inductance (L), and a capacitance (C) all in series, the reactance (X) is equal to $(2\pi fL) - (1/2\pi fC)$. The total opposition to the current, called the *impedance (Z), is then equal to the ratio of the r.m.s. applied p.d. to the r.m.s. current and is given by $\sqrt{(R^2 + X^2)}$.

alternation of generations The occurrence within the *life cycle of an organism of two or more distinct forms (generations), which differ from each other in appearance, habit, and method of reproduction. The phenomenon occurs in certain lower animals (e.g. coelenterates and parasitic protozoans and flatworms) and in plants. The malaria parasite (*Plasmodium*), for example, has a complex life cycle involving the alternation of sexually and asexually reproducing generations. In plants the generation with sexual reproduction is called the *gametophyte and the asexual generation is the *sporophyte, either of which may dominate the life cycle, and there is also alternation of the haploid and diploid states. Thus in ferns the dominant plant is the diploid sporophyte; it produces spores that germinate into small haploid gametophytes bearing sex organs. In mosses the gametophyte is the dominant plant and the sporophyte is the spore-bearing capsule.

alternator An *alternating-current generator consisting of a coil or coils that rotate in the magnetic field produced by one or more permanent magnets or electromagnets. The electromagnets are supplied by an independent direct-current source. The frequency of the alternating current produced depends on the speed at which the coil rotates and the number of pairs of magnetic poles. In the large alternators of power stations the electromagnets rotate inside fixed coils; many bicycle dynamos are alternators with rotating permanent magnets inside fixed coils.

altimeter A device used to measure height above sea level. It usually consists of an aneroid *barometer measuring atmospheric pressure. Aircraft are fitted with altimeters, which are set to the atmospheric pressure at a convenient level, usually sea level, before take off. The height of the aircraft can then be read off the instrument as the aircraft climbs and the pressure falls.

altruism Behaviour by an animal that decreases its chances of survival or reproduction while increasing those of another member of the same species. For example, a lapwing puts itself at risk by luring a predator away from the nest through feigning injury, but by so doing saves its offspring. Altruism in its biological sense does not imply any conscious benevolence on the part of the performer. Altruism can evolve through *kin selection, if the recipients of altruistic acts tend on average to be more closely related to the altruist than the population as a whole. *See also* inclusive fitness.

alum *See* aluminium potassium sulphate; alums.

alumina *See* aluminium oxide; aluminium hydroxide.

aluminate A salt formed when aluminium hydroxide or γ-alumina is dissolved in solutions of strong bases, such as sodium hydroxide. Aluminates exist in solutions containing the aluminate ion, commonly written $[Al(OH)_4]^-$. In fact the ion probably is a complex hydrated ion and can be regarded as formed from a hydrated Al^{3+} ion by removal of four hydrogen ions:

$$[Al(H_2O)_6]^{3+} + 4OH^- \rightarrow 4H_2O + [Al(OH)_4(H_2O)_2]^-$$

Other aluminates and polyaluminates, such as $[Al(OH)_6]^{3-}$ and $[(HO)_3AlOAl(OH)_3]^{2-}$, are also present. *See also* aluminium hydroxide.

aluminium Symbol Al. A silvery-white lustrous metallic element belonging to *group III of the periodic table; a.n. 13; r.a.m. 26.98; r.d. 2.7; m.p. 660°C; b.p. 2467°C. The metal itself is highly reactive but is protected by a thin transparent layer of the oxide, which forms quickly in air. Aluminium and its oxide are amphoteric. The metal is extracted from purified bauxite (Al_2O_3) by electrolysis; the main process uses a *Hall–Heroult cell but other electrolytic methods are under development, including conversion of bauxite with chlorine and electrolysis of the molten chloride. Pure aluminium is soft and ductile but its strength can be increased by work-hardening. A large number of alloys are manufactured; alloying elements include copper, manganese, silicon, zinc, and magnesium. Its lightness, strength (when alloyed), corrosion resistance, and electrical conductivity (62% of that of copper) make it suitable for a variety of uses, including vehicle and aircraft construction, building (window and door frames), and overhead power cables. Although it is the third most abundant element in the earth's crust (8.1% by weight) it was not isolated until 1825 by H. C. Oersted (1777–1851).

aluminium acetate *See* aluminium ethanoate.

aluminium chloride A whitish solid, $AlCl_3$, which fumes in moist air and reacts violently with water (to give

Structure of aluminium trichloride dimer

hydrogen chloride). It is known as the anhydrous salt (hexagonal; r.d. 2.44 (fused solid); m.p. 190°C (2.5 atm.); sublimes at · 178°C) or the hexahydrate $AlCl_3.6H_2O$ (rhombic; r.d. 2.398; loses water at 100°C), both of which are deliquescent. Aluminium chloride may be prepared by passing hydrogen chloride or chlorine over hot aluminium or (industrially) by passing chlorine over heated aluminium oxide and carbon. The chloride ion is polarized by the small positive aluminium ion and the bonding in the solid is intermediate between covalent and ionic. In the liquid and vapour phases dimer molecules exist, Al_2Cl_6, in which there are chlorine bridges making coordinate bonds to aluminium atoms (see formula). The $AlCl_3$ molecule can also form compounds with other molecules that donate pairs of electrons (e.g. amines or hydrogen sulphide); i.e. it acts as a Lewis *acid. At high temperatures the Al_2Cl_6 molecules in the vapour dissociate to (planar) $AlCl_3$ molecules. Aluminium chloride is used commercially as a catalyst in the cracking of oils. It is also a catalyst in certain other organic reactions, especially the Friedel–Crafts reaction.

aluminium ethanoate (aluminium acetate) A white solid, $Al(OOCCH_3)_3$, which decomposes on heating, is very slightly soluble in cold water, and decomposes in warm water. The normal salt, $Al(OOCCH_3)_3$, can only be made in the absence of water (e.g. ethanoic anhydride and aluminium chloride at 180°C); in water it forms the basic salts $Al(OH)(OOCCH_3)_2$ and $Al_2(OH)_2(OOCCH_3)_4$. The reaction of aluminium hydroxide with ethanoic acid gives these basic salts directly. The compound is used extensively in dyeing as a mordant, particularly in combination with aluminium sulphate (known as *red liquor*); in the paper and board industry for sizing and hardening; and in tanning. It was previously used as an antiseptic and astringent.

aluminium hydroxide A white crystalline compound, $Al(OH)_3$; r.d. 2.42–2.52. The compound occurs naturally as the mineral *gibbsite* (monoclinic). In the laboratory it can be prepared by precipitation from solutions of aluminium salts. Such solutions contain the hexaquoaluminium(III) ion with six water molecules coordinated, $[Al(H_2O)_6]^{3+}$. In neutral solution this ionizes:

$$[Al(H_2O)_6]^{3+} \leftrightarrows H^+ + [Al(H_2O)_5OH]^{2+}$$

The presence of a weak base such as S^{2-} or CO_3^{2-} (by bubbling hydrogen sulphide or carbon dioxide through the solution) causes further ionization with precipitation of aluminium hydroxide

$$[Al(H_2O)_6]^{3+}(aq) \rightarrow$$
$$Al(H_2O)_3(OH)_3(s) + 3H^+(aq)$$

The substance contains coordinated water molecules and is more correctly termed *hydrated aluminium hydroxide*. In addition, the precipitate has water molecules trapped in it and has a characteristic gelatinous form. The substance is amphoteric. In strong bases the *aluminate ion is produced by loss of a further proton:

$$Al(H_2O)_3(OH)_3(s) + OH^-(aq) \leftrightarrows$$
$$[Al(H_2O)_2(OH)_4]^-(aq) + H_2O(l)$$

On heating, the hydroxide transforms to a mixed oxide hydroxide, AlO.OH (rhombic; r.d. 3.01). This substance occurs naturally as *diaspore* and *boehmite*. Above 450°C it transforms to γ-alumina.

In practice various substances can be produced that are mixed crystalline forms of $Al(OH)_3$, AlO.OH, and aluminium oxide (Al_2O_3) with water molecules. These are known as *hydrated alumina*. Heating the hydrated hydroxide causes loss of water, and produces various *activated aluminas*, which differ in porosity, number of remaining –OH groups, and particle size. These are used as catalysts (particularly for organic dehydration reactions), as catalyst supports, and in chromatography. Gelatinous freshly precipitated aluminium hydroxide was formerly widely used as a mordant for dyeing and calico printing because of its ability to form insoluble coloured *lakes with vegetable dyes. *See also* aluminium oxide.

aluminium oxide (alumina) A white or colourless oxide of aluminium occurring in two main forms. The stable form α-alumina (r.d. 3.97; m.p. 2020°C; b.p. 2980 $\pm$ 60°C) has colourless hexagonal or rhombic crystals; γ-alumina (r.d. 3.5–3.9) transforms to the α-form on heating and is a white microcrystalline solid. The compound occurs naturally as *corundum* or *emery* in the α-form with a hexagonal-close-packed structure of oxide ions with aluminium ions in the octahedral interstices. The gemstones ruby and sapphire are aluminium oxide coloured by minute traces of chromium and cobalt respectively. A number of other forms of aluminium oxide have been described (β-, δ-, and ζ-alumina) but these contain alkali-metal ions. There is also a short-lived spectroscopic suboxide AlO. The highly protective film of oxide formed on the surface of aluminium metal is yet another structural variation, being a defective rock-salt form (every third Al missing).

Pure aluminium oxide is obtained by dissolving the ore bauxite in sodium hydroxide solution; impurities such as iron oxides remain insoluble because they are not amphoteric. The hydrated oxide is precipitated by seeding with material from a previous batch and this is then roasted at 1150 –1200°C to give pure α-alumina, or at 500–800°C to give γ-alumina. The bonding in aluminium hydroxide is not purely ionic due to polarization of the oxide ion. Although the compound might be expected to be amphoteric, α-alumina is weakly acidic, dissolving in alkalis to give solutions containing aluminate ions; it is resistant to acid attack. In contrast γ-alumina is typically amphoteric dissolving both in acids to give aluminium salts and in bases to give aluminates. α-alumina is one of the hardest materials known (silicon carbide and diamond are harder) and is widely used as an abrasive in both natural (corundum) and synthetic forms. Its refractory nature makes alumina brick an ideal material for furnace linings and alumina is also used in cements for high-temperature conditions. *See also* aluminium hydroxide.

aluminium potassium sulphate (potash alum; alum) A white or colourless crystalline compound, $Al_2(SO_4)_3.K_2SO_4.24H_2O$; r.d. 1.757; loses $18H_2O$ at 92.5°C; becomes anhydrous at 200°C. It forms cubic or octahedral crystals that are soluble in cold water, very soluble in hot water, and insoluble in ethanol and acetone. The compound occurs naturally as the mineral *kalinite*. It is a double salt and can be prepared by recrystallization from a solution containing equimolar quantities of potassium sulphate and aluminium sulphate. It is used as a mordant for dyeing and in the tanning and finishing of leather goods (for white leather). *See also* alums.

aluminium sulphate A white or colourless crystalline compound, $Al_2(SO_4)_3$, known as the anhydrous compound (r.d. 2.71; decomposes at 770°C) or as the hydrate $Al_2(SO)_3.18H_2O$ (monoclinic; r.d. 1.69; loses water at 86.5°C). The anhydrous salt is soluble in water and slightly soluble in ethanol; the hydrate is very soluble in water and insoluble in ethanol. The compound occurs naturally in the rare mineral *alunogenite* ($Al_2(SO)_3.18H_2O$). It may be prepared by dissolving aluminium hydroxide or china clays (aluminosilicates) in sulphuric acid. It decomposes on heating to sulphur dioxide, sulphur trioxide, and aluminium oxide. Its solutions are acidic because of hydrolysis.

Aluminium sulphate is commercially one of the most important aluminium compounds; it is used in sewage treatment (as a flocculating agent) and in the purification of drinking water, the paper industry, and in the preparation of mordants. It is also a fire-proofing agent. Aluminium sulphate is often wrongly called *alum* in these industries.

aluminium trimethyl See trimethyl-aluminium.

alums A group of double salts with the formula $A_2SO_4.B_2(SO_4)_3.24H_2O$, where A is a monovalent metal and B a trivalent metal. The original example contains potassium and aluminium (called *potash alum* or simply *alum*); its formula is often written $AlK(SO_4)_2.12H_2O$ (aluminium potassium sulphate-12-water). *Ammonium alum* is $AlNH_4(SO_4)_2.12H_2O$, *chrome alum* is $KCr(SO_4)_2.12H_2O$ (see potassium chromium sulphate), etc. The alums are isomorphous and can be made by dissolving equivalent amounts of the two salts in water and recrystallizing. See also aluminium sulphate.

alunogenite A mineral form of hydrated *aluminium sulphate, $Al_2(SO_4)_3.18H_2O$.

alveolus 1. The tiny air sac in the *lung of mammals and reptiles at the end of each *bronchiole. It is lined by a delicate moist membrane, has many blood capillaries, and is the site of exchange of respiratory gases (carbon dioxide and oxygen). **2.** The socket in the jawbone in which a tooth is rooted by means of the *periodontal membrane.

AM (amplitude modulation) See modulation.

amalgam An alloy of mercury with one or more other metals. Most metals form amalgams (iron and platinum are exceptions), which may be liquid or solid. Some contain definite intermetallic compounds, such as $NaHg_2$.

amber A yellow or reddish-brown fossil resin. The resin was exuded by certain trees and other plants and often contains preserved insects, flowers, or leaves that were trapped by its sticky surface before the resin hardened. Amber is used for jewellery and ornaments. It also has the property of acquiring an electrical charge when rubbed (the term electricity is derived from *electron*, the Greek name for amber). It occurs throughout the world in rock strata from the Cretaceous to the Pleistocene, but most commonly in Cretaceous and Tertiary rocks.

americium Symbol Am. A radioactive metallic transuranic element belonging to the *actinoids; a.n. 95; mass number of most stable isotope 243 (half-life 7.95×10^3 years); r.d. 13.67 (20°C); m.p. 994 ± 4°C; b.p. 2607°C. Ten isotopes are known. The element was discovered by G. T. Seaborg and associates in 1945, who obtained it by bombarding uranium–238 with alpha particles.

amethyst The purple variety of the mineral *quartz. It is found chiefly in Brazil, the Urals (Soviet Union), Arizona (USA), and Uruguay. The colour is due to impurities, especially iron oxide. It is used as a gemstone.

Amide structure

amides 1. Organic compounds containing the group $-CO.NH_2$ (the *amide group*). Amides are volatile solids; examples are ethanamide,

CH_3CONH_2, and propanamide, $C_2H_5CONH_2$. They are made by heating the ammonium salt of the corresponding carboxylic acid. **2.** Inorganic compounds containing the ion NH_2^-, e.g. KNH_2 and $Cd(NH_2)_2$. They are formed by the reaction of ammonia with electropositive metals.

amination A chemical reaction in which an amino group ($-NH_2$) is introduced into a molecule. Examples of amination reaction include the reaction of halogenated hydrocarbons with ammonia (high pressure and temperature) and the reduction of nitro compounds and nitriles.

amines Organic compounds derived by replacing one or more of the hydrogen atoms in ammonia by organic groups (see illustration). *Primary amines* have one hydrogen replaced, e.g. methylamine, CH_3NH_2. They contain the functional group $-NH_2$ (the *amino group*). *Secondary amines* have two hydrogens replaced, e.g. methylethylamine, $CH_3(C_2H_5)NH$. The group $=NH$ is the *imino group*. *Tertiary amines* have all three hy-

drogens replaced, e.g. trimethylamine, $(CH_3)_3N$. Amines are produced by the decomposition of organic matter. They can be made by reducing nitro compounds or amides.

amine salts Salts similar to ammonium salts in which the hydrogen atoms attached to the nitrogen are replaced by one or more organic groups. Amines readily form salts by reaction with acids, gaining a proton to form a positive ammonium ion. They are named as if they were substituted derivatives of ammonium compounds; for example, dimethylamine ($(CH_3)_2NH$) will react with hydrogen chloride to give dimethylammonium chloride, which is an ionic compound $[(CH_3)_2NH_2]^+Cl^-$. When the amine has a common nonsystematic name the suffix *-ium* can be used; for example, phenylamine (aniline) would give $[C_6H_5NH_3]^+Cl^-$, known as anilinium chloride. Formerly, such compounds were sometimes called *hydrochlorides*, e.g. aniline hydrochloride with the formula $C_6H_5NH_2.HCl$.

Salts formed by amines are crystalline substances that are readily soluble in water. Many insoluble *alkaloids (e.g. quinine and atropine) are used medicinally in the form of soluble salts ('hydrochlorides'). If alkali (sodium hydroxide) is added to solutions of such salts the free amine is liberated.

If all four hydrogen atoms of an ammonium salt are replaced by organic groups a *quaternary ammonium compound* is formed. Such compounds are made by reacting tertiary amines with halogen compounds; for example, trimethylamine ($(CH_3)_3N$) with chloromethane (CH_3Cl) gives tetramethylammonium chloride, $(CH_3)_4N^+Cl^-$. Salts of this type do not liberate the free amine when alkali is added, and quaternary hydroxides (such as $(CH_3)_4N^+OH^-$) can be isolated. Such compounds are

ammonia

amino group

primary (methylamine)

imino group

secondary amine (dimethylamine)

tertiary amine (trimethylamine)

Examples of amines

strong alkalis, comparable to sodium hydroxide.

amino acid Any of a group of water-soluble organic compounds that possess both a carboxyl (–COOH) and an amino (–NH₂) group attached to the α-carbon atom. Amino acids can be represented by the general formula R–CH(NH₂)COOH. R may be hydrogen or an organic group and determines the properties of any particular amino acid. Through the formation of peptide bonds, amino acids join together to form short chains (*peptides) or much longer chains (*polypeptides). Proteins are composed of various proportions of about 20 commonly occurring amino acids (see table). The sequence of these amino acids in the protein polypeptides determines the shape, properties, and hence biological role of the protein. Some amino acids that never occur in proteins are nevertheless important, e.g. *ornithine and citrulline, which are intermediates in the urea cycle.

Plants and many microorganisms can synthesize amino acids from simple inorganic compounds, but animals rely on adequate supplies in their diet. The *essential amino acids must be present in the diet whereas others can be manufactured from them.

aminobenzene See phenylamine.

amino group See amines.

amino sugar Any sugar containing an amino group in place of a hydroxyl group. The *hexosamines* are amino derivatives of hexose sugars and include *glucosamine* (based on glucose) and *galactosamine* (based on galactose). The former is a constituent of *chitin and the latter occurs in cartilage.

ammeter An instrument that measures electric current. The main types are the *moving-coil* ammeter, the *moving-iron* ammeter, and the *thermoammeter*. The moving-coil instrument is a moving-coil *galva-

nometer fitted with a *shunt to reduce its sensitivity. It can only be used for d.c., but can be adapted for a.c. by using a *rectifier. In moving-iron instruments, a piece of soft iron moves in the magnetic field created when the current to be measured flows through a fixed coil. They can be used with a.c. or d.c. but are less accurate (though more robust) than the moving-coil instruments. In thermoammeters, which can also be used with a.c. or d.c., the current is passed through a resistor, which heats up as the current passes. This is in contact with a thermocouple, which is connected to a galvanometer. This indirect system is mainly used for measuring high frequency a.c. In the *hot-wire* instrument the wire is clamped at its ends and its elongation as it is heated causes a pointer to move over a scale.

ammine A coordination *complex in which the ligands are ammonia molecules. An example of an ammine is the tetraamminecopper(II) ion [Cu(NH₃)₄]²⁺.

ammonia A colourless gas, NH₃, with a strong pungent odour; r.d. 0.59 (relative to air); m.p. –74°C; b.p. –30.9°C. It is very soluble in water and soluble in alcohol. The compound may be prepared in the laboratory by reacting ammonium salts with bases such as calcium hydroxide, or by the hydrolysis of a nitride. Industrially it is made by the *Haber process and over 80 million tonnes per year are used either directly or in combination. Major uses are the manufacture of nitric acid, ammonium nitrate, ammonium phosphate, and urea (the last three as fertilizers), explosives, dyestuffs and resins.

Liquid ammonia has some similarity to water as it is hydrogen bonded and has a moderate dielectric constant, which permits it to act as an ionizing solvent. It is weakly self-ion-

amino acid	abbreviation	formula
alanine	ala	$CH_3 - \underset{\underset{NH_2}{\vert}}{\overset{\overset{H}{\vert}}{C}} - COOH$
*arginine	arg	$H_2N - \underset{\underset{NH}{\vert}}{C} - NH - CH_2 - CH_2 - CH_2 - \underset{\underset{NH_2}{\vert}}{\overset{\overset{H}{\vert}}{C}} - COOH$
asparagine	asn	$H_2N - \underset{\underset{O}{\vert\vert}}{C} - CH_2 - \underset{\underset{NH_2}{\vert}}{\overset{\overset{H}{\vert}}{C}} - COOH$
aspartic acid	asp	$HOOC - CH_2 - \underset{\underset{NH_2}{\vert}}{\overset{\overset{H}{\vert}}{C}} - COOH$
cysteine	cys	$HS - CH_2 - \underset{\underset{NH_2}{\vert}}{\overset{\overset{H}{\vert}}{C}} - COOH$
glutamic acid	glu	$HOOC - CH_2 - CH_2 - \underset{\underset{NH_2}{\vert}}{\overset{\overset{H}{\vert}}{C}} - COOH$
glutamine	gln	$\underset{O}{\overset{H_2N}{\diagdown}}C - CH_2 - CH_2 - \underset{\underset{NH_2}{\vert}}{\overset{\overset{H}{\vert}}{C}} - COOH$
glycine	gly	$H - \underset{\underset{NH_2}{\vert}}{\overset{\overset{H}{\vert}}{C}} - COOH$
*histidine	his	$HC = C - CH_2 - \underset{\underset{NH_2}{\vert}}{\overset{\overset{H}{\vert}}{C}} - COOH$ (imidazole ring: $N \; NH$, $\underset{H}{C}$)
*isoleucine	ile	$CH_3 - CH_2 - \underset{\underset{CH_3}{\vert}}{CH} - \underset{\underset{NH_2}{\vert}}{\overset{\overset{H}{\vert}}{C}} - COOH$
*leucine	leu	$\underset{H_3C}{\overset{H_3C}{\diagup}}CH - CH_2 - \underset{\underset{NH_2}{\vert}}{\overset{\overset{H}{\vert}}{C}} - COOH$
*lysine	lys	$H_3N - CH_2 - CH_2 - CH_2 - CH_2 - \underset{\underset{NH_2}{\vert}}{\overset{\overset{H}{\vert}}{C}} - COOH$

*methionine	met	$CH_3-S-CH_2-CH_2-\overset{\overset{H}{\mid}}{\underset{\underset{NH_2}{\mid}}{C}}-COOH$
*phenylalanine	phe	$\langle\!\!\bigcirc\!\!\rangle-CH_2-\overset{\overset{H}{\mid}}{\underset{\underset{NH_2}{\mid}}{C}}-COOH$
proline	pro	![proline to 4-hydroxyproline structure] 4–hydroxyproline
serine	ser	$HO-CH_2-\overset{\overset{H}{\mid}}{\underset{\underset{NH_2}{\mid}}{C}}-COOH$
*threonine	thr	$CH_3-\underset{\underset{OH}{\mid}}{CH}-\overset{\overset{H}{\mid}}{\underset{\underset{NH_2}{\mid}}{C}}-COOH$
*tryptophan	trp	indole ring $-C-CH_2-\overset{\overset{H}{\mid}}{\underset{\underset{NH_2}{\mid}}{C}}-COOH$
tyrosine	tyr	$HO-\langle\!\!\bigcirc\!\!\rangle-CH_2-\overset{\overset{H}{\mid}}{\underset{\underset{NH_2}{\mid}}{C}}-COOH$
*valine	val	$\begin{matrix}H_3C\\H_3C\end{matrix}\!\!>\!CH-\overset{\overset{H}{\mid}}{\underset{\underset{NH_2}{\mid}}{C}}-COOH$

* an essential amino acid

The amino acids occurring in proteins

ized to give ammonium ions, NH_4^+ and amide ions, NH_2^-. It also dissolves electropositive metals to give blue solutions, which are believed to contain solvated electrons. Ammonia is extremely soluble in water giving basic solutions that contain solvated NH_3 molecules and small amounts of the ions NH_4^+ and OH^-. The combustion of ammonia in air yields nitrogen and water. In the presence of catalysts NO, NO_2, and water are formed; this last reaction is the basis for the industrial production of nitric acid. Ammonia is a good proton acceptor (i.e. it is a base) and gives rise to a series of ammonium salts, e.g. $NH_3 + HCl \rightarrow NH_4^+ + Cl^-$. It is also a reducing agent.

The participation of ammonia in the

*nitrogen cycle is a most important natural process. Nitrogen-fixing bacteria are able to achieve similar reactions to those of the Haber process, but under normal conditions of temperature and pressure. These release ammonium ions, which are converted by nitrifying bacteria into nitrite and nitrate ions.

ammoniacal Describing a solution in which the solvent is aqueous ammonia.

ammonia clock A form of atomic clock in which the frequency of a quartz oscillator is controlled by the vibrations of excited ammonia molecules (*see* excitation). The ammonia molecule (NH_3) consists of a pyramid with a nitrogen atom at the apex and one hydrogen atom at each corner of the triangular base. When the molecule is excited, once every 20.9 microseconds the nitrogen atom passes through the base and forms a pyramid the other side: 20.9 microseconds later it returns to its original position. This vibration back and forth has a frequency of 23 870 hertz and ammonia gas will only absorb excitation energy at exactly this frequency. By using a *crystal oscillator to feed energy to the gas and a suitable feedback mechanism, the oscillator can be locked to exactly this frequency.

ammonia–soda process *See* Solvay process.

ammonite An extinct aquatic mollusc of the class *Cephalopoda. Ammonites were abundant in the Mesozoic era (225–65 million years ago) and are commonly found as fossils in rock strata of that time, being used as *index fossils for the Jurassic period. They were characterized by a coiled shell divided into many chambers, which acted as a buoyancy aid. The external suture lines on these shells increased in complexity with the advance of the group.

ammonium alum *See* alums.

ammonium carbonate A colourless or white crystalline solid, $(NH_4)_2CO_3$, usually encountered as the monohydrate. It is very soluble in cold water. The compound decomposes slowly to give ammonia, water, and carbon dioxide. Commercial 'ammonium carbonate' is a double salt of ammonium hydrogencarbonate and ammonium aminomethanoate (carbamate), $NH_4HCO_3.NH_2COONH_4$. This material is manufactured by heating a mixture of ammonium chloride and calcium carbonate and recovering the product as a sublimed solid. It readily releases ammonia and is the basis of sal volatile. It is also used in dyeing and wool preparation and in baking powders.

ammonium chloride (sal ammoniac) A white or colourless cubic solid, NH_4Cl; r.d. 1.53; sublimes at 340°C. It is very soluble in water and slightly soluble in ethanol but insoluble in ether. It may be prepared by fractional crystallization from a solution containing ammonium sulphate and sodium chloride or ammonium carbonate and calcium chloride. Pure samples may be made directly by the gas-phase reaction of ammonia and hydrogen chloride. Because of its ease of preparation it can be manufactured industrially alongside any plant that uses or produces ammonia. The compound is used in dry cells, metal finishing, and in the preparation of cotton for dyeing and printing.

ammonium ion The monovalent cation NH_4^+. It may be regarded as the product of the reaction of ammonia (a Lewis base) with a hydrogen ion. The ion has tetrahedral symmetry. The chemical properties of ammonium salts are frequently very similar to those of equivalent alkali-metal salts.

ammonium nitrate A colourless crystalline solid, NH_4NO_3; r.d. 1.72; m.p. 169.6°C; b.p. 210°C. It is very soluble in water and soluble in ethanol. The crystals are rhombic when obtained below 32°C and monoclinic above 32°C. It may be readily prepared in the laboratory by the reaction of nitric acid with aqueous ammonia. Industrially, it is manufactured by the same reaction using ammonia gas. Vast quantities of ammonium nitrate are used as fertilizers (over 20 million tonnes per year) and it is also a component of some explosives.

ammonium sulphate A white rhombic solid, $(NH_4)_2SO_4$; r.d. 1.67; decomposes at 235°C. It is very soluble in water and insoluble in ethanol. It occurs naturally as the mineral *mascagnite*. Ammonium sulphate was formerly manufactured from the 'ammoniacal liquors' produced during coal-gas manufacture but is now produced by the direct reaction between ammonia gas and sulphuric acid. It is decomposed by heating to release ammonia (and ammonium hydrogensulphate) and eventually water, sulphur dioxide, and ammonia. Vast quantities of ammonium sulphate are used as fertilizers.

amniocentesis The taking of a sample of amniotic fluid from a pregnant woman to determine the condition of an unborn baby. A hollow needle is inserted through the woman's abdomen and wall of the uterus and the fluid drawn off. Chemical and microscopical examination of cells shed from the embryo's skin into the fluid are used to detect spina bifida, *Down's syndrome, or other serious biochemical or chromosomal abnormalities.

amnion A membrane that encloses the embryo of reptiles, birds, and mammals within the *amniotic cavity*. This cavity is filled with *amniotic fluid*, in which the embryo is protected from desiccation and from external pressure. *See also* extraembryonic membranes.

amniote A vertebrate whose embryos are totally enclosed in a fluid-filled sac – the *amnion. The evolution of the amnion provided the necessary fluid environment for the developing embryo and therefore allowed animals to breed away from water. Amniotes comprise the reptiles, birds, and mammals. *Compare* anamniote.

Amoeba A genus of *Protozoa, members of which have temporary body projections called *pseudopodia. These are used for locomotion and feeding and result in a constantly changing body shape. Most species are free-living in soil, mud, or water, where they feed on smaller protozoans and single-celled plants, but a few are parasitic. The best known species is the much studied *A. proteus*.

amoebocyte An animal cell whose location is not fixed and is therefore able to wander through the body tissues. Amoebocytes are named after their resemblance, especially in their movement, to *Amoeba* and they feed on foreign particles (including invading bacteria). They occur, for example, in sponges and mammalian blood (e.g. some *leucocytes).

amorphous Describing a solid that is not crystalline; i.e. one that has no long-range order in its lattice. Many powders that are described as 'amorphous' in fact are composed of microscopic crystals, as can be demonstrated by X-ray diffraction. *Glasses are examples of true amorphous solids.

amount of substance Symbol n. A measure of the number of entities present in a substance. The specified entity may be an atom, molecule, ion, electron, photon, etc., or any specified group of such entities. The amount of substance of an element, for example, is proportional to the number of atoms present. For all entities, the constant of proportionality is the

AMP

*Avogadro constant. The SI unit of amount of substance is the *mole.

AMP *See* ATP; cyclic AMP.

ampere Symbol A. The SI unit of electric current. The constant current that, maintained in two straight parallel infinite conductors of negligible cross section placed one metre apart in a vacuum, would produce a force between the conductors of 2×10^{-7} N m^{-1}. This definition replaced the earlier international ampere defined as the current required to deposit 0.001 118 00 gram of silver from a solution of silver nitrate in one second. The unit is named after A. M. Ampère (1775–1836).

ampere-hour A practical unit of electric charge equal to the charge flowing in one hour through a conductor passing one ampere. It is equal to 3600 coulombs.

ampere-turn The SI unit of *magnetomotive force equal to the magnetomotive force produced when a current of one ampere flows through one turn of a magnetizing coil.

Amphibia The class of vertebrates that contains the frogs, toads, newts, and salamanders. The amphibians evolved in the Devonian period (about 370 million years ago) as the first vertebrates to occupy the land, and many of their characteristics are adaptations to terrestrial life. All adult amphibians have a passage linking the roof of the mouth with the nostrils so they may breathe air and keep the mouth closed. The moist scaleless skin is used to supplement the lungs in gas exchange. They have no diaphragm, and therefore the muscles of the mouth and pharynx provide the pumping action for breathing. Fertilization is usually external and the eggs are soft and prone to desiccation, therefore reproduction commonly occurs in water. Amphibian larvae are aquatic, having gills for respiration; they undergo metamorphosis to the adult form.

amphiboles A large group of rock-forming metasilicate minerals. They have a structure of silicate tetrahedra linked to form double endless chains, in contrast to the single chains of the *pyroxenes, to which they are closely related. They are present in many igneous and metamorphic rocks. The amphiboles show a wide range of compositional variation but all of them conform to the general formula: $X_{2-3}Y_5Z_8O_{22}(OH)_2$, where X = Ca, Na, K, Mg, or Fe^{2+}; Y = Mg, Fe^{2+}, Fe^{3+}, Al, Ti, or Mn; and Z = Si or Al. The hydroxyl ions may be replaced by F, Cl, or O. Most amphiboles are monoclinic, including: cummingtonite,
$$(Mg,Fe^{2+})_7(Si_8O_{22})(OH)_2;$$
tremolite, $Ca_2Mg_5(Si_8O_{22})(OH,F)_2$; actinolite,
$$Ca_2(Mg,Fe^{2+})_5(Si_8O_{22})(OH,F)_2;$$
*hornblende,
$$NaCa_2(Mg,Fe^{2+},Fe^{3+},Al)_5((Si,Al)_8O_{22})-$$
$$(OH,F)_2;$$
edenite,
$$NaCa_2(Mg,Fe^{2+})_5(Si_7AlO_{22})(OH,F)_2;$$
riebeckite, $Na_2,Fe_3^{2+}(Si_8O_{22})(OH,F)_2$. The minerals anthophyllite,
$$(Mg,Fe^{2+})_7(Si_8O_{22})(OH,F)_2,$$
and gedrite,
$$(Mg,Fe^{2+})_6Al(Si,Al)_8O_{22})(OH,F)_2,$$
are orthorhombic amphiboles.

amphimixis True sexual reproduction, involving the fusion of male and female gametes and the formation of a zygote. *Compare* apomixis.

amphioxus *See* Chordata.

amphiprotic *See* solvent.

ampholyte A substance that can act as either an acid, in the presence of a strong base, or a base, when in the presence of a strong acid.

ampholyte ion *See* zwitterion.

amphoteric Describing a compound that can act as both an acid and a base (in the traditional sense of the term). For instance, aluminium hydroxide is amphoteric: as a base $Al(OH)_3$ it reacts with acids to form aluminium salts; as an acid H_3AlO_3 it reacts with alkalis to give *aluminates. Oxides of metals are typically basic and oxides of nonmetals tend to be acidic. The existence of amphoteric oxides is sometimes regarded as evidence that an element is a *metalloid.

amplifier A device that increases the strength of an electrical signal by drawing energy from a separate source to that of the signal. The original device used in electronic amplifiers was the *triode valve, in which the cathode–anode current is varied in accordance with the low-voltage signal applied to the valve's control grid. In the more recent *transistor, the emitter–collector current is controlled in much the same way by the signal applied to the transistor's base region. In the most modern devices the complete amplifier circuit is manufactured as a single *integrated circuit. The ratio of the output amplitude (of p.d. or current) of an amplifier (or stage of an amplifier) to the corresponding input amplitude is called the *gain* of the amplifier.

amplitude *See* wave.

amplitude modulation (AM) *See* modulation; radio.

a.m.u. *See* atomic mass unit.

amylase Any of a group of closely related enzymes that degrade starch, glycogen, and other polysaccharides. Plants contain both α- and β-amylases; the name *diastase* is given to the component of malt containing β-amylase, important in the brewing industry. Animals possess only α-amylases, found in pancreatic juice and also (in humans and some other species) in saliva (as *ptyalin*). Amylases cleave the long polysaccharide chains, producing a mixture of glucose and maltose.

amyl group Formerly, any of several isomeric groups with the formula $C_5H_{11}-$.

amylopectin A *polysaccharide comprising highly branched chains of glucose molecules. It is one of the constituents (the other being amylose) of *starch.

amylose A *polysaccharide consisting of linear chains of between 100 and 1000 linked glucose molecules. Amylose is a constituent of *starch. In water, amylose reacts with iodine to give a characteristic blue colour.

anabolic steroid Any steroid compound that promotes tissue growth, especially of muscles. Naturally occurring anabolic steroids include the male sex hormones (*androgens). Synthetic forms of these are used medically to help weight gain after debilitating diseases; their use by athletes to build up body muscles can cause liver damage and is banned by most athletic authorities.

anabolism The metabolic synthesis of proteins, fats, and other constituents of living organisms from molecules or simple precursors. This process requires energy in the form of ATP. Drugs that promote such metabolic activity are described as *anabolic* (*see* anabolic steroid). *See* metabolism. *Compare* catabolism.

anaerobe *See* anaerobic respiration.

anaerobic respiration A type of *respiration in which foodstuffs (usually carbohydrates) are partially oxidized, with the release of chemical energy, in a process not involving atmospheric oxygen. Since the substrate is never completely oxidized the energy yield of this type of respiration is lower than that of *aerobic respiration. It occurs in some yeasts and bacteria and in muscle tissue when oxygen is absent (*see* oxygen debt). *Obligate anaerobes* are organisms that cannot use free oxygen for respiration; *facultative anaerobes* are normally aerobic but can respire anaerobically during periods of oxygen shortage. Alcoholic *fermentation is a type of anaerobic respiration in which one of the end products is ethanol.

analogous Describing features of organisms that are superficially similar but have evolved in different ways. The wings of butterflies and birds are analogous organs.

analysis The determination of the components in a chemical sample. *Qualitative analysis* involves determining the nature of a pure unknown compound or the compounds present in a mixture. Various chemical tests exist for different elements or types of compound, and systematic analytical procedures can be used for mixtures. *Quantitative analysis* involves measuring the proportions of known components in a mixture. Chemical techniques for this fall into two main classes: *volumetric analysis and *gravimetric analysis. In addition, there are numerous physical methods of qualitative and quantitative analysis, including spectroscopic techniques, mass spectrometry, polarography, chromatography, activation analysis, etc.

analytical geometry (coordinate geometry) A form of geometry in which points are located in a two-dimensional or three-dimensional space by means of a system of coordinates. Curves are represented by an equation for a set of such points. The geometry of figures can thus be analysed by algebraic methods. *See* Cartesian coordinates; polar coordinates.

anamniote A vertebrate that lacks an *amnion and whose embryos and larvae must therefore develop in water. Anamniotes comprise the agnathans, fishes, and amphibians. *Compare* amniote.

anaphase The third stage of cell division. In *mitosis the chromatids of each chromosome move apart to opposite ends of the spindle. In the first anaphase of *meiosis, the paired homologous chromosomes separate and move to opposite ends; in the second anaphase the chromatids move apart, as in mitosis.

anaphylaxis An abnormal *immune response that occurs when an individual previously exposed to a particular *antigen is re-exposed to the same antigen. Anaphylaxis may follow an insect bite or the injection of a drug (such as penicillin). It is caused by the release of *histamine and similar substances and may produce a localized reaction or a more generalized and severe one, with difficulty in breathing, pallor, or drop in blood pressure, unconsciousness, and possibly heart failure and death. *See also* allergy.

anastigmatic lens 1. An objective lens for an optical instrument in which all *aberrations, including *astigmatism, are reduced greatly. **2.** A spectacle lens designed to correct astigmatism. It has different radii of curvature in the vertical and horizontal planes.

anatomy The study of the structure of living organisms, especially of their internal parts by means of dissection and microscopical examination. *Compare* morphology.

anchor ring *See* torus.

and circuit *See* logic circuits.

androecium The male sex organs (*stamens) of a flower. *Compare* gynoecium.

androgen One of a group of male sex hormones that stimulate development of the testes and of male *secondary sexual characteristics (such as growth of facial and pubic hair in men). *Testosterone* is the most important. Androgens are produced principally by the testes when stimulated with *luteinizing hormone but they are also secreted in smaller amounts by the adrenal glands and the ovaries. Injections of natural or synthetic androgens are used to treat hormonal disorders of the testes and breast cancer and to build up body tissue (*see* anabolic steroid).

anechoic Having a low degree of reverberation with little or no reflection of sound. An *anechoic* chamber is one designed for experiments in acoustics. The walls are covered with small pyramids to avoid the formation of stationary waves between facing surfaces and the whole of the interior surface is covered with an absorbent material to avoid reflections.

anemometer An instrument for measuring the speed of the wind or any other flowing fluid. The simple *vane anemometer* consists of a number of cups or blades attached to a central spindle so that the air, or other fluid, causes the spindle to rotate. The instrument is calibrated to give a wind speed directly from a dial. The instrument can be mounted to rotate about a vertical axis and in this form it also gives an indication of the direction of the wind. A *hot-wire anemometer* consists of an electrically heated wire that is cooled by the flow of fluid passing round it. The faster the flow the lower the temperature of the wire and the lower its resistance. Thus the rate of flow can be calculated by measuring the resistance of the wire.

anemophily Pollination of a flower in which the pollen is carried by the wind. Examples of anemophilous flowers are those of grasses and conifers. *Compare* entomophily; hydrophily.

aneroid barometer *See* barometer.

Angiospermae The flowering plants: a subdivision of the *Spermatophyta or a class of the *Pteropsida. Angiosperm gametes are produced within *flowers and the ovules (and the seeds into which they develop) are enclosed in a carpel (*compare* Gymnospermae). The angiosperms are the dominant plant forms of the present day. They show the most advanced structural organization in the plant kingdom, enabling them to inhabit a very diverse range of habitats. There are two classes (or subclasses) within this group: the *Monocotyledonae with one seed leaf (cotyledon) in the seed, and the *Dicotyledonae with two seed leaves.

angiotensin Either of two related peptide hormones that raise blood pressure. Angiotensin I is derived, by the action of the enzyme *renin, from a protein (α-globulin) secreted by the liver into the bloodstream. As blood passes through the lungs, another enzyme splits angiotensin I, forming angiotensin II. This causes constriction of blood vessels and stimulates the release of the hormones *vasopressin and *aldosterone, which increase blood pressure.

angle modulation *See* modulation.

angle of incidence 1. The angle between a ray falling on a surface and the perpendicular (normal) to the surface at the point at which the ray strikes the surface. 2. The angle between a wavefront and a surface that it strikes.

angle of reflection 1. The angle between a ray leaving a reflecting surface and the perpendicular (normal) to the surface at the point at which the ray leaves the surface. 2. The angle between a wavefront and a surface that it leaves.

angle of refraction 1. The angle between a ray that is refracted at a surface between two different media and the perpendicular (normal) to the surface at the point of refraction. 2. The angle between a wavefront and a surface at which it has been refracted.

anglesite A mineral form of *lead(II) sulphate, $PbSO_4$.

angstrom Symbol Å. A unit of length equal to 10^{-10} metre. It was formerly used to measure wavelengths and intermolecular distances but has now been replaced by the nanometre. 1 Å = 0.1 nanometre. The unit is named after the Swedish pioneer of spectroscopy A. J. Ångstrom (1814–74).

angular displacement, velocity, and acceleration See rotational motion.

angular frequency (pulsatance) A quantity proportional to the *frequency of a periodic phenomenon but having the dimensions of angular velocity. The angular frequency in radians per second = frequency in hertz × 2π radians per cycle.

angular magnification (magnifying power) See magnification.

angular momentum Symbol L. The product of the angular velocity of a body and its *moment of inertia about the axis of rotation, i.e. $L = I\omega$.

anhydride A compound that produces a given compound on reaction with water. For instance, sulphur trioxide is the (acid) anhydride of sulphuric acid
$$SO_3 + H_2O \rightarrow H_2SO_4$$
See also acid anhydrides.

anhydrite An important rock-forming anhydrous mineral form of calcium sulphate, $CaSO_4$. It is chemically similar to *gypsum but is harder and heavier and crystallizes in the rhombic form (gypsum is monoclinic). Under natural conditions anhydrite slowly hydrates to form gypsum. It occurs chiefly in white and greyish granular masses and is often found in the caprock of certain salt domes. It is used as a raw material in the chemical industry and in the manufacture of cement and fertilizers.

anhydrous Denoting a chemical compound lacking water: applied particularly to salts lacking their water of crystallization.

aniline See phenylamine.

anilinium ion The ion $C_6H_5NH_3^+$, derived from *phenylamine.

animal Any living organism of the kingdom Animalia, characterized by an inability to manufacture its own food, so that it feeds on other organisms or organic matter (holozoic nutrition; see heterotrophism). Animals are therefore typically mobile (to search for food) and have evolved specialized sense organs for detecting changes in the environment; a *nervous system coordinates information received by the sense organs and enables rapid responses to environmental stimuli. Animal *cells lack the cellulose cells walls of *plant cells. For a classification of the animal kingdom, see Appendix.

animal behaviour The activities that constitute an animal's response to its external environment. Certain categories of behaviour are seen in all animals (e.g. feeding, reproduction) but these activities involve different movements in different species and develop in different ways. Some movements are highly characteristic of a species (see instinct) whereas others are more variable and depend on the interaction between innate tendencies and *learning during the individual's lifetime. Physiologists study how changes in the body (e.g. hormone levels) affect behaviour, psychologists study the mechanisms of learning, and ethologists study the behaviour of the whole animal: how this develops during the individual's lifetime and how it evolved through natural selection (see ethology).

animal charcoal See charcoal.

animal starch See glycogen.

anion A negatively charged *ion, i.e. an ion that is attracted to the *anode in *electrolysis. Compare cation.

anionic detergent See detergent.

anionic resin See ion exchange.

anisogamy Sexual reproduction involving the fusion of gametes that differ in size and sometimes also in form. *See also* oogamy. *Compare* isogamy.

anisotropic Denoting a medium in which certain physical properties are different in different directions. Wood, for instance, is an anisotropic material: its strength along the grain differs from that perpendicular to the grain. Single crystals that are not cubic are anisotropic with respect to some physical properties, such as the transmission of electromagnetic radiation. *Compare* isotropic.

annealing A form of heat treatment applied to a metal to soften it, relieve internal stresses and instabilities, and make it easier to work or machine. It consists of heating the metal to a specified temperature for a specified time, both of which depend on the metal involved, and then allowing it to cool slowly It is applied to both ferrous and nonferrous metals and a similar process can be applied to other materials, such as glass.

Annelida A phylum of invertebrates comprising the segmented worms (e.g. the earthworm). Annelids have cylindrical soft bodies showing *metameric segmentation, obvious externally as a series of rings separating the segments. Each segment is internally separated from the next by a membrane and bears stiff bristles (*see* chaeta). Between the gut and other body organs there is a fluid-filled cavity called the *coelom, which acts as a hydrostatic skeleton. Movement is by alternate contraction of circular and longitudinal muscles in the body wall. The phylum contains three classes: *Polychaeta, *Oligochaeta, and *Hirudinea.

annihilation The destruction of a particle and its *antiparticle as a result of a collision between them. The *annihilation radiation* produced is carried away by *photons or

*mesons. For example, in a collision between an electron and a positron the energy produced is carried away by two photons, each having an energy of 0.511 MeV, which is equivalent to the rest-mass energies of the annihilated particles plus their kinetic energies. When nucleons annihilate each other the energy is carried away by mesons.

annual A plant that completes its life cycle in one year, during which time it germinates, flowers, produces seeds, and dies. Examples are the sunflower and marigold. *Compare* biennial; ephemeral; perennial.

annual ring *See* growth ring.

annulus 1. (in mathematics) The plane figure formed between two concentric circles of different radii, R and r. Its area is $\pi(R^2 - r^2)$. 2. (in botany) a. A ragged ring of tissue that remains on the stalk of a mushroom or toadstool. Also called a *velum*, it is formed from the ruptured membrane that originally covered the lower surface of the cap. b. The region of the wall of a fern sporangium that is specialized for spore dispersal. It consists of cells that are thickened except on their outer walls. On drying out, the cells contract and the sporangium ruptures, releasing the spores. The annulus springs back into position when the residual water in the cells vaporizes and any remaining spores are dispersed. 3. (in zoology) Any of various ring-shaped structures in animals, such as any of the segments of an earthworm or other annelid.

anode A positive electrode. In *electrolysis anions are attracted to the anode. In an electronic vacuum tube it attracts electrons from the *cathode and it is therefore from the anode that electrons flow out of the device. In these instances the anode is made positive by external means; however in a *voltaic cell the anode is the electrode that spontaneously becomes positive and therefore attracts electrons to it from the external circuit.

anode sludge *See* electrolytic refining.

anodizing A method of coating objects made of aluminium with a protective oxide film, by making them the anode in an electrolytic bath containing an oxidizing electrolyte. Anodizing can also be used to produce a decorative finish by formation of an oxide layer that can absorb a coloured dye.

anomaly An angle used to fix the position of a body, such as a planet, in an elliptical orbit. The *true anomaly* of a planet is the angle between the *perihelion, the sun, and the planet in the direction of the planet's motion. The *mean anomaly* is the angle between the perihelion, the sun, and an imaginary planet having the same period as the real planet, but assumed to be moving at constant speed.

Anoplura *See* Siphunculata.

ANS *See* autonomic nervous system.

antagonism 1. The interaction of two substances (e.g. drugs, hormones, or enzymes) having opposing effects in a system in such a way that the action of one partially or completely inhibits the effects of the other. For example, one group of anti-cancer drugs acts by antagonizing the effects of certain enzymes controlling the activities of the cancer cells. **2.** An interaction between two muscles in which contraction of one (the *antagonist*) prevents that of the other (the *agonist*). The antagonist must relax to enable the agonist to contract and effect movement. **3.** An interaction between two organisms (e.g. moulds or bacteria) in which the growth of one is inhibited by the other. *Compare* synergism.

antenna 1. (in zoology) A long whiplike jointed mobile paired appendage on the head of many arthropods, usually concerned with the senses of smell, touch, etc. In insects, millipedes, and centipedes they are the first pair of head appendages and are specialized and modified in many insects. In crustaceans they are the second pair of head appendages, the first pair (the *antennules*) having the sensory function, while the antennae are modified for swimming and for attachment. **2.** (in radio) *See* aerial.

antennule *See* antenna.

anterior 1. Designating the part of an animal that faces to the front, i.e. that leads when the animal is moving. In man and bipedal animals the anterior surface corresponds to the *ventral surface. **2.** Designating the side of a flower or axillary bud that faces away from the flower stalk or main stem, respectively. *Compare* posterior.

anther The upper two-lobed part of a plant *stamen, usually yellow in colour. Each lobe contains two pollen sacs within which are numerous pollen grains, which are released when the anther ruptures.

antheridium The male sex organ of algae, fungi, bryophytes, and pteridophytes. It produces the male gametes (*antherozoids*). It may consist of a single cell or it may have a wall that is made up of one or several layers forming a sterile jacket around the developing gametes. *Compare* archegonium.

antherozoid (spermatozoid) The motile male gamete of algae, fungi, bryophytes, pteridophytes, and certain gymnosperms. Antherozoids usually develop in an *antheridium but in certain gymnosperms, such as *Ginkgo* and *Cycas*, they develop from a cell in the pollen tube.

anthocyanin One of a group of *flavonoid pigments. Anthocyanins occur in the cell vacuoles of various plant organs and are responsible for many of the blue, red, and purple colours in plants (particularly in flowers).

Anthracene

anthracene A white crystalline solid, $C_{14}H_{10}$; r.d. 1.25; m.p. 286°C; b.p. 379.8°C. It is an aromatic hydrocarbon with three fused rings (see formula), and is obtained by the distillation of crude oils. The main use is in the manufacture of dyes.

anthracite *See* coal.

antiaromatic *See* pseudoaromatic.

antibiotics Substances obtained from microorganisms, especially moulds, that destroy or inhibit the growth of other microorganisms, particularly disease-producing bacteria and fungi. Common antibiotics include *penicillin, streptomycin, and tetracyclines. They are used to treat various infections but tend to weaken the body's natural defence mechanisms and can cause allergies. Overuse of antibiotics can lead to the development of resistant strains of microorganisms.

antibody A protein (*see* immunoglobulin) produced by certain white blood cells (*lymphocytes) in response to entry into the body of a foreign substance (*antigen) in order to render it harmless. An antibody–antigen reaction is highly specific. Antibody production is one aspect of the *immune response and is stimulated by such antigens as invading bacteria, foreign red blood cells (*see* ABO system), inhaled pollen grains or dust, and foreign tissue *grafts. *See also* immunity; monoclonal antibody.

antibonding orbital *See* orbital.

anticline *See* fold.

anticoagulant A substance that prevents the formation of blood clots. *Heparin is a natural anticoagulant, which is extracted to treat such conditions as thrombosis and embolism. Synthetic anticoagulants include *warfarin.

anticyclone (high) An area of the atmosphere that is at a higher pressure than the surrounding air. It appears on synoptic charts with a roughly circular closed *isobar at its centre. Winds, which are generally light, circulate in a clockwise direction around the high-pressure centre in the northern hemisphere and in an anticlockwise direction in the southern hemisphere. An anticyclone is characteristically slow moving and is associated with settled weather.

antidiuretic hormone (ADH) *See* vasopressin.

antiferromagnetism *See* magnetism.

antigen Any substance that the body regards as foreign and that therefore elicits an *immune response. Antigens may be formed in, or introduced into, the body. They are usually proteins. *Histocompatibility antigens* are associated with the tissues and are involved in the rejection of tissue or organ *grafts (*see* histocompatibility); an example is the group of antigens encoded by the *HLA system. A graft will be rejected if the recipient's body regards such antigens on the donor's tissues as foreign. *See also* antibody.

antigorite *See* serpentine.

antihistamine Any drug that inhibits the effects of *histamine in the body and is therefore used to relieve and prevent the symptoms associated with allergic reactions, such as hay fever. Since one of the side-effects produced by antihistamines is sleepiness, some are used to prevent motion sickness and induce sleep.

***anti*-isomer** *See* isomerism.

antiknock agent A petrol additive that inhibits preignition ('knocking') in internal-combustion engines. Antiknock agents work by retarding combustion chain reactions. The commonest, lead(IV) tetraethyl, causes environmental pollution, and its use is being discouraged.

antilogarithm *See* logarithm.

antimatter *See* antiparticle.

antimonic compounds Compounds of antimony in its +5 oxidation state; e.g. antimonic chloride is antimony(V) chloride ($SbCl_5$).

antimonous compounds Compounds of antimony in its +3 oxidation state; e.g. antimonous chloride is antimony(III) chloride ($SbCl_3$).

antimony Symbol Sb. An element belonging to *group VA of the periodic table; a.n. 51; r.a.m. 121.75; r.d. 6.73; m.p. 630.5°C; b.p. 1380°C. Antimony has several allotropes. The stable form is a bluish-white metal. Yellow antimony and black antimony are unstable nonmetallic allotropes made at low temperatures. The main source is stibnite (Sb_2S_3), from which antimony is extracted by reduction with iron metal or by roasting (to give the oxide) followed by reduction with carbon and sodium carbonate. The main use of the metal is as an alloying agent in lead-accumulator plates, type metals, bearing alloys, solders, Britannia metal, and pewter. It is also an agent for producing pearlitic cast iron. Its compounds are used in flame-proofing, paints, ceramics, enamels, glass dyestuffs, and rubber technology. The element will burn in air but is unaffected by water or dilute acids. It is attacked by oxidizing acids and by halogens. It was first reported by Tholden in 1450.

antinode *See* stationary wave.

antiparallel vectors Vectors directed along the same line but in opposite directions.

antiparticle A subatomic particle that has the same mass as another particle and equal but opposite values of some other property or properties. For example, the antiparticle of the electron is the positron, which has a positive charge equal in magnitude to the electron's negative charge. The antiproton has a negative charge equal to the proton's positive charge. The neutron and the antineutron have *magnetic moments opposite in sign relative to their *spins. The existence of antiparticles is predicted by relativistic *quantum mechanics. When a particle and its corresponding antiparticle collide *annihilation takes place. *Antimatter* is postulated to consist of matter made up of antiparticles. For example, antihydrogen would consist of an antiproton with an orbiting positron. It appears that the universe consists overwhelmingly of (normal) matter, and explanations of the absence of large amounts of antimatter have been incorporated into cosmological models that involve the use of *grand unified theories of elementary particles.

antiseptic Any substance that kills or inhibits the growth of disease-causing microorganisms but is essentially nontoxic to cells of the body. Common antiseptics include hydrogen peroxide, the detergent cetrimide, and ethanol. They are used to treat minor wounds. *Compare* disinfectant.

anus The terminal opening of the *alimentary canal in most animals, through which indigestible material (*faeces) is expelled.

anyon *See* quantum statistics.

aorta The major blood vessel in higher vertebrates through which oxygenated blood leaves the *heart from the left ventricle. The aorta branches to form many smaller arteries, which in turn branch many times to supply oxygen and essential nutrients to all living cells in the body.

apatite A highly complex mineral form of *calcium phosphate, $Ca_5(PO_4)_3(OH,F,Cl)$; the commonest of the phosphate minerals. It has a hexagonal structure and occurs widely as an accessory mineral in igneous rocks (e.g. pegmatite) and often in regional and contact metamorphic rocks, especially limestone. Large deposits occur in the Kola Peninsula, USSR. It is used in the production of fertilizers and is a major source of phosphorus. The enamel of teeth is composed chiefly of apatite.

aperture The effective diameter of a lens or mirror. The ratio of the effective diameter to the focal length is called the *relative aperture*, which is commonly known as the aperture, especially in photographic usage. The reciprocal of the relative aperture is called the *focal ratio*. The numerical value of the focal ratio is known as the *f-number* of a lens. For example, a camera lens with a 40 mm focal length and a 10 mm aperture has a relative aperture of 0.25 and a focal ratio of 4. Its f-number would be f/4, often written f4.
The light-gathering power of a telescope depends on the area of the lens, i.e. it is related to the square of the aperture. However, the larger the relative aperture the greater the *aberrations. In microscopy large-aperture objectives (corrected for aberrations) are preferred, since they reduce the blurring caused by *diffraction of light waves.

aperture synthesis See radio telescope.

aphelion The point in the orbit of a planet, comet, or artificial satellite in solar orbit at which it is farthest from the sun. The earth is at aphelion on about July 3. *Compare* perihelion.

apical dominance Inhibition of the growth of lateral buds in a plant by the presence of a growing apical bud. It is brought about by the action of auxins (produced by the apical bud) and abscisic acid.

apical meristem A region at the tip of each shoot and root of a plant in which cell divisions are continually occurring to produce new stem and root tissue, respectively. The new tissues produced are known collectively as the *primary tissues* of the plant. *See also* meristem. *Compare* cambium.

aplanatic lens A lens that reduces both spherical *aberration and *coma.

apocarpy The condition in which the female reproductive organs (*carpels) of a flower are not joined to each other. It occurs, for example, in the buttercup. *Compare* syncarpy.

apochromatic lens See achromatic lens.

apocrine secretion See secretion.

apocynthion The point in the orbit around the moon of a satellite launched from the earth that is furthest from the moon. For a satellite launched from the moon the equivalent point is the *apolune*. *Compare* pericynthion.

apoenzyme An inactive enzyme that must associate with a specific *cofactor molecule or ion in order to function. *Compare* holoenzyme.

apogee The point in the orbit of the moon, or an artificial earth satellite, at which it is furthest from the earth. At apogee the moon is 406 700 km from the earth, some 42 000 km further away than at *perigee*, the nearest point to the earth.

apolune See apocynthion.

apomixis (agamospermy) A reproductive process in plants that superficially resembles normal sexual reproduction but in which there is no fusion of gametes. In apomictic flowering plants there is no fertilization by pollen, and the embryos develop simply by division of a *diploid cell of the ovule. *See also* parthenocarpy; parthenogenesis.

aposematic coloration *See* warning coloration.

apparent expansivity *See* expansivity.

appeasement Behaviour that inhibits aggression from another animal of the same species, frequently taking the form of a special posture or *display emphasizing the weakness of the performer. Threatening structures (e.g. antlers) and markings are covered or turned away, and vulnerable parts of the body may be exposed. Appeasement is seen in *courtship, in greeting ceremonies, and often (from the loser) after a fight.

appendix (vermiform appendix) An outgrowth of the *caecum in the alimentary canal. In humans it is a *vestigial organ containing lymphatic tissue and serves no function in normal digestive processes. Appendicitis is caused by inflammation of the appendix.

Appleton layer *See* earth's atmosphere.

aprotic *See* solvent.

apsides The two points in an astronomical orbit that lie closest to (*periapsis*) and farthest from (*apoapsis*) the centre of gravitational attraction. The *line of apsides* is the straight line that joins the two apsides. If the orbit is elliptical the line of apsides is the major axis of the ellipse.

aqua regia A mixture of concentrated nitric acid and concentrated hydrochloric acid in the ratio 1:3 respectively. It is a very powerful oxidizing mixture and will dissolve all metals (except silver, which forms an insoluble chloride) including such noble metals as gold and platinum, hence its name ('royal water'). Nitrosyl chloride (NOCl) is believed to be one of the active constituents.

aqueous Describing a solution in water.

aqueous humour The fluid that fills the space between the cornea and the lens of the vertebrate eye. In addition to supplying the cornea and lens with nutrients, the aqueous humour helps to maintain the shape of the eye. It is produced and renewed every four hours by the *ciliary body.

aquifer A deposit of rock that yields economic supplies of water to wells or springs as a result of its porosity or permeability. It may be, for example, a zone of sandstone, unconsolidated gravels, or jointed limestone.

Arachnida A class of terrestrial arthropods comprising about 65 000 species and including spiders, scorpions, harvestmen, ticks, and mites. An arachnid's body is divided into an anterior *cephalothorax (*prosoma*) and a posterior abdomen (*opisthosoma*). The prosoma bears a pair of grasping or piercing appendages (the *chelicerae*), a pair of *pedipalps* used for manipulation or as sensory structures, and four pairs of walking legs. The opisthosoma may bear various sensory or silk-spinning appendages (*see* spinneret). Arachnids are generally carnivorous, feeding on the body fluids of their prey or secreting enzymes to digest prey externally. Spiders immobilize their prey with poison injected by the fanglike chelicerae, while scorpions grasp their prey in large clawed pedipalps and may poison it using the posterior stinging organ. Ticks and some mites are parasitic but most arachnids are free-living. They breathe either via *tracheae (like insects) or by means of thin highly folded regions of the body wall called *lung-books*.

arachnoid membrane One of the three membranes (*meninges) that surround the brain and spinal cord of vertebrates. It lies between the *pia mater and the *dura mater. The arachnoid membrane is very delicate and carries *cerebrospinal fluid, which sustains and cushions the nervous tissue.

aragonite A rock-forming anhydrous mineral form of calcium carbonate, $CaCO_3$. It is much less stable than *calcite, the commoner form of calcium carbonate, from which it may be distinguished by its greater hardness and specific gravity. Over time aragonite undergoes recrystallization to calcite. Aragonite occurs in cavities in limestone, as a deposit in limestone caverns, as a precipitate around hot springs and geysers, and in high-pressure low-temperature metamorphic rocks; it is also found in the shells of a number of molluscs and corals and is the main constituent of pearls. It is white or colourless when pure but the presence of impurities may tint it grey, blue, green, or pink.

arbovirus Obsolete name for any RNA-containing virus that is transmitted from animals to man through the bite of mosquitoes and ticks (i.e. arthropods, hence *ar*thropod-*bo*rne viruses). They cause various forms of encephalitis (inflammation of the brain) and serious fevers, such as dengue and yellow fever.

arccos, arcsin, arctan *See* inverse functions.

archegonium The multicellular flask-shaped female sex organ of bryophytes, pteridophytes, and many gymnosperms. Such plants are described as *archegoniate* to distinguish them from algae, which do not possess archegonia. The dilated base, the *venter*, contains the oosphere (female gamete). The cells of the narrow neck liquefy to allow the male gametes to swim towards the oosphere. The archegonium is thus an adaptation to the terrestrial environment as it provides a means for the male gametes to reach the female gamete. *Compare* antheridium.

archenteron (gastrocoel) A cavity within an animal embryo at the *gastrula stage of development. All or part of the archenteron eventually forms the cavity of the gut. It is connected to the outside by an opening (the *blastopore*), which becomes either the mouth, the mouth and anus, or the anal opening of the animal.

Archimedes' principle The weight of the liquid displaced by a floating body is equal to the weight of the body. The principle was not in fact stated by the Greek mathematician Archimedes (287–212 BC), though it has some connection with his discoveries. The principle is often stated in the form: when a body is (partially or totally) immersed in a fluid, the upthrust on the body is equal to the weight of fluid displaced.

arc lamp *See* electric lighting.

arcosh, arsinh, artanh *See* inverse functions.

arenes Aromatic hydrocarbons, such as benzene, toluene, and naphthalene.

Argand diagram *See* complex number.

argentic compounds Compounds of silver in its higher ($+2$) oxidation state; e.g. argentic oxide is silver(II) oxide (AgO).

argentite A sulphide ore of silver, Ag_2S. It crystallizes in the cubic system but most commonly occurs in massive form. It is dull grey-black in colour but bright when first cut and occurs in veins associated with other silver minerals. Important deposits occur in Mexico, Peru, Chile, Bolivia, Norway, and Czechoslovakia.

argentous compounds Compounds of silver in its lower ($+1$) oxidation state; e.g. argentous chloride is silver(I) chloride.

arginine *See* amino acid.

argon Symbol Ar. A monatomic noble gas present in air (0.93%); a.n. 18; r.a.m. 39.948; d. $0.00178\,g\,cm^{-3}$; m.p. $-189°C$; b.p. $-185°C$. Argon is separated from liquid air by fractional distillation. It is slightly soluble in water, colourless, and has no smell. Its uses include inert atmospheres in welding and special-metal manufacture (Ti and Zr), and (when mixed with 20% nitrogen) in gas-filled electric-light bulbs. The element is inert and has no true compounds. Lord Rayleigh and Sir William Ramsey identified argon in 1894.

argument 1. A sequence of logical propositions based on a set of premisses and leading to a conclusion. **2.** *See* complex number.

aril An outgrowth that grows around and may completely enclose the testa (seed coat) of a seed. It develops from the placenta, funicle, or micropyle of an ovule. The aril surrounding the nutmeg seed forms the spice mace. *See also* caruncle.

arithmetic average (arithmetic mean) *See* average.

arithmetic series (arithmetic progression) A series or progression of numbers in which there is a common difference between terms, e.g. 3, 9, 15, 21, ... is an arithmetic series with a common difference of 6. The general formula for the nth term is

$$[a + (n - 1)d]$$

and the sum of n terms is

$$n[2a + (n - 1)d]/2.$$

Compare geometric series.

armature Any moving part in an electrical machine in which a voltage is induced by a magnetic field, especially the rotating coils in an electric motor or generator and the ferromagnetic bar attracted by an electromagnet in a *relay.

aromatic compound An organic compound that contains a benzene ring in its molecules or that has chemical properties similar to benzene. Aromatic compounds are unsaturated compounds, yet they do not easily partake in addition reactions. Instead they undergo electrophilic substitution.

Benzene, the archetypal aromatic compound, has an hexagonal ring of carbon atoms and the classical formula (the Kekulé structure) would have alternating double and single bonds. In fact all the bonds in benzene are the same length intermediate between double and single $C-C$ bonds. The properties arise because the electrons in the π-orbitals are delocalized over the ring, giving an extra stabilization energy of 150 kJ mol^{-1} over the energy of a Kekulé structure. The condition for such delocalization is that a compound should have a planar ring with $(4n + 2)$ pi electrons – this is known as the *Hückel rule*. Aromatic behaviour is also found in heterocyclic compounds such as pyridine. *See also* nonbenzenoid aromatic; pseudoaromatic.

aromaticity The property characteristic of *aromatic compounds.

arousal A level of physiological and behavioural responsiveness in an animal, which tends to vary between sleep and full alertness. It is controlled by a particular part of the brain (the *reticular activating system*) and can be detected by changes in brain electrical activity, heart rate and muscle tone, responsiveness to new stimuli, and general activity.

Arrhenius equation An equation of the form

$$k = A\exp(-E_a/RT)$$

where k is the rate constant of a given reaction and E_a the *activation energy. A is a constant for a given reaction, called the *pre-exponential factor*. Often the equation is written in logarithmic form

$$\ln k = \ln A - E_a/RT$$

A graph of $\ln k$ against $1/T$ is a straight line with a gradient $-E_a/R$ and an intercept on the $\ln k$ axis of $\ln A$.

Arrhenius theory *See* acid.

arsenate(III) *See* arsenic(III) oxide.

arsenate(V) *See* arsenic(V) oxide.

arsenic Symbol As. A metalloid element of *group V of the periodic table; a.n. 33; r.a.m. 74.92; r.d. 5.7; sublimes at 613°C. It has three allotropes – yellow, black, and grey. The grey metallic form is the stable and most common one. Over 150 minerals contain arsenic but the main sources are as impurities in sulphide ores and in the minerals orpiment (As_2S_3) and realgar (As_4S_4). Ores are roasted in air to form arsenic oxide and then reduced by hydrogen or carbon to metallic arsenic. Arsenic compounds are used in insecticides and as doping agents in semiconductors. The element is included in some lead-based alloys to promote hardening. Confusion can arise because As_4O_6 is often sold as white arsenic. Arsenic compounds are accumulative poisons. The element will react with halogens, concentrated oxidizing acids, and hot alkalis. Albertus Magnus is believed to have been the first to isolate the element in 1250.

arsenic acid *See* arsenic(V) oxide.

arsenic(III) acid *See* arsenic(III) oxide.

arsenic hydride *See* arsine.

arsenic(III) oxide (arsenic trioxide; arsenious oxide; white arsenic) A white or colourless compound, As_4O_6, existing in three solid forms. The commonest has cubic or octahedral crystals (r.d. 3.85; sublimes at 193°C) and is soluble in water, ethanol, and alkali solutions. It occurs naturally as *arsenolite*. A vitreous form can be prepared by slow condensation of the vapour (r.d. 3.74); its solubility in cold water is more than double that of the cubic form. The third modification, which occurs naturally as *claudetite*, has monoclinic crystals (r.d. 4.15). Arsenic(III) oxide is obtained commercially as a byproduct from the smelting of non-ferrous sulphide ores; it may be produced in the laboratory by burning elemental arsenic in air. The structure of the molecule is similar to that of P_4O_6, with a tetrahedral arrangement of As atoms edge linked by oxygen bridges. Arsenic(III) oxide is acidic; its solutions were formerly called *arsenious acid* (technically, *arsenate(III) acid*). It forms *arsenate(III)* salts (formerly called *arsenites*). Arsenic(III) oxide is extremely toxic and is used as a poison for vermin; trace doses are used for a variety of medicinal purposes. It is also used for producing opalescent glasses and enamels.

arsenic(V) oxide (arsenic oxide) A white amorphous deliquescent solid, As_2O_5; r.d. 4.32; decomposes at 315°C. It is soluble in water and ethanol. Arsenic(V) oxide cannot be obtained by direct combination of arsenic and oxygen; it is usually prepared by the reaction of arsenic with nitric acid followed by dehydration of the arsenic acid thus formed. It readily loses oxygen on heating to give arsenic(III) oxide. Arsenic(V) oxide is acidic, dissolving in water to give arsenic(V) acid (formerly called *arsenic acid*), H_3AsO_4; the acid is tribasic and slightly weaker than phosphoric acid and should be visualized as $(HO)_3AsO$. It gives *arsenate(V)* salts (formerly called *arsenates*).

arsenic trioxide *See* arsenic(III) oxide.

arsenious acid *See* arsenic(III) oxide.

arsenious oxide *See* arsenic(III) oxide.

arsenite *See* arsenic(III) oxide.

arsenolite A mineral form of *arsenic(III) oxide, As_4O_6.

arsine (arsenic hydride) A colourless gas, AsH_3; m.p. $-116.3°C$; b.p. $-55°C$. It is soluble in water, chloroform, and benzene. Liquid arsine has a relative density of 1.69. Arsine is produced by the reaction of mineral acids with arsenides of electropositive metals or by the reduction of many

Artesian basin

arsenic compounds using nascent hydrogen. It is extremely poisonous and, like the hydrides of the heavier members of group V, is readily decomposed at elevated temperatures (around 260–300°C). Like ammonia and phosphine, arsine has a pyramidal structure.

Arsine gas has a very important commercial application in the production of modern microelectronic components. It is used in a dilute gas mixture with an inert gas and its ready thermal decomposition is exploited to enable other growing crystals to be doped with minute traces of arsenic to give n-type semiconductors.

arteriole A small muscular blood vessel that receives blood from the arteries and carries it to the capillaries.

artery A blood vessel that carries blood away from the heart towards the other body tissues. Most arteries carry oxygenated blood (the *pulmonary artery is an exception). The large arteries branch to form smaller ones, which in turn branch into *arterioles. All arteries have muscular walls, whose contraction aids in pumping blood around the body. The accumulation of fatty deposits in the walls of the arteries leads to *atherosclerosis*, which limits and may eventually block the flow of blood. *Compare* vein.

artesian basin A structural basin in the earth's crust in which a zone of water-bearing rock (an *aquifer) is

confined between impermeable beds. Water enters the aquifer where it reaches the surface and becomes trapped. As the point of intake of the water is above the level of the ground surface a well sunk into the aquifer through the overlying impermeable rock will result in water being forced up to the surface by hydrostatic pressure. However, the number of wells sunk into artesian basins often considerably lowers the level of water so that pumping is frequently required. Artesian wells were named after the province of Artois in France, where they were first observed. Other examples include the London Basin where chalk and Lower Eocene sandstones lie confined below the London clay and above the Gault clay.

Arthropoda A phylum of invertebrate animals comprising over one million species – the largest in the animal kingdom. Arthropods inhabit marine, freshwater, and terrestrial habitats worldwide. Characteristically, they possess an outer body layer – the *cuticle – that functions as a rigid protective exoskeleton; growth is thus possible only by periodic moults (*see* ecdysis). The arthropod body is composed of segments (*see* metameric segmentation) usually forming distinct specialized body regions, e.g. head, thorax, and abdomen. These segments may possess hardened jointed appendages, modified variously as *mouthparts, limbs, wings, reproductive organs, or sense organs. The main body cavity, containing the internal organs, is a continuous blood-filled *haemocoel*, within which lies the heart. The origins and relationships of the various groups of arthropods remain uncertain. The major classes are: *Arachnida (spiders, scorpions, mites, and ticks), *Crustacea (shrimps, barnacles, crabs, etc.), *Insecta (insects), *Myriapoda (centipedes and millipedes), and the wormlike Onychophora (e.g. *Peripatus*).

articulation The attachment of two bones, usually by means of a *joint. The thigh bone (femur), for instance, articulates with the pelvic girdle.

artificial insemination The deposition of semen, using a syringe, at the mouth of the uterus to make conception possible. It is used for selected breeding of domestic animals and in humans in some cases of impotence and infertility. It is timed to coincide with ovulation in the female.

artinite A mineral form of basic *magnesium carbonate,
$$MgCO_3.Mg(OH)_2.3H_2O.$$

Artiodactyla An order of hooved mammals comprising the even-toed ungulates, in which the third and fourth digits are equally developed and bear the weight of the body. The order includes cattle and other ruminants (*see* Ruminantia), camels, hippopotamuses, and pigs. All except the latter are herbivorous, having an elongated gut and teeth with enamel ridges for grinding tough grasses. *Compare* Perissodactyla.

aryl group A group obtained by removing a hydrogen atom from an aromatic compound, e.g. phenyl group, C_6H_5-, derived from benzene.

asbestos Any one of a group of fibrous amphibole minerals (amosite, crocidolite (blue asbestos), tremolite, anthophyllite, and actinolite) or the fibrous serpentine mineral chrysotile. Asbestos has widespread commercial uses because of its resistance to heat, chemical inertness, and high electrical resistance. The fibres may be spun and woven into fireproof cloth for use in protective clothing, curtains, brake linings, etc., or moulded into blocks. Exposure to large amounts of asbestos can cause the respiratory disease asbestosis. Canada and the USSR are the largest producers of asbestos;

others include South Africa, Zimbabwe, and China.

ascomycetes A group of fungi classified as a class (Ascomycetes) or a subdivision (Ascomycotina); they include many yeasts. Sexual reproduction is by means of *ascospores*, eight of which are characteristically produced within a spherical or cylindrical cell, the *ascus*. The asci are usually grouped together in an *ascocarp*.

ascorbic acid *See* vitamin C.

asexual reproduction Reproduction in which new individuals are produced from a single parent without the formation of gametes. It occurs chiefly in lower animals, microorganisms, and plants. In lower animals the chief methods are *fission (e.g. in protozoans), *fragmentation (e.g. in some aquatic annelid worms), and *budding (e.g. in coelenterates). The principal methods of asexual reproduction in plants are by *vegetative propagation (e.g. bulbs, corms, tubers) and by the formation of sexual *spores. Spore formation occurs in mosses, ferns, and other plants showing alternation of generations, as a dormant stage between sporophyte and gametophyte, and in some algae and fungi, to produce replicas of the plant. *Compare* sexual reproduction.

asparagine *See* amino acid.

aspartic acid *See* amino acid.

aspirin (acetylsalicylic acid) A drug that reduces inflammation, combats fever, and alleviates pain. Aspirin works by inhibiting the formation of *prostaglandins, which are major factors in the inflammation process. It also reduces the aggregation of blood platelets, hence its use in maintaining blood flow following heart and circulatory disorders.

assimilation The utilization by a living organism of absorbed food materials in the processes of growth, reproduction, or repair.

association 1. (in ecology) An ecological unit in which two or more species occur in closer proximity to one another than would be expected on the basis of chance. Early plant ecologists recognized associations of fixed composition on the basis of the *dominant species present (e.g. a coniferous forest association). Associations now tend to be detected by using more objective statistical sampling methods. *See also* consociation. **2.** (in chemistry) The combination of molecules of one substance with those of another to form chemical species that are held together by forces weaker than normal chemical bonds. For example, ethanol and water form a mixture (an *associated liquid*) in which hydrogen bonding holds the different molecules together.

associative law The mathematical law stating that the value of an expression is independent of the grouping of the numbers, symbols, or terms in the expression. The *associative law for addition* states that numbers may be added in any order, e.g. $(x + y) + z = (x + z) + y$. The *associative law for multiplication* states that numbers can be multiplied in any order, e.g. $x(yz) = (xy)z$. Subtraction and division are not associative. *Compare* commutative law; distributive law.

astatic galvanometer A sensitive form of moving-magnet *galvanometer in which any effects of the earth's magnetic field are cancelled out. Two small oppositely directed magnets are suspended at the centres of two oppositely wound coils. As its resultant moment on the magnets is zero, the earth's field has no effect and the only restoring torque on the magnets is that provided by the suspending fibre. This makes a sensitive but delicate instrument.

astatine Symbol At. A radioactive *halogen element; a.n. 85; r.a.m. 211; m.p. 302°C; b.p. 377°C. It occurs naturally by radioactive decay from uranium and thorium isotopes. Astatine forms at least 20 isotopes, the most stable astatine−210 has a half-life of 8.3 hours. It can also be produced by alpha bombardment of bismuth−200. Astatine is stated to be more metallic than iodine; at least 5 oxidation states are known in aqueous solutions. It will form interhalogen compounds, such as AtI and AtCl. The existence of At_2 has not yet been established. The element was synthesized by nuclear bombardment in 1940 by D. R. Corson, K. R. MacKenzie, and E. Segré at the University of California.

aster A starlike arrangement of small fibres radiating from a *centriole. Asters become conspicuous in animal cells at the ends of the spindle when cell division starts.

asteroids (minor planets; planetoids) A number of small bodies that revolve around the sun between the orbits of Mars and Jupiter (in the *asteroid belt*). The size of the bodies varies from the largest, Ceres (with a diameter of about 1000 km), to objects less than 1 km in diameter. It is estimated that there are about 10 bodies with diameters in excess of 300 km and some 200 bodies with diameters over 100 km.

asthenosphere A layer of the earth's mantle (*see* earth) that underlies the lithosphere at a depth of about 70 km. The velocity of seismic waves is considerably reduced in the asthenosphere and it is thought to be a zone of partial melting. It extends to a depth of about 250 km where rocks again become solid.

astigmatism A lens defect in which when rays in one plane are in focus those in another plane are not. In lenses and mirrors it occurs with objects not on the axis and is best controlled by reducing the *aperture to restrict the use of the lens or mirror to its central portion. The eye can also suffer from astigmatism, usually when the cornea is not spherical. It is corrected by using an *anastigmatic lens.

astrometry The branch of astronomy concerned with the measurement of the positions of the celestial bodies on the *celestial sphere.

astronomical telescope *See* telescope.

astronomical unit (AU) The mean distance between the sun and the earth. It is equal to 149 597 870 km (499 light seconds).

astronomy The study of the universe beyond the earth's atmosphere. The main branches are *astrometry, *celestial mechanics, and *astrophysics.

astrophysics The study of the physical and chemical processes involving astronomical phenomena. Astrophysics deals with stellar structure and evolution (including the generation and transport of energy within stars), the properties of the interstellar medium and its interactions with stellar systems, and the structure and dynamics of systems of stars (such as *clusters and *galaxies), and of systems of galaxies. *See also* cosmology.

asymmetric atom *See* optical activity.

asymptote A line that a curve approaches but only touches at infinity.

asymptotic freedom The consequence of certain *gauge theories, particularly *quantum chromodynamics, that the forces between such particles as quarks become weaker at shorter distances (i.e. higher energies) and vanish as the distance between particles tends to zero. Only non-Abelian gauge theories with unbroken gauge symmetries can have asymptotic freedom (*see* group). In contrast, *quantum electrodynamics implies that the interaction between particles decreases as a result of dielectric

screening; asymptotic freedom for quarks implies that antiscreening occurs. Physically, asymptotic freedom postulates that the *vacuum state for gluons is a medium that has colour paramagnetism, i.e. the vacuum antiscreen colour charges. Asymptotic freedom explains the successes of the *parton model of point-like objects inside hadrons and enables systematic corrections to the parton model to be calculated using perturbation theory. That the interaction between quarks increases as the distance between them increases has given rise to the hypothesis of *quark confinement. It appears that if a theory requires the presence of Higgs bosons, asymptotic freedom is destroyed. Thus, *electroweak theory does not have asymptotic freedom.

asymptotic series A series formed by the expansion of a function in the form $a_0 + a_1/x + a_2/x^2 + \ldots + a_n/x^n + \ldots$, such that the error resulting from terminating the series at the term a_n/x^n tends to zero more rapidly than $1/x^n$ as x tends to infinity. An asymptotic series expansion is not necessarily a *convergent series.

atactic polymer *See* polymer.

atlas The first *cervical vertebra, a ringlike bone that joins the skull to the vertebral column in terrestrial vertebrates. In advanced vertebrates articulation between the skull and atlas permits nodding movements of the head. *See also* axis.

atm. *See* atmosphere.

atmolysis The separation of a mixture of gases by means of their different rates of diffusion. Usually, separation is effected by allowing the gases to diffuse through the walls of a porous partition or membrane.

atmosphere 1. (atm.) A unit of pressure equal to 101 325 pascals. This is equal to 760.0 mmHg. The actual *atmospheric pressure fluctuates around this value. The unit is usually used for expressing pressures well in excess of standard atmospheric pressure, e.g. in high-pressure chemical processes. **2.** *See* earth's atmosphere.

atmospheric pressure The pressure exerted by the weight of the air above it at any point on the earth's surface. At sea level the atmosphere will support a column of mercury about 760 mm high. This decreases with increasing altitude. The standard value for the atmospheric pressure at sea level in SI units is 101 325 pascals.

atom The smallest part of an element that can exist. Atoms consist of a small dense nucleus of protons and neutrons surrounded by moving electrons. The number of electrons equals the number of protons so the overall charge is zero. The electrons may be thought of as moving in circular or elliptical orbits (*see* Bohr theory) or, more accurately, in regions of space around the nucleus (*see* orbital).

The *electronic structure* of an atom refers to the way in which the electrons are arranged about the nucleus, and in particular the *energy levels that they occupy. Each electron can be characterized by a set of four quantum numbers, as follows:

(1) The *principal quantum number n* gives the main energy level and has values 1, 2, 3, etc. (the higher the number, the further the electron from the nucleus). Traditionally, these levels, or the orbits corresponding to them, are referred to as *shells* and given letters K, L, M, etc. The K-shell is the one nearest the nucleus. The maximum number of electrons in a given shell is $2n^2$.

(2) The *orbital quantum number l*, which governs the angular momentum of the electron. The possible values of l are $(n - 1)$, $(n - 2)$, ..., 1, 0. Thus, in the first shell ($n = 1$) the electrons can only have angular momentum zero ($l = 0$). In the second shell ($n = 2$), the values of l

can be 1 or 0, giving rise to two *subshells* of slightly different energy. In the third shell ($n = 3$) there are three subshells, with $l = 2$, 1, or 0. The subshells are denoted by letters s ($l = 0$), p ($l = 1$), d ($l = 2$), f ($l = 3$). The number of electrons in each subshell is written as a superscript numeral to the subshell symbol, and the maximum number of electrons in each subshell is s^2, p^6, d^{10}, and f^{14}. The orbital quantum number is sometimes called the *azimuthal quantum number*.

(3) The *magnetic quantum number m*, which governs the energies of electrons in an external magnetic field. This can take values of $+l$, $+(l - 1)$, ..., 1, 0, -1, ..., $-(l - 1)$, $-l$. In an s-subshell (i.e. $l = 0$) the value of $m = 0$. In a p-subshell ($l = 1$), m can have values $+1$, 0, and -1; i.e. there are three p-orbitals in the p-subshell, usually designated p_x, p_y, and p_z. Under normal circumstances, these all have the same energy level.

(4) The *spin quantum number m_s*, which gives the spin of the individual electrons and can have the values $+\frac{1}{2}$ or $-\frac{1}{2}$.

According to the *Pauli exclusion principle, no two electrons in the atom can have the same set of quantum numbers. The numbers define the *quantum state* of the electron, and explain how the electronic structures of atoms occur.

atomic absorption spectroscopy (AAS) An analytical technique in which a sample is vaporized and the nonexcited atoms absorb electromagnetic radiation at characteristic wavelengths.

atomic bomb *See* nuclear weapons.

atomic clock An apparatus for measuring or standardizing time that is based on periodic phenomena within atoms or molecules. *See* ammonia clock; caesium clock.

atomic emission spectroscopy An analytical technique in which a sample is vaporized and the atoms present are detected by their emission of electromagnetic radiation at characteristic wavelengths.

atomic energy *See* nuclear energy.

atomic force microscope (AFM) A type of microscope in which a small probe, consisting of a tiny chip of diamond, is held on a spring-loaded cantilever in contact with the surface of the sample. The probe is moved slowly across the surface and the tracking force between the tip and the surface is monitored. The probe is raised and lowered so as to keep this force constant, and a profile of the surface is produced. Scanning the probe over the sample gives a computer-generated contour map of the surface. The instrument is similar to the *scanning tunnelling microscope, but uses mechanical forces rather than electrical signals. It can resolve individual molecules and, unlike the scanning tunnelling microscope, can be used with nonconducting samples, such as biological specimens.

atomicity The number of atoms in a given molecule. For example, oxygen (O_2) has an atomicity of 2, ozone (O_3) an atomicity of 3, benzene (C_6H_6) an atomicity of 12, etc.

atomic mass unit (a.m.u.) A unit of mass used to express *relative atomic masses. It is 1/12 of the mass of an atom of the isotope carbon–12 and is equal to $1.660\,33 \times 10^{-27}$ kg. This unit superseded both the physical and chemical mass units based on oxygen –16 and is sometimes called the *unified mass unit* or the *dalton*.

atomic number (proton number) Symbol Z. The number of protons in the nucleus of an atom. The atomic number is equal to the number of electrons orbiting the nucleus in a neutral atom.

atomic orbital *See* orbital.

atomic pile An early form of *nuclear reactor using graphite as a *moderator.

atomic volume The relative atomic mass of an element divided by its density.

atomic weight *See* relative atomic mass.

ATP (adenosine triphosphate) A nucleotide that is of fundamental importance as a carrier of chemical energy in all living organisms. It consists of adenine linked to D-ribose (i.e. adenosine); the D-ribose component bears three phosphate groups, linearly linked together by covalent bonds (see formula). These bonds can undergo hydrolysis to yield either a

ATP

53

atrium

molecule of *ADP* (*adenosine diphosphate*) and inorganic phosphate or a molecule of *AMP* (*adenosine monophosphate*) and pyrophosphate. Both these reactions yield a large amount of energy (about 30.6 kJ mol^{-1}) that is used to bring about such biological processes as muscle contraction, the active transport of ions and molecules across cell membranes, and the synthesis of biomolecules. The reactions bringing about these processes often involve the enzyme-catalysed transfer of the phosphate group to intermediate substrates. Most ATP-mediated reactions require Mg^{2+} ions as *cofactors.

ATP is regenerated by the rephosphorylation of AMP and ADP using the chemical energy obtained from the oxidation of food. This takes place during *glycolysis and the *Krebs cycle but, most significantly, is also a result of the reduction–oxidation reactions of the *electron transport chain, which ultimately reduces molecular oxygen to water (oxidative phosphorylation).

atrium 1. (*or* **auricle**) A chamber of the *heart that receives blood from the veins and forces it by powerful muscular contraction into the *ventricle(s). Fish have a single atrium but all other vertebrates have two. **2.** Any of various cavities or chambers in animals, such as the chamber surrounding the gill slits of *Amphioxus* and other invertebrate chordates.

atrophy The degeneration or withering of an organ or part of the body.

atropine A poisonous crystalline alkaloid, $C_{17}H_{23}NO_3$; m.p. 114–16°C. It can be extracted from deadly nightshade and other solanaceous plants and is used in medicine to treat colic, to reduce secretions, and to dilate the pupil of the eye.

attenuation 1. (in physics) **a.** A loss of intensity suffered by sound, radiation, etc., as it passes through a medium. It may be caused by absorp-

tion or scattering. **b.** The drop in voltage or current experienced by a signal as it passes through a circuit. **2.** (in medicine) A process of reducing the disease-producing ability of a microorganism. It can be achieved by chemical treatment, heating, drying, irradiation, by growing the organism under adverse conditions, or by serial passage through another organism. Attenuated bacteria or viruses are used for some *vaccines. **3.** (in mycology) The conversion by yeasts of carbohydrates to alcohol, as in brewing and wine and spirit production.

atto- Symbol a. A prefix used in the metric system to denote 10^{-18}. For example, 10^{-18} second = 1 attosecond (as).

attractor The set of points in *phase space to which the representative point of a dissipative system (i.e. one with internal friction) tends as the system evolves. The attractor can be: a single point; a closed curve (a *limit cycle*), which describes a system with periodic behaviour; or a *fractal (or *strange attractor*), in which case the system exhibits *chaos.

AU *See* astronomical unit.

audibility The limits of audibility of the human ear are between about 20 hertz (a low rumble) and 20 000 hertz (a shrill whistle). With increased age the upper limit falls considerably.

audiofrequency A frequency that is audible to the human ear. *See* audibility.

audiometer An instrument that generates a sound of known frequency and intensity in order to measure an individual's hearing ability.

Aufbau principle A principle that gives the order in which orbitals are filled in successive elements in the periodic table. The order of filling is 1*s*, 2*s*, 2*p*, 3*s*, 3*p*, 4*s*, 3*d*, 4*p*, 5*s*, 4*d*, 5*p*, 6*s*, 4*f*, 5*d*, 6*p*, 7*s*, 5*f*, 6*d*. *See* atom.

Auger effect The ejection of an electron from an atom without the emission of an X- or gamma-ray photon, as a result of the de-excitation of an excited electron within the atom. This type of transition occurs in the X-ray region of the emission spectrum. The kinetic energy of the ejected electron, called an *Auger electron*, is equal to the energy of the corresponding X-ray photon minus the binding energy of the Auger electron. The effect was discovered by Pierre Auger (1899–) in 1925.

auric compounds Compounds of gold in its higher (+3) oxidation state; e.g. auric chloride is gold(III) chloride ($AuCl_3$).

auricle 1. *See* atrium. **2.** *See* pinna.

aurora The luminous phenomena seen in the night sky in high latitudes, occurring most frequently near the earth's geomagnetic poles. The displays of aurora appear as coloured arcs, rays, bands, streamers, and curtains, usually green or red. The aurora is caused by the interaction of the atoms (mainly atomic oxygen) and molecules in the upper atmosphere (above about 100 km) with charged particles streaming from the sun, attracted to the auroral regions by the earth's magnetic field. The aurora is known as the *aurora borealis* (or northern lights) in the northern hemisphere and as the *aurora australis* (or southern lights) in the southern hemisphere.

aurous compounds Compounds of gold in its lower (+1) oxidation state; e.g. aurous chloride is gold(I) chloride (AuCl).

austenite *See* steel.

Australopithecus A genus of fossil primates that lived 1.3–4 million years ago, coexisting for some of this time with early forms of men (*see Homo*). They walked erect and had teeth resembling those of modern man, but the brain capacity was less than half that of a human. Various finds have been made, chiefly in East and South Africa (hence the name, which means 'southern ape'). *Australopithecus* and related genera are known as *australopithecines*.

autocatalysis *Catalysis in which one of the products of the reaction is a catalyst for the reaction. Reactions in which autocatalysis occurs have a characteristic S-shaped curve for reaction rate against time – the reaction starts slowly and increases as the amount of catalyst builds up, falling off again as the products are used up.

autoclave A strong steel vessel used for carrying out chemical reactions, sterilizations, etc., at high temperature and pressure.

autogamy 1. A type of reproduction that occurs in single isolated individuals of ciliate protozoan animals of the genus *Paramecium*. The nucleus divides into two genetically identical haploid nuclei, which then fuse to form a diploid zygote. The onset of autogamy is associated with changing environmental conditions and may be necessary to maintain cell vitality. **2.** Self-fertilization in plants. *See* fertilization.

autograft *See* graft.

autoimmunity A disorder of the body's defence mechanism in which an *immune response is elicited against its own tissues. Rheumatoid arthritis is basically an autoimmune disease.

autolysis The process of self-destruction of a cell, cell organelle, or tissue. It occurs by the action of enzymes within or released by *lysosomes. *See also* lysis.

autonomic nervous system (ANS) The part of the vertebrate *peripheral nervous system that supplies stimulation via motor nerves to the sm and cardiac muscles (the in muscles) and to the gl body. It is divided i pathetic and the *sympathetic nervous systems, which tend to work antagonistically on the same organs. The activity of the ANS is controlled principally by the *medulla oblongata and *hypothalamus of the brain.

autoradiography An experimental technique in which a radioactive specimen is placed in contact with (or close to) a photographic plate, so as to produce a record of the distribution of radioactivity in the specimen. The film is darkened by the ionizing radiation from radioactive parts of the sample. Autoradiography has a number of applications, being used particularly to study the distribution of particular substances in living tissues and cells. A radioactive isotope of the substance is introduced into the organism or tissue, which is killed, sectioned, and examined after enough time has elapsed for the isotope to be incorporated into the substance.

autosome Any of the chromosomes in a cell other than the *sex chromosomes.

autotomy The shedding by an animal of part of its body followed by the regeneration of the lost part. Autotomy is achieved by the contraction of muscles at specialized regions in the body. It serves as a protective mechanism if the animal is damaged or attacked (e.g. tail loss in certain reptiles) and is common as a method of asexual reproduction in polychaete worms   h both new head and tai'   be regenerated.

vpe of nutrition in
synthesize the
v require from
ef sources of
carbon diox-
tively. All
and use
for the
rophic
teria
are

chemoautotrophic, using energy derived from chemical processes. *Compare* heterotrophism.

auxanometer Any mechanical instrument or measuring device used to study the growth or movement of plant organs. One type of auxanometer consists of a recording device that translates any increase in stem height into movement of a needle across a scale.

auxin Any of a group of plant *growth substances responsible for such processes as the promotion of growth by cell enlargement, the maintenance of *apical dominance, and the initiation of root formation in cuttings. Naturally occurring auxins, such as *indoleacetic acid* (*IAA*), are synthesized in the shoot tips. Synthetic auxins, such as *2,4-D* and *2,4,5-T*, are used as weedkillers.

auxochrome A group in a dye molecule that influences the colour due to the *chromophore. Auxochromes are groups, such as $-OH$ and $-NH_2$, containing lone pairs of electrons that can be delocalized along with the delocalized electrons of the chromophore. The auxochrome intensifies the colour of the dye. Formerly, the term was also used of such groups as $-SO_2O^-$, which make the molecule soluble and affect its application.

avalanche A shower of ionized particles created by a single *ionization as a result of secondary ionizations caused by the original electron and ion being accelerated in an electric field. Each ionization leads to the formation of more electrons and ions, which themselves cause further ionizations. Such avalanches occur in a *Geiger counter.

average (mean) 1. The *arithmetic average* of a set of n numbers is the sum of the numbers divided by n. 2. The *geometric average* of a set of n numbers is the nth root of their product. *See also* root-mean-square value.

Aves The birds: a class of bipedal vertebrates with *feathers, wings, and a beak. They evolved from reptilian ancestors in the Jurassic period (190–136 million years ago) and modern birds still have scaly legs, like reptiles. Birds are warm-blooded (*see* homoiothermy). The skin is dry and loose and has no sweat glands, so cooling is effected by panting. Their efficient lungs and four-chambered heart (which completely separates oxygenated and deoxygenated blood) ensure a good supply of oxygen to the tissues. Birds can therefore sustain a high body temperature and level of activity necessary for flight. The breastbone bears a keel for the attachment of flight muscles. The skeleton is very light; many of the bones are tubular, having internal struts to provide strength and air sacs to reduce weight and provide extra oxygen in flight. Their feathers are vital for flight, streamlining the body, and insulation against heat loss.

Many birds show a high degree of social behaviour in forming large flocks and pair bonding for nesting, egg incubation, and rearing young. Fertilization is internal and the female lays hard-shelled eggs. *See also* Ratitae.

Avogadro constant Symbol N_A or L. The number of atoms or molecules in one *mole of substance. It has the value $6.022\ 1367(36) \times 10^{23}$. Formerly it was called *Avogadro's number*.

Avogadro's law Equal volumes of all gases contain equal numbers of molecules at the same pressure and temperature. The law, often called *Avogadro's hypothesis*, is true only for ideal gases. It was first proposed in 1811 by Count Amadeo Avogadro (1776–1856).

axenic culture A *culture medium in which only one type of microorganism is growing. Such cultures are widely used in microbiology to determine the basic growth requirements

or degree of inhibition by antibiotics or other chemicals of a particular species.

axil The angle between a branch or leaf and the stem it grows from. *Axillary* (or *lateral*) buds develop in the axil of a leaf. The presence of axillary buds distinguishes a leaf from a leaflet.

axion A hypothetical elementary particle postulated to explain why there is no observed CP violation (*see* CP invariance) in the strong interaction (*see* fundamental interactions). Axions have not been detected experimentally, although it has been possible to put limits on their mass and other properties from the effects that they would have on some astrophysical phenomena (e.g. the cooling of stars). It has also been suggested that they may account for some or all of the *missing matter in the universe.

axis 1. (in mathematics) **a.** One of a set of reference lines used to locate points on a graph or in a coordinate system. *See* Cartesian coordinates; polar coordinates. **b.** A line about which a figure, curve, or body is symmetrical (*axis of symmetry*) or about which it rotates (*axis of rotation*). **2.** (in anatomy) The second *cervical vertebra, which articulates with the *atlas (the first cervical vertebra, which articulates with the skull). The articulation between the axis and atlas in reptiles, birds, and mammals permits side-to-side movement of the head. The body of the axis is elongated to form a peg (the *odontoid process*), which extends into the ring of the atlas and acts as a pivot on which the atlas (and skull) can turn.

axon The long threadlike part of a nerve cell (*neurone). It carries the nerve impulse (in the form of an *action potential) away from the body of a neurone towards the effector organ or the next nerve fibre.

azeotrope (azeotropic mixture; constant-boiling mixture) A mixture of two liquids that boils at constant composition; i.e. the composition of the vapour is the same as that of the liquid. Azeotropes occur because of deviations in Raoult's law leading to a maximum or minimum in the *boiling-point–composition diagram. When the mixture is boiled, the vapour initially has a higher proportion of one component than is present in the liquid, so the proportion of this in the liquid falls with time. Eventually, the maximum and minimum point is reached, at which the two liquids distil together without change in composition. The composition of an azeotrope depends on the pressure.

azeotropic distillation A technique for separating components of an azeotrope by adding a third liquid to form a new azeotrope with one of the original components. It is most commonly used to separate ethanol from water, adding benzene to associate with the ethanol.

azides Compounds containing the ion N_3^- or the group $-N_3$.

azimuth *See* polar coordinates.

azimuthal quantum number *See* atom.

azine An organic heterocyclic compound containing a six-membered ring formed from carbon and nitrogen atoms. Pyridine is an example containing one nitrogen atom (C_5... have two nitrogen ...(e.g. $C_4H_4N_2$), and ...ling on the rela-...nitrogen atoms. ...ree nitrogen

...ompounds
...N– link-
...can be
...nium

azoimide *See* hydrogen azide.

azurite A secondary mineral consisting of hydrated basic copper carbonate, $Cu_3(OH)_2(CO_3)_2$, in monoclinic crystalline form. It is generally formed in the upper zone of copper ore deposits and often occurs with *malachite. Its intense azure-blue colour made it formerly important as a pigment. It is a minor ore of copper and is used as a gemstone.

B

Babbit metal Any of a group of related alloys used for making bearings. They consist of tin containing antimony (about 10%) and copper (1–2%), and often lead. The original alloy was invented in 1839 by the US inventor Isaac Babbit (1799–1862).

Babo's law The vapour pressure of a liquid is decreased when a solute is added, the amount of the decrease being proportional to the amount of solute dissolved. The law was discovered in 1847 by the German chemist Lambert Babo (1818–99). *See also* Raoult's law.

Bacillariophyta A division of *algae comprising the diatoms. These marine or freshwater unicellular algae have cell walls composed of pectin impregnated with silica and consisting of two halves, one overlapping the other. Diatoms are found in huge numbers in plankton and are important in the food chains of seas and rivers. Past deposition has resulted in diatomaceous earths (kieselguhr) and the oil reserves of these species have contributed to oil deposits.

bacillus Any rod-shaped bacterium. Generally, bacilli are large, Gram-positive, spore-bearing, and have a tendency to form chains and produce a *capsule. Some are motile, bearing flagella. They are ubiquitous in soil and air and many are responsible for food spoilage. The group also includes *Bacillus anthracis*, which causes anthrax.

backbone *See* vertebral column.

back cross (test cross) A mating made to identify hidden *recessive alleles. If an organism displays a *dominant characteristic, it may possess two dominant alleles (i.e. it is homozygous) or a dominant and a recessive allele for that characteristic (i.e. it is heterozygous). To find out which is the case, the organism is crossed with one displaying the recessive characteristic. If all the offspring show the dominant characteristic then the organism is homozygous, but if half show the recessive characteristic, then the organism is heterozygous.

back donation A form of chemical bonding in which a *ligand forms a sigma bond to an atom or ion by donating a pair of electrons, and the central atom donates electrons back by overlap of its *d*-orbitals with empty *p*- or *d*-orbitals on the ligand.

back e.m.f. An electromotive force that opposes the main current flow in a circuit. For example, when the coils of the armature in an electric motor are rotated a back e.m.f is generated in these coils by their interaction with the field magnet (*see* inductance). Also, in an electric cell, *polarization causes a back e.m.f. to be set up, in this case by chemical means.

background radiation Low intensity *ionizing radiation present on the surface of the earth and in the atmosphere as a result of *cosmic radiation and the presence of radioisotopes in the earth's rocks, soil, and atmosphere. The radioisotopes are either natural or the result of nuclear fallout or waste gas from power stations. Background counts must be taken into account when measuring the radiation produced by a specified source. *See also* microwave background radiation.

A generalized bacterial cell

bacteria A diverse group of ubiquitous microorganisms all of which consist of only a single *cell that lacks a distinct nuclear membrane and has a *cell wall of a unique composition (see illustration). Bacteria are usually classified by means of *Gram's stain, whether or not they require oxygen (*see* aerobic respiration, anaerobic respiration), and on the basis of shape. A bacterial cell may be spherical (*see* coccus), rodlike (*see* bacillus), spiral (*see* spirillum), comma-shaped (*see* vibrio), corkscrew-shaped (*see* spirochaete), or filamentous, resembling a fungal cell. The majority of bacteria range in size from 0.5 to 5 μm. Many are motile, bearing flagella, possess an outer slimy *capsule, and produce resistant spores (*see* endospore). In general bacteria reproduce only asexually, by simple division of cells, but a few groups undergo a form of sexual reproduction (*see* conjugation). Bacteria are largely responsible for decay and decomposition of organic matter, producing a cycling of such chemicals as carbon (*see* carbon cycle), oxygen (*see* oxygen cycle), nitrogen (*see* nitrogen cycle), and sulphur. A few bacteria obtain their food by means of *photosynthesis (including the *blue-green algae, now regarded as bacteria), some are saprophytes, and others are parasites, causing disease. The symptoms of bacterial infections are produced by *toxins.

bactericidal Capable of killing bacteria. Common bactericides are some *antibiotics, *antiseptics, and *disinfectants. *Compare* bacteriostatic.

bacteriophage (phage) A virus that is parasitic within a bacterium. Each phage is specific for only one type of bacterium. Most phages (*virulent phages*) infect, quickly multiply within, and destroy (lyse) their host cells. However, some (*temperate phages*) remain dormant in their hosts after initial infection: their nucleic acid becomes integrated into that of the host and multiplies with it, producing infected daughter cells. Lysis may eventually be triggered by environmental factors Phages are used experimentally to identify bacteria, to control manufacturing processes (such as cheese production) that depend on bacteria, and, because they can alter the genetic make-up of bacterial cells, they are important tools in genetic engineering as cloning *vectors.

bacteriostatic Capable of inhibiting or slowing down the growth and reproduction of bacteria. Some *antibiotics are bacteriostatic. *Compare* bactericidal.

Bakelite A trade name for certain phenol–formaldehyde resins, first introduced in 1909 by the Belgian–American chemist Leo Hendrik Baekeland (1863–1944).

baking soda *See* sodium hydrogencarbonate.

balance An accurate weighing device. The simple *beam balance* consists of two pans suspended from a centrally pivoted beam. Known masses are placed on one pan and the substance or body to be weighed is placed in the other. When the beam is exactly horizontal the two masses are equal. An accurate laboratory balance weighs to the nearest hundredth of a milligram. Specially designed balances can be accurate to a millionth of a milligram. More modern *substitution balances* use the substitution principle.

In this calibrated weights are removed from the single lever arm to bring the single pan suspended from it into equilibrium with a fixed counter weight. The substitution balance is more accurate than the two-pan device and enables weighing to be carried out more rapidly. In automatic electronic balances, mass is determined not by mechanical deflection but by electronically controlled compensation of an electric force. A scanner monitors the displacement of the pan support generating a current proportional to the displacement. This current flows through a coil forcing the pan support to return to its original position by means of a magnetic force. The signal generated enables the mass to be read from a digital display. The mass of the empty container can be stored in the balance's computer memory and automatically deducted from the mass of the container plus its contents. *See also* spring balance.

baleen *See* whalebone.

ballistic galvanometer A moving-coil *galvanometer designed for measuring charge by detecting a surge of current. It has a heavy coil with minimal damping. When a surge of current is passed through the coil, the initial maximum deflection (the 'throw') is proportional to the total charge that has passed.

ballistic pendulum A device used to measure the velocity of a projectile, such as a bullet. A large mass of relatively soft material is suspended from a horizontal bar and the angle through which this mass is displaced when it is struck by the projectile in flight enables the momentum and hence the velocity of the projectile to be calculated by successive application of the laws of conservation of linear momentum and of energy.

ballistics The study of the flight of projectiles, especially those that have a parabolic flight path from one point on the earth's surface to another.

Balmer series *See* hydrogen spectrum.

banana bond Informal name for the type of electron-deficient bond holding the B–H–B bridges in *boranes and similar compounds.

band spectrum *See* spectrum.

band theory *See* energy bands.

bandwidth The frequency range over which a radio signal of specified frequency spreads. For example, in a *modulation system it is the range of frequencies occupied by the modulating signal on either side of the carrier wave. In an amplifier, it is the range of frequencies over which the power amplification falls within a specified fraction of the maximum value. In an aerial it is the range of frequencies that an aerial system can handle without mismatch.

bar A c.g.s. unit of pressure equal to 10^6 dynes per square centimetre or 10^5 pascals (approximately 750 mmHg or 0.987 atmosphere). The *millibar* (100 Pa) is commonly used in meteorology.

rachis of feather — barbule — barbicels — barb

Interlocking barbs of a contour feather

barb 1. (in zoology) Any one of the stiff filaments forming a row on each side of the longitudinal shaft of a feather (see illustration). Together the barbs form the expanded part (*vane*) of the feather. *See* barbule. **2.** (in botany) A hooked hair.

barbule Any of the minute filaments forming a row on each side of the *barb of a feather. In a *contour feather adjacent barbules interlock by means of hooks (*barbicels*) and grooves, forming a firm vane. Down feathers have no barbicels.

Barfoed's test A biochemical test to detect monosaccharide (reducing) sugars in solution, devised by the Swedish physician C. T. Barfoed (1815–99). *Barfoed's reagent*, a mixture of ethanoic (acetic) acid and copper(II) acetate, is added to the test solution and boiled. If any reducing sugars are present a red precipitate of copper(II) oxide is formed. The reaction will be negative in the presence of disaccharide sugars as they are weaker reducing agents.

barite *See* barytes.

barium Symbol Ba. A silvery-white reactive element belonging to *group II of the periodic table; a.n. 56; r.a.m. 137.34; r.d. 3.51; m.p. 725°C; b.p. 1640°C. It occurs as the minerals barytes ($BaSO_4$) and witherite. Extraction is by high-temperature reduction of barium oxide with aluminium or silicon in a vacuum, or by electrolysis of fused barium chloride. The metal is used as a getter in vacuum systems. It oxidizes readily in air and reacts with ethanol and water. Soluble barium compounds are extremely poisonous. It was first identified in 1774 by Karl Scheele, and was extracted by Humphry Davy in 1808.

barium bicarbonate *See* barium hydrogencarbonate.

barium carbonate A white insoluble compound, $BaCO_3$; r.d. 4.43. It decomposes on heating to give barium oxide and carbon dioxide:
$$BaCO_3(s) \rightarrow BaO(s) + CO_2(g)$$
The compound occurs naturally as the mineral *witherite* and can be prepared by adding an alkaline solution of a carbonate to a solution of a barium salt. It is used as a raw material for making other barium

salts, as a flux for ceramics, and as a raw material in the manufacture of certain types of optical glass.

barium chloride A white compound, $BaCl_2$. The anhydrous compound has two crystalline forms: an α form (monoclinic; r.d. 3.856), which transforms at 962°C to a β form (cubic; r.d. 3.917; m.p. 963°C; b.p. 1560°C). There is also a dihydrate, $BaCl_2.2H_2O$ (cubic; r.d. 3.1), which loses water at 113°C. It is prepared by dissolving barium carbonate (witherite) in hydrochloric acid and crystallizing out the dihydrate. The compound is used in the extraction of barium by electrolysis.

barium hydrogencarbonate (barium bicarbonate) A compound, $Ba(HCO_3)_2$, which is only stable in solution. It can be formed by the action of carbon dioxide on a suspension of barium carbonate in cold water:

$$BaCO_3(s) + CO_2(g) + H_2O(l) \rightarrow$$
$$Ba(HCO_3)_2(aq)$$

On heating, this reaction is reversed.

barium hydroxide (baryta) A white solid, $Ba(OH)_2$, sparingly soluble in water. The common form is the octahydrate, $Ba(OH)_2.8H_2O$; monoclinic; r.d. 2.18; m.p. 78°C. It can be produced by adding water to barium monoxide or by the action of sodium hydroxide on soluble barium compounds and is used as a weak alkali in volumetric analysis.

barium oxide A white or yellowish solid, BaO, obtained by heating barium in oxygen or by the thermal decomposition of barium carbonate or nitrate; cubic; r.d. 5.72; m.p. 1920°C; b.p. 2000°C. When barium oxide is heated in oxygen the peroxide, BaO_2, is formed in a reversible reaction that was once used as a method for obtaining oxygen (the *Brin process*). Barium oxide is now used in the manufacture of lubricating-oil additives.

barium peroxide A dense off-white solid, BaO_2, prepared by carefully heating *barium oxide in oxygen; r.d. 4.96; m.p. 450°C. It is used as a bleaching agent. With acids, hydrogen peroxide is formed and the reaction is used in the laboratory preparation of hydrogen peroxide.

barium sulphate An insoluble white solid, $BaSO_4$, that occurs naturally as the mineral *barytes (or *heavy spar*) and can be prepared as a precipitate by adding sulphuric acid to barium chloride solution; r.d. 4.50; m.p. 1580°C. The rhombic form changes to a monoclinic form at 1149°C. It is used as a raw material for making other barium salts, as a pigment extender in surface coating materials (called *blanc fixe*), and in the glass and rubber industries. Barium compounds are opaque to X-rays, and a suspension of the sulphate in water is used in medicine to provide a contrast medium for X-rays of the stomach and intestine. Although barium compounds are extremely poisonous, the sulphate is safe to use because it is very insoluble.

bark The protective layer of mostly dead cells that covers the outside of woody stems and roots. It includes the living and dead tissues external to the xylem, including the phloem and periderm. The term can be used more specifically to describe the periderm together with other tissues isolated by the activity of the *cork cambium. In some species, such as birch, there is one persistent cork cambium but in the older stems of certain other species a second cork cambium becomes active beneath the periderm and further periderm layers are formed every few years. The result is a composite tissue called *rhytidome*, composed of cork, dead cortex, and dead phloem cells.

Barkhausen effect The magnetization of a ferromagnetic substance by an increasing magnetic field takes place in discontinuous steps rather than continuously. The effect results from the orientation of magnetic domains (*see* magnetism). It was first observed by H. Barkhausen (1881–1956) in 1919.

barn A unit of area sometimes used to measure *cross sections in nuclear interactions involving incident particles. It is equal to 10^{-28} square metre. The name comes from the phrase 'side of a barn' (something easy to hit).

barograph A meteorological instrument that records on paper variations in atmospheric pressure over a period. It often consists of an aneroid barometer operating a pen that rests lightly on a rotating drum to which the recording paper is attached.

barometer A device for measuring *atmospheric pressure. The *mercury barometer* in its simplest form consists of a glass tube about 80 cm long sealed at one end and filled with mercury. The tube is then inverted and the open end is submerged in a reservoir of mercury; the mercury column is held up by the pressure of the atmosphere acting on the surface of mercury in the reservoir. This type of device was invented by the Italian scientist Evangelista Torricelli (1608–47), who first noticed the variation in height from day to day, and constructed a barometer in 1644.

In such a device, the force exerted by the atmosphere balanced the weight of the mercury column. If the height of the column is h and the cross-sectional area of the tube is A, then the volume of the mercury in the column is hA and its weight is $hA\rho$, where ρ is the density of mercury. The force is thus $hA\rho g$, where g is the acceleration of free fall and the pressure exerted is this force divided by the area of the tube; i.e. $h\rho g$. Note that the height of the mercury is independent of the diameter of the tube. At standard atmospheric pressure the column is 760 mm high. The pressure is then expressed as 760 mmHg (101 325 pascals).

Mercury barometers of this type, with a reservoir of mercury, are known as *cistern barometers*. A common type is the *Fortin barometer*, in which the mercury is held in a leather bag so that the level in the reservoir can be adjusted. The height is read from a scale along the side of the tube in conjunction with a vernier scale that can be moved up and down. Corrections are made for temperature.

The second main type of barometer is the *aneroid barometer*, in which the cumbersome mercury column is replaced by a metal box with a thin corrugated lid. The air is removed from the box and the lid is supported by a spring. Variations in atmospheric pressure cause the lid to move against the spring. This movement is magnified by a system of delicate levers and made to move a needle around a scale. The aneroid barometer is less accurate than the mercury type but much more robust and convenient, hence its use in *altimeters.

baroreceptor A *receptor that responds to changes in pressure. The *carotid sinus in the carotid artery contains baroreceptors that respond to changes in arterial pressure and are therefore involved in the regulation of blood pressure and heart beat.

Barr body A structure consisting of a condensed X chromosome (*see* sex chromosome) that is found in nondividing nuclei of female mammals. The presence of a Barr body is used to confirm the sex of athletes in sex determination tests. It is named after the Canadian anatomist M. L. Barr (1908–), who identified it in 1949.

barrel A measurement of volume, widely used in the chemical industry, equal to 35 UK gallons (approximately 159 litres).

barycentre The *centre of mass of a system.

barye A c.g.s. unit of pressure equal to one dyne per square centimetre (0.1 pascal).

baryon A *hadron with half-integral spin. Nucleons comprise a subclass of baryons. According to currently accepted theory, baryons are made up of three *quarks (*antibaryons* are made up of three antiquarks) held together by gluons (*see* quantum chromodynamics). Baryons possess a quantum number, called the *baryon number*, which is +1 for baryons, −1 for antibaryons, and 0 for all other particles. Baryon number has always appeared to have been conserved experimentally, but *grand unified theories postulate interactions at very high energies that allow it not to be conserved. *See* proton decay.

baryta *See* barium hydroxide.

barytes (barite) An orthorhombic mineral form of *barium sulphate, $BaSO_4$; the chief ore of barium. It is usually white but may also be yellow, grey, or brown. Large deposits occur in Andalusia, Spain, and in the USA.

basal ganglia Small masses of nervous tissue within the brain that connect the *cerebrum with other parts of the nervous system. They are involved with the subconscious regulation of voluntary movements.

basal metabolic rate (BMR) The rate of energy metabolism required to maintain an animal at rest. BMR is measured in terms of heat production per unit time: it indicates the energy consumed in order to sustain such vital functions as heartbeat, breathing, nervous activity, active transport, and secretion. Different tissues have different metabolic rates (e.g. the BMR of brain tissue is much greater than that of bone tissue) and therefore the tissue composition of an animal determines its overall BMR. For any comparable group of animals (such as mammals) BMR is proportional to body weight according to the allometric equation (*see* allometric growth); small animals tend to have a higher metabolic rate per unit weight than large ones.

basalt A fine-grained basic igneous rock. It is composed chiefly of calcium-rich plagioclase feldspar and pyroxene; other minerals present may be olivine, magnetite, and apatite. Basalt is the commonest type of lava.

base 1. (in chemistry) A compound that reacts with a protonic acid to give water (and a salt). The definition comes from the Arrhenius theory of acids and bases. Typically, bases are metal oxides, hydroxides, or compounds (such as ammonia) that give hydroxide ions in aqueous solution. Thus, a base may be either: (1) An insoluble oxide or hydroxide that reacts with an acid, e.g.

$$CuO(s) + 2HCl(aq) \rightarrow CuCl_2(aq) + H_2O(l)$$

Here the reaction involves hydrogen ions from the acid

$$CuO(s) + 2H^+(aq) \rightarrow H_2O(l) + Cu^{2+}(aq)$$

(2) A soluble hydroxide, in which case the solution contains hydroxide ions. The reaction with acids is a reaction between hydrogen ions and hydroxide ions:

$$H^+ + OH^- \rightarrow H_2O$$

(3) A compound that dissolves in water to produce hydroxide ions. For example, ammonia reacts as follows:

$$NH_3(g) + H_2O(l) \rightleftharpoons NH_4^+(aq) + {}^-OH$$

Similar reactions occur with organic *amines (*see also* nitrogenous base; amine salts). A base that dissolves in water to give hydroxide ions is called an *alkali*. Ammonia and sodium hydroxide are common examples.

The original Arrhenius definition of a base has been extended by the Lowry –Brønsted theory and by the Lewis theory. *See* acid.

2. (in mathematics) **a.** The number of different symbols in a number system. In the decimal system the base is 10; in *binary notation it is 2. **b.** The number that when raised to a certain power has a *logarithm equal to that power. For example if 10 is raised to the power of 3 it is equal to 1000; 3 is then the (common) logarithm of 1000 to the base 10. In natural or Napierian logarithms the base is e. To change the base from common to natural logarithms the formula used is: $\log_{10}y = \log_e y \times \log_{10}e = 0.43429\log_e y$.
3. (in electronics) *See* transistor.

base dissociation constant *See* dissociation.

base metal A common relatively inexpensive metal, such as iron or lead, that corrodes, oxidizes, or tarnishes on exposure to air, moisture, or heat, as distinguished from precious metals, such as gold and silver.

base pairing The chemical linking of two complementary nitrogenous bases in *DNA and in certain types of *RNA molecules. Of the four such bases in DNA, adenine pairs with thymine and cytosine with guanine. In RNA, thymine is replaced by uracil. Base pairing is responsible for holding together the two strands of a DNA molecule to form a double helix and for faithful reproduction and reading of the *genetic code. The links between bases take the form of *hydrogen bonds.

base unit A unit that is defined arbitrarily rather than being defined by simple combinations of other units. For example, the ampere is a base unit in the SI system defined in terms of the force produced between two current-carrying conductors, whereas the coulomb is a *derived unit*, defined as the quantity of charge transferred by one ampere in one second.

basic 1. Describing a compound that is a base. **2.** Describing a solution containing an excess of hydroxide ions; alkaline.

basic dye *See* dyes.

basicity constant *See* dissociation.

basic-oxygen process (BOP process) A high-speed method of making high-grade steel. It originated in the *Linnz–Donnewitz (L–D) process*. Molten pig iron and scrap are charged into a tilting furnace, similar to the Bessemer furnace except that it has no tuyeres. The charge is converted to steel by blowing high-pressure oxygen onto the surface of the metal through a water-cooled lance. The excess heat produced enables up to 30% of scrap to be incorporated into the charge. The process has largely replaced the Bessemer and open-hearth processes.

basic salt A compound that can be regarded as being formed by replacing some of the oxide or hydroxide ions in a base by other negative ions. Basic salts are thus mixed salt–oxides (e.g. bismuth(III) chloride oxide, BiOCl) or salt–hydroxides (e.g. lead(II) chloride hydroxide, Pb(OH)Cl).

basic slag *Slag formed from a basic flux (e.g. calcium oxide) in a blast furnace. The basic flux is used to remove acid impurities in the ore and contains calcium silicate, phosphate, and sulphide. If the phosphorus content is high the slag can be used as a fertilizer.

basic stains *See* staining.

basidiomycetes A group of fungi classified as a class (Basidiomycetes) or a subdivision (Basidiomycotina). Sexual reproduction is by means of *basidiospores* (spores produced externally on a club-shaped or cylindrical cell, the *basidium*). Basidia are often grouped together forming fruiting structures, such as mushrooms, puffballs, and bracket fungi. Exceptions are the *rusts and *smuts,

bast

which do not produce obvious fruiting bodies.

bast An old name for *phloem.

bats *See* Chiroptera.

battery A number of electric cells joined together. The common car battery, or *accumulator, usually consists of six secondary cells connected in series to give a total e.m.f. of 12 volts. A torch battery is usually a dry version of the *Leclanché primary cell, two of which are often connected in series. Batteries may also have cells connected in parallel, in which case they have the same e.m.f. as a single cell, but their capacity is increased, i.e. they will provide more total charge. The capacity of a battery is usually specified in ampere-hours, the ability to supply 1 A for 1 hr, or the equivalent.

bauxite The chief ore of aluminium, consisting of hydrous aluminium oxides and aluminous laterite. It is a claylike amorphous material formed by the weathering of silicate rocks under tropical conditions. The chief producers are Australia, Guinea, Jamaica, USSR, Brazil, and Surinam.

b.c.c. Body-centred cubic. *See* cubic crystal.

BCS theory *See* superconductivity.

beam A group of rays moving in an organized manner. It may consist of particles (e.g. an electron beam) or of electromagnetic radiation (e.g. a radar beam).

beam balance *See* balance.

beam hole A hole through the shielding of a *nuclear reactor to enable a beam of neutrons or other particles to escape for experimental purposes.

beats A periodic increase and decrease in loudness heard when two notes of slightly different frequency are sounded at the same time. If a note of frequency n is heard at the same time as a note of frequency m,

the resulting note will have a frequency of about $(n + m)/2$. However the amplitude of this note will vary from the difference to the sum of the amplitudes of the m and n notes and the frequency (called the *beat frequency*) of this variation will be $(m - n)$. The beating sound produced occurs as the waves successively reinforce and oppose each other as they move in and out of phase. *See also* interference.

Beaufort wind scale A scale of wind speed that was devised in the early 19th century by Admiral Francis Beaufort (1774–1857). Originally based on observations of the effect of various wind speeds on the sails of a full-rigged frigate, it has since been modified and is now based on observations of the sea surface or, on land, such easily observable indicators as smoke and tree movement. The scale ranges from 0 (calm) to 12 (hurricane).

Beckmann thermometer A thermometer for measuring small changes of temperature (see illustration). It

Beckmann thermometer

Beaufort number	description of wind	wind speed	
		(knots)	(metres per second)
0	calm	<1	0.0– 0.2
1	light air	1– 3	0.3– 1.5
2	light breeze	4– 6	1.6– 3.3
3	gentle breeze	7–10	3.4– 5.4
4	moderate breeze	11–16	5.5– 7.9
5	fresh breeze	17–21	8.0–10.7
6	strong breeze	22–27	10.8–13.8
7	near gale	28–33	13.9–17.1
8	gale	34–40	17.2–20.7
9	strong gale	41–47	20.8–24.4
10	storm	48–55	24.5–28.4
11	violent storm	56–63	28.5–32.6
12	hurricane	≥64	≥32.7

Beaufort wind scale

consists of a mercury-in-glass thermometer with a scale covering only 5 or 6°C calibrated in hundredths of a degree. It has two mercury bulbs, the range of temperature to be measured is varied by running mercury from the upper bulb into the larger lower bulb. It is used particularly for measuring *depression of freezing point or *elevation of boiling point of liquids when solute is added, in order to find relative molecular masses. The instrument was invented by the German chemist E. O. Beckmann (1853–1923).

becquerel Symbol Bq. The SI unit of activity (*see* radiation units). The unit is named after the discoverer of radioactivity A. H. Becquerel (1852–1908).

bees *See* Hymenoptera.

beetles *See* Coleoptera.

beet sugar *See* sucrose.

behavioural genetics The branch of genetics concerned with determining the relative importance of the genetic constitution of animals as compared to environmental factors in influencing animal behaviour.

bel Ten *decibels.

bell metal A type of *bronze used in casting bells. It consists of 60–85% copper alloyed with tin, often with some zinc and lead included.

Belousov–Zhabotinskii reaction *See* B–Z reaction.

bending moment (about any point or section of a horizontal beam under load) The algebraic sum of the *moments of all the vertical forces to either side of that point or section.

Benedict's test A biochemical test to detect reducing sugars in solution, devised by the US chemist S. R. Benedict (1884–1936). *Benedict's reagent* – a mixture of copper(II) sulphate and a filtered mixture of hydrated sodium citrate and hydrated sodium carbonate – is added to the test solution and boiled. A high concentration of reducing sugars induces the formation of a red precipitate; a lower concentration produces a yellow precipitate. Benedict's test is a more sensitive alternative to *Fehling's test.

beneficiation (ore dressing) The separation of an ore into the valuable components and the waste material (gangue). This may be achieved by a number of processes, including crushing, grinding, magnetic separation, froth flotation, etc. The dressed ore, consisting of a high proportion of valuable components, is then ready for smelting or some other refining process.

benthos Flora and fauna occurring on the bottom of a sea or lake. Benthic organisms may crawl, burrow, or remain attached to a substrate. *Compare* pelagic.

benzaldehyde *See* benzenecarbaldehyde.

Kekulé structures   Dewar structures
                           (3 in all)

Benzene

benzene A colourless liquid hydrocarbon, C_6H_6; r.d. 0.88; m.p. 5.5°C; b.p. 80.1°C. It is now made from gasoline from petroleum by catalytic reforming (formerly obtained from coal tar). Benzene is the archetypal *aromatic compound. It has an unsaturated molecule, yet will not readily undergo addition reactions. On the other hand, it does undergo substitution reactions in which hydrogen atoms are replaced by other atoms or groups. This behaviour occurs because of delocalization of p-electrons over the benzene ring, and all the C–C bonds in benzene are equivalent and intermediate in length between single and double bonds. It can be regarded as a resonance hybrid of Kekulé and Dewar structures (see formulae). In formulae it can be represented by a hexagon with a ring inside it.

benzenecarbaldehyde (benzaldehyde) A yellowish volatile oily liquid, C_6H_5CHO; r.d. 1.04; m.p. −26°C; b.p. 178.1°C. The compound occurs in almond kernels and has an almond-like smell. It is made from methylbenzene (by conversion to dichloromethyl benzene, $C_6H_5CHCl_2$, followed by hydrolysis). Benzenecarbaldehyde is used in flavourings, perfumery, and the dyestuffs industry.

benzenecarbonyl chloride (benzoyl chloride) A colourless liquid, C_6H_5COCl; r.d. 1.21; m.p. 0°C; b.p. 197.2°C. It is an *acyl halide, used to

introduce benzenecarbonyl groups into molecules. *See* acylation.

benzenecarbonyl group (benzoyl group) The organic group $C_6H_5CO–$.

benzenecarboxylate (benzoate) A salt or ester of benzenecarboxylic acid.

benzenecarboxylic acid (benzoic acid) A white crystalline compound, C_6H_5COOH; r.d. 1.27; m.p. 122.4°C; b.p. 249°C. It occurs naturally in some plants and is used as a food preservative. Benzenecarboxylic acid has a carboxyl group bound directly to a benzene ring. It is a weak carboxylic acid ($K_a = 6.4 \times 10^{-5}$ at 25°C), which is slightly soluble in water. It also undergoes substitution reactions on the benzene ring.

benzene-1,4-diol (hydroquinone; quinol) A white crystalline solid, $C_6H_4(OH)_2$; r.d. 1.33; m.p. 170°C; b.p. 285°C. It is used in making dyes. *See also* quinhydrone electrode.

benzene hexachloride (BHC) A crystalline substance, $C_6H_6Cl_6$, made by adding chlorine to benzene. It is used as a pesticide and, like *DDT, concern has been expressed at its environmental effects.

benzoate *See* benzenecarboxylate.

benzoic acid *See* benzenecarboxylic acid.

benzoquinone *See* cyclohexadiene-1,4-dione.

benzoylation A chemical reaction in which a benzoyl group (benzenecarbonyl group, C_6H_5CO) is introduced into a molecule. *See* acylation.

benzoyl chloride *See* benzenecarbonyl chloride.

benzoyl group *See* benzenecarbonyl group.

benzyl alcohol *See* phenylmethanol.

benzyne A compound, C_6H_4, having a hexagonal ring of carbon atoms containing two double bonds and one triple bond. The compound is highly reactive and cannot be isolated, although benzyne and its derivatives occur as short-lived intermediates in organic reactions.

Bergius process A process for making hydrocarbon mixtures (for fuels) from coal by heating powdered coal mixed with tar and iron(III) oxide catalyst at 450°C under hydrogen at a pressure of about 200 atmospheres. In later developments of the process, the coal was suspended in liquid hydrocarbons and other catalysts were used. The process was developed by the German chemist Friederich Bergius (1884–1949) during World War I as a source of motor fuel.

berkelium Symbol Bk. A radioactive metallic transuranic element belonging to the *actinoids; a.n. 97; mass number of the most stable isotope 247 (half-life 1.4×10^3 years); r.d. (calculated) 14. There are eight known isotopes. It was first produced by G. T. Seborg and associates in 1949 by bombarding americium–241 with alpha particles.

Bernoulli theorem At any point in a pipe through which a fluid is flowing the sum of the pressure energy, the kinetic energy, and the potential energy of a given mass of the fluid is constant. This is equivalent to a statement of the law of the conservation of energy. The law was published in 1738 by the Swiss physicist Daniel Bernoulli (1700–82).

berry A fleshy fruit formed from either one carpel or from several fused together and containing many seeds. The fruit wall may have two or three layers but the inner layer is never hard and stony (as in some drupes). Examples of berries are grapes and tomatoes. A berry, such as a cucumber, that develops a hard outer rind is called a *pepo*. One that is segmented and has a leathery rind, such as a citrus fruit, is called a *hesperidium*. The rind contains oil glands

and is lined by the white mesocarp, commonly called *pith*.

Berthollide compound *See* nonstoichiometric compound.

beryl A hexagonal mineral form of beryllium aluminium silicate, $Be_3Al_2Si_6O_{18}$; the chief ore of beryllium. It may be green, blue, yellow, or white and has long been used as a gemstone. Beryl occurs throughout the world in granite and pegmatites. *Emerald, the green gem variety, occurs more rarely and is of great value. Important sources of beryl are found in Brazil, Madagascar, and the USA.

beryllate A compound formed in solution when beryllium metal, or the oxide or hydroxide, dissolves in strong alkali. The reaction (for the metal) is often written

$$Be + 2OH^-(aq) \rightarrow BeO_2^{2-}(aq) + H_2(g)$$

The ion BeO_2^{2-} is the beryllate ion. In fact, as with the *aluminates, the ions present are probably hydroxy ions of the type $Be(OH)_4^{2-}$ (the *tetrahydroxoberyllate(II)* ion) together with polymeric ions.

beryllia *See* beryllium oxide.

beryllium Symbol Be. A grey metallic element of *group II of the periodic table; a.n. 4; r.a.m. 9.012; r.d. 1.85; m.p. 1285°C; b.p. 2970°C. Beryllium occurs as beryl $(3BeO.Al_2O_3.6SiO_2)$ and chrysoberyl $(BeO.Al_2O_3)$. The metal is extracted from a fused mixture of BeF_2/NaF by electrolysis or by magnesium reduction of BeF_2. It is used to manufacture Be–Cu alloys, which are used in nuclear reactors as reflectors and moderators because of their low absorption cross section. Beryllium oxide is used in ceramics and in nuclear reactors. Beryllium and its compounds are toxic and can cause serious lung diseases and dermatitis. The metal is resistant to oxidation by air because of the formation of an oxide layer, but will react with dilute hydrochloric and sulphuric acids. Beryllium compounds show high covalent character. The element was isolated independently by F. Wohler and A. A. Bussy in 1828.

beryllium hydroxide A white crystalline compound, $Be(OH)_2$, precipitated from solutions of beryllium salts by adding alkali. Like the oxide, it is amphoteric and dissolves in excess alkali to give *beryllates.

beryllium oxide (beryllia) An insoluble solid compound, BeO; hexagonal; r.d. 3.01; m.p. 2550°C; b.p. 4120°C. It occurs naturally as *bromellite*, and can be made by burning beryllium in oxygen or by the decomposition of beryllium carbonate or hydroxide. It is an important amphoteric oxide, reacting with acids to form salts and with alkalis to form compounds known as *beryllates. Beryllium oxide is used in the production of beryllium and beryllium—copper refractories, transistors, and integrated circuits.

Bessemer process A process for converting *pig iron from a *blast furnace into *steel. The molten pig iron is loaded into a refractory-lined tilting furnace (*Bessemer converter*) at about 1250°C. Air is blown into the furnace from the base and *spiegel is added to introduce the correct amount of carbon. Impurities (especially silicon, phosphorus, and manganese) are removed by the converter lining to form a slag. Finally the furnace is tilted so that the molten steel can be poured off. In the modern VLN (very low nitrogen) version of this process, oxygen and steam are blown into the furnace in place of air to minimize the absorption of nitrogen from the air by the steel. The process is named after the British engineer Sir Henry Bessemer (1813–98), who announced it in 1856. *See also* basic-oxygen process.

beta decay A type of weak interaction (*see* fundamental interactions) in which an unstable atomic nucleus changes into a nucleus of the same mass number but different proton number. The change involves the conversion of a neutron into a proton with the emission of an electron and an electron antineutrino:

$$n \rightarrow p + e^- + \bar{\nu}_e$$

or of a proton into a neutron with the emission of a positron and an electron neutrino:

$$p \rightarrow n + e^+ + \nu_e.$$

An example is the decay of carbon–14:

$$^{14}_6C \rightarrow {}^{14}_7N + e^- + \bar{\nu}_e.$$

The electrons or positrons emitted are called *beta particles* and streams of beta particles are known as *beta radiation*.

beta-iron A nonmagnetic allotrope of iron that exists between 768°C and 900°C.

beta particle *See* beta decay.

betatron A particle *accelerator for producing high-energy electrons (up to 340 MeV) for research purposes, including the production of high-energy X-rays. The electrons are accelerated by electromagnetic induction in a doughnut-shaped (toroidal) ring from which the air has been removed. This type of accelerator was first developed by D. W. Kerst (1911–) in 1939; the largest such machine, at the University of Illinois, was completed in 1950.

Bevatron A colloquial name for the proton *synchrotron at the Berkeley campus of the University of California. It produces energies up to 6 GeV.

BHC *See* benzene hexachloride.

biaxial crystal *See* double refraction.

bicarbonate *See* hydrogencarbonate.

bicarbonate of soda *See* sodium hydrogencarbonate.

biconcave *See* concave.

bicuspid valve *See* mitral valve.

biennial A plant that requires two growing seasons to complete its life cycle. During the first year it builds up food reserves, which are used during the second year in the production of flowers and seeds. Examples are carrot and parsnip.

big-bang theory The cosmological theory that all the matter and energy in the universe originated from a state of enormous density and temperature that exploded at a finite moment in the past. This theory successfully explains the *expansion of the universe; the observed *microwave background radiation, characteristic of *black-body radiation at a temperature of 3 K; and the observed abundance of helium in the universe, formed in the first 100 seconds after the explosion from deuterium at a temperature of 10^9 K. It is now generally considered to be more satisfactory than the rival *steady-state theory. The big-bang theory was first developed in 1927 by A. G. E. Lemaitre (1894–1966) and revived and revised in 1946 by George Gamow (1904–68). Several variants of it have been proposed. *See* early universe.

bilateral symmetry A type of arrangement of the parts and organs of an animal in which the body can be divided into two halves that are mirror images of each other along one plane only (usually passing through the midline at right angles to the dorsal and ventral surfaces). Bilaterally symmetrical animals are characterized by a type of movement in which one end of the body always leads. In botany this type of symmetry is usually called *zygomorphy* when applied to flowers (e.g. foxglove and antirrhinum flowers are zygomorphic). *Compare* radial symmetry.

bile (gall) A bitter-tasting greenish-yellow alkaline fluid produced by the *liver, stored in the *gall bladder, and secreted into the *duodenum of vertebrates. It assists the digestion and absorption of fats by the action of *bile salts*, which chemically reduce fatty substances and decrease the surface tension of fat droplets so that they are broken down and emulsified. Bile may also stimulate gut muscle contraction (*peristalsis). Bile also contains the pigments *bilirubin* and *biliverdin*, which are produced by the breakdown of the blood pigment *haemoglobin.

bile duct The tube through which bile passes from the *liver or (when present) the *gall bladder to the duodenum.

bilirubin *See* bile.

biliverdin *See* bile.

billion 1. (in the UK and Germany) One million million, 10^{12}. **2.** (in the USA and France) One thousand million, 10^9.

bimetallic strip A strip consisting of two metals of different *expansivity riveted or welded together so that the strip bends on heating. If one end is fixed the other end can be made to open and close an electric circuit, as in a *thermostat.

bimolecular reaction A step in a chemical reaction that involves two molecules. *See* molecularity.

bimorph cell A device consisting of two plates of piezoelectric material, such as Rochelle salt, joined together so that one expands on the application of a potential difference and the other contracts. The cell thus bends as a result of the applied p.d. The opposite effect is also used, in which the mechanical bending of the cell is used to produce a p.d., as in the crystal microphone and some types of record-player pickups.

binary Describing a compound or alloy formed from two elements.

binary acid An *acid in which the acidic hydrogen atom(s) are bound directly to an atom other than oxygen. Examples are hydrogen chloride (HCl) and hydrogen sulphide (H_2S).

Such compounds are sometimes called *hydracids. Compare* oxo acid.

binary notation A number system using only two different digits, 0 and 1. Instead of units, tens, hundreds, etc., as used in the decimal system, digits in the binary notation represent units, twos, fours, eights, etc. Thus one in decimal notation is represented by 0001, two by 0010, four by 0100, and eight by 1000. Because 0 and 1 can be made to correspond to off and on conditions in an electric circuit, the binary notation is widely used in computers.

binary stars A pair of stars revolving about a common centre of mass. In a *visual binary* the stars are far enough apart to be seen separately by an optical telescope. In an *astrometric binary* one component is too faint to be seen and its presence is inferred from the perturbations in the motion of the other. In a *spectrosopic binary* the stars cannot usually be resolved by a telescope, but the motions can be detected by different Doppler shifts in the spectrum at each side of the binary, according to whether the components are approaching or receding from the observer.

binding energy The energy equivalent to the *mass defect when nucleons bind together to form an atomic nucleus. When a nucleus is formed some energy is released by the nucleons, since they are entering a more stable lower-energy state. Therefore the energy of a nucleus consists of the energy equivalent of the mass of its individual nucleons minus the binding energy. The binding energy per nucleon plotted against the mass number provides a useful graph showing that up to a mass number of 50–60, the binding energy per nucleon increases rapidly, thereafter it falls slowly. Energy is released both by fission of heavy elements and by fusion of light elements because both processes entail a rearrangement of nuclei in the lower part of the graph to form nuclei in the higher part of the graph.

binoculars Any optical instrument designed to serve both the observer's eyes at once. *Binocular field glasses* consist of two refracting astronomical *telescopes inside each of which is a pair of prisms to increase the effective length and produce an upright image. Simpler binoculars, such as *opera glasses*, consist of two Galilean telescopes which produce upright images without prisms. Commonly, binoculars are specified by a pair of numbers, such as 10 × 50. The first number indicates the angular *magnification produced. The second is the diameter of the objective lens in millimetres, and indicates the amount of light gathered by the instrument. *Binocular microscopes* are used in biology and surgery to enable the observer to obtain a stereoscopic view of small objects or parts.

binocular vision The ability, found only in animals with forward-facing eyes, to produce a focused image of the same object simultaneously on the retinas of both eyes. This permits three-dimensional vision and contributes to distance judgment.

binomial nomenclature The system of naming organisms using a two-part Latinized (or scientific) name that was devised by the Swedish botanist Linnaeus (Carl Linné; 1707–78); it is also known as the *Linnaean system*. The first part is the generic name (*see* genus), the second is the specific epithet or name (*see* species). The Latin name is usually printed in italics, starting with a capital letter. For example, in the scientific name of the common frog, *Rana temporaria, Rana* is the generic name and *temporaria* the specific name. The name of the species may be followed by an abbreviated form of the name of its discoverer; for example, the common daisy is *Bellis perennis* L. (for Linnaeus). There are several International Codes of Taxonomic Nomenclature that lay down the rules for naming organisms. *See also* classification; taxonomy.

binomial theorem (binomial expansion) A rule for the expansion of a binomial expression (expression consisting of the sum of two variables raised to a given power). The general binomial expression $(x + y)^n$ expands to:

$$x^n + nx^{n-1}y + [n(n - 1)/2!]x^{n-2}y^2 + \ldots y^n.$$

bioassay (biological assay) A controlled experiment for the quantitative estimation of a substance by measuring its effect in a living organism. For example, the amount of the plant hormone auxin can be estimated by observing its effect on the curvature of oat coleoptiles – the concentration of the hormone is proportional to the curvature of the coleoptile.

biochemical oxygen demand (BOD) The amount of oxygen taken up by microorganisms that decompose organic waste matter in water. It is therefore used as a measure of the amount of certain types of organic pollutant in water. BOD is calculated by keeping a sample of water containing a known amount of oxygen for five days at 20°C. The oxygen content is measured again after this time. A high BOD indicates the presence of a large number of microorganisms, which suggests a high level of pollution.

biochemical taxonomy *See* taxonomy.

biochemistry The study of the chemistry of living organisms, especially the structure and function of their chemical components (principally proteins, carbohydrates, lipids, and nucleic acids). Biochemistry has advanced rapidly with the development, from the mid-20th century, of such techniques as chromatography, X-ray diffraction, radioisotopic labelling, and electron microscopy. Using these techniques to separate and analyse biologically important molecules, the steps of the metabolic pathways in which they are involved (e.g. *glycolysis and the *Krebs cycle) have been determined. This has provided some knowledge of how organisms obtain and store energy, how they manufacture and degrade their biomolecules, how they sense and respond to their environment, and how all this information is carried and expressed by their genetic material. Biochemistry forms an important part of many other disciplines, especially physiology, nutrition, and genetics, and its discoveries have made a profound impact in medicine, agriculture, industry, and many other areas of human activity.

biodegradable *See* pollution.

bioenergetics The study of the flow and the transformations of energy that occur in living organisms. Typically, the amount of energy that an organism takes in (from food or sunlight) is measured and divided into the amount used for growth of new tissues; that lost through death, wastes, and (in plants) transpiration; and that lost to the environment as heat (through respiration).

bioengineering 1. The use of artificial tissues, organs, and organ components to replace parts of the body that are damaged, lost, or malfunctioning, e.g. artificial limbs, heart valves, and heart pacemakers. **2.** The application of engineering knowledge to medicine and zoology.

biofeedback The technique whereby a subject can learn to control certain body functions, such as heart rate or blood pressure, that are usually unconsciously regulated by the autonomic nervous system. It is facilitated by the use of monitoring devices, such as pulse monitors, electroencephalographs, and electromyographs, and can be useful in treating high blood pressure, migraine, epilepsy, and other disorders.

biogenesis The principle that a living organism can only arise from other living organisms similar to itself (i.e. that like gives rise to like) and can never originate from nonliving material. *Compare* spontaneous generation.

biological clock The mechanism, presumed to exist within many animals and plants, that produces regular periodic changes in behaviour or physiology. Biological clocks underlie many of the *circadian, tidal, and annual rhythms seen in organisms (e.g. hibernation in animals). They continue to run even when conditions are kept artificially constant, but eventually drift out of step with the natural environment without the specific signals that normally keep them synchronized.

biological control The control of pests by biological (rather than chemical) means. This may be achieved, for example, by breeding disease-resistant crops or by introducing a natural enemy of the pest, such as a predator or a parasite. This technique, which may offer substantial advantages over the use of pesticides or herbicides, has been employed successfully on a number of occasions. One notable example was the control of the prickly pear cactus (*Opuntia*) in Australia by introducing the cactus moth (*Cactoblastis cactorum*), whose caterpillars feed on the plant's growing shoots. Insect pests have been controlled by releasing large numbers of males of the pest species that have been sterilized by radiation: infertile matings subsequently cause a decline in the pest population.

biology The study of living organisms, which includes their structure (gross and microscopical), functioning, origin and evolution, classification, interrelationships, and distribution.

bioluminescence The emission of light without heat (*see* luminescence) by living organisms. The phenomenon occurs in glow-worms and fireflies, bacteria and fungi, and in many deep-sea fish (among others); in animals it may serve as a means of protection (e.g. by disguising the shape of a fish) or species recognition or it may provide mating signals. The light is produced during the oxidation of a compound called *luciferin* (the composition of which varies according to the species), the reaction being catalysed by an enzyme, *luciferase*. Bioluminescence may be continuous (e.g. in bacteria) or intermittent (e.g. in fireflies).

biomass The total mass of all the organisms of a given type and/or in a given area; for example, the world biomass of trees, or the biomass of elephants in the Serengeti National Park.

biome A large naturally occurring assemblage of plant and animal species that are of the same general type, being adapted to the particular conditions in which they occur. Examples of biomes are tundra and tropical rain forest.

biophysics The study of the physical aspects of biology, including the application of physical laws and the techniques of physics to study biological phenomena. *See also* physics.

biopoiesis The development of living matter from complex organic molecules that are themselves nonliving but self-replicating. It is the process by which life is assumed to have begun. *See* origin of life.

biorhythm A roughly periodic change in the behaviour or physiology of an organism that is generated and maintained by a *biological clock. A well-known example is the circadian rhythm occurring in many animals and plants. The term biorhythm has unfortunately also become associated with a pseudoscientific cult postulating that man's physical, emotional, and intellectual behaviour is influenced by three biological cycles with periodicities of exactly 23 days, 28 days, and 33 days respectively. There is no reliable evidence to support this theory.

biosphere The whole of the region of the earth's surface, the sea, and the air that is inhabited by living organisms.

biostratigraphy See stratigraphy.

biosynthesis The production of molecules by a living cell, which is the essential feature of *anabolism.

biosystematics See systematics.

biotechnology The development of techniques for the application of biological processes to the production of materials of use in medicine and industry. For example, the production of antibiotics, cheese, and wine rely on the activity of various fungi and bacteria. *Genetic engineering can modify bacterial cells to synthesize completely new substances, e.g. hormones.

biotic environment The part of an organism's environment that consists of other living organisms. These may affect it in many ways; for example, as competitors, predators, parasites, prey, or symbionts. In time, the distribution and abundance of the organism will be affected by its inter-relationships with the biotic environment. See also abiotic factor.

biotin A vitamin in the *vitamin B complex. It is the *coenzyme for various enzymes that catalyse the incorporation of carbon dioxide into various compounds. Adequate amounts are normally produced by the intestinal bacteria in animals although deficiency can be induced by consuming large amounts of raw egg white. This contains a protein, avidin, that specifically binds biotin, preventing its absorption from the gut. Other

sources of biotin include cereals, vegetables, milk, and liver.

biotite An important rock-forming silicate mineral, a member of the *mica group of minerals, in common with which it has a sheetlike crystal structure. It is usually black, dark brown, or green in colour.

biprism A glass prism with an obtuse angle that functions as two acute-angle prisms placed base-to-base. A double image of a single object is thus formed; the device was used by Fresnel to produce two coherent beams for interference experiments.

bipyramid See complex.

birds See Aves.

birefringence See double refraction.

Birkeland–Eyde process A process for the fixation of nitrogen by passing air through an electric arc to produce nitrogen oxides. It was introduced in 1903 by the Norwegian chemists Kristian Birkeland (1867–1913) and Samuel Eyde (1866–1940). The process is economic only if cheap hydroelectricity is available.

bisexual (in biology) See hermaphrodite.

bismuth Symbol Bi. A white crystalline metal with a pinkish tinge belonging to *group V of the periodic table; a.n. 83; r.a.m. 208.98; r.d. 9.78; m.p. 271.3°C; b.p. 1560°C. The most important ores are bismuthinite (Bi_2S_3) and bismite (Bi_2O_3). Peru, Japan, Mexico, Bolivia, and Canada are major producers. The metal is extracted by carbon reduction of its oxide. Bismuth is the most diamagnetic of all metals and its thermal conductivity is lower than any metal except mercury. The metal has a high electrical resistance and a high Hall effect when placed in magnetic fields. It is used to make low-melting-point casting alloys with tin and cadmium. These alloys expand on solidification to give clear replication of intricate features. It is also used to make thermally activated safety devices for fire-detection and sprinkler systems. More recent applications include its use as a catalyst for making acrylic fibres, as a constituent of malleable iron, as a carrier of uranium–235 fuel in nuclear reactors, and as a specialized thermocouple material. Bismuth compounds (when lead-free) are used for cosmetics and medical preparations. It is attacked by oxidizing acids, steam (at high temperatures), and by moist halogens. It burns in air with a blue flame to produce yellow oxide fumes. C. G. Junine first demonstrated that it was different from lead in 1753.

bistable circuit *See* flip-flop.

bisulphate *See* hydrogensulphate.

bisulphite *See* hydrogensulphite; aldehydes.

bit (binary digit) Either of the digits 0 or 1 as used in the *binary notation. Bits are therefore the basic unit of information in a computer system.

bittern The solution of salts remaining when sodium chloride is crystallized from sea water.

Bitter pattern A microscopic pattern that forms on the surface of a ferromagnetic material that has been painted with a colloidal suspension of small iron particles. The patterns outline the boundaries of the magnetic domains (*see* magnetism). They were first observed by F. Bitter in 1931.

bitumen *See* petroleum.

bituminous coal *See* coal.

bituminous sand *See* oil sand.

biuret test A biochemical test to detect proteins in solution, named after the substance *biuret* ($H_2NCONHCONH_2$), which is formed when urea is heated. Sodium hydroxide is mixed with the test solution and drops of 1% copper(II) sulphate solution are then added slowly. A positive result is indicated by a violet ring, caused by the reaction of *pep-

tide bonds in the proteins or peptides. Such a result will not occur in the presence of free amino acids.

bivalent 1. (in chemistry) (*or* **divalent**) Having a valency of two. **2.** (in genetics) *See* pairing.

bivalves *See* Lamellibranchia.

black body A hypothetical body that absorbs all the radiation falling on it. It thus has an *absorptance and an *emissivity of 1. While a true black body is an imaginary concept, a small hole in the wall of an enclosure at uniform temperature is the nearest approach that can be made to it in practice.

Black-body radiation is the electromagnetic radiation emitted by a black body. It extends over the whole range of wavelengths and the distribution of energy over this range has a characteristic form with a maximum at a certain wavelength. The position of the maximum depends on temperature, moving to shorter wavelengths with increasing temperature. *See* Stefan's law; Wien's displacement law.

blackdamp (choke damp) Air left depleted in oxygen following the explosion of firedamp in a mine.

black earth *See* chernozem.

black hole An object in space that has collapsed under its own gravitational forces to such an extent that its *escape velocity is equal to the speed of light. Black holes are believed to be formed in the gravitational collapse of massive stars at the ends of their lives (*see* stellar evolution; supernova). If the mass of an evolved stellar core is greater than the Chandrasekhar limit for neutron stars then neutron degeneracy pressure is unable to prevent contraction until the gravitational field is sufficiently strong to prevent the escape of electromagnetic radiation. The boundary of the black hole, which is known as the *event horizon,* is the surface in space at which the gravitational field reaches this critical value. Events occurring within this horizon (i.e. in the interior of the black hole) cannot be observed from outside.

The theoretical study of black holes involves the use of general *relativity. It has been shown that a black hole can be characterized uniquely by just three properties: its mass, angular momentum, and electrical charge (this is known as the no-hair theorem). Mathematical expressions have been derived for describing black holes; these are the *Schwarzschild solution* (uncharged nonrotating hole), the *Reissner–Nordstrøm solution* (charged nonrotating hole), the *Kerr solution* (uncharged rotating hole), and the *Kerr–Newman solution* (charged rotating hole).

The ultimate fate of matter inside the black hole's event horizon is as yet unknown. General relativity predicts that at the centre of the hole there is a *singularity*, a point at which the density becomes infinite and the presently understood laws of physics break down. It is possible that a successful quantum theory of gravity could resolve this problem. However, since any singularity is hidden within the event horizon, it cannot influence the outside universe, so the normal laws of physics, including general relativity, can be used to describe processes outside the black hole.

Observational evidence of objects thought to be black holes comes from their effect on surrounding matter. Thus, if a black hole is part of a binary system with another star it will attract and capture matter from this star. The material leaving the star first forms a rotating *accretion disc* around the black hole, in which the matter becomes compressed and heated to such an extent that it emits X-rays. In the constellation Cygnus there is an X-ray source, Cygnus X-1, which consists of a supergiant star revolving around a small invisible companion with a mass of about ten times that of the sun, and therefore well above the Chandrasekhar limit. The companion is thought to be a black hole. Black holes have also been postulated as the power sources of *quasars and as possible generators of *gravitational waves.

Theoreticians have also postulated the existence of 'mini' black holes (with masses of about 10^{12} kilograms and radii about 10^{-15} metre). Such entities might have been formed shortly after the big bang when the universe was created. Quantum-mechanical effects are important for mini black holes, which emit Hawking radiation (*see* Hawking process). *See also* Schwarzschild radius.

black lead *See* carbon.

bladder 1. (in anatomy) **a.** A hollow muscular organ in most vertebrates, also known as the *urinary bladder*, in which urine is stored before being discharged. In mammals urine is conveyed from the *kidneys to the bladder by the *ureters and is discharged to the outside through the *urethra. **b.** Any of various other saclike organs in animals for the storage of liquid or gas. *See* gall bladder; swim bladder. **2.** (in botany) **a.** A modified submerged leaf of certain aquatic insectivorous plants, such as the bladderwort (*Utricularia*). It forms a hollow with a single opening that is sealed by a valve to trap small aquatic invertebrates after they have been sucked in. **b.** An air-filled cavity in the thallus of certain seaweeds, such as the bladderwrack (*Fucus vesiculosus*).

blanc fixe *See* barium sulphate.

blast furnace A furnace for smelting iron ores, such as haematite (Fe_2O_3) or magnetite (Fe_3O_4), to make *pig iron. The furnace is a tall refractory-lined cylindrical structure that is charged at the top with the dressed ore (*see* beneficiation), coke, and a flux, usually limestone. The conversion of the iron oxides to metallic iron is a reduction process in which carbon monoxide and hydrogen are the reducing agents. The overall reaction can be summarized thus:

$$Fe_3O_4 + 2CO + 2H_2 \rightarrow 3Fe + 2CO_2 + 2H_2O$$

The CO is obtained within the furnace by blasting the coke with hot air from a ring of tuyeres about two-thirds of the way down the furnace. The reaction producing the CO is:

$$2C + O_2 \rightarrow 2CO$$

In most blast furnaces hydrocarbons (oil, gas, tar, etc.) are added to the blast to provide a source of hydrogen. In the modern *direct-reduction process* the CO and H_2 may be produced separately so that the reduction process can proceed at a lower temperature. The pig iron produced by a blast furnace contains about 4% carbon and further refining is usually required to produce steel or cast iron.

blasting gelatin A high explosive made from nitroglycerine and gun cotton (cellulose nitrate).

blastocoel *See* blastula.

blastocyst *See* blastula; implantation.

blastula The first stage of development of an animal embryo, formed as a result of *cleavage of a fertilized egg. This stage generally resembles a hollow ball with the dividing cells (*blastomeres*) of the embryo forming a layer (*blastoderm*) around a central cavity (*blastocoel*). Insect eggs have no blastula. In vertebrates the blastula forms a disc (*blastodisc*) on the surface of the yolk. In mammals the blastula stage is known as a *blastocyst*. *See also* gastrula.

bleaching powder A white solid regarded as a mixture of calcium chlorate(I), calcium chloride, and calcium hydroxide. It is prepared on a large scale by passing chlorine gas through a solution of calcium hydroxide. Bleaching powder is sold on the basis of available chlorine, which is liberated when it is treated with a

dilute acid. It is used for bleaching paper pulps and fabrics and for sterilizing water.

blende A naturally occurring metal sulphide, e.g. zinc blende ZnS.

blending inheritance The early theory that assumed that hereditary substances from parents merge together in their offspring. Mendel showed that this does not occur (*see* Mendel's laws). In breeding experiments an appearance of blending may result from co-dominant alleles (*see* co-dominance) and *polygenes but close study shows that the alleles retain their identity through successive generations. *Compare* particulate inheritance.

blind spot The portion of the retina at which blood vessels and nerve fibres enter the optic nerve. There are no rods or cones in this area, so no visual image can be transmitted from it.

block *See* periodic table.

block copolymer *See* polymer.

blood A fluid body tissue that acts as a transport medium within an animal. It is contained within a blood *vascular system and in vertebrates is circulated by means of contractions of the *heart. Oxygen and food are carried to tissues, and carbon dioxide and chemical (nitrogenous) waste are transported from tissues to excretory organs for disposal (*excretion). In addition blood carries *hormones and also acts as a defence system. Blood consists of a liquid (*see* blood plasma) containing blood cells (*see* erythrocyte; leucocyte) and *platelets (see illustration).

blood–brain barrier The mechanism that controls the passage of substances from the blood to the cerebrospinal fluid bathing the brain and spinal cord. It takes the form of a semipermeable lipid membrane permitting the passage of solutions but excluding particles and large mol-

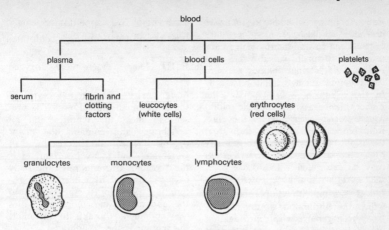

blood

plasma | blood cells | platelets

serum | fibrin and clotting factors | leucocytes (white cells) | erythrocytes (red cells)

granulocytes | monocytes | lymphocytes

Composition of mammalian blood

ecules. This barrier provides the central nervous system with a constant environment, while not interfering with the transport of essential metabolites.

blood cell (blood corpuscle) Any of the cells that are normally found in the blood plasma. These include red cells (*see* erythrocyte) and white cells (*see* leucocyte).

blood clotting (blood coagulation) The production of a mass of semi-solid material at the site of an injury that closes the wound, helping to prevent further blood loss and bacterial invasion. The clot is formed by the action of *clotting factors and *platelets. Calcium ions and platelets trigger the release of the enzyme *thromboplastin* from injured tissues. This converts *prothrombin* in the blood to its enzymically active form *thrombin*. Thrombin catalyses the formation of the insoluble protein *fibrin* from soluble *fibrinogen*; the fibrin forms a fibrous network in which blood cells become enmeshed, producing a clot.

blood groups The many types into which an individual's blood may be classified, based on the presence or absence of certain proteins (*antigens) on the surface of the red blood cells. Blood of one group contains *antibodies in the serum that react against the antigens on the cells of other groups. *Incompatibility* between groups results in clumping of cells (*agglutination), so knowledge of blood groups is important for blood transfusions. In man, the two most important blood group systems are the *ABO system and the system involving the *rhesus factor.

blood plasma The liquid part of the *blood (i.e. excluding blood cells). It consists of water containing a large number of dissolved substances, including proteins, salts (especially sodium and potassium chlorides and bicarbonates), food materials (glucose, amino acids, fats), hormones, vitamins, and excretory materials. *See also* blood serum; lymph.

blood pressure The pressure exerted by the flow of blood through the major arteries of the body. This pressure is greatest during the contraction of the ventricles of the heart (*systolic pressure*; *see* systole), which forces blood into the arterial system. Pressure falls to its lowest level when the heart is filling with blood (*diastolic pressure*; *see* diastole). Blood pressure is measured in millimetres of mercury using an instrument called a *sphygmomanometer*. Normal blood pressure for a young average adult human is in the region of 120/80 mmHg (the higher number is the systolic blood pressure; the lower number the diastolic blood pressure), but individual variations are common. Abnormally high blood pressure (*hypertension*) may be associated with disease or it may occur without an apparent cause.

blood serum Blood plasma from which the fibrin and clotting factors have been removed by centrifugation or vigorous stirring, so that it cannot clot. Serum containing a specific antibody or antitoxin may be used in the treatment or prevention of certain infections. Such serum is generally derived from a nonhuman mammal (e.g. a horse).

blood vascular system The tissues and organs of an animal that transport blood through the body. In vertebrates it consists of the heart and blood vessels. *See* vascular system.

blood vessel A tubular structure through which the blood of an animal flows. *See* artery; arteriole; capillary; venule; vein.

blue-green algae Prokaryotic organisms (*see* prokaryote) formerly classified as algae (division Cyanophyta) but now regarded as bacteria (*cyanobacteria*); 'blue-green algae' is still used as the common name, along with 'blue-green bacteria'. The colour is due to the presence of the pigment phycocyanin. Blue-green algae are unicellular but sometimes become joined in colonies or filaments by a sheath of mucilage. They occur in all aquatic habitats. A few species fix atmospheric nitrogen and thus contribute to soil fertility (*see* nitrogen fixation). Others exhibit *symbiosis (*see* lichens).

blue vitriol *See* copper(II) sulphate.

boat conformation *See* conformation.

BOD *See* biochemical oxygen demand.

body cavity The internal cavity of the body of an animal, which is present in most invertebrates and all vertebrates and contains the major organs. The body cavity of vertebrates and many invertebrates is the *coelom. In vertebrates the body cavity is divided by a transverse septum just posterior to the heart into the abdominal and thoracic cavities (*see* abdomen; thorax). In mammals the septum is the *diaphragm.

body-centred cubic (b.c.c.) *See* cubic crystal.

boehmite A mineral form of a mixed aluminium oxide and hydroxide, AlO.OH. It is named after the German scientist J. Böhm. *See* aluminium hydroxide.

Bohr theory The theory published in 1913 by the Danish physicist Niels Bohr (1885–1962) to explain the line spectrum of hydrogen. He assumed that a single electron of mass m travelled in a circular orbit of radius r, at a velocity v, around a positively charged nucleus. The *angular momentum of the electron would then be mvr. Bohr proposed that electrons could only occupy orbits in which this angular momentum had certain fixed values, $h/2\pi$, $2h/2\pi$, $3h/2\pi, \ldots nh/2\pi$, where h is the Planck constant. This means that the angular momentum is quantized, i.e. can only have certain values, each of which is a multiple of n. Each permitted value of n is associated with an orbit of different radius and Bohr assumed that when the atom emitted or absorbed radiation of frequency v, the electron jumped from one orbit to another; the energy emitted or absorbed by each jump is equal to hv. This theory gave good results in predicting the lines observed in the spectrum of hydrogen and simple ions such as He$^+$, Li^{2+}, etc. The idea of quantized values of angular momentum was later explained by the wave nature of the electron. Each orbit has to have a whole number of wavelengths around it; i.e. $n\lambda = 2\pi r$, where λ is the wavelength and n a whole number. The wavelength of a particle is given by h/mv, so $nh/mv = 2\pi r$, which leads to $mvr = nh/2\pi$. Modern atomic theory does not allow subatomic particles to be treated in the same way as large objects, and Bohr's reasoning is somewhat discredited. However, the idea of quantized angular momentum has been retained.

boiling point (b.p.) The temperature at which the saturated vapour pressure of a liquid equals the external atmospheric pressure. As a consequence, bubbles form in the liquid and the temperature remains constant until all the liquid has evaporated. As the boiling point of a liquid depends on the external atmospheric pressure, boiling points are usually quoted for standard atmospheric pressure (760 mmHg = 101 325 Pa).

boiling-point–composition diagram A graph showing how the boiling point and vapour composition of a mixture of two liquids depends on the composition of the mixture. The abscissa shows the range of compositions from 100% A at one end to 100% B at the other. The diagram has two curves: the lower one gives the boiling points (at a fixed pressure) for the different compositions. The upper one is plotted by taking the composition of vapour at each temperature on the boiling-point curve. The two curves would coincide for an ideal mixture, but generally they are different because of deviations from *Raoult's law. In some cases, they may show a maximum or minimum and coincide at some intermediate

composition, explaining the formation of *azeotropes.

boiling-water reactor *See* nuclear reactor.

bolometer A sensitive instrument used to measure radiant heat. The original form consists of two elements, each comprising blackened platinum strips (about 10^{-3} mm thick) arranged in series on an insulated frame to form a zigzag. The two elements are connected into the adjacent arms of a *Wheatstone bridge; one element is exposed to the radiation, the other is shielded from it. The change in the resistance of the exposed element, as detected by the bridge galvanometer, enables the heat reaching it to be calculated.

Modern semiconductor bolometers are now common, in which the platinum is replaced by a strip of semiconductor: this has a much greater (though usually negative) *temperature coefficient of resistance, and makes the system more sensitive.

Boltzmann constant Symbol k or k_B. The ratio of the universal gas constant (R) to the Avogadro constant (N_A). It may be thought of therefore as the gas constant per molecule:
$$k = R/N_A = 1.380\ 658(12) \times 10^{-23}\ \text{J K}^{-1}$$
It is named after the Austrian physicist Ludwig Boltzmann (1844–1906).

bomb calorimeter An apparatus used for measuring heats of combustion (e.g. calorific values of fuels and foods). It consists of a strong container in which the sample is sealed with excess oxygen and ignited electrically. The heat of combustion at constant volume can be calculated from the resulting rise in temperature.

bond *See* chemical bond.

bond energy An amount of energy associated with a bond in a chemical compound. It is obtained from the heat of atomization. For instance, in methane the bond energy of the C–H bond is one quarter of the enthalpy of the process

$$CH_4(g) \rightarrow C(g) + 4H(g)$$

Bond energies can be calculated from the standard enthalpy of formation of the compound and from the enthalpies of atomization of the elements. Energies calculated in this way are called *average bond energies* or *bond–energy terms*. They depend to some extent on the molecule chosen; the C–H bond energy in methane will differ slightly from that in ethane. The *bond dissociation energy* is a different measurement, being the energy required to break a particular bond; e.g. the energy for the process:

$$CH_4(g) \rightarrow CH_3 \cdot (g) + H \cdot (g)$$

bonding orbital *See* orbital.

bone The hard connective tissue of which the *skeleton of most vertebrates is formed. It comprises a matrix of *collagen fibres (30%) impregnated with bone salts (70%), mostly calcium phosphate, in which are embedded bone cells (*osteoblasts* and *osteocytes*), which secrete the matrix. Bone generally replaces embryonic *cartilage and is of two sorts – compact bone and spongy bone. The outer *compact bone* is formed as concentric layers (*lamellae*) that surround small holes (*Haversian canals): see illustration. The inner

single Haversian
system

Structure of compact bone

spongy bone is chemically similar but forms a network of bony bars. The spaces between the bars may contain bone marrow or (in birds) air for lightness. *See also* cartilage bone; membrane bone.

bone black *See* charcoal.

bone marrow A soft tissue contained within the central cavity and internal spaces of a bone. At birth and in young animals the marrow of all bones is concerned with the formation of blood cells: it contains *myeloid tissue and is known as *red marrow*. In mature animals the marrow of the long bones ceases producing blood cells and is replaced by fat, being known as *yellow marrow*.

bony fishes *See* Osteichthyes.

Boolean algebra A form of symbolic logic, devised by George Boole (1815–64) in the middle of the 19th century, which provides a mathematical procedure for manipulating logical relationships in symbolic form. For example in Boolean algebra $a + b$ means a or b, while ab means a and b. It makes use of *set theory and is extensively used by the designers of computers to enable the bits 0 and 1, as used in the binary notation, to relate to the logical functions the computer needs in carrying out its calculations.

borane (boron hydride) Any of a group of compounds of boron and hydrogen, many of which can be prepared by the action of acid on magnesium boride (MgB_2). Others are made by pyrolysis of the products of this reaction in the presence of hydrogen and other reagents. They are all volatile, reactive, and oxidize readily in air, some explosively so. The boranes are a remarkable group

Diborane, the simplest of the boranes

of compounds in that their structures cannot be described using the conventional two-electron covalent bond model (see electron-deficient compound). The simplest example is *diborane* (B_2H_6): see formula. Other boranes include B_4H_{10}, B_5H_9, B_5H_{11}, B_6H_{10}, and $B_{10}H_4$. The larger borane molecules have open or closed polyhedra of boron atoms. In addition, there is a wide range of borane derivatives containing atoms of other elements, such as carbon and phosphorus.

borate Any of a wide range of ionic compounds that have negative ions containing boron and oxygen (see formulae). Lithium borate, for example, contains the simple anion $B(OH)_4^-$. Most borates, however, are inorganic polymers with rings, chains, or other networks based on the planar BO_3 group or the tetrahedral $BO_3(OH)$ group. 'Hydrated' borates are ones containing –OH groups; many examples occur naturally. Anhydrous borates, which contain BO_3 groups, can be made by melting together boric acid and metal oxides.

borax (disodium tetraborate-10-water) A colourless monoclinic solid, $Na_2B_4O_7.10H_2O$, soluble in water and very slightly soluble in ethanol, monoclinic; r.d. 1.73; loses $8H_2O$ at 75°C; loses $10H_2O$ at 320°C. The formula gives a misleading impression of the structure. The compound contains the ion $[B_4O_5(OH)_4]^{2-}$ (see borate). Attempts to recrystallize this compound above 60.8°C yield the pentahydrate. The main sources of borax are the borate minerals *kernite* ($Na_2B_4O_7.4H_2O$) and *tincal* ($Na_2B_4O_7.10H_2O$). The ores are purified by carefully controlled dissolution and recrystallization. On treatment with mineral acids borax gives boric acid.

Borax is a very important substance in the glass and ceramics industries as a raw material for making borosilicates. It is also important as a metallurgical flux because of the ability of molten borates to dissolve metal oxides. In solution it partially hydrolyses to boric acid and can thus act as a buffer. For this reason it is used as a laundry pre-soak. It is used medicinally as a mild alkaline antiseptic and astringent for the skin and mucous membranes.

Disodium tetraborate is the source of many industrially important boron compounds, such as barium borate (fungicidal paints), zinc borate (fire-retardant additive in plastics), and boron phosphate (heterogeneous acid catalyst in the petrochemicals industry).

borax–bead test A simple laboratory test for certain metal ions in salts. A small amount of the salt is mixed with borax and a molten bead formed on the end of a piece of platinum wire. Certain metals can be identified

$B_3O_6^{3-}$ as in $Na_3B_3O_6$

$(BO_2)_n^{n-}$ as in CaB_2O_4

$[B_4O_5(OH)_4]^{2-}$ as in borax $Na_2B_4O_7.10H_2O$

Structure of some typical borate ions

by the colour of the bead produced in the oxidizing and reducing parts of a Bunsen flame. For example, iron gives a bead that is red when hot and yellow when cold in the oxidizing flame and a green bead in the reducing flame.

borax carmine A red dye, used in optical microscopy, that stains nuclei and cytoplasm pink. It is frequently used to stain large pieces of animal tissue.

borazon *See* boron nitride.

boric acid Any of a number of acids containing boron and oxygen. Used without qualification the term applies to the compound H_3BO_3 (which is also called *orthoboric acid* or, technically, *trioxoboric(III) acid*). This is a white or colourless solid that is soluble in water and ethanol; triclinic; r.d. 1.435; m.p. 169°C. It occurs naturally in the condensate from volcanic steam vents (suffioni). Commercially, it is made by treating borate minerals (e.g. kernite, $Na_2B_4O_7.4H_2O$) with sulphuric acid followed by recrystallization.

In the solid there is considerable hydrogen bonding between H_3BO_3 molecules resulting in a layer structure, which accounts for the easy cleavage of the crystals. H_3BO_3 molecules also exist in dilute solutions but in more concentrated solutions polymeric acids and ions are formed (e.g. $H_4B_2O_7$; *pyroboric acid* or *tetrahydroxomonoxodiboric(III) acid*). The compound is a very weak acid but also acts as a Lewis *acid in accepting hydroxide ions:

$$B(OH)_3 + H_2O \leftrightharpoons B(OH)_4^- + H^+$$

If solid boric acid is heated it loses water and transforms to another acid at 300°C. This is given the formula HBO_2 but is in fact a polymer $(HBO_2)_n$. It is called *metaboric acid* or, technically, *polydioxoboric(III) acid*.

Boric acid is used in the manufacture of glass (borosilicate glass), glazes and enamels, leather, paper, adhesives, and explosives. It is widely used (particularly in the USA) in detergents, and because of the ability of fused boric acid to dissolve other metal oxides it is used as a flux in brazing and welding. Because of its mild antiseptic properties it is used in the pharmaceutical industry and as a food preservative.

boride A compound of boron with a metal. Most metals form at least one boride of the type MB, MB_2, MB_4, MB_6, or MB_{12}. The compounds have a variety of structures; in particular, the hexaborides contain clusters of B_6 atoms. The borides are all hard high-melting materials with metal-like conductivity. They can be made by direct combination of the elements at high temperatures (over 2000°C) or, more usually, by high-temperature reduction of a mixture of the metal oxide and boron oxide using carbon or aluminium. Chemically, they are stable to nonoxidizing acids but are attacked by strong oxidizing agents and by strong alkalis. Magnesium boride (MgB_2) is unusual in that it can be hydrolysed to boranes. Industrially, metal borides are used as refractory materials. The most important are CrB, CrB_2, TiB_2, and ZnB_2. Generally, they are fabricated using high-temperature powder metallurgy, in which the article is produced in a graphite die at over 2000°C and at very high pressure. Items are pressed as near to final shape as possible as machining requires diamond cutters and is extremely expensive.

Born–Haber cycle A cycle of reactions used for calculating the lattice energies of ionic crystalline solids. For a compound MX, the lattice energy is the enthalpy of the reaction

$$M^+(g) + X^-(g) \rightarrow M^+X^-(s) \quad \Delta H_L$$

The standard enthalpy of formation of the ionic solid is the enthalpy of the reaction

$$M(s) + \tfrac{1}{2}X_2(g) \rightarrow M^+X^-(s) \quad \Delta H_f$$

The cycle involves equating this enthalpy (which can be measured) to the sum of the enthalpies of a number of steps proceeding from the elements to the ionic solid. The steps are:

(1) Atomization of the metal:
$$M(s) \rightarrow M(g) \; \Delta H_1$$
(2) Atomization of the nonmetal:
$$\tfrac{1}{2}X_2(g) \rightarrow X(g) \; \Delta H_2$$
(3) Ionization of the metal:
$$M(g) \rightarrow M^+(g) + e \; \Delta H_3$$
This is obtained from the ionization potential.
(4) Ionization of the nonmetal:
$$X(g) + e \rightarrow X^-(g) \; \Delta H_4$$
This is the electron affinity.
(5) Formation of the ionic solids:
$$M^+(g) + X^-(g) \rightarrow M^+X^-(s) \; \Delta H_L$$
Equating the enthalpies gives:
$$\Delta H_f = \Delta H_1 + \Delta H_2 + \Delta H_3 + \Delta H_4 + \Delta H_L$$
from which ΔH_L can be found. It is named after the German physicist Max Born (1882–1970) and the chemist Fritz Haber (1868–1934).

bornite An important ore of copper composed of a mixed copper–iron sulphide, Cu_5FeS_4. Freshly exposed surfaces of the mineral are a metallic reddish-brown but a purplish iridescent tarnish soon develops – hence it is popularly known as *peacock ore*. Bornite is mined in Chile, Peru, Bolivia, Mexico, and the USA.

boron Symbol B. An element of *group III of the periodic table; a.n. 5; r.a.m. 10.81; r.d. 2.35; m.p. 2079°C; b.p. 2550°C. It forms two allotropes; amorphous boron is a brown powder but metallic boron is black. The metallic form is very hard (9.3 on Mohs' scale) and is a poor electrical conductor at room temperature. At least three crystalline forms are possible; two are rhombohedral and the other tetragonal. The element is never found free in nature. It occurs as orthoboric acid in volcanic springs in Tuscany, as borates in kernite ($Na_2B_4O_7.4H_2O$), and as coleman-

ite ($Ca_2B_6O_{11}.5H_2O$) in California. Samples usually contain isotopes in the ratio of 19.78% boron–10 to 80.22% boron–11. Extraction is achieved by vapour-phase reduction of boron trichloride with hydrogen on electrically heated filaments. Amorphous boron can be obtained by reducing the trioxide with magnesium powder. Boron when heated reacts with oxygen, halogens, oxidizing acids, and hot alkalis. It is used in semiconductors and in filaments for specialized aerospace applications. Amorphous boron is used in flares, giving a green coloration. The isotope boron–10 is used in nuclear reactor control rods and shields. The element was discovered in 1808 by Sir Humphry Davy and by J. L. Gay-Lussac and L. J. Thenard.

boron carbide A black solid, B_4C, soluble only in fused alkali, it is extremely hard, over $9\tfrac{1}{2}$ on Mohs' scale; rhombohedral; r.d. 2.52; m.p. 2350°C; b.p. >3500°C. Boron carbide is manufactured by the reduction of boric oxide with petroleum coke in an electric furnace. It is used largely as an abrasive, but objects can also be fabricated using high-temperature powder metallurgy. Boron nitride is also used as a neutron absorber because of its high proportion of boron–10.

boron counter A *counter tube containing a *boron chamber*, used for counting slow neutrons. The boron chamber is lined with boron or a boron compound or is filled with the gas boron trifluoride (BF_3). As natural boron contains about 18% of the isotope boron–10, and as this isotope absorbs neutrons with the emission of an alpha particle, the chamber can be coupled with a scaler to count the alpha particles emitted when neutrons enter the chamber.

boron hydride *See* borane.

boron nitride A solid, BN, insoluble in cold water and slowly decomposed

boron trichloride

by hot water; r.d. 2.25 (hexagonal); sublimes above 3000°C. Boron nitride is manufactured by heating boron oxide to 800°C on an acid-soluble carrier, such as calcium phosphate, in the presence of nitrogen or ammonia. It is isoelectronic with carbon and, like carbon, it has a very hard cubic form (*borazon*) and a softer hexagonal form; unlike graphite this is a nonconductor. It is used in the electrical industries where its high thermal conductivity and high resistance are of especial value.

boron trichloride A colourless fuming liquid, BCl_3, which reacts with water to give hydrogen chloride and boric acid; r.d. 1.349; m.p. −107°C; b.p. 12.5°C. Boron trichloride is prepared industrially by the exothermic chlorination of boron carbide at above 700°C, followed by fractional distillation. An alternative, but more expensive, laboratory method is the reaction of dry chlorine with boron at high temperature. Boron trichloride is a Lewis *acid, forming stable addition compounds with such donors as ammonia and the amines and is used in the laboratory to promote reactions that liberate these donors. The compound is important industrially as a source of pure boron (reduction with hydrogen) for the electronics industry. It is also used for the preparation of boranes by reaction with metal hydrides.

borosilicate Any of a large number of substances in which BO_3 and SiO_4 units are linked to form networks with a wide range of structures. Borosilicate glasses are particularly important; the addition of boron to the silicate network enables the glass to be fused at lower temperatures than pure silica and also extends the plastic range of the glass. Thus such glasses as Pyrex have a wider range of applications than soda glasses (narrow plastic range, higher thermal expansion) or silica (much higher melting point). Borosilicates are also used in glazes and enamels and in the production of glass wools.

Bosch process *See* Haber process.

Bose condensation (Bose–Einstein condensation) A phenomenon occurring in a macroscopic system consisting of a large number of *bosons at a sufficiently low temperature, in which a significant fraction of the particles occupy a single quantum state of lowest energy (the ground state). Bose condensation can only take place for bosons whose total number is conserved in collisions. Because of the Pauli exclusion principle, it is impossible for two or more fermions to occupy the same quantum state, and so there is no analogous condensation phenomenon for such particles. Bose condensation is of fundamental importance in explaining the phenomenon of *superfluidity.

Bose–Einstein statistics *See* quantum statistics.

boson An *elementary particle (or bound state of an elementary particle, e.g. an atomic nucleus or an atom) with integral spin; i.e. a particle that conforms to Bose–Einstein statistics (*see* quantum statistics), from which it derives its name. *Compare* fermion.

botany The scientific study of plants, including their anatomy, morphology, physiology, biochemistry, taxonomy, cytology, genetics, ecology, evolution, and geographical distribution.

boundary layer The thin layer of fluid formed around a solid body or surface relative to which the fluid is flowing. Adhesion between the molecules of the fluid and that of the body or surface causes the molecules of the fluid closest to the solid to be stationary relative to it. The transfer of heat or mass between a solid and a fluid flowing over it is largely controlled by the nature of the boundary layer.

Bourdon gauge A pressure gauge consisting essentially of a C-shaped or spiral tube with an oval cross section. One end of the tube is connected to the fluid whose pressure is to be measured and the other end is sealed. As the pressure inside the tube is increased, the oval tube tends to become circular and this causes the tube to straighten. The movement of the end of the tube is transferred by a simple mechanism to a needle moving round a dial or to a digital display. With suitable design, Bourdon gauges can be used for high-pressure measurement and also for low pressures.

Bowman's capsule The cup-shaped end of a kidney *nephron. It is named after its discoverer, the British physician Sir William Bowman (1816 –92).

Boyle's law The volume (V) of a given mass of gas at a constant temperature is inversely proportional to its pressure (p), i.e. pV = constant. This is true only for an *ideal gas. This law was discovered in 1662 by the Irish physicist Robert Boyle (1627 –91). On the continent of Europe it is known as *Mariotte's law* after E. Mariotte (1620–84), who discovered it independently in 1676. *See also* gas laws.

Brackett series *See* hydrogen spectrum.

bract A modified leaf with a flower or inflorescence in its axil. Bracts are often brightly coloured and may be mistaken for the petals of a flower. For example the showy 'flowers' of poinsettia and *Bougainvillea* are composed of bracts; the true flowers are comparatively inconspicuous. *See also* involucre.

bracteole A reduced leaf that arises from the stalk of an individual flower.

bradykinin *See* kinin.

Bragg's law When a beam of X-rays (wavelength λ) strikes a crystal surface in which the layers of atoms or ions are separated by a distance d, the maximum intensity of the reflected ray occurs when $\sin\theta = n\lambda/2d$, where θ (known as the *Bragg angle*) is the complement of the angle of incidence and n is an integer. The law enables the structure of many crystals to be determined. It was discovered in 1912 by Sir Lawrence Bragg (1890–1971).

brain 1. The enlarged anterior part of the vertebrate central nervous system, which is encased within the cranium of the skull. Continuous with the spinal cord, the brain is surrounded by three membranes (*see* meninges) and bathed in cerebrospinal fluid, which fills internal cavities (*ventricles). It functions as the main coordinating centre for nervous activity, receiving information (in the form of nerve impulses) from sense organs, interpreting it, and transmitting 'instructions' to muscles and other *effectors. It is also the seat of intelligence and memory. The embryonic vertebrate brain is in three sections (*see* forebrain; hindbrain; midbrain), which become further differentiated during development into specialized regions.

The human brain

The main parts of the adult human brain are a highly developed *cerebrum in the form of two cerebral hemispheres, a *cerebellum, *medulla oblongata, and *hypothalamus (see illustration). **2.** A concentration of nerve *ganglia at the anterior end of an invertebrate animal.

brain death The permanent absence of vital functions of the brain, which is marked by cessation of breathing and other reflexes controlled by the *brainstem and by a zero reading on an *electroencephalogram. Organs may be removed for transplantation when brain death is established, which may not necessarily be associated with permanent absence of heart beat.

brainstem The part of the brain comprising the *medulla oblongata, the *midbrain, and the *pons. It resembles and is continuous with the spinal cord. The midbrain controls and integrates reflex activities (such as respiration) that originate in higher centres of the brain via a network of nerve pathways (the *reticular formation*).

branched chain *See* chain.

brass A group of alloys consisting of copper and zinc. A typical yellow brass might contain about 67% copper and 33% zinc.

breakdown The sudden passage of a current through an insulator. The voltage at which this occurs is the *breakdown voltage*.

breaking stress *See* elasticity.

breathing *See* respiratory movement.

breed A domesticated *variety of an animal or, rarely, a cultivated variety of plant. Cultivated plants are more often simply called varieties or, more correctly, *cultivars. Examples of animal breeds are Friesian cattle and Shetland sheepdogs.

breeder reactor *See* nuclear reactor.

Bremsstrahlung (German: braking radiation) The X-rays emitted when a charged particle, especially a fast electron, is rapidly slowed down, as when it passes through the electric field around an atomic nucleus. The X-rays cover a whole continuous range of wavelengths down to a minimum value, which depends on the energy of the incident particles. Bremsstrahlung are produced by a metal target when it is bombarded by electrons. *See also* X-ray spectrum.

Brewster's law The extent of the polarization of light reflected from a transparent surface is a maximum when the reflected ray is at right angles to the refracted ray. The angle of incidence (and reflection) at which this maximum polarization occurs is called the *Brewster angle* or *polarizing angle*. For this angle i_B, the condition is that $\tan i_B = n$, where n is the refractive index of the transparent medium. The law was discovered in 1811 by the British physicist David Brewster (1781–1868).

bridge An atom joining two other atoms in a molecule. *See* aluminium chloride; borane.

Brinell hardness A scale for measuring the hardness of metals introduced around 1900 by the Swedish metallurgist J. A. Brinell (1849–1925). A small chromium-steel ball is pressed into the surface of the metal by a load of known weight. The ratio of the mass of the load in kilograms to the area of the depression formed in square millimetres is the *Brinell number*.

Brin process A process formerly used for making oxygen by heating barium oxide in air to form the peroxide and then heating the peroxide at higher temperature (>800°C) to produce oxygen
$$2BaO_2 \rightarrow 2BaO + O_2$$
Britannia metal A silvery alloy consisting of 80–90% tin, 5–15% antimony, and sometimes small

percentages of copper, lead, and zinc. It is used in bearings and some domestic articles.

British thermal unit (Btu) The Imperial unit of heat, being originally the heat required to raise the temperature of 1lb of water by 1°F. 1 Btu is now defined as 1055.06 joules.

broken symmetry A situation in which the ground state of a many-body system or *vacuum state of a relativistic *quantum field theory has a lower symmetry than the Hamiltonian or Langrangian defining the system. Examples in solid-state physics include antiferromagnetism and superconductivity. In particle physics, the Weinberg–Salam model (*see* electroweak theory) is an important example of a relativistic quantum field theory with broken symmetry.

A result associated with broken symmetry is *Goldstone's theorem*. This states that a relativistic quantum field theory having continuous symmetry that is broken must include the existence of massless particles called *Goldstone bosons*. In many-body theory Goldstone bosons are *collective excitations. An exception to Goldstone's theorem is provided in the case of broken *gauge theories, such as the Weinberg–Salam model, in which the Goldstone bosons become massive bosons known as *Higgs bosons. In many-body theory, long-range forces provide the analogous exception to Goldstone's theorem, with the Higgs bosons being excitations with a nonzero gap.

bromate A salt or ester of a bromic acid.

bromic(I) acid (hypobromous acid) A yellow liquid, HBrO. It is a weak acid but a strong oxidizing agent.

bromic(V) acid A colourless liquid, $HBrO_3$, made by adding sulphuric acid to barium bromate. It is a strong acid.

bromide *See* halide.

bromination A chemical reaction in which a bromine atom is introduced into a molecule. *See also* halogenation.

bromine Symbol Br. A *halogen element; a.n. 35; r.a.m. 79.909; r.d. 3.13; m.p. −7.2°C; b.p. 58.78°C. It is a red volatile liquid at room temperature, having a red-brown vapour. Bromine is obtained from brines in the USA (displacement with chlorine); a small amount is obtained from sea water in Anglesey. Large quantities are used to make 1,2-dibromoethane as a petrol additive. It is also used in the manufacture of many other compounds. Chemically, it is intermediate in reactivity between chlorine and iodine. It forms compounds in which it has oxidation states of 1, 3, 5, or 7. The liquid is harmful to human tissue and the vapour irritates the eyes and throat. The element was discovered in 1826 by Antoine Balard.

bromoethane (ethyl bromide) A colourless flammable liquid, C_2H_5Br; r.d. 1.43; m.p. −119°C; b.p. 38.4°C. It is a typical *haloalkane, which can be prepared from ethene and hydrogen bromide. Bromoethane is used as a refrigerant.

bromoform *See* tribromomethane; haloforms.

bromomethane (methyl bromide) A colourless volatile nonflammable liquid, CH_3Br; r.d. 1.73; m.p. −93°C; b.p. 4.5°C. It is a typical *haloalkane.

bromothymol blue An acid–base *indicator that is yellow in acid solutions and blue in alkaline solutions. It changes colour over the pH range 6–8.

bronchiole A fine respiratory tube in the lungs of reptiles, birds, and mammals. It is formed by the subdivision of a *bronchus and in reptiles and mammals it terminates in a number of *alveoli.

bronchus (bronchial tube) One of the major air tubes in the *lung. The

bronze

deciduous woodland vegetation

litter layer — horizon A

grey-brown porous crumb structure

B

transition to parent material — C

Brown earth profile

*trachea divides into two main bronchi, one for each lung, which split into smaller bronchi and then into *bronchioles. The walls of the bronchi are stiffened by rings of cartilage.

bronze Any of a group of alloys of copper and tin, sometimes with lead and zinc present. The amount of tin varies from 1% to 30%. The alloy is hard and easily cast and extensively used in bearings, valves, and other machine parts. Various improved bronzes are produced by adding other elements; for instance, *phosphor bronzes* contain up to 1% phosphorus. In addition certain alloys of copper and metals other than tin are called bronzes – *aluminium bronze* is a mixture of copper and aluminium. Other special bronzes include *bell metal and *gun metal.

brown algae *See* Phaeophyta.

brown dwarf An astronomical object with a mass intermediate between the mass of a planet and that of a small star. The mass of a brown dwarf is large enough to generate energy by gravitational pressure, but not large enough to sustain nuclear fusion. The masses of brown dwarfs lie between a few times the mass of Jupiter and 80 times the mass of Jupiter. Brown dwarfs are faint objects; it has been suggested that they may contribute to the *missing mass of the universe.

brown earth (brown forest soil) A type of soil that is characteristic of those mid-latitude parts of the world that were originally covered with deciduous woodland. It is rich in organic matter derived from the annual leaf fall of deciduous trees and from associated shrubs, herbs, and grasses. Brown earths occur in the NE USA, N China, central Japan, and NW and central Europe. They are important agriculturally, possessing a good crumb structure, mild acidity, and free drainage, and consequently most of the original forest has long since been cleared for agricultural use.

brown fat A darker coloured region of *adipose tissue found in newborn and hibernating animals (in which it may also be called the *hibernating gland*). Compared to normal white *fat, deposits of brown fat are more richly supplied with blood vessels and have a higher proportion of unsaturated fatty acids. They can also be more rapidly converted to heat energy, especially during arousal from hibernation and during cold stress in young animals. Since the deposits are strategically placed near major blood vessels, the heat they generate warms the blood returning to the heart.

It has been said that some types of obesity in man may be linked to a lack of brown fat in affected individuals: in normal individuals, an intake of excess food is converted into brown fat and rapidly metabolized.

Brownian movement The continuous random movement of microscopic solid particles (of about 1 micrometre in diameter) when suspended in a

90

fluid medium. First observed by the botanist Robert Brown (1773–1858) in 1827 when studying pollen particles, it was originally thought to be the manifestation of some vital force. It was later recognized to be a consequence of bombardment of the particles by the continually moving molecules of the liquid. The smaller the particles the more extensive is the motion. The effect is also visible in particles of smoke suspended in a still gas and in the protoplasm of dead cells.

brown-ring test A test for ionic nitrates. The sample is dissolved and iron(II) sulphate solution added in a test tube. Concentrated sulphuric acid is then added slowly so that it forms a separate layer. A brown ring (of $Fe(NO)SO_4$) at the junction of the liquids indicates a positive result.

brucite A mineral form of *magnesium hydroxide, $Mg(OH)_2$.

brush An electrical contact to a moving commutator on a motor or generator. It is made of a specially prepared form of carbon and is kept in contact with the moving part by means of a spring.

brush discharge A luminous discharge from a conductor that takes the form of luminous branching threads that penetrate into the surrounding gas. It is a form of *corona and it occurs when the electric field near the surface of the conductor exceeds a certain value but is not sufficiently high for a spark to appear.

Bryophyta A division of simple plants possessing no vascular tissue and rudimentary rootlike organs (rhizoids). It includes the classes *Musci (mosses) and *Hepaticae (liverworts). Bryophytes grow in a variety of damp habitats, from fresh water to rock surfaces. Some use other plants for support. Bryophytes show a marked *alternation of generations between gamete-bearing forms (gametophytes) and spore-bearing forms (sporo-

phytes), the latter being dependent on the former for water and nutrients. The leaves, stems, and roots of the gametophyte generation are not equivalent to those of higher (vascular) plants since the whole structure is *haploid.

Bryozoa (Polyzoa) *See* Ectoprocta.

bubble chamber A device for detecting ionizing radiation. It consists of a chamber containing a liquid, often hydrogen, kept at slightly above its boiling point at its preliminary pressure. Immediately before the passage of the ionizing particles the pressure is reduced, and the particles then act as centres for the formation of bubbles, which can be photographed to obtain a record of the particles' tracks. The device was invented in 1952 by the US physicist D. A. Glaser (1926–). *Compare* cloud chamber.

buccal cavity The mouth cavity: the beginning of the *alimentary canal, which leads to the pharynx and (in vertebrates) to the oesophagus. In mammals it contains the tongue and teeth, which assist in the mechanical breakdown of food, and the openings of the *salivary gland ducts.

Buchner funnel A type of funnel with an internal perforated tray on which a flat circular filter paper can be placed, used for filtering by suction. It is named after the German chemist Eduard Buchner (1860–1917).

bud 1. (in botany) A condensed immature shoot with a short stem bearing small folded or rolled leaves. The outer leaves of a bud are often scalelike and protect the delicate inner leaves. A *terminal* (or *apical*) *bud* exists at the tip of a stem or branch while *axillary* (or *lateral*) *buds* develop in the *axils of leaves. However, in certain circumstances buds can be produced anywhere on the surface of a plant. Some buds remain dormant, but may become active if the terminal bud is removed. It is common gardening practice to remove the terminal buds of some shoots to induce the development of lateral shoots from axillary buds. *See also* apical dominance. **2.** (in biology) An outgrowth from a parent organism that breaks away and develops into a new individual in the process of *budding.

budding 1. (in biology) A method of asexual reproduction in which a new individual is derived from an outgrowth (*bud*) that becomes detached from the body of the parent. In animals the process is also called *gemmation*; it is common in coelenterates (e.g. *Hydra*) and also occurs in some sponges and other invertebrates. Among plants, budding is characteristic of the yeasts and other unicellular fungi. **2.** (in horticulture) A method of grafting in which a bud of the scion is inserted onto the stock, usually beneath the bark.

buffer A solution that resists change in pH when an acid or alkali is added or when the solution is diluted. Acidic buffers consist of a weak acid with a salt of the acid. The salt provides the negative ion A^-, which is the conjugate base of the acid HA. An example is carbonic acid and sodium hydrogencarbonate. Basic buffers have a weak base and a salt of the base (to provide the conjugate acid). An example is ammonia solution with ammonium chloride.

In an acidic buffer, for example, molecules HA and ions A^- are present. When acid is added most of the extra protons are removed by the base:

$$A^- + H^+ \rightarrow HA$$

When base is added, most of the extra hydroxide ions are removed by reaction with undissociated acid:

$$OH^- + HA \rightarrow A^- + H_2O$$

Thus, the addition of acid or base changes the pH very little. The hydrogen-ion concentration in a buffer is given by the expression

$$K_a = [H^+] = [A^-]/[HA]$$

i.e. it depends on the ratio of conjugate base to acid. As this is not altered by dilution, the hydrogen-ion concentration for a buffer does not change much during dilution.

In the laboratory, buffers are used to prepare solutions of known stable pH. Natural buffers occur in living organisms, where the biochemical reactions are very sensitive to change in pH (*see* acid–base balance). The main natural buffers are $H_2CO_3/$ HCO_3^- and $H_2PO_4^-/HPO_4^{2-}$. Buffer solutions are also used in medicine (e.g. in intravenous injections), in agriculture, and in many industrial processes (e.g. dyeing, fermentation processes, and the food industry).

bugs *See* Hemiptera.

bulb An underground plant organ that enables a plant to survive from one growing season to the next. It is a modified shoot with a short flattened stem. A terminal bud develops at the centre of its upper surface, surrounded by swollen leaf bases that contain food stored from the previous growing season. Papery brown scale leaves cover the outside of the bulb. The stored food is used in the growing season when the terminal bud produces foliage leaves and flowers. The new leaves photosynthesize and some of the manufactured food passes into the leaf bases forming a new bulb (see illustration). If more than one bud develops, then additional bulbs form, resulting in vegetative propagation. Examples of bulb-forming plants are daffodil, onion, and tulip. *Compare* corm.

bulbil A small bulblike organ that may develop in place of a flower, from an axillary bud, or at the base of a stem in certain plants. If it becomes detached it develops into a new plant.

bulk modulus *See* elastic modulus.

terminal bud

scale leaf

swollen leaf bases

flattened stem

winter

foliage leaves and flowers develop from terminal bud

axillary bud

leaf bases shrivel as food used for growing plant

spring

withered foliage leaves

swelling leaf bases forming new bulb

new bud

autumn

Development of a bulb

bulla The rounded hollow projection of bone from the skull that encloses the *middle ear in mammals.

bumping Violent boiling of a liquid caused by superheating so that bubbles form at a pressure above atmospheric pressure. It can be prevented by putting pieces of porous pot in the liquid to enable bubbles of vapour to form at the normal boiling point.

Bunsen burner A laboratory gas burner having a vertical metal tube into which the gas is led, with a hole

in the side of the base of the tube to admit air. The amount of air can be regulated by a sleeve on the tube. When no air is admitted the flame is luminous and smoky. With air, it has a faintly visible hot outer part (the oxidizing part) and an inner blue cone where combustion is incomplete (the cooler reducing part of the flame). The device is named after the German chemist Robert Bunsen (1811 –99), who used a similar device (without a regulating sleeve) in 1855.

Bunsen cell A *primary cell consisting of a zinc cathode immersed in dilute sulphuric acid and a carbon anode immersed in concentrated nitric acid. The electrolytes are separated by a porous pot. The cell gives an e.m.f. of about 1.9 volts.

buoyancy The upward thrust on a body immersed in a fluid. This force is equal to the weight of the fluid displaced (*see* Archimedes' principle).

burette A graduated glass tube with a tap at one end leading to a fine outlet tube, used for delivering known volumes of a liquid (e.g. in titration).

buta-1,3-diene (butadiene) A colourless gaseous hydrocarbon, CH_2:CHCH:CH_2; m.p. 109°C; b.p. −4.5°C. It is made by catalytic dehydrogenation of butane (from petroleum or natural gas) and polymerized in the production of synthetic rubbers. The compound is a conjugated *diene in which the electrons in the pi orbitals are partially delocalized over the whole molecule. It can have trans and cis forms, the latter taking part in *Diels–Alder reactions.

butanal (butyraldehyde) A colourless flammable liquid aldehyde, C_3H_9CHO; r.d. 0.8; m.p. −109°C; b.p. 75.7°C.

butane A gaseous hydrocarbon, C_4H_{10}; d. 0.58 g cm^{-3}; m.p. −135°C; b.p. 0°C. Butane is obtained from petroleum (from refinery gas or by cracking higher hydrocarbons). The fourth member of the *alkane series, it has a straight chain of carbon atoms and is isomeric with 2-methyl-propane ($CH_3CH(CH_3)CH_3$, formerly called *isobutane*). It can easily be liquefied under pressure and is supplied in cylinders for use as a fuel gas. It is also a raw material for making buta-1,3-diene (for synthetic rubber).

butanedioic acid (succinic acid) A colourless crystalline fatty acid, $(CH_2)_2(COOH)_2$; r.d. 1.5; m.p. 185°C; b.p. 235°C. A weak carboxylic acid, it is produced by fermentation of sugar or ammonium tartrate and used as a sequestrant and in making dyes. It occurs in living organisms as an intermediate in metabolism, especially in the *Krebs cycle.

butanoic acid (butyric acid) A colourless liquid water-soluble acid, C_3H_7COOH; r.d. 0.96; b.p. 163°C. It is a weak acid ($K_a = 1.5 \times 10^{-5}$ mol dm^{-3} at 25°C) with a rancid odour. Its esters are present in butter and in human perspiration. The acid is used to make esters for flavourings and perfumery.

butanol Either of two aliphatic alcohols with the formula C_4H_9OH. *Butan-1-ol*, $CH_3(CH_2)_3OH$, is a primary alcohol; r.d. 0.81; m.p. −89.5°C; b.p. 117.3°C. *Butan-2-ol*, $CH_3CH(OH)C_2H_5$, is a secondary alcohol; r.d. 0.81; m.p. −114.7°C; b.p. 100°C. Both are colourless volatile liquids obtained from butane and used as solvents.

butanone (methyl ethyl ketone) A colourless flammable water-soluble liquid, $CH_3COC_2H_5$; r.d. 0.8; m.p. −86.4°C; b.p. 77.6°C. It can be made by the catalytic oxidation of butane and is used as a solvent.

butenedioic acid Either of two isomers with the formula HCOOHC:CHCOOH. Both compounds can be regarded as derivatives of ethene in which a hydrogen atom on each carbon has been replaced by a –COOH group. The compounds show cis–trans isomerism. The trans form is *fumaric acid* (r.d. 1.64; sublimes at 200°C) and the cis form is *maleic acid* (r.d. 1.59; m.p. 130°C). Both are colourless crystalline compounds used in making synthetic resins. The cis form is rather less stable than the trans form and converts to the trans form at 120°C. Unlike the trans form it can eliminate water on heating to form a cyclic anhydride containing a –CO.O.CO– group (*maleic anhydride*). Fumaric acid is an intermediate in the *Krebs cycle.

butterflies *See* Lepidoptera.

buttress root *See* prop root.

butyl group The organic group $CH_3(CH_2)_3-$.

butyl rubber A type of synthetic rubber obtained by copolymerizing 2-methylpropene (CH_2:$C(CH_3)CH_3$; isobutylene) and methylbuta-1,3-diene (CH_2:$C(CH_3)CH$:CH_2, isoprene). Only small amounts of isoprene (about 2 mole %) are used. The rubber can be vulcanized. Large amounts were once used for tyre inner tubes.

butyraldehyde *See* butanal.

butyric acid *See* butanoic acid.

Buys Ballot's law A law relating to winds that states that if an observer stands with his back to the wind, pressure is lower on his left than on his right in the northern hemisphere, and lower on his right than on his left in the southern hemisphere. The law was propounded by the Dutch meteorologist C. H. D. Buys Ballot (1817–90) in 1857.

by-product A compound formed during a chemical reaction at the same time as the main product. Commercially useful by-products are obtained from a number of industrial processes. For example, calcium chloride is a by-product of the *Solvay process for making sodium carbonate. Propanone is a by-product in the manufacture of *phenol.

byte A subdivision of a *word in a computer, it usually consists of eight *bits.

B–Z reaction (Belousov–Zhabotinskii reaction) A chemical reaction that shows a periodic colour change between magenta and blue with a period of about one minute. It occurs with a mixture of sulphuric acid, potassium bromate(V), cerium sulphate, and propanedioic acid. The colour change is caused by alternating oxidation–reductions in which cerium changes its oxidation state (Ce^{3+} gives a magenta solution while Ce^{4+} gives a blue solution). The B–Z reaction is an example of a chemical *clock reaction* – a reaction in which there is a regular periodic change in the concentration of one or more reactants. The phenomenon can be explained by *chaos theory.

C

cadmium Symbol Cd. A soft bluish metal belonging to *group IIB of the periodic table; a.n. 48; r.a.m. 112.41; r.d. 8.65; m.p. 320.9°C; b.p. 765°C. The element's name is derived from the ancient name for calamine, zinc carbonate $ZnCO_3$, and it is usually found associated with zinc ores, such as sphalerite (ZnS), but does occur as the mineral greenockite (CdS). Cadmium is usually produced as an associate product when zinc, copper, and lead ores are reduced. Cadmium is used in low-melting-point alloys to make solders, in Ni–Cd batteries, in bearing alloys, and in electroplating (over 50%). Cadmium compounds are used as phosphorescent coatings in TV tubes. Cadmium and its compounds are extremely toxic at low concentrations; great care is essential where solders are used or where fumes are emitted. It has similar chemical properties to zinc but shows a greater tendency towards complex formation. The element was discovered in 1817 by F. Stromeyer.

cadmium cell *See* Weston cell.

cadmium sulphide A water-insoluble compound, CdS; r.d. 4.82. It occurs naturally as the mineral *greenockite* and is used as a pigment and in semiconductors and fluorescent materials.

caecum A pouch in the alimentary canal of vertebrates between the *small intestine and *colon. The caecum (and its *appendix) is large and highly developed in herbivorous animals (e.g. rabbits and cows), in which it contains a large population of bacteria essential for the breakdown of cellulose. In humans the caecum is a *vestigial organ and is poorly developed.

caesium Symbol Cs. A soft silvery-white metallic element belonging to *group I of the periodic table; a.n. 55; r.a.m. 132.905; r.d. 1.88; m.p. 28.4°C; b.p. 678°C. It occurs in small amounts in a number of minerals, the main source being carnallite ($KCl.MgCl_2.6H_2O$). It is obtained by electrolysis of molten caesium cyanide. The natural isotope is caesium–133. There are 15 other radioactive isotopes. Caesium–137 (half-life 33 years) is used as a gamma source. As the heaviest alkali metal, caesium has the lowest ionization potential of all elements, hence its use in photoelectric cells, etc.

caesium clock An *atomic clock that depends on the energy difference between two states of the caesium–133 nucleus when it is in a magnetic field. In one type, atoms of caesium–133 are irradiated with *radio-frequency radiation, whose frequency is chosen to correspond to the energy difference between the two states. Some caesium nuclei absorb this radiation and are excited to the higher state. These atoms are deflected by a further magnetic field, which causes them to hit a detector. A signal from this detector is fed back to the radio-frequency oscillator to prevent it drifting from the resonant frequency of 9 192 631 770 hertz. In this way the device is locked to this frequency with an accuracy better than 1 part in 10^{13}. The caesium clock is used in the *SI unit definition of the second.

cage compound *See* clathrate.

Cainozoic *See* Cenozoic.

calciferol *See* vitamin D.

calcination The formation of a calcium carbonate deposit from hard water. *See* hardness of water.

calcinite A mineral form of *potassium hydrogencarbonate, $KHCO_3$.

calcite One of the most common and widespread minerals, consisting of crystalline calcium carbonate, $CaCO_3$. Calcite crystallizes in the rhombohedral system; it is usually colourless or white and has a hardness of 3 on the Mohs' scale. It has the property of double refraction, which is apparent in Iceland spar – the transparent variety of calcite. It is an important rock-forming mineral and is a major constituent in limestones, marbles, and carbonatites.

calcitonin (thyrocalcitonin) A hormone, produced by the mammalian *thyroid gland, that lowers the concentration of calcium (and phosphate) in the blood. It operates in opposition to parathyroid hormone (*see* parathyroid glands).

calcium Symbol Ca. A soft grey metallic element belonging to *group II of the periodic table; a.n. 20; r.a.m. 40.08; r.d. 1.55; m.p. 840°C; b.p. 1484°C. Calcium compounds are common in the earth's crust; e.g. limestone and marble ($CaCO_3$), gypsum ($CaSO_4.2H_2O$), and fluorite (CaF_2). The element is extracted by electrolysis of fused calcium chloride and is used as a getter in vacuum systems and a deoxidizer in produc-

ing nonferrous alloys. It is also used as a reducing agent in the extraction of such metals as thorium, zirconium, and uranium.

Calcium is an *essential element for living organisms, being required for normal growth and development. In animals it is an important constituent of bones and teeth and is present in the blood, being required for muscle contraction and other metabolic processes. In plants it is a constituent (in the form of calcium pectate) of the *middle lamella.

calcium acetylide *See* calcium dicarbide.

calcium bicarbonate *See* calcium hydrogencarbonate.

calcium carbide *See* calcium dicarbide.

calcium carbonate A white solid, $CaCO_3$, which is only sparingly soluble in water. Calcium carbonate decomposes on heating to give *calcium oxide (quicklime) and carbon dioxide. It occurs naturally as the minerals *calcite (rhombohedral; r.d. 2.71) and *aragonite (rhombic; r.d. 2.93). Rocks containing calcium carbonate dissolve slowly in acidified rainwater (containing dissolved CO_2) to cause temporary hardness. In the laboratory, calcium carbonate is precipitated from *limewater by carbon dioxide. Calcium carbonate is used in making lime (calcium oxide) and is the main raw material for the *Solvay process.

calcium chloride A white deliquescent compound, $CaCl_2$, which is soluble in water; r.d. 2.15; m.p. 772°C; b.p. 7600°C. There are a number of hydrated forms, including the monohydrate, $CaCl_2.H_2O$, the dihydrate, $CaCl_2.2H_2O$ (r.d. 0.84), and the hexahydrate, $CaCl_2.6H_2O$ (trigonal; r.d. 1.71; the hexahydrate loses $4H_2O$ at 30°C and the remaining $2H_2O$ at 200°C). Large quantities of it are formed as a byproduct of the *Solvay process and it can be prepared by dissolving calcium carbonate or calcium oxide in hydrochloric acid. Crystals of the anhydrous salt can only be obtained if the hydrated salt is heated in a stream of hydrogen chloride. Solid calcium chloride is used in mines and on roads to reduce dust problems, whilst the molten salt is the electrolyte in the extraction of calcium. An aqueous solution of calcium chloride is used in refrigeration plants.

calcium cyanamide A colourless solid, $CaCN_2$, which sublimes at 1150°C. It is prepared by heating calcium dicarbide at 800°C in a stream of nitrogen:

$$CaC_2(s) + N_2(g) \rightarrow CaCN_2(s) + C(s)$$

The reaction has been used as a method of fixing nitrogen in countries in which cheap electricity is available to make the calcium dicarbide (the *cyanamide process*). Calcium cyanamide can be used as a fertilizer because it reacts with water to give ammonia and calcium carbonate:

$$CaCN_2(s) + 3H_2O(l) \rightarrow CaCO_3(s) + 2NH_3(g)$$

It is also used in the production of melamine, urea, and certain cyanide salts.

calcium cyclamate An organic calcium salt, $(C_6H_{11}NHSO_3)Ca.2H_2O$, that has a very sweet taste and was formerly used as a sweetening agent. Its use was banned in the UK in 1969 as it was suspected of causing cancer.

calcium dicarbide (calcium acetylide; calcium carbide; carbide) A colourless solid compound, CaC_2; tetragonal; r.d. 2.22; m.p. 450°C; b.p. 2300°C. In countries in which electricity is cheap it is manufactured by heating calcium oxide with either coke or ethyne at temperatures above 2000°C in an electric arc furnace. The crystals consist of Ca^{2+} and C_2^- ions arranged in a similar way to the ions in sodium chloride. When water

is added to calcium dicarbide, the important organic raw material ethyne (acetylene) is produced:

$$CaC_2(s) + 2H_2O(l) \rightarrow Ca(OH)_2(s) + C_2H_2(g)$$

calcium fluoride A white crystalline solid, CaF_2; r.d. 3.2; m.p. 1360°C; b.p. 2500°C. It occurs naturally as the mineral *fluorite (or fluorspar) and is the main source of fluorine. The *calcium fluoride structure (fluorite structure)* is a crystal structure in which the calcium ions are each surrounded by eight fluoride ions arranged at the corners of a cube. Each fluoride ion is surrounded by four calcium ions at the corners of a tetrahedron.

calcium hydrogencarbonate (calcium bicarbonate) A compound, $Ca(HCO_3)_2$, that is stable only in solution and is formed when water containing carbon dioxide dissolves calcium carbonate:

$$CaCO_3(s) + H_2O(l) + CO_2(g) \rightarrow Ca(HCO_3)_2(aq)$$

It is the cause of temporary *hardness in water, because the calcium ions react with soap to give scum. Calcium hydrogencarbonate is unstable when heated and decomposes to give solid calcium carbonate. This explains why temporary hardness is removed by boiling and the formation of 'scale' in kettles and boilers.

calcium hydroxide (slaked lime) A white solid, $Ca(OH)_2$, which dissolves sparingly in water (*see* limewater); hexagonal; r.d. 2.24. It is manufactured by adding water to calcium oxide, a process that evolves much heat and is known as slaking. It is used as a cheap alkali to neutralize the acidity in certain soils and in the manufacture of mortar, whitewash, bleaching powder, and glass.

calcium nitrate A white deliquescent compound, $Ca(NO_3)_2$, that is very soluble in water; cubic; r.d. 2.50; m.p. 561°C. It can be prepared by neutralizing nitric acid with calcium carbonate and crystallizing it from solution as the tetrahydrate $Ca(NO_3)_2.4H_2O$, which exists in two monoclinic crystalline forms (α, r.d. 1.9; β, r.d. 1.82). There is also a trihydrate, $Ca(NO_3)_2.3H_2O$. The anhydrous salt can be obtained from the hydrate by heating but it decomposes on strong heating to the oxide, nitrogen dioxide, and oxygen. Calcium nitrate is sometimes used as a nitrogenous fertilizer.

calcium octadecanoate (calcium stearate) An insoluble white salt, $Ca(CH_3(CH_2)_{16}COO)_2$, which is formed when soap is mixed with water containing calcium ions and is the scum produced in hard-water regions.

calcium oxide (quicklime) A white solid compound, CaO, formed by heating calcium in oxygen or by the thermal decomposition of calcium carbonate; cubic; r.d. 3.35; m.p. 2600°C; b.p. 2850°C. On a large scale, calcium carbonate in the form of limestone is heated in a tall tower (lime kiln) to a temperature above 550°C:

$$CaCO_3(s) \leftrightharpoons CaO(s) + CO_2(g)$$

Although the reaction is reversible, the carbon dioxide is carried away by the upward current through the kiln and all the limestone decomposes. Calcium oxide is used to make calcium hydroxide, as a cheap alkali for treating acid soil, and in extractive metallurgy to produce a slag with the impurities (especially sand) present in metal ores.

calcium phosphate(V) A white insoluble powder, $Ca_3(PO_4)_2$; r.d. 3.14. It is found naturally in the mineral *apatite ($Ca_5(PO_4)_3(OH,F,Cl)$) and as rock phosphate. It is also the main constituent of animal bones. Calcium phosphate can be prepared by mixing solutions containing calcium ions and hydrogenphosphate ions in the presence of an alkali:

$$HPO_4{}^{2-} + OH^- \rightarrow PO_4{}^{3-} + H_2O$$

$$3Ca^{2+} + 2PO_4{}^{3-} \rightarrow Ca_3(PO_4)_2$$

It is used extensively as a fertilizer. The compound was formerly called *calcium orthophosphate* (*see* phosphate).

calcium stearate *See* calcium octadecanoate.

calcium sulphate A white solid compound, $CaSO_4$; r.d. 2.96; 1450°C. It occurs naturally as the mineral *anhydrite, which has a rhombic structure, transforming to a monoclinic form at 200°C. More commonly, it is found as the dihydrate, *gypsum, $CaSO_4.2H_2O$ (monoclinic; r.d. 2.32). When heated, gypsum loses water at 128°C to give the hemihydrate, $2CaSO_4.H_2O$, better known as *plaster of Paris. Calcium sulphate is sparingly soluble in water and is a cause of permanent *hardness of water. It is used in the manufacture of certain paints, ceramics, and paper. The naturally occurring forms are used in the manufacture of sulphuric acid.

calculus A series of mathematical techniques developed independently by Isaac Newton (1642–1727) and Gottfried Leibniz (1646–1716). *Differential calculus* treats a continuously varying quantity as if it consisted of an infinitely large number of infinitely small changes. For example, the velocity v of a body at a particular instant can be regarded as the infinitesimal distance, written ds, that it travels in the vanishingly small time interval, dt; the instantaneous velocity v is then ds/dt, which is called the *derivative* of s with respect to t. If s is a known function of t, v at any instant can be calculated by the process of *differentiation. The differential calculus is a powerful technique for solving many problems concerned with rate processes, maxima and minima, and similar problems.

Integral calculus is the opposite technique. For example, if the velocity of a body is a known function of time, the infinitesimal distance ds travelled in the brief instant dt is given by ds = vdt. The measurable distance s travelled between two instants t_1 and t_2 can then be found by a process of summation, called *integration, i.e.

$$s = \int_{t_1}^{t_2} v \, dt$$

The technique is used for finding areas and volumes and other problems involving the summation of infinitesimals.

Calgon Trade name for a water-softening agent. *See* hardness of water.

caliche A mixture of salts found in deposits between gravel beds in the Atacama and Tarapaca regions of Chile. They vary from 4 m to 15 cm thick and were formed by periodic leaching of soluble salts during wet geological epochs, followed by drying out of inland seas in dry periods. They are economically important as a source of nitrates. A typical composition is $NaNO_3$ 17.6%, NaCl 16.1%, Na_2SO_4 6.5%, $CaSO_4$ 5.5%, $MgSO_4$ 3.0%, KNO_3 1.3%, $Na_2B_4O_7$ 0.94%, $KClO_3$ 0.23%, $NaIO_3$ 0.11%, sand and gravel to 100%.

californium Symbol Cf. A radioactive metallic transuranic element belonging to the *actinoids; a.n. 98; mass number of the most stable isotope 251 (half-life about 700 years). Nine isotopes are known; californium –252 is an intense neutron source, which makes it useful in neutron *activation analysis and potentially useful as a radiation source in medicine. The element was first produced by G. T. Seaborg and associates in 1950.

callus 1. (in botany) A protective tissue, consisting of parenchyma cells, that develops over a cut or damaged plant surface. Callus tissue can also be induced to form in cell cultures by hormone treatment. **2.** (in pathology) A thick hard area of skin that commonly forms on the palms of the hands and soles of the feet as a result of continuous pressure or friction. **3.** (in physiology) Hard tissue formed round bone ends following a fracture, which is gradually converted to new bone.

calomel *See* mercury(I) chloride.

calomel half cell (calomel electrode) A type of half cell in which the electrode is mercury coated with calomel (HgCl) and the electrolyte is a solution of potassium chloride and saturated calomel. The standard electrode potential is –0.2415 volt (25°C). In the calomel half cell the reactions are

$$HgCl(s) \leftrightarrows Hg^+(aq) + Cl^-(aq)$$
$$Hg^+(aq) + e \leftrightarrows Hg(s)$$

The overall reaction is

$$HgCl(s) + e \leftrightarrows Hg(s) + Cl^-(aq)$$

This is equivalent to a $Cl_2(g)|Cl^-(aq)$ half cell in which the pressure is the dissociation pressure of HgCl.

caloric theory A former theory concerning the nature of heat, which was regarded as a weightless fluid (called *caloric*). It was unable to account for the fact that friction could produce an unlimited quantity of heat and it was abandoned when Joule showed that heat is a form of energy.

calorie The quantity of heat required to raise the temperature of 1 gram of water by 1°C (1 K). The calorie, a c.g.s. unit, is now largely replaced by the *joule, an *SI unit. 1 calorie = 4.186 8 joules.

Calorie (kilogram calorie; kilocalorie) 1000 calories. This unit is still in limited use in estimating the energy value of foods, but is obsolescent.

calorific value The heat per unit mass produced by complete combustion of a given substance. Calorific values are used to express the energy values of fuels; usually these are expressed in megajoules per kilogram ($MJ \, kg^{-1}$). They are also used to measure the energy content of foodstuffs; i.e. the energy produced when the food is oxidized in the body. The units here are kilojoules per gram ($kJ\,g^{-1}$), although Calories (kilocalories) are often still used in nontechnical contexts. Calorific values are measured using a *bomb calorimeter.

calorimeter Any of various devices used to measure thermal properties, such as *calorific value, *specific heat capacity, *specific latent heat, etc. *See* bomb calorimeter.

calx A metal oxide formed by heating an ore in air.

calyptra 1. A layer of cells that covers the developing sporophyte of bryophytes and pteridophytes. In mosses it forms a hood over the *capsule and in liverworts it forms a sheath at the base of the capsule stalk. 2. *See* root cap.

calyptrogen The region within the root *apical meristem that divides to produce the *root cap (calyptra).

calyx The *sepals of a flower, collectively, forming the outer whorl of the *perianth. It encloses the petals, stamens, and carpels and protects the flower in bud. *See also* pappus.

cambium (lateral meristem) A plant tissue consisting of actively dividing cells (*see* meristem) that is responsible for increasing the girth of the plant, i.e. it causes secondary growth. The two most important cambia are the *vascular cambium* and the *cork cambium. The vascular cambium occurs in the stem and root; it divides to produce secondary *xylem and secondary *phloem (new food- and water-conducting tissues). *Compare* apical meristem.

Cambrian The earliest geological period of the Palaeozoic era. It is estimated to have begun about 570 million years ago and lasted for some 100 million years. During this period marine animals with mineralized shells made their first appearance and Cambrian rocks are the first to contain an abundance of fossils. Cambrian fossils are all of marine animals; they include *trilobites, which dominated the Cambrian seas, echinoderms, brachiopods, molluscs, and primitive *graptolites (from the mid Cambrian). Trace *fossils also provide evidence for a variety of worms.

camera 1. An optical device for obtaining still photographs or for exposing cinematic film. It consists of a light-proof box with a lens at one end and a plate or film at the other. To make an exposure the shutter is opened and an image of the object to be photographed is formed on the light-sensitive film. The length of the exposure is determined by the intensity of light available, the film speed, and the *aperture of the lens. In the simpler cameras the shutter speed and aperture are controlled manually, but in automatic cameras the iris over the lens or the shutter is adjusted on the basis of information provided by a built-in *exposure meter. In ciné cameras the shutter automatically opens as the film comes to rest behind the lens for each frame; the film passes through the camera so that a set number (commonly 16, 18, or 24) of frames are exposed every second. 2. The part of a television system that converts optical images into electronic signals. It consists of a lens system, which focuses the image to be televised on the photosensitive mosaic of the camera tube, causing localized discharge of those of its elements that are illuminated. This mosaic is scanned from behind by an electron beam so that the beam current is varied as it passes over areas of light and shade. The signal so picked up by the scanning beam is preamplified in the camera and passed to the transmitter with sound and synchronization signals. In *colour television three separate camera tubes are used, one for each *primary colour.

camphor A white crystalline cyclic ketone, $C_{10}H_{16}O$; r.d. 0.99; m.p. 179°C; b.p. 204°C. It was formerly obtained from the wood of the Formosan camphor tree, but can now be synthesized. The compound has a characteristic odour associated with its use in mothballs. It is a plasticizer in celluloid.

Canada balsam A yellow-tinted resin used for mounting specimens in optical microscopy. It has similar optical properties to glass.

canal rays Streams of positive ions produced in a *discharge tube by boring holes (canals) in the cathode. The positive ions attracted to the cathode pass through the holes and emerge on the other side as positive rays.

cancer Any disorder of cell growth that results in invasion and destruction of surrounding healthy tissue by abnormal cells. Cancer cells arise from normal cells whose nature is permanently changed. They multiply more rapidly than healthy body cells and do not seem subject to normal control by nerves and hormones. They may spread via the bloodstream or lymphatic system to other parts of the body, where they produce further tissue damage (*metastases*). *Malignant tumour* is another name for cancer; *leukaemia* is cancer of white blood cells. Causative agents (*carcinogens*) include various chemicals (including those in tobacco smoke), ionizing radiation, silica and asbestos particles, and *oncogenic viruses. Hereditary factors and stress may also play a role.

candela Symbol Cd. The *SI unit of luminous intensity equal to the luminous intensity in a given direction of a source that emits monochromatic radiation of frequency 540×10^{12} Hz and has a radiant intensity in that direction of 1/683 watt per steradian.

candle power Luminous intensity as formerly expressed in terms of the international candle but now expressed in candela.

cane sugar *See* sucrose.

canine tooth A sharp conical *tooth in mammals that is large and highly developed in carnivores (e.g. dogs) for tearing meat. There are two canines in each jaw, each situated between the second *incisor and the first *premolar. In some animals (e.g. herbivores, such as giraffes and rabbits) canine teeth are absent.

Cannizzaro reaction A reaction of aldehydes to give carboxylic acids and alcohols. It occurs in the presence of strong bases with aldehydes that do not have alpha hydrogen atoms. For example, benzene-carbaldehyde gives benzenecarboxylic acid and benzyl alcohol:

$$2C_6H_5CHO \rightarrow C_6H_5COOH + C_6H_5CH_2OH$$

Aldehydes that have alpha hydrogen atoms undergo the *aldol reaction instead. The Cannizzaro reaction is an example of a *disproportionation. It was discovered in 1853 by the Italian chemist Stanislao Cannizzaro (1826–1910).

canonical form One of the possible structures of a molecule that together form a *resonance hybrid.

capacitance The property of a conductor or system of conductors that describes its ability to store electric charge. The capacitance (C) is given by Q/V, where Q is stored charge on one conductor and V the potential difference between the two conductors (or between a single conductor and earth); it is measured in farads.

An isolated sphere has a capacitance of $4\pi\varepsilon r$, where r is the radius and ε the *permittivity of the medium surrounding it. Capacitance is more commonly applied to systems of conductors (or semiconductors) separated by insulators (*see* capacitor).

capacitor An arrangement of conductors separated by an insulator (dielectric) used to store charge or introduce *reactance into an alternating-current circuit. The earliest form was the *Leyden jar. Capacitors used as circuit elements have two conducting plates separated by the dielectric. The dielectric may be air, paper impregnated with oil or wax, plastic film, or ceramic. The simplest form has two parallel rectangular conducting plates (area A) separated by a dielectric (thickness d, permittivity ε). The capacitance of such a capacitor is $A\varepsilon/d$. *Electrolytic capacitors* are devices in which the dielectric is formed by electrolysis; for example, it may be an oxide film on one of the electrodes. In these the electrolysis commonly continues during the charging process and is reversed during discharge.

capacitor microphone A microphone consisting of a *capacitor with a steady voltage applied across its parallel plates. One plate is fixed, the other is a thin diaphragm that is moved by the pressure of the sound waves. The movements of the diaphragm cause a variation in the spacing and therefore in the *capacitance of the device. This variation in capacitance is, in turn, reflected in a similar variation in the charge carried by each plate. The consequent current to and from one plate is carried by a resistor, the varying potential difference across which constitutes the device's output signal. It was formerly known as a *condenser microphone*.

capillarity *See* surface tension.

capillary 1. A tube of small diameter. **2.** The narrowest type of blood vessel in the vertebrate circulatory system. Capillaries conduct blood from *arterioles to all living cells: their walls are only one cell layer thick, so that oxygen and nutrients can pass through them into the surrounding tissues. Capillaries also transport waste material (e.g. urea and carbon dioxide) to venules for ultimate excretion. Capillaries can be constricted or dilated, according to local tissue requirements.

capitulum A type of flowering shoot (*see* racemose inflorescence) characteristic of plants of the family Compositae, e.g. daisy and dandelion. The tip of the shoot is flattened and bears many small stalkless flowers (*florets*) surrounded by an involucre (ring) of bracts. This arrangement gives the appearance of a single flower.

capric acid *See* decanoic acid.

caproic acid *See* hexanoic acid.

caprolactam A white crystalline substance, $C_6H_{11}NO$; r.d. 1.02; m.p. 170°C; b.p. 150°C. It is a *lactam containing the $-NH.CO-$ group with five CH_2 groups making up the rest of the seven-membered ring. Caprolactam is used in making *nylon.

caprylic acid *See* octanoic acid.

capsid The protein coat of a *virus. The chemical nature of the capsid is important in stimulating the body's *immune response against the invading virus.

capsule 1. (in botany) **a.** A dry fruit that releases its seeds when ripe; it is formed from several fused carpels and contains many seeds. The seeds may be dispersed through pores (as in the poppy), through a lid (as in plantain), or by the splitting and separation of the individual carpels (as in the crocus). Various other forms of capsules include the *silicula and *siliqua. **b.** The part of the sporophyte of mosses and liverworts in which the haploid spores are produced. It is borne on a long stalk (*seta*) and sheds its spores when mature (*see* peristome). **2.** (in microbiology) A thick gelatinous layer completely surrounding the cell wall of certain bacteria. It appears to have a protective function, making ingestion of the bacterial cell by *phagocytes more difficult and preventing desiccation. **3.** (in animal anatomy) **a.** The membranous or fibrous envelope that surrounds certain organs, e.g. the kidneys, spleen, and lymph nodes. **b.** The ligamentous sheath of connective tissue that surrounds various skeletal joints.

capture Any of various processes in which a system of particles absorbs an extra particle. There are several examples in atomic and nuclear physics. For instance, a positive ion may capture an electron to give a neutral atom or molecule. Similarly, a neutral atom or molecule capturing an electron becomes a negative ion. An atomic nucleus may capture a neutron to produce a different (often unstable) nucleus. Another type of nuclear capture is the process in which the nucleus of an atom absorbs an electron from the innermost orbit (the K shell) to transform into a different nucleus. In this process (called *K capture*) the atom is left in an excited state and generally decays by emission of an X-ray photon.

Radiative capture is any such process in which the capture results in an excited state that decays by emission of photons. A common example is neutron capture to yield an excited nucleus, which decays by emission of a gamma ray.

carapace 1. The dorsal part of the *exoskeleton of some crustaceans (e.g. crabs), which spreads like a shield over several segments of the head and thorax. **2.** The domed dorsal part of the shell of tortoises and turtles, formed of bony plates fused with the ribs and vertebrae and covered by a horny epidermal layer. The ventral part of the shell (*plastron*) is similar but flatter.

carat 1. A measure of fineness (purity) of gold. Pure gold is described as 24-carat gold. 14-carat gold contains 14 parts in 24 of gold, the remainder usually being copper. **2.** A unit of mass equal to 0.200

gram, used to measure the masses of diamonds and other gemstones.

carbamide *See* urea.

carbanion An organic ion with a negative charge on a carbon atom; i.e. an ion of the type R_3C^-. Carbanions are intermediates in certain types of organic reaction (e.g. the *aldol reaction).

carbene A species of the type $R_2C:$, in which the carbon atom has two electrons that do not form bonds. *Methylene*, $:CH_2$, is the simplest example. Carbenes are highly reactive and exist only as transient intermediates in certain organic reactions. They attack double bonds to give cyclopropane derivatives. They also cause insertion reactions, in which the carbene group is inserted between the carbon and hydrogen atoms of a C–H bond:

$$C–H + :CR_2 \rightarrow C–CR_2–H$$

carbide Any of various compounds of carbon with metals or other more electropositive elements. True carbides contain the ion C^{4-} as in Al_4C_3. These are saltlike compounds giving methane on hydrolysis, and were formerly called *methanides*. Compounds containing the ion C_2^{2-} are also saltlike and are known as *dicarbides*. They yield ethyne (acetylene) on hydrolysis and were formerly called *acetylides*. The above types of compound are ionic but have partially covalent bond character, but boron and silicon form true covalent carbides, with giant molecular structures. In addition, the transition metals form a range of interstitial carbides in which the carbon atoms occupy interstitial positions in the metal lattice. These substances are generally hard materials with metallic conductivity. Some transition metals (e.g. Cr, Mn, Fe, Co, and Ni) have atomic radii that are too small to allow individual carbon atoms in the interstitial holes. These form carbides in which the metal lattice is distorted and chains of carbon atoms exist (e.g. Cr_3C_2, Fe_3C). Such compounds are intermediate in character between interstitial carbides and ionic carbides. They give mixtures of hydrocarbons on hydrolysis with water or acids.

carbocyclic See cyclic.

carbohydrate One of a group of organic compounds based on the general formula $C_x(H_2O)_y$. The simplest carbohydrates are the *sugars (saccharides), including glucose and sucrose. *Polysaccharides are carbohydrates of much greater molecular weight and complexity; examples are starch, glycogen, and cellulose. Carbohydrates perform many vital roles in living organisms. Sugars, notably glucose, and their derivatives are essential intermediates in the conversion of food to energy. Starch and other polysaccharides serve as energy stores in plants, particularly in seeds, tubers, etc., which provide a major energy source for animals, including man. Cellulose, lignin, and others form the supporting cell walls and woody tissue of plants. Chitin is a structural polysaccharide found in the body shells of many invertebrate animals. Carbohydrates also occur in the surface coat of animal cells and in bacterial cell walls.

carbolic acid See phenol.

carbon Symbol C. A nonmetallic element belonging to *group IV of the periodic table; a.n. 6; r.a.m. 12.011; m.p. $\sim$3550°C; b.p. $\sim$4289°C. Carbon has two main allotropic forms (see allotropy).
*Diamond (r.d. 3.52) occurs naturally and small amounts can be produced synthetically. It is extremely hard and has highly refractive crystals. The hardness of diamond results from the covalent crystal structure, in which each carbon atom is linked by covalent bonds to four others situated at the corners of a tetrahedron. The C–

C bond length is 0.154 nm and the bond angle is 109.5°.
Graphite (r.d. 2.25), the other allotrope, is a soft black slippery substance (sometimes called *black lead* or *plumbago*). It occurs naturally and can also be made by the *Acheson process. In graphite the carbon atoms are arranged in layers, in which each carbon atom is surrounded by three others to which it is bound by single or double bonds. The layers are held together by much weaker van der Waals' forces. The carbon–carbon bond length in the layers is 0.142 nm and the layers are 0.34 nm apart. Graphite is a good conductor of heat and electricity. It has a variety of uses including electrical contacts, high-temperature equipment, and as a solid lubricant. Graphite mixed with clay is the 'lead' in pencils (hence its alternative name). There are also several amorphous forms of carbon, such as *carbon black and *charcoal.
There are two stable isotopes of carbon (proton numbers 12 and 13) and four radioactive ones (10, 11, 14, 15). Carbon–14 is used in *carbon dating. Carbon occurs in all organic compounds and is therefore basic to the structure of all living organisms. It is an *essential element for plants and animals, being ultimately derived from atmospheric carbon dioxide assimilated by plants during photosynthesis (see carbon cycle). The ubiquitous nature of carbon in living organisms is due to its unique ability to form stable covalent bonds with other carbon atoms and also with hydrogen, oxygen, nitrogen, and sulphur atoms, resulting in the formation of a variety of compounds containing chains and rings of carbon atoms.

carbonate A salt of carbonic acid containing the carbonate ion, CO_3^{2-}. The free ion has a plane triangular structure. Metal carbonates may be ionic or may contain covalent metal–

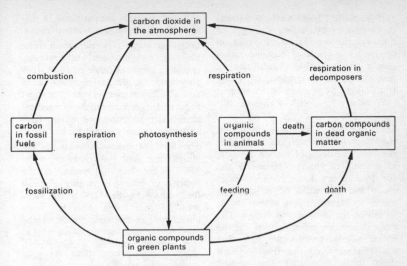

The carbon cycle in nature

carbonate bonds (complex carbonates) via one or two oxygen atoms. The carbonates of the alkali metals are all soluble but other carbonates are insoluble; they all react with mineral acids to release carbon dioxide.

carbonate minerals A group of common rock-forming minerals containing the anion CO_3^{2-} as the fundamental unit in their structure. The most important carbonate minerals are *calcite, *dolomite, and *magnesite. *See also* aragonite.

carbonation The solution of carbon dioxide in a liquid under pressure.

carbon bisulphide *See* carbon disulphide.

carbon black A fine carbon powder made by burning hydrocarbons in insufficient air. It is used as a pigment and a filler (e.g. for rubber).

carbon cycle 1. (in biology) One of the major cycles of chemical elements in the environment (see illustration). Carbon (as carbon dioxide) is taken up from the atmosphere and incorporated into the tissues of plants in *photosynthesis. It may then pass into the bodies of animals as the plants are eaten (*see* food chain). During the respiration of plants, animals, and organisms that bring about decomposition, carbon dioxide is returned to the atmosphere. The combustion of fossil fuels (e.g. coal and peat) also releases carbon dioxide into the atmosphere.

2. (in physics) A series of nuclear reactions in which four hydrogen nuclei combine to form a helium nucleus with the liberation of energy, two positrons, and two neutrinos. The process is believed to be the source of energy in many stars and to take place in six stages. In this series carbon–12 acts as if it were a catalyst, being reformed at the end of the series:

$$^{12}_{6}C + {}^{1}_{1}H \rightarrow {}^{13}_{7}N + \gamma$$
$$^{13}_{7}N \rightarrow {}^{13}_{6}C + e^{+} + \nu_e$$
$$^{13}_{6}C + {}^{1}_{1}H \rightarrow {}^{14}_{7}N + \gamma$$
$$^{14}_{7}N + {}^{1}_{1}H \rightarrow {}^{15}_{8}O + \gamma$$
$$^{15}_{8}O \rightarrow {}^{15}_{7}N + e^{+} + \nu_e$$
$$^{15}_{7}N + {}^{1}_{1}H \rightarrow {}^{12}_{6}C + {}^{4}_{2}He.$$

See stellar evolution.

carbon dating (radiocarbon dating)
A method of estimating the ages of archaeological specimens of biological origin. As a result of *cosmic radiation a small number of atmospheric nitrogen nuclei are continuously being transformed by neutron bombardment into radioactive nuclei of carbon–14:

$$^{14}_{7}N + n \rightarrow ^{14}_{6}C + p$$

Some of these radiocarbon atoms find their way into living trees and other plants in the form of carbon dioxide, as a result of *photosynthesis. When the tree is cut down photosynthesis stops and the ratio of radiocarbon atoms to stable carbon atoms begins to fall as the radiocarbon decays. The ratio $^{14}C/^{12}C$ in the specimen can be measured and enables the time that has elapsed since the tree was cut down to be calculated. The method has been shown to give consistent results for specimens up to some 40 000 years old, though its accuracy depends upon assumptions concerning the past intensity of the cosmic radiation. The technique was developed by Willard F. Libby (1908–80) and his coworkers in 1946–47.

carbon dioxide A colourless odourless gas, CO_2, soluble in water, ethanol, and acetone; d. 1.977 g dm^{-3} (0°C); m.p. –56.6°C; b.p. –78.5°C. It occurs in the atmosphere (0.03% by volume) but has a short residence time in this phase as it is both consumed by plants during *photosynthesis and produced by *respiration and by combustion. It is readily prepared in the laboratory by the action of dilute acids on metal carbonates or of heat on heavy-metal carbonates. Carbon dioxide is a by-product from the manufacture of lime and from fermentation processes. Carbon dioxide has a small liquid range and liquid carbon dioxide is produced only at high pressures. The molecule CO_2 is linear with each oxygen making a double bond to the carbon. Chemically, it is unreactive and will not support combustion. It dissolves in water to give *carbonic acid.

Large quantities of solid carbon dioxide (*dry ice*) are used in processes requiring large-scale refrigeration. It is also used in fire extinguishers as a desirable alternative to water for most fires, and as a constituent of medical gases as it promotes exhalation. It is also used in carbonated drinks.

The level of carbon dioxide in the atmosphere has increased by some 12% in the last 100 years, mainly because of extensive burning of fossil fuels and the destruction of large areas of rain forest. This has been postulated as the main cause of the average increase of 0.5°C in global temperatures over the same period, through the *greenhouse effect; environmentalists urge that measures be taken to prevent further increases in atmospheric CO_2 concentration and subsequent global warming.

carbon disulphide (carbon bisulphide) A colourless highly refractive liquid, CS_2, slightly soluble in water and soluble in ethanol and ether; r.d. 1.261; m.p. –110°C; b.p. 46.3°C. Pure carbon disulphide has an ethereal odour but the commercial product is contaminated with a variety of other sulphur compounds and has a most unpleasant smell. It was previously manufactured by heating a mixture of wood, sulphur, and charcoal; modern processes use natural gas and sulphur. Carbon disulphide is an excellent solvent for oils, waxes, rubber, sulphur, and phosphorus, but its use is decreasing because of its high toxicity and its flammability. It is used for the preparation of xanthates in the manufacture of viscose yarns.

carbon fibres Fibres of carbon in which the carbon has an oriented crystal structure. Carbon fibres are made by heating textile fibres and are used in strong composite materials for use at high temperatures.

carbonic acid A dibasic acid, H_2CO_3, formed in solution when carbon dioxide is dissolved in water:

$$CO_2(aq) + H_2O(l) \rightleftharpoons H_2CO_3(aq)$$

The acid is in equilibrium with dissolved carbon dioxide, and also dissociates as follows:

$$H_2CO_3 \rightleftharpoons H^+ + HCO_3^-$$
$$K_a = 4.5 \times 10^{-7} \text{ mol dm}^{-3}$$
$$HCO_3^- \rightleftharpoons CO_3^{2-} + H^+$$
$$K_a = 4.8 \times 10^{-11} \text{ mol dm}^{-3}$$

The pure acid cannot be isolated, although it can be produced in ether solution at $-30°C$. Carbonic acid gives rise to two series of salts: the *carbonates and the *hydrogencarbonates.

Carboniferous A geological period in the Palaeozoic era. It began about 345 million years ago, following the Devonian period, and extended until the beginning of the Permian period, about 280 million years ago. In Europe the period is divided into the Lower and Upper Carboniferous, which roughly correspond to the Mississippian and Pennsylvanian periods, respectively, of North America. During the Lower Carboniferous a marine transgression occurred and the characteristic rock of this division – the Carboniferous limestone – was laid down in the shallow seas. Fauna included foraminiferans, corals, bryozoans, brachiopods, blastoids, and other invertebrates. The Upper Carboniferous saw the deposition of the millstone grit, a mixture of shale and sandstone formed in deltaic conditions, followed by the coal measures, alternating beds of coal, sandstone, shale, and clay. The coal was formed from the vast swamp forests composed of seed ferns, lycopsids, and other plants. During the period fishes continued to diversify and amphibians became more common.

carbonium ion An organic ion with a positive charge on a carbon atom; i.e. an ion of the type R_3C^+. Carbonium ions are intermediates in certain types of organic reaction (e.g. *Williamson's synthesis).

carbonize (carburize) To change an organic compound into carbon by heating, or to coat something with carbon in this way.

$C \equiv O$	carbon monoxide
$O - C = O$	carbon dioxide
$O = C = C = C = O$	tricarbon dioxide (carbon suboxide)

Oxides of carbon

carbon monoxide A colourless odourless gas, CO, sparingly soluble in water and soluble in ethanol and benzene; d. 1.25 g dm^{-3}(0°C); m.p. $-199°C$; b.p. $-191.5°C$. It is flammable and highly toxic. In the laboratory it can be made by the dehydration of methanoic acid (formic acid) using concentrated sulphuric acid. Industrially it is produced by the oxidation of natural gas (methane) or (formerly) by the water-gas reaction. It is formed by the incomplete combustion of carbon and is present in car-exhaust gases.

It is a neutral oxide, which burns in air to give carbon dioxide, and is a good reducing agent, used in a number of metallurgical processes. It has the interesting chemical property of forming a range of transition metal carbonyls, e.g. $Ni(CO)_4$. Carbon monoxide is able to use vacant p-orbitals in bonding with metals; the stabilization of low oxidation states, including the zero state, is a consequence of this. This also accounts for its toxicity, which is due to the binding of the CO to the iron in haemoglobin, thereby blocking the uptake of oxygen.

carbon suboxide See tricarbon dioxide.

carbon tetrachloride See tetrachloromethane.

carbonyl chloride (phosgene) A colourless gas, $COCl_2$, with an odour of freshly cut hay. It is used in organic chemistry as a chlorinating agent, and was formerly used as a war gas.

carbonyl compound A compound containing the carbonyl group $>C=O$. Aldehydes, ketones, and carboxylic acids are examples of organic carbonyl compounds. Inorganic carbonyls are complexes in which carbon monoxide has coordinated to a metal atom or ion, as in *nickel carbonyl, $Ni(CO)_4$. *See also* ligand.

carbonyl group The group $>C=O$, found in aldehydes, ketones, carboxylic acids, amides, etc., and in inorganic carbonyl complexes (*see* carbonyl compound).

carborundum *See* silicon carbide.

R—C with =O and O—H — carboxyl group

Carboxylic acid structure

carboxyl group The organic group −COOH, present in *carboxylic acids.

carboxylic acids Organic compounds containing the group −COOH (the *carboxyl group*; i.e. a carbonyl group attached to a hydroxyl group). In systematic chemical nomenclature carboxylic-acid names end in the suffix *-oic*, e.g. ethanoic acid, CH_3COOH. They are generally weak acids. Many long-chain carboxylic acids occur naturally as esters in fats and oils and are therefore also known as *fatty acids. *See also* glycerides.

carburize *See* carbonize.

carbylamine reaction *See* isocyanide test.

carcerulus A dry fruit that is a type of *schizocarp. It consists of a number of one-seeded fragments (*mericarps*) that adhere to a central axis. It is characteristic of mallow.

carcinogen Any agent that produces *cancer, e.g. tobacco smoke, certain industrial chemicals, and *ionizing radiation (such as X-rays and ultraviolet rays).

cardiac 1. Relating to the heart. **2.** Relating to the part of the stomach nearest to the oesophagus.

cardiac muscle A specialized form of *muscle that is peculiar to the vertebrate heart. The muscle fibres are branched and interlock, and the muscle itself shows spontaneous contraction and does not need nervous stimulation (*see* pacemaker). The vagus nerve to the heart can, however, affect the rate of contraction.

Carius method A method of determining the amount of sulphur and halogens in an organic compound, by heating the compound in a sealed tube with silver nitrate in concentrated nitric acid. The compound is decomposed and silver sulphide and halides are precipitated, separated, and weighed.

carnallite A mineral consisting of a hydrated mixed chloride of potassium and magnesium, $KCl.MgCl_2.6H_2O$.

carnassial teeth Molar and premolar teeth modified for shearing flesh by having cusps with sharp cutting edges. They are typical of animals of the order *Carnivora (e.g. tigers, wolves), in which they are the first molars in the lower jaw and the last premolars in the upper.

Carnivora An order of mainly flesh-eating mammals that includes the dogs, wolves, bears, badgers, weasels, and cats. Carnivores typically have very keen sight, smell, and hearing. The hinge joint between the lower jaw and skull is very tight, allowing no lateral movement of the lower jaw. This − together with the arrangement of jaw muscles − enables a very powerful bite. The teeth are specialized for stabbing and tearing flesh: canines are large and pointed and some of the cheek teeth are modified for shearing (*see* carnassial teeth).

carnivore An animal that eats meat, especially a member of the order *Carnivora (e.g. tigers, wolves). Carnivores are specialized by having strong powerful jaws and well-developed canine teeth. They may be predators or carrion eaters. *Compare* herbivore; omnivore.

carnivorous plant (insectivorous plant) Any plant that supplements its supply of nitrates in conditions of nitrate deficiency by digesting small animals, especially insects. Such plants are adapted in various ways to attract and trap the insects and produce proteolytic enzymes to digest them. Venus' fly trap (*Dionaea*), for example, has spiny-margined hinged leaves that snap shut on an alighting insect. Sundews (*Drosera*) trap and digest insects by means of glandular leaves that secrete a sticky substance, and pitcher plants (families Nepenthaceae and Sarraceniaceae) have leaves modified as pitchers into which insects fall, drowning in the water and digestive enzymes at the bottom.

Carnot cycle The most efficient cycle of operations for a reversible *heat engine. Published in 1824 by the French physicist N. L. S. Carnot (1746–1832), it consists of four operations on the working substance in the engine (see illustration):

Carnot cycle

a. Isothermal expansion at thermodynamic temperature T_1 with heat Q_1 taken in.
b. Adiabatic expansion with a fall of temperature to T_2.
c. Isothermal compression at temperature T_2 with heat Q_2 given out.
d. Adiabatic compression with a rise of temperature back to T_1.

According to the *Carnot principle*, the efficiency of any reversible heat engine depends only on the temperature range through which it works, rather than the properties of the working substances. In any reversible engine, the efficiency (η) is the ratio of the work done (W) to the heat input (Q_1), i.e. $\eta = W/Q_1$. As, according to the first law of *thermodynamics, $W = Q_1 - Q_2$, it follows that $\eta = (Q_1 - Q_2)/Q_1$. For the Kelvin temperature scale, $Q_1/Q_2 = T_1/T_2$ and $\eta = (T_1 - T_2)/T_1$. For maximum efficiency T_1 should be as high as possible and T_2 as low as possible.

carnotite A radioactive mineral consisting of hydrated uranium potassium vanadate, $K_2(UO_2)_2(VO_4)_2.nH_2O$. It varies in colour from bright yellow to lemon- or greenish-yellow. It is a source of uranium, radium, and vanadium. The chief occurrences are in the Colorado Plateau, USA; Radium Hill, Australia; and Katanga, Zaïre.

Caro's acid *See* peroxosulphuric(VI) acid.

carotene A member of a class of *carotenoid pigments. Examples are β-carotene and lycopene, which colour carrot roots and ripe tomato fruits respectively. α- and β-carotene yield vitamin A when they are broken down during animal digestion.

carotenoid Any of a group of yellow, orange, red, or brown plant pigments chemically related to terpenes. Carotenoids are responsible for the characteristic colour of many plant organs, such as ripe tomatoes, carrots, and autumn leaves. They also function in the light reactions of *photosynthesis. *See* carotene; xanthophyll.

carotid artery The major artery that supplies blood to the head. A pair of *common carotid arteries* arise from the aorta (on the left) and the innominate artery (on the right) and run up the neck; each branches into an *external* and an *internal carotid artery*, which supply the head.

carotid body One of a pair of tissue masses adjacent to the *carotid sinus. Each contains receptors that are sensitive to oxygen and pH levels (acidity) in the blood. High levels of carbon dioxide in the blood lower the pH (i.e. increase the acidity). By responding to fluctuations in pH, the carotid body coordinates reflex changes in respiration rate.

carotid sinus An enlarged region of the *carotid artery at its major branching point in the neck. Its walls contain many receptors that are sensitive to changes in pressure and it regulates blood pressure by initiating reflex changes in heart rate and dilation of blood vessels.

carpal (carpal bone) One of the bones that form the wrist (*see* carpus) in terrestrial vertebrates.

carpel The female reproductive organ of a flower. Typically it consists of a *stigma, *style, and *ovary. It is thought to have evolved by the fusion of the two edges of a flattened megasporophyll (*see* sporophyll). Each flower may have one carpel (*monocarpellary*) or many (*polycarpellary*), either free (*apocarpous*) or fused together (*syncarpous*). *See also* pistil.

carpus The wrist (or corresponding part of the forelimb) in terrestrial vertebrates, consisting of a number of small bones (*carpals*). The number of carpal bones varies with the species. The rabbit, for example, has two rows of carpals, the first (proximal) row containing three bones and the second (distal) row five. In humans

there are also eight carpals. This large number of bones enables flexibility at the wrist joint, between the hand and forelimb. *See also* pentadactyl limb.

carrier 1. (in radio) *See* carrier wave. **2.** (in physics) *See* charge carrier. **3.** (in medicine) An individual who harbours a particular disease-causing microorganism without ill-effects and who can transmit the microorganism to others. *Compare* vector. **4.** (in genetics) An individual with an *allele for some defective condition that is masked by a normal *dominant allele. Such individuals therefore do not suffer from the condition themselves but they may pass on the defective allele to their offspring. In humans, women may be carriers of such conditions as red–green colour blindness and haemophilia, the alleles for which are carried on the X chromosomes (*see* sex linkage).

carrier gas The gas that carries the sample in *gas chromatography.

carrier wave An electromagnetic wave of specified frequency and amplitude that is emitted by a radio transmitter in order to carry information. The information is superimposed onto the carrier by making small changes in either its amplitude or its frequency (*see* modulation).

Cartesian coordinates A system used in analytical geometry to locate a point P, with reference to two or three *axes* (see graphs). In a two-dimensional system the vertical axis is the y-axis and the horizontal axis is the x-axis. The point at which the axes intersect each other is called the *origin*, O. Values of y <0 fall on the y-axis below the origin, values of x <0 fall on the x-axis to the left of the origin. Any point P is located by its perpendicular distances from the two axes. The distance from the x-axis is called the *ordinate*; the distance from the y-axis is the *abscissa*. The position is indicated numerically by enclosing the values of the

two-dimensional system

three-dimensional system

Cartesian coordinates

abscissa and the ordinate in parentheses and separating the two by means of a comma, e.g. (x,y). In three dimensions the system can be used to locate a point with reference to a third, z-axis. It is named after René Descartes (1596–1650).

cartilage (gristle) A firm flexible connective tissue that forms the adult skeleton of cartilaginous fish (e.g. sharks). It comprises a matrix consisting chiefly of a mucopolysaccharide called *chondroitin sulphate* secreted by cells (*chondroblasts*) that become embedded in the matrix as *chondrocytes*. In other vertebrates cartilage

forms the skeleton of the embryo, being largely replaced by *bone in mature animals. It persists in areas of high wear (e.g. bone ends, joints, and intervertebral discs) and in the nose and pinna of the ear.

cartilage bone (replacing bone) *Bone that is formed by replacing the cartilage of an embryo skeleton. The process, called *ossification*, is brought about by the cells (osteoblasts) that secrete bone. *Compare* membrane bone.

cartilaginous fishes *See* Elasmobranchii.

cartography The science of the production of maps and charts. Maps may be based on original surveys, aerial photographs (photogrammetry), or compiled from existing maps and records. *See also* map projections.

caruncle A small outgrowth from the testa of a seed that develops from the placenta, funicle, or micropyle. Examples include the warty outgrowth from the castor-oil seed and the tuft of hairs on the testa of the seed of willowherb. *See also* aril.

caryopsis A dry single-seeded indehiscent fruit that differs from an *achene in that the fruit wall is fused to the testa of the seed. It is the grain of cereals and grasses.

cascade liquefier An apparatus for liquefying a gas of low *critical temperature. Another gas, already below its critical temperature, is liquified and evaporated at a reduced pressure in order to cool the first gas to below its critical temperature. In practice a series of steps is often used, each step enabling the critical temperature of the next gas to be reached.

cascade process Any process that takes place in a number of steps, usually because the single step is too inefficient to produce the desired result. For example, in some uranium-enrichment processes the separation of the desired isotope is only poorly achieved in a single stage; to achieve better separation the process has to be repeated a number of times, in a series, with the enriched fraction of one stage being fed to the succeeding stage for further enrichment. Another example of cascade process is that operating in a *cascade liquefier.

case hardening The hardening of the surface layer of steel, used for tools and certain mechanical components. The commonest method is to carburize the surface layer by heating the metal in a hydrocarbon or by dipping the red hot metal into molten sodium cyanide. Diffusion of nitrogen into the surface layer to form nitrides is also used.

casein One of a group of phosphate-containing proteins (phosphoproteins) found in milk. Caseins are easily digested by the enzymes of young mammals and represent a major source of phosphorus. *See* rennin.

Cassegrainian telescope *See* telescope.

cassiterite A yellow, brown, or black form of tin(IV) oxide, SnO_2, that forms tetragonal, often twinned, crystals; the principal ore of tin. It occurs in hydrothermal veins and metasomatic deposits associated with acid igneous rocks and in alluvial (placer) deposits. The chief producers are Malaysia, Indonesia, Zaïre, and Nigeria.

caste A division found in social insects, such as the *Hymenoptera (ants, bees, wasps) and the Isoptera (termites), in which the individuals are structurally and physiologically specialized to perform a particular function. For example, in honeybees there are queens (fertile females), workers (sterile females), and drones (males). There are several different castes of workers (all sterile females) among ants.

cast iron A group of iron alloys containing 1.8 to 4.5% of carbon. It is usually cast into specific shapes ready for machining, heat treatment, or assembly. It is sometimes produced direct from the *blast furnace or it may be made from remelted *pig iron.

catabolism The metabolic breakdown of large molecules in living organisms to smaller ones, with the release of energy. Respiration is an example of a catabolic series of reactions. *See* metabolism. *Compare* anabolism.

catalysis The process of changing the rate of a chemical reaction by use of a *catalyst.

catalyst A substance that increases the rate of a chemical reaction without itself undergoing any permanent chemical change. Catalysts that have the same phase as the reactants are *homogeneous catalysts* (e.g. *enzymes in biochemical reactions or transition-metal complexes used in the liquid phase for catalysing organic reactions). Those that have a different phase are *hetereogeneous catalysts* (e.g. metals or oxides used in many industrial gas reactions). The catalyst provides an alternative pathway by which the reaction can proceed, in which the activation energy is lower. It thus increases the rate at which the reaction comes to equilibrium, although it does not alter the position of the equilibrium. The catalyst itself takes part in the reaction and consequently may undergo physical change (e.g. conversion into powder). In certain circumstances, very small quantities of catalyst can speed up very large reactions. Some catalysts are also highly specific in the type of reaction they catalyse, particularly enzymes in biochemical reactions. Generally, the term is used for a substance that increases reaction rate (a *positive catalyst*). Some reactions can be slowed down by *negative catalysts* (*see* inhibition).

catalytic cracking *See* cracking.

cataphoresis *See* electrophoresis.

catechol *See* 1,2-dihydroxybenzene.

catecholamine Any of a class of amines (including *dopamine, *adrenaline, and *noradrenaline) that function as *neurotransmitters and/or hormones.

category (in taxonomy) *See* rank.

catenary A curve formed when a chain or rope of uniform density hangs from two fixed points. If the lowest point on the curve passes through the origin, the equation is $y = c(\cosh x/c)$, where c is the distance between the x-axis and the directrix.

catenation The formation of chains of atoms in chemical compounds.

cathetometer A telescope or microscope fitted with crosswires in the eyepiece and mounted so that it can slide along a graduated scale. Cathetometers are used for accurate measurement of lengths without mechanical contact. The microscope type is often called a *travelling microscope*.

cathode A negative electrode. In *electrolysis cations are attracted to the cathode. In vacuum electronic devices electrons are emitted by the cathode and flow to the *anode. It is therefore from the cathode that electrons flow into these devices. However, in a primary or secondary cell the cathode is the electrode that spontaneously becomes negative during discharge, and from which therefore electrons emerge.

cathode-ray oscilloscope (CRO) An instrument based on the *cathode-ray tube that provides a visual image of electrical signals. The horizontal deflection is usually provided by an internal time base, which causes the beam to sweep across the screen at a specified rate. The signal to be investigated is fed to the vertical deflection plates after amplifica-

tion. Thus the beam traces a graph of the signal against time.

cathode rays Streams of electrons emitted at the cathode in an evacuated tube containing a cathode and an anode. They were first observed in gas *discharge tubes operated at low pressure. Under suitable conditions electrons produced by secondary emission at the cathode are accelerated down the tube to the anode. In such devices as the *cathode-ray tube the electrons are produced by *thermionic emission from a hot cathode in a vacuum.

cathode-ray tube (CRT) The device that provides the viewing screen in the television tube, the radar viewer, and the *cathode-ray oscilloscope. The cathode-ray tube consists of an evacuated tube containing a heated cathode and two or more ring-shaped anodes through which the cathode rays can pass so that they strike the enlarged end of the tube (see illustration). This end of the tube is coated with fluorescent material so that it provides a screen. Any point on the screen that is struck by the cathode ray becomes luminous. A *control grid between the cathode and the anode enables the intensity of the beam to be varied, thus controlling the brightness of the illumination on the screen. The assembly of cathode, control grid, and anode is called the *electron gun. The beam emerging from the electron gun is focused and

Cathode-ray tube

113

deflected by means of plates providing an electric field or coils providing a magnetic field. This enables the beam to be focused to a small point of light and deflected to produce the illusion of an illuminated line as this point sweeps across the tube.

The television tube is a form of cathode-ray tube in which the beam is made to scan the screen 625 times to form a frame, with 25 new frames being produced every second. (These are the figures for standard television tubes in the UK). Each frame creates a picture by variations in the intensity of the beam as it forms each line.

cathodic protection See sacrificial protection.

cation A positively charged ion, i.e. an ion that is attracted to the cathode in *electrolysis. Compare anion.

cationic detergent See detergent.

cationic dye See dyes.

cationic resin See ion exchange.

catkin A type of flowering shoot (see racemose inflorescence) in which the axis, which is often long, bears many small stalkless unisexual flowers. Usually the male catkins hang down from the stem; the female catkins are shorter and often erect. Examples include birch and hazel. Most plants with catkins are adapted for wind pollination, the male flowers producing large quantities of pollen; willows are an exception, having nectar-secreting flowers and being pollinated by insects.

caudal vertebrae The bones (see vertebra) of the tail, which articulate with the *sacral vertebrae. The number of caudal vertebrae varies with the species. Rabbits, for example, have 15 caudal vertebrae, while in man these vertebrae are fused to form a single bone, the *coccyx.

caustic 1. (in chemistry) Describing a substance that is strongly alkaline (e.g. caustic soda). **2.** (in optics) The curve or surface formed by the reflection of parallel rays of light in a large-aperture concave mirror. The apex of the caustic lies at the principal focus of the mirror. Such a curve can sometimes be seen on the surface of the liquid in a cup as a result of reflection by the curved walls of the cup. A similar curve is formed by a convex lens with spherical surfaces refracting parallel rays of light.

caustic potash See potassium hydroxide.

caustic soda See sodium hydroxide.

cavitation The formation of gas- or vapour-filled cavities in liquids in motion when the pressure is reduced to a critical value while the ambient temperature remains constant. If the velocity of the flowing liquid exceeds a certain value, the pressure can be reduced to such an extent that *Bernoulli's theorem breaks down. It is at this point that cavitation occurs, causing a restriction on the speed at which hydraulic machinery can be run without noise, vibration, erosion of metal parts, or loss of efficiency.

cavity resonator See resonant cavity.

c.c.p. Cubic close packing. See close packing.

CD (cluster of differentiation) Any group of antigens that is associated with a specific subpopulation of human *T cells. The differentiation antigens expressed by a T cell vary with its stage of development and thus with its role in the immune response. Hence, for example, CD4 antigens are expressed by helper T cells, whereas CD8 antigens are expressed by cytotoxic and suppressor T cells. The antigens are characterized using *monoclonal antibodies.

cDNA See complementary DNA.

celestial equator See equator.

celestial mechanics The study of the motions of and forces between the celestial bodies. It is based on *Newton's laws of motion and *Newton's law of gravitation. Refinements

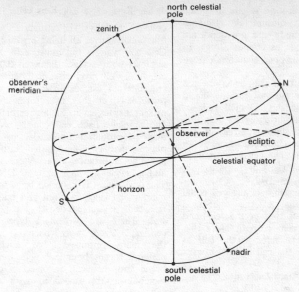

Celestial sphere

based on the general theory of *relativity are also included in the study, although the differences between the two theories are only important in a few cases.

celestial sphere The imaginary sphere of infinite radius within which celestial bodies appear to lie. The earth, and the observer, are visualized as being at the centre of the sphere and the sphere as rotating once every sidereal *day (see illustration). The sphere is used to describe the position of celestial bodies with respect to the earth.

celestine A mineral form of strontium sulphate, $SrSO_4$.

cell (in physical chemistry) **1.** A system in which two electrodes are in contact with an electrolyte. The electrodes are metal or carbon plates or rods or, in some cases, liquid metals (e.g. mercury). In an *electrolytic cell a current from an outside source is passed through the electrolyte to produce chemical change (see electrolysis). In a *voltaic cell, spontaneous reactions between the electrodes and electrolyte(s) produce a potential difference between the two electrodes.

Voltaic cells can be regarded as made up of two *half cells, each composed of an electrode in contact with an electrolyte. For instance, a zinc rod dipped in zinc sulphate solution is a $Zn|Zn^{2+}$ half cell. In such a system zinc atoms dissolve as zinc ions, leaving a negative charge on the electrode

$$Zn(s) \rightarrow Zn^{2+}(aq) + 2e$$

The solution of zinc continues until the charge build-up is sufficient to prevent further ionization. There is then a potential difference between the zinc rod and its solution. This cannot be measured directly, since measurement would involve making contact with the electrolyte, thereby introducing another half cell (see electrode potential). A rod of copper in copper sulphate solution comprises another half cell. In this case the spontaneous reaction is one in which copper ions in solution take electrons from the electrode and are deposited on the electrode as copper atoms. In this case, the copper acquires a positive charge.

The two half cells can be connected by using a porous pot for the liquid junction (as in the *Daniell cell) or by using a salt bridge. The resulting cell can then supply current if the electrodes are connected through an external circuit. The cell is written

$$Zn(s)|Zn^{2+}(aq)|Cu^{2+}(aq)|Cu$$
$$E = 1.10 \text{ V}$$

Here, E is the e.m.f. of the cell equal to the potential of the right-hand electrode minus that of the left-hand electrode for zero current. Note that 'right' and 'left' refer to the cell as written. Thus, the cell could be written

$$Cu(s)|Cu^{2+}(aq)|Zn^{2+}(aq)|Zn(s)$$
$$E = -1.10 \text{ V}$$

The overall reaction for the cell is

$$Zn(s) + Cu^{2+}(aq) \rightarrow Cu(s) + Zn^{2+}(aq)$$

This is the direction in which the cell reaction occurs for a positive e.m.f.

The cell above is a simple example of a *chemical cell*; i.e. one in which the e.m.f. is produced by a chemical difference. *Concentration cells* are cells in which the e.m.f. is caused by a difference of concentration. This may be a difference in concentration of the electrolyte in the two half cells. Alternatively, it may be an electrode concentration difference (e.g. different concentrations of metal in an amalgam, or different pressures of gas in two gas electrodes). Cells are also classified into cells *without transport* (having a single electrolyte) and *with transport* (having a liquid junction across which ions are transferred). Various types of voltaic cell exist, used as sources of current, standards of potential, and experimental set-ups for studying electrochemical reactions.

2. *See* photoelectric cell.

3. *See* solar cell.

4. *See* Kerr effect (for Kerr cell).

cell (in biology) The structural and functional unit of all living organisms. Cell size varies, but most cells are microscopic (average diameter 0.01– 0.1 mm). Cells may exist as independent units of life, as in bacteria and protozoans, or they may form colonies or tissues, as in all higher plants and animals. Each cell consists of a mass of protein material (*see* protoplasm) that is differentiated into a jelly-like substance (*see* cytoplasm) and a *nucleus, which contains DNA. The protoplasm is bounded by a *cell membrane, which in plant and bacterial cells is surrounded by a *cell wall. There are two main types of cell. *Prokaryotic cells* (bacteria) are the more primitive. The nuclear material is not bounded by a membrane and chemicals involved in cell metabolism are associated with the cell membrane. Reproduction is asexual and involves simple cell cleavage. In *eukaryotic cells* the nucleus is bounded by a nuclear membrane and the cytoplasm is divided by membranes into a system of interconnected cavities and separate compartments (*organelles*), e.g. *mitochondria, *endoplasmic reticulum, *Golgi apparatus, *lysosomes, and *ribosomes (see illustration). Reproduction can be either asexual (*see* mitosis) or sexual (*see* meiosis). Plants and animals consist of eukaryotic cells but plant cells possess *chloroplasts and other *plastids and bear a rigid cellulose cell wall.

cell body (perikaryon) The part of a *neurone that contains the nucleus. The cell processes that are involved in the transmission and reception of nervous impulses (the axon and the dendrites respectively) develop as extensions from the cell body.

cell division The formation of two daughter cells from a single mother cell. The nucleus divides first and this is followed by the formation of a cell

Generalized eukaryotic cells

membrane between the daughter nuclei. In *mitosis the daughter nuclei are identical to the original nucleus; *meiosis results in daughter nuclei each with half the number of chromosomes in the mother cell nucleus.

cell membrane (plasmalemma; plasma membrane) The semipermeable *membrane forming the outer limit of a *cell. It consists mostly of protein and lipid. The membrane regulates the flow of materials in and out of the cell and plays a role in the production or assembly of the *cell wall or *capsule (when this is present).

celluloid A transparent highly flammable substance made from cellulose nitrate with a camphor plasticizer. It was formerly widely used as a thermoplastic material, especially for film, (a use now discontinued owing to the inflammability of celluloid).

cellulose A polysaccharide that consists of a long unbranched chain of glucose units. It is the main constituent of the cell walls of all higher plants, many algae, and some fungi and is responsible for providing the rigidity of the cell wall. It is an important constituent of dietary *fibre. The fibrous nature of extracted cellulose has led to its use in the textile industry for the production of cotton, artificial silk, etc.

cellulose acetate *See* cellulose ethanoate.

cellulose ethanoate (cellulose acetate) A compound prepared by treating cellulose (cotton linters or wood pulp) with a mixture of ethanoic anhydride, ethanoic acid, and concentrated sulphuric acid. Cellulose in the cotton is ethanoylated and when the resulting solution is treated with water, cellulose ethanoate forms as a flocculent white mass. It is used in lacquers, nonshatterable glass, varnishes, and as a fibre (*see also* rayon).

cellulose nitrate A highly flammable material made by treating cellulose (wood pulp) with concentrated nitric acid. Despite the alternative name *nitrocellulose*, the compound is in fact an ester (containing $CONO_2$ groups), not a nitro compound (which would contain $C-NO_2$). It is used in explosives (as *guncotton*) and celluloid.

cell wall A rigid outer layer that surrounds the cell membrane of plant and bacterial (but not animal) cells.

It protects and/or gives shape to a cell, and in herbaceous plants provides mechanical support for the plant body. Most plant cell walls are composed of the polysaccharide *cellulose and may be strengthened by the addition of *lignin. The cell walls of fungi consist mainly of *chitin. Bacterial cell walls consist of complex polymers of polysaccharides and amino acids.

Celsius scale A *temperature scale in which the fixed points are the temperatures at standard pressure of ice in equilibrium with water (0°C) and water in equilibrium with steam (100°C). The scale, between these two temperatures, is divided in 100 degrees. The degree Celsius (°C) is equal in magnitude to the *kelvin. This scale was formerly known as the *centigrade scale*; the name was officially changed in 1948 to avoid confusion with a hundredth part of a grade. It is named after the Swedish astronomer Anders Celsius (1701–44), who devised the inverted form of this scale (ice point 100°, steam point 0°) in 1742.

cement 1. Any of various substances used for bonding or setting to a hard material. Portland cement is a mixture of calcium silicates and aluminates made by heating limestone ($CaCO_3$) with clay (containing aluminosilicates) in a kiln. The product is ground to a fine powder. When mixed with water it sets in a few hours and then hardens over a longer period of time due to the formation of hydrated aluminates and silicates. **2.** (or **cementum**) A thin layer of bony material that fixes teeth to the jaw. It covers the dentine of the root of a *tooth, below the level of the gum, and is attached to the *periodontal membrane lining the tooth socket in the jawbone.

cementation Any metallurgical process in which the surface of a metal is impregnated by some other substance, especially an obsolete process for making steel by heating bars of wrought iron to red heat for several days in a bed of charcoal. *See also* case hardening.

cementite *See* steel.

Cenozoic (Cainozoic; Kainozoic) The geological era that began about 65 million years ago and extends to the present. It followed the *Mesozoic era and is subdivided into the *Tertiary and *Quaternary periods. The Cenozoic is often known as the *Age of Mammals* as these animals evolved to become an abundant, diverse, and dominant group. Birds and flowering plants also flourished. The era saw the formation of the major mountain ranges of the Himalayas and the Alps.

centi- Symbol c. A prefix used in the metric system to denote one hundredth. For example, 0.01 metre = 1 centimetre (cm).

centigrade scale *See* Celsius scale.

centipedes *See* Myriapoda.

Central Dogma The basic belief originally held by molecular geneticists, that flow of genetic information can only occur from *DNA to *RNA to proteins. It is now known, however, that information contained within RNA molecules of viruses can also flow back to DNA. *See also* genetic code.

central nervous system (CNS) The part of the nervous system that coordinates all neural functions. In invertebrates it may comprise simply a few *nerve cords and their associated *ganglia. In vertebrates it consists of the *brain and the *spinal cord. The vertebrate CNS contains *reflex arcs, which produce automatic and rapid responses to particular stimuli.

central processing unit (CPU) *See* computer.

centre (in neurology) A part of the nervous system, consisting of a group of nerve cells, that coordinates a particular process. An example is the respiratory centre in the vertebrate brainstem, which controls breathing movements. The stimulation of a centre will initiate the process, while destruction of the centre will prevent or impair it.

centre of curvature The centre of the sphere of which a *lens surface or curved *mirror forms a part. The *radius of curvature* is the radius of this sphere.

centre of gravity *See* centre of mass.

centre of mass The point at which the whole mass of a body may be considered to be concentrated. This is the same as the *centre of gravity*, the point at which the whole weight of a body may be considered to act, if the body is situated in a uniform gravitational field.

centrifugal force *See* centripetal force.

centrifugal pump *See* pump.

centrifuge A device in which solid or liquid particles of different densities are separated by rotating them in a tube in a horizontal circle. The denser particles tend to move along the length of the tube to a greater radius of rotation, displacing the lighter particles to the other end.

centriole A cylindrical structure in a cell that has a role in cell division. Centrioles are not normally found in plant cells but are commonly present in pairs in animal cells. During cell division they move to opposite sides of the nucleus to form the ends of the *spindle. The spindle fibres develop from them.

centripetal force A force acting on a body causing it to move in a circular path. If the mass of the body is m, its constant speed v, and the radius of the circle r, the magnitude of the force is mv^2/r and it is directed towards the centre of the circle. Even though the body is moving with a

constant speed v, its velocity is changing, because its direction is constantly changing. There is therefore an acceleration v^2/r towards the centre of the circle. For example, when an object is tied to a string and swung in a horizontal circle there is a tension in the string equal to mv^2/r. If the string breaks, this restraining force disappears and the object will move off in a straight line along the tangent to the circle in which it was previously moving.

In the case of a satellite (mass m) orbiting the earth (mass M), the centripetal force holding the satellite in orbit is the *gravitational force, GmM/d^2, where G is the gravitational constant and d is the height of the satellite above the centre of the earth. Therefore $GmM/d^2 = mv^2/d$. This equation enables the height of the orbit to be calculated for a given orbital velocity.

Another way of looking at this situation, which was once popular, is to assume that the centripetal force is balanced by an equal and opposite force, acting away from the centre of the circle, called the *centrifugal force*. One could then say that the satellite stays in orbit when the centrifugal force balances the gravitational force. This is, however, a confusing argument because the centrifugal force is fictitious − it does not exist. The gravitational force is not balanced by the centrifugal force: it *is* the centripetal force.

Another example is that of a car rounding a bend. To an observer in the car, a tennis ball lying on the back shelf will roll across the shelf as if it was acted on by an outward centrifugal force. However, to an observer outside the car it can be seen that the ball, because of its almost frictionless contact with the car, is continuing in its straight line motion, uninfluenced by the centripetal force. Occasionally the concept of a centrifugal force can be useful, as long as it is recognized as a fictitious force. A true centrifugal force is exerted, as a *reaction, by the rotating object on whatever is providing its centripetal force.

centroid The point within an area or volume at which the centre of mass would be if the surface or body had a uniform density. For a symmetrical area or volume it coincides with the centre of mass. For a nonsymmetrical area or volume it has to be found by integration.

centromere (kinomere; spindle attachment) The part of a *chromosome that contains no genes. It usually appears as a constriction when chromosomes contract during cell division (*see* meiosis; mitosis). The position of the centromere is a distinguishing feature of individual chromosomes and it attaches them to the *spindle fibres. *See also* kinetochore.

centrum *See* vertebra.

cephalization The tendency among animal groups for the major sense organs, mouth, and brain to be grouped together at the front (anterior) end of the body. These are usually contained in a specialized cephalic region – the head.

Cephalopoda The most advanced class of molluscs, containing the squids, cuttlefishes, octopuses, and the extinct *ammonites. Cephalopods have a highly concentrated central nervous system within a protective cartilaginous case. The eye has a well-developed retina and is comparable to that of vertebrates. All cephalopods are predacious carnivores capable of swimming by jet propulsion; they have highly mobile tentacles for catching and holding prey.

cephalothorax The fused head and thorax of crustaceans and arachnids (spiders, scorpions), which is connected to the abdomen.

ceramics Inorganic materials, such as pottery, enamels, and refractories. Ceramics are metal silicates, oxides, nitrides, etc.

cerebellum The part of the vertebrate brain concerned with the coordination and regulation of muscle activity and the maintenance of muscle tone and balance. In mammals it consists of two connected hemispheres, composed of a core of white matter and a much-folded outer layer of grey matter, and is situated above the medulla oblongata and partly beneath the cerebrum.

cerebral cortex (pallium) The layer of *grey matter that forms the outer layer of the hemispheres of the *cerebrum in many vertebrates. It is most highly developed in mammals. The cortex is responsible for the control and integration of voluntary movement and the senses of vision, hearing, touch, etc.; it also contains centres concerned with memory, language, thought, and intellect.

cerebral hemisphere Either of the two halves of the vertebrate *cerebrum.

cerebrospinal fluid (CSF) The fluid, similar in composition to *lymph, that bathes the central nervous system of vertebrates. It is secreted by the *choroid plexus into the *ventricles of the brain, filling these and other cavities in the brain and spinal cord, and is reabsorbed by veins on the brain surface. Its function is to protect the central nervous system from mechanical injury.

cerebrum The largest part of the vertebrate brain. It consists of two *cerebral hemispheres*, which develop from the embryonic *forebrain. The hemispheres have an outer convoluted layer of grey matter – the *cerebral cortex – which contains an estimated ten billion nerve cells. Underneath this is *white matter. The two halves of the cerebrum are linked by several neural structures. The function of the cerebrum is to integrate complex sensory and neural functions. The cerebrum is thought to play a critical role in the process of learning, which involves both short-term and long-term memory.

Cerenkov counter (Cerenkov detector) A type of *counter for detecting and counting high-energy charged particles. The particles pass through a liquid and the light emitted as *Cerenkov radiation is registered by a *photomultiplier tube.

Cerenkov radiation Electromagnetic radiation, usually bluish light, emitted by a beam of high-energy charged particles passing through a transparent medium at a speed greater than the speed of light in that medium. It was discovered in 1934 by the Russian physicist Pavel Cerenkov (1904–). The effect is similar to that of a *sonic boom when an object moves faster than the speed of sound; in this case the radiation is a shock wave set up in the electromagnetic field. Cerenkov radiation is used in the *Cerenkov counter.

cerium Symbol Ce. A silvery metallic element belonging to the *lanthanoids; a.n. 58; r.a.m. 140.12; r.d. 6.77 (20°C); m.p. 798°C; b.p. 3433°C. It occurs in allanite, bastnasite, cerite, and monazite. Four isotopes occur naturally: cerium–136, –138, –140, and –142; fifteen radioisotopes have been identified. Cerium is used in mischmetal, a rare-earth metal containing 25% cerium, for use in lighter flints. The oxide is used in the glass industry. It was discovered by M. H. Klaproth in 1803.

cermet A composite material consisting of a ceramic in combination with a sintered metal, used when a high resistance to temperature, corrosion, and abrasion is needed.

CERN (Conseil Européen pour la Recherche Nucléaire) The European Laboratory for Particle Physics, formerly known as the European Organization for Nuclear Research, which is situated close to Geneva in Switzerland and is supported by a number of European nations. It runs the *Super Proton Synchrotron* (SPS), which has a 7-kilometre underground tunnel enabling protons to be accelerated to 400 GeV, and the *Large Electron-Positron Collider* (LEP), in which 50 GeV electron and positron beams are collided.

cerussite An ore of lead consisting of lead carbonate, $PbCO_3$. It is usually of secondary origin, formed by the weathering of *galena. Pure cerussite is white but the mineral may be grey due to the presence of impurities. It forms well-shaped orthorhombic crystals. It occurs in the USA, Spain, and SW Africa.

cervical vertebrae The *vertebrae of the neck. The number of cervical vertebrae varies: for example, rabbits have 7 and humans have 12. Their main functions are to support the head and to provide articulating surfaces against which it can move relative to the backbone. See atlas; axis.

cervix A narrow or necklike part of an organ. The cervix of the uterus (*cervix uteri*) leads to the vagina. Glands in its walls produce mucus, whose viscosity changes according to the oestrous cycle. During labour, the cervix enlarges greatly to allow passage of the fetus.

Cestoda A class of flatworms (*see* Platyhelminthes) comprising the tapeworms – ribbon-like parasites within the gut of vertebrates. Tapeworms are surrounded by partially digested food in the host gut so they are able to absorb nutrients through their whole body surface. The body consists of a head (*scolex*), bearing suckers and hooks for attachment, and a series of segments (*proglottides*), containing male and female reproductive systems. The life cycle of a tapeworm requires two hosts, the primary host usually being a predator of the secondary host. *Taenia solium* has man for its primary host and the pig as its secondary host. Mature segments, containing thousands of fertilized eggs, leave the primary host with its faeces and develop into embryos and then larvae that continue the life cycle in the gut of a secondary host.

Cetacea An order of marine mammals comprising the whales, which includes what is probably the largest known animal – the blue whale (*Balaenoptera musculus*), over 30 m long and over 150 tonnes in weight. The forelimbs of whales are modified as short stabilizing flippers and the skin is very thin and almost hairless. A thick layer of blubber insulates the body against heat loss and is an important food store. Whales breathe through a dorsal blowhole, which is closed when the animal is submerged. The toothed whales (suborder Odontoceti), such as the dolphins and killer whale, are carnivorous; whalebone whales (suborder Mysticeti), such as the blue whale, feed on plankton filtered by *whalebone plates.

cetane *See* hexadecane.

cetane number A number that provides a measure of the ignition characteristics of a Diesel fuel when it is burnt in a standard Diesel engine. It is the percentage of cetane (hexadecane) in a mixture of cetane and 1-methylnaphthalene that has the same ignition characteristics as the fuel being tested. *Compare* octane number.

CFC *See* chlorofluorocarbon.

c.g.s. units A system of *units based on the centimetre, gram, and second. Derived from the metric system, it was badly adapted to use with thermal quantities (based on the inconsistently defined *calorie) and with electrical quantities (in which two systems, based respectively on unit permittivity and unit permeability of free space, were used). For many scientific purposes c.g.s. units have now been replaced by *SI units.

chaeta A bristle, made of *chitin, occurring in annelid worms. In the earthworm they occur in small groups projecting from the skin in each segment and function in locomotion. The chaetae of polychaete worms (e.g. ragworm) are borne in larger groups on paddle-like appendages (*parapodia*).

chain A line of atoms of the same type in a molecule. In a *straight chain* the atoms are attached only to single atoms, not to groups. Propane, for instance, is a straight-chain alkane, $CH_3CH_2CH_3$, with a chain of three carbon atoms. A *branched chain* is one in which there are side groups attached to the chain. Thus, 3-ethyloctane, $CH_3CH_2CH(C_2H_5)C_5H_{11}$, is a branched-chain alkane in which there is a *side chain* (C_2H_5) attached to the third carbon atom. A *closed chain* is a *ring of atoms in a molecule; otherwise the molecule has an *open chain*.

chain reaction A reaction that is self-sustaining as a result of the products of one step initiating a subsequent step.

In nuclear chain reactions the succession depends on production and capture of neutrons. Thus, one nucleus of the isotope uranium–235 can disintegrate with the production of two or three neutrons, which cause similar fission of adjacent nuclei. These in turn produce more neutrons. If the total amount of material exceeds a *critical mass, the chain reaction may cause an explosion.

Chemical chain reactions usually involve free radicals as intermediates. An example is the reaction of chlorine with hydrogen initiated by ultraviolet radiation. A chlorine molecule is first split into atoms:

$$Cl_2 \rightarrow Cl \cdot + Cl \cdot$$

These react with hydrogen as follows

$$Cl\cdot + H_2 \rightarrow HCl + H\cdot$$
$$H\cdot + Cl_2 \rightarrow HCl + Cl\cdot \text{ etc.}$$
Combustion and explosion reactions involve similar free-radical chain reactions.

chair conformation See conformation.

chalaza 1. A twisted strand of fibrous albumen in a bird's egg that is attached to the membrane at either end of the yolk and thus holds the yolk in position in the albumen. **2.** The part of a plant *ovule where the nucellus and integuments merge.

chalcedony A mineral consisting of a microcrystalline variety of *quartz. It occurs in several forms, including a large number of semiprecious gemstones; for example, sard, carnelian, jasper, onyx, chrysoprase, agate, and tiger's-eye.

chalcogens See group VI elements.

chalconides Binary compounds formed between metals and group VI elements; i.e. oxides, sulphides, selenides, and tellurides.

chalcopyrite (copper pyrites) A brassy yellow mineral consisting of a mixed copper–iron sulphide, $CuFeS_2$, crystallizing in the tetragonal system; the principal ore of copper. It is similar in appearance to pyrite and gold. It crystallizes in igneous rocks and hydrothermal veins associated with the upper parts of acid igneous intrusions. Chalcopyrite is the most widespread of the copper ores, occurring, for example, in Cornwall (UK), Sudbury (Canada), Chile, Tasmania (Australia), and Rio Tinto (Spain).

chalk A very fine-grained white rock composed of the fossilized skeletal remains of marine plankton known as *coccoliths* and consisting largely of *calcium carbonate ($CaCO_3$). It is used in toothpaste and cosmetics and is the characteristic rock of the *Cretaceous period. It should not be confused with blackboard 'chalk', which is made from calcium sulphate.

Chandrasekhar limit The maximum possible mass of a star that is prevented from collapsing under its own gravity by the *degeneracy pressure of either electrons (a *white dwarf) or neutrons (a *neutron star). For white dwarfs the *Chandrasekhar mass* is about 1.4 times the mass of the sun. For neutron stars its value is less precisely known because of uncertainties regarding the equation of state of neutron matter, but it is generally taken to be in the range of 1.5 to 3 (and almost certainly no more than 5) times the mass of the sun.

change of phase (change of state) A change of matter in one physical *phase (solid, liquid, or gas) into another. The change is invariably accompanied by the evolution or absorption of energy, even if it takes place at constant temperature (see latent heat).

channel 1. The region between the source and the drain in a field-effect *transistor. The conductivity of the channel is controlled by the voltage applied to the gate. **2.** A path, or a specified frequency band, along which signals, information, or data flow.

chaos Unpredictable and seemingly random behaviour occurring in a system that should be governed by deterministic laws. In such systems, the equations that describe the way the system changes with time are nonlinear and involve several variables. Consequently, they are very sensitive to the initial conditions, and a very small initial difference may make an enormous change to the future state of the system. Originally, the theory was introduced to describe unpredictability in meteorology, as exemplified by the *butterfly effect*. It has been suggested that the dynamical equations governing the weather are so sensitive to the initial data that whether or not a butterfly flaps its wings in one part of the world may make the difference between a tornado occurring or not occurring in some other part of the world. Chaos theory has subsequently been extended to other branches of science; for example to turbulent flow, planetary dynamics, and electrical oscillations in physics, and to combustion processes and oscillatory (clock) reactions, such as the *B–Z reaction, in chemistry. *See also* attractor; fractal.

character A distinctive inherited feature of an organism. Organisms in a population may display different aspects of a particular character, e.g. the A, B, and O human blood groups (*see* ABO system) are different aspects of the blood group character.

characteristic *See* logarithm.

charcoal A porous form of carbon produced by the destructive distillation of organic material. Charcoal from wood is used as a fuel. All forms of charcoal are porous and are used for adsorbing gases and purifying and clarifying liquids. There are several types depending on the source. Charcoal from coconut shells is a particularly good gas adsorbent. *Animal charcoal* (or *bone black*) is made by heating bones and dissolving out the calcium phosphates and other mineral salts with acid. It is used in sugar refining. *Activated charcoal* is charcoal that has been activated for adsorption by steaming or by heating in a vacuum.

charge A property of some *elementary particles that gives rise to an interaction between them and consequently to the host of material phenomena described as electrical. Charge occurs in nature in two forms, conventionally described as *positive* and *negative* in order to distinguish between the two kinds of interaction between particles. Two particles that have similar charges (both negative or both positive) interact by repelling each other; two particles that have

dissimilar charges (one positive, one negative) interact by attracting each other. The size of the interaction is determined by *Coulomb's law.

The natural unit of negative charge is the charge on an *electron, which is equal but opposite in effect to the positive charge on the proton. Large-scale matter that consists of equal numbers of electrons and protons is electrically neutral. If there is an excess of electrons the body is negatively charged; an excess of protons results in a positive charge. A flow of charged particles, especially a flow of electrons, constitutes an electric current. Charge is measured in coulombs, the charge on an electron being 1.602×10^{-19} coulombs.

charge carrier The entity that transports electric charge in an electric current. The nature of the carrier depends on the type of conductor: in metals, the charge carriers are electrons; in *semiconductors the carriers are electrons (n-type) or positive *holes (p-type); in gases the carriers are positive ions and electrons; in electrolytes they are positive and negative ions.

charge conjugation Symbol C. A property of elementary particles that determines the difference between a particle and its *antiparticle. The property is not restricted to electrically charged particles (i.e. it applies to neutral particles such as the neutron). *See* CP invariance.

charge density 1. The electric charge per unit volume of a medium or body (*volume charge density*). **2.** The electric charge per unit surface area of a body (*surface charge density*).

charge-transfer complex A complex in which there is weak coordination involving the transfer of charge. An example is phenoquinone, in which the phenol and quinone molecules are not held together by formal chemical bonds but are associated by transfer of charge between the compounds' aromatic ring systems.

Charles' law The volume of a fixed mass of gas at constant pressure expands by a constant fraction of its volume at 0°C for each Celsius degree or kelvin its temperature is raised. For any *ideal gas the fraction is approximately 1/273. This can be expressed by the equation

$$V = V_0(1 + t/273),$$

where V_0 is the volume at 0°C and V is its volume at t°C. This is equivalent to the statement that the volume of a fixed mass of gas at constant pressure is proportional to its thermodynamic temperature, $V = kT$, where k is a constant. The law resulted from experiments begun around 1787 by the French scientist J. A. C. Charles (1746–1823) but was properly established only by the more accurate results published in 1802 by the French scientist Joseph Gay-Lussac (1778–1850). Thus the law is also known as *Gay-Lussac's law*. An equation similar to that given above applies to pressures for ideal gases:

$$p = p_0(1 + t/273),$$

a relationship known as *Charles' law of pressures*. See also gas laws.

charm A property of certain *elementary particles that is expressed as a *quantum number and is used in the quark model. It was originally suggested to account for the unusually long lifetime of the *psi particle. In this theory the three original quark –antiquark pairs were supplemented by a fourth pair – the charmed quark and its antiquark. The psi particle itself is a meson having zero charm as it consists of the charmed pair. However, charmed *hadrons do exist; they are said to possess *naked charm*. Charm is thought to be conserved in strong and electromagnetic interactions.

cheddite Any of a group of high explosives made from nitro compounds mixed with sodium or potassium chlorate.

Chelate formed by coordination of two molecules of $H_2N(CH_2)_2NH_2$

chelate An inorganic complex in which a *ligand is coordinated to a metal ion at two (or more) points, so that there is a ring of atoms including the metal (see formula). The process is known as *chelation*. The bidentate ligand diaminoethane forms chelates with many ions. See also sequestration.

chemical bond A strong force of attraction holding atoms together in a molecule or crystal. Typically chemical bonds have energies of about 1000 $kJ\,mol^{-1}$ and are distinguished from the much weaker forces between molecules (see van der Waals' forces). There are various types.

Ionic (or *electrovalent*) bonds can be formed by transfer of electrons. For instance, the calcium atom has an electron configuration of $[Ar]4s^2$, i.e. it has two electrons in its outer shell. The chlorine atom is $[Ne]3s^23p^5$, with seven outer electrons. If the calcium atom transfers two electrons, one to each chlorine atom, it becomes a Ca^{2+} ion with the stable configuration of an inert gas $[Ar]$. At the same time each chlorine, having gained one electron, becomes a Cl^- ion, also with an inert-gas configuration $[Ar]$. The bonding in calcium chloride is the electrostatic attraction between the ions.

Covalent bonds are formed by sharing of valence electrons rather than by transfer. For instance, hydrogen atoms have one outer electron ($1s^1$).

chemical cell

In the hydrogen molecule, H_2, each atom contributes 1 electron to the bond. Consequently, each hydrogen atom has control of 2 electrons – one of its own and the second from the other atom – giving it the electron configuration of an inert gas [He]. In the water molecule, H_2O, the oxygen atom, with six outer electrons, gains control of an extra two electrons supplied by the two hydrogen atoms. This gives it the configuration [Ne]. Similarly, each hydrogen atom gains control of an extra electron from the oxygen, and has the [He] electron configuration.

A particular type of covalent bond is one in which one of the atoms supplies both the electrons. These are known as *coordinate* (*semipolar* or *dative*) bonds, and written A→B, where the direction of the arrow denotes the direction in which electrons are donated.

Covalent or coordinate bonds in which one pair of electrons is shared are *electron-pair bonds* and are known as *single bonds*. Atoms can also share two pairs of electrons to form *double bonds* or three pairs in *triple bonds*. See orbital.

In a compound such as sodium chloride, Na^+Cl^-, there is probably complete transfer of electrons in forming the ionic bond (the bond is said to be *heteropolar*). Alternatively, in the hydrogen molecule H–H, the pair of electrons is equally shared between the two atoms (the bond is *homopolar*). Between these two extremes, there is a whole range of *intermediate bonds*, which have both ionic and covalent contributions. Thus, in hydrogen chloride, H–Cl, the bonding is predominantly covalent with one pair of electrons shared between the two atoms. However, the chlorine atom is more electronegative than the hydrogen and has more control over the electron pair; i.e. the molecule is polarized with a positive charge on the hydrogen and a negative charge

on the chlorine. *See also* banana bond; hydrogen bond; metallic bond; multicentre bond; multiple bond.

chemical cell *See* cell.

chemical combination The combination of elements to give compounds. There are three laws of chemical combination.
(1) The *law of constant composition* states that the proportions of the elements in a compound are always the same, no matter how the compound is made. It is also called the *law of constant proportions* or *definite proportions*.
(2) The *law of multiple proportions* states that when two elements A and B combine to form more than one compound, then the masses of B that combine with a fixed mass of A are in simple ratio to one another. For example, carbon forms two oxides. In one, 12 grams of carbon is combined with 16 grams of oxygen (CO); in the other 12 g of carbon is combined with 32 grams of oxygen (CO_2). The oxygen masses combining with a fixed mass of carbon are in the ratio 16:32, i.e. 1:2.
(3) The *law of equivalent proportions* states that if two elements A and B each form a compound with a third element C, then a compound of A and B will contain A and B in the relative proportions in which they react with C. For example, sulphur and carbon both form compounds with hydrogen. In methane 12 g of carbon react with 4 g of hydrogen. In hydrogen sulphide, 32 g of sulphur react with 2 g of hydrogen (i.e. 64 g of S for 4 g of hydrogen). Sulphur and carbon form a compound in which the C:S ratio is 12:64 (i.e. CS_2). The law is sometimes called the law of *reciprocal proportions*.

chemical dating An absolute *dating technique that depends on measuring the chemical composition of a specimen. Chemical dating can be used when the specimen is known to

126

undergo slow chemical change at a known rate. For instance, phosphate in buried bones is slowly replaced by fluoride ions from the ground water. Measurement of the proportion of fluorine present gives a rough estimate of the time that the bones have been in the ground. Another, more accurate, method depends on the fact that amino acids in living organisms are L-optical isomers. After death, these racemize and the age of bones can be estimated by measuring the relative amounts of D- and L-amino acids present.

chemical engineering The study of the design, manufacture, and operation of plant and machinery in industrial chemical processes.

chemical equation A way of denoting a chemical reaction using the symbols for the participating particles (atoms, molecules, ions, etc.); for example,

$$x\text{A} + y\text{B} \rightarrow z\text{C} + w\text{D}$$

The single arrow is used for an irreversible reaction; double arrows ($\rightleftharpoons$) are used for reversible reactions. When reactions involve different phases it is usual to put the phase in brackets after the symbol (s = solid; l = liquid; g = gas; aq = aqueous). The numbers x, y, z, and w, showing the relative numbers of molecules reacting, are called the *stoichiometric coefficients*. The sum of the coefficients of the reactants minus the sum of the coefficients of the products ($x + y - z - w$ in the example) is the *stoichiometric sum*. If this is zero the equation is balanced. Sometimes a generalized chemical equation is considered

$$\nu_1\text{A}_1 + \nu_2\text{A}_2 + \ldots \rightarrow \ldots \nu_n\text{A}_n + \nu_{n+1}\text{A}_{n+1} \ldots$$

In this case the reaction can be written $\Sigma\nu_i\text{A}_i = 0$, where the convention is that stoichiometric coefficients are positive for reactants and negative for products. The stoichiometric sum is $\Sigma\nu_i$.

chemical equilibrium A reversible chemical reaction in which the concentrations of reactants and products are not changing with time because the system is in thermodynamic equilibrium. For example, the reversible reaction

$$3\text{H}_2 + \text{N}_2 \rightleftharpoons 2\text{NH}_3$$

is in chemical equilibrium when the rate of the *forward reaction*

$$3\text{H}_2 + \text{N}_2 \rightarrow 2\text{NH}_3$$

is equal to the rate of the *back reaction*

$$2\text{NH}_3 \rightarrow 3\text{H}_2 + \text{N}_2$$

See also equilibrium constant.

chemical equivalent *See* equivalent weight.

chemical fossil Any of various organic compounds found in ancient geological strata that appear to be biological in origin and are assumed to indicate that life existed when the rocks were formed. The presence of chemical fossils in Precambrian strata indicates that life existed over 3000 million years ago.

chemical potential Symbol: μ. For a given component in a mixture, the coefficient $\partial G/\partial n$, where G is the Gibbs free energy and n the amount of substance of the component. The chemical potential is the change in Gibbs free energy with respect to change in amount of the component, with pressure, temperature, and amounts of other components being constant. Components are in equilibrium if their chemical potentials are equal.

chemical reaction A change in which one or more chemical elements or compounds (the *reactants*) form new compounds (the *products*). All reactions are to some extent *reversible*; i.e. the products can also react to give the original reactants. However, in many cases the extent of this back reaction is negligibly small, and the reaction is regarded as *irreversible*.

chemical shift A change in the normal wavelength of absorption or

Chernozem soil profile

emission of electromagnetic wavelength in a process in which there is a nuclear energy change (as in the *Mössbauer effect and *nuclear magnetic resonance) or a change in electron energy levels in the inner shells of an atom (as in X-ray *photoelectron spectroscopy).

chemiluminescence *See* luminescence.

chemisorption *See* adsorption.

chemistry The study of the elements and the compounds they form. Chemistry is mainly concerned with effects that depend on the outer electrons in atoms. *See* biochemistry; geochemistry; inorganic chemistry; organic chemistry; physical chemistry.

chemoreceptor A *receptor that detects the presence of particular chemicals and (in multicellular organisms) transmits this information to sensory nerves. Examples include the *taste buds and the receptors in the *carotid body.

chemosystematics *See* systematics.

chemotaxis *See* taxis.

chemotaxonomy The *classification of plants and microorganisms based on similarities and differences in their natural products and the biochemical pathways involved in their manufacture. *See also* taxonomy.

chemotropism The growth or movement of a plant or plant part in response to a chemical stimulus. An example is the growth of a pollen tube down the style during fertilization in response to the presence of sugars in the style.

chernozem (black earth) A type of soil that is characteristic of the continental interiors of the mid-latitudes, in which grassland formed the natural vegetation. Chernozems occur across the Russian steppes and parts of Romania and Hungary; these soils also occur in North America. The deep surface layer (A horizon) of a chernozem is black and rich in alkaline humus derived from the decomposition of the natural grassland vegetation. The underlying horizon contains calcium carbonate concretions. Chernozems are important agriculturally and most have been ploughed up for cereal production.

chert *See* flint.

chiasma The point at which paired *homologous chromosomes remain in contact as they begin to separate during the first prophase of *meiosis, forming a cross shape. A number of chiasmata can usually be identified and at these points *crossing over occurs.

Chile saltpetre A commercial mineral largely composed of *sodium nitrate from the caliche deposits in Chile. Before the ammonia-oxidation process for nitrates most imported Chilean saltpetre was used by the chemical industry; its principal use today is as an agricultural source of nitrogen.

chimaera An organism composed of tissues that are genetically different.

Chimaeras can develop if a *mutation occurs in a cell of a developing embryo. All the cells arising from it have the mutation and therefore produce tissue that is genetically different from adjacent tissue, e.g. brown patches in otherwise blue eyes in humans. *Graft hybrids are examples of plant chimaeras.

china clay See kaolin.

Chinese white See zinc oxide.

chip See silicon chip.

chirality The property of existing in left- and right-handed structural forms. See optical activity.

Chiron A minor planet discovered in 1977. It has an orbit of 50.68 years that, unlike other known minor planets, lies almost entirely outside that of Saturn. Its diameter is uncertain, but appears to be of the order of 300 km.

Chiroptera An order of flying mammals comprising the bats. Their membranous wings are supported by very elongated forelimbs and digits and stretch along the sides of the body to the hindlimbs and tail. Whenever bats rest they allow their body temperature to fall, hibernating in winter when food is scarce. Most bats are nocturnal; their ears are enlarged and specialized for *echolocation, which they use to hunt prey and avoid obstacles. Bats feed variously on insects, fruit, nectar, or blood.

chitin A *polysaccharide comprising chains of N-acetyl-D-glucosamine, a derivative of glucose. Chitin is structurally very similar to cellulose and serves to strengthen the supporting structures of various invertebrates. It also occurs in fungi.

chloral See trichloroethanal.

chloral hydrate See 2,2,2-trichloroethanediol.

chlorates Salts of the chloric acids; i.e. salts containing the ions ClO^- (chlorate(I) or hypochlorite), ClO_2^- (chlorate(III) or chlorite), ClO_3^- (chlorate(V)), or ClO_4^- (chlorate(VII) or perchlorate). When used without specification of an oxidation state the term 'chlorate' refers to a chlorate(V) salt.

chloric acid Any of the oxoacids of chlorine: *chloric(I) acid, *chloric(III) acid, *chloric(V) acid, and *chloric(VII) acid. The term is commonly used without specification of the oxidation state of chlorine to mean chloric(V) acid, $HClO_3$.

chloric(I) acid (hypochlorous acid) A liquid acid that is stable only in solution, HOCl. It may be prepared by the reaction of chlorine with an agitated suspension of mercury(II) oxide. Because the disproportionation of the ion ClO^- is slow at low temperatures chloric(I) acid may be produced, along with chloride ions by the reaction of chlorine with water at 0°C. At higher temperatures disproportionation to the chlorate(V) ion, ClO_3^-, takes place. Chloric(I) acid is a very weak acid but is a mild oxidizing agent and is widely used as a bleaching agent.

chloric(III) acid (chlorous acid) A pale-yellow acid known only in solution, $HClO_2$. It is formed by the reaction of chlorine dioxide and water and is a weak acid and an oxidizing agent.

chloric(V) acid (chloric acid) A colourless unstable liquid, $HClO_3$; r.d. 1.2; m.p. $< -20°C$; decomposes at 40°C. It is best prepared by the reaction of barium chlorate with sulphuric acid although chloric(V) acid is also formed by the disproportionation of chloric(I) acid in hot solutions. It is both a strong acid and a powerful oxidizing agent; hot solutions of the acid or its salts have been known to detonate in contact with readily oxidized organic material.

chloric(VII) acid (perchloric acid) An unstable liquid acid, $HClO_4$; r.d. 1.76; m.p. $-112°C$; b.p. 39°C (50

chloride

mmHg); explodes at about 90°C at atmospheric pressure. There is also a monohydrate (r.d. 1.88 (solid), 1.77 (liquid); m.p. 48°C; explodes at about 110°C) and a dihydrate (r.d. 1.65; m.p. −17.8°C; b.p. 200°C). Commercial chloric(VII) acid is a water azeotrope, which is 72.5% $HClO_4$, boiling at 203°C. The anhydrous acid may be prepared by vacuum distillation of the concentrated acid in the presence of magnesium perchlorate as a dehydrating agent. Chloric(VII) acid is both a strong acid and a strong oxidizing agent. It is widely used to decompose organic materials prior to analysis, e.g. samples of animal or vegetable matter requiring heavy-metal analysis.

chloride *See* halide.

chlorination 1. A chemical reaction in which a chlorine atom is introduced into a compound. *See* halogenation. 2. The treatment of water with chlorine to disinfect it.

chlorine Symbol Cl. A *halogen element; a.n. 17; r.a.m. 35.453; d. 3.214 g dm⁻³; m.p. −100.98°C; b.p. −34.6°C. It is a poisonous greenish-yellow gas and occurs widely in nature as sodium chloride in seawater and as halite (NaCl), carnallite ($KCl.MgCl_2.6H_2O$), and sylvite (KCl). It is manufactured by the electrolysis of brine and also obtained in the *Downs process for making sodium. It has many applications, including the chlorination of drinking water, bleaching, and the manufacture of a large number of organic chemicals.

It reacts directly with many elements and compounds and is a strong oxidizing agent. Chlorine compounds contain the element in the 1, 3, 5, and 7 oxidation states. It was discovered by Karl Scheele in 1774 and Humphry Davy confirmed it as an element in 1810.

chlorine dioxide A yellowish-red explosive gas, ClO_2; d. 3.09 g dm⁻³; m.p. −59.5°C; b.p. 9.9°C. It is sol-

uble in cold water but decomposed by hot water to give chloric(VII) acid, chlorine, and oxygen. Because of its high reactivity, chlorine dioxide is best prepared by the reaction of sodium chlorate and moist oxalic acid at 90°–100°C, as the product is then diluted by liberated carbon dioxide. Commercially the gas is produced by the reaction of sulphuric acid containing chloride ions with sulphur dioxide. Chlorine dioxide is widely used as a bleach in flour milling and in wood pulping and also finds application in water purification.

chlorine monoxide *See* dichlorine oxide.

chlorite 1. *See* chlorates. 2. A group of layered silicate minerals, usually green or white in colour, that are similar to the micas in structure and crystallize in the monoclinic system. Chlorites are composed of complex silicates of aluminium, magnesium, and iron in combination with water, with the formula $(Mg,Al,Fe)_{12}$-$(Si,Al)_8O_{20}(OH)_{16}$. They are most common in low-grade metamorphic rocks and also occur as secondary minerals in igneous rocks as alteration products of pyroxenes, amphiboles, and micas. The term is derived from *chloros*, the Greek word for green.

chloroacetic acids *See* chloroethanoic acids.

chlorobenzene A colourless highly inflammable liquid, C_6H_5Cl; r.d. 1.106; m.p. −45.43°C; b.p. 131.85°C. It is prepared by the direct chlorination of benzene using a halogen carrier (*see* Friedel–Crafts reaction), or manufactured by the *Raschig process. It is used mainly as an industrial solvent.

2-chlorobuta-1,3-diene (chloroprene) A colourless liquid chlorinated diene, $CH_2:CCICH:CH_2$; r.d. 0.96; b.p. 59°C. It is polymerized to make synthetic rubbers (e.g. neoprene).

chloroethane (ethyl chloride) A colourless flammable gas, C_2H_5Cl; m.p. $-45°C$; b.p. $132°C$. It is made by reaction of ethene and hydrogen chloride and used in making lead tetraethyl for petrol.

chloroethanoic acids (chloroacetic acids) Three acids in which hydrogen atoms in the methyl group of ethanoic acid have been replaced by chlorine atoms. They are: *monochloroethanoic acid* ($CH_2ClCOOH$); *dichloroethanoic acid* ($CHCl_2COOH$); *trichloroethanoic acid* (CCl_3COOH). The presence of chlorine atoms in the methyl group causes electron withdrawal from the COOH group and makes the chloroethanoic acids stronger acids than ethanoic acid itself. The K_a values (in moles dm^{-3} at $25°C$) are

$$CH_3COOH \ 1.7 \times 10^{-5}$$
$$CH_2ClCOOH \ 1.3 \times 10^{-3}$$
$$CHCl_2COOH \ 5.0 \times 10^{-2}$$
$$CCl_3COOH \ 2.3 \times 10^{-1}$$

chloroethene (vinyl chloride) A gaseous compound, $CH_2{:}CHCl$; r.d. 0.911; m.p. $-153.8°C$; b.p. $-13.37°C$. It is made by chlorinating ethene to give dichloroethane, then removing HCl:

$$C_2H_4 + Cl_2 \rightarrow CH_2ClCH_2Cl \rightarrow CH_2CHCl$$

The compound is used in making PVC.

chlorofluorocarbon (CFC) A type of compound in which some or all of the hydrogen atoms of a hydrocarbon (usually an alkane) have been replaced by chlorine and fluorine atoms. Most chlorofluorocarbons are chemically unreactive and are stable at high temperatures. They are used as aerosol propellants, refrigerants, and solvents, and in the manufacture of rigid packaging foam. A commonly encountered commercial name for these compounds is *freon*; freon 12, for example, is dichlorodifluoromethane (CCl_2F_2). Chlorofluorocarbons, because of their chemical inertness, can diffuse unchanged into the upper atmosphere. Here, photochemical reactions cause them to break down and react with ozone (*see* ozone layer). For this reason, their use has been discouraged. *See also* pollution.

chloroform *See* trichloromethane.

chloromethane (methyl chloride) A colourless flammable gas, CH_3Cl; r.d. 0.916; m.p. $-97°C$; b.p. $-24°C$. It is a *haloalkane, made by direct chlorination of methane and used as a local anaesthetic and refrigerant.

chlorophyll One of two pigments (*chlorophyll a* and *chlorophyll b*) responsible for the green colour of most plants. Chlorophyll molecules are the principal sites of light absorption in the light reactions of photosynthesis. They are magnesium-containing *porphyrins, chemically related to *cytochrome and *haemoglobin.

Chlorophyta (green algae) A large division of *algae, the members of which possess chlorophylls *a* and *b*, store food reserves as starch, and have cellulose cell walls. In these respects they resemble higher plants more closely than do any of the other algal divisions. The Chlorophyta are widely distributed and diverse in form. Unicellular forms may occur singly (sometimes with *flagella for motility) or in colonies, while multicellular forms may be filamentous (e.g. *Spirogyra*) or platelike (e.g. *Ulva*).

chloroplast Any of the chlorophyll-containing organelles (*see* plastid) that are found in large numbers in those plant cells undergoing *photosynthesis. Chloroplasts are typically lens-shaped and made up of stacks of membranes (*see* granum) enclosed in a gel-like matrix. The light reactions of photosynthesis occur on the membranes while the dark reactions take place in the matrix.

chloroplatinic acid A reddish crystalline compound, H_2PtCl_6, made by dissolving platinum in aqua regia.

chloroprene *See* 2-chlorobuta-1,3-diene.

chlorosis The abnormal condition in plant stems and leaves in which synthesis of the green pigment chlorophyll is inhibited, resulting in a pale yellow coloration. This may be caused by lack of light, mineral deficiency, infection (particularly by viruses), or genetic factors.

chlorosulphanes *See* disulphur dichloride.

chlorous acid *See* chloric(III) acid.

choanae (internal nares) *See* nares.

choke A coil of wire with high inductance and low resistance. It is used in radio circuits to impede the passage of audio-frequency or radio-frequency currents or to smooth the output of a rectifying circuit.

choke damp *See* blackdamp.

cholecalciferol *See* vitamin D.

cholecystokinin (pancreozymin) A hormone, produced by the duodenal region of the small intestine, that induces the gall bladder to contract and eject bile into the intestine and stimulates the pancreas to secrete its digestive enzymes. Cholecystokinin output is stimulated by contact with the contents of the stomach.

cholesteric crystal *See* liquid crystal.

cholesterol A *sterol occurring widely in animal tissues and also in some higher plants and algae. It can exist as a free sterol or esterified with a long-chain fatty acid. Cholesterol is absorbed through the intestine or manufactured in the liver. It serves principally as a constituent of blood plasma *lipoproteins and of the lipid –protein complexes that form cell membranes. It is also important as a precursor of various steroids, especially the bile acids, sex hormones, and adrenocorticoid hormones. The

derivative 7-dehydrocholesterol is converted to vitamin D_3 by the action of sunlight on skin. Increased levels of dietary and blood cholesterol have been associated with *atherosclerosis*, a condition in which lipids accumulate on the inner walls of arteries and eventually obstruct blood flow.

choline An amino alcohol, $CH_2OHCH_2N(CH_3)_3OH$. It occurs widely in living organisms as a constituent of certain types of phospholipids – the *lecithins and sphingomyelins. It is sometimes classified as a member of the *vitamin B complex.

cholinergic Describing a nerve fibre that either releases *acetylcholine when stimulated or is itself stimulated by acetylcholine. *Compare* adrenergic.

Chondrichthyes *See* Elasmobranchii.

Chordata A phylum of animals characterized by a hollow dorsal nerve cord and, at some stage in their development, a flexible skeletal rod (the *notochord) and *gill slits opening from the pharynx. There are four subphyla, the largest of which is the *Vertebrata, in which the notochord is only present in the embryo or larva and becomes replaced by the backbone before birth or metamorphosis. The other three subphyla are often grouped together as the Protochordata. They are relatively primitive chordates including the lancelet, or amphioxus (*Branchiostoma*) and the sea squirts.

chorion 1. A membrane enclosing the embryo yolk sac and allantois of reptiles, birds, and mammals. *See* extraembryonic membranes. **2.** The protective shell of an insect egg, produced by the ovary. It is pierced by a small pore (*micropyle*) that allows the entry of spermatozoa for fertilization. *See also* egg membrane.

chorionic gonadotrophin *See* gonadotrophin.

choroid A pigmented layer, rich in blood vessels, that lies between the retina and the sclerotic of the vertebrate eye. At the front of the eye the choroid is modified to form the *ciliary body and the *iris.

choroid plexus A membrane rich in blood vessels that lines the *ventricles of the brain. It is an extension of the *pia mater and secretes *cerebrospinal fluid into the ventricles; it also controls exchange of materials between the blood and cerebrospinal fluid.

chromate A salt containing the ion $CrO_4{}^{2-}$.

chromatic aberration *See* aberration.

chromaticity An objective description of the colour quality of a visual stimulus that does not depend on its luminance but which, together with its luminance, completely specifies the colour. The colour quality is defined in terms of *chromaticity coordinates*, x, y, and z, where

$$x = X/(X + Y + Z)$$
$$y = Y/(X + Y + Z)$$
$$\text{and } z = Z/(X + Y + Z)$$

X, Y, and Z are the *tristimulus values* of a light, i.e. they are the amounts of three reference stimuli needed to match exactly the light under consideration in a trichromatic system.

chromatid A threadlike strand formed from a *chromosome during the early stages of cell division. Each chromosome divides along its length into two chromatids, which are at first held together at the centromere. They separate completely at a later stage. The DNA of the chromosome reproduces itself exactly so that each chromatid has the complete amount of DNA and becomes a daughter chromosome with exactly the same genes as the original chromosome from which it was formed.

chromatin The substance of which eukaryotic *chromosomes are composed. It consists of proteins (principally histones), DNA, and small amounts of RNA. In a metabolically inactive nucleus, chromatin is mainly in a condensed form, *heterochromatin*, but in an active nucleus most of the chromatin is in an expanded form, *euchromatin*. Messenger RNA molecules are formed in euchromatic regions.

chromatogram A record obtained by chromatography. The term is applied to the developed records of *paper chromatography and *thin-layer chromatography and also to the graphical record produced in *gas chromatography.

chromatography A technique for analysing or separating mixtures of gases, liquids, or dissolved substances. The original technique (invented by the Russian botanist Mikhail Tsvet in 1906) is a good example of *column chromatography*. A vertical glass tube is packed with an adsorbing material, such as alumina. The sample is poured into the column and continuously washed through with a solvent (a process known as *elution*). Different components of the sample are adsorbed to different extents and move down the column at different rates. In Tsvet's original application, plant pigments were used and these separated into coloured bands in passing down the column (hence the name chromatography). The usual method is to collect the liquid (the *eluate*) as it passes out from the column in fractions.

In general, all types of chromatography involve two distinct phases – the *stationary phase* (the adsorbent material in the column in the example above) and the *moving phase* (the solution in the example). The separation depends on competition for molecules of sample between the moving phase and the stationary phase. The form of column chromatography above is an example of

chromatophore

adsorption chromatography, in which the sample molecules are adsorbed on the alumina. In *partition chromatography*, a liquid (e.g. water) is first absorbed by the stationary phase and the moving phase is an immiscible liquid. The separation is then by *partition between the two liquids. In ion-exchange chromatography (*see* ion exchange), the process involves competition between different ions for ionic sites on the stationary phase. *Gel filtration is another chromatographic technique in which the size of the sample molecules is important.

See also gas chromatography; high-performance liquid chromatography; paper chromatography; R_F value; thin-layer chromatography.

chromatophore 1. A pigment-containing cell found in the skin of many lower vertebrates (e.g. chameleon) and in the integument of crustaceans. Concentration or dispersion of the pigment granules in the cytoplasm of the cell causes the colour of the animal to alter to match its surroundings. A common type of chromatophore is the *melanophore*, which contains the pigment *melanin. **2.** *See* chromoplast.

chrome alum *See* potassium chromium sulphate.

chrome iron ore A mixed iron–chromium oxide, $FeO.Cr_2O_3$, used to make ferrochromium for chromium steels.

chrome red A basic lead chromate, $PbO.PbCrO_4$, used as a red pigment.

chrome yellow Lead chromate, $PbCrO_4$, used as a pigment.

chromic acid A hypothetical acid, H_2CrO_4, known only in chromate salts.

chromic anhydride *See* chromium(VI) oxide.

chromic compounds Compounds containing chromium in a higher (+3 or +6) oxidation state; e.g. chromic oxide is chromium(VI) oxide (CrO_3).

chromite A spinel mineral, $FeCr_2O_4$; the principal ore of chromium. It is black with a metallic lustre and usually occurs in massive form. It is a common constituent of peridotites and serpentines. The chief producing countries are Turkey, South Africa, the USSR, the Philippines, and Zimbabwe.

chromium Symbol Cr. A hard silvery *transition element; a.n. 24; r.a.m. 52.00; r.d. 7.19; m.p. 1900°C; b.p. 2640°C. The main ore is chromite ($FeCr_2O_4$). The metal is extracted by heating chromite with sodium chromate, from which chromium can be obtained by electrolysis. Alternatively, chromite can be heated with carbon in an electric furnace to give ferrochrome, which is used in making alloy steels. The metal is also used as a shiny decorative electroplated coating and in the manufacture of certain chromium compounds.

At normal temperatures the metal is corrosion-resistant. It reacts with dilute hydrochloric and sulphuric acids to give chromium(II) salts. These readily oxidize to the more stable chromium(III) salts. Chromium also forms compounds with the +6 oxidation state, as in chromates, which contain the CrO_4^{2-} ion. The element was discovered in 1797 by Vauquelin.

chromium(II) oxide A black insoluble powder, CrO. Chromium(II) oxide is prepared by oxidizing chromium amalgam with air. At high temperatures hydrogen reduces it to the metal.

chromium(III) oxide A green crystalline water-insoluble salt, Cr_2O_3; r.d. 5.21; m.p. 2266°C; b.p. 4000°C. It is obtained by heating chromium in a stream of oxygen or by heating ammonium dichromate. The industrial preparation is by reduction of sodium dichromate with carbon. Chromium(III) oxide is amphoteric, dissolving in acids to give chromium(III) ions and in concentrated solutions of alkalis to give *chromites*. It is used as a green pigment in glass, porcelain, and oil paint.

chromium(IV) oxide (chromium dioxide) A black insoluble powder, CrO_2; m.p. 300°C. It is prepared by the action of oxygen on chromium(VI) oxide or chromium(III) oxide at 420–450°C and 200–300 atmospheres. The compound is unstable.

chromium(VI) oxide (chromium trioxide; chromic anhydride) A red compound, CrO_3; rhombic; r.d. 2.70; m.p. 196°C. It can be made by careful addition of concentrated sulphuric acid to an ice-cooled strong aqueous solution of sodium dichromate with stirring. The mixture is then filtered through sintered glass, washed with nitric acid, then dried out 120°C in a desiccator.

Chromium(VI) oxide is an extremely powerful oxidizing agent, especially to organic matter; it immediately inflames ethanol. It is an acidic oxide and dissolves in water to form 'chromic acid', a powerful oxidizing agent and cleansing fluid for glassware. At 400°C, chromium(VI) oxide loses oxygen to give chromium(III) oxide.

chromium potassium sulphate A red crystalline solid, $K_2SO_4.Cr_2(SO_4)_3.24H_2O$; r.d. 1.91. It is used as a mordant. *See also* alums.

chromium steel Any of a group of *stainless steels containing 8–25% of chromium. A typical chromium steel might contain 18% of chromium, 8% of nickel, and 0.15% of carbon. Chromium steels are highly resistant to corrosion and are used for cutlery, chemical plant, ball bearings, etc.

chromophore A group causing coloration in a *dye. Chromophores are generally groups of atoms having delocalized electrons.

chromoplast (chromatophore) Any of various organelles in plant cells

that contain pigments. Red, orange, and yellow chromoplasts contain carotenoid pigments, while green chromoplasts (called *chloroplasts) contain the pigment chlorophyll and provide the green coloration of most plants. *Compare* leucoplast. *See* plastid.

chromosome A threadlike structure several to many of which are found in the nucleus of plant and animal (eukaryotic) cells. Chromosomes are composed of *chromatin and carry the *genes, which determine the individual characteristics of an organism. When the nucleus is not dividing, individual chromosomes cannot be identified with a light microscope. During the first stage of nuclear division, however, the chromosomes contract and, when stained, can be clearly seen under a microscope. Each consists of two *chromatids held together at the *centromere (*see also* meiosis; mitosis). The number of chromosomes in each cell is constant for and characteristic of the species concerned. In the normal body cells of *diploid organisms the chromosomes occur in pairs (*see* homologous chromosomes); in the gamete-forming germ cells, however, the diploid number is halved and each cell contains only one member of each chromosome pair. Thus in man each body cell contains 46 chromosomes (22 matched pairs and one pair of *sex chromosomes) and each germ cell 23. Abnormalities in the number or structure of chromosomes may give rise to abnormalities in the individual; *Down's syndrome is the result of one such abnormality.

Bacterial and viral cells contain only a single chromosome; it differs from the eukaryotic chromosome in being much simpler, lacking histones and consisting simply of a single or double strand of DNA or (in some viruses) RNA.

chromosome map (linkage map) A plan showing the relative positions of *genes along the length of the chromosomes of an organism. It is constructed by making crosses and observing whether certain characteristics tend to be inherited together. The closer together two allele pairs are situated on *homologous chromosomes, the less often will they be separated and rearranged as the reproductive cells are formed (see chiasma; crossing over). The proportion of offspring that show *recombination of the alleles concerned thus reflects their spacing and is used as a unit of length in mapping chromosomes. See also restriction mapping.

chromosphere The layer of the *sun's atmosphere immediately above the *photosphere. The chromosphere is normally only visible when the photosphere is totally eclipsed by the moon. The chromosphere is about 10 000 kilometres thick and the temperature in it rises from 4000 K, where it merges with the photosphere, to about 50 000 K, where it reaches the transition region below the *corona.

chromous compounds Compounds containing chromium in its lower (+2) oxidation state; e.g. chromous chloride is chromium(II) chloride $(CrCl_2)$.

chromyl chloride (chromium oxychloride) A dark red liquid, CrO_2Cl_2; r.d. 1.911; m.p. $-96.5°C$; b.p. $117°C$. It is evolved as a dark-red vapour on addition of concentrated sulphuric acid to a mixture of solid potassium dichromate and sodium chloride; it condenses to a dark-red covalent liquid, which is immediately hydrolysed by solutions of alkalis to give the yellow chromate. Since bromides and iodides do not give analogous compounds this is a specific test for chloride ions. The compound is a powerful oxidizing agent, exploding on contact with phosphorus and inflaming

sulphur and many organic compounds.

chrysalis See pupa.

chrysotile See serpentine.

chyle A milky fluid consisting of *lymph that contains absorbed food materials (especially emulsified fats). Most chyle occurs in the lymphatic ducts (*lacteals) in the *villi of the small intestine during the absorption of fat.

chyme The semisolid and partly digested food that is discharged from the stomach into the duodenum.

ciliary body The circular band of tissue surrounding and supporting the *lens of the vertebrate eye. It contains the ciliary muscles, which bring about changes in the shape of the lens (see also accommodation). The ciliary body produces the *aqueous humour.

ciliary feeding A method of feeding used by amphioxus and many aquatic invertebrates. The movement of cilia causes a current of water to be drawn towards and through the animal, and microorganisms in the water are filtered out by the cilia.

cilium A short minute hairlike structure (up to 10 μm long) present on the surface of many animal and a few plant cells, notably in certain protozoans and some types of vertebrate epithelium. Cilia usually occur in large groups. Beating of cilia can produce cell movement or create a current in fluid surrounding a cell. Compare flagellum.

cinnabar A bright red mineral form of mercury(II) sulphide, HgS, crystallizing in the hexagonal system; the principal ore of mercury. It is deposited in veins and impregnations near recent volcanic rocks and hot springs. The chief sources include Spain, Italy, and Yugoslavia.

cinnamic acid (3-phenylpropenoic acid) A white crystalline aromatic *carboxylic acid, $C_6H_5CH:CHCOOH$;

r.d. 1.28 (trans isomer); m.p. 133°C; b.p. 300°C. Esters of cinnamic acid occur in some essential oils.

circadian rhythm (diurnal rhythm) Any 24-hour periodicity in the behaviour or physiology of animals or plants. Examples are the sleep/activity cycle in many animals and the growth movements of plants. Circadian rhythms are generally controlled by *biological clocks.

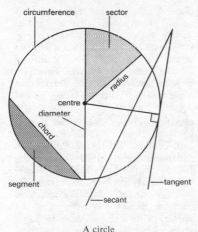

A circle

circle A closed curve every point on which is a fixed distance (the *radius*) from a point (the *centre*) within the curve (see illustration). The *diameter* is a line that joins two points on the *circumference* and passes through the centre: the diameter is twice the radius (*r*). The circumference of a circle is equal to $2\pi r$; the area is πr^2, where π is a constant with the value 3.141 592. In analytical geometry the equation of a circle, centred at the origin, is $x^2 + y^2 = r^2$.

circular measure A method of measuring angles by treating them as the angle formed by a sector of a circle at the circle's centre. The unit of measure is the *radian*, the angle subtended at the centre of a circle by an arc of equal length to the radius.

Since an arc of length *r* subtends an angle of 1 radian, the whole circumference, length $2\pi r$, will subtend an angle of $2\pi r/r = 2\pi$ radians. Thus, $360° = 2\pi$ radians; 1 radian = 57.296°.

circular polarization *See* polarization of light.

circulatory system The heart, blood vessels, blood, lymphatic vessels, and lymph, which together serve to transport materials throughout the body. *See also* vascular system.

circumnutation *See* nutation.

cis-isomer *See* isomerism.

cis–trans isomerism *See* isomerism.

citrate A salt or ester of citric acid.

citric acid A white crystalline hydroxycarboxylic acid, HOOC-CH$_2$C(OH)(COOH)CH$_2$COOH; r.d. 1.67; m.p. 153°C. It is present in citrus fruits and is an intermediate in the *Krebs cycle in plant and animal cells.

citric-acid cycle *See* Krebs cycle.

cladding 1. A thin coating of an expensive metal rolled on to a cheaper one. **2.** A thin covering of a metal around a fuel element in a nuclear reactor to prevent corrosion of the fuel elements by the coolant.

clade A group of organisms that share a common ancestor. *See* cladistics.

A cladogram showing the relationships of six species (A–F)

cladistics A controversial method of classification in which animals and plants are placed into taxonomic groups called *clades* when they share characteristics (known as *homologies*) that are thought to indicate common ancestry. It is based on the assumption that two new species are formed suddenly, by splitting from a common ancestor, and not by gradual evolutionary change. A diagram indicating these relationships (called a *cladogram*) therefore consists of a system of dichotomous branches: each point of branching represents divergence from a common ancestor, as shown in the diagram. Thus the species A to F form a clade as they share the common ancestor X, and species A to D form a clade of a different taxonomic rank, sharing the ancestor X_2. Species C to F do not form a clade, since the latter must include *all* the descendants of a common ancestor.

Cladistics has been used as evidence by the opponents of neo-Darwinism in the debate on the nature of evolution. Cladists argue that the major branching patterns of cladograms correspond to large-scale evolutionary events that cannot be explained by orthodox neo-Darwinism, which postulates a series of small changes occurring over a long period of time.

cladode A flattened stem or internode that resembles and functions as a leaf. It is an adaptation to reduce water loss, since it contains fewer *stomata than a leaf. An example of a plant with cladodes is asparagus.

cladogram *See* cladistics.

Claisen condensation A reaction of esters in which two molecules of the ester react to give a keto ester, e.g.

$$2CH_3COOR \rightarrow CH_3COCH_2COOR + ROH$$

The reaction is catalysed by sodium ethoxide, the mechanism being similar to that of the *aldol reaction.

Clark cell A type of *voltaic cell consisting of an anode made of zinc amalgam and a cathode of mercury both immersed in a saturated solution of zinc sulphate. The Clark cell was formerly used as a standard of e.m.f.; the e.m.f. at 15°C is 1.4345 volts. It is named after the British scientist Hosiah Clark (d. 1898).

Clark process *See* hardness of water.

class A category used in the *classification of organisms that consists of similar or closely related orders. Similar classes are grouped into a phylum in the case of animals and a division in the case of plants. Examples include, among animals, Mammalia (mammals), Aves (birds), and Agnatha (jawless fish).

classical physics Theoretical physics up to approximately the end of the 19th century, before the concepts of *quantum theory (1900) and special *relativity (1905). Classical physics relied largely on Newton's mechanics and James Clerk Maxwell's theory of electromagnetism. It may still be applied with high precision to large-scale phenomena involving no very rapid relative motion.

classification The arrangement of organisms into a series of groups based on physiological, biochemical, anatomical, or other relationships. An *artificial classification* is based on one or a few characters simply for ease of identification or for a specific purpose; for example, birds are often arranged according to habit and habitat (seabirds, songbirds, birds of prey, etc.) while fungi may be classified as edible or poisonous. Such systems do not reflect evolutionary relationships. A *natural classification* is based on resemblances and is a hierarchical arrangement. The smallest group commonly used is the *species. In plant classification, species are grouped into genera (*see* genus), the hierarchy continuing up through *tribes, *families, *orders, *classes, and *divisions to *kingdom. Animal classification follows the same system except that the *phylum replaces the division. Higher up in the hierarchy the similarities between members of a group become fewer. Present-day natural classifications try to take into account as many features as possible and in so doing aim to reflect evolutionary relationships (*see* cladistics). Natural classifications are also predictive. Thus if an organism is placed in a particular genus because it shows certain features characteristic of the genus, then it can be assumed it is very likely to possess most (if not all) of the other features of that genus. *See also* binomial nomenclature; taxonomy.

clathrate A solid mixture in which small molecules of one compound or element are trapped in holes in the crystal lattice of another substance. Clathrates are sometimes called *enclosure compounds* or *cage compounds*, but they are not true compounds (the molecules are not held by chemical bonds). Quinol and ice both form clathrates with such substances as sulphur dioxide and xenon.

Claude process A process for liquefying air on a commercial basis. Air under pressure is used as the working substance in a piston engine, where it does external work and cools adiabatically. This cool air is fed to a counter-current *heat exchanger, where it reduces the temperature of the next intake of high-pressure air. The same air is re-compressed and used again, and after several cycles eventually liquefies. The process was perfected in 1902 by the French scientist Georges Claude (1870–1960).

claudetite A mineral form of *arsenic(III) oxide, As_4O_6.

clavicle A bone that forms part of the *pectoral (shoulder) girdle, linking the *scapula (shoulder blade) to the sternum (breastbone). In man it forms the collar bone and serves as a brace for the shoulders.

clay A fine-grained deposit consisting chiefly of *clay minerals. It is characteristically plastic and virtually impermeable when wet and cracks when it dries out. In geology the size of the constituent particles is usually taken to be less than 1/256 mm. In soil science clay is regarded as a soil with particles less than 0.002 mm in size.

clay minerals Very small particles, chiefly hydrous silicates of aluminium, sometimes with magnesium and/or iron substituting for all or part of the aluminium, that are the major constituents of clay materials. The particles are essentially crystalline (either platy or fibrous) with a layered structure, but may be amorphous or metalloidal. The clay minerals are responsible for the plastic properties of clay; the particles have the property of being able to hold water. The chief groups of clay minerals are:
kaolinite, $Al_4Si_4O_{10}(OH)_8$, the chief constituent of *kaolin;
halloysite, $Al_4Si_4(OH)_8O_{10}.4H_2O$;
illite, $KAl_4(Si,Al)_8O_{18}.2H_2O$;
montmorillonite,
$(Na,Ca)_{0.33}(Al,Mg)_2Si_4O_{10}(OH)_2.nH_2O$, formed chiefly through alteration of volcanic ash;
vermiculite,
$(Mg,Fe,Al)_3(Al,Si)_4O_{10}(OH)_2.4H_2O$, used as an insulating material and potting soil.

cleavage 1. (in embryology) The series of cell divisions by which a single fertilized egg cell is transformed into a multicellular body, the *blastula. Characteristically no growth occurs during cleavage, the shape of the embryo is unchanged except for the formation of a central cavity (the blastocoel), and the ratio of nuclear material (DNA) to cytoplasm increases. **2.** (in crystallography) The splitting of a crystal along planes of atoms in the lattice.

climate The characteristic pattern of weather elements in an area over a period. The weather elements include temperature, rainfall, humidity, solar insolation, wind, etc. The climate of a large area is determined by several climatic controls: (1) the latitude of the area, which accounts for the amount of solar radiation it receives; (2) the distribution of land and sea masses; (3) the altitude and topography of the area; and (4) the location of the area in relation to the ocean currents. Climates may be classified into groups; the classification devised by the meteorologist Vladimir Köppen (1846–1940) in 1918, with subsequent modifications, is the most frequently used today. The scientific study of climate is *climatology*.

climax community A relatively stable ecological *community that is achieved at the end of a *succession.

clinostat A mechanical device that rotates whole plants (usually seedlings), so removing the effect of any stimulus that normally acts in one particular direction. It is most often used to study the growth of plant organs when the influence of gravity has been removed.

clitoris An erectile rod of tissue in female mammals (and also some reptiles and birds) that is the equivalent of the male penis. It lies in front of the *urethra and *vagina.

cloaca The cavity in the pelvic region into which the terminal parts of the alimentary canal and the urinogenital ducts open in most vertebrates. Placental mammals, however, have a separate anus and urinogenital opening.

clock reaction See B–Z reaction.

clone 1. A group of organisms or cells that have arisen from a single individual by asexual reproduction and are therefore all genetically identical, e.g. a group of plants propagated by bulbs or cuttings. **2. (gene**

clone) An exact replica of a gene. *See* gene cloning.

cloning vector *See* vector.

closed chain *See* chain; ring.

close packing The packing of spheres so as to occupy the minimum amount of space. In a single plane, each sphere is surrounded by six close neighbours in a hexagonal arrangement. The spheres in the second plane fit into depressions in the first layer, and so on. Each sphere has 12 other touching spheres. There are two types of close packing. In *hexagonal close packing* the spheres in the third layer are directly over those in the first, etc., and the arrangement of planes is ABAB.... In *cubic close packing* the spheres in the third layer occupy a different set of depressions than those in the first. The arrangement is ABCABC.... *See also* cubic crystal.

clotting factors (coagulation factors) A group of substances present in blood plasma that, under certain circumstances, undergo a series of chemical reactions leading to the conversion of blood from a liquid to a solid state (*see* blood clotting). Although they have specific names, most coagulation factors are referred to by an agreed set of Roman numerals (e.g. *Factor VIII, Factor IX). Lack of any of these factors in the blood results in the inability of the blood to clot. *See also* haemophilia.

cloud A mass of minute water droplets or ice crystals held in suspension in the atmosphere that appears as an opaque drifting body. The droplets or crystals are formed by the condensation of water vapour in the atmosphere in the presence of condensation nuclei – minute particles, such as smoke or salt. A number of cloud classifications have been devised; that most commonly used is based on cloud appearance and height. The high clouds comprise cirrocumulus, cirrostratus, and cirrus (mares' tails) clouds; the medium clouds comprise altocumulus and altostratus; and the low clouds are nimbostratus, stratocumulus, and stratus. Some clouds with great vertical development cannot be confined to these height categories; these are cumulus and cumulonimbus clouds.

cloud chamber A device for making visible the paths of particles of *ionizing radiation. The *Wilson (expansion) cloud chamber* consists of a container containing air and ethanol vapour, which is cooled suddenly by adiabatic expansion, causing the vapour to become supersaturated. The excess moisture in the vapour is then deposited in drops on the tracks of ions created by the passage of the ionizing radiation. The resulting row of droplets can be photographed. If the original moving particle was being deflected by electric or magnetic fields, the extent of the deflection provides information on its mass and charge. This device was invented in 1911 by C. T. R. Wilson (1869–1961). A simpler version of this apparatus is the *diffusion cloud chamber*, developed by Cowan, Needels, and Nielsen in 1950, in which supersaturation is achieved by placing a row of felt strips soaked in a suitable alcohol at the top of the chamber. The lower part of the chamber is cooled by solid carbon dioxide. The vapour continuously diffuses downwards, and that in the centre (where it becomes supersaturated) is almost continuously sensitive to the presence of ions created by the radiation.

clubmoss *See* Lycopsida.

Clusius column A device for separating isotopes by *thermal diffusion. One form consists of a vertical column some 30 metres high with a heated electric wire running along its axis. The lighter isotopes in a gaseous mixture of isotopes diffuse faster than the heavier isotopes. Heated by the axial wire, and assisted by natural convection, the lighter atoms are carried to the top of the column, where a fraction rich in lighter isotopes can be removed for further enrichment.

cluster *See* galaxy cluster; star cluster.

cluster compound A compound in which groups of metal atoms are joined together by metal–metal bonds. The formation of such compounds is a feature of the chemistry of certain transition elements, particularly molybdenum and tungsten, but also vanadium, tantalum, niobium, and uranium. *Isopoly compounds* are ones in which the cluster contains atoms of the same element; *heteropoly compounds* contain a mixture of different elements.

cluster of differentiation *See* CD.

Cnidaria *See* Coelenterata.

cnidoblast *See* thread cell.

CNS *See* central nervous system.

CoA *See* coenzyme A.

coagulation The process in which colloidal particles come together to form larger masses. Coagulation can be brought about by adding ions to neutralize the charges stabilizing the colloid. Ions with a high charge are particularly effective (e.g. alum, containing Al^{3+}, is used in styptics to coagulate blood). Another example of ionic coagulation is in the formation of river deltas, which occurs when colloidal silt particles in rivers are coagulated by ions in sea water. Heating is another way of coagulating certain colloids (e.g. boiling an egg coagulates the albumin). *See also* blood clotting.

coal A brown or black carbonaceous deposit derived from the accumulation and alteration of ancient vegetation, which originated largely in swamps or other moist environments. As the vegetation decomposed it formed layers of peat, which were subsequently buried (for example, by marine sediments following a rise in sea level or subsidence of the land). Under the increased pressure and resulting higher temperatures the peat was transformed into coal. Two types of coal are recognized: *humic* (or *woody*) *coals*, derived from plant remains; and *sapropelic coals*, which are derived from algae, spores, and finely divided plant material.

As the processes of coalification (i.e. the transformation resulting from the high temperatures and pressures) continue, there is a progressive transformation of the deposit: the proportion of carbon relative to oxygen rises and volatile substances and water are driven out. The various stages in this process are referred to as the *ranks* of the coal. In ascending order, the main ranks of coal are: *lignite* (or *brown coal*), which is soft, brown, and has a high moisture content; *subbituminous coal*, which is used chiefly by generating stations; *bituminous coal*, which is the most abundant rank of coal; *semibituminous coal*; *semianthracite coal*, which has a fixed carbon content of between 86% and 92%; and *anthracite coal*, which is hard and black with a fixed carbon content of between 92% and 98%.

Most deposits of coal were formed during the Carboniferous and Permian periods. More recent periods of coal formation occurred during the early Jurassic and Tertiary periods. Coal deposits occur in all the major continents; the leading producers include the USA, China, USSR, Poland, UK, South Africa, India, Australia, and West Germany. Coal is used as a fuel and in the chemical industry; by-products include coke and coal tar.

coal gas A fuel gas produced by the destructive distillation of coal. In the late-19th and early-20th centuries coal gas was a major source of energy and was made by heating coal in the absence of air in local gas works. Typically, it contained hydrogen (50%), methane (35%), and carbon monoxide (8%). By-products of the process were *coal tar and coke. The use of this type of gas declined with the increasing availability of natural gas, although since the early 1970s interest has developed in using coal in making *SNG.

coal tar A tar obtained from the destructive distillation of coal. Formerly, coal tar was obtained as a by-product in manufacturing *coal gas. Now it is produced in making coke for steel making. The crude tar contains a large number of organic compounds, such as benzene, naphthalene, methylbenzene, phenols, etc., which can be obtained by distillation. The residue is *pitch*. At one time coal tar was the major source of organic chemicals, most of which are now derived from petroleum and natural gas.

coaxial cable A cable consisting of a central conductor surrounded by an insulator, which is in turn contained in an earthed sheath of another conductor. The central conductor and the outer conductor are coaxial (i.e. have the same axis). They are used to transmit high-frequency signals as they produce no external fields and are not influenced by them.

cobalt Symbol Co. A light-grey *transition element; a.n. 27; r.a.m. 58.933; r.d. 8.9; m.p. 1495°C; b.p. 2870°C. Cobalt is ferromagnetic below its Curie point of 1150°C. Small amounts of metallic cobalt are present in meteorites but it is usually extracted from ore deposits worked in Canada, Morocco, and Zaïre. It is present in the minerals cobaltite, smaltite, and erythrite but also associated with copper and nickel as sulphides and arsenides. Cobalt ores are usually roasted to the oxide and then reduced with carbon or water gas. Cobalt is usually alloyed for use. Alnico is a well-known magnetic alloy and cobalt is also used to make stainless steels and in high-strength alloys that are resistant to oxidation at high temperatures (for turbine blades and cutting tools).

The metal is oxidized by hot air and also reacts with carbon, phosphorus, sulphur, and dilute mineral acids. Cobalt salts, usual oxidation states II and III, are used to give a brilliant blue colour in glass, tiles, and pottery. Anhydrous cobalt(II) chloride paper is used as a qualitative test for water and as a heat-sensitive ink. Small amounts of cobalt salts are essential in a balanced diet for mammals (see essential element). Artificially produced cobalt–60 is an important radioactive tracer and cancer-treatment agent. The element was discovered by G. Brandt in 1737.

cobalt(II) oxide A pink solid, CoO; cubic; r.d. 6.45; m.p. 1795°C. The addition of potassium hydroxide to a solution of cobalt(II) nitrate gives a bluish-violet precipitate, which on boiling is converted to pink impure cobalt(II) hydroxide. On heating this in the absence of air, cobalt(II) oxide is formed. The compound is readily oxidized in air to form tricobalt tetroxide, Co_3O_4, and is readily reduced by hydrogen to the metal.

cobalt(III) oxide (cobalt sesquioxide) A black grey insoluble solid, Co_2O_3; hexagonal or rhombic; r.d. 5.18; decomposes at 895°C. It is produced by the ignition of cobalt nitrate; the product however never has the composition corresponding exactly to cobalt(III) oxide. On heating it readily forms Co_3O_4, which contains both Co(II) and Co(III), and is easily reduced to the metal by hydrogen. Cobalt(III) oxide dissolves in strong acid to give unstable brown solutions of trivalent cobalt salts. With dilute acids cobalt(II) salts are formed.

cobalt steel Any of a group of *alloy steels containing 5–12% of cobalt, 14–20% of tungsten, usually with 4% of chromium and 1–2% of vanadium. They are very hard but somewhat brittle. Their main use is in high-speed tools.

coccus Any spherical bacterium. Cocci may occur singly, in pairs, in groups of four or more, in cubical packets, in grapelike clusters (*Staphylococcus), or in chains (*Streptococcus). Staphylococci and streptococci include pathogenic species. They are generally nonmotile and do not form spores.

coccyx The last bone in the *vertebral column in apes and man (i.e. tailless primates). It is formed by the fusion of 3–5 *caudal vertebrae.

cochlea Part of the *inner ear of mammals, birds, and some reptiles that transforms sound waves into nerve impulses. In mammals it is coiled, resembling a snail shell. The cochlea is lined with sensitive cells that bear tiny hairs; it is filled with fluid (endolymph) and surrounded by fluid (perilymph). Sound-induced vibrations of the *fenestra ovalis are transmitted through the perilymph and endolymph and stimulate the cells that line the cochlea. These in turn stimulate nerve cells that transmit information, via the *auditory nerve, to the brain for interpretation of the sounds.

Cockcroft–Walton generator The first proton accelerator; a simple *linear accelerator producing a potential difference of some 800 kV (d.c.) from a circuit of rectifiers and capacitors fed by a lower (a.c.) voltage. The experimenters, Sir John Cockcroft (1897–1967) and E. T. S. Walton (1903–), used this device in 1932 to achieve the first artificially induced nuclear reaction by bombarding lithium with protons to produce helium:

$$^1_1H + ^7_3Li = ^4_2He + ^4_2He$$

cockroaches See Dictyoptera.

cocoon A protective covering for eggs and/or larvae produced by many

invertebrates. For example, the larvae
of many insects spin a cocoon in
which the pupae develop (that of the
silkworm moth produces silk), and
earthworms secrete a cocoon for the
developing eggs.

co-dominance The condition that
arises when both alleles in a *heterozygous organism are dominant and
are fully expressed. For example, the
human blood group AB is the result
of two alleles, A and B, both being
expressed. A is not dominant to B,
nor vice versa. *Compare* incomplete
dominance.

codon A triplet of nitrogen-containing bases within a molecule of *DNA
or messenger *RNA that codes for a
particular amino acid during the
synthesis of proteins in a cell (*see*
translation). *See also* genetic code.

coefficient 1. (in mathematics) A
number or other known factor by
which a variable quantity is multiplied, e.g. in $ax^2 + bx + c = 0$, a is
the coefficient of x^2 and b is the coefficient of x. 2. (in physics) A measure
of a specified property of a particular
substance under specified conditions,
e.g. the coefficient of *friction of a
substance.

coefficient of expansion *See* expansivity.

coefficient of friction *See* friction.

coelacanth A bony fish, *Latimeria
chalumnae*, that was believed to have
been extinct until 1938, when the first
specimen of modern times was discovered. The coelacanth belongs to
the same order (Crossopterygii –
lobe-finned fishes) as the ancestors of
the amphibians. It lives at a depth of
200–300 m in the Indian Ocean. The
coelacanth is a large fish, 1–2 m long
and weighing 80 kg or more, with a
three-lobed tail fin. The body is
covered in rough heavy scales and the
pectoral fins can be used like crutches
to help movement across the sea bed.
The young are born alive. Fossil coe-

lacanths are most abundant in
deposits about 400 million years old
and no fossils less than 70 million
years old have been found.

Coelenterata (Cnidaria) A phylum
of aquatic invertebrates that includes
Hydra, jellyfish, sea anemones, and
*corals. A coelenterate's body is *dip-
loblastic, with two cell layers of the
body wall separated by *mesoglea,
and shows *radial symmetry. The
body cavity (*coelenteron*) is sac-
shaped, with one opening acting as
both mouth and anus. This opening
is surrounded by tentacles bearing
*thread cells. Coelenterates exist both
as free-swimming *medusae (e.g. jellyfish) and as sedentary *polyps. The
latter may be colonial (e.g. corals) or
solitary (e.g. sea anemones and
Hydra). In many coelenterates the life
cycle alternates between these two
forms (*see* alternation of generations).

coelom A fluid-filled cavity that
forms the main *body cavity of vertebrate and most invertebrate animals.
It is formed by the splitting of the
*mesoderm. Ciliated ducts
(*coelomoducts*) connect the coelom to
the exterior allowing the exit of waste
products and gametes; in higher
animals these are specialized as
oviducts, etc. The coelom is large and
often subdivided in annelid worms (in
which it functions as a hydrostatic
skeleton) and vertebrates. In arthropods it is restricted to the cavities of
the gonads and excretory organs, the
body cavity being a blood-filled
haemocoel.

coelomoduct *See* coelom.

coelostat A device that enables light
from the same area of the sky to be
continuously reflected into the field of
view of an astronomical telescope or
other instrument. It consists of a
plane mirror driven by a clockwork
or electrical mechanism so that it
rotates from east to west to compensate for the apparent west-to-east
rotation of the *celestial sphere.

coenocyte A mass of protoplasm containing many nuclei and enclosed by a cell wall. It is found in certain algae and fungi. *Compare* cell; plasmodium; syncytium.

coenzyme An organic nonprotein molecule that associates with an enzyme molecule in catalysing biochemical reactions. Coenzymes usually participate in the substrate–enzyme interaction by donating or accepting certain chemical groups. Many vitamins are precursors of coenzymes. *See also* cofactor.

coenzyme A (CoA) A complex organic compound that acts in conjunction with enzymes involved in various biochemical reactions, notably the oxidation of pyruvate via the *Krebs cycle and fatty-acid oxidation and synthesis. It comprises principally the B vitamin *pantothenic acid, the nucleotide *adenine, and a ribose–phosphate group.

coenzyme Q (ubiquinone) Any of a group of related quinone-derived compounds that serve as electron carriers in the *electron transport chain reactions of cellular respiration. Coenzyme Q molecules have side chains of different lengths in different types of organisms but function in similar ways.

coercive force (coercivity) The magnetizing force necessary to reduce the flux density in a magnetic material to zero. *See* hysteresis.

cofactor A nonprotein component essential for the normal catalytic activity of an enzyme. Cofactors may be organic molecules (*coenzymes) or inorganic ions. They may activate the enzyme by altering its shape or they may actually participate in the chemical reaction.

coherent radiation Electromagnetic radiation in which two or more sets of waves have a constant phase relationship, i.e. with peaks and troughs always similarly spaced.

coherent units A system of *units of measurement in which derived units are obtained by multiplying or dividing base units without the use of numerical factors. *SI units form a coherent system; for example the unit of force is the newton, which is equal to 1 kilogram metre per second squared ($kg\ m\ s^{-2}$), the kilogram, metre, and second all being base units of the system.

coinage metals A group of three malleable ductile metals forming subgroup IB of the *periodic table: copper (Cu), silver (Ag), and gold (Au). Their outer electronic configurations have the form $nd^{10}(n+1)s^1$. Although this is similar to that of alkali metals, the coinage metals all have much higher ionization energies and higher (and positive) standard electrode potentials. Thus, they are much more difficult to oxidize and are more resistant to corrosion. In addition, the fact that they have *d*-electrons makes them show variable valency (Cu^I, Cu^{II}, and Cu^{III}; Ag^I and Ag^{II}; Au^I and Au^{III}) and form a wide range of coordination compounds. They are generally classified with the *transition elements.

coincidence circuit An electronic logic device that gives an output only if two input signals are fed to it simultaneously or within a specified time of each other. A *coincidence counter* is an electronic counter incorporating such a device.

coke A form of carbon made by the destructive distillation of coal. Coke is used for blast-furnaces and other metallurgical and chemical processes requiring a source of carbon. Lower-grade cokes, made by heating the coal to a lower temperature, are used as smokeless fuels for domestic heating.

colchicine An *alkaloid derived from the autumn crocus, *Colchicum autumnale*. It inhibits *spindle formation in cells, so preventing their division and inducing multiple sets of chromosomes (*polyploidy). Colchicine is used in genetics, cytology, and plant breeding research and also in cancer therapy to inhibit cell division.

cold-bloodedness *See* poikilothermy.

cold emission The emission of electrons by a solid without the use of high temperature (thermal emission), either as a result of field emission (*see* field-emission microscope) or *secondary emission.

cold fusion *See* nuclear fusion.

Coleoptera An order of insects comprising the beetles and weevils and containing about 330 000 known species – the largest order in the animal kingdom. The forewings are hardened and thickened to form *elytra*, which meet at a precise mid-dorsal line and protect the underlying pair of hindwings and abdomen. The mouthparts are generally modified for biting and in some species assume antler-like proportions. Beetles occur in a wide variety of terrestrial and aquatic habitats; many feed on decaying organic matter, some eat living vegetation, while others prey on other arthropods. A number of beetles and weevils are economically important pests of stored grain, timber, and crops. The young emerge as larvae and generally undergo metamorphosis via a pupal stage to form the adult beetle.

coleoptile A protective sheath that covers the young shoot of the embryo in plants of the grass family. It bursts open when the first leaves develop. Experiments investigating growth movements of the oat coleoptile led to the discovery of the plant growth substance indoleacetic acid (IAA).

coleorhiza A protective sheath that covers the young root of the embryo in plants of the grass family.

collagen An insoluble fibrous protein found extensively in the connective tissue of skin, tendons, and bone. The polypeptide chains of collagen (containing the amino acids glycine and proline predominantly) form triple-stranded helical coils that are bound together to form fibrils, which have great strength and limited elasticity. Collagen accounts for over 30% of the total body protein of mammals.

collective excitation A quantized mode in a many-body system, occurring because of cooperative motion of the whole system as a result of interactions between particles. *Plasmons and *phonons in solids are examples of collective excitations. Collective excitations obey Bose–Einstein statistics (*see* quantum statistics).

collector *See* transistor.

collenchyma *See* ground tissues.

colligative properties Properties that depend on the concentration of particles (molecules, ions, etc.) present in a solution, and not on the nature of the particles. Examples of colligative properties are osmotic pressure (*see* osmosis), *lowering of vapour pressure, *depression of freezing point, and *elevation of boiling point.

collimator 1. Any device for producing a parallel beam of radiation. A common arrangement used for light consists of a convex achromatic lens fitted to one end of a tube with an adjustable slit at the other end, the slit being at the principal focus of the lens. Light rays entering the slit leave the lens as a parallel beam. Collimators for particle beams and other types of electromagnetic radiation utilize a system of slits or apertures. **2.** A small fixed telescope attached to a large astronomical telescope to assist in lining up the large one onto the desired celestial body.

collision density The number of collisions that occur in unit volume in unit time when a given neutron flux passes through matter.

collodion A thin film of cellulose nitrate made by dissolving the cellulose nitrate in ethanol or ethoxyethane, coating the surface, and evaporating the solvent.

colloids Colloids were originally defined by Thomas Graham in 1861 as substances, such as starch or gelatin, which will not diffuse through a membrane. He distinguished them from *crystalloids* (e.g. inorganic salts), which would pass through membranes. Later it was recognized that colloids were distinguished from true solutions by the presence of particles that were too small to be observed with a normal microscope yet were much larger than normal molecules. Colloids are now regarded as systems in which there are two or more phases, with one (the *dispersed phase*) distributed in the other (the *continuous phase*). Moreover, at least one of the phases has small dimensions (in the range 10^{-9}–10^{-6} m). Colloids are classified in various ways.

Sols are dispersions of small solid particles in a liquid. The particles may be macromolecules or may be clusters of small molecules. *Lyophobic sols* are those in which there is no affinity between the dispersed phase and the liquid. An example is silver chloride dispersed in water. In such colloids the solid particles have a surface charge, which tends to stop them coming together. Lyophobic sols are inherently unstable and in time the particles aggregate and form a precipitate. *Lyophilic sols*, on the other hand, are more like true solutions in which the solute molecules are large and have an affinity for the solvent. Starch in water is an example of such a system. *Association colloids* are systems in which the dispersed phase consists of clusters of molecules that have lyophobic and lyophilic parts. Soap in water is an association colloid (*see* micelle).

Emulsions are colloidal systems in which the dispersed and continuous phases are both liquids, e.g. oil-in-water or water-in-oil. Such systems require an emulsifying agent to stabilize the dispersed particles.

Gels are colloids in which both dispersed and continuous phases have a three-dimensional network throughout the material, so that it forms a jelly-like mass. Gelatin is a common example. One component may sometimes be removed (e.g. by heating) to leave a rigid gel (e.g. silica gel).

Other types of colloid include *aerosols (dispersions of liquid or solid particles in a gas, as in a mist or smoke) and foams (dispersions of gases in liquids or solids).

cologarithm The logarithm of the reciprocal of a number.

colon The section of the vertebrate *large intestine that lies between the *caecum and the *rectum. Its prime function is to absorb water and minerals from indigestible food residues passing from the small intestine, which results in the formation of *faeces.

colony 1. (in zoology) A group of animals of the same species living together and dependent upon each other. Some, such as the corals and sponges, are physically connected and function as a single unit. Others, such as insect colonies, are not physically joined but show a high level of social organization with members specialized for different functions (*see* caste). **2.** (in microbiology) A group of microorganisms, usually bacteria or yeasts, that are considered to have developed from a single parent cell. Colonies that grow on *agar plates differ in shape, colour, surface texture, and translucency and can therefore be used as a means of identification.

colony-stimulating factor (CSF) One of several molecules that regulate the development of blood cells. *GM-CSF* is a glycoprotein manufactured in all tissues, large amounts being found in the lungs. It induces haemopoietic stem cells (the precursors of mature blood cells) to develop into granulocytes or monocytes (hence GM). *G-CSF* is a protein that induces stem cells to form granulocytes only. *M-CSF*, a glycoprotein, causes stem cells to form monocytes only.

colorimeter Any instrument for comparing or reproducing colours. Monochromatic colorimeters match a *colour with a mixture of monochromatic and white lights. Trichromatic colorimeters use a mixture of three *primary colours.

colorimetric analysis Quantitative analysis of solutions by estimating their colour, e.g. by comparing it with the colours of standard solutions.

colostrum A liquid with a high content of nitrogen, antibodies, and vitamins that is secreted from the mammary glands before and just after giving birth. The change of secretion from colostrum to proper milk takes place gradually during the days after birth.

colour The sensation produced when light of different wavelengths falls on the human eye. Although the visible spectrum covers a continuously varying range of colours from red to violet it is usually split into seven colours (the *visible spectrum*) with the following approximate wavelength ranges:
red 740–620 nm
orange 620–585 nm
yellow 585–575 nm
green 575–500 nm
blue 500–445 nm
indigo 445–425 nm
violet 425–390 nm
A mixture of all these colours in the proportions encountered in daylight gives white light; other colours are produced by varying the proportions or omitting components.

A coloured light has three attributes: its *hue*, depending on its wavelength; its *saturation*, depending on the degree to which it departs from white light; and its *luminosity. Coloured

objects that owe their colour to pigments or dyes absorb some components of white light and reflect the rest. For example, a red book seen in white light absorbs all the components except the red, which it reflects. This is called a *subtractive process* as the final colour is that remaining after absorption of the rest. This is the basis of the process used in *colour photography. Combining coloured lights, on the other hand, is an *additive process* and this is the method used in *colour television. *See also* primary colours.

colour blindness Any disorder of vision in which colours are confused. The most common type is red–green colour blindness. This is due to a recessive gene carried on the X chromosome (*see* sex linkage), and therefore men are more likely to show the defect although women may be *carriers. It results in absence or malfunctioning of nerve cells sensitive to coloured light (*cones) in the retina of the eyes.

colour charge *See* elementary particles.

colour photography Any of various methods of forming coloured images on film or paper by photographic means. One common process is a subtractive reversal system that utilizes a film with three layers of light-sensitive emulsion, one responding to each of the three *primary colours. On development a black image is formed where the scene is blue. The white areas are dyed yellow, the *complementary colour of blue, and the blackened areas are bleached clean. A yellow filter between this emulsion layer and the next keeps blue light from the second emulsion, which is green-sensitive. This is dyed magenta where no green light has fallen. The final emulsion is red-sensitive and is given a cyan (blue-green) image on the negative after dying. When white light shines through the three dye layers the cyan dye subtracts red where it does not occur in the scene, the magenta subtracts green, and the yellow subtracts blue. The light projected by the negative therefore reconstructs the original scene either as a transparency or for use with printing paper.

colour television A television system in which the camera filters the light from the scene into the three component *primary colours, red, blue, and green, which are detected by separate camera tubes. The separate information so obtained relating to the colour of the image is combined with the sound and synchronization signals and transmitted using one of three systems, the American, British, or French. At the receiver, the signal is split again into red, blue, and green components, each being fed to a separate *electron gun in the cathode-ray tube of the receiver. By an additive process (*see* colour) the picture is reconstituted by the beam from each gun activating a set of phosphor dots of that colour on the screen.

colour temperature The temperature of a non-black body as indicated by the temperature of a black body having approximately the same spectral distribution.

columbium A former name for the element *niobium.

column chromatography *See* chromatography.

coma 1. A nebulous cloud of gas and dust that surrounds the nucleus of a *comet. **2.** An *aberration of a lens or mirror in which the image of a point lying off the axis has a comet-shaped appearance.

combinations *See* permutations and combinations.

combustion A chemical reaction in which a substance reacts rapidly with oxygen with the production of heat and light. Such reactions are often free-radical chain reactions, which can usually be summarized as the oxidation of carbon to form its oxides and the oxidation of hydrogen to form water. *See also* flame.

comet A small body that travels around the sun in an eccentric orbit. *Short-period comets* have orbital periods of less than 150 years. The others have very long periods, some exceeding 100 000 years. Typical comets have three components: the *nucleus* of ice and dust, the *coma of gas and dust, and the *comet tail*, which only appears when the comet is near the sun (it, too, consists of gas and dust). The nuclei of most comets are thought to be 'dirty snowballs' about one kilometre in diameter, although the solar system has a few comets with nuclei exceeding 10 km in diameter. The coma may be 10^4-10^5 km in diameter, and the tail can be 10^7 km in length. *See also* Halley's comet.

commensalism An interaction between two animal or plant species that habitually live together in which one species (the *commensal*) benefits from the association while the other is not significantly affected. For example, the burrows of many marine worms contain commensals that take advantage of the shelter provided but do not affect the worm.

common logarithm *See* logarithm.

common salt *See* sodium chloride.

communication satellite An unmanned artificial satellite sent by rocket into a geostationary orbit (*see* synchronous orbit) around the earth to enable television broadcasts and telephone communications to be made between points on the earth's surface that could not otherwise communicate by radio owing to the earth's curvature. Modulated *microwaves are transmitted to the satellite, which amplifies them and retransmits them at a different frequency to the receiving station. The satellites are powered by *solar cells. Three or more satellites in equatorial orbits can provide a world-wide communications linkage. The satellites are placed well above the *ionosphere and therefore the carrier waves used have to be in the microwave region of the spectrum in order to pass through the ionosphere.

community A naturally occurring assemblage of plant and animal species living within a defined area or habitat. Communities are named after one of their *dominant species (e.g. a pine community) or the major physical characteristics of the area (e.g. a freshwater pond community). Members of a community interact in various ways (e.g. through *food chains and *competition). Large communities may be divided into smaller component communities. *See* association.

commutative law The mathematical law stating that the value of an expression is independent of the order of combination of the numbers, symbols, or terms in the expression. The *commutative law* for addition applies if $x + y = y + x$. The commutative law of multiplication applies if $x \times y = y \times x$. Subtraction and division are not commutative. *Compare* associative law; distributive law.

commutator The part of the armature of an electrical motor or generator through which connections are made to external circuits. It consists of a cylindrical assembly of insulated copper conductors, each of which is connected to one point in the armature winding. Spring-loaded carbon brushes are positioned around the commutator to carry the current to or from it.

companion cell A type of cell found within the *phloem of flowering plants. Each companion cell is usually closely associated with a *sieve element. Its function is uncertain, though it appears to regulate the activity of the adjacent sieve element.

In gymnosperms a similar function is attributed to *albuminous cells*, which are found closely associated with gymnosperm sieve elements.

compass A small magnet pivoted at its central point to revolve in a horizontal plane. In the earth's magnetic field the magnet (called the compass needle) aligns itself so that its north-seeking end points to the earth's magnetic north pole. A scale (called a compass card) is placed below the needle for use in navigation. In some navigation compasses the entire card is pivoted, indicating direction by a fixed mark on the casing. Such compasses are often filled with alcohol to provide damping. Magnetic compasses suffer from being affected by magnetic metals in their vicinity and to a large extent they have been replaced by *gyrocompasses.

compass plant A plant that has its leaves permanently orientated in a north–south direction. Such an arrangement enables the plant to take full advantage of morning and evening sun, while avoiding the stronger midday sunlight. An example is the compass plant of the prairies (*Silphium laciniatum*).

competent Describing embryonic tissue that is capable of developing into a specialized tissue when suitably stimulated. *See* induction; evocation.

competition The interaction that occurs between two or more organisms, populations, or species that share some environmental resource when this is in short supply. Competition is an important force in evolution: plants, for example, become tall to compete for light, and animals evolve various foraging methods to compete for food. There may be a direct confrontation between competitors, as occurs between barnacles competing for space on a rock, or the numbers or fecundity of the competitors are indirectly reduced through joint dependence on limited resources.

Competition occurs both between members of a species (*intraspecific competition*) and between different species (*interspecific competition*). Since competition ultimately results in the displacement by one competitor of the others, it is to the advantage of the competitors to avoid one another wherever possible. Thus in time the competitors become separated from each other geographically or ecologically, which promotes evolutionary change. Competition for mates may lead to *sexual selection.

complement A group of proteins present in blood plasma and tissue fluid that aids the body's defences following an *immune response; the genes encoding it form part of the *major histocompatibility complex. Following an antibody–antigen reaction, complement is activated chemically and becomes bound to the antibody–antigen complex (*complement fixation*); it becomes involved in destroying foreign cells and attracting scavenging white blood cells (*phagocytes) to the area of conflict in the body.

complemental males The small males of certain animals that live in or on the females and are usually more or less degenerate apart from the reproductive organs They occur in certain crustaceans (e.g. some barnacles), in which the normal individuals are hermaphrodite but the complemental males have suppressed ovaries, lose their alimentary canal, and lead a semiparasitic existence in the mantle cavity of the larger partner. This may ensure that cross fertilization occurs.

complementarity The concept that a single model may not be adequate to explain all the observations made of atomic or subatomic systems in different experiments. For example, *electron diffraction is best explained by assuming that the electron is a wave (*see* de Broglie wavelength), whereas

the *photoelectric effect is described by assuming that it is a particle. The idea of two different but complementary concepts to treat quantum phenomena was first put forward by the Danish physicist Niels Bohr (1855–1962) in 1927. *See also* light.

complementary colours A pair of coloured lights of specific hue (*see* colour) that produce the sensation of white when mixed in appropriate intensities. There is an infinite number of such pairs, an example (with wavelengths) is orange (608 nm) and blue (490 nm).

complementary DNA (cDNA) A form of DNA prepared in the laboratory using messenger *RNA (mRNA) as template, i.e. the reverse of the usual process of *transcription in cells; the synthesis is catalysed by *reverse transcriptase. cDNA thus has a base sequence that is complementary to that of the mRNA template; unlike genomic DNA, it contains no noncoding sequences (*introns). cDNA is used in *gene cloning for the expression of eukaryote genes in prokaryote host cells.

complex A compound in which molecules or ions form coordinate bonds to a metal atom or ion (see illustration). The complex may be a positive ion (e.g. $[Cu(H_2O)_6]^{2+}$), a negative ion (e.g. $Fe[(CN)_6]^{3-}$), or a neutral molecule (e.g. $PtCl_2(NH_3)_2$). The formation of such coordination complexes is typical behaviour of transition metals. The complexes formed are often coloured and have unpaired electrons (i.e. are paramagnetic). *See also* ligand; chelate.

complex number A number that has a real part, x, and an imaginary part, iy, where $i = \sqrt{-1}$ and x and y are real (x can also equal 0). The complex number therefore has the form $x + iy$, which can also be written in the polar form $r\cos\theta + ir\sin\theta$, where r is the *modulus* and θ is the *argument* (or *amplitude*). A complex numoctahedral

square-planar

tetrahedral

trigonal-bipyramid

Some common shapes of coordination complexes

ber can be represented on an *Argand diagram*, devised by J. R. Argand (1768–1822), in which the horizontal axis represents the real part of the number and the vertical axis the imaginary part (see illustration). In the polar form the modulus is the line joining the origin to the point representing the complex number and the argument is the angle between the modulus and the *x*-axis.

Argand diagram

complexometric analysis A type of volumetric analysis in which the reaction involves the formation of an inorganic *complex.

component A distinct chemical species in a mixture. If there are no reactions taking place, the number of components is the number of separate chemical species. A mixture of water and ethanol, for instance, has two components (but is a single phase). A mixture of ice and water has two phases but one component (H_2O). If an equilibrium reaction occurs, the number of components is taken to be the number of chemical species minus the number of reactions. Thus, in

$$H_2 + I_2 \leftrightarrows 2HI$$

there are two components. *See also* phase rule.

component vectors Two or more vectors that produce the same effect as a given vector; the vectors that combine to produce the effect of a resultant vector. A component vector in a given direction is the projection of the given vector (*V*) along that direction, i.e. $V\cos\theta$, where θ is the angle between the given vector and the direction.

composite fruit A type of fruit that develops from an inflorescence rather than from a single flower. *See* pseudocarp; sorosis; strobilus; syconus.

compound A substance formed by the combination of elements in fixed proportions. The formation of a compound involves a chemical reaction; i.e. there is a change in the configuration of the valence electrons of the atoms. Compounds, unlike mixtures, cannot be separated by physical means. *See also* molecule.

compound eye The eye of insects and crustaceans, which consists of numerous visual units, the *ommatidia*. Each ommatidium consists of a lens beneath which is a crystalline cone and light-sensitive cells. The eye is convex, with the apices of the cones converging onto the optic nerve. There are two types of compound eye. In *apposition eyes*, typical of diurnal insects, each ommatidium focuses rays parallel to its long axis so that each gives an image of a minute part of the visual field, producing a detailed mosaic image. In *superposition eyes*, typical of nocturnal insects, each ommatidium receives light from a larger part of the visual field and the image may overlap with those received by many neighbouring ommatidia. This produces an image that is bright but lacks sharpness of detail.

compound microscope *See* microscope.

compressibility The reciprocal of bulk modulus (*see* elastic modulus). The compressibility (*k*) is given by $-V^{-1}dV/dp$, where dV/dp is the rate of change of volume (*V*) with pressure.

compression ratio The ratio of the total volume enclosed in the cylinder of an *internal-combustion engine at the beginning of the compression stroke to the volume enclosed at the end of the compression stroke. For petrol engines the compression ratio is 8.5–9:1, with a recent tendency to the lower end of the range in order to make use of unleaded petrols. For Diesel engines the compression ratio is in the range 12–18:1.

Compton effect The reduction in the energy of high-energy (X-ray or gamma-ray) photons when they are scattered by free electrons, which thereby gain energy. The phenomenon, first observed in 1923 by the US physicist A. H. Compton (1892–1902), occurs when the photon collides with an electron; some of the photon's energy is transferred to the electron and consequently the photon loses energy $h(\nu_1 - \nu_2)$, where h is the *Planck constant and ν_1 and ν_2 are the frequencies before and after collision. As $\nu_1 > \nu_2$, the wavelength of the radiation increases after the collision. This type of inelastic scattering is known as *Compton scattering* and is similar to the *Raman effect. *See also* inverse Compton effect.

Compton wavelength The length below which a particle's quantum-mechanical properties become relevant in relativistic *quantum mechanics. For a particle of rest mass m the Compton wavelength is $\hbar/mc$, where $\hbar$ is the rationalized Planck constant and c is the speed of light. The Compton wavelength is so named because of its occurrence in the theory of the *Compton effect, where its value for the electron is 3.8616×10^{-13} m. The Compton wavelength is sometimes defined as h/mc, with h

being the Planck constant, in which case the electron value is 2.4263 × 10^{-12} m.

computer An electronic device that processes information according to a set of instructions, called the *program*. The most versatile type of computer is the *digital computer*, in which the input is in the form of characters, represented within the machine in *binary notation. The three basic components of a digital computer are the peripheral input and output devices, the memory, and the *central processing unit* (CPU). The input usually consists of a keyboard, tape drive, or a punched-card or tape reader. The output is usually a printer providing information on paper, a tape drive that provides it on magnetic tape, or a visual-display unit (VDU) providing it on a cathode-ray screen. The memory consists of one or more storage devices, such as magnetic disks, on which both data and programs are stored. The CPU controls the whole system and performs arithmetical and logical operations on the data. Computers range in size from the *microprocessor* with a few thousand logic elements, to the large *mainframe computer* with millions of logic circuits.

The *analog computer* is used in scientific experiments, industrial control, etc. In this type of device the input and output are continuously varying quantities, such as a voltage, rather than the discrete digits of the more commercially useful digital device. *Hybrid computers* combine the properties of both digital and analog devices. Input is usually in analog form, but processing is carried out digitally in a CPU.

Computer *hardware* consists of the actual electronic or mechanical devices used in the system; the *software* consists of the programs and data.

concave Curving inwards. A *concave mirror* is one in which the reflecting surface is formed from the interior surface of a sphere or paraboloid. A *concave lens* has at least one face formed from the interior surface of a sphere. A *biconcave lens* has both faces concave and is therefore thinnest at its centre. The *plano-concave lens* has one plane face and one concave face. The *concavo-convex lens* (also called a *meniscus*) has one concave face and one *convex face. See* lens.

concavo-convex *See* concave.

concentrated Describing a solution that has a relatively high concentration of solute.

concentration The quantity of dissolved substance per unit quantity of solvent in a solution. Concentration is measured in various ways. The amount of substance dissolved per unit volume (symbol c) has units of mol dm^{-3} or mol l^{-1}. It is now called 'concentration' (formerly *molarity*). The *mass concentration* (symbol ρ) is the mass of solute per unit volume of solvent. It has units of kg dm^{-3}, g cm^{-3}, etc. The *molal concentration* (or *molality*; symbol m) is the amount of substance per unit mass of solvent, commonly given in units of mol kg^{-1}. *See also* mole fraction.

concentration cell *See* cell.

conceptacle A flask-shaped cavity with a small opening (the *ostiole*) that is found in the swollen tip of certain brown algae, such as *Fucus*. It contains the sex organs.

conchoidal fracture Fracture of a solid in which the surface of the material is curved and marked by concentric rings. It occurs particularly in amorphous materials.

condensation The change of a vapour or gas into a liquid. The change of phase is accompanied by the evolution of heat (*see* latent heat).

condensation polymerization *See* polymer.

condensation pump *See* diffusion pump.

condensation reaction A chemical reaction in which two molecules combine to form a larger molecule with elimination of a small molecule (e.g. H_2O). *See* aldehydes; ketones.

condensed-matter physics *See* solid-state physics.

condenser 1. A mirror or set of lenses used in optical instruments, such as a microscope or film projector, to concentrate the light diverging from a compact source. A common form consists of two plano-convex lenses with the plane faces pointing outwards. **2.** A device used to cool a vapour to cause it to condense to a liquid. In a steam engine the condenser acts as a reservoir that collects the part of the steam's internal energy that has not been used in doing work on the piston. The cooling water passed through the condenser is warmed and is used as fresh feedwater for the boiler. *See also* Liebig condenser. **3.** *See* capacitor.

condenser microphone *See* capacitor microphone.

conditional response (conditioned reflex) A learned response that develops to an initially ineffective stimulus in classical *conditioning.

conditioning A process by which animals learn about a relation between two events. In *classical* (or *Pavlovian*) *conditioning*, repeated presentations of a neutral stimulus (e.g. the sound of a bell or buzzer) are followed each time by a biologically important stimulus (such as food or electric shock), which elicits a response (e.g. salivation). Eventually the neutral stimulus presented by itself produces a response (the *conditional response*, or *conditioned reflex*) similar to that originally evoked by the biologically important stimulus.

For example, Pavlov's dogs learned to salivate in response to the sound of a metronome that preceded the presentation of food. In *instrumental* (or *operant*) *conditioning* the animal is rewarded (or punished) each time it makes a particular response; this eventually causes the frequency of the response to increase (or decrease). For example, a rat will learn to press a lever in order to obtain food. *See* reinforcement.

conductance The reciprocal of electrical resistance in a direct-current circuit. The ratio of the resistance to the square of the *impedance in an alternating-current circuit. The SI unit is the siemens, formerly called the mho or reciprocal ohm.

conducting polymer An organic polymer that conducts electricity. Conducting polymers have a crystalline structure in which chains of conjugated unsaturated carbon–carbon bonds are aligned. Examples are polyacetylene and polypyrrole. There has been considerable interest in the development of such materials because they would be cheaper and lighter than metallic conductors. They do, however, tend to be chemically unstable and, so far, no commercial conducting polymers have been developed.

conductiometric titration A type of titration in which the electrical conductivity of the reaction mixture is continuously monitored as one reactant is added. The equivalence point is the point at which this undergoes a sudden change. The method is used for titrating coloured solutions, which cannot be used with normal indicators.

conduction 1. (thermal conduction) The transmission of heat through a substance from a region of high temperature to a region of lower temperature. In gases and most liquids, the energy is transmitted mainly by collisions between atoms and molecules with those possessing lower kinetic energy. In solid and liquid metals, heat conduction is predominantly by migration of fast-moving electrons, followed by collisions between these electrons and ions. In solid insulators the absence of *free electrons restricts heat transfer to the vibrations of atoms and molecules within crystal lattices. See conductivity. 2. (electrical conduction) The passage of electric charge through a substance under the influence of an electric field. See also charge carrier; energy bands.

conduction band See energy bands.

conductivity 1. (thermal conductivity) A measure of the ability of a substance to conduct heat. For a block of material of cross section A, the energy transferred per unit time E/t, between faces a distance, l, apart is given by $E/t = \lambda A(T_2 - T_1)/l$, where λ is the conductivity and T_2 and T_1 are the temperatures of the faces. This equation assumes that the opposite faces are parallel and that there is no heat loss through the sides of the block. The SI unit is therefore $J\,s^{-1}\,m^{-1}\,K^{-1}$. 2. (electrical conductivity) The reciprocal of the *resistivity of a material. It is measured in siemens per metre in SI units. When a fluid is involved the electrolytic conductivity is given by the ratio of the current density to the electric field strength.

conductivity water See distilled water.

conductor 1. A substance that has a high thermal *conductivity. Metals are good conductors on account of the high concentration of *free electrons they contain. Energy is transmitted through a metal predominantly by means of collisions between electrons and ions. Most nonmetals are poor conductors (good *thermal insulators*) because there are relatively few free electrons. 2. A substance that has a high electrical conductivity. Again

conduction results from the movement of free electrons. See energy bands.

condyle A smooth round knob of bone that fits into a socket on an adjoining bone, forming a *joint. Such a joint permits up-and-down or side-to-side movement but does not allow rotation. There are condyles where the lower jawbone (mandible) is attached to the skull, which permits chewing movements. See also occipital condyle.

Condy's fluid A mixture of calcium and potassium permanganates (manganate(VII)) used as an antiseptic.

cone 1. (in botany) A reproductive structure occurring in gymnosperms and some pteridophytes, known technically as a *strobilus*. It consists of *sporophylls bearing the spore-producing sporangia. Gymnosperms produce different male and female cones. The large woody female cones of pines, firs, and other conifers are made up of structures called *ovuliferous scales*, which bear the ovules. 2. (in animal anatomy) A type of light-sensitive receptor cell, found in the *retinas of all diurnal vertebrates. Cones are specialized to transmit information about colour and they function best in bright light. They are not evenly distributed on the retina, being concentrated in the *fovea and absent on the margin of the retina. *Compare* rod. 3. (in mathematics) A solid figure generated by a line (the *generator*) joining a point on the perimeter of a closed plane curve (the *directrix*) to a point (the *vertex*) outside this plane, as the line moves round the directrix. If the directrix is a circle, the figure is a *circular cone* standing on a circular *base*. If the line joining the vertex to the centre of the base (the *axis*) is perpendicular to the base the figure is a *right circular cone*, which has a volume $\pi r^2 h/3$, where r is the radius of the base and h is the height of the vertex above the base. If the axis of the cone is not perpendicular to the base, the figure is an *oblique cone*. In general, the volume of any cone is one third of its base area multiplied by the perpendicular distance of the vertex from the base.

configuration 1. The arrangement of atoms or groups in a molecule. **2.** The arrangement of electrons in atomic *orbitals in an atom.

configuration space The n-dimensional space with coordinates $(q_1, q_2, \ldots, q_n)$ associated with a system that has n *degrees of freedom, where the values q describe the degrees of freedom. For example, in a gas of N atoms each atom has three positional coordinates, so the configuration space is $3N$-dimensional. If the particles also have internal degrees of freedom, such as those caused by vibration and rotation in a molecule, then these must be included in the configuration space, which is consequently of a higher dimension. *See also* phase space.

confinement *See* quantum chromodynamics; quark confinement.

conformation Any of the large number of possible shapes of a molecule resulting from rotation of one part of the molecule about a single bond. See illustration.

congeners Elements that belong to the same group in the periodic table.

congenital Present at birth. Congenital disorders of the body may be due to genetic factors, e.g. *Down's syndrome, or caused by injury or environmental factors, e.g. drugs (such as thalidomide), chemicals (such as dioxin), and infections (such as those caused by *Listeria* and *cytomegalovirus).

conic A figure formed by the intersection of a plane and a *cone. If the intersecting plane is perpendicular to the axis of a right circular cone, the figure formed is a *circle. If the intersecting plane is inclined to the axis at an angle in excess of half the apex angle of the cone it is an *ellipse. If the plane is parallel to the sloping side of the cone, the figure is a *parabola. If the plane cuts both halves of the cone a *hyperbola is formed.

A conic can be defined as a plane curve in which for all points on the curve the ratio of the distance from a fixed point (the *focus*) to the perpendicular distance from a straight line (the *directrix*) is a constant called the *eccentricity e*. For a parabola $e = 1$, for an ellipse $e < 1$, and for a hyperbola $e > 1$.

conidiospore *See* conidium.

conidium (conidiospore) A spore of certain fungi, such as mildews and moulds, that is produced by the constriction of the tip of a specialized hypha, the *conidiophore*. Chains of conidia may be cut off in this way.

Coniferales An order within the *Gymnospermae containing some extinct species and most of the present-day gymnosperms, such as the pines, firs, and spruces. They are typically evergreen trees inhabiting cool temperate regions and have leaves reduced to needles or scales. The

| eclipsed | staggered | gauche |

Conformation for rotation about a single bond

chair conformation   boat conformation

Conformations of cyclohexane ring

wood of conifers, which is called *softwood* in contrast to the *hardwood* of angiosperm trees, is widely used for timber and pulp.

conjugate acid *or* **base** *See* acids.

conjugated Describing double or triple bonds in a molecule that are separated by one single bond. For example, the organic compound buta-1,3-diene, $H_2C=CH-CH=CH_2$, has conjugated double bonds. In such molecules, there is some delocalization of electrons in the pi orbitals between the carbon atoms linked by the single bond.

conjugate points Two points in the vicinity of a *lens or *mirror such that a bright object placed at one will form an image at the other.

conjugation 1. The fusion of two reproductive cells, particularly when these are both the same size (*see* isogamy). **2.** A form of sexual reproduction seen in some algae (e.g. *Spirogyra*), some bacteria (e.g. *Escherichia coli*), and ciliate protozoans. Two individuals are united by a tube formed by outgrowths from one or both of the cells. Genetic material from one cell (designated the male) then passes through the tube to unite with that in the other (female) cell.

conjunction The alignment of two celestial bodies within the solar system so that they have the same longitude as seen from the earth. A planet that orbits between the sun and the earth (Venus and Mercury) is in *superior conjunction* when it is in line with the sun and the earth but on the opposite side of the sun to the earth. It is in *inferior conjunction* when it lies between the earth and the sun. Conjunction may also occur between two planets or a moon and a planet.

conjunctiva The delicate membrane that covers the cornea and lines the inside of the eyelid of a vertebrate eye. It is kept clean by secretions of the *lacrimal (tear) gland and the reflex blink mechanism.

connective tissue An animal tissue consisting of a small number of cells (e.g. *fibroblasts and *mast cells) and fibres and a large amount of intercellular material (*matrix*). It is widely distributed and has many functions, including support, packing, defence, and repair. The individual constituents vary, depending on the function of the tissue. Different types of connective tissue include *mesenchyme* in the embryo, *adipose tissue, loose *areolar tissue* for packing and support, blood, lymph, cartilage, and bone.

conservation The sensible use of the earth's natural resources in order to avoid excessive degradation and impoverishment of the environment. It should include the search for alternative food and fuel supplies when these are endangered (as by deforestation and overfishing); an awareness of the dangers of *pollution; and the maintenance and preservation of natural habitats and the creation of new ones (e.g. nature reserves and national parks).

conservation law A law stating that the total magnitude of a certain physical property of a system, such as its mass, energy, or charge, remains unchanged even though there may be exchanges of that property between components of the system. For example, imagine a table with a bottle of salt solution (NaCl), a bottle of silver nitrate solution ($AgNO_3$), and a beaker standing on it. The mass of this table and its contents will not change even when some of the contents of the bottles are poured into the beaker. As a result of the reaction between the chemicals two new substances (silver chloride and sodium nitrate) will appear in the beaker: $NaCl + AgNO_3 \rightarrow AgCl + NaNO_3$, but the total mass of the table and its contents will not change. This *conser-*

vation of mass is a law of wide and general applicability, which is true for the universe as a whole, provided that the universe can be considered a closed system (nothing escaping from it, nothing being added to it). According to Einstein's mass–energy relationship, every quantity of energy (E) has a mass (m), which is given by E/c^2, where c is the speed of light. Therefore if mass is conserved, the law of *conservation of energy* must be of equally wide application. The laws of *conservation* of *linear momentum* and *angular momentum* also are believed to be universally true.

Because no way is known of either creating or destroying electric charge, the law of *conservation of charge* is also a law of universal application. Other quantities are also conserved in reactions between elementary particles.

conservative field A field of force in which the work done in moving a body from one point to another is independent of the path taken. The force required to move the body between these points in a conservative field is called a *conservative force*.

consociation A climax plant *community that is dominated by one particular species, e.g. a pine forest. *See* dominant. *Compare* association.

consolute temperature The temperature at which two partially miscible liquids become fully miscible as the temperature is increased.

constant 1. A component of a relationship between variables that does not change its value, e.g. in $y = ax + b$, b is a constant. **2.** A fixed value that has to be added to an indefinite integral. Known as the *constant of integration*, it depends on the limits between which the integration has been performed. **3.** *See* fundamental constants.

constantan An alloy having an electrical resistance that varies only very

slightly with temperature (over a limited range around normal room temperatures). It consists of copper (50–60%) and nickel (40–50%) and is used in resistance wire, thermocouples, etc.

constant-boiling mixture *See* azeotrope.

constant proportions *See* chemical combination.

consumer An organism that feeds upon those below it in a *food chain (i.e. at the preceding *trophic level). Herbivores, which feed upon green plants, are *primary consumers*; a carnivore that feeds only upon herbivores is a *secondary consumer*. *Compare* producer.

contact potential difference The potential difference that occurs between two electrically connected metals or between the base regions of two semiconductors. If two metals with work functions ϕ_1 and ϕ_2 are brought into contact, their Fermi levels will coincide. If $\phi_1 > \phi_2$ the first metal will acquire a positive surface charge with respect to the other at the area of contact. As a result, a contact potential difference occurs between the two metals or semiconductors.

contact process A process for making sulphuric acid from sulphur dioxide (SO_2), which is made by burning sulphur or by roasting sulphide ores. A mixture of sulphur dioxide and air is passed over a hot catalyst

$$2SO_2 + O_2 \rightarrow 2SO_3$$

The reaction is exothermic and the conditions are controlled to keep the temperature at an optimum 450°C. Formerly, platinum catalysts were used but vanadium–vanadium oxide catalysts are now mainly employed (although less efficient, they are less susceptible to poisoning). The sulphur trioxide is dissolved in sulphuric acid

$$H_2SO_4 + SO_3 \rightarrow H_2S_2O_7$$

and the oleum is then diluted.

(a) 200 million years ago

(b) 135 million years ago

(c) 65 million years ago

Continental drift

containment 1. The prevention of the escape of radioactive materials from a *nuclear reactor. 2. The process of preventing the plasma in a *thermonuclear reactor from touching the walls of the vessel by means of magnetic fields.

continent A large landmass that rises above the deep ocean floor. Geologically, the boundary of a continent lies offshore at the edge of the gentle slope of the continental shelf. The British Isles and other offshore islands consequently are parts of the nearby continents. It is generally accepted that there are seven continents: Asia, Africa, North America,

South America, Europe, Australia, and Antarctica.

continental drift The theory that the earth's continents once formed a single mass and have since moved relative to each other. It was first postulated by A. Snider in 1858 and greatly developed by Alfred Wegener in 1912. He used evidence, such as the fit of South America into Africa and the distribution of rock types, flora, fauna, and geological structures, to suggest that the present distribution of the continents results from the breaking up of one or two greater land masses. The original land mass was named Pangaea and it was suggested that this broke up into the northerly Laurasia and the southerly Gondwanaland (see illustration). The theory was not accepted for about 50 years by the majority of geologists but during the early 1960s, H. H. Hess's seafloor-spreading hypothesis and the subsequent development of *plate tectonics produced a mechanism to explain the drift of the continents.

continuous function A function $f(x)$ is continuous at $x = a$ if the limit of $f(x)$ as x approaches a is $f(a)$. A function that does not satisfy this condition is said to be a *discontinuous function*.

continuous phase *See* colloid.

continuous spectrum *See* spectrum.

continuous variation (quantitative variation) The range of differences that can be observed in many characteristics in a population. Characteristics resulting from *multifactorial inheritance show continuous variation, e.g. the wide range of foot sizes in an adult human population. *Compare* discontinuous variation.

continuous wave A wave that is transmitted continuously rather than in pulses.

continuum A system of axes that form a *frame of reference. The three dimensions of space and the dimension of time together can be taken to form a four-dimensional continuum; this was suggested by Minkowski in connection with special *relativity.

contour A line drawn on a map or chart that joins points with equal elevation above (or below) a level (usually mean sea level). Contours thus show the relief of the land surface. The difference in height between two consecutive contours is the *contour interval.*

contour feathers *Feathers that are arranged in regular rows on a bird's body, giving the body its streamlined shape. Each has a central horny shaft (the *rachis*) with a flattened *vane* on each side. Each vane is composed of two rows of filament-like *barbs, which are connected to each other by means of hooked *barbules to form a smooth surface. There is often a small second vane, the *aftershaft*, near the base of the feather.

contractile root Any of the modified adventitious roots that develop from the base of the stem of a bulb or corm. The new bulb or corm develops at a higher level in the soil than the old one. The contractile roots shorten and pull it down to a suitable level.

contractile vacuole A membrane-surrounded cavity in a cell that periodically expands, filling with water, and then suddenly contracts, expelling its contents to the cell's exterior. It is thus an organ of *osmoregulation and excretion. Contractile vacuoles are common in freshwater sponges and typical of freshwater protozoans, such as *Amoeba* (which has one spherical vacuole) and *Euglena* and *Paramecium* (in which a number of accessory vacuoles are attached to a main vacuole).

control grid A wire-mesh electrode placed between the cathode and anode in a *thermionic valve or a *cathode-ray tube to control the flow of electrons from one to the other. A fluctuating potential signal fed to the control grid produces at the anode a current signal with similar but amplified fluctuations. It thus forms the basis of the electronic valve amplifier. In a cathode-ray tube the grid controls the intensity of the electron beam and hence the brightness of the image on the screen.

control rod One of a number of rods of a material, such as boron or cadmium, that absorbs neutrons. Control rods can be moved into or out of the core of a *nuclear reactor to control the rate of the reaction taking place within it.

convection A process by which heat is transferred from one part of a fluid to another by movement of the fluid itself. In *natural convection* the movement occurs as a result of gravity; the hot part of the fluid expands, becomes less dense, and is displaced by the colder denser part of the fluid as this drops below it. This is the process that occurs in most domestic hot-water systems between the boiler and the hot-water cylinder. A natural convection current is set up transferring the hot water from the boiler up to the cylinder (always placed above the boiler) so that the cold water from the cylinder can move down into the boiler to be heated. In some modern systems, where small-bore pipes are used or it is inconvenient to place the cylinder above the boiler, the circulation between boiler and hot-water cylinder relies upon a pump. This is an example of *forced convection*, where hot fluid is transferred from one region to another by a pump or fan.

conventional current A 19th-century convention, still in use, that treats any electrical current as a flow of positive charge from a region of positive potential to one of negative potential. The real motion, however, in the case of electrons flowing through a metal conductor, is in the opposite direction, from negative to positive. In semiconductors *hole conduction is in the direction of the conventional current; electron conduction is in the opposite direction.

convergent evolution The development of superficially similar structures in unrelated organisms, usually because the organisms live in the same kind of environment. Examples are the wings of insects and birds and the streamlined bodies of whales and fish. *Compare* adaptive radiation.

convergent series A series $a_1 + a_2 + \ldots + a_i + \ldots$, for which a partial sum $S_n = a_1 + a_2 + \ldots + a_n$ tends to a finite (or zero) limit as n tends to infinity. This limit is the *sum* of the series. For example, the series $1 + 1/2 + 1/3 + 1/4 + \ldots$ (with the general term a_i equal to $(1/2)^{i-1}$) tends to the limit 2. A series that is not convergent is said to be a *divergent series*. In such a series the partial sum tends to plus or minus infinity or may oscillate. For example, the series $1 + 1/2 + 1/3 + 1/4 + \ldots$ (with a_i equal to $1/i$) is divergent. As can be seen from this latter example, a series may be divergent even if the individual terms a_i tend to zero as i tends to infinity.

converging lens or mirror A lens or mirror that can refract or reflect a parallel beam of light so that it converges at a point (the principal focus). Such a mirror is concave; a converging lens is thicker at its centre than at its edges (i.e. it is biconvex, plano-convex, or convexo-concave). *Compare* diverging lens or mirror.

conversion electron *See* internal conversion.

converter 1. An electrical machine for converting alternating current into direct current, or less frequently, vice versa. **2.** The reaction vessel in the *Bessemer process or some similar steel-making process. **3.** A computer device for converting information

coded in one form into some other form.

converter reactor A *nuclear reactor that converts fertile material (e.g. thorium–232) into *fissile material (e.g. uranium–233). A converter reactor can also be used to produce electrical power.

convex Curving outwards. A *convex mirror* is one in which the reflecting surface is formed from the exterior surface of a sphere or paraboloid. A *convex lens* has at least one face formed from the exterior surface of a sphere. A *biconvex lens* has both faces convex and is therefore thickest at its centre. The *plano-convex lens* has one plane face and one convex face. The *convexo-concave lens* (also called a *meniscus*) has one convex face and one *concave face. *See* lens.

coolant A fluid used to remove heat from a system by *convection (usually forced), either to control the temperature or to extract energy. In a water-cooled car engine the coolant is water (or water and antifreeze), which is pumped around the engine and cooled in the radiator. In a *nuclear reactor the coolant is used to transfer the heat of the reaction from the core to a heat exchanger or to the steam-raising plant. In gas-cooled reactors the coolant is usually carbon dioxide. Pressurized water or boiling water is used as both coolant and *moderator in several types of reactor. In fast reactors, liquid sodium is used as the coolant.

Cooper pairs *See* superconductivity.

coordinate *See* Cartesian coordinates; polar coordinates.

coordinate bond *See* chemical bond.

coordinate geometry *See* analytical geometry.

coordination compound A compound in which coordinate bonds are formed (*see* chemical bond). The term is used especially for inorganic *complexes.

coordination number The number of groups, molecules, atoms, or ions surrounding a given atom or ion in a complex or crystal. For instance, in a square-planar complex the central ion has a coordination number of four. In a close-packed crystal (*see* close packing) the coordination number is twelve.

Copepoda A subclass of crustaceans occurring in marine and freshwater habitats. Copepods are usually 0.5–2 mm long and lack both a carapace and compound eyes. Copepods are important members of plankton: some are free-living, feeding on microscopic organisms; others are parasitic. A familiar freshwater genus is *Cyclops*, so named because the members have a single median eye.

Copernican astronomy The system of astronomy that was proposed by the Polish astronomer Nicolaus Copernicus (1473–1543) in his book *De revolutionibus orbium coelestium*, which was published in the month of his death and first seen by him on his deathbed. It used some elements of *Ptolemaic astronomy, but rejected the notion, then current, that the earth was a stationary body at the centre of the universe. Instead, Copernicus proposed the apparently unlikely concept that the sun was at the centre of the universe and that the earth was hurtling through space in a circular orbit about it. Galileo's attempts, some 70 years later, to convince the Catholic church that in spite of scriptural authority to the contrary, the Copernican system was correct, resulted in *De revolutionibus* being placed on the Index of forbidden books, where it remained until 1835.

copolymer *See* polymer.

copper Symbol Cu. A red-brown *transition element; a.n. 29; r.a.m. 63.546; r.d. 8.93; m.p. 1083.4°C; b.p. 2582°C. Copper has been extracted for thousands of years; it was known to the Romans as cuprum, a name linked with the island of Cyprus. The metal is malleable and ductile and an excellent conductor of heat and electricity. Copper-containing minerals include cuprite (Cu_2O) as well as azurite ($2CuCO_3.Cu(OH)_2$), chalcopyrite ($CuFeS_2$), and malachite ($CuCO_3.Cu(OH)_2$). Native copper appears in isolated pockets in some parts of the world. The large mines in the USA, Chile, Canada, Zambia, Zaïre, and Peru extract ores containing sulphides, oxides, and carbonates. They are usually worked by smelting, leaching, and electrolysis. Copper metal is used to make electric cables and wires. Its alloys, brass (copper–zinc) and bronze (copper–tin), are used extensively.

Water does not attack copper but in moist atmospheres it slowly forms a characteristic green surface layer (patina). The metal will not react with dilute sulphuric or hydrochloric acids, but with nitric acid oxides of nitrogen are formed. Copper compounds contain the element in the +1 and +2 oxidation states. Copper(I) compounds are mostly white (the oxide is red). Copper(II) salts are blue in solution. The metal also forms a large number of coordination complexes.

copperas *See* iron(II) sulphate.

copper(I) chloride A white solid compound, CuCl; cubic; r.d. 4.14; m.p. 430°C; b.p. 1490°C. It is obtained by boiling a solution containing copper(II) chloride, excess copper turnings, and hydrochloric acid. Copper(I) is present as the $[CuCl_2]^-$ complex ion. On pouring the solution into air-free distilled water copper(I) chloride precipitates. It must be kept free of air and moisture since it oxidizes to copper(II) chloride under those conditions.

Copper(I) chloride is essentially covalent and its structure is similar to that of diamond; i.e. each copper atom is surrounded tetrahedrally by four chlorine atoms and vice versa. In the vapour phase, dimeric and trimeric species are present. Copper(I) chloride is used in conjunction with ammonium chloride as a catalyst in the dimerization of ethyne to but-1-ene-3-yne (vinyl acetylene), which is used in the production of synthetic rubber. In the laboratory a mixture of copper(I) chloride and hydrochloric acid is used for converting benzene diazonium chloride to chlorobenzene – the Sandmeyer reaction.

copper(II) chloride A brown-yellow powder, $CuCl_2$; r.d. 3.386; m.p. 620°C. It exists as a blue-green dihydrate (rhombic; r.d. 2.54; loses H_2O at 100°C). The anhydrous solid is obtained by passing chlorine over heated copper. It is predominantly covalent and adopts a layer structure in which each copper atom is surrounded by four chlorine atoms at a distance of 0.23 nm and two more at a distance of 0.295 nm. A concentrated aqueous solution is dark brown in colour due to the presence of complex ions such as $[CuCl_4]^{2-}$. On dilution the colour changes to green and then blue because of successive replacement of chloride ions by water molecules, the final colour being that of the $[Cu(H_2O)_6]^{2+}$ ion. The dihydrate can be obtained by crystallizing the solution.

copper glance A mineral form of copper(I) sulphide, Cu_2S.

copper(II) nitrate A blue deliquescent solid, $Cu(NO_3)_2.3H_2O$; r.d. 2.32; m.p. 114.5°C. It may be obtained by reacting either copper(II) oxide or copper(II) carbonate with dilute nitric acid and crystallizing the resulting solution. Other hydrates containing 6 or 9 molecules of water are known. On heating it readily decomposes to give copper(II) oxide, nitrogen dioxide, and oxygen. The anhydrous form can be obtained by reacting copper with a solution of nitrogen dioxide in

ethyl ethanoate. It sublimes on heating suggesting that it is appreciably covalent.

copper(I) oxide A red insoluble solid, Cu_2O; r.d. 6.0; m.p. 1235°C. It is obtained by reduction of an alkaline solution of copper(II) sulphate. Since the addition of alkalis to a solution of copper(II) salt results in the precipitation of copper(II) hydroxide the copper(II) ions are complexed with tartrate ions; under such conditions the concentration of copper(II) ions is so low that the solubility product of copper(II) hydroxide is not exceeded.

When copper(I) oxide reacts with dilute sulphuric acid a solution of copper(II) sulphate and a deposit of copper results, i.e. disproportionation occurs.

$$Cu_2O + 2H^+ \rightarrow Cu^{2+} + Cu + H_2O$$

When dissolved in concentrated hydrochloric acid the $[CuCl_2]^-$ complex ion is formed. Copper(I) oxide is used in the manufacture of rectifiers and the production of red glass.

copper(II) oxide A black insoluble solid, CuO; monoclinic; r.d. 6.3; m.p. 1326°C. It is obtained by heating either copper(II) carbonate or copper(II) nitrate. It decomposes on heating above 800°C to copper(I) oxide and oxygen. Copper(II) oxide reacts readily with mineral acids on warming, with the formation of copper(II) salts; it is also readily reduced to copper on heating in a stream of hydrogen. Copper(II) oxide is soluble in dilute acids forming blue solutions of cupric salts.

copper pyrites *See* chalcopyrite.

copper(II) sulphate A blue crystalline solid, $CuSO_4.5H_2O$; triclinic; r.d. 2.284. The pentahydrate loses $4H_2O$ at 110°C and the fifth H_2O at 150°C to form the white anhydrous compound (rhombic; r.d. 3.6; decomposes above 200°C). The pentahydrate is prepared either by reacting copper(II) oxide or copper(II) carbonate with dilute sulphuric acid; the solution is heated to saturation and the blue pentahydrate crystallizes out on cooling (a few drops of dilute sulphuric acid are generally added to prevent hydrolysis). It is obtained on an industrial scale by forcing air through a hot mixture of copper and dilute sulphuric acid. In the pentahydrate each copper(II) ion is surrounded by four water molecules at the corner of a square, the fifth and sixth octahedral positions are occupied by oxygen atoms from the sulphate anions, and the fifth water molecule is held in place by hydrogen bonding. Copper(II) sulphate has many industrial uses, including the preparation of the Bordeaux mixture (a fungicide) and the preparation of other copper compounds. It is also used in electroplating and textile dying and as a timber preservative. The anhydrous form is used in the detection of traces of moisture.

Copper(II) sulphate pentahydrate is also known as *blue vitriol*.

coral Any of a group of sedentary colonial marine invertebrates belonging to the class Anthozoa of the phylum *Coelenterata. A coral colony consists of individual *polyps within a protective skeleton that they secrete. This skeleton may be soft and jelly-like, horny, or stony. The horny skeleton secreted by corals of the genus *Corallium*, especially *C. rubrum*, constitutes the red, or precious, coral used as a gemstone. The skeleton of stony, or true, corals consists of almost pure calcium carbonate and forms the coral reefs common in tropical seas.

cordite An explosive mixture of cellulose nitrate and nitroglycerin, with added plasticizers and stabilizers, used as a propellant for guns.

core 1. A rod or frame of magnetic material that increases the inductance of a coil through which it passes.

Cores are used in transformers, electromagnets, and the rotors and stators of electrical machines. It may consist of laminated metal, ferrite, or compressed ferromagnetic particles in a matrix of an insulating binder (*dust core*). **2.** The inner part of a *nuclear reactor in which the nuclear reaction takes place. **3.** The devices that make up the memory in certain types of computer. **4.** The central region of a star or planet.

Coriolis force A fictitious force sometimes used to simplify calculations involving rotating systems, such as the movement of air, water, and projectiles over the surface of the rotating earth. The concept was first used in 1835 by Gaspard de Coriolis (1792–1843), a French physicist. The daily rotation of the earth means that in 24 hours a point on its equator moves a distance of some 40 000 kilometres, giving it a tangential velocity of about 1670 kilometres per hour. A point at the latitude of, say, Rome, travels a shorter distance in the same time and therefore has a lower tangential velocity – about 1340 km/hr. Air over the equator has the full tangential velocity of 1670 km/hr and as it travels north, say, it will retain this velocity; to an observer outside the earth this would be clear. However, to an observer in Rome it appears to be moving eastwards, because the earth at that point is moving eastwards more slowly than the air. The Coriolis force (which is quite fictitious) is the force that a naive observer thinks is needed to push the air eastwards.

cork (phellem) A protective waterproof plant tissue produced by the *cork cambium. It develops in plants undergoing *secondary growth and replaces the epidermis. Its cells, whose walls are impregnated with *suberin, are arranged in radial rows and fit closely together except where the cork is interrupted by *lenticels.

Development of a corm

Some cork cells become air-filled while others contain deposits of lignin, tannins, and fatty acids, which give the cork a particular colour. The cork oak (*Quercus suber*) produces cork that can be used commercially.

cork cambium (phellogen) A type of *cambium arising within the outer layers of the stems of woody plants, usually as a complete ring surrounding the inner tissues. The cells of the cork cambium divide to produce an outer corky tissue (*cork or *phellem*) and an inner secondary cortex (*phelloderm*). Cork, cork cambium, and

phelloderm together make up the *periderm*, an impermeable outer layer that protects the inner stem tissues if the outer tissues split as the stem girth increases with age. It thus takes over the functions of the epidermis.

corm An underground organ formed by certain plants, e.g. crocus and gladiolus, that enables them to survive from one growing season to the next (see illustration). It consists of a short swollen food-storing stem surrounded by protective scale leaves. One or more buds in the axils of scale leaves produce new foliage leaves and flowers in the subsequent season, using up the food stored in the stem. *Compare* bulb.

cornea A transparent layer of tissue, continuous with the *sclerotic, that forms the front part of the vertebrate eye, over the iris and lens. The cornea refracts light entering the eye onto the lens, thus assisting in the focusing of images onto the *retina. *See also* astigmatism.

cornification *See* keratinization.

CORN rule *See* absolute configuration.

corolla The *petals of a flower, collectively, forming the inner whorl of the *perianth. It encircles the stamens and carpels. The form of the corolla is very variable. The petals may either be free (*polypetalous*) or united to form a tube (*gamopetalous*).

corona 1. The outer part of the sun's atmosphere. Its two main components are the K-corona (or inner corona), with a temperature of about 2×10^6 K at a height of some 75 000 km, and the F-corona (or outer corona), which is considerably cooler and extends for several million kilometres into space. **2.** A glowing region of the air surrounding a conductor when the potential gradient near it exceeds a critical value. It is caused by ionization of the air and may be accompanied by hissing sounds. *Corona dis-*

166

charge (or *point discharge*) occurs at sharp points where the surface charge density is high by the attraction, charging, and consequent repulsion of air molecules.

coronary vessels Two pairs of blood vessels (the coronary arteries and coronary veins) that supply the muscles of the heart itself. The coronary arteries arise from the aorta and divide into branches that encircle the heart. A blood clot in a coronary artery (*coronary thrombosis*) is one of the causes of a 'heart attack'.

corpuscular theory *See* light.

corpus luteum The yellowish body that forms in the cavity of a *Graafian follicle in the ovary of a mammal after the release of the egg cell. It secretes the hormone *progesterone. Some species of sharks, reptiles, and birds have similar structures in their ovaries but the function of these is less well understood.

corrosion Chemical or electrochemical attack on the surface of a metal. *See also* electrolytic corrosion; rusting.

cortex 1. (in botany) The tissue between the epidermis and the vascular system in plant stems and roots. It is composed of *parenchyma cells and shows little or no structural differentiation. Cortex is produced by activity of the *apical meristem. *See also* endodermis. **2.** (in zoology) The outermost layer of tissue of various organs, including the adrenal glands (*adrenal cortex*), kidneys (*renal cortex*), and cerebral hemispheres (*see* cerebral cortex).

corticosteroid Any of several hormones produced by the cortex of the *adrenal glands. *Glucocorticoids* regulate the use of carbohydrates, proteins, and fats in the body and include *cortisol and *cortisone. *Mineralocorticoids* regulate salt and water balance (*see* aldosterone).

corticotrophin *See* ACTH.

cortisol (hydrocortisone) A hormone (*see* corticosteroid), produced by the adrenal glands, that promotes the synthesis and storage of glucose and is therefore important in the normal response to stress, suppresses or prevents inflammation, and regulates deposition of fat in the body. It is used as treatment for various allergies and for rheumatic fever, certain skin conditions, and adrenal failure (Addison's disease).

cortisone A hormone (*see* corticosteroid), produced by the adrenal glands, that is closely related to, and has an action similar to that of, *cortisol. Cortisone is also the name for a synthetic version of cortisol.

corundum A mineral form of aluminium oxide, Al_2O_3. It crystallizes in the trigonal system and occurs as well-developed hexagonal crystals. It is colourless and transparent when pure but the presence of other elements gives rise to a variety of colours. *Ruby is a red variety containing chromium; *sapphire is a blue variety containing iron and titanium. Corundum occurs as a rock-forming mineral in both metamorphic and igneous rocks. It is chemically resistant to weathering processes and so also occurs in alluvial (placer) deposits. The second hardest mineral after diamond (it has a hardness of 9 on the Mohs' scale), it is used as an abrasive.

corymb A type of flowering shoot (*see* racemose inflorescence) in which the lower flower stalks are longer than the higher ones, resulting in a flat-topped cluster of flowers. Examples are candytuft and wallflower.

cos *See* trigonometric functions.

cosecant *See* trigonometric functions.

cosech *See* hyperbolic functions.

cosh *See* hyperbolic functions.

cosine *See* trigonometric functions.

cosine rule In any triangle, with sides of length a, b, and c, $c^2 = a^2 +$

$b^2 - 2ab\cos\theta$, where θ is the angle between sides a and b.

cosmic radiation High-energy particles that fall on the earth from space. *Primary cosmic rays* consist of nuclei of the most abundant elements, with *protons (hydrogen nuclei) forming by far the highest proportion; electrons, positrons, neutrinos, and gamma-ray photons are also present. The particle energies range from 10^{-11} J to 10 J (10^8 to 10^{20} eV) and as they enter the earth's atmosphere they collide with oxygen and nitrogen nuclei producing *secondary cosmic rays*. The secondary rays consist of elementary particles and gamma-ray photons. A single high-energy primary particle can produce a large *shower* of secondary particles. The sources of the primary radiation are not all known, although the sun is believed to be the principal source of particles with energies up to about 10^{10} eV. It is believed that all particles with energies of less than 10^{18} eV originate within the Galaxy.

cosmic string *See* string.

cosmid A hybrid *vector, used in *gene cloning, that includes the *cos* gene (from the lambda bacteriophage). It also contains drug resistance *marker genes and other plasmid genes. Cosmids can incorporate larger DNA fragments than either phage or plasmid vectors alone and are especially suitable for cloning large mammalian genes or multigene fragments.

cosmoid scale *See* scales.

cosmology The study of the nature, origin, and evolution of the universe. Various theories concerning the origin and evolution of the universe exist. *See* big-bang theory; steady-state theory.

cotangent *See* trigonometric functions.

coth *See* hyperbolic functions.

Cottrell precipitator An electrostatic precipitator used to remove dust particles from industrial waste gases, by attracting them to charged grids or wires.

cotyledon (seed leaf) A part of the embryo in a seed plant. The number of cotyledons is an important feature in classifying plants. Among the flowering plants, the class known as *Monocotyledonae have a single cotyledon and *Dicotyledonae have two. Gymnosperms have either two cotyledons, as in *Taxus* (yews) and *Cycas*, or five to ten, as in *Pinus* (pines). In seeds without an *endosperm, e.g. garden pea and broad bean, the cotyledons store food, which is used in germination. In seeds showing *epigeal germination, e.g. runner bean, they emerge above the soil surface and become the first photosynthetic leaves.

coudé system *See* telescope.

coulomb Symbol C. The *SI unit of electric charge. It is equal to the charge transferred by a current of one ampere in one second. The unit is named after Charles de Coulomb (1736–1806), a French physicist.

coulomb explosion The sudden disruption of a molecule from which the electrons have been stripped to leave only the nuclei, which repel each other because of their electric charge. The technique of *coulomb explosion imaging* uses this effect to investigate the shape of molecules. A beam of high-energy neutral molecules is produced by first adding electrons, accelerating the ions in an electric field, and then removing the electrons. The beam collides with a thin metal foil having a thickness of about 30 atoms. As the molecules pass through this foil their electrons are scattered and only the nuclei of the molecules emerge. The process occurs within a very short period of time, shorter than the time required for a complete molecular vibration, and consequently the nuclei retain the molecular shape until they are suddenly repulsed by the like charges. The nuclei then impinge on a detector that records their velocity and direction, thus enabling the spatial arrangement of the original molecule to be derived.

Coulomb's law The force (sometimes called the *Coulomb force*) between two charged particles, regarded as point charges Q_1 and Q_2 a distance d apart, is proportional to the product of the charges and inversely proportional to the square of the distance between them. The law is now usually stated in the form $F = Q_1 Q_2 / 4\pi\varepsilon d^2$, where ε is the absolute *permittivity of the intervening medium. $\varepsilon = \varepsilon_r \varepsilon_0$, where ε_r is the relative permittivity (the dielectric constant) and ε_0 is the electric constant. The electric field surrounding a point charge is called the *Coulomb field* and the scattering of charged particles by the Coulomb field surrounding an atomic nucleus is called *Coulomb scattering*. The law was first published by Charles de Coulomb in 1785.

counter Any device for detecting and counting objects or events, often incident charged particles or photons. The latter devices usually work by allowing the particle to cause ionization, which creates a current or voltage pulse. The pulses are then counted electronically. *See* Cerenkov counter; crystal counter; Geiger counter; proportional counter; scintillation counter; semiconductor counter; spark counter. These names are often applied merely to the actual detectors; the ancillary counting mechanism is then called a *scaler.

couple Two equal and opposite parallel forces applied to the same body that do not act in the same line. The forces create a torque, the *moment of which is equal to the product of the force and the perpendicular distance between them.

coupling A type of chemical reaction in which two molecules join together; for example, the formation of an *azo compound by coupling of a diazonium ion with a benzene ring.

courtship Behaviour in animals that plays a part in the initial attraction of a mate or as a prelude to copulation. Courtship often takes the form of *displays that have evolved through *ritualization; some are derived from other contexts (e.g. food begging in some birds). Chemical stimuli (*see* pheromone) are also important in many mammals and insects.

As well as ensuring that the prospective mate is of the same species, the male's courtship performance allows females to choose between different males. The later stages of courtship may involve both partners in an alternating series of displays that inhibit *aggression and fear responses and ensure synchrony of sexual arousal.

covalent bond See chemical bond.

covalent crystal A crystal in which the atoms are held together by covalent bonds. Covalent crystals are sometimes called *macromolecular* or *giant-molecular crystals*. They are hard high-melting substances. Examples are diamond and boron nitride.

covalent radius An effective radius assigned to an atom in a covalent compound. In the case of a simple diatomic molecule, the covalent radius is half the distance between the nuclei. Thus, in Cl_2 the internuclear distance is 0.198 nm so the covalent radius is taken to be 0.099 nm. Covalent radii can also be calculated for multiple bonds; for instance, in the case of carbon the values are 0.077 nm for single bonds, 0.0665 nm for double bonds, and 0.0605 nm for triple bonds. The values of different covalent radii can sometimes be added to give internuclear distances. For example, the length of the bond in interhalogens (e.g. ClBr) is nearly equal to the sum of the covalent radii of the halogens involved. This, however, is not always true because of other effects (e.g. ionic contributions to the bonding).

coxa The first segment, attached to the thorax, of an insect's leg. See also femur; trochanter.

CP invariance The symmetry generated by the combined operation of changing *charge conjugation (C) and *parity (P). CP violation occurs in weak interactions in kaon decay. See also CPT theorem; time reversal.

CPT theorem The theorem that the combined operation of changing *charge conjugation C, *parity P, and *time reversal T, denoted CPT, is a fundamental *symmetry of relativistic *quantum field theory. No violation of the CPT theorem is known experimentally. When C, P, and T (or any two of them) are violated, the principles of relativistic quantum field theory are not affected; however, violation of CPT invariance would drastically alter the fundamentals of relativistic quantum field theory. It is not known whether *superstrings or *supermembranes obey versions of the CPT theorem.

cracking The process of breaking down chemical compounds by heat. The term is applied particularly to the cracking of hydrocarbons in the kerosine fraction obtained from *petroleum refining to give smaller hydrocarbons and alkenes. It is an important process, both as a source of branched-chain hydrocarbons suitable for gasoline (for motor fuel) and as a source of ethene and other alkenes. Catalytic cracking is a similar process in which a catalyst is used to lower the temperature required and to modify the products obtained.

cranial nerves Ten to twelve pairs of nerves in vertebrates that emerge directly from the brain. They supply the sense organs and muscles of the head, neck, and viscera. Examples of cranial nerves include the *optic nerve (II) and the *vagus nerve (X). With the *spinal nerves, the cranial nerves form an important part of the *peripheral nervous system.

Craniata See Vertebrata.

cranium (brain case) The part of the vertebrate *skull that encloses and protects the brain. It is formed by the fusion of several flattened bones, which have immovable joints (sutures) between them.

cream of tartar See potassium hydrogentartrate.

creatine A compound, synthesized from the amino acids arginine, glycine, and methionine, that occurs in muscle. In the form of creatine phosphate (or phosphocreatine), it is an important reserve of energy for muscle contraction, which is released when creatine phosphate loses its phosphate and is converted to creatinine, which is excreted in the urine. See also phosphagen.

creationist A proponent of the theory of *special creation.

creep The continuous deformation of a solid material, usually a metal, under a constant stress that is well below its *yield point. It usually only occurs at high temperatures and the creep characteristics of any material destined to be used under conditions of high stress at high temperatures must be investigated.

cremocarp A dry fruit that is a type of *schizocarp formed from two one-seeded carpels. The carpels remain separate and form indehiscent mericarps that are attached to a central supporting strand (carpophore) for some time before dispersal. It is characteristic of the Umbelliferae (carrot family).

creosote 1. (wood creosote) An almost colourless liquid mixture of phenols obtained by distilling tar obtained by the destructive distillation of wood. It is used medically as an antiseptic and expectorant. **2.** (**coal-tar creosote**) A dark liquid mixture of phenols and cresols obtained by distilling coal tar. It is used for preserving timber.

cresols *See* methylphenols.

Cretaceous The final geological period of the Mesozoic era. It extended from about 136 million years ago, following the Jurassic, to about 65 million years ago, when it was succeeded by the Palaeocene epoch. The name of the period is derived from *creta* (Latin: chalk) and the Cretaceous was characterized by the deposition of large amounts of *chalk in western Europe. The Cretaceous was the time of greatest flooding in the Mesozoic. Angiosperm plants made their first appearance on land and in the early Cretaceous Mesozoic reptiles reached their peak. At the end of the period there was a widespread extinction of the dinosaurs, flying reptiles, and ammonites, the cause of which is not definitely known, although it may be partly related to changing environmental conditions at the time.

crista 1. *See* semicircular canals. **2.** *See* mitochondrion.

cristobalite A mineral form of *silicon(IV) oxide. SiO_2

critical angle *See* total internal reflection.

critical damping *See* damping.

critical group A large group of related organisms that, although variations exist between them, cannot be divided into smaller groups of equivalent taxonomic rank to the parent group.

critical mass The minimum mass of fissile material that will sustain a nuclear *chain reaction. For example, when a nucleus of uranium–235 disintegrates two or three neutrons are released in the process, each of which is capable of causing another nucleus

to disintegrate, so creating a chain reaction. However, in a mass of U–235 less than the critical mass, too many neutrons escape from the surface of the material for the chain reaction to proceed. In the atom bomb, therefore, two or more subcritical masses have to be brought together to make a mass in excess of the critical mass before the bomb will explode.

critical pressure The pressure of a fluid in its *critical state; i.e. when it is at its critical temperature and critical volume.

critical reaction A nuclear *chain reaction in which, on average, one transformation causes exactly one other transformation so that the chain reaction is self-sustaining. If the average number of transformations caused by one transformation falls below one, the reaction is *subcritical* and the chain reaction ceases; if it exceeds one the reaction is *supercritical* and proceeds explosively.

critical state The state of a fluid in which the liquid and gas phases both have the same density. The fluid is then at its *critical temperature, *critical pressure, and *critical volume.

critical temperature 1. The temperature above which a gas cannot be liquefied by an increase of pressure. *See also* critical state. **2.** *See* transition point.

critical volume The volume of a fixed mass of a fluid in its *critical state; i.e. when it is at its critical temperature and critical pressure. The *critical specific volume* is its volume per unit mass in this state: in the past this has often been called the critical volume.

CRO *See* cathode-ray oscilloscope.

Cromagnon man The earliest form of modern man (*Homo sapiens*), which is believed to have appeared in Europe about 35 000 years ago and

171

Crossing over at two chiasmata in a pair of homologous chromosomes

possibly earlier in Africa and Asia. Fossils indicate that these hominids were taller and more delicate than *Neanderthal man, which they replaced. They used intricately worked tools of stone and bone and left the famous cave drawings at Lascaux in the Dordogne. The name is derived from the site at Cromagnon, France, where the first fossils were found in 1868.

crop An enlarged portion of the anterior section of the alimentary canal in some animals, in which food may be stored and/or undergo preliminary digestion. The term is most commonly applied to the thin-walled sac in birds between the oesophagus and the *proventriculus. In female pigeons the crop contains glands that secrete *crop milk*, used to feed nestlings.

cross 1. A mating between two selected individuals. Controlled crosses are made for many reasons, e.g. to investigate the inheritance of a particular characteristic or to improve a livestock or crop variety. *See also* back cross; reciprocal cross. **2.** An organism resulting from such a mating.

crossing over An exchange of portions of chromatids between *homologous chromosomes. This occurs as the chromosomes begin to move apart at the end of the first prophase of *meiosis; they remain in contact at a number of points (*see* chiasma). At these points the chromatids break and rejoin in such a way that sections are exchanged (see illustration). Crossing over thus alters the pattern of genes in the chromosomes. *See* recombination.

cross linkage A short side chain of atoms linking two longer chains in a polymeric material.

cross product *See* vector product.

cross section 1. A plane surface formed by cutting a solid, especially by cutting at right angles to its longest axis. **2.** The area of such a surface. **3.** A measure of the probability that a collision will occur between a beam of radiation and a particular particle, expressed as the effective area presented by the particle in that particular process. It is measured in square metres or *barns.

crucible A dish or other vessel in which substances can be heated to a high temperature.

crude oil *See* petroleum.

Crustacea A class of arthropods containing over 35 000 species distributed worldwide, mainly in freshwater and marine habitats, where they constitute a major component of plankton. Crustaceans include shrimps, crabs, and lobsters (*see* Decapoda), barnacles, water fleas (*see* Daphnia), copepods (*see* Copepoda), and the terrestrial woodlice. The segmented body usually has a distinct head (bearing *compound eyes, two pairs of *antennae, and various mouthparts), thorax, and abdomen, and is protected by a shell-like carapace. Each body segment may bear a pair of branched (*biramous*) appendages used for locomotion, as gills, and for filtering food particles from the water. Appendages in the head region are modified to form jaws and in the abdominal region are often reduced or absent. Typically, the eggs hatch to produce a free-swimming *nauplius* larva. This develops either by a series of moults or undergoes metamorphosis to the adult form.

cryogenic pump A *vacuum pump in which pressure is reduced by condensing gases on surfaces maintained at about 20 K by means of liquid hydrogen or at 4 K by means of liquid helium. Pressures down to 10^{-8} mmHg (10^{-6} Pa) can be maintained; if they are used in conjunction with a *diffusion pump, pressures as low as 10^{-15} mmHg (10^{-13} Pa) can be reached.

cryogenics The study of very low temperatures and the techniques for producing them. Objects are most simply cooled by placing them in a bath containing liquefied gas maintained at a constant pressure. In general, a liquefied gas can provide a constant bath temperature from its triple point to its critical temperature and the bath temperature can be varied by changing the pressure above the liquid. The lowest practical temperature for a liquid bath is 0.3 K. Refrigerators (*see* refrigeration) consist essentially of devices operating on a repeated cycle, in which a low-temperature reservoir is a continuously replenished liquid bath. Above 1 K they work by compressing and expanding suitable gases. Below this temperature liquids or solids are used and by *adiabatic demagnetization it is possible to reach 10^{-6} K.

cryohydrate A eutectic mixture of ice and some other substance (e.g. an ionic salt) obtained by freezing a solution.

cryolite A rare mineral form of sodium aluminofluoride, Na_3AlF_6, which crystallizes in the monoclinic

system. It is usually white but may also be colourless. The only important occurrence of the mineral is in Greenland. It is used chiefly as a flux in the production of aluminium from bauxite.

cryometer A thermometer designed to measure low temperatures. *Thermocouples can be used down to about 1 K and *resistance thermometers can be used at 0.01 K. Below this magnetic thermometers (0.001 K) and nuclear-resonance thermometers (3 × 10^{-7} K) are required.

cryoscopic constant *See* depression of freezing point.

cryostat A vessel enabling a sample to be maintained at a very low temperature. The *Dewar flask is the most satisfactory vessel for controlling heat leaking in by radiation, conduction, or convection. Cryostats usually consist of two or more Dewar flasks nesting in each other. For example, a liquid nitrogen bath is often used to cool a Dewar flask containing a liquid helium bath.

cryotron A switch that relies on *superconductivity. It consists of a coil of wire of one superconducting material surrounding a straight wire of another superconducting material; both are immersed in a liquid-helium bath. A current passed through the coil creates a magnetic field, which alters the superconducting properties of the central wire, switching its resistance from zero to a finite value. Cryotron switches can be made very small and take very little current.

cryptic coloration The type of colouring or marking of an animal that helps to camouflage it in its natural environment. It may enable the animal to blend with its background or, like the stripes of zebras and tigers, help to break up the outline of its body.

crystal A solid with a regular polyhedral shape. All crystals of the same substance grow so that they have the same angles between their faces. However, they may not have the same external appearance because different faces can grow at different rates, depending on the conditions. The external form of the crystal is referred to as the *crystal habit*. The atoms, ions, or molecules forming the crystal have a regular arrangement and this is the *crystal structure*.

crystal counter A type of solid-state *counter in which a potential difference is applied across a crystal; when the crystal is struck by an elementary particle or photon, the electron–ion pairs created cause a transient increase in conductivity. The resulting current pulses are counted electronically.

crystal-field theory A theory of the electronic structures of inorganic *complexes, in which the complex is assumed to consist of a central metal atom or ion surrounded by ligands that are ions. For example, the complex $[PtCl_4]^{2-}$ is thought of as a Pt^{2+} ion surrounded by four Cl^- ions at the corners of a square. The presence of these ions affects the energies of the *d*-orbitals, causing a splitting of energy levels. The theory can be used to explain the spectra of complexes and their magnetic properties. *Ligand-field theory* is a development of crystal-field theory in which the overlap of orbitals is taken into account.

crystal habit *See* crystal.

crystal lattice The regular pattern of atoms, ions, or molecules in a crystalline substance. A crystal lattice can be regarded as produced by repeated translations of a *unit cell* of the lattice. *See also* crystal system.

crystalline Having the regular internal arrangement of atoms, ions, or molecules characteristic of crystals. Crystalline materials need not necessarily exist as crystals; all metals, for example, are crystalline although they are not usually seen as regular geometric crystals.

crystallite A small crystal, e.g. one of the small crystals forming part of a microcrystalline substance.

crystallization The process of forming crystals from a liquid or gas.

crystallography The study of crystal form and structure. *See also* X-ray crystallography.

crystalloids *See* colloids.

crystal microphone A microphone in which the sound waves fall on a plate of Rochelle salt or similar material with piezoelectric properties, the variation in pressure being converted into a varying electric field by the *piezoelectric effect. Crystal microphones have strong high-frequency response and are nondirectional; they are now rarely used except when their cheapness in important.

crystal oscillator An oscillator of precisely determined frequency in which a piezoelectric crystal is used to produce the oscillation in a tuned circuit or is coupled to a tuned circuit to control its frequency. An alternating electric field applied to two metallic films sputtered onto the parallel faces of a crystal, usually of quartz, causes it to vibrate at its natural frequency; this frequency can be in the kilohertz or megahertz range, depending on how the crystal is cut. The mechanical vibrations in turn create an alternating electric field across the crystal that does not suffer from frequency drift. The device is widely used in *quartz clocks and watches.

crystal pick-up A pick-up in a record player in which the mechanical vibrations produced by undulations in the record groove are transmitted to a piezoelectric crystal, which produces a varying electric field of the same frequency as the sound. This signal is body-centred   simple cubic   face-centred

Cubic crystal structures

amplified and fed to loudspeakers in order to recreate the sound.

crystal structure *See* crystal.

crystal system A method of classifying crystalline substances on the basis of their unit cell. There are seven crystal systems. If the cell is a parallelopiped with sides a, b, and c and if α is the angle between b and c, β the angle between a and c, and γ the angle between a and b, the systems are:

(1) *cubic* $a=b=c$ and $\alpha=\beta=\gamma=90°$

(2) *tetragonal* $a=b\neq c$ and $\alpha=\beta=\gamma=90°$

(3) *rhombic* (or *orthorhombic*) $a\neq b\neq c$ and $\alpha=\beta=\gamma=90°$

(4) *hexagonal* $a=b\neq c$ and $\alpha=\beta=\gamma=90°$

(5) *trigonal* $a=b\neq c$ and $\alpha=\beta=\gamma\neq 90°$

(6) *monoclinic* $a\neq b\neq c$ and $\alpha=\gamma=90°\neq\beta$

(7) *triclinic* $a=b=c$ and $\alpha\neq\beta\neq\gamma$

CSF 1. *See* cerebrospinal fluid. **2.** *See* colony-stimulating factor.

CS gas The vapour from a white solid, $C_6H_4(Cl)CH:C(CN)_2$, causing tears and choking, used in 'crowd control'.

CT scanner (computerized tomography scanner) *See* tomography.

cubic close packing *See* close packing.

cubic crystal A crystal in which the unit cell is a cube (*see* crystal system). There are three possible packings for cubic crystals: *simple cubic*, *face-centred cubic*, and *body-centred cubic*. See illustration.

cubic equation An equation in which the highest power of the variable is three. It has the general form $ax^3 + bx^2 + cx + d = 0$ and, in general, is satisfied by three values of x.

cubic expansivity *See* expansivity.

cultivar A plant that has been developed and maintained by cultivation as a result of agricultural or horticultural practices. The term is derived from *culti*vated *var*iety.

culture medium A nutrient material, either solid or liquid, used to support the growth and reproduction of microorganisms or to maintain tissue or organ cultures. *See also* agar.

cumene process An industrial process for making phenol from benzene. A mixture of benzene vapour and propene is passed over a phosphoric acid catalyst at 250°C and high pressure

$$C_6H_6 + CH_3CH:CH_2 \rightarrow$$
$$C_6H_5CH(CH_3)_2$$

The product is called *cumene*, and it can be oxidized in air to a peroxide, $C_6H_5C(CH_3)_2O_2H$. This reacts with dilute acid to give phenol (C_6H_5OH)

and propanone (acetone, CH_3OCH_3), which is a valuable by-product.

cupellation A method of separating noble metals (e.g. gold or silver) from base metals (e.g. lead) by melting the mixture with a blast of hot air in a shallow porous dish (the *cupel*). The base metals are oxidized, the oxide being carried away by the blast of air or absorbed by the porous container.

cuprammonium ion The tetraamminecopper(II) ion $[Cu(NH_3)_4]^{2+}$. *See* ammine.

cupric compounds Compounds containing copper in its higher ($+2$) oxidation state; e.g. cupric chloride is copper(II) chloride ($CuCl_2$).

cuprite A red mineral cubic form of copper(I) oxide, Cu_2O; an important ore of copper. It occurs where deposits of copper have been subjected to oxidation. The mineral has been mined as a copper ore in Chile, Zaïre, Bolivia, Australia, the USSR, and the USA.

cupronickel A type of corrosion-resistant alloy of copper and nickel containing up to 45% nickel.

cuprous compounds Compounds containing copper in its lower ($+1$) oxidation state; e.g. cuprous chloride is copper(I) chloride ($CuCl$).

cupule 1. A hard or membranous cup-shaped structure formed from bracts and enclosing various fruits, such as the hazelnut and acorn. **2.** A structure in club mosses (*Lycopodium* species) that protects the gemma (resting bud) during its development. It is composed of six leaflike structures. **3.** The bright red tissue around the seed of yew (*Taxus*), forming the yew 'berry'.

curie The former unit of *activity (*see* radiation units). It is named after the French physicist Pierre Curie (1859–1906).

Curie point (Curie temperature) The temperature at which a ferromagnetic substance loses its ferromagnetism and becomes only paramagnetic. For iron the Curie point is 760°C and for nickel 356°C.

Curie's law The susceptibility (χ) of a paramagnetic substance is proportional to the thermodynamic temperature (T), i.e. $\chi = C/T$, where C is the Curie constant. A modification of this law, the *Curie–Weiss law*, is more generally applicable. It states that $\chi = C/(T - \theta)$, where θ is the Weiss constant, a characteristic of the material. The law was first proposed by the French physicist Pierre Curie (1859–1906) and modified by another French physicist, Pierre-Ernest Weiss (1865–1940).

curium Symbol Cm. A radioactive metallic transuranic element belonging to the *actinoids; a.n. 96; mass number of the most stable isotope 247 (half-life 1.64×10^7 years); r.d. (calculated) 13.51; m.p. $1340\pm40°C$. There are nine known isotopes. The element was first identified by G. T. Seaborg and associates in 1944 and first produced by L. B. Werner and I. Perlman in 1947 by bombarding americium–241 with neutrons.

curl (rot) The *vector product of the *gradient operator with a vector. For a vector u that has components u_1, u_2, and u_3 in the x, y, and z directions (with respective unit vectors i, j, and k), and is a function of x, y, and z, the curl is given by:

$$\text{curl}\,u = \nabla \times u$$
$$= (\partial u_3/\partial y - \partial u_2/\partial z)i +$$
$$(\partial u_1/\partial z - \partial u_3/\partial x)j +$$
$$(\partial u_2/\partial x - \partial u_1/\partial y)k.$$

See also divergence.

current Symbol I. A flow of electric charge through a conductor. The current at a particular cross section is the rate of flow of charge. The charge may be carried by electrons, ions, or positive holes (*see* charge carrier). The unit of current is the ampere. *See also* conventional current.

current balance An instrument used to measure a current absolutely, on the basis of the definition of the ampere. An accurate form consists of a beam balance with similar coils attached to the ends of the balance arms. Fixed coils are situated above and below these two coils. The six coils are then connected in series so that a current passing through them creates a torque on the beam, which is restored to the horizontal by means of a rider. From the position and weight of the rider, and the geometry of the system, the current can be calculated.

current density 1. The current flowing through a conductor per unit cross-sectional area, measured in amperes per square metre. **2.** The current flowing through an electrolyte per unit area of electrode.

cusp 1. (in dentistry) A sharp raised protuberance on the surface of a *molar tooth. The cusps of opposing molars (i.e. on opposite jaws) are complementary to each other, which increases the efficiency of grinding food during chewing. **2.** (in anatomy) A flap forming part of a *valve. **3.** (in mathematics) A point at which two arcs of a curve intersect.

cuticle 1. (in botany) The continuous waxy layer that covers the aerial parts of a plant. Composed of *cutin, it is secreted by the *epidermis and its primary function is to prevent water loss. **2.** (in zoology) A layer of horny noncellular material covering, and secreted by, the epidermis of many invertebrates. It is usually made of a collagen-like protein or of *chitin and its main function is protection. In arthropods it is also strong enough to act as a skeleton (see exoskeleton) and in insects it reduces water loss. Growth is allowed by moulting of the cuticle (see ecdysis).

cuticularization The secretion by the outer (epidermal) layer of cells of plants and many invertebrates of substances that then harden to form a *cuticle.

cutin A mixture of waxy substances that impregnates the cell wall and covers the outer surface of mature epidermal plant cells, forming a *cuticle. The deposition of cutin (*cutinization*) reduces water loss by the plant and helps prevent the entry of pathogens. See also suberin.

cutinization The deposition of *cutin in plant cell walls, principally in the outermost layers of leaves and young stems.

cutis See dermis.

cutting A part of a plant, such as a bud, leaf, or a portion of a root or shoot, that, when detached from the plant and inserted in soil, can take root and give rise to a new daughter plant. Taking or striking cuttings is a horticultural method for propagating plants. See also vegetative propagation.

cyanamide 1. An inorganic salt containing the ion CN_2^{2-}. See calcium cyanamide. **2.** A colourless crystalline solid, H_2NCN, made by the action of carbon dioxide on hot sodamide. It is a weakly acidic compound (the parent acid of cyanamide salts) that is soluble in water and ethanol. It is hydrolysed to urea in acidic solutions.

cyanamide process See calcium cyanamide.

cyanate See cyanic acid.

cyanic acid An unstable explosive acid, HOCN. The compound has the structure $H-O-C\equiv N$, and is also called *fulminic acid*. Its salts and esters are *cyanates* (or *fulminates*). The compound is a volatile liquid, which readily polymerizes. In water it hydrolyses to ammonia and carbon dioxide. It is isomeric with another acid, $H-N=C=O$, which is known as *isocyanic acid*. Its salts and esters are *isocyanates*.

cyanide 1. An inorganic salt containing the cyanide ion CN^-. Cyanides are extremely poisonous because of the ability of the CN^- ion to coordinate with the iron in haemoglobin, thereby blocking the uptake of oxygen by the blood. **2.** A metal coordination complex formed with cyanide ions.

cyanide process A method of extracting gold by dissolving it in potassium cyanide (to form the complex ion $[Au(CN)_2]^-$). The ion can be reduced back to gold with zinc.

cyanine dyes A class of dyes that contain a $-CH=$ group linking two nitrogen-containing heterocyclic rings. They are used as sensitizers in photography.

cyanobacteria *See* blue-green algae.

cyanocobalamin *See* vitamin B complex.

cyanogen A colourless gas, $(CN)_2$, with a pungent odour; soluble in water, ethanol, and ether; d. 2.335 $g\,dm^{-3}$; m.p. $-27.9°C$; b.p. $-20.7°C$. The compound is very toxic. It may be prepared in the laboratory by heating mercury(II) cyanide; industrially it is made by gas-phase oxidation of hydrogen cyanide using air over a silver catalyst, chlorine over activated silicon(IV) oxide, or nitrogen dioxide over a copper(II) salt. Cyanogen is an important intermediate in the preparation of various fertilizers and is also used as a stabilizer in making nitrocellulose. It is an example of a *pseudohalogen.

cyano group The group $-CN$ in a chemical compound. *See* nitrile.

cyanohydrins Organic compounds formed by the addition of hydrogen cyanide to aldehydes or ketones (in the presence of a base). The first step is attack by a CN^- ion on the carbonyl carbon atom. The final product is a compound in which a $-CN$ and $-OH$ group are attached to the same carbon atom. For example, ethanal reacts as follows

$$CH_3CHO + HCN \rightarrow$$
$$CH_3CH(OH)(CN)$$

The product is 2-hydroxypropanonitrile. Cyanohydrins of this type can be oxidized to α-hydroxy carboxylic acids.

Cyanophyta *See* blue-green algae.

cyanuric acid A white crystalline water-soluble trimer of cyanic acid, $(HNCO)_3$. It is a cyclic compound having a six-membered ring made of alternating imide (NH) and carbonyl (CO) groups.

Cycadales An order within the *Gymnospermae that contains many extinct species and a few of the present-day gymnosperms. Cycads inhabit tropical and subtropical regions, sometimes growing to a height of 20 m. The stem bears a crown of fernlike leaves. These species are among the most primitive of living seed plants.

cyclamates Salts of the acid, $C_6H_{11}.NH.SO_3H$, where $C_6H_{11}-$ is a cyclohexyl group. Sodium and calcium cyclamates were formerly used as sweetening agents in soft drinks, etc, until their use was banned when they were suspected of causing cancer.

cycle A regularly repeated set of changes to a system that brings back all its parameters to their original values once in every set of changes. The duration of one cycle is called its *period and the rate of repetition of cycle, called the *frequency, is measured in *hertz. *See* simple harmonic motion.

cyclic Describing a compound that has a ring of atoms in its molecules. In *homocyclic* compounds all the atoms in the ring are the same type, e.g. benzene (C_6H_6) and cyclohexane (C_6H_{12}). These two examples are also examples of *carbocyclic* compounds; i.e. the rings are of carbon atoms. If different atoms occur in the ring, as in pyridine (C_5H_5N), the compound is said to be *heterocyclic*.

cyclic AMP A derivative of *ATP that is widespread in animal cells as an intermediate messenger in many biochemical reactions induced by hormones. Upon reaching their target cells, the hormones activate the enzyme that catalyses cyclic AMP production. Cyclic AMP ultimately activates the enzymes of the reaction induced by the hormone concerned. Cyclic AMP is also involved in controlling gene expression and cell division, in immune responses, and in nervous transmission.

cyclization The formation of a cyclic compound from an open-chain compound. *See* ring.

cyclo- Prefix designating a cyclic compound, e.g. a cycloalkane or a cyclosilicate.

cycloalkanes Cyclic saturated hydrocarbons containing a ring of carbon atoms joined by single bonds. They have the general formula C_nH_{2n}, for example cyclohexane, C_6H_{12}, etc. In general they behave like the *alkanes but are rather less reactive.

cyclohexadiene-1,4-dione (benzoquinone; quinone) A yellow solid, $C_6H_4O_2$; r.d. 1.3; m.p. 116°C. It has a six-membered ring of carbon atoms with two opposite carbon atoms linked to oxygen atoms (C=O) and the other two pairs of carbon atoms linked by double bonds (HC=CH). The compound is used in making dyes. *See also* quinhydrone electrode.

cyclohexane A colourless liquid *cycloalkane, C_6H_{12}; r.d. 0.78; m.p. 6.5°C; b.p. 81°C. It occurs in petroleum and is made by passing benzene and hydrogen under pressure over a heated Raney nickel catalyst at 150°C, or by the reduction of cyclohexanone. It is used as a solvent and paint remover and can be oxidized using hot concentrated nitric acid to hexanedioic acid (adipic acid). The cyclohexane ring is not planar and can adopt boat and chain

*conformations; in formulae it is represented by a single hexagon.

cycloid The curve traced by a point on the circumference of a circle as it rolls without slipping along a straight line. The length of the arc formed by one revolution of the circle is $8r$, where r is the radius of the circle. The horizontal distance between cusps is $2\pi r$.

cyclone An area of low pressure in the atmosphere. Winds rotate about the low-pressure centre in an anticlockwise direction in the northern hemisphere and in a clockwise direction in the southern hemisphere. In the mid- and high-latitudes these low-pressure systems are now commonly referred to as *depressions, or lows, and the term cyclone is avoided. *Tropical cyclones* only develop in oceanic areas in which sea temperatures are above 27°C and at least 5° of latitude away from the equator. They are accompanied by strong winds and can develop into intense vortical storms known as *hurricanes or *typhoons.

cyclonite (RDX) A highly explosive nitro compound, $(CH_2N.NO_2)_3$. It has a cyclic structure with a six-membered ring of alternating CH_2 groups and nitrogen atoms, with each nitrogen being attached to a NO_2 group. It is made by nitrating hexamine, $C_6H_{12}N_4$, which is obtained from ammonia and methanal. Cyclonite is a very powerful explosive used mainly for military purposes.

cyclopentadiene A colourless liquid cyclic *alkene, C_5H_6; r.d. 0.8047; b.p. 42.5°C. It is prepared as a by-product during the fractional distillation of crude benzene from coal tar. It undergoes condensation reactions with ketones to give highly coloured compounds (fulvenes) and readily undergoes polymerization at room temperature to give the dimer, dicyclopentadiene. The compound itself is not *aromatic because it does

cyclopentadienyl ion

buttercup   forget-me-not

monochasial cymes

dichasial cyme

1 = oldest flower

Types of cymose inflorescence

not have the required number of pi electrons. However, removal of a hydrogen atom produces the stable *cyclopentadienyl ion*, $C_5H_5^-$, which does have aromatic properties. In particular, the ring can coordinate to positive ions in such compounds as *ferrocene.

cyclopentadienyl ion *See* cyclopentadiene.

Cyclostomata *See* Agnatha.

cyclotron A cyclic particle *accelerator in which charged particles fed into the centre of the device are accelerated in an outward spiral path inside two hollow D-shaped conductors placed to form a split circle. A magnetic field is applied at right-angles to the plane of the dees and an alternating potential difference is

applied between them. The frequency of the alternating p.d. is arranged so that the particles are accelerated each time they reach the evacuated gap between the dees. The magnetic field makes them follow curved paths. After several thousand revolutions inside the dees the particles reach the perimeter of the dees, where a deflecting field directs them onto the target. In this device protons can achieve an energy of 10^{-12} J (10 MeV). The first working cyclotron was produced in 1931 by the US physicist E. O. Lawrence (1901–58). *See also* synchrocyclotron.

cylindrical polar coordinates *See* polar coordinates.

cyme *See* cymose inflorescence.

cymose inflorescence (cyme; definite inflorescence) A type of flowering shoot (*see* inflorescence) in which the first-formed flower develops from the growing region at the top of the flower stalk (see illustration). Thus no new flower buds can be produced at the tip and other flowers are produced from lateral buds beneath. In a *monochasial cyme* (or *monochasium*), the development of the flower at the tip is followed by a new flower axis growing from a single lateral bud. Subsequent new flowers may develop from the same side of the lateral shoots, as in the buttercup, or alternately on opposite sides, as in forget-me-not. In a *dichasial cyme* (or *dichasium*), the development of the flower at the apex is followed by two new flower axes developing from buds opposite one another, as in plants of the family Caryophyllaceae (such as stitchwort). *Compare* racemose inflorescence.

cypsela A dry single-seeded fruit that does not split open during seed dispersal and is formed from a double ovary in which only one ovule develops into a seed. It is similar to an *achene and characteristic of members of the family Compositae, such as the dandelion. *See also* pappus.

cysteine *See* amino acid.

cystine A molecule resulting from the oxidation reaction between the sulphydryl (–SH) groups of two cysteine molecules (*see* amino acid). This often occurs between adjacent cysteine residues in polypeptides. The resultant *disulphide bonds* (–S–S–) are important in stabilizing the structure of protein molecules.

cytidine A nucleoside comprising one cytosine molecule linked to a D-ribose sugar molecule. The derived nucleotides, cytidine mono-, di-, and triphosphate (CMP, CDP, and CTP respectively), participate in various biochemical reactions, notably in phospholipid synthesis.

cytochrome Any of a group of proteins, each with an iron-containing *haem group, that form part of the *electron transport chain in mitochondria and chloroplasts. Electrons are transferred by reversible changes in the iron atom between the reduced Fe(II) and oxidized Fe(III) states.

cytogenetics The study of inheritance in relation to the structure and function of cells. For example, the results of breeding experiments can be explained in terms of the behaviour of chromosomes during the formation of the reproductive cells.

cytokine Any soluble factor secreted by cells of the lymphoid system that acts as a signal to other lymphoid cells. There are two categories: *lymphokines, secreted by lymphocytes; and *monokines*, secreted by macrophages. However, certain cytokines, notably *interferons and *interleukins, are secreted by both lymphocytes and macrophages.

cytokinin (kinin) Any of a group of plant *growth substances chemically related to the purine adenine. Cytokinins stimulate cell division in the presence of *auxin and have also

been found to delay senescence, overcome *apical dominance, and promote cell expansion. Zeatin is a naturally occurring cytokinin.

cytology The study of the structure and function of cells. The development of the light and electron microscopes has enabled the detailed structure of the nucleus (including the chromosomes) and other organelles to be elucidated. Microscopic examination of cells, either live or as stained sections on a slide, is also used in the detection and diagnosis of various diseases, especially *cancer.

cytolysis The breakdown of cells, usually as a result of destruction or dissolution of their outer membranes. Certain drugs (*cytotoxic drugs*) have this effect and are used in the treatment of some forms of cancer.

cytomegalovirus A virus belonging to the herpes group (*see* herpesvirus). In man it normally causes symptoms that are milder than the common cold, but it can produce more serious symptoms in those whose *immune response is disturbed (e.g. cancer patients). Infection in pregnant women may cause congenital handicap in their children.

cytoplasm The jelly-like material surrounding the nucleus of a *cell. It can be differentiated into dense outer *ectoplasm*, which is concerned primarily with cell movement, and less dense *endoplasm*, which contains most of the cell's structures.

cytoplasmic inheritance The inheritance of genes contained in the cytoplasm of a cell, rather than the nucleus. Only a very small number of genes are inherited in this way. The phenomenon occurs because certain organelles, the *mitochondria and (in plants) the *chloroplasts, contain their own genes and can reproduce independently. The female reproductive cell (the egg) has a large amount of cytoplasm containing many such organelles, which are consequently incorporated into the cytoplasm of all the cells of the embryo. The male reproductive cells (sperm or pollen), however, consist almost solely of a nucleus. Cytoplasmic organelles are thus not inherited from the male parent. In plants, male sterility can be inherited via the cytoplasm. The inheritance of any such factors does not follow Mendelian laws.

cytosine A *pyrimidine derivative. It is one of the principal component bases of *nucleotides and the nucleic acids *DNA and *RNA.

cytotaxonomy *See* taxonomy.

cytotoxic Destructive to living cells. The term is applied particularly to a class of drugs that inhibit cell division and are therefore used to destroy cancer cells and to a group of *T cells that destroy virus-infected cells.

D

2,4-D 2,4-dichlorophenoxyacetic acid: a synthetic *auxin frequently used as a weedkiller of broad-leaved weeds.

dalton *See* atomic mass unit.

Dalton's atomic theory A theory of *chemical combination, first stated by the British chemist John Dalton (1766 –1844) in 1803. It involves the following postulates:
(1) Elements consist of indivisible small particles (atoms).
(2) All atoms of the same element are identical; different elements have different types of atom.
(3) Atoms can neither be created nor destroyed.
(4) 'Compound elements' (i.e. compounds) are formed when atoms of different elements join in simple ratios to form 'compound atoms' (i.e. molecules).
Dalton also proposed symbols for atoms of different elements (later

replaced by the present notation using letters).

Dalton's law The total pressure of a mixture of gases or vapours is equal to the sum of the partial pressures of its components, i.e. the sum of the pressures that each component would exert if it were present alone and occupied the same volume as the mixture of gases. Strictly speaking, the principle is true only for ideal gases.

dam A structure built across a river to stop the flow of water. A dam may be constructed for one or a number of purposes. It may raise the water level to increase the depth for navigation purposes; divert water; provide a head of water for the generation of hydroelectric power; or store water for industrial or domestic use, irrigation, flood control, or power production. An additional use of water stored by dams is often for recreational purposes. Most dams are constructed of either concrete or of earth and rock. *Gravity dams* depend on the weight of their bulk for stability and usually have a flat vertical face upstream. *Arch dams* are curved concrete structures with a convex curve upstream. Pressure is transmitted to the sides of the dam.

damping A decrease in the amplitude of an oscillation as a result of energy being drained from the oscillating system to overcome frictional or other resistive forces. For example, a pendulum soon comes to rest unless it is supplied with energy from an outside source; in a pendulum clock, energy is supplied through an *escapement from a wound spring or a falling mass to compensate for the energy lost through friction. Damping is introduced intentionally in measuring instruments of various kinds to overcome the problem of taking a reading from an oscillating needle. A measuring instrument is said to be *critically damped* if the system just fails to oscillate and the system comes to rest in the shortest possible time. If it is *underdamped* it will oscillate repeatedly before coming to rest; if it is *overdamped* it will not oscillate but it will take longer to come to rest than it would if it was critically damped. An instrument, such as a galvanometer, that is critically damped is often called a *deadbeat* instrument.

dance of the bees A celebrated example of communication in animals, first investigated by Karl von Frisch (1886–1982). Honeybee workers on returning to the hive after a successful foraging expedition perform a 'dance' on the comb that contains coded information about the distance and direction of the food source. Other workers, sensing vibrations from the dance, follow the instructions to find the food source.

Daniell cell A type of primary *voltaic cell with a copper positive electrode and a negative electrode of a zinc amalgam. The zinc-amalgam electrode is placed in an electrolyte of dilute sulphuric acid or zinc sulphate solution in a porous pot, which stands in a solution of copper sulphate in which the copper electrode is immersed. While the reaction takes place ions move through the porous pot, but when it is not in use the cell should be dismantled to prevent the diffusion of one electrolyte into the other. The e.m.f. of the cell is 1.08 volts with sulphuric acid and 1.10 volts with zinc sulphate. It was invented in 1836 by the British chemist John Daniell (1790–1845).

Daphnia A genus of crustaceans belonging to the suborder Cladocera (water fleas). *Daphnia* species have a transparent carapace and a protruding head with a pair of highly branched antennae for swimming and a single median compound eye. The five pairs of thoracic appendages form an efficient filter-feeding mechanism. Reproduction can take place without mating, i.e. by *parthenogenesis.

dark matter *See* missing mass.
dark reactions *See* photosynthesis.

Darwinism The theory of *evolution proposed by Charles Darwin (1809–82) in *On the Origin of Species* (1859), which postulated that present-day species have evolved from simpler ancestral types by the process of *natural selection acting on the variability found within populations. *On the Origin of Species* caused a furore when it was first published because it suggested that species are not immutable nor were they specially created – a view directly opposed to the doctrine of *special creation. However the wealth of evidence presented by Darwin gradually convinced most people and the only major unresolved problem was to explain how the variations in populations arose and were maintained from one generation to the next. This became clear with the rediscovery of Mendel's work on classical genetics in the 1900s and led to the present theory known as *neo-Darwinism.

Darwin's finches (Galapagos finches) The 14 species of finch, unique to the Galapagos Islands, that Charles Darwin studied during his journey on HMS *Beagle*. Each is adapted to exploit a different food source. They are not found on the mainland because competition there for these food sources from other birds is fiercer. Darwin believed all the Galapagos finches to be descendants of a few that strayed from the mainland, and this provided important evidence for his theory of evolution. *See also* adaptive radiation.

database A large collection of information that has been coded and stored in a computer in such a way that it can be extracted under a number of different category headings.

dating techniques Methods of estimating the age of rocks, palaeontological specimens, archaeological sites, etc. *Relative dating techniques* date

dative bond

specimens in relation to one another; for example, *stratigraphy is used to establish the succession of fossils. *Absolute* (or *chronometric*) *techniques* give an absolute estimate of the age and fall into two main groups. The first depends on the existence of something that develops at a seasonally varying rate, as in *dendrochronology and *varve dating. The other uses some measurable change that occurs at a known rate, as in *chemical dating, *radioactive* (or *radiometric*) *dating* (*see* carbon dating; fission-track dating; potassium–argon dating; rubidium–strontium dating; uranium–lead dating), and *thermoluminescence.

dative bond *See* chemical bond.

daughter 1. A nuclide produced by radioactive *decay of some other nuclide (the *parent*). **2.** An ion or free radical produced by dissociation or reaction of some other (*parent*) ion or radical.

Davy lamp An oil-burning miner's safety lamp invented by Sir Humphry Davy (1778–1829) in 1816 when investigating firedamp (methane) explosions in coal mines. The lamp has a metal gauze surrounding the flame, which cools the hot gases by conduction and prevents ignition of gas outside the gauze. If firedamp is present it burns within the gauze cage, and lamps of this type are still used for testing for gas.

day The time taken for the earth to complete one revolution on its axis. The *solar day* is the interval between two successive returns of the sun to the *meridian. The *mean solar day* of 24 hours is the average value of the solar day for one year. The *sidereal day* is measured with respect to the fixed stars and is 4.09 minutes shorter than the mean solar day as a result of the imposition of the earth's orbital motion on its rotational motion.

day-neutral plant A plant in which flowering can occur irrespective of the day length. Examples are cucumber and maize. *See* photoperiodism. *Compare* long-day plant; short-day plant.

***d*-block elements** The block of elements in the *periodic table consisting of scandium, yttrium, and lanthanum together with the three periods of transition elements: titanium to zinc, zirconium to cadmium, and hafnium to mercury. These elements all have two outer *s*-electrons and have *d*-electrons in their penultimate shell; i.e. an outer electron configuration of the form $(n-1)d^x ns^2$, where x is 1 to 10. *See also* transition elements.

d.c. *See* direct current.

DDT Dichlorodiphenyltrichloroethane; a colourless organic crystalline compound, $(ClC_6H_4)_2CH(CCl_3)$, made by the reaction of trichloromethanal with chlorobenzene. DDT is the best known of a number of chlorine-containing pesticides used extensively in agriculture in the 1940s and 1950s. The compound is stable, accumulates in the soil, and concentrates in fatty tissue, reaching dangerous levels in carnivores high in the food chain. Restrictions are now placed on the use of DDT and similar pesticides.

Deacon process A former process for making chlorine by oxidizing hydrogen chloride in air at 450°C using a copper chloride catalyst. It was patented in 1870 by Henry Deacon (1822–76).

deactivation A partial or complete reduction in the reactivity of a substance, as in the poisoning of a catalyst.

deadbeat *See* damping.

deamination The removal of an amino group ($-NH_2$) from a compound. Enzymatic deamination occurs in the liver and is important in amino-acid metabolism, especially in their degradation and subsequent oxidation. The amino group is removed as ammonia and excreted, either unchanged or as urea or uric acid.

de Broglie wavelength The wavelength of the wave associated with a moving particle. The wavelength (λ) is given by $\lambda = h/mv$, where h is the Planck constant, m is the mass of the particle, and v its velocity. The *de Broglie wave* was first suggested by the French physicist Louis de Broglie (1892–1987) in 1924 on the grounds that electromagnetic waves can be treated as particles (*photons) and one could therefore expect particles to behave in some circumstances like waves (*see* complementarity). The subsequent observation of *electron diffraction substantiated this argument and the de Broglie wave became the basis of *wave mechanics.

debye A unit of electric *dipole moment in the electrostatic system, used to express dipole moments of molecules. It is the dipole moment produced by two charges of opposite sign, each of 1 statcoulomb and placed 10^{-18} cm apart, and has the value $3.335\,64 \times 10^{-30}$ coulomb metre.

Debye–Hückel theory A theory to explain the nonideal behaviour of electrolytes, published in 1923 by Peter Debye (1884–1966) and Erich Hückel (1896–). It assumes that electrolytes in solution are fully dissociated and that nonideal behaviour arises because of electrostatic interactions between the ions. The theory shows how to calculate the extra free energy per ion resulting from such interactions, and consequently the activity coefficient. It gives a good description of nonideal electrolyte behaviour for very dilute solutions, but cannot be used for more concentrated electrolytes.

deca- Symbol da. A prefix used in the metric system to denote ten times. For example, 10 coulombs = 1 decacoulomb (daC).

decahydrate A crystalline hydrate containing ten molecules of water per molecule of compound.

decanoic acid (capric acid) A white crystalline straight-chain saturated *carboxylic acid, $CH_3(CH_2)_8COOH$; m.p. 31.5°C. Its esters are used in perfumes and flavourings.

Decapoda An order of crustaceans distributed worldwide, mainly in marine habitats. Decapods comprise swimming forms (shrimps and prawns) and crawling forms (crabs, lobsters, and crayfish). All are characterized by five pairs of walking legs, the first pair of which are highly modified in crawling forms to form powerful grasping pincers. The carapace is fused with the thorax and head forming a *cephalothorax. The antennae are especially long in shrimps and prawns, which also possess several pairs of well-developed swimming appendages (*pleopods*) posterior to the walking legs. Following fertilization by the male, females usually carry the eggs until they hatch. The larvae undergo several transformations before attaining adult form.

decay The spontaneous transformation of one radioactive nuclide into a daughter nuclide, which may be radioactive or may not, with the emission of one or more particles or photons. The decay of N_0 nuclides to give N nuclides after time t is given by $N = N_0\exp(-\gamma t)$, where γ is called the *decay constant* or the *disintegration constant*. The reciprocal of the decay constant is the *mean life*. The time required for half the original nuclides to decay (i.e. $N = \frac{1}{2}N_0$) is called the *half-life* of the nuclide. The same terms are applied to elementary particles that spontaneously transform into other particles. For example, a free neutron decays into a proton and an electron (*see* beta decay). *See also* alpha particle.

deci- Symbol d. A prefix used in the metric system to denote one tenth. For example, 0.1 coulomb = 1 decicoulomb (dC).

decibel A unit used to compare two power levels, usually applied to sound or electrical signals. Although the decibel is one tenth of a *bel*, it is the decibel, not the bel, that is invariably used. Two power levels P and P_0 differ by n decibels when $n = 10\log_{10}P/P_0$. If P is the level of sound intensity to be measured, P_0 is a reference level, usually the intensity of a note of the same frequency at the threshold of audibility.

The logarithmic scale is convenient as human audibility has a range of 1 (just audible) to 10^{12} (just causing pain) and one decibel, representing an increase of some 26%, is about the smallest change the ear can detect.

deciduous Describing plants in which all the leaves are shed at the end of each growing season, usually the autumn in temperate regions or at the beginning of a dry season in the tropics. This seasonal leaf fall helps the plant retain water that would otherwise be lost by transpiration from the leaves. Examples of deciduous plants are rose and horse chestnut. *Compare* evergreen.

deciduous teeth (milk teeth) The first of two sets of teeth of a mammal. These teeth are smaller than those that replace them (the *permanent teeth) and fewer in number, since there are no deciduous *molars. *See also* diphyodont.

decimal system A number system based on the number 10; the number system in common use. All rational numbers can be written as a *finite decimal* (e.g. $\frac{1}{4}$ = 0.25) or a *repeating decimal* (e.g. $5/27 = 0.185\,185\,185\ldots$). An *irrational number can be written to any number of decimal places, but can never be given exactly (e.g. $\sqrt{3} = 1.732\,050\,8\ldots$).

declination 1. The angle between the magnetic meridian and the geographic meridian at a point on the surface of the earth. *See* geomagnetism. **2.** The angular distance of a celestial body north (positive) or south (negative) of the celestial *equator.

decomposer An organism that obtains energy from the chemical breakdown of dead organisms or animal or plant wastes. Examples are earthworms and many bacteria and fungi. Many decomposers (e.g. nitrifying bacteria) are specialized to break down organic materials that are difficult for other organisms to digest. Decomposers fulfil a vital role in the *ecosystem, returning the constituents of organic matter to the environment in inorganic form so that they can again be assimilated by plants. *See also* carbon cycle; nitrogen cycle.

decomposition Chemical reaction in which a compound breaks down into simpler compounds or into elements.

deconfinement temperature *See* quark confinement.

decrepitation A crackling noise produced when certain crystals are heated, caused by changes in structure resulting from loss of water of crystallization.

defect 1. A discontinuity in a crystal lattice. A *point defect* consists either of a missing atom or ion creating a *vacancy* in the lattice (a vacancy is sometimes called a *Schottky defect*) or an extra atom or ion between two normal lattice points creating an *interstitial*. A *Frenkel defect* consists of a vacancy in which the missing atom or ion has moved to an interstitial position. If more than one adjacent point defect occurs in a crystal there may be a slip along a surface causing a *line defect* (or *dislocation*). Defects are caused by strain or, in some cases, by irradiation. All crystalline solids contain an equilibrium number of point defects above absolute zero; this number increases with temperature. The existence of defects in crystals is important in the conducting properties of *semiconductors. 2. *See* mass defect.

deficiency disease Any disease caused by an inadequate intake of an essential nutrient in the diet, primarily vitamins, minerals, and amino acids. Examples are scurvy (lack of vitamin C), rickets (lack of vitamin D), and iron deficiency anaemia.

definite inflorescence *See* cymose inflorescence.

definite integral *See* integration.

definite proportions *See* chemical combination.

degassing The removal of dissolved or absorbed gases from a liquid or solid. Degassing is important in vacuum systems, where gas absorbed in the walls of the vacuum vessel starts to desorb as the pressure is lowered.

degaussing The process of neutralizing the magnetization in an object that has inadvertently become magnetized. For example, ferromagnetic components of TV sets may become magnetized and misdirect the electron beams. A degaussing coil is often provided and fed with a diminishing alternating current each time the set is switched on. Ships can be degaussed by surrounding them with current-carrying cables that set up an equal and opposite field. This prevents the ships from detonating magnetic mines. Degaussing is used to protect scientific and other electronic devices from strong magnetic fields; usually a system of coils is designed to neutralize such fields over the important region or the equipment is surrounded by a shield of suitable alloy (e.g. Mumetal).

degeneracy pressure The pressure in a *degenerate gas of fermions caused by the Pauli exclusion principle and the Heisenberg uncertainty principle. Because of the exclusion principle, fermions at a high density, with small interparticle spacing, must have different momenta. Because of the uncertainty principle, the momentum difference must be inversely proportional to the spacing. Consequently, in a high-density gas (small spacing) the particles have high relative momenta, which leads to a degeneracy pressure much greater than the thermal pressure. *White dwarfs and *neutron stars are supported against collapse under their own gravitational fields by the degeneracy pressure of electrons and neutrons respectively.

degenerate 1. Having quantum states with the same energy. For example, the five *d*-orbitals in an isolated transition-metal atom have the same energy (although they have different spatial arrangements) and are thus *degenerate levels*. The application of a magnetic or electric field may cause the quantum states to have different energies (*see* crystal-field theory). In this case, the degeneracy is said to be 'lifted'. **2.** Having a high particle concentration such that the *Maxwell–Boltzmann distribution does not apply and the behaviour of the particles is governed by *quantum statistics. Systems in which the particles are degenerate are called *degenerate gases*; examples are the conduction electrons in a metal (or *degenerate semiconductor), the electrons in a *white dwarf, and the neutrons in a *neutron star. *See also* degeneracy pressure.

degenerate semiconductor A heavily doped *semiconductor in which the *Fermi level is located in either the valence band or the conduction band (*see* energy levels) causing the material to behave as a metal.

degenerate states *Quantum states of a system that have the same energy.

degeneration 1. Changes in cells, tissues, or organs due to disease, etc., that result in an impairment or loss of function and possibly death and breakdown of the affected part. **2.** The reduction in size or complete loss of organs during evolution. The human appendix has undergone this process and performs no function in man. Degeneration of external organs may cause animals to appear to be more primitive than they really are; for example, early zoologists believed whales were fish rather than mammals because of the degeneration of their limbs. *See also* vestigial organ.

deglutition (swallowing) A reflex action initiated by the presence of food in the pharynx. During deglutition, the soft *palate is raised, which prevents food from entering the nasal cavity; the *epiglottis closes, which blocks the entrance to the windpipe; and the oesophagus starts to contract (*see* peristalsis), which ensures that food is conveyed to the stomach.

degradation A type of organic chemical reaction in which a compound is converted into a simpler compound in stages.

degree 1. A unit of plane angle equal to 1/360th of a complete revolution. **2.** A division on a *temperature scale. **3.** The power to which a variable is raised. If one expression contains several variables the overall degree of the expression is the sum of the powers. For example, the expression $p^2q^3r^4$ has a degree of 9 overall (it is a second-degree expression in p). The degree of a polynomial is the degree of the variable with the highest power, e.g. $ax^5 + bx^4 + c$ has a degree of 5. **4.** The highest power to which the derivative of the highest order is raised in a *differential equation. For example, $(d^2y/dx^2)^3 + dy/dx = c$ is a differential equation of the third degree (but second order).

degrees of freedom 1. The number of independent parameters required to specify the configuration of a system. This concept is applied in the *kinetic theory to specify the number of independent ways in which an atom or molecule can take up energy. There are however various sets of parameters that may be chosen, and the details of the consequent theory vary with the choice. For example, in a monatomic gas each atom may be allotted three degrees of freedom, corresponding to the three coordinates in space required to specify its position. The mean energy per atom for each degree of freedom is the same, according to the principle of the *equipartition of energy, and is equal to $kT/2$ for each degree of freedom (where k is the *Boltzmann constant and T is the thermodynamic temperature). Thus for a monatomic gas the total molar energy is $3LkT/2$, where L is the Avogadro constant (the number of atoms per mole). As $k = R/L$, where R is the molar gas constant, the total molar energy is $3RT/2$.

In a diatomic gas the two atoms require six coordinates between them, giving six degrees of freedom. Commonly these are interpreted as six independent ways of storing energy: on this basis the molecule has three degrees of freedom for different directions of *translational motion, and in addition there are two degrees of freedom for rotation of the molecular axis and one vibrational degree of freedom along the bond between the atoms. The rotational degrees of freedom each contribute $kT/2$, to the total energy; similarly the vibrational degree of freedom has an equal share of kinetic energy and must on average have as much potential energy (*see* simple harmonic motion). The total energy per molecule for a diatomic gas is therefore $3kT/2$ (for translational energy of the whole molecule) plus $2kT/2$ (for rotational energy of each atom) plus $2kT/2$ (for vibrational energy), i.e. a total of $7kT/2$.

2. The least number of independent variables required to define the state of a system in the *phase rule. In this sense a gas has two degrees of freedom (e.g. temperature and pressure).

dehiscence The spontaneous and often violent opening of a fruit, seed pod, or anther to release and disperse the seeds or pollen. Examples are the splitting of laburnum pods and primrose capsules; such structures are described as *dehiscent* (*compare* indehiscent).

dehydration 1. Removal of water from a substance. **2.** A chemical reaction in which a compound loses hydrogen and oxygen in the ratio 2:1. For instance, ethanol passed over hot pumice undergoes dehydration to ethene:

$$C_2H_5OH - H_2O \rightarrow CH_2{:}CH_2$$

Substances such as concentrated sulphuric acid, which can remove H_2O in this way, are known as *dehydrating agents*. For example, with sulphuric acid, methanoic acid gives carbon monoxide:

$$HCOOH - H_2O \rightarrow CO$$

dehydrogenase Any enzyme that catalyses the removal of hydrogen atoms in biological reactions. Dehydrogenases occur in many biochemical pathways but are particularly important in driving the *electron-transport-chain reactions of cell respiration. They work in conjunction with the hydrogen-accepting coenzymes *NAD and *FAD.

dekatron A neon-filled tube with a central anode surrounded by ten cathodes and associated transfer electrodes. As voltage pulses are received by the tube a glow discharge moves from one set of electrodes to the next, enabling the device to be used as a visual counting tube in the decimal system. The tube can also be used for switching.

delayed neutrons The small proportion of neutrons that are emitted with a measurable time delay in a nuclear fission process. *Compare* prompt neutrons.

delay line A component in an electronic circuit that is introduced to provide a specified delay in transmitting the signal. Coaxial cable or inductor-capacitor networks can be used to provide a short delay but for longer delays an *acoustic delay line* is required. In this device the signal is converted by the *piezoelectric effect into an acoustic wave, which is passed through a liquid or solid medium, before reconversion to an electronic signal.

deliquescence The absorption of water from the atmosphere by a hygroscopic solid to such an extent that a concentrated solution of the solid eventually forms.

delocalization In certain chemical compounds the valence electrons cannot be regarded as restricted to definite bonds between the atoms but are 'spread' over several atoms in the molecule. Such electrons are said to be *delocalized*. Delocalization occurs particularly when the compound contains alternating (conjugated) double or triple bonds, the delocalized electrons being those in the pi *orbitals. The molecule is then more stable than it would be if the electrons were localized, an effect accounting for the properties of benzene and other aromatic compounds. Another example is in the ions of carboxylic acids, containing the carboxylate group $-COO^-$. In terms of a simple model of chemical bonding, this group would have the carbon joined to one oxygen by a double bond (i.e. $C{=}O$) and the other joined to O^- by a single bond ($C{-}O^-$). In fact, the two C $-O$ bonds are identical because the extra electron on the O^- and the electrons in the pi bond of $C{=}O$ are delocalized over the three atoms.

delta-iron *See* iron.

demagnetization The removal of the ferromagnetic properties of a body by

189

disordering the domain structure (*see* magnetism). One method of achieving this is to insert the body within a coil through which an alternating current is flowing; as the magnitude of the current is reduced to zero, the domains are left with no predominant direction of magnetization.

deme A group of organisms in the same *taxon. The term is used with various prefixes that denote how the group differs from other groups. For example, an *ecodeme* occurs in a particular ecological habitat, *cytodemes* differ from each other cytologically, and *genodemes* differ genetically.

demodulation The process of extracting the information from a modulated carrier wave (*see* modulation; radio). The device used is called a *demodulator* or a *detector*.

denature 1. To add a poisonous or unpleasant substance to ethanol to make it unsuitable for human consumption (*see* methylated spirits). **2.** To produce a structural change in a protein or nucleic acid that results in the reduction or loss of its biological properties. Denaturation is caused by heat, chemicals, and extremes of pH. The differences between raw and boiled eggs are largely a result of denaturation. **3.** To add another isotope to a fissile material to make it unsuitable for use in a nuclear weapon.

dendrite 1. (in chemistry) A crystal that has branched in growth into two parts. Crystals that grow in this way (*dendritic growth*) have a branching treelike appearance. **2.** (in neurology) A slender branching process of the cell body of a *neurone. It forms connections (*see* synapse) with the axons of other neurones and transmits nerve impulses from these to the cell body.

dendrochronology An absolute *dating technique using the *growth rings of trees. It depends on the fact that trees in the same locality show a characteristic pattern of growth rings resulting from climatic conditions. Thus it is possible to assign a definite date for each growth ring in living trees, and to use the ring patterns to date fossil trees or specimens of wood (e.g. used for buildings or objects on archaeological sites) with lifespans that overlap those of living trees. The bristlecone pine (*Pinus aristata*), which lives for up to 5000 years, has been used to date specimens over 8000 years old. Fossil specimens accurately dated by dendrochronology have been used to make corrections to the *carbon-dating technique. Dendrochronology is also helpful in studying past climatic conditions. Analysis of trace elements in sections of rings can also provide information on past atmospheric pollution.

denitrification A chemical process in which nitrates in the soil are reduced to molecular nitrogen, which is released into the atmosphere. This process is effected by denitrifying bacteria (e.g. *Pseudomonas denitrificans*), which use nitrates as a source of energy for other chemical reactions in a manner similar to respiration in other organisms. *Compare* nitrification. *See* nitrogen cycle.

densitometer An instrument used to measure the *photographic density of an image on a film or photographic print. Densitometers work by letting the specimen transmit or reflect a beam of light and monitoring the transmitted or reflected intensity. They originally consisted of visual *photometers but most instruments are now photoelectric. The simplest transmission densitometer consists of a light source, a photosensitive cell, and a microammeter: the density is measured in terms of the meter readings with and without the sample in place. They have a variety of uses, including detecting the sound track on a cinematic film, measuring intensities in spectrographic records, and checking photographic prints.

density 1. The mass of a substance per unit of volume. In *SI units it is measured in kg m^{-3}. *See also* relative density; vapour density *2. See* charge density. 3. *See* photographic density.

2	1	2	3	man
2	1	2	3	(32 teeth)

2	0	3	3	rabbit
1	0	2	3	(28 teeth)

3	1	4	2	bear
3	1	4	3	(42 teeth)

Representative dental formulas

dental formula A representation of the dentition of an animal. A dental formula consists of eight numbers, four above and four below a horizontal line. The numbers represent (from left to right) the numbers of incisors, canines, premolars, and molars in either half of the upper and lower jaws. The total number of teeth in both jaws is therefore obtained by adding up all the numbers in the dental formula and multiplying by 2. Representative dental formulas are shown in the illustration.

denticle (placoid scale) *See* scales.

dentine The bony material that forms the bulk of a *tooth. Dentine is similar in composition to bone but is perforated with many tiny canals for nerve fibres, blood capillaries, and processes of the dentine-forming cells (*odontoblasts). Ivory, the material that forms elephant tusks, is made of dentine.

dentition The type, number, and arrangement of teeth in a species. This can be represented concisely by a *dental formula. *See also* diphyodont; monophyodont; polyphyodont; heterodont; homodont.

deoxyribonuclease *See* DNAase.

deoxyribonucleic acid *See* DNA.

depleted Denoting a material that contains less of a particular isotope than it normally contains, especially a residue from a nuclear reactor or isotope-separation plant containing fewer fissile atoms than natural uranium.

depletion layer A region in a *semiconductor that has a lower-than-usual number of mobile charge carriers. A depletion layer forms at the interface between two dissimilar regions of conductivity (e.g. a $p-n$ junction). *See* semiconductor diode.

depolarization 1. (in physics) The prevention of *polarization in a *primary cell. For example, maganese(IV) oxide (the *depolarizer*) is placed around the positive electrode of a *Leclanché cell to oxidize the hydrogen released at this electrode. 2. (in neurophysiology) A reduction in the difference of electrical potential that exists across the cell membrane of a nerve or muscle cell. Depolarization of a nerve-cell membrane occurs during the passage of an *action potential along the axon when the nerve is transmitting an impulse.

deposition The laying down of material on the earth's surface. This includes materials that have been eroded elsewhere and transported by natural agents, such as rivers, wind, ice (e.g. glaciers), and the tides and currents in the sea. It also includes the deposition of materials resulting from chemical precipitation, the formation of crusts on the earth's surface through evaporation, and the growth, accumulation, and decay of natural organisms.

depression (low; disturbance) An area of low atmospheric pressure in the mid- and high-latitudes that is surrounded by several closed isobars. Depressions generally move towards the northeast in the northern hemisphere and towards the southeast in the southern hemisphere. Unsettled weather conditions are associated with

depression

191

depressions and they form the main source of precipitation in most lowland parts of the mid-latitudes.

depression of freezing point The reduction in the freezing point of a pure liquid when another substance is dissolved in it. It is a *colligative property – i.e. the lowering of the freezing point is proportional to the number of dissolved particles (molecules or ions), and does not depend on their nature. It is given by $\Delta t = K_f C_m$, where C_m is the molar concentration of dissolved solute and K_f is a constant (the *cryoscopic constant*) for the solvent used. Measurements of freezing-point depression (using a Beckmann thermometer) can be used for finding relative molecular masses of unknown substances.

depth of field The range of distance in front of and behind an object that is being focused by an optical instrument, such as a microscope or camera, within which other objects will be in focus. The *depth of focus* is the amount by which the distance between the camera and the film can be changed without upsetting the sharpness of the image.

derivative 1. (in chemistry) A compound that is derived from some other compound and usually maintains its general structure, e.g. trichloromethane (chloroform) is a derivative of methane. **2.** (in mathematics) *See* differentiation; calculus.

derived unit *See* base unit.

dermal bone *See* membrane bone.

Dermaptera An order of insects comprising the earwigs. Earwigs typically have long thin cylindrical bodies with biting mouthparts and a stout pair of curved forceps at the tip of the abdomen, used for catching prey and in courtship. Some species have a single pair of wings, which at rest are folded back over the abdomen like a fan; others are wingless. Most earwigs are nocturnal and omnivorous.

dermis (corium; cutis) The thicker and innermost layer of the *skin of vertebrates, the other layer being the *epidermis. The dermis consists of fibrous connective tissue in which are embedded blood vessels, sensory nerve endings, and (in mammals) hair follicles, sebaceous glands, and sweat ducts. Beneath the dermis lies the *subcutaneous tissue.

desalination The removal of salt from sea water for irrigation of the land or to provide drinking water. The process is normally only economic if a cheap source of energy, such as the waste heat from a nuclear power station, can be used. Desalination using solar energy has the greatest economic potential since shortage of fresh water is most acute in hot regions. The methods employed include evaporation, often under reduced pressure (flash evaporation); freezing (pure ice forms from freezing brine); *reverse osmosis; *electrodialysis; and *ion exchange.

desiccator A container for drying substances or for keeping them free from moisture. Simple laboratory desiccators are glass vessels containing a drying agent, such as silica gel. They can be evacuated through a tap in the lid.

desmids A group of unicellular algae that belongs to the division *Chlorophyta. Like *Spirogyra*, they have an elaborate chloroplast. The cells of desmids are characteristically split into two halves joined by a narrow neck, each half being a mirror image of the other. The outer wall of the cell is patterned with various protuberances. Desmids are an important component of the plankton.

desorption The removal of adsorbed atoms, molecules, or ions from a surface.

destructive distillation The process of heating complex organic substances in the absence of air so that they break down into a mixture of volatile

Devonian

products, which are condensed and collected. At one time the destructive distillation of coal (to give coke, coal tar, and coal gas) was the principal source of industrial organic chemicals.

detector 1. *See* demodulation. **2.** *See* counter.

detergent A substance added to water to improve its cleaning properties. Although water is a powerful solvent for many compounds, it will not dissolve grease and natural oils. Detergents are compounds that cause such nonpolar substances to go into solution in water. *Soap is the original example, owing its action to the presence of ions formed from long-chain fatty acids (e.g. the octadecanoate (stearate) ion, $CH_3(CH_2)_{16}COO^-$). These have two parts: a nonpolar part (the hydrocarbon chain), which attaches to the grease; and a polar part (the $-COO^-$ group), which is attracted to the water. A disadvantage of soap is that it forms a scum with hard water (*see* hardness of water) and is relatively expensive to make. Various synthetic ('soapless') detergents have been developed from petrochemicals. The commonest, used in washing powders, is sodium dodecylbenzenesulphonate, which contains $CH_3(CH_2)_{11}C_6H_4SO_2O^-$ ions. This, like soap, is an example of an *anionic detergent*, i.e. one in which the active part is a negative ion. *Cationic detergents* have a long hydrocarbon chain connected to a positive ion. Usually they are amine salts, as in $CH_3(CH_2)_{15}N(CH_3)_3{}^+Br^-$, in which the polar part is the $-N(CH_3)_3{}^+$ group. *Nonionic detergents* have nonionic polar groups of the type $-C_2H_4-O-C_2H_4-OH$, which form hydrogen bonds with the water. Synthetic detergents are also used as wetting agents, emulsifiers, and stabilizers for foam.

determined Describing embryonic tissue at a stage when it can develop only as a certain kind of tissue (rather than as any kind).

deuterated compound A compound in which some or all of the hydrogen −1 atoms have been replaced by deuterium atoms.

deuterium (heavy hydrogen) Symbol D. The isotope of hydrogen that has a mass number 2 (r.a.m. 2.0144). Its nucleus contains one proton and one neutron. The abundance of deuterium in natural hydrogen is about 0.015%. It is present in water as the oxide HDO (*see also* heavy water), from which it is usually obtained by electrolysis or fractional distillation. Its chemical behaviour is almost identical to hydrogen although deuterium compounds tend to react rather more slowly than the corresponding hydrogen compounds. Its physical properties are slightly different from those of hydrogen, e.g. b.p. 23.6 K (hydrogen 20.4 K).

deuterium oxide *See* heavy water.

deuteron A nucleus of a deuterium atom, consisting of a proton and a neutron bound together.

deviation 1. (angle of deviation) The angle formed between a ray of light falling on a surface or transparent body and the ray leaving it. **2.** The difference between one of an observed set of values and the true value, usually represented by the mean of all the observed values. The *mean deviation* is the mean of all the individual deviations of the set. *See* standard deviation.

devitrification Loss of the amorphous nature of glass as a result of crystallization.

Devonian A geological period in the Palaeozoic era that extended from the end of the Silurian (about 395 million years ago) to the beginning of the Carboniferous (about 345 million years ago). It was named by Adam Sedgwick (1785–1873) and Roderick Murchison (1792–1871) in 1839. The

193

Devonian is divided into seven stages based on invertebrate fossil remains, such as corals, brachiopods, ammonoids, and crinoids, found in marine deposits. There were also extensive continental deposits consisting of conglomerates, red silts, and sandstones, forming the Old Red Sandstone facies. Fossils in the Old Red Sandstone include fishes and the earliest land plants. Graptolites became extinct early in the Devonian and the trilobites declined.

dew *See* precipitation.

Dewar flask A vessel for storing hot or cold liquids so that they maintain their temperature independently of the surroundings. Heat transfer to the surroundings is reduced to a minimum: the walls of the vessel consist of two thin layers of glass (or, in large vessels, steel) separated by a vacuum to reduce conduction and convection; the inner surface of a glass vessel is silvered to reduce radiation; and the vessel is stoppered to prevent evaporation. It was devised around 1872 by the British physicist Sir James Dewar (1842–1923) and is also known by its first trade name *Thermos flask*. *See also* cryostat.

Dewar structure A proposed structure of *benzene, having a hexagonal ring of six carbon atoms with two opposite atoms joined by a long single bond across the ring and with two double C–C bonds, one on each side of the hexagon. Dewar structures contribute to the resonance hybrid of benzene.

dew point The temperature at which the water vapour in the air is saturated. As the temperature falls the dew point is the point at which the vapour begins to condense as droplets of water.

dew-point hygrometer *See* hygrometer.

dextrorotatory Denoting a chemical compound that rotates the plane of polarization of plane-polarized light to the right (clockwise as observed by someone facing the oncoming radiation). *See* optical activity.

dextrose *See* glucose.

diabetes *See* insulin; vasopressin.

diagonal relationship A relationship within the periodic table by which certain elements in the second period have a close chemical similarity to their diagonal neighbours in the next group of the third period. This is particularly noticeable with the following pairs.

Lithium and magnesium:
(1) both form chlorides and bromides that hydrolyse slowly and are soluble in ethanol;
(2) both form colourless or slightly coloured crystalline nitrides by direct reaction with nitrogen at high temperatures;
(3) both burn in air to give the normal oxide only;
(4) both form carbonates that decompose on heating.

Beryllium and aluminium:
(1) both form highly refractory oxides with polymorphs;
(2) both form crystalline nitrides that are hydrolysed in water;
(3) addition of hydroxide ion to solutions of the salts gives an amphoteric hydroxide, which is soluble in excess hydroxide giving beryllate or aluminate ions $[Be(OH)_4]^{2-}$ and $[Al(OH)_4]^-$;
(4) both form covalent halides and covalent alkyl compounds that display bridging structures;
(5) both metals dissolve in alkalis.

Boron and silicon:
(1) both display semiconductor properties;
(2) both form hydrides that are unstable in air and chlorides that hydrolyse in moist air;
(3) both form acidic oxides with covalent crystal structures, which are readily incorporated along with other

oxides into a wide range of glassy materials.

The reason for this relationship is a combination of the trends to increase size down a group and to decrease size along a period, and a similar, but reversed, effect in electronegativity, i.e. decrease down a group and increase along a period.

diakinesis The period at the end of the first prophase of *meiosis when the separation of *homologous chromosomes is almost complete and *crossing over has occurred.

dialysis A method by which large molecules (such as starch or protein) and small molecules (such as glucose or amino acids) in solution may be separated by selective diffusion through a semipermeable membrane. For example, if a mixed solution of starch and glucose is placed in a closed container made of a semipermeable substance (such as Cellophane), which is then immersed in a beaker of water, the smaller glucose molecules will pass through the membrane into the water while the starch molecules remain behind. The cell membranes of living organisms are semipermeable, and dialysis takes place naturally in the kidneys for the excretion of nitrogenous waste. An artificial kidney (*dialyser*) utilizes the principle of dialysis by taking over the functions of diseased kidneys.

diamagnetism *See* magnetism.

1,6-diaminohexane (hexamethyl-enediamine) A solid colourless amine, $H_2N(CH_2)_6NH_2$; m.p. 41°C; b.p. 204°C. It is made by oxidizing cyclohexane to hexanedioic acid, reacting this with ammonia to give the ammonium salt, and dehydrating the salt to give hexanedionitrile ($NC(CH_2)_6CN$). This is reduced with hydrogen to the diamine. The compound is used, with hexanedioic acid, for producing *nylon 6,6.

diamond The hardest known mineral (with a hardness of 10 on Mohs'

scale). It is an allotropic form of pure *carbon that has crystallized in the cubic system, usually as octahedra or cubes, under great pressure. Diamond crystals may be colourless and transparent or yellow, brown, or black They are highly prized as gemstones but also have extensive uses in industry, mainly for cutting and grinding tools. Diamonds occur in ancient volcanic pipes of kimberlite; the most important deposits are in South Africa but others are found in Tanzania, the USA, USSR, and Australia. Diamonds also occur in river deposits that have been derived from weathered kimberlite, notably in Brazil, Zaïre, Sierra Leone, and India. Industrial diamonds are increasingly being produced synthetically.

diapause A period of suspended development or growth occurring in some insects during which metabolism is greatly decreased. It is often triggered by seasonal changes and regulated by an inborn rhythm and enables the insect to survive unfavourable environmental conditions so that its offspring may be produced in more favourable ones.

diaphragm 1. (in optics) An opaque disc with a circular aperture at its centre. Diaphragms of different sizes are used to control the total light flux passing through an optical system or to reduce aberration by restricting the light passing through a system to the central portion. An *iris diaphragm* consists of a number of overlapping crescent-shaped discs arranged so that the central aperture can be continuously varied in diameter. **2.** (in anatomy) The muscular membrane that divides the thorax (chest) from the abdomen in mammals. It plays an essential role in breathing (*see* respiratory movement), being depressed during inhalation and raised during exhalation.

diaphysis The shaft of a limb bone, which in immature animals is separated from the ends of the bone (*see* epiphysis) by cartilage.

diaspore A mineral form of a mixed aluminium oxide and hydroxide, AlO.OH. *See* aluminium hydroxide.

diastase *See* amylase.

diastereoisomers Stereoisomers that are not identical and yet not mirror images. For instance, the *d*-isomer of tartaric acid and the meso-isomer constitute a pair of diastereoisomers. *See* optical activity.

diastole The phase of a heart beat that occurs between two contractions of the heart, during which the heart muscles relax and the ventricles fill with blood. *Compare* systole. *See* blood pressure.

diatomic molecule A molecule formed from two atoms (e.g. H_2 or HCl).

diatoms *See* Bacillariophyta.

diazine *See* azine.

diazo compounds Organic compounds containing two linked nitrogen compounds. The term includes *azo compounds, diazonium compounds, and also such compounds as diazomethane, CH_2N_2.

Structure of diazonium ion $C_6H_5N_2^+$

diazonium salts Unstable salts containing the ion $C_6H_5N_2^+$ (the *diazonium ion*: see formula). They are formed by *diazotization reactions.

diazotization The formation of a *diazonium salt by reaction of an aromatic amine with nitrous acid at low temperature (below 5°C). The nitrous acid is produced in the reaction mixture from sodium nitrite and hydrochloric acid:

$$ArNH_2 + NaNO_2 + HCl \rightarrow$$
$$ArN^+N + Cl^- + Na^+ + OH^- + H_2O$$

dibasic acid An *acid that has two acidic hydrogen atoms in its molecules. Sulphuric (H_2SO_4) and carbonic (H_2CO_3) acids are common examples.

1,2-dibromoethane A colourless liquid *haloalkane, $BrCH_2CH_2Br$; r.d. 2.2; m.p. 9°C; b.p. 121°C. It is made by addition of bromine to ethene and used as an additive in petrol to remove lead as the volatile lead bromide.

dicarbide *See* carbide.

dicarboxylic acid A *carboxylic acid having two carboxyl groups in its molecules. In systematic chemical nomenclature, dicarboxylic acids are denoted by the suffix -*dioic*; e.g. hexanedioic acid, $HOOC(CH_2)_4COOH$.

dichasium *See* cymose inflorescence.

dichlorine oxide (chlorine monoxide) A strongly oxidizing orange gas, Cl_2O, made by oxidation of chlorine using mercury(II) oxide. It is the acid anhydride of chloric(I) acid.

dichloroethanoic acid *See* chloroethanoic acids.

2,4-dichlorophenoxyacetic acid *See* 2,4-D.

dichogamy The condition in which the male and female reproductive organs of a flower mature at different times, thereby ensuring that self-fertilization does not occur. *Compare* homogamy. *See also* protandry; protogyny.

dichotomous Describing the type of branching in plants that results when the growing point (apical bud) divides into two equal growing points, which in turn divide in a similar manner after a period of growth, and so on. Dichotomous branching is common is ferns and mosses.

dichroism The property of some crystals, such as tourmaline, of selectively absorbing light vibrations in one plane while allowing light vibrations at right angles to this plane to pass through. Polaroid is a synthetic dichroic material. *See* polarization.

dichromate(VI) A salt containing the ion $Cr_2O_7^-$. Solutions containing dichromate(VI) ions are strongly oxidizing.

Dicotyledonae One of the two classes or subclasses of plants within the *Angiospermae, distinguished by having two seed leaves (*cotyledons) within the seed. The dicotyledons usually have leaf veins in the form of a net, a ring of vascular bundles in the stem, and flower parts in fours or fives or multiples of these. Dicotyledons include many food plants (e.g. potatoes, peas, beans), ornamentals (e.g. roses, ivies, honeysuckles), and hardwood trees (e.g. oaks, limes, beeches). *Compare* Monocotyledonae.

Dictyoptera An order of insects comprising the cockroaches (suborder Blattaria) and the mantids (suborder Mantodea), occurring mainly in tropical regions. Cockroaches are oval and flattened in shape; some have a single well-developed pair of wings, folded back over the abdomen at rest, while in others the wings may be reduced or absent. They are usually found in forest litter, feeding on dead organic matter, but some species, e.g. the American cockroach (*Periplaneta americana*), are major household pests, scavenging on starchy foods, fruits, etc. In most species the females produce capsules (*oothecae*) containing 16–40 eggs. These are either deposited or carried by the female during incubation.

dielectric A nonconductor of electric charge in which an applied electric field causes a *displacement of charge but not a flow of charge. Electrons within the atoms of a dielectric are, on average, displaced by an applied field with respect to the nucleus, giving rise to a dipole that has an electric moment in the direction of the field. The resulting stress within the dielectric is known as the *electric polarization* (*P*) and is defined by *P*

$= D - E\varepsilon_0$, where D is the displacement, E is the electric field strength, and ε_0 is the electric constant.

The *dielectric constant* is now called the relative *permittivity. The *dielectric strength* is the maximum potential gradient that can be applied to a material without causing it to break down. It is usually expressed in volts per millimetre. *See also* capacitor.

dielectric constant *See* permittivity.

dielectric heating The heating of a dielectric material, such as a plastic, by applying a radio-frequency electric field to it. The most common method is to treat the material as the dielectric between the plates of a capacitor. The heat produced is proportional to $V^2 f A \phi / t$, where V is the applied potential difference, f its frequency, A is the area of the dielectric, t its thickness, and ϕ is the loss factor of the material (related to its *permittivity).

Diels–Alder reaction A type of chemical reaction in which a compound containing two double bonds separated by a single bond (i.e. a conjugated *diene) adds to a suitable compound containing one double bond (known as the *dienophile*) to give a ring compound. In the dienophile, the double bond must have a carbonyl group on each side. It is named after the German chemists Otto Diels (1876–1954) and Kurt Alder (1902–58), who discovered it in 1928.

diene An *alkene that has two double bonds in its molecule. If the two bonds are separated by one single bond, as in buta-1,3-diene $CH_2{:}CHCH{:}CH_2$, the compound is a *conjugated diene*.

dienophile *See* Diels–Alder reaction.

Diesel engine *See* internal-combustion engine.

dietary fibre *See* fibre.

diethyl ether *See* ethoxyethane.

differential calculus *See* calculus.

differential equation An equation in which a derivative of y with respect to x appears as well as the variables x and y. The *order* of a differential equation is the order of its highest derivative. The *degree* of the equation is the highest power present of the highest order derivative. There are many types of differential equation, each having its own method of solution. The simplest type has separable variables, enabling each side of the equation to be integrated separately.

differential scanning calorimetry (DSC) *See* thermal analysis.

differential thermal analysis (DTA) *See* thermal analysis.

differentiation 1. (in mathematics) The process of finding the *derivative* of a function in differential *calculus. If $y = f(x)$, the derivative of y, written dy/dx or $f'(x)$, is equal to the limit as $\Delta x \to 0$ of $[f(x + \Delta x) - f(x)]/\Delta x$. In general, if $y = x^n$, then $dy/dx = nx^{n-1}$. On a graph of $y = f(x)$, the derivative dy/dx is the gradient of the tangent to the curve at the point x. **2.** (in biology) The changes from simple to more complex forms undergone by developing tissues and organs so that they become specialized for particular functions. Differentiation occurs during embryonic development, *regeneration, and (in plants) meristematic activity (*see* meristem).

diffraction The spreading or bending of waves as they pass through an aperture or round the edge of a barrier. The diffracted waves subsequently interfere with each other (*see* interference) producing regions of reinforcement and weakening. First noticed as occurring with light by Francesco Grimaldi (1618–63), the phenomenon gave considerable support to the wave theory of light. Diffraction also occurs with streams of particles. *See also* Fresnel diffraction; Fraunhofer diffraction; electron diffraction.

diffraction grating A device for producing spectra by diffraction and interference. The usual grating consists of a glass or speculum-metal sheet with a very large number of equidistant parallel lines ruled on it (usually of the order of 1000 per mm). Diffracted light after transmission through the glass or reflection by the speculum produces maxima of illumination (spectral lines) according to the equation $m\lambda = d(\sin i + \sin\theta)$, where d is the distance between grating lines, λ is the wavelength of the light, i is the angle of incidence, θ the direction of the diffracted maximum, and m is the 'order' of the spectral line. Reflection gratings are also used to produce spectra in the ultraviolet region of the electromagnetic spectrum.

diffusion 1. The process by which different substances mix as a result of the random motions of their component atoms, molecules, and ions. In gases, all the components are perfectly miscible with each other and mixing ultimately becomes nearly uniform, though slightly affected by gravity (*see also* Graham's law). The diffusion of a solute through a solvent to produce a solution of uniform concentration is slower, but otherwise very similar to the process of gaseous diffusion. In solids, however, diffusion occurs very slowly at normal temperatures. **2.** The scattering of a beam of light by reflection at a rough surface or by transmission through a translucent (rather than transparent) medium, such as frosted glass. **3.** The passage of elementary particles through matter when there is a high probability of scattering and a low probability of capture.

diffusion cloud chamber *See* cloud chamber.

diffusion pump (condensation pump) A *vacuum pump in which oil or mercury vapour is diffused through a jet, which entrains the gas molecules from the container in which the pressure is to be reduced. The diffused vapour and entrained gas molecules are condensed on the cooled walls of the pump. Pressures down to 10^{-7} Pa can be reached by sophisticated forms of the diffusion pump.

digestion The breakdown by a living organism of ingested food material into chemically simpler forms that can be readily absorbed and assimilated by the body. This process requires the action of digestive enzymes and may take place extracellularly (i.e. in the *alimentary canal), as is the case in most animals; or intracellularly (e.g. by engulfing phagocytic cells), as occurs in protozoans and coelenterates.

digit 1. (in mathematics) A symbol used to represent a single number. For example, the number 479 consists of three digits. **2.** (in anatomy) A finger or toe. In the basic limb structure of terrestrial vertebrates there are five digits (see pentadactyl limb). This number is retained in man and other primates, but in some other species the number of digits is reduced. Frogs, for example, have four fingers and five toes, and in ungulate (hooved) mammals, the digits are reduced and their tips are enclosed in horn, forming hooves.

digital computer See computer.

digital display A method of indicating a reading of a measuring instrument, clock, etc., in which the appropriate numbers are generated on a fixed display unit by the varying parameter being measured rather than fixed numbers on a scale being indicated by a moving pointer or hand. See digitron; light-emitting diode; liquid-crystal display.

digital recording A method of recording or transmitting sound in which the sound itself is not transmitted or recorded. Instead the pressure in the sound wave is sampled at least 30 000 times per second and the successive values represented by numbers, which are then transmitted or recorded. Afterwards they are restored to analogue form in the receiver or player. This method is used for very high fidelity recordings as no distortion or interference occurs during transmission or in the recording process.

digitigrade Describing the gait of most fast-running animals, such as dogs and cats, in which only the toes are on the ground and the rest of the foot is raised off the ground. Compare plantigrade; unguligrade.

digitron An electronic gas-discharge tube that provides a *digital display in calculators, counters, etc. It usually has 10 cold cathodes shaped into the form of the digits 0–9. The cathode selected receives a voltage pulse causing a glow discharge to illuminate the digit. It has now largely been superseded by *light-emitting diodes and *liquid-crystal displays.

dihedral (dihedron) An angle formed by the intersection of two planes (e.g. two faces of a polyhedron). The dihedral angle is the angle formed by taking a point on the line of intersection and drawing two lines from this point, one in each plane, perpendicular to the line of intersection.

dihydrate A crystalline hydrate containing two molecules of water per molecule of compound.

dihydric alcohol See diol.

1,2-dihydroxybenzene (catechol) A colourless crystalline phenol, $C_6H_4(OH)_2$; r.d. 1.4; m.p. 105°C; b.p. 240°C. It is used as a photographic developer.

2,3-dihydroxybutanedioic acid See tartaric acid.

dilatancy See Newtonian fluid.

dilation (or dilatation) 1. An increase in volume. See also vasodilation. **2.** See time dilation.

dilatometer

dilatometer A device for measuring the cubic *expansivities of liquids. It consists of a bulb of known volume joined to a graduated capillary tube, which is closed at the top to prevent evaporation. A known mass of liquid is introduced into the device, which is submerged in a bath maintained at different temperatures t_1 and t_2. The two volumes corresponding to these temperatures, V_1 and V_2, are read off the calibrated stem. The value of the cubic expansivity (γ) is then given by
$$\gamma = (V_2 - V_1)/V_1(t_2 - t_1).$$

dilead(II) lead(IV) oxide A red amorphous powder, Pb_3O_4; r.d. 9.1; decomposes at 500°C to lead(II) oxide. It is prepared by heating lead(II) oxide to 400°C and has the unusual property of being black when hot and red-orange when cold. The compound is nonstoichiometric, generally containing less oxygen than implied by the formula. It is largely covalent and has $Pb(IV)O_6$ octahedral groups linked together by $Pb(II)$ atoms, each joined to three oxygen atoms. It is used in glass making but its use in the paint industry has largely been discontinued because of the toxicity of lead. Dilead(II) lead(IV) oxide is commonly called *red lead* or, more accurately, *red lead oxide*.

dilute Describing a solution that has a relatively low concentration of solute.

dilution The volume of solvent in which a given amount of solute is dissolved.

dilution law See Ostwald's dilution law.

dimensional analysis A method of checking an equation or a solution to a problem by analysing the dimensions in which it is expressed. It is also useful for establishing the form, but not the numerical coefficients, of an empirical relationship. If the two sides of an equation do not have the same dimensions, the equation is wrong. If they do have the same dimensions, the equation may still be wrong, but the error is likely to be in the arithmetic rather than the method of solution.

dimensions The product or quotient of the basic physical quantities, raised to the appropriate powers, in a derived physical quantity. The basic physical quantities of a mechanical system are usually taken to be mass (M), length (L), and time (T). Using these dimensions, the derived physical quantity velocity will have the dimensions L/T and acceleration will have the dimensions L/T^2. As force is the product of a mass and an acceleration (see Newton's law of motion), force has the dimensions MLT^{-2}. In electrical work in *SI units, current, I, can be regarded as dimensionally independent and the dimensions of other electrical units can be found from standard relationships. Charge, for example, is measured as the product of current and time. It therefore has the dimension IT. Potential difference is given by the relationship $P = VI$, where P is power. As power is force × distance ÷ time ($MLT^{-2} \times L \times T^{-1} = ML^2T^{-3}$), voltage V is given by $V = ML^2T^{-3}I^{-1}$.

dimer An association of two identical molecules linked together. The molecules may react to form a larger molecule, as in the formation of dinitrogen tetroxide (N_2O_4) from nitrogen dioxide (NO_2), or the formation of an *aluminium chloride dimer (Al_2Cl_6) in the vapour. Alternatively, they may be held by hydrogen bonds. For example, carboxylic acids form dimers in organic solvents, in which hydrogen bonds exist between the O of the C=O group and the H of the −O−H group.

dimethylbenzenes (xylenes) Three compounds with the formula $(CH_3)_2C_6H_4$, each having two methyl groups substituted on the benzene ring. 1,2-dimethylbenzene is *o*-xylene, etc. A mixture of the isomers (b.p. 135–145°C) is obtained from petroleum and is used as a clearing agent in preparing specimens for optical microscopy.

dimethylformamide (DMF) A colourless liquid, $HCON(CH_3)_2$; m.p. −61°C; b.p. 153°C. It is widely used as a solvent for organic compounds.

dimethyl sulphoxide (DMSO) A colourless solid, $(CH_3)_2SO$; m.p. 18°C; b.p. 189°C. It is used as a solvent and as a reagent in organic synthesis.

dimorphism 1. (in biology) The existence of two distinctly different types of individual within a species. An obvious example is *sexual dimorphism* in certain animals, in which the two sexes differ in colouring, size, etc. Dimorphism also occurs in some lower plants, such as mosses and ferns, that show an *alternation of generations. **2.** (in chemistry) *See* polymorphism.

dinitrogen oxide (nitrous oxide) A colourless gas, N_2O, d. 1.97 g dm⁻³; m.p. −90.8°C; b.p. −88.5°C. It is soluble in water, ethanol, and sulphuric acid. It may be prepared by the controlled heating of ammonium nitrate (chloride free) to 250°C and passing the gas produced through solutions of iron(II) sulphate to remove impurities of nitrogen monoxide. It is relatively unreactive, being inert to halogens, alkali metals, and ozone at normal temperatures. It is decomposed on heating above 520°C to nitrogen and oxygen and will support the combustion of many compounds. Dinitrogen oxide is used as an anaesthetic gas ('laughing gas') and as an aerosol propellant.

dinitrogen tetroxide A colourless to pale yellow liquid or a brown gas, N_2O_4; r.d. 1.45 (liquid); m.p. −11.2°C; b.p. 21.2°C. It dissolves in water with reaction to give a mixture of nitric acid and nitrous acid. It may be readily prepared in the laboratory by the reaction of copper with concentrated nitric acid; mixed nitrogen oxides containing dinitrogen oxide may also be produced by heating metal nitrates. The solid compound is wholly N_2O_4 and the liquid is about 99% N_2O_4 at the boiling point; N_2O_4 is diamagnetic. In the gas phase it dissociates to give *nitrogen dioxide*

$$N_2O_4 \leftrightharpoons 2NO_2$$

Because of the unpaired electron this is paramagnetic and brown. Liquid N_2O_4 has been widely studied as a nonaqueous solvent (self-ionizes to NO^+ and NO_3^-). Dinitrogen tetroxide, along with other nitrogen oxides, is a product of combustion engines and is thought to be involved in the depletion of stratospheric ozone.

dinosaur An extinct terrestrial reptile belonging to a group that constituted the dominant land animals of the Jurassic and Cretaceous periods, 190–65 million years ago. There were two orders. The Ornithischia were typically quadrupedal herbivores, many with heavily armoured bodies, and included *Stegosaurus*, *Triceratops*, and *Iguanodon*. They were all characterized by birdlike pelvic girdles. The Saurischia included many bipedal carnivorous forms, such as *Tyrannosaurus* (the largest known carnivore), and some quadrupedal herbivorous forms, such as *Apatosaurus* (*Brontosaurus*) and *Diplodocus*. They all had lizard-like pelvic girdles. Many of the herbivorous dinosaurs were amphibious or semiaquatic.

diode An electronic device with two electrodes. In the obsolescent *thermionic diode* a heated cathode emits electrons, which flow across the intervening vacuum to the anode when a positive potential is applied to it. The device permits flow of current in one direction only as a negative potential applied to the anode repels the electrons. This property of diodes was made use of in the first

dioecious

thermionic radios, in which the diode was used to demodulate the transmitted signal (*see* modulation). In the *semiconductor diode*, a $p-n$ junction performs a similar function. The forward current increases with increasing potential difference whereas the reverse current is very small indeed. *See* semiconductor; transistor.

dioecious Describing plant species that have male and female flowers on separate plants. Examples of dioecious plants are willows. *Compare* monoecious.

diol (dihydric alcohol) An *alcohol containing two hydroxyl groups per molecule.

dioptre A unit for expressing the power of a lens or mirror equal to the reciprocal of its focal length in metres. Thus a lens with a focal length of 0.5 metre has a power of $1/0.5 = 2$ dioptres. The power of a converging lens is usually taken to be positive and that of a diverging lens negative. Because the power of a lens is a measure of its ability to cause a beam to converge, the dioptre is now sometimes called the radian per metre.

dioxan A colourless toxic liquid, $C_4H_8O_2$; r.d. 1.03; m.p. 11°C; b.p. 101.5°C. The molecule has a six-membered ring containing four CH_2 groups and two oxygen atoms at opposite corners. It can be made from ethane-1,2-diol and is used as a solvent.

dioxonitric(III) acid *See* nitrous acid.

dioxygenyl compounds Compounds containing the positive ion O_2^+, as in dioxygenyl hexafluoroplatinate O_2PtF_6 – an orange solid that sublimes in vacuum at 100°C. Other ionic compounds of the type $O_2^+[MF_6]^-$ can be prepared, where M is P, As, or Sb.

dip *See* geomagnetism.

diphosphane (diphosphine) A yellow liquid, P_2H_4, which is sponta-neously flammable in air. It is obtained by hydrolysis of calcium phosphide. Many of the references to the spontaneous flammability of phosphine (PH_3) are in fact due to traces of P_2H_4 as impurities.

diphosphine *See* diphosphane.

diphyodont Describing a type of dentition that is characterized by two successive sets of teeth: the *deciduous (milk) teeth, which are followed by the *permanent (adult) teeth. Mammals have a diphyodont dentition. *Compare* monophyodont; polyphyodont.

diploblastic Describing an animal with a body wall composed only of two layers, *ectoderm and *endoderm, sometimes with a noncellular *mesoglea between them. The Coelenterata are diploblastic; all other Metazoa are *triploblastic.

diploid Describing a nucleus, cell, or organism with twice the *haploid number of chromosomes characteristic of the species. The diploid number is designated as $2n$. Two sets of chromosomes are present, one set being derived from the female parent and the other from the male. In animals, all the cells except the reproductive cells are diploid.

diplotene The period in the first prophase of *meiosis when paired *homologous chromosomes begin to move apart. They remain attached at a number of points (*see* chiasma).

Dipnoi A subclass of bony fishes that contains the lungfishes, which have lungs and breathe air. They are found in Africa, Australia, and South America, where they live in freshwater lakes and marshes that tend to become stagnant or even dry up in summer. They survive in these conditions by burrowing into the mud, leaving a small hole for breathing air, and entering a state of *aestivation, in which they can remain for six months or more. The Dipnoi date from the Devonian era (395–345 million years ago) and share many features with the modern *Amphibia.

dipole 1. A pair of separated opposite electric charges. The *dipole moment* (symbol μ) is the product of the positive charge and the distance between the charges. Dipole moments are often stated in *debyes; the SI unit is the coulomb metre. In a diatomic molecule, such as HCl, the dipole moment is a measure of the polar nature of the bond (*see* polar molecule); i.e. the extent to which the average electron charge is displaced towards one atom (in the case of HCl, the electrons are attracted towards the more electronegative chlorine atom). In a polyatomic molecule, the dipole moment is the vector sum of the dipole moments of the individual bonds. In a symmetrical molecule, such as tetrachloromethane (CCl$_4$), there is no overall dipole moment, although the individual C–Cl bonds are polar. **2.** An aerial commonly used for frequencies below 30 megahertz. It consists of a horizontal rod, fed or tapped at its centre. It may be half a wavelength or a full wavelength long.

Diptera An order of insects comprising the true, or two-winged, flies. Flies possess only one pair of wings – the forewings; the hindwings are modified to form small clublike *halteres* that function as balancing organs. Typically fluid feeders, flies have mouthparts adapted for piercing and sucking or for lapping; the diet includes nectar, sap, decaying organic matter, and blood. Some species prey on insects; others are parasitic. Dipteran larvae (*maggots*) are typically wormlike with an inconspicuous head. They undergo metamorphosis via a pupal stage to the adult form. Many flies or their larvae are serious pests, either by feeding on crops (e.g. fruit flies) or as vectors of disease organisms (e.g. the house fly (*Musca domestica*) and certain mosquitoes).

direct current (d.c.) An electric current in which the net flow of charge is in one direction only. *Compare* alternating current.

direct-current motor *See* electric motor.

direct dye *See* dyes.

direct motion 1. The apparent motion of a planet from west to east as seen from the earth against the background of the stars. **2.** The anticlockwise rotation of a planet, as seen from its north pole. *Compare* retrograde motion.

directrix 1. A plane curve defining the base of a *cone. **2.** A straight line from which the distance to any point on a *conic is in a constant ratio to the distance from that point to the focus.

disaccharide A sugar consisting of two linked *monosaccharide molecules. For example, sucrose comprises one glucose molecule and one fructose molecule bonded together.

discharge 1. The conversion of the chemical energy stored in a *secondary cell into electrical energy. **2.** The release of electric charge from a capacitor in an external circuit. **3.** The passage of charge carriers through a gas at low pressure in a *discharge tube*. A potential difference applied between cathode and anode creates an electric field that accelerates any free electrons and ions to their appropriate electrodes. Collisions between electrons and gas molecules create more ions. Collisions also produce excited ions and molecules (*see* excitation), which decay with emission of light in certain parts of the tube.

discontinuous function *See* continuous function.

discontinuous variation (qualitative variation) Clearly defined differences in a characteristic that can be observed in a population. Characteristics that are determined by different *alleles at a single locus show discontinuous variation, e.g. garden peas are either wrinkled or smooth. *Compare* continuous variation.

disilane *See* silane.

disinfectant Any substance that kills or inhibits the growth of disease-producing microorganisms and is in general toxic to human tissues. Disinfectants include cresol, bleaching powder, and phenol. They are used to cleanse surgical apparatus, sick-rooms, and household drains and if sufficiently diluted can be used as *antiseptics.

disintegration Any process in which an atomic nucleus breaks up spontaneously into two or more fragments in a radioactive decay process or breaks up as a result of a collision with a high-energy particle or nuclear fragment.

disintegration constant *See* decay.

dislocation *See* defect.

disodium hydrogenphosphate(V) (disodium orthophosphate) A colourless crystalline solid, Na_2HPO_4, soluble in water and insoluble in ethanol. It is known as the dihydrate (r.d. 2.066), heptahydrate (r.d. 1.68), and dodecahydrate (r.d. 1.52). It may be prepared by titrating phosphoric acid with sodium hydroxide to an alkaline end point (phenolphthalein) and is used in treating boiler feed water and in the textile industry.

disodium orthophosphate *See* disodium hydrogenphosphate(V).

disodium tetraborate-10-water *See* borax.

***d*-isomer** *See* optical activity.

D-isomer *See* absolute configuration.

disperse dye *See* dye.

disperse phase *See* colloids.

dispersion The splitting up of a ray of light of mixed wavelengths by refraction into its components. Dispersion occurs because the *deviation

for each wavelength is different on account of the different speeds at which waves of different wavelengths pass through the refracting medium. If a ray of white light strikes one face of a prism and passes out of another face, the white light will be split into its components and the full visible spectrum will be formed. The *dispersive power* of a prism (or other medium) for white light is defined by

$$(n_b - n_r)/(n_y - 1),$$

where n_b, n_r, and n_y are the *refractive indexes for blue, red, and yellow light respectively. The term is sometimes applied to the separation of wavelengths produced by a *diffraction grating.

dispersion forces *See* van der Waals' force.

dispersive power *See* dispersion.

displacement (electric flux density) Symbol D. The ratio of the charge displaced in a medium subjected to an electric field and the area of an imagined plane perpendicular to the potential gradient.

displacement activity An activity shown by an animal that appears to be irrelevant to its situation. Displacement activities are frequently observed when there is conflict between opposing tendencies. For example, birds in aggressive situations, in which there are simultaneous tendencies to attack and to flee, may preen their feathers as a displacement activity.

displacement reaction *See* substitution reaction.

display behaviour Stereotyped movement or posture that serves to influence the behaviour of another animal. Many displays in *courtship and *aggression are conspicuous and characteristic of the species; special markings or parts of the body may be prominently exhibited (for example, the male peacock spreads its tail in courtship). Other displays are cryptic and make it harder for a predator to recognize the displaying animal as potential prey. For example, geometer moth caterpillars, which look like twigs, hold themselves on plant stems with one end sticking into the air.

disproportionation A type of chemical reaction in which the same compound is simultaneously reduced and oxidized. For example, copper(I) chloride disproportionates thus:

$$2CuCl \rightarrow Cu + CuCl_2$$

The reaction involves oxidation of one molecule

$$Cu^I \rightarrow Cu^{II} + e$$

and reduction of the other

$$Cu^I + e \rightarrow Cu$$

The reaction of halogens with hydroxide ions is another example of a disproportionation reaction, for example

$$Cl_2(g) + 2OH^-(aq) \leftrightarrows$$
$$Cl^-(aq) + ClO^-(aq) + H_2O(l)$$

dissociation The breakdown of a molecule, ion, etc., into smaller molecules, ions, etc. An example of dissociation is the reversible reaction of hydrogen iodide at high temperatures

$$2HI(g) \leftrightarrows H_2(g) + I_2(g)$$

The *equilibrium constant of a reversible dissociation is called the *dissociation constant*. The term 'dissociation' is also applied to ionization reactions of *acids and *bases in water; for example

$$HCN + H_2O \leftrightarrows H_3O^+ + CN^-$$

which is often regarded as a straightforward dissociation into ions

$$HCN \leftrightarrows H^+ + CN^-$$

The equilibrium constant of such a dissociation is called the *acid dissociation constant* or *acidity constant*, given by

$$K_a = [H^+][A^-]/[HA]$$

for an acid HA (the concentration of water [H_2O] can be taken as constant). K_a is a measure of the strength of the acid. Similarly, for a nitrogenous base B, the equilibrium

$$B + H_2O \leftrightarrows BH^+ + OH^-$$

is also a dissociation; with the *base dissociation constant*, or *basicity constant*, given by

$$K_b = [BH^+][OH^-]/[B]$$

For a hydroxide MOH,

$$K_b = [M^+][OH^-]/[MOH]$$

dissociation pressure When a solid compound dissociates to give one or more gaseous products, the dissociation pressure is the pressure of gas in equilibrium with the solid at a given temperature. For example, when calcium carbonate is maintained at a constant high temperature in a closed container, the dissociation pressure at that temperature is the pressure of carbon dioxide from the equilibrium

$$CaCO_3(s) \leftrightarrows CaO(s) + CO_2(g)$$

distal Describing the part of an organ that is farthest from the organ's point of attachment to the rest of the body. For example, hands and feet are at the distal ends of arms and legs, respectively. *Compare* proximal.

distillation The process of boiling a liquid and condensing and collecting the vapour. The liquid collected is the *distillate*. It is used to purify liquids and to separate liquid mixtures (*see* fractional distillation; steam distillation). *See also* destructive distillation.

distilled water Water purified by distillation so as to free it from dissolved salts and other compounds. Distilled water in equilibrium with the carbon dioxide in the air has a conductivity of about 0.8×10^{-6} siemens cm^{-1}. Repeated distillation in a vacuum can bring the conductivity down to 0.043×10^{-6} siemens cm^{-1} at 18°C (sometimes called *conductivity water*). The limiting conductivity is due to self-ionization: $H_2O \leftrightarrows H^+ + OH^-$.

distortion The extent to which a system fails to reproduce the characteristics of its input in its output. It is most commonly applied to electronic amplifiers and to optical systems. *See* aberration.

distributive law The mathematical law stating that one operation is independent of being carried out before or after another operation. For example, multiplication is distributive with respect to addition and subtraction, i.e. $x(y + z) = xy + xz$. Compare associative law; commutative law.

disulphur dichloride (sulphur monochloride) An orange–red liquid, S_2Cl_2, which is readily hydrolysed by water and is soluble in benzene and ether; r.d. 1.678; m.p. $-80°C$; b.p. $136°C$. It may be prepared by passing chlorine over molten sulphur; in the presence of iodine or metal chlorides *sulphur dichloride*, SCl_2, is also formed. In the vapour phase S_2Cl_2 molecules have $Cl-S-S-Cl$ chains. The compound is used as a solvent for sulphur and can form higher *chlorosulphanes* of the type $Cl-(S)_n-Cl$ ($n < 100$), which are of great value in *vulcanization processes.

disulphuric(VI) acid (pyrosulphuric acid) A colourless hygroscopic crystalline solid, $H_2S_2O_7$; r.d. 1.9; m.p. $35°C$. It is commonly encountered mixed with sulphuric acid as it is formed by dissolving sulphur trioxide in concentrated sulphuric acid. The resulting fuming liquid, called *oleum* or *Nordhausen sulphuric acid*, is produced during the *contact process and is also widely used in the *sulphonation of organic compounds. *See also* sulphuric acid.

dithionate A salt of dithionic acid, containing the ion $S_2O_6^{2-}$, usually formed by the oxidation of a sulphite using manganese(IV) oxide. The ion has neither pronounced oxidizing nor reducing properties.

dithionic acid An acid, $H_2S_2O_6$, known in the form of its salts (dithionates).

dithionite *See* sulphinate.

dithionous acid *See* sulphinic acid.

diurnal Daily; denoting an event that happens once every 24 hours.

diurnal rhythm *See* circadian rhythm.

divalent (bivalent) Having a valency of two.

divergence (div) The *scalar product of the *gradient operator ∇ with a vector. For a vector u that has components u_1, u_2, and u_3 in the x, y, and z directions, and is a function of x, y, and z, the divergence is given by:

$$\text{div}u = \nabla.u$$
$$= \partial u_1/\partial x + \partial u_2/\partial y + \partial u_3/\partial z.$$

The divergence of a vector at a given point represents the flux of the vector per unit volume in the neighbourhood of that point. *See also* curl; Laplace equation.

divergent evolution *See* adaptive radiation.

divergent series *See* convergent series.

diverging lens or mirror A lens or mirror that can refract or reflect a parallel beam of light into a diverging beam. A diverging lens is predominantly concave; a diverging mirror is convex. *Compare* converging lens or mirror.

diverticulum A saclike or tubular outgrowth from a tubular or hollow internal organ. Diverticula may occur as normal structures (e.g. the *caecum and *appendix in the alimentary canal) or abnormally, from a weakened area of the organ.

division A category used in the *classification of plants that consists of one or several similar classes. Division names end in -phyta. Examples are the Bryophyta (mosses and liverworts) and the Tracheophyta (vascular plants). The divisions are grouped in the kingdom Plantae. In animal classification, the term *phylum is substituted for division.

dizygotic twins *See* fraternal twins.

D-lines Two close lines in the yellow region of the visible spectrum of sodium, having wavelengths 589.0 and 589.6 nm. As they are prominent and

Detail of molecular structure of sugar–phosphate backbone. Each deoxyribose unit is attached to a phosphate group and a base, forming a nucleotide

The four bases of DNA, showing the hydrogen bonding between base pairs

Double helical structure of DNA

Molecular structure of DNA

easily recognized they are used as a standard in spectroscopy.

dl-isomer *See* optical activity; racemic mixture.

DMF *See* dimethylformamide.

DMSO *See* dimethyl sulphoxide.

DNA (deoxyribonucleic acid) The genetic material of most living organisms, which is a major constituent of the *chromosomes within the cell nucleus and plays a central role in the determination of hereditary characteristics by controlling *protein synthesis in cells (*see also* genetic

code). DNA is a nucleic acid composed of two chains of *nucleotides in which the sugar is *deoxyribose* and the bases are *adenine, *cytosine, *guanine, and *thymine (*compare* RNA). The two chains are wound round each other and linked together by hydrogen bonds between specific complementary bases (*see* base pairing) to form a spiral ladder-shaped molecule (*double helix*). See illustration.

When the cell divides, its DNA also replicates in such a way that each of the two daughter molecules is identical to the parent molecule. The hydrogen bonds between the complementary bases on the two strands of the parent molecule break and the strands unwind. Using as building bricks nucleotides present in the nucleus, each strand directs the synthesis of a new one complementary to itself. Replication is initiated, controlled, and stopped by means of *polymerase enzymes. *See also* complementary DNA.

DNAase (DNase; deoxyribonuclease) An enzyme that catalyses the cleavage of DNA. DNAase I is a digestive enzyme, secreted by the pancreas, that degrades DNA into shorter nucleotide fragments. Many other *endonucleases and *exonucleases cleave DNA, including the *restriction enzymes and enzymes involved in DNA repair and replication.

DNA hybridization A method of determining the similarity of DNA from different sources. Single strands of DNA from two sources, e.g. different bacterial species, are put together and the extent to which double hybrid strands are formed is estimated. The greater the tendency to form these hybrid molecules, the greater the extent of complementary base sequences, i.e. gene similarity. The method is one way of determining the genetic relationships of species.

DNA polymerase *See* polymerase.

Döbereiner's triads A set of triads of chemically similar elements noted by J. W. Döbereiner in 1817. Even with the inaccurate atomic mass data of the day it was observed that when each triad was arranged in order of increasing atomic mass, then the mass of the central member was approximately the average of the values for the other two. The chemical and physical properties were similarly related. The triads are now recognized as consecutive members of the groups

of the periodic table. Examples are: lithium, sodium, and potassium; calcium, strontium, and barium; and chlorine, bromine, and iodine.

dodecanoic acid (lauric acid) A white crystalline *fatty acid, $CH_3(CH_2)_{10}COOH$; r.d. 0.8; m.p. 44°C; b.p. 225°C. Glycerides of the acid are present in natural fats and oils (e.g. coconut and palm-kernel oil).

dodecene A straight-chain alkene, $CH_3(CH_2)_9CH:CH_2$, obtained from petroleum and used in making *dodecylbenzene.

dodecylbenzene A hydrocarbon, $CH_3(CH_2)_{11}C_6H_5$, manufactured by a Friedel–Crafts reaction between dodecene $(CH_3(CH_2)_9CH:CH_2)$ and benzene. It can be sulphonated, and the sodium salt of the sulphonic acid is the basis of common *detergents.

dolomite A carbonate mineral consisting of a mixed calcium–magnesium carbonate, $CaCO_3.MgCO_3$, crystallizing in the rhombohedral system. It is usually white or colourless. The term is also used to denote a rock with a high ratio of magnesium to calcium carbonate. *See* limestone.

domain *See* magnetism.

dominance hierarchy *See* dominant.

dominant 1. (in genetics) Describing the *allele that functions when two different alleles of a gene are present in the cells of an organism. For example, the height of garden peas is controlled by two alleles, 'tall' (*T*) and 'dwarf' (*t*). When both are present (*Tt*), i.e. when the cells are *heterozygous, the plant is tall since *T* is dominant and *t* is *recessive. *See also* incomplete dominance. **2.** (in ecology) Describing the most conspicuously abundant and characteristic species in a *community. The term is usually used of a plant species in plant ecology; for example, pine trees in a pine forest. **3.** (in animal behaviour) Describing an animal that is allowed priority in access to food, mates, etc., by others of its species because of its success in previous aggressive encounters. Less dominant animals frequently show *appeasement behaviour towards a more dominant individual, so overt *aggression is minimized. In a stable group there may be a linear *dominance hierarchy* or *peak order* (so called because it was first observed in domestic fowl), with each animal being subservient to those above it in the hierarchy and taking precedence over those below it.

donor 1. (in chemistry) An atom, ion, or molecule that provides a pair of electrons in forming a coordinate bond. **2.** (in electronics) *See* semiconductor. **3.** (in medicine) An individual whose tissues or organs are transferred to another (the *recipient*). Donors may provide blood for transfusion or a kidney or heart for transplantation.

dopa (dihydroxyphenylalanine) A derivative of the amino acid tyrosine. It is found in particularly high levels in the adrenal glands and is a precursor in the synthesis of *dopamine, *noradrenaline, and *adrenaline. The laevorotatory form, *L-dopa*, is administered in the treatment of Parkinson's disease, in which brain levels of dopamine are reduced.

dopamine A *catecholamine that is a precursor in the synthesis of *noradrenaline and *adrenaline. It is also believed to function as a neurotransmitter in the brain.

doping *See* semiconductor.

Doppler effect The apparent change in the observed frequency of a wave as a result of relative motion between the source and the observer. For example, the sound made by a low-flying aircraft as it approaches appears to fall in pitch as it passes and flies away. In fact, the frequency of the aircraft engine remains constant but as it is approaching more

sound waves per second impinge on the ear and as it recedes fewer sound waves per second impinge on the ear. The apparent frequency, F, is given by

$$F = f(c - u_o)/(c - u_s),$$

where f is the true frequency, c is the speed of sound, and u_o and u_s are the speeds of the observer and the source, respectively.

Although the example of sound is most commonly experienced, the effect was suggested by Christian Johann Doppler (1803–53), an Austrian physicist, as an attempt to explain the coloration of stars. In fact the Doppler effect cannot be observed visually in relation to the stars, although the effect does occur with electromagnetic radiation and the *red shift of light from receding stars can be observed spectroscopically. The Doppler effect is also used in radar to distinguish between stationary and moving targets and to provide information regarding the speed of moving targets by measuring the frequency shift between the emitted and reflected radiation.

For electromagnetic radiation, the speed of light, c, features in the calculation and as there is no fixed medium to provide a frame of reference, relativity has to be taken into account, so that

$$F = f\sqrt{[(1 - v/c)/(1 + v/c)]},$$

where v is the speed at which source and observer are moving apart. If v^2/c^2 is small compared to 1, i.e. if the speed of separation is small compared to the speed of light, this equation simplifies to

$$F = f(1 - v/c).$$

d-orbital *See* orbital.

dormancy An inactive period in the life of an animal or plant during which growth slows or completely ceases. Physiological changes associated with dormancy help the organism survive adverse environmental conditions. Annual plants survive the winter as dormant seeds while many perennial plants survive as dormant tubers, rhizomes, or bulbs. *Hibernation and *aestivation in animals help them survive extremes of cold and heat, respectively.

dorsal Describing the surface of a plant or animal that is farthest from the ground or other support, i.e. the upper surface. In vertebrates, the dorsal surface is that down which the backbone runs. Thus in upright (bipedal) mammals, such as man and kangaroos, it is the backward-directed (*posterior) surface. *Compare* ventral.

dose A measure of the extent to which matter has been exposed to *ionizing radiation. The *absorbed dose* is the energy per unit mass absorbed by matter as a result of such exposure. The SI unit is the gray, although it is often measured in rads (1 rad = 0.01 gray; *see* radiation units). The *maximum permissible dose* is the recommended upper limit of absorbed dose that a person or organ should receive in a specified period according to the International Commission on Radiological Protection. *See also* linear energy transfer.

dosimeter Any device used to measure absorbed *dose of ionizing radiation. Methods used include the *ionization chamber, photographic film, or the rate at which certain chemical reactions occur in the presence of ionizing radiation.

dot product *See* scalar product.

double bond *See* chemical bond.

double decomposition (metathesis) A chemical reaction involving exchange of radicals, e.g.

$$AgNO_3(aq) + KCl(aq) \rightarrow$$
$$KNO_3(aq) + AgCl(s)$$

double helix *See* DNA.

double recessive An organism with two *recessive alleles for a particular characteristic.

double refraction The property, possessed by certain crystals (notably calcite), of forming two refracted rays from a single incident ray. The *ordinary ray* obeys the normal laws of refraction. The other refracted ray, called the *extraordinary ray*, follows different laws. The light in the ordinary ray is polarized at right angles to the light in the extraordinary ray. Along an *optic axis the ordinary and extraordinary rays travel with the same speed. Some crystals, such as calcite, quartz, and tourmaline, have only one optic axis; they are *uniaxial crystals*. Others, such as mica and selenite, have two optic axes; they are *biaxial crystals*. The phenomenon is also known as *birefringence* and the double-refracting crystal as a *birefringent crystal*. *See also* polarization.

double salt A crystalline salt in which there are two different anions and/or cations. An example is the mineral dolomite, $CaCO_3.MgCO_3$, which contains a regular arrangement of Ca^{2+} and Mg^{2+} ions in its crystal lattice. *Alums are double sulphates. Double salts only exist in the solid; when dissolved they act as a mixture of the two separate salts. *Double oxides* are similar.

doublet 1. A pair of optical lenses of different shapes and made of different materials used together so that the chromatic aberration produced by one is largely cancelled by the reverse aberration of the other. **2.** A pair of associated lines in certain spectra, e.g. the two lines that make up the sodium D-lines.

down feathers (plumules) Small soft feathers that cover and insulate the whole body of a bird. In nestlings they are the only feathers; in adults they lie beneath and beneath the *contour feathers. Down feathers have a fluffy appearance as their *barbs are not joined together to form a smooth vane.

Downs process A process for extracting sodium by the electrolysis of molten sodium chloride. The

Downs cell has a central graphite anode surrounded by a cylindrical steel cathode. Chlorine released is led away through a hood over the anode. Molten sodium is formed at the cathode and collected through another hood around the top of the cathode cylinder (it is less dense than the sodium chloride). The two hoods and electrodes are separated by a coaxial cylindrical steel gauze. A small amount of calcium chloride is added to the sodium chloride to lower its melting point. The sodium chloride is melted electrically and kept molten by the current through the cell. More sodium chloride is added as the electrolysis proceeds.

Down's syndrome A congenital form of mental retardation due to a chromosome defect in which there are three copies of chromosome no. 21 instead of the usual two. The affected individual has a short broad face and slanted eyes (as in the Mongolian races), short fingers, and weak muscles. Down's syndrome can be detected before birth by *amniocentesis. It is named after the British physician John Down (1828–96), who first studied the incidence of the disorder.

dragonflies *See* Odonata.

drain *See* transistor.

drupe (pyrenocarp) A fleshy fruit that develops from either one or several fused carpels and contains one or many seeds. The seeds are enclosed by the hard protective endocarp (*see* pericarp) of the fruit. Thus the stone of a peach is the endocarp containing the seed. Plums, cherries, coconuts, and almonds are other examples of one-seeded drupes; holly and elder fruits are examples of many-seeded drupes. *See also* etaerio.

dry cell A primary or secondary cell in which the electrolytes are in the form of a paste. Many torch, radio, and calculator batteries are *Leclanché cells in which the electro-lyte is an ammonium chloride paste and the container is the negative zinc electrode (with an outer plastic wrapping).

dry ice Solid carbon dioxide used as a refrigerant. It is convenient because it sublimes at $-78°C$ (195 K) at standard pressure rather than melting.

drying oil A natural oil, such as linseed oil, that hardens on exposure to the air. Drying oils contain unsaturated fatty acids, such as linoleic and linolenic acids, which polymerize on oxidation. They are used in paints, varnishes, etc.

Dryopithecus A genus of extinct apes, fossils of which have been found in India and dated as 30–40 million years old. Fossils of *Dryopithecus* and of the similar genus *Proconsul* are often referred to as *dryopithecines*. Dryopithecines are believed to be either the ancestors of the fossil hominids or ancestral to both the human and ape lines.

DSC Differential scanning calorimetry. *See* thermal analysis.

D-series *See* absolute configuration.

DTA Differential thermal analysis. *See* thermal analysis.

ductility The ability of certain metals, such as copper, to retain their strength when their shape is changed, especially the ability of such metals to be drawn into a thin wire without cracking or breaking.

ductless gland *See* endocrine gland.

Dulong and Petit's law For a solid element the product of the relative atomic mass and the specific heat capacity is a constant equal to about $25 \, J \, mol^{-1} \, K^{-1}$. Formulated in these terms in 1819 by the French scientists Pierre Dulong (1785–1838) and A. T. Petit (1791–1820), the law in modern terms states: the molar heat capacity of a solid element is approximately equal to $3R$, where R is the *gas constant. The law is only approximate but applies with fair accuracy at normal temperatures to elements with a simple crystal structure.

Dumas' method 1. A method of finding the amount of nitrogen in an organic compound. The sample is weighed, mixed with copper(II) oxide, and heated in a tube. Any nitrogen present in the compound is converted into oxides of nitrogen, which are led over hot copper to reduce them to nitrogen gas. This is collected and the volume measured, from which the mass of nitrogen in a known mass of sample can be found. 2. A method of finding the relative molecular masses of volatile liquids by weighing. A thin-glass bulb with a long narrow neck is used. This is weighed full of air at known temperature, then a small amount of sample is introduced and the bulb heated (in a bath) so that the liquid is vaporized and the air is driven out. The tip of the neck is sealed and the bulb cooled and weighed at known (room) temperature. The volume of the bulb is found by filling it with water and weighing again. If the density of air is known, the mass of vapour in a known volume can be calculated.

The techniques are named after the French chemist Jean Baptiste André Dumas (1800–84).

duodenum The first section of the *small intestine of vertebrates. It is the site where food from the stomach is subjected to the action of bile (from the bile duct) and pancreatic enzymes (from the pancreatic duct) as well as the enzymes secreted by digestive glands in the duodenum itself (*see* succus entericus), which are required in the breakdown of proteins, carbohydrates, and fats. By neutralizing the acidic secretions of the stomach, the duodenum provides an alkaline environment necessary for the action of the intestinal enzymes.

duplet A pair of electrons in a covalent chemical bond.

duplex Describing a biological molecule comprising two cross-linked polymeric chains; the term is applied particularly to the double-stranded structure of *DNA.

Duralumin Trade name for a class of strong lightweight aluminium alloys containing copper, magnesium, manganese, and sometimes silicon. Duralumin alloys combine strength with lightness and are extensively used in aircraft, racing cars, etc.

dura mater The outermost and toughest of the three membranes (*meninges) that surround the central nervous system in vertebrates. It lies adjacent to the skull and its purpose is to protect the delicate inner meninges (the *arachnoid membrane and the *pia mater).

duramen *See* heartwood.

dust core *See* core.

dwarf star A star, such as the sun, that lies on the main sequence in a *Hertzsprung–Russell diagram. *See also* white dwarf.

dyes Substances used to impart colour to textiles, leather, paper, etc. Compounds used for dyeing (*dyestuffs*) are generally organic compounds containing conjugated double bonds. The group producing the colour is the *chromophore; other noncoloured groups that influence or intensify the colour are called *auxochromes. Dyes can be classified according to the chemical structure of the dye molecule. For example, *azo dyes* contain the $-N=N-$ group (*see* azo compounds). In practice, they are classified according to the way in which the dye is applied or is held on the substrate.

Acid dyes are compounds in which the chromophore is part of a negative ion (usually an organic sulphonate RSO_2O^-). They can be used for protein fibres (e.g. wool and silk) and for polyamide and acrylic fibres. Originally, they were applied from an acidic bath. *Metallized dyes* are forms of acid dyes in which the negative ion contains a chelated metal atom. *Basic dyes* have chromophores that are part of a positive ion (usually an amine salt or ionized imino group). They are used for acrylic fibres and also for wool and silk, although they have only moderate fastness with these materials.

Direct dyes are dyes that have a high affinity for cotton, rayon, and other cellulose fibres. They are applied directly from a neutral bath containing sodium chloride or sodium sulphate. Like acid dyes, they are usually sulphonic acid salts but are distinguished by their greater substantivity (affinity for the substrate), hence the alternative name *substantive dyes*.

Vat dyes are insoluble substances used for cotton dyeing. They usually contain keto groups, $C=O$, which are reduced to $C-OH$ groups, rendering the dye soluble (the *leuco form* of the dye). The dye is applied in this form, then oxidized by air or oxidizing agents to precipitate the pigment in the fibres. Indigo and anthroquinone dyes are examples of vat dyes. *Sulphur dyes* are dyes applied by this technique using sodium sulphide solution to reduce and dissolve the dye. Sulphur dyes are used for cellulose fibres.

Disperse dyes are insoluble dyes applied in the form of a fine dispersion in water. They are used for cellulose acetate and other synthetic fibres.

Reactive dyes are compounds that contain groups capable of reacting with the substrate to form covalent bonds. They have high substantivity and are used particularly for cellulose fibres.

dynamical meteorology The branch of meteorology concerned with motions in the atmosphere. It is based on hydrodynamics and thermodynamics.

dynamic equilibrium *See* equilibrium.

dynamics The branch of mechanics concerned with the motion of bodies under the action of forces. Time intervals, distances, and masses are regarded as fundamental and bodies are assumed to possess *inertia. Bodies in motion have an attribute called *momentum (*see* Newton's laws of motion), which can only be changed by the application of a force. *Compare* kinetics; statics.

dynamite Any of a class of high explosives based on nitroglycerin. The original form, invented in 1867 by Alfred Nobel, consisted of nitroglycerin absorbed in kieselguhr. Modern dynamites, which are used for blasting, contain sodium or ammonium nitrate sensitized with nitroglycerin and use other absorbers (e.g. wood pulp).

dynamo An electric *generator, especially one designed to provide *direct current. Alternating-current generators can be called dynamos but are more often called alternators.

dynamo action The generation of electrical current and magnetic field by the motion of an electrically conducting fluid. It is generally believed that the magnetic fields of the earth and the sun are produced by dynamo action in the molten iron–nickel core of the earth and in the plasma of the solar interior.

dynamometer 1. An instrument used to measure a force, often a spring balance. **2.** A device used to measure the output power of an engine or motor. **3. (current dynamometer)** A variety of *current balance, for measuring electric current.

dyne The unit of force in the *c.g.s. system; the force required to give a mass of one gram an acceleration of 1 cm s^{-2}. 1 dyne $= 10^{-5}$ newton.

dysprosium Symbol Dy. A soft silvery metallic element belonging to the

Structure of the mammalian ear

*lanthanoids; a.n. 66; r.a.m. 162.50; r.d. 8.551 (20°C); m.p. 1412°C; b.p. 2567°C. It occurs in apatite, gadolinite, and xenotime, from which it is extracted by an ion-exchange process. There are seven natural isotopes and twelve artificial isotopes have been identified. It finds limited use in some alloys as a neutron absorber, particularly in nuclear technology. It was discovered by François Lecoq de Boisbaudran in 1886.

dystectic mixture A mixture of substances that has a constant maximum melting point.

E

e The irrational number defined as the limit as n tends to infinity of $(1 + 1/n)^n$. It has the value 2.718 28 .... It is used as the base of natural *logarithms and occurs in the *exponential function, e^x.

ear The sense organ in vertebrates that is specialized for the detection of sound and the maintenance of balance. It can be divided into the *outer ear and *middle ear, which collect and transmit sound waves, and the *inner ear, which contains the organs of balance and (except in fish) hearing (see illustration). The term ear is often used for the *pinna of the mammalian outer ear.

eardrum *See* tympanum.

early universe The study of *cosmology at the time very soon after the *big bang. Theories of the early universe have led to a mutually beneficial interaction between cosmology and the theory of *elementary particles, particularly *grand unified theories.

Because there were very high temperatures in the early universe many of the *broken symmetries in *gauge theories become unbroken symmetries at these temperatures. As the universe cools after the big bang there is thought to be a sequence of transitions to broken symmetry states.

Combining cosmology with grand unified theories helps to explain why the observed universe appears to consist of matter with no *antimatter. This means that one has a nonzero *baryon number for the universe. This solution relies on the fact that there were nonequilibrium conditions in the early universe due to its rapid expansion after the big bang.

An important idea in the theory of the early universe is that of *inflation* – the idea that the nature of the *vacuum state gave rise, after the big bang, to an exponential expansion of the universe. The hypothesis of the *inflationary universe* solves several long-standing problems in cosmology, such as the flatness and homogeneity of the universe.

ear ossicles Three small bones – the *incus* (*anvil*), *malleus* (*hammer*), and *stapes* (*stirrup*) – that lie in the mammalian *middle ear, forming a bridge between the tympanum (eardrum) and the fenestra ovalis. The function of the ossicles is to transmit (and amplify) vibrations of the tympanum across the middle ear to the fenestra ovalis, which transfers them to the *inner ear. Muscles of the middle ear constrict the movement of the ossicles. This serves to safeguard the ear from damage caused by excessively loud noise.

earth The planet that orbits the sun between the planets Venus and Mars at a mean distance from the sun of 149 600 000 km. It has a mass of about 5.976×10^{24} kg and an equatorial diameter of 12 756 km. The earth consists of three layers: the gaseous atmosphere (*see* earth's atmosphere), the liquid *hydrosphere, and the solid *lithosphere. The solid part of the earth also consists of three layers: the *crust* with a mean thickness of about 32 km under the land and 10 km under the seas; the *mantle*, which extends some 2900 km below the crust; and the *core*, part of which is believed to be liquid. The crust has a relative density of about 3 and consists largely of sedimentary rocks

overlaying igneous rocks. The composition of the crust is: oxygen 47%, silicon 28%, aluminium 8%, iron 4.5%, calcium 3.5%, sodium and potassium 2.5% each, and magnesium 2.2%. Hydrogen, carbon, phosphorus, and sulphur are all present to an extent of less than 1%. The mantle reaches a relative density of about 5.5 at its maximum depth and is believed to consist mainly of silicate rocks. The core is believed to have a maximum relative density of 13 and a maximum temperature of 6400 K. *See also* geomagnetism; plate tectonics.

earthquake A sudden movement or fracturing within the earth's lithosphere, followed by the series of shocks generated by this movement. This may range from a mild tremor to a large-scale earth movement causing extensive damage over a wide area. The point at which the earthquake originates is known as the *seismic focus*; the point on the earth's surface directly above this is the *epicentre* (or *hypocentre*). Earthquakes result from a build-up of stresses within the rocks until they are strained to the point beyond which they will fracture. They occur in narrow continuous belts of activity, which correspond with the junction of lithospheric plates, including the circum-Pacific belt, the Alpine–Himalayan belt, and mid-ocean ridges. The scale of the shock of an earthquake is known as the magnitude; the most commonly used scale for comparing the magnitude of earthquakes is the logarithmic *Richter scale (8.9 is the highest recorded magnitude on the scale).

earth's atmosphere The gas that surrounds the earth. The composition of dry air at sea level is: nitrogen 78.08%, oxygen 20.95%, argon 0.93%, carbon dioxide 0.03%, neon 0.0018%, helium 0.0005%, krypton 0.0001%, and xenon 0.00001%. In addition to water vapour, air in some localities

earth's atmosphere

Earth's atmosphere

contains sulphur compounds, hydrogen peroxide, hydrocarbons, and dust particles.

The lowest level of the atmosphere, in which most of the weather occurs, is called the *troposphere*. Its thickness varies from about 7 km at the poles to 28 km at the equator and in this layer temperature falls with increasing height. The next layer is the *stratosphere*, which goes up to about 50 km. Here the temperature remains approximately constant. Above this is the *ionosphere*, which extends to about 1000 km, with the temperature rising and the composition changing substantially. At about 100 km and above most of the oxygen has dissociated into atoms; at above 150 km the percentage of nitrogen has dropped to nil. In the ionosphere the gases are ionized by the absorption of solar radiation. This enables radio transmissions to be made round the curved surface of the earth as the ionized gas acts as a reflector for cer-

tain wavelengths. The ionosphere is divided into three layers. The D-layer (50–90 km) contains a low concentration of free electrons and reflects low-frequency radio waves. The E-layer (90–150 km) is also called the *Heaviside layer* or *Heaviside–Kennelly layer* as its existence was predicted independently by Oliver Heaviside (1850–1925) and Arthur E. Kennelly (1861–1939). This layer reflects medium-frequency waves. The F-layer (150–1000 km) is also called the *Appleton layer* after its discoverer Sir Edward Appleton (1892–1965). It has the highest concentration of free electrons and is the most useful for radio transmission. Wavelengths between 8 mm and 20 m are not reflected by the ionosphere but escape into space. Therefore television transmissions, which utilize this range, require artificial *satellites for reflection (or reception, amplification, and retransmission). From about 400 km, the outermost region of the atmosphere is also called the *exosphere*. See illustration.

earth sciences A group of sciences concerned with the study of the earth. The chief earth sciences are geology, geography, oceanography, meteorology, geophysics, and geochemistry.

earthshine Sunlight reflected from the surface of the earth. An observer in space may see nearby objects dimly illuminated by earthshine, as things on earth may be illuminated by moonlight. Under certain conditions near new moon the dark disc of the moon can be seen faintly illuminated by earthshine – a phenomenon called 'the old moon in the new moon's arms'.

earth's magnetic field *See* geomagnetism.

earwigs *See* Dermaptera.

ebonite *See* vulcanite.

ebullioscopic constant *See* elevation of boiling point.

eccentricity *See* conic.

ecdysis (moulting) 1. The periodic loss of the outer cuticle of arthropods. It starts with the reabsorption of some materials in the inner part of the old cuticle and the formation of a new soft cuticle. The remains of the old cuticle then split; the animal emerges and absorbs water or swallows air and increases in size while the new cuticle is still soft. This cuticle is then hardened with chitin and lime salts. In insects and crustaceans ecdysis is controlled by the hormone *ecdysone. **2.** The periodic shedding of the outer layer of the epidermis of reptiles (except crocodiles) to allow growth to occur.

ecdysone A steroid hormone, produced by insects and crustaceans, that stimulates moulting (*see* ecdysis) and metamorphosis. It acts on specific gene loci, stimulating the synthesis of proteins involved in these bodily changes.

ECG *See* electrocardiogram.

echelon A form of *interferometer consisting of a stack of glass plates arranged stepwise with a constant offset. It gives a high resolution and is used in spectroscopy to study hyperfine line structure. In the *transmission echelon* the plates are made equal in optical thickness to introduce a constant delay between adjacent parts of the wavefront. The *reflecting echelon* has the exposed steps metallized and acts like an exaggerated *diffraction grating.

Echinodermata A phylum of marine invertebrates that includes the sea urchins, starfish, brittlestars, and sea cucumbers. Echinoderms have an exoskeleton (*test*) of calcareous plates embedded in the skin. In many species (e.g. sea urchins) spines protrude from the test. A system of water-filled canals (the *water vascular system*) provides hydraulic power for thousands of *tube feet*: saclike protrusions of the body wall used for locomotion, feeding, and respiration. Echinoderms have a long history: fossils of primitive echinoderms are known from rocks 500 million years old.

echo The reflection of a wave by a surface or object so that a weaker version of it is detected shortly after the original. The delay between the two is an indication of the distance of the reflecting surface. An *echo sounder* is an apparatus for determining the depth of water under a ship. The ship sends out a sound wave and measures the time taken for the echo to return after reflection by the sea bottom. *Sonar* (*sound navigation ranging*) is a technique for locating underwater objects by a similar method. Echoes also occur with radio waves; reflection of waves causes an echo in radio transmission and ghosts in television pictures. *See also* radar.

echolocation 1. *See* radar; echo. **2.** A method used by some animals (such as bats, dolphins, and certain

eclipse

Solar and lunar eclipses

Moon's shadow in solar eclipse

birds) to detect objects in the dark. The animal emits a series of high-pitched sounds that echo back from the object and are detected by the ear or some other sensory receptor. From the direction of the echo and from the time between emission and reception of the sounds the object is located, often very accurately.

eclipse The total (*total eclipse*) or partial (*partial eclipse*) obscuring of light from a celestial body as it passes behind or through the shadow of another body. A *lunar eclipse* occurs when the sun, earth, and moon are in a straight line and the shadow of the earth falls on the moon. A *solar eclipse* occurs when the shadow of the moon falls on the earth. See illustrations.

eclipsed conformation *See* conformation.

ecliptic The *great circle in which the plane of the earth's orbit round the sun intersects the *celestial sphere. It is thus the sun's apparent annual path across the sky.

E. coli *See Escherichia coli.*

ecological niche The status or role of an organism in its environment. An organism's niche is defined by the types of food it consumes, its predators, temperature tolerances, etc. Two species cannot coexist stably if they occupy identical niches.

ecology The study of the inter-relationships between organisms and their natural environment, both living and nonliving. For this purpose, ecologists study organisms in the context of the *populations and *communities in which they can be grouped and the *ecosystems of which they form a part. The study of ecological interactions provides important information on the nature and mechanisms of evolutionary change. Advances made in ecology over the last 25 years have led to increased concern about the effects of man's activities on the environment (notably the effects of *pollution), which has resulted in a greater awareness of the importance of *conservation.

ecosystem A biological *community and the physical environment associated with it. Nutrients pass between the different organisms in an ecosystem in definite pathways; for example, nutrients in the soil are taken up by plants, which are then eaten by herbivores, which in turn may be eaten by carnivores (*see* food chain). Organisms are classified on the basis of their position in an eco-system into various *trophic levels. Nutrients and energy move round ecosystems in loops or cycles (in the case above, for example, nutrients are returned to the soil via animal wastes and decomposition). *See* carbon cycle; nitrogen cycle.

ectoderm The external layer of cells of the *gastrula, which will develop into the epidermis and the nervous system in the adult. *See also* germ layers.

ectoparasite A parasite that lives on the outside of its host's body. *See* parasitism.

ectoplasm *See* cytoplasm.

Ectoprocta (Bryozoa; Polyzoa) A phylum of aquatic, mainly marine, invertebrates – the moss animals and sea mats. They live in colonies, 50 cm or more across, which are attached to rocks, seaweeds, or shells. The individuals making up the colonies are about 1 mm long and superficially resemble coelenterate *polyps, with a mouth surrounded by ciliated tentacles that trap minute particles of organic matter in the water. Some have a horny or calcareous outer skeleton into which the body can be withdrawn.

edaphic factor A factor relating to the physical or chemical composition of the soil found in a particular area. For example, very alkaline soil may be an edaphic factor limiting the variety of plants growing in a region.

eddy current A current induced in a conductor situated in a changing magnetic field or moving in a fixed one. Any imagined circuit within the conductor will change its magnetic flux linkage, and the consequent induced e.m.f. will drive current around the circuit. In a substantial block of metal the resistance will be small and the current therefore large. Eddy currents occur in the cores of transformers and other electrical machines and represent a loss of useful energy (the *eddy-current loss*). To reduce this loss to a minimum metal cores are made of insulated sheets of metal, the resistance between these laminations reducing the current. In high-frequency circuits *ferrite cores can be used. Eddy currents in a moving conductor interact with the magnetic field producing them to retard the motion of the conductor. This enables some electrical instruments

(moving-coil type) to utilize eddy currents to create damping. Eddy currents are also used in *induction heating.

Edison cell *See* nickel–iron accumulator.

EDTA Ethylenediaminetetracetic acid, $(HOOCCH_2)_2N(CH_2)_2N(CH_2COOH)_2$

A compound that acts as a chelating agent, reversibly binding with iron, magnesium, and other metal ions. It is used in certain culture media bound with iron, which it slowly releases into the medium.

EEG *See* electroencephalogram.

effective temperature *See* luminosity.

effective value *See* root-mean-square value.

effector A cell or organ that produces a physiological response when stimulated by a nerve impulse. Examples include muscles and glands.

efferent Carrying (nerve impulses, blood, etc.) away from the centre of a body or organ towards peripheral regions. The term is usually applied to types of nerve fibres or blood vessels. *Compare* afferent.

effervescence The formation of gas bubbles in a liquid by chemical reaction.

efficiency A measure of the performance of a machine, engine, etc., being the ratio of the energy or power it delivers to the energy or power fed to it. In general, the efficiency of a machine varies with the conditions under which it operates and there is usually a load at which it operates with the highest efficiency. The *thermal efficiency* of a heat engine is the ratio of the work done by the engine to the heat supplied by the fuel. For a *reversible heat engine this efficiency equals $(T_1 - T_2)/T_1$, where T_1 is the thermodynamic temperature at which all the heat is taken up and T_2 is the thermodynamic temperature at which it is given out (*see* Carnot cycle). For real engines it is always less than this.

efflorescence The process in which a crystalline hydrate loses water, forming a powdery deposit on the crystals.

effusion The flow of a gas through a small aperture. The relative rates at which gases effuse, under the same conditions, is approximately inversely proportional to the square roots of their densities.

egg 1. The fertilized ovum (*zygote) in egg-laying animals, e.g. birds and insects, after it emerges from the body. The egg is covered by *egg membranes that protect it from environmental damage, such as drying. **2.** (*or* **egg cell**) The mature female reproductive cell in animals and plants. *See* oosphere; ovum.

egg membrane The layer of material that covers an animal egg cell. *Primary membranes* develop in the ovary and cover the egg surface in addition to the normal cell membrane. The primary membrane is called the *vitelline membrane* in insects, molluscs, birds, and amphibians, the *chorion* in tunicates and fish, and the *zona pellucida* in mammals. Insects have a second thicker membrane, also called the chorion. *Secondary membranes* are secreted by the oviducts and parts of the genital system while the egg is passing to the outside. They include the jelly coat of frogs' eggs and the albumen and shell of birds' eggs.

eigenfunction An allowed *wave function of a system in quantum mechanics. The associated energies are *eigenvalues*.

Einstein equation 1. The mass–energy relationship announced by Einstein in 1905 in the form $E = mc^2$, where E is a quantity of energy, m its mass, and c is the speed of light. It presents the concept that energy has mass. *See also* relativity. **2.** The relationship $E_{max} = hf - W$, where E_{max} is the maximum kinetic energy of the electrons emitted in the photoemissive effect, h is the Planck constant, f the frequency of the incident radiation, and W the *work function of the emitter. This is also written $E_{max} = hf - \phi e$, where e is the electronic charge and ϕ a potential difference, also called the work function. (Sometimes W and ϕ are distinguished as *work function energy* and *work function potential*.) The equation can also be applied to photoemission from gases, when it has the form: $E = hf - I$, where I is the ionization potential of the gas.

einsteinium Symbol Es. A radioactive metallic transuranic element belonging to the *actinoids; a.n. 99; mass number of the most stable isotope 254 (half-life 270 days). Eleven isotopes are known. The element was first identified by A. Ghiorso and associates in debris from the first hydrogen bomb explosion in 1952. Microgram quantities of the element did not become available until 1961.

Einstein shift *See* redshift.

Elasmobranchii (Chondrichthyes) A class of vertebrates comprising the fishes with cartilaginous skeletons. The majority belong to the order Selachi (skates, rays, and sharks). Most cartilaginous fishes are marine carnivores with powerful jaws. Unlike bony fishes, they have no swim bladder, and therefore avoid sinking only by constant swimming with the aid of an asymmetrical (*heterocercal*) tail. There is no operculum covering the gill slits, the first of which is modified as a *spiracle. Fertilization is internal so the few eggs produced are consequently yolky, large, and well-protected. Some elasmobranchs show viviparous development of the young (*see* viviparity).

elastance The reciprocal of *capacitance. It is measured in $farad^{-1}$ (sometimes called a 'daraf').

Elasticity

elastic collision A collision in which the total kinetic energy of the colliding bodies after collision is equal to their total kinetic energy before collision. Elastic collisions occur only if there is no conversion of kinetic energy into other forms, as in the collision of atoms. In the case of macroscopic bodies this will not be the case as some of the energy will become heat. In a collision between polyatomic molecules, some kinetic energy may be converted into vibrational and rotational energy of the molecules, but otherwise molecular collisions appear to be elastic.

elasticity The property of certain materials that enables them to return to their original dimensions after an applied *stress has been removed. In general, if a stress is applied to a wire, the *strain will increase in proportion (see *OA* on the illustration) until a certain point called the *limit of proportionality* is reached. This is in accordance with *Hooke's law. Thereafter there is at first a slight increase in strain with increased load until a point *L* is reached. This is the *elastic limit*; up to this point the deformation of the specimen is elastic, i.e. when the stress is removed the specimen returns to its original length. Beyond the point *L* there is permanent deformation when the stress is removed, i.e. the material has ceased to be *elastic* and has become *plastic*. In the plastic stages individual materials vary somewhat; in general, how-

ever, at a point *B* there is a sudden increase in strain with further increases of stress – this is the *yield point*. Beyond the point *C*, the *breaking stress*, the wire will snap (which occurs at point *D*).

elastic modulus The ratio of the *stress applied to a body to the *strain produced. The *Young modulus of elasticity*, named after the British physicist Thomas Young (1773–1829), refers to longitudinal stress and strain. The *bulk modulus* is the ratio of the pressure on a body to its fractional decrease in volume. The *shear* (or *rigidity*) *modulus* is the tangential force per unit area divided by the angular deformation in radians.

elastomer A natural or synthetic rubber or rubberoid material, which has the ability to undergo deformation under the influence of a force and regain its original shape once the force has been removed.

electret A permanently electrified substance or body that has opposite charges at its extremities. Electrets resemble permanent magnets in many ways. An electret can be made by cooling certain waxes in a strong electric field.

electrical energy A form of energy related to the position of an electric charge in an electric field. For a body with charge Q and an electric potential V, its electrical energy is QV. If V is a potential difference, the same expression gives the energy transformed when the charge moves through the p.d.

electric arc A luminous discharge between two electrodes. The discharge raises the electrodes to incandescence, the resulting thermal ionization largely providing the carriers to maintain the high current between the electrodes.

electric-arc furnace A furnace used in melting metals to make alloys, especially in steel manufacture, in which the heat source is an electric arc. In the direct-arc furnace, such as the Héroult furnace, an arc is formed between the metal and an electrode. In the indirect-arc furnace, such as the Stassano furnace, the arc is formed between two electrodes and the heat is radiated onto the metal.

bell

hammer

light iron trembler

soft
iron core

coil

spring

switch

cell

Electric bell

electric bell A device in which an electromagnetically operated hammer strikes a bell (see illustration). Pressing the bell-push closes a circuit, causing current to flow from a battery or mains step-down transformer through an electromagnet. The electromagnet attracts a piece of soft iron attached to the hammer, which strikes the bell and at the same time breaks the circuit. The hammer springs back into its original position again, closing the circuit and causing the magnet to attract the soft iron. This process continues until the bell-push is released.

electric charge *See* charge.

electric constant *See* permittivity.

electric current *See* current.

electric displacement (electric flux density) Symbol D. The charge per unit area that would be displaced across a layer of conductor placed across an *electric field. This describes also the charge density on an extended surface that could be causing the field.

electric field A region in which an electric charge experiences a force usually because of a distribution of other charges. The *electric field strength* or *electric intensity* (E) at any point in an electric field is defined as the force per unit charge experienced by a small charge placed at that point. This is equivalent to a potential gradient along the field and is measured in volts per metre. The strength of the field can alternatively be described by its *electric displacement D. The ratio D/E for measurements in a vacuum is the electric constant ε_0. In a substance the observed potential gradient is reduced by electron movement so that D/E appears to increase: the new ratio (ε) is called the *permittivity of the substance. An electric field can be created by an isolated electric charge, in which case the field strength at a distance r from a point charge Q is given by $E = Q/4\pi r^2\varepsilon$, where ε is the permittivity of the intervening medium (*see* Coulomb's law). An electric field can also be created by a changing magnetic field.

electric flux Symbol Ψ. In an *electric field, the product of the electric flux density and the relevant area. *See* electric displacement.

electric flux density *See* electric displacement.

electricity Any effect resulting from the existence of stationary or moving electric charges.

electric lighting Illumination provided by electric currents. The devices used are the *arc lamp*, the *light bulb* (incandescent filament lamp), and the *fluorescent tube*. In the arc lamp, which is no longer used as a general means of illumination, an electric current flows through a gap between two carbon electrodes, between which a high potential difference is maintained. The current is carried by electrons and ions in the vapour produced by the electrodes and a mechanism is required to bring the electrodes closer together as they are vaporized. The device produces a strong white light but has many practical disadvantages. However, arcs enclosed in an inert gas (usually xenon) are increasingly used for such purposes as cinema projectors. The common light bulb is a glass bulb containing a tungsten filament and usually an inert gas. The passage of an electric current through the filament heats it to a white heat. Inert gas is used in the bulb to minimize blackening of the glass by evaporation of tungsten. In the fluorescent tube a glass tube containing mercury vapour (or some other gas) at a low pressure has its inner surface coated with a fluorescent substance. A discharge is created within the tube between two electrodes. Electrons emitted by the cathode collide with gas atoms or molecules and raise them to an excited state (*see* excitation). When they fall back to the *ground state they emit photons of ultraviolet radiation, which is converted to visible light by the coating of phosphor on the inner walls of the tube. In some lamps, such as the *sodium-vapour and *mercury-vapour lamps used in street lighting, no fluorescent substance is used, the light being emitted directly by the excited atoms of sodium or mercury. Vapour lights are more efficient than filament lights as less of the energy is converted into heat.

electric motor A machine for converting electrical energy into mechanical energy. They are quiet, clean, and have a high efficiency (75–95%). They

work on the principle that a current passing through a coil within a magnetic field will experience forces that can be used to rotate the coil. In the *induction motor*, alternating current is fed to a stationary coil (the *stator*), which both creates the magnetic field and induces a current in the rotating coil (*rotor*), which it surrounds. The advantage of this kind of motor is that current does not have to be fed through a commutator to a moving part. In the *synchronous motor*, alternating current fed to the stator produces a magnetic field that rotates and locks with the field of the rotor, in this case an independent magnet, causing the rotor to rotate at the same speed as the stator field rotates. The rotor is either a permanent magnet or an electromagnet fed by a direct current through slip rings. In the *universal motor*, current is fed to the stator and, through a commutator, to the rotor. In the series-wound motor the two are in series; in the shunt-wound motor they are in parallel. These motors can be used with either a.c. or d.c. but some small motors use a permanent magnet as the stator and require d.c. for the rotor (via the commutator). *See also* linear motor.

electric organ An organ occurring on the body or tail of certain fish, such as the electric ray (*Torpedo*) and electric eel (*Electrophorus electricus*). It gives an electric shock when touched and is used either to stun prey or predators or, in some species, to maintain a weak electric field in the surrounding water that is used in navigation. The organ is composed of modified muscle cells (*electroplate cells*), nervous stimulation of which greatly increases the potential difference across the cell. The electroplates are in series so a high overall voltage can be achieved.

electric polarization *See* dielectric.

electric potential Symbol *V*. The energy required to bring unit electric charge from infinity to the point in an electric field at which the potential is being specified. The unit of electric potential is the volt. The *potential difference* (*p.d.*) between two points in an electric field or circuit is the difference in the values of the electric potentials at the two points, i.e. it is the work done in moving unit charge from one point to the other.

electric power The rate of expending energy or doing work in an electrical system. For a direct-current circuit, it is given by the product of the current passing through a system and the potential difference across it. In alternating-current circuits, the power is given by *VI*cosϕ, where *V* and *I* are the RMS values and ϕ is the *phase angle. Cosϕ is called the *power factor* of the circuit.

electric spark The transient passage of an electric current through a gas between two points of high opposite potential, with the emission of light and sound. *Lightning consists of a spark between a cloud and earth or between two oppositely charged parts of the same cloud.

electric susceptibility *See* susceptibility.

electrocardiogram (ECG) A tracing or graph of the electrical activity of the heart. Recordings are made from electrodes fastened over the heart and usually on both arms and a leg. Changes in the normal pattern of an ECG may indicate heart irregularities or disease.

electrochemical cell *See* cell.

electrochemical equivalent Symbol *z*. The mass of a given element liberated from a solution of its ions in electrolysis by one coulomb of charge. *See* Faraday's laws (of electrolysis).

electrochemical series *See* electromotive series.

electrochemistry The study of chemical properties and reactions involving ions in solution, including electrolysis and electric cells.

electrochromatography *See* electrophoresis.

electrode 1. A conductor that emits or collects electrons in a cell, thermionic valve, semiconductor device, etc. The *anode* is the positive electrode and the *cathode* is the negative electrode. **2.** *See* half cell.

electrodeposition The process of depositing one metal on another by electrolysis, as in *electroforming and *electroplating.

electrode potential The potential difference produced between the electrode and the solution in a *half cell. It is not possible to measure this directly since any measurement involves completing the circuit with the electrolyte, thereby introducing another half cell. *Standard electrode potentials* $E^{\ominus}$ are defined by measuring the potential relative to a standard *hydrogen half cell using 1.0 molar solution at 25°C. The convention is to designate the cell so that the oxidized form is written first. For example,

$$Pt(s)|H_2(g)H^+(aq)|Zn^{2+}(aq)|Zn(s)$$

The e.m.f. of this cell is −0.76 volt (i.e. the zinc electrode is negative). Thus the standard electrode potential of the $Zn^{2+}|Zn$ half cell is −0.76 V. Electrode potentials are also called *reduction potentials*. *See also* electromotive series.

electrodialysis A method of obtaining pure water from water containing a salt, as in *desalination. The water to be purified is fed into a cell containing two electrodes. Between the electrodes is placed an array of *semipermeable membranes alternately semipermeable to positive ions and negative ions. The ions tend to segregate between alternate pairs of membranes, leaving pure water in the other gaps between membranes. In this way, the feed water is separated into two streams: one of pure water and the other of more concentrated solution.

electrodynamics The study of electric charges in motion, the forces created by electric and magnetic fields, and the relationship between them. *Compare* electrostatics.

electroencephalogram (EEG) A tracing or graph of the electrical activity of the brain. Electrodes taped to the scalp record electrical waves from different parts of the brain. The pattern of an EEG reflects an individual's level of consciousness and can be used to detect such disorders as epilepsy, tumours, or brain damage. *See also* brain death.

electroforming A method of forming intricate metal articles or parts by *electrodeposition of the metal on a removable conductive mould.

electroluminescence *See* luminescence.

electrolysis The production of a chemical reaction by passing an electric current through an electrolyte. In electrolysis, positive ions migrate to the cathode and negative ions to the anode. The reactions occurring depend on electron transfer at the electrodes and are therefore redox reactions. At the anode, negative ions in solution may lose electrons to form neutral species. Alternatively, atoms of the electrode can lose electrons and go into solution as positive ions. In either case the reaction is an oxidation. At the cathode, positive ions in solution can gain electrons to form neutral species. Thus cathode reactions are reductions.

electrolyte A liquid that conducts electricity as a result of the presence of positive or negative ions. Electrolytes are molten ionic compounds or solutions containing ions, i.e. solutions of ionic salts or of compounds that ionize in solution. Liquid metals,

in which the conduction is by free electrons, are not usually regarded as electrolytes.

electrolytic capacitor *See* capacitor.

electrolytic cell A cell in which electrolysis occurs; i.e. one in which current is passed through the electrolyte from an external source.

electrolytic corrosion Corrosion that occurs through an electrochemical reaction. *See* rusting.

electrolytic gas The highly explosive gas formed by the electrolysis of water. It consists of two parts hydrogen and one part oxygen by volume.

electrolytic rectifier A *rectifier consisting of two dissimilar electrodes immersed in an electrolyte. By suitable choice of electrodes and electrolyte the cell can be made to pass current easily in one direction but hardly at all in the other. Examples include a lead–aluminium cell with ammonium phosphate(V) electrolyte and a tantalum–lead cell with sulphuric acid as the electrolyte.

electrolytic refining The purification of metals by electrolysis. It is commonly applied to copper. A large piece of impure copper is used as the anode with a thin strip of pure copper as the cathode. Copper(II) sulphate solution is the electrolyte. Copper dissolves at the anode: $Cu \rightarrow Cu^{2+} + 2e$, and is deposited at the cathode. The net result is transfer of pure copper from anode to cathode. Gold and silver in the impure copper form a so-called *anode sludge* at the bottom of the cell, which is recovered.

electrolytic separation A method of separating isotopes by exploiting the different rates at which they are released in electrolysis. It was formerly used for separating deuterium and hydrogen. On electrolysis of water, hydrogen is formed at the cathode more readily than deuterium, thus the water becomes enriched with deuterium oxide.

electromagnet A magnet consisting of a soft ferromagnetic core with a coil of insulated wire wound round it. When a current flows through the wire the core becomes magnetized; when the current ceases to flow the core loses its magnetization. Electromagnets are used in switches, solenoids, electric bells, metal-lifting cranes, and many other applications.

electromagnetic induction The production of an electromotive force in a conductor when there is a change of magnetic flux linkage with the conductor or when there is relative motion of the conductor across a magnetic field. The magnitude of the e.m.f. is proportional (and in modern systems of units equal) to the rate of change of the flux linkage or the rate of cutting flux $d\Phi/dt$; the sense of the induced e.m.f. is such that any induced current opposes the change causing the induction, i.e. $E = -d\Phi/dt$. *See* Faraday's laws; Lenz's law; Neumann's law; inductance.

electromagnetic interaction *See* fundamental interactions.

electromagnetic pump A pump used for moving liquid metals, such as the liquid-sodium coolant in a fast nuclear reactor. The liquid is passed through a flattened pipe over two electrodes between which a direct current flows. A magnetic field at right angles to the current causes a force to be created directly on the liquid, along the axis of the tube. The pump has no moving parts and is therefore safe and trouble free.

electromagnetic radiation Energy resulting from the acceleration of electric charge and the associated electric fields and magnetic fields. The energy can be regarded as waves propagated through space (requiring no supporting medium) involving oscillating electric and magnetic fields at right angles to each other and to the direction of propagation. In a vacuum the waves travel with a constant speed (the speed of light) of 2.9979×10^8 metres per second; if material is present they are slower. Alternatively, the energy can be regarded as a stream of *photons travelling at the speed of light, each photon having an energy hc/λ, where h is the Planck constant, c is the speed of light, and λ is the wavelength of the associated wave. A fusion of these apparently conflicting concepts is possible using the methods of *quantum mechanics or *wave mechanics. The characteristics of the radiation depend on its wavelength. *See* electromagnetic spectrum.

electromagnetic spectrum The range of wavelengths over which *electromagnetic radiation extends. The longest waves $(10^5 - 10^{-3}$ metres) are radio waves, the next longest $(10^{-3} - 10^{-6}$ m) are infrared waves, then comes the narrow band $(4 - 7 \times 10^{-7}$ m) of visible light, followed by ultraviolet waves $(10^{-7} - 10^{-9}$ m), X-rays $(10^{-9} - 10^{-11}$ m), and gamma rays $(10^{-11} - 10^{-14}$ m).

electromagnetic units (e.m.u.) A system of electrical units formerly used in the *c.g.s. system. The e.m.u. of electric current is the *abampere* (all e.m.u. have the prefix *ab-* attached to the names of practical units). The abampere is the current that, flowing in an arc of a circle (1 centimetre in diameter), exerts a force of 1 dyne on unit magnetic pole at the centre of the circle. In e.m.u. the magnetic constant is of unit magnitude. The system has now been replaced by *SI units for most purposes. *Compare* electrostatic units; Gaussian units; Heaviside–Lorentz units.

electromagnetic wave *See* electromagnetic radiation; wave.

electrometallurgy The uses of electrical processes in the separation of metals from their ores, the refining of metals, or the forming or plating of metals.

electrometer A measuring instrument for determining a voltage difference without drawing an appreciable current from the source. Originally electrostatic instruments based on the electroscope, they are now usually based on operational amplifiers, solid-state devices with high input impedances. Electrometers are also used to measure low currents (nanoamperes), by passing the current through a high resistance.

electromotive force (e.m.f.) The greatest potential difference that can be generated by a particular source of electric current. In practice this may be observable only when the source is not supplying current, because of its *internal resistance.

electromotive series (electrochemical series) A series of chemical elements arranged in order of their *electrode potentials. The hydrogen electrode ($H^+ + e \rightarrow \frac{1}{2}H_2$) is taken as having zero electrode potential. Elements that have a greater tendency than hydrogen to lose electrons to their solution are taken as *electropositive*; those that gain electrons from their solution are below hydrogen in the series and are called *electronegative*. The series shows the order in which metals replace one another from their salts; electropositive metals will replace hydrogen from acids. The chief metals and hydrogen, placed in order in the series, are: potassium, calcium, sodium, magnesium, aluminium, zinc, cadmium, iron, nickel, tin, lead, hydrogen, copper, mercury, silver, platinum, gold.

electron An *elementary particle, classed as a *lepton, with a rest mass (symbol m_e) of 9.109 3897(54) × 10^{-31} kg and a negative charge of 1.602 177 33(49) × 10^{-19} coulomb. Electrons are present in all atoms in groupings called shells around the nucleus; when they are detached from the atom they are called *free electrons*. The antiparticle of the electron is the *positron*.

electron affinity Symbol A. The energy change occurring when an atom or molecule gains an electron to form a negative ion. For an atom or molecule X, it is the energy released for the electron-attachment reaction
$$X(g) + e \rightarrow X^-(g)$$
Often this is measured in electronvolts. Alternatively, the molar enthalpy change, ΔH, can be used.

electron biprism An arrangement of fields that splits a beam of electrons or other charged particles in an analogous way to an optical biprism.

electron capture 1. The formation of a negative ion by an atom or molecule when it acquires an extra free electron. **2.** A radioactive transformation in which a nucleus acquires an electron from an inner orbit of the atom, thereby transforming, initially, into a nucleus with the same mass number but an atomic number one less than that of the original nucleus (capture of the electron transforms a proton into a neutron). This type of capture is accompanied by emission of an X-ray photon or Auger electron as the vacancy in the inner orbit is filled by an outer electron.

electron-deficient compound A compound in which there are fewer electrons forming the chemical bonds than required in normal electron-pair bonds. *See* borane.

electron diffraction *Diffraction of a beam of electrons by atoms or molecules. The fact that electrons can be diffracted in a similar way to light and X-rays shows that particles can act as waves (*see* de Broglie wavelength). An electron (mass m, charge e) accelerated through a potential difference V acquires a kinetic energy $mv^2/2 = eV$, where v is the velocity of the electron. The (nonrelativistic) momentum (p) of the electron is $\sqrt{(2eVm)}$. The de Broglie wavelength

(λ) of an electron is given by h/p, where h is the Planck constant, thus $\lambda = h/\sqrt{(2eVm)}$. For an accelerating voltage of 3600 V, the wavelength of the electron beam is 0.02 nanometre, some 3×10^4 times shorter than visible radiation.

Electrons then, like X-rays, show diffraction effects with molecules and crystals in which the interatomic spacing is comparable to the wavelength of the beam. They have the advantage that their wavelength can be set by adjusting the voltage. Unlike X-rays they have very low penetrating power. The first observation of electron diffraction was by George Thomson (1892–1975) in 1927, in an experiment in which he passed a beam of electrons in a vacuum through a very thin gold foil onto a photographic plate. Concentric circles were produced by diffraction of electrons by the lattice. The same year Clinton J. Davisson (1881–1958) and Lester Germer (1896–1971) performed a classic experiment in which they obtained diffraction patterns by glancing an electron beam off the surface of a nickel crystal. Both experiments were important verifications of de Broglie's theory and the new quantum theory.

Electron diffraction, because of the low penetration, cannot easily be used to investigate crystal structure. It is, however, employed to measure bond lengths and angles of molecules in gases. Moreover, it is extensively used in the study of solid surfaces and absorption. The main techniques are low-energy electron diffraction (*LEED*) in which the electron beam is reflected onto a fluorescent screen, and high-energy electron diffraction (*HEED*) used either with reflection or transmission in investigating thin films.

electronegative Describing elements that tend to gain electrons and form negative ions. The halogens are typical electronegative elements. For example, in hydrogen chloride, the chlorine atom is more electronegative than the hydrogen and the molecule is polar, with negative charge on the chlorine atom. There are various ways of assigning values for the *electronegativity* of an element. *Mulliken electronegativities* are calculated from $E = (I + A)/2$, where I is ionization potential and A is electron affinity. More commonly, *Pauling electronegativities* are used. These are based on bond dissociation energies using a scale in which fluorine, the most electronegative element, has a value 4. Some other values on this scale are B 2, C 2.5, N 3.0, O 3.5, Si 1.8, P 2.1, S 2.5, Cl 3.0, Br 2.8.

electron gun A device used in *cathode-ray tubes (including television tubes), electron microscopes, etc., to produce a steady narrow beam of electrons. It usually consists of a heated cathode, control grid, and two or more annular anodes inserted in an evacuated tube. The electrons emitted by the cathode are attracted to the final anode, through which they pass. The intensity of the beam is regulated by the control grid and potential differences between the anodes create electric fields that focus the diverging electrons into a narrow beam.

electronics The study and design of control, communication, and computing devices that rely on the movement of electrons in circuits containing semiconductors, thermionic valves, resistors, capacitors, and inductors.

electron lens A device used to focus an electron beam. It is analogous to an optical lens but instead of using a refracting material, such as glass, it uses a coil or coils to produce a magnetic field or an arrangement of electrodes between which an electric field is created. Electron lenses are used in *electron microscopes and *cathode-ray tubes.

Principle of transmission
electron microscope

electron microscope A form of microscope that uses a beam of electrons instead of a beam of light (as in the optical microscope) to form a large image of a very small object. In optical microscopes the resolution is limited by the wavelength of the light. High-energy electrons, however, can be associated with a considerably shorter wavelength than light; for example, electrons accelerated to an energy of 10^5 electronvolts have a wavelength of 0.004 nanometre (*see* de Broglie wavelength) enabling a resolution of 0.2–0.5 nm to be achieved. The *transmission electron microscope* (see illustration) has an electron beam, sharply focused by *electron lenses, passing through a very thin metallized specimen (less than 50 nanometres thick) onto a fluorescent screen, where a visual image is formed. This image can be photographed. The *scanning electron microscope* can be used with thicker specimens and forms a perspective image, although the resolution and magnification are lower. In this type

of instrument a beam of primary electrons scans the specimen and those that are reflected, together with any secondary electrons emitted, are collected. This current is used to modulate a separate electron beam in a TV monitor, which scans the screen at the same frequency, consequently building up a picture of the specimen. The resolution is limited to about 10–20 nm. *See also* field-emission microscope; field-ionization microscope.

electron optics The study of the use of *electron lenses in the *electron microscope, *cathode-ray tubes, and other similar devices. The focusing of beams of positive or negative ions also relies on these methods.

electron probe microanalysis (EPM) A method of analysing a very small quantity of a substance (as little as 10^{-13} gram). The method consists of directing a very finely focused beam of electrons on to the sample to produce the characteristic X-ray spectrum of the elements present. It can be used quantitatively for elements with atomic numbers in excess of 11.

electron-spin resonance (ESR) A spectroscopic method of locating electrons within the molecules of a paramagnetic substance (*see* magnetism) in order to provide information regarding its bonds and structure. The spin of an unpaired electron is associated with a *magnetic moment that is able to align itself in one of two ways with an applied external magnetic field. These two alignments correspond to different *energy levels, with a statistical probability, at normal temperatures, that there will be slightly more in the lower state than in the higher. By applying microwave radiation to the sample a transition to the higher state can be achieved. The precise energy difference between the two states of an electron depends on the surrounding electrons in the atom or molecule. In this way the position of unpaired electrons can be investigated. The technique is used particularly in studying free radicals and paramagnetic substances such as inorganic complexes. *See also* nuclear magnetic resonance.

electron transport chain (respiratory chain) A sequence of biochemical reduction–oxidation reactions that forms the final stage of *aerobic respiration. It results in the transfer of electrons or hydrogen atoms derived from the *Krebs cycle to molecular oxygen, with the formation of water. At the same time it conserves energy from food or light in the form of *ATP. The chain comprises a series of electron carriers that undergo reversible reduction–oxidation reactions, accepting electrons and then donating them to the next carrier in the chain. In the mitochondria, NADH and $FADH_2$, generated by the Krebs cycle, transfer their electrons to a chain comprising flavin mononucleotide (FMN), *coenzyme Q, and a series of *cytochromes. This process is coupled to the formation of ATP at three sites along the chain (*see* oxidative phosphorylation). The ATP is then carried across the mitochondrial membrane in exchange for ADP. An electron transport chain also occurs in *photosynthesis.

electronvolt Symbol eV. A unit of energy equal to the work done on an electron in moving it through a potential difference of one volt. It is used as a measure of particle energies although it is not an *SI unit. 1 eV = 1.602×10^{-19} joule.

electroorganic reaction An organic reaction produced in an electrolytic cell. Electroorganic reactions are used to synthesize compounds that are difficult to produce by conventional techniques.

electrophile An ion or molecule that is electron deficient and can accept electrons. Electrophiles are often reducing agents and Lewis *acids.

They are either positive ions (e.g. NO_2^+) or molecules that have a positive charge on a particular atom (e.g. SO_3, which has an electron-deficient sulphur atom). In organic reactions they tend to attack negatively charged parts of a molecule. *Compare* nucleophile.

electrophilic addition An *addition reaction in which the first step is attack by an electrophile (e.g. a positive ion) on an electron-rich part of the molecule. An example is addition to the double bonds in alkenes.

electrophilic substitution A *substitution reaction in which the first step is attack by an electrophile. Electrophilic substitution is a feature of reactions of benzene (and its compounds) in which a positive ion approaches the delocalized pi electrons on the benzene ring.

electrophoresis (cataphoresis) A technique for the analysis and separation of colloids, based on the movement of charged colloidal particles in an electric field. There are various experimental methods. In one the sample is placed in a U-tube and a buffer solution added to each arm, so that there are sharp boundaries between buffer and sample. An electrode is placed in each arm, a voltage applied, and the motion of the boundaries under the influence of the field is observed. The rate of migration of the particles depends on the field, the charge on the particles, and on other factors, such as the size and shape of the particles. More simply, electrophoresis can be carried out using an adsorbent, such as a strip of filter paper, soaked in a buffer with two electrodes making contact. The sample is placed between the electrodes and a voltage applied. Different components of the mixture migrate at different rates, so the sample separates into zones. The components can be identified by the rate at which they move. This technique has also been called *electrochromatography*.

Electrophoresis is used extensively in studying mixtures of proteins, nucleic acids, carbohydrates, enzymes, etc. In clinical medicine it is used for determining the protein content of body fluids.

electrophorus An early form of *electrostatic generator. It consists of a flat dielectric plate and a metal plate with an insulated handle. The dielectric plate is charged by friction and the metal plate is placed on it and momentarily earthed, which leaves the metal plate with an induced charge of opposite polarity to that of the dielectric plate. The process can be repeated until all of the original charge has leaked away.

electroplating A method of plating one metal with another by *electrodeposition. The articles to be plated are made the cathode of an electrolytic cell and a rod or bar of the plating metal is made the anode. Electroplating is used for covering metal with a decorative, more expensive, or corrosion-resistant layer of another metal.

electropositive Describing elements that tend to lose electrons and form positive ions. The alkali metals are typical electropositive elements.

electroscope A device for detecting electric charge and for identifying its polarity. In the *gold-leaf electroscope* two rectangular gold leaves are attached to the end of a conducting rod held in an insulated frame. When a charge is applied to a plate attached to the other end of the conducting rod, the leaves move apart owing to the mutual repulsion of the like charges they have received.

electrostatic field The *electric field that surrounds a stationary charged body.

electrostatic generator A device used to build up electric charge to an

extreme potential usually for experimental purposes. The *electrophorus and the *Wimshurst machine were early examples; a more usual device now is the *Van de Graaff generator.

electrostatic precipitation A method of removing solid and liquid particles from suspension in a gas. The gas is exposed to an electric field so that the particles are attracted to and deposited on a suitably placed electrode. Electrostatic precipitation is widely used to remove dust and other pollutants from waste gases and from air. *See also* Cottrell precipitator.

electrostatics The study of electric charges at rest, the forces between them (*see* Coulomb's law), and the electric fields associated with them. *Compare* electrodynamics.

electrostatic units (e.s.u.) A system of electrical units in the *c.g.s. system. The e.s.u. of electric charge is the *statcoulomb* (all e.s.u. have the prefix *stat-* attached to the names of practical units). The statcoulomb is the quantity of electric charge that will repel an equal quantity 1 centimetre distant with a force of 1 dyne. In e.s.u. the electric constant is of unit magnitude. The system has now been replaced for most purposes by *SI units. *Compare* electromagnetic units, Gaussian units, Heaviside–Lorentz units.

electrostriction A change in the dimensions of a body as a result of reorientation of its molecules when it is placed in an electric field. If the field is not homogeneous the body will tend to move; if its relative permittivity is higher than that of its surroundings it will tend to move into a region of higher field strength. *Compare* magnetostriction.

electrovalent bond *See* chemical bond.

electroweak theory A *gauge theory (sometimes called *quantum flavourdynamics*, or *QFD*) that gives a unified description of the electromagnetic and weak interactions (*see* fundamental interactions). A successful electroweak theory was proposed in 1967 by Steven Weinberg and Abdul Salam, known as the *Weinberg–Salam model* or *WS model*. Because similar ideas were put forward by Sheldon Glashow, it is sometimes known as the *Glashow–Weinberg–Salam model* or *GWS model*. In this electroweak theory the gauge group is non-Abelian and the gauge symmetry is a *broken symmetry. The electroweak interaction is mediated by photons and by intermediate vector bosons, called the *W boson and the *Z boson. The observation of these particles in 1983/84, with their predicted energies, was a major success of the theory. The theory successfully accounts for existing data for electroweak processes and also predicts the existence of a heavy particle with spin 0, the *Higgs boson.

electrum 1. An alloy of gold and silver containing 55–88% of gold. **2.** A *German silver alloy containing 52% copper, 26% nickel, and 22% zinc.

element A substance that cannot be decomposed into simpler substances. In an element, all the atoms have the same number of protons or electrons, although the number of neutrons may vary. There are 92 naturally occurring elements. *See also* periodic table; transuranic element.

elementary particles The fundamental constituents of all the matter in the universe. Until the discovery of the electron by J. J. Thomson in 1897, it was assumed that atoms were the fundamental constituents of matter. This discovery, and Rutherford's discovery of the atomic nucleus and the proton in 1911, made it apparent that atoms were not themselves elementary, in the sense that they have an internal structure. Chadwick's discovery of the neutron in 1932 completed the atomic model based on an atomic nucleus consisting of protons and neutrons surrounded by sufficient electrons to balance the nuclear charge. It did not, however, explain the great stability of the nucleus, which clearly could not be held together by an electromagnetic interaction, as the neutron has no electric charge. In 1935 Yukawa suggested that the *exchange forces* that held them together involved short-lived particles, called *mesons*, which jumped from proton to neutron and back again. This concept led to the discovery of strong interactions, and weak interactions, bringing the total of *fundamental interactions to four. It also led to the discovery of some 200 short-lived 'elementary' particles, some of which were clearly more elementary than others. In the current classification two main classes of particles are recognized: *leptons (electron, muon, neutrinos, tau particle), which interact either by the electromagnetic interaction or the weak interaction and have no apparent internal structure; and *hadrons (nucleons, pions, etc.), which interact by the strong interaction and have a complex internal structure.

Hadron structure is currently based on Murray Gell-Mann's concept of the *quark*, introduced in 1964. In this model, hadrons are divided into *baryons* (which decay into protons) and *mesons* (which decay into leptons and photons). Baryons consist of three quarks and mesons consist of two quarks (a quark and antiquark). In the quark theory, therefore, the only truly elementary particles are leptons and quarks. Unlike electrons and protons, which have exactly equal but opposite charges, quarks have charges that are fractions of the electronic charge ($+2/3$ or $-1/3$ of the electronic charge). Quarks occur in six *flavours* (no connection with taste): up (u; $+2/3$ charge), down (d; $-1/3$), charmed (c; $+2/3$), strange (s; $-1/3$), top (t; $+2/3$), and bottom (b; −1/3). The evidence for the existence of top is not totally established. The proton, being a baryon, consists of three quarks uud (2/3 + 2/3 − 1/3 = 1) and the neutron consists of udd (2/3 − 1/3 − 1/3 = 0). For each flavour there are equivalent antiquarks (ū, d̄, etc.), which have opposite values of electric charge (ū = −2/3, etc.).

In order to avoid conflict with the *Pauli exclusion principle it has proved necessary to add the concept of *colour charge* to the six flavours. Each flavour of quark occurs in the three primary colours red, green, and blue; each antiquark has the complementary anticolours cyan, magenta, and yellow. There are therefore a total of 18 quarks and 18 antiquarks. The rules of combination to form hadrons are that the combinations of colours should always give white, either by mixing three primary colours to make baryons or by mixing one primary colour with its complementary anticolour to make mesons. The use of the word 'colour' in this context implies an analogy to visual colours, not that the particles are coloured. The theory governing these combinations is modelled on quantum electrodynamics and is called *quantum chromodynamics.

The strong interaction between quarks is visualized as occurring by the exchange of eight types of chargeless particles, with no rest mass, called *gluons* (because they glue the quarks together). Although gluons, like the photons that perform an analogous function between leptons, have no electrical charge, they do have a colour charge. Each gluon carries one colour and one anticolour. There are nine colour–anticolour pairs but one is excluded as it is equivalent to white. In an interaction a quark can change its colour, but all colour changes are accompanied by the emission of a gluon. This gluon is then absorbed by another quark, whose colour changes to compensate for the original change. For example, if a red quark changes to green it emits a gluon bearing the colours red + antigreen. When this is absorbed by a green quark the green of the quark and the antigreen of the gluon annihilate each other leaving the second quark with the colour red, acquired from the gluon. The net change is nil, because as a result of the interaction there is still one green quark and one red quark. All hadrons therefore remain white, even though quark colours move from point to point. The strong force can be envisaged as the system of interactions needed to maintain this condition.

The whole elaborate quark theory is now well established by circumstantial evidence, but as neither quarks nor gluons have ever been identified in experiments the theory does not claim to have been directly verified. Individual quarks would have the curious property of being much more massive than the hadrons they usually form (because of the enormous potential energy they would have when separated) and some theorists believe it is in consequence fundamentally impossible for them to exist in isolation. However some experimenters have reported results consistent with the presence of the fractional charges that unattached quarks would have. *See also* grand unified theory; superstring theory; supersymmetry.

elements of an orbit Six parameters that can be used to define the path of a celestial body. The shape of the orbit is defined by its eccentricity (*see* conic) and semimajor axis. The orientation of the orbit is specified by the *inclination of the orbital plane to the reference plane (usually the *ecliptic) and by the longitude of the ascending *node (the angular distance from the vernal equinox to the ascending node). The position of the body in its orbit is defined by its eccentric *anomaly and the position as a function of time is calculated from the time of periapsis passage (*see* apsides).

elevation of boiling point An increase in the boiling point of a liquid when a solid is dissolved in it. The elevation is proportional to the number of particles dissolved (molecules or ions) and is given by $\Delta t = k_B C$, where C is the molal concentration of solute. The constant k_B is the *ebullioscopic constant* of the solvent and if this is known, the molecular weight of the solute can be calculated from the measured value of Δt. The elevation is measured by a Beckmann thermometer. *See also* colligative property.

elimination reaction A reaction in which a molecule decomposes to two molecules, one smaller than the other.

Elinvar Trade name for a nickel–chromium steel containing about 36% nickel, 12% chromium, and smaller proportions of tungsten and manganese. Its elasticity does not vary with temperature and it is therefore used to make hairsprings for watches.

ELISA (enzyme-linked immunosorbent assay) A sensitive technique (*see* immunoassay) for accurately determining the amount of protein or other antigen in a given sample by means of an enzyme-catalysed colour change. Antibody specific to the test protein is adsorbed onto a solid substrate, such as a PVC sheet, and a measured amount of the sample is added; all molecules of the test protein in the sample are bound by the antibody. A second antibody specific for a second site on the test protein is added; this is conjugated with an enzyme; which catalyses a colour change in the fourth reagent, added finally to the sheet. The colour change can be measured photometrically and compared against a standard curve to give the concentration of protein in the sample. ELISA is widely used for diagnostic and other purposes.

An ellipse

ellipse A *conic formed by the intersection of a plane with a right circular cone, so that the plane is inclined to the axis of the cone at an angle in excess of half the apex angle of the cone. The ellipse has two vertices, which are joined by a line called the *major axis*. The centre of the ellipse falls on this line, midway between the vertices. The *minor axis* is the line perpendicular to the major axis that passes through the centre and joins two points on the ellipse. The *foci* of an ellipse are two points on the major axis so placed that for any point on the ellipse the sum of the distances from that point to each focus is constant. (See illustration.) The area of an ellipse is πab, where a and b are half the major and minor axes, respectively. For an ellipse centred at the origin, the equation in Cartesian coordinates is $x^2/a^2 + y^2/b^2 = 1$. The foci are at $(ea, 0)$ and $(-ea, 0)$, where e is the eccentricity. Each of the two chords of the ellipse passing through a focus and parallel to the minor axis is called a *latus rectum* and has a length equal to $2b^2/a$.

ellipsoid A solid body formed when an *ellipse is rotated about an axis. If it is rotated about its major axis it is a *prolate ellipsoid*; if it is rotated about its minor axis it is an *oblate ellipsoid*. For an ellipsoid centred at the origin the equation in Cartesian coordinates is:

$$x^2/a^2 + y^2/b^2 + z^2/c^2 = 1.$$

uterus wall

fallopian tube

yolk sac

placenta

blood vessel

umbilical cord

amnion

embryo

chorion

amniotic fluid

vagina

A developing human embryo

elliptical galaxy *See* galaxy.

elliptical polarization *See* polarization of light.

eluate *See* chromatography; elution.

eluent *See* chromatography; elution.

elution The process of removing an adsorbed material (*adsorbate*) from an adsorbent by washing it in a liquid (*eluent*). The solution consisting of the adsorbate dissolved in the eluent is the *eluate*. Elution is the process used to wash components of a mixture through a *chromatography column.

elutriation The process of suspending finely divided particles in an upward flowing stream of air or water to wash and separate them into sized fractions.

elytra The thickened horny forewings of the *Coleoptera (beetles), which cover and protect the membranous hindwings when the insect is at rest.

emanation The former name for the gas radon, of which there are three isotopes: Rn-222 (radium emanation), Rn-220 (thoron emanation), and Rn-219 (actinium emanation).

emasculation The removal of the anthers of a flower in order to prevent self-pollination or the undesirable pollination of neighbouring plants.

Embden–Meyerhof pathway *See* glycolysis.

embryo 1. An animal in the earliest stages of its development, from the time when the fertilized ovum starts to divide (*see* cleavage), while it is contained within the egg or reproductive organs of the mother, until hatching or birth. A human embryo (see illustration) is called a *fetus after the first eight weeks of pregnancy. **2.** The structure in bryophytes, pteridophytes, and seed plants that develops from the zygote prior to germination. In seed plants the zygote is situated in the *embryo sac of the ovule. It divides by mitosis to form the embryonic cell and a structure called the *suspensor*, which embeds the embryo in the surrounding nutritive tissue. The embryonic cell divides continuously and eventually gives rise to the *radicle (young root), *plumule (young shoot), and one or two *cotyledons (seed leaves). Changes also take place in the surrounding tissues of the ovule, which becomes the *seed enclosing the embryo plant.

embryology The study of the development of animals from the fertilized egg to the new adult organism. It is sometimes limited to the period between fertilization of the egg and hatching or birth (see embryo).

embryo sac A large cell that develops in the *ovule of flowering plants. It is equivalent to the female *gametophyte of lower plants, although it is very much reduced. Typically, it contains eight nuclei formed by division of the original female gamete. One, the *oosphere (egg nucleus), is fertilized by a male nucleus and becomes the *embryo. Two of the remaining nuclei fuse with a second male nucleus to form a triploid nucleus that gives rise to the *endosperm.

emerald The green gem variety of *beryl: one of the most highly prized gemstones. The finest specimens occur in the Muzo mines, Colombia. Other occurrences include the Ural Mountains, the Transvaal in South Africa, and Kaligunan in India. Emeralds can also be successfully synthesized.

emery A rock composed of corundum (natural aluminium oxide, Al_2O_3) with magnetite, haematite, or spinel. It occurs on the island of Naxos (Greece) and in Turkey. Emery is used as an abrasive and polishing material and in the manufacture of certain concrete floors.

e.m.f. See electromotive force.

emission spectrum See spectrum.

emissivity Symbol ε. The ratio of the power per unit area radiated by a surface to that radiated by a *black body at the same temperature. A black body therefore has an emissivity of 1 and a perfect reflector has an emissivity of 0. The emissivity of a surface is equal to its *absorptance.

emittance See exitance.

emitter See transistor.

empirical Denoting a result that is obtained by experiment or observation rather than from theory.

empirical formula See formula.

emulsion A *colloid in which small particles of one liquid are dispersed in another liquid. Usually emulsions involve a dispersion of water in an oil or a dispersion of oil in water, and are stabilized by an *emulsifier*. Commonly emulsifiers are substances, such as *detergents, that have lyophobic and lyophilic parts in their molecules. Dietary fats are reduced to an emulsion in the duodenum to facilitate their subsequent digestion.

enamel The material that forms a covering over the crown of a *tooth (i.e. the part that projects above the gum). Enamel is smooth, white, and extremely hard, being rich in minerals containing calcium. It is produced by certain cells (*ameloblasts*) of the oral epithelium and protects the underlying dentine of the tooth. Enamel may also be found in the placoid *scales of certain fish, which demonstrates the common developmental origin of scales and teeth.

enantiomers See optical activity.

enantiomorphism See optical activity.

enantiotropy See allotropy.

encephalin (enkephalin) A peptide *neurotransmitter found principally in the brain. Two encephalins have so far been isolated. They bind to opiate receptors in the brain and their release is thought to control levels of pain and other sensations. Encephalins have also been found in spinal cord nerve cells and in gut epithelial cells. Compare endorphin.

endocarp See pericarp.

endocrine gland (ductless gland) Any gland in an animal that manufactures *hormones and secretes them directly into the bloodstream to act at distant sites in the body (known as *target organs* or *cells*). Endocrine glands tend to control slow long-term activities in the body, such as growth and sexual development. In mammals they include the *pituitary, *adrenal, *thyroid, and *parathyroid glands, the *ovary and *testis, the *placenta, and part of the pancreas (*see* islets of Langerhans). The activity of endocrine glands is controlled by negative feedback, i.e. a rise in output of hormone inhibits a further increase in its production, either directly or indirectly via the target organ or cell. *See also* neuroendocrine system. *Compare* exocrine gland.

endocrinology The study of the structure and functions of the *endocrine glands and of the *hormones they produce.

endoderm (entoderm) The internal layer of cells of the *gastrula, which will develop into the alimentary canal (gut) and digestive glands of the adult. *See also* germ layers.

endodermis The innermost layer of the root *cortex of a plant, lying immediately outside the vascular tissue. Various modifications of the endodermal cell walls indicate that they regulate the passage of materials both into and out of the vascular system. An endodermis may also be seen in the stems of some plants.

endoergic Denoting a nuclear process that absorbs energy. *Compare* exoergic.

endogamy The fusion of reproductive cells from closely related parents, i.e. *inbreeding. *Compare* exogamy.

endogenous Describing a substance, stimulus, organ, etc., that originates from within an organism. For example, growth rhythms not directed by environmental stimuli are termed endogenous rhythms. Lateral roots, which always grow from inside the main root rather than from its

surface, are said to arise endogenously. *Compare* exogenous.

endolymph The fluid that fills the membranous labyrinth of the vertebrate *inner ear. *See* cochlea; semicircular canals. *Compare* perilymph.

endonuclease An enzyme that catalyses the internal cleavage of nucleic acids. *See also* restriction enzyme. *Compare* exonuclease.

endoparasite A parasite that lives inside its host's body. *See* parasitism.

endoplasm *See* cytoplasm.

endoplasmic reticulum (ER) A system of membranes within the cytoplasm of plant and animal *cells. It forms a link between the cell and nuclear membranes and is the site of protein synthesis. It is also concerned with the transport of proteins and lipids within the cell. *Rough ER* has *ribosomes attached to its surface; in *smooth ER* ribosomes are absent.

end organ The structure at the end of a peripheral nerve. Examples of end organs are the muscle *end plate at the end of a motor neurone and the *receptor at the end of a sensory neurone.

endorphin A peptide neurotransmitter found in the pituitary gland and hypothalamus that has pain-relieving properties similar to the *encephalins.

endoskeleton A supporting framework that lies entirely within the body of an animal, such as the bony *skeleton of vertebrates or the spicules of a sponge. The function of an endoskeleton is to support the body and in vertebrates it also protects the organs and provides a system of levers on which the muscles can act to produce movement. *Compare* exoskeleton.

endosperm A nutritive tissue, characteristic of flowering plants, that surrounds the developing embryo in a seed. It develops from nuclei in the *embryo sac and its cells are triploid. In *endospermic* seeds it remains and increases in size; in *nonendospermic* seeds it disappears as the food is absorbed by the embryo, particularly the *cotyledons. Many plants with endospermic seeds, such as cereals and oil crops, are cultivated for the rich food reserves in the endosperm.

endospore The resting stage of certain bacteria, formed in response to adverse conditions. The bacterial cell becomes enclosed in a tough resistant spore coat. On return to favourable conditions the spore changes back to the normal vegetative form of the organism.

endothelium A single layer of thin platelike cells that line the inner surfaces of blood and lymph vessels and the heart. Endothelium is derived from the *mesoderm. *Compare* epithelium; mesothelium.

endothermic Denoting a chemical reaction that takes heat from its surroundings. *Compare* exothermic.

endotoxin *See* toxin.

end plate The area of muscle cell membrane that lies immediately beneath a motor nerve ending. Release of a *neurotransmitter at the end plate induces contraction of the muscle fibre.

end point The point in a titration at which reaction is complete as shown by the *indicator.

energy A measure of a system's ability to do work. Like work itself, it is measured in joules. Energy is conveniently classified into two forms: *potential energy* is the energy stored in a body or system as a consequence of its position, shape, or state (this includes gravitational energy, electrical energy, nuclear energy, and chemical energy); *kinetic energy* is energy of motion and is usually defined as the work that will be done by the body possessing the energy when it is brought to rest. For a body of mass m having a speed v, the kinetic energy is $mv^2/2$ (classical) or $(m - m_0)c^2$ (relativistic). The rotational kinetic energy of a body having an angular velocity ω is $I\omega^2/2$, where I is its moment of inertia.

The *internal energy of a body is the sum of the potential energy and the kinetic energy of its component atoms and molecules.

energy band A range of energies that electrons can have in a solid. In a single atom, electrons exist in discrete *energy levels. In a crystal, in which large numbers of atoms are held closely together in a lattice, electrons are influenced by a number of adjacent nuclei and the sharply defined levels of the atoms become bands of allowed energy (see illustration); this approach to energy levels in solids is often known as the *band theory*. Each band represents a large number of allowed quantum states. Between the bands are *forbidden bands*. The outermost electrons of the atoms (i.e. the ones responsible for chemical bonding) form the *valence band* of the solid. This is the band, of those occupied, that has the highest energy.

The band structure of solids accounts for their electrical properties. In order to move through the solid, the electrons have to change from one quantum state to another. This can only occur if there are empty quantum states with the same energy. In general, if the valence band is full, electrons cannot change to new quantum states in the same band. For conduction to occur, the electrons have to be in an unfilled band – the *conduction band*. Metals are good conductors either because the valence band and the conduction band are only half-filled or because the conduction band overlaps with the valence band; in either case vacant states are available. In insulators the conduction band and valence band are separated by a wide forbidden band and electrons do

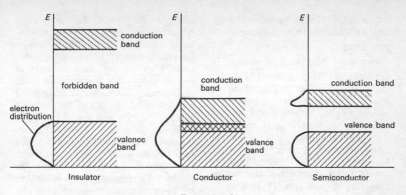

Energy bands

not have enough energy to 'jump' from one to the other.

In intrinsic *semiconductors the forbidden gap is narrow and, at normal temperatures, electrons at the top of the valence band can move by thermal agitation into the conduction band (at absolute zero, a semiconductor would act as an insulator). Doped semiconductors have extra bands in the forbidden gap.

energy level A definite fixed energy that a system described by *quantum mechanics, such as a molecule, atom, electron, or nucleus, can have. In an atom, for example, the atom has a fixed energy corresponding to the *orbitals in which its electrons move around the nucleus. The atom can accept a quantum of energy to become an excited atom (see excitation) if that extra energy will raise an electron to a permitted orbital. Between the *ground state*, which is the lowest possible energy level for a particular system, and the first excited state there are no permissible energy levels. According to the *quantum theory, only certain energy levels are possible. An atom passes from one energy level to the next without passing through fractions of that energy transition. These levels are usually described by the energies associated

with the individual electrons in the atoms, which are always lower than an arbitrary level for a free electron. The energy levels of molecules also involve quantized vibrational and rotational motion.

Engel's salt *See* potassium carbonate.

engine Any device for converting some forms of energy by mechanical work. *See* heat engine; Carnot cycle; internal-combustion engine; steam engine.

enols Compounds containing the group $-CH=C(OH)-$ in their molecules. *See also* keto–enol tautomerism.

enrichment The process of increasing the abundance of a specified isotope in a mixture of isotopes. It is usually applied to an increase in the proportion of $U-235$, or the addition of Pu -239 to natural uranium for use in a nuclear reactor or weapon.

enthalpy Symbol H. A thermodynamic property of a system defined by $H = U + pV$, where H is the enthalpy, U is the internal energy of the system, p its pressure, and V its volume. In a chemical reaction carried out in the atmosphere the pressure remains constant and the enthalpy of reaction, ΔH, is equal to

239

$\Delta U + p\Delta V$. For an exothermic reaction ΔH is taken to be negative.

entoderm *See* endoderm.

entomology The study of insects.

entomophily Pollination of a flower in which the pollen is carried on an insect. Entomophilous flowers are usually brightly coloured and scented and often secrete nectar. In some species (e.g. primulas) there are structural differences between the flowers to ensure that cross-pollination occurs. Other examples of entomophilous flowers are orchids and antirrhinums. *Compare* anemophily; hydrophily.

entropy Symbol S. A measure of the unavailability of a system's energy to do work; in a closed system an increase in entropy is accompanied by a decrease in energy availability. When a system undergoes a reversible change the entropy (S) changes by an amount equal to the energy (Q) transferred to the system by heat divided by the thermodynamic temperature (T) at which this occurs, i.e. $\Delta S = \Delta Q/T$. However, all real processes are to a certain extent irreversible changes and in any closed system an irreversible change is always accompanied by an increase in entropy.

In a wider sense entropy can be interpreted as a measure of disorder; the higher the entropy the greater the disorder. As any real change to a closed system tends towards higher entropy, and therefore higher disorder, it follows that the entropy of the universe (if it can be considered a closed system) is increasing and its available energy is decreasing (*see* heat death of the universe). This increase in the entropy of the universe is one way of stating the second law of *thermodynamics.

environment (in ecology) The physical, chemical, and biological conditions of the region in which an organism lives. *See also* ecology; ecosystem.

enzyme A protein that acts as a *catalyst in biochemical reactions. Each enzyme is specific to a particular reaction or group of similar reactions. Many require the association of certain nonprotein *cofactors in order to function. The molecule undergoing reaction (the *substrate*) binds to a specific *active site on the enzyme molecule to form a short-lived intermediate: this greatly increases (by a factor of up to 10^{20}) the rate at which the reaction proceeds to form the product. Enzyme activity is influenced by substrate concentration and by temperature and pH, which must lie within a certain range. Other molecules may compete for the active site, causing *inhibition of the enzyme or even irreversible destruction of its catalytic properties.

Enzyme production is governed by a cell's genes. Enzyme activity is further controlled by pH changes, alterations in the concentrations of essential cofactors, feedback inhibition by the products of the reaction, and activation by another enzyme, either from a less active form or an inactive precursor (*zymogen). Such changes may themselves be under the control of hormones or the nervous system.

The names of most enzymes end in *-ase*, which is added to the names of the substrates on which they act. Thus *lactase* is the enzyme that acts to break down lactose.

enzyme-linked immunosorbent assay *See* ELISA.

Eocene The second geological epoch of the *Tertiary period. It extended from the end of the Palaeocene epoch, about 54 million years ago, to the beginning of the Oligocene epoch, about 38 million years ago. The term was first proposed by Sir Charles Lyell (1797–1875) in 1833. In some classifications of geological time the *Palaeocene is included as part of the Eocene. Mammals were dominant in the Eocene: rodents, artiodactyls, carnivores, perissodactyls (including early horses), and whales were among the groups to make their first appearance.

eosin One of a series of acidic dyes, used in optical microscopy, that colours cytoplasm pink and cellulose red. It is frequently used as a counterstain with *haematoxylin for colouring tissue smears and sections of animal tissue.

ephemeral 1. (in botany) An *annual plant that completes its life cycle in considerably less than one growing season. A number of generations can therefore occur in one year. Many troublesome weeds, such as groundsel and willowherb, are ephemerals. Certain desert plants are also ephemerals, completing their life cycles in a short period following rain. **2.** (in zoology) A short-lived animal, such as a mayfly.

ephemeris A tabulation showing the calculated future positions of the sun, moon, and planets, together with other useful information for astronomers and navigators. It is published at regular intervals.

ephemeris time (ET) A time system that has a constant uniform rate as opposed to other systems that depend on the earth's rate of rotation, which has inherent irregularities. It is reckoned from an instant in 1900 (Jan 0d 12h) when the sun's mean longitude was 279.696 677 8°. The unit by which ephemeris time is measured is the tropical year, which contains 31 556 925.9747 *ephemeris seconds*. This fundamental definition of the *second was replaced in 1964 by the caesium second of atomic time.

epicalyx A ring of bracts below a flower that resembles a calyx. It is seen, for example, in the strawberry flower.

epicarp *See* pericarp.

epicentre The point on the surface of the earth directly above the focus of an earthquake or directly above or below a nuclear explosion.

epicotyl The region of a seedling stem above the stalks of the seed leaves (*cotyledons) of an embryo plant. It grows rapidly in seeds showing *hypogeal germination and lifts the stem above the soil surface. *Compare* hypocotyl.

epicycle A small circle whose centre rolls around the circumference of a larger fixed circle. The curve traced out by a point on the epicycle is called an *epicycloid*.

epidermis 1. (in zoology) The outermost layer of cells of the body of an animal. In invertebrates the epidermis is normally only one cell thick and is covered by an impermeable *cuticle. In vertebrates the epidermis is the thinner of the two layers of *skin (*compare* dermis). It consists of a basal layer of actively dividing cells (*see* Malpighian layer), covered by layers of cells that become impregnated with keratin (*see* keratinization). The outermost layers of epidermal cells (the *stratum corneum) form a water-resistant protective layer. The epidermis may bear a variety of specialized structures (e.g. *feathers, *hairs). **2.** (in botany) The outermost layer of cells covering a plant. It is overlaid by a *cuticle and its functions are principally to protect the plant from injury and to reduce water loss. Some epidermal cells are modified to form guard cells (*see* stoma) or hairs of various types (*see* piliferous layer). In woody plants the functions of the shoot epidermis are taken over by the periderm tissues (*see* cork cambium) and in mature roots the epidermis is sloughed off and replaced by the *hypodermis.

epidiascope An optical instrument used by lecturers, etc., for projecting an enlarged image of either a translucent object (such as a slide or transparency) or an opaque object (such as a diagram or printed page) onto a screen.

epididymis A long coiled tube in which spermatozoa are stored in vertebrates. In reptiles, birds, and mammals it is attached at one end to the *testis and opens into the sperm duct (*vas deferens) at the other.

epigamic Serving to attract a mate. Epigamic characters include the bright plumage of some male birds.

epigeal Describing seed germination in which the seed leaves (cotyledons) emerge from the ground and function as true leaves. Examples of epigeal germination are seen in sycamore and sunflower. *Compare* hypogeal.

epiglottis A flexible flap of cartilage in mammals that is attached to the wall of the pharynx near the base of the tongue. During swallowing (*see* deglutition) it covers the *glottis (the opening to the respiratory tract), so preventing food from entering the trachea (windpipe).

epimerism A type of optical isomerism in which a molecule has two chiral centres; two optical isomers (*epimers*) differ in the arrangement about one of these centres. *See also* optical activity.

epinephrine *See* adrenaline.

epiphysis The terminal section of a growing bone (especially a long limb bone). It is separated from the bone shaft (*diaphysis*) by cartilage. New bone is produced on the side of the cartilage facing the diaphysis, while new cartilage is produced on the other side of the cartilage disc. When the bone reaches adult length the epiphysis merges with the diaphysis.

epiphyte A plant that grows upon another plant but is neither parasitic on it nor rooted in the ground. Epiphytes include many mosses and lichens and some tropical orchids.

episome A genetic element that can exist and replicate either independently of its host cell's chromosomes

or as an integrated part of the chromosomes. Examples include certain bacterial *plasmids.

epitaxy (epitaxial growth) Growth of a layer of one substance on a single crystal of another, such that the crystal structure in the layer is the same as that in the substrate. It is used in making semiconductor devices.

epithelium A tissue in vertebrates consisting of closely packed cells in a sheet with little intercellular material. It forms a membrane over the outer surfaces of the body and walls of the internal cavities (coeloms). It also forms glands and parts of sense organs. Its functions are protective, absorptive, secretory, and sensory. The types of cell vary, giving rise to *squamous, cuboidal, columnar,* and *ciliated epithelia*. Epithelium is derived from *ectoderm and *endoderm. *Compare* endothelium; mesothelium.

epithermal neutron A neutron with an energy in excess of that associated with a thermal neutron (*see* moderator) but less than that of a *fast neutron, i.e. a neutron having an energy in the range 0.1 to 100 eV.

EPM *See* electron probe microanalysis.

$$R_1 \diagdown C - C \diagup R_3$$
$$R_2 \diagup \diagdown O \diagdown \diagup R_4$$

The functional group in epoxides

epoxides Compounds that contain oxygen atoms in their molecules as part of a three-membered ring (see formula). Epoxides are thus *cyclic ethers*.

epoxyethane (ethylene oxide) A colourless flammable gas, C_2H_4O; m.p. $-111°C$; b.p. $11°C$. It is a cyclic ether (*see* epoxides), made by the catalytic oxidation of ethene. It can be hydrolysed to ethane-1,2-diol and also polymerizes to:

... $-O-C_2H_4-O-C_2H_4-$ ..., which is used for lowering the viscosity of water (e.g. in fire fighting).

epoxy resins Synthetic resins produced by copolymerizing epoxide compounds with phenols. They contain $-O-$ linkages and epoxide groups and are usually viscous liquids. They can be hardened by addition of agents, such as polyamines, that form cross-linkages. Alternatively, catalysts may be used to induce further polymerization of the resin. Epoxy resins are used in electrical equipment and in the chemical industry (because of resistance to chemical attack). They are also used as adhesives.

epsomite A mineral form of *magnesium sulphate heptahydrate, $MgSO_4$.$7H_2O$.

Epsom salt *See* magnesium sulphate.

Epstein–Barr virus *See* herpesvirus.

equation of motion (kinematic equation) Any of four equations that apply to bodies moving linearly with uniform acceleration (a). The equations, which relate distance covered (s) to the time taken (t), are:

$$v = u + at$$
$$s = (u + v)t/2$$
$$s = ut + at^2/2$$
$$v^2 = u^2 + 2as$$

where u is the initial velocity of the body and v is its final velocity.

equation of state An equation that relates the pressure p, volume V, and thermodynamic temperature T of an amount of substance n. The simplest is the ideal *gas law:

$$pV = nRT,$$

where R is the universal gas constant. Applying only to ideal gases, this equation takes no account of the volume occupied by the gas molecules (according to this law if the pressure is infinitely great the volume becomes zero), nor does it take into account any forces between molecules. A more

accurate equation of state would therefore be

$$(p + k)(V - nb) = nRT,$$

where k is a factor that reflects the decreased pressure on the walls of the container as a result of the attractive forces between particles, and nb is the volume occupied by the particles themselves when the pressure is infinitely high. In the *van der Waals equation of state*, proposed by the Dutch physicist J. D. van der Waals (1837–1923),

$$k = n^2a/V^2,$$

where a is a constant. This equation more accurately reflects the behaviour of real gases; several others have done better but are more complicated.

equation of time The length of time that must be added to the mean solar time, as shown on a clock, to give the apparent solar time, as shown by a sundial. The amount varies during the year, being a minimum of -14.2 minutes in February and a maximum of $+16.4$ minutes in October. It is zero on four days (April 15/16, June 14/15, Sept. 1/2, Dec. 25/26). The difference arises as a result of two factors: the eccentricity of the earth's orbit and the inclination of the ecliptic to the celestial equator.

equator 1. The great circle around the earth that lies in a plane perpendicular to the earth's axis. It is equidistant from the two geographical poles. 2. The *magnetic equator* is a line of zero magnetic dip (*see* geomagnetism) that is close to the geographical equator but lies north of it in Africa and south of it in America. 3. The *celestial equator* is the circle formed on the *celestial sphere by the extension of the earth's equatorial plane.

equilibrium A state in which a system has its energy distributed in the statistically most probable manner; a state of a system in which forces, influences, reactions, etc., balance

equilibrium constant

each other out so that there is no net change.

A body is in *static equilibrium* if the resultants of all forces and all couples acting on it are both zero; it may be at rest and will certainly not be accelerated. Such a body at rest is in *stable equilibrium* if after a slight displacement it returns to its original position – for a body whose weight is the only downward force this will be the case if the vertical line through its centre of gravity always passes through its base. If a slight displacement causes the body to move to a new position, then the body is in *unstable equilibrium*.

A body is said to be in *thermal equilibrium* if no net heat exchange is taking place within it or between it and its surroundings. A system is in *chemical equilibrium when a reaction and its reverse are proceeding at equal rates (*see also* equilibrium constant). These are examples of *dynamic equilibrium*, in which activity in one sense or direction is in aggregate balanced by comparable reverse activity.

equilibrium constant For a reversible reaction of the type

$$xA + yB \rightleftharpoons zC + wD$$

chemical equilibrium occurs when the rate of the forward reaction equals the rate of the back reaction, so that the concentrations of products and reactants reach steady-state values. It can be shown that at equilibrium the ratio of concentrations

$$[C]^z[D]^w/[A]^x[B]^y$$

is a constant for a given reaction and fixed temperature, called the equilibrium constant K_c (where the c indicates concentrations have been used). Note that, by convention, the products on the right-hand side of the reaction are used on the top line of the expression for equilibrium constant. This form of the equilibrium constant was originally introduced in 1863 by C. M. Guldberg and P.

Waage using the law of *mass action. They derived the expression by taking the rate of the forward reaction

$$k_f[A]^x[B]^y$$

and that of the back reaction

$$k_b[C]^z[D]^w$$

Since the two rates are equal at equilibrium, the equilibrium constant K_c is the ratio of the rate constants k_f/k_b. The principle that the expression is a constant is known as the *equilibrium law* or *law of chemical equilibrium*.

The equilibrium constant shows the *position* of equilibrium. A low value of K_c indicates that [C] and [D] are small compared to [A] and [B]; i.e. that the back reaction predominates. It also indicates how the equilibrium shifts if concentration changes. For example, if [A] is increased (by adding A) the equilibrium shifts towards the right so that [C] and [D] increase, and K_c remains constant.

For gas reactions, partial pressures are used rather than concentrations. The symbol K_p is then used. Thus, in the example above

$$K_p = p_C^z p_D^w / p_A^x p_B^y$$

It can be shown that, for a given reaction $K_p = K_c(RT)^{\Delta\nu}$, where $\Delta\nu$ is the difference in stoichiometric coefficients for the reaction (i.e. $z + w - x - y$). Note that the units of K_p and K_c depend on the numbers of molecules appearing in the stoichiometric equation. The value of the equilibrium constant depends on the temperature. If the forward reaction is exothermic, the equilibrium constant decreases as the temperature rises; if endothermic it increases (*see also* van't Hoff's isochore).

The expression for the equilibrium constant can also be obtained by thermodynamics; it can be shown that the standard equilibrium constant $K^\ominus$ is given by $\exp(-\Delta G^\ominus/RT)$, where $\Delta G^\ominus$ is the standard Gibbs free energy change for the complete reaction. Strictly, the expressions above for equilibrium constants are true only for ideal gases (pressure) or infinite dilution (concentration). For accurate work *activities are used.

equilibrium law See equilibrium constant.

equinox 1. Either of the two points on the *celestial sphere at which the *ecliptic intersects the celestial equator. The sun appears to cross the celestial equator from south to north at the *vernal equinox* and from north to south at the *autumnal equinox*. **2.** Either of the two instants at which the centre of the sun appears to cross the celestial equator. In the northern hemisphere the vernal equinox occurs on or about March 21 and the autumnal equinox on or about Sept. 23. In the southern hemisphere the dates are reversed. See precession of the equinoxes.

equipartition of energy The theory, proposed by Ludwig Boltzmann (1844 –1906) and given some theoretical support by James Clerk Maxwell (1831–79), that the energy of gas molecules in a large sample under thermal *equilibrium is equally divided among their available *degrees of freedom, the average energy for each degree of freedom being $kT/2$, where k is the *Boltzmann constant and T is the thermodynamic temperature. The proposition is not generally true if *quantum considerations are important, but is frequently a good approximation.

equivalence point The point in a titration at which reaction is complete. See indicator.

equivalent proportions See chemical combination.

equivalent weight The mass of an element or compound that could combine with or displace one gram of hydrogen (or eight grams of oxygen or 35.5 grams of chlorine) in a chemical reaction. The equivalent weight represents the 'combining power' of

the substance. For an element it is the relative atomic mass divided by the valency. For a compound it depends on the reaction considered.

erbium Symbol Er. A soft silvery metallic element belonging to the *lanthanoids; a.n. 68; r.a.m. 167.26; r.d. 9.066 (20°C); m.p. 1529°C; b.p. 2868°C. It occurs in apatite, gadolinite, and xenotine from certain sources. There are six natural isotopes, which are stable, and twelve artificial isotopes are known. It has been used in alloys for nuclear technology as it is a neutron absorber; it is being investigated for other potential uses. It was discovered by C. G. Mosander in 1843.

erecting prism A glass prism used in optical instruments to convert an inverted image into an erect image, as in prismatic binoculars.

erepsin An obsolete name for a mixture of protein-degrading enzymes secreted by the small intestine.

erg A unit of work or energy used in the c.g.s. system and defined as the work done by a force of 1 dyne when it acts through a distance of 1 centimetre. 1 erg = 10^{-7} joule.

ergocalciferol See vitamin D.

ergonomics The study of the engineering aspects of the relationship between workers and their working environment.

ergosterol A *sterol occurring in fungi, bacteria, algae, and higher plants. It is converted into vitamin D_2 by the action of ultraviolet light.

erosion The wearing away of the land surface by natural agents that involves the transport of rock debris. These natural agents include moving waters (e.g. rivers, ocean waves), ice (e.g. glaciers), wind, organisms, and gravity.

erythroblast Any of the cells in the *myeloid tissue of red bone marrow that develop into erythrocytes (red blood cells). Erythroblasts have a

erythrocyte

nucleus and are at first colourless, but fill with *haemoglobin as they develop. In mammals the nucleus disappears.

erythrocyte (red blood cell) The most numerous type of blood cell, which contains the red pigment *haemoglobin and is responsible for oxygen transport. Mammalian erythrocytes are disc-shaped and lack a nucleus; those of other vertebrates are oval and nucleated. In man the number of erythrocytes in the blood varies between 4.5 and 5.5 million per cubic millimetre. They survive for about four months and are then destroyed in the spleen and liver. *See also* erythroblast. *Compare* leucocyte.

Esaki diode *See* tunnel diode.

ESCA *See* photoelectron spectroscopy.

escapement A device in a clock or watch that controls the transmission of power from the spring or falling weight to the hands. It is usually based on a balance wheel or pendulum. It thus allows energy to enter the mechanism in order to move the hands round the face, overcome friction in the gear trains, and maintain the balance wheel or pendulum in continuous motion.

escape velocity The minimum speed needed by a space vehicle, rocket, etc., to escape from the gravitational field of the earth, moon, or other celestial body. The gravitational force between a rocket of mass m and a celestial body of mass M and radius r is MmG/r^2 (*see* Newton's law of gravitation). Therefore the gravitational potential energy of the rocket with respect to its possible position very far from the celestial body on which it is resting can be shown to be $-GmM/r$, assuming (by convention) that the potential energy is zero at an infinite distance from the celestial body. If the rocket is to escape from the gravitational field it must have a kinetic energy that exceeds

this potential energy, i.e. the kinetic energy $mv^2/2$ must be greater than MmG/r, or $v > \sqrt{(2MG/r)}$. This is the value of the escape velocity. Inserting numerical values for the earth and moon into this relationship gives an escape velocity from the earth of 11 200 m s^{-1} and from the moon of 2370 m s^{-1}.

Escherichia coli (E. coli) A species of Gram-negative aerobic bacteria that is found in the intestine and is also widely used in microbiological and genetics research. The motile rod-shaped cells ferment lactose and are usually harmless commensals, although certain strains are pathogenic. Studies of *E. coli* laboratory cultures have revealed much about the genetics of prokaryotes; the species is also frequently used in genetic engineering, particularly as a host for *gene cloning and the expression of recombinant foreign genes in culture.

ESR *See* electron-spin resonance.

essential amino acid An *amino acid that an organism is unable to synthesize in sufficient quantities. It must therefore be present in the diet. In man the essential amino acids are arginine, histidine, lysine, threonine, methionine, isoleucine, leucine, valine, phenylalanine, and tryptophan. These are required for protein synthesis and deficiency leads to retarded growth and other symptoms. Most of the amino acids required by man are also essential for all other multicellular animals and for most protozoans.

essential element Any of a number of elements required by living organisms to ensure normal growth, development, and maintenance. Apart from the elements found in organic compounds (i.e. carbon, hydrogen, oxygen, and nitrogen), plants, animals, and microorganisms all require a range of elements in inorganic forms in varying amounts, depending on the type of organism.

The *major elements*, present in tissues in relatively large amounts (greater than 0.005%), are calcium, phosphorus, potassium, sodium, chlorine, sulphur, and magnesium. The *trace elements* occur at much lower concentrations and thus requirements are much less. The most important are iron, manganese, zinc, copper, iodine, cobalt, selenium, molybdenum, chromium, and silicon. Each element may fulfil one or more of a variety of metabolic roles. Sodium, potassium, and chloride ions are the chief electrolytic components of cells and body fluids and thus determine their electrical and osmotic status. Calcium, phosphorus, and magnesium are all present in bone. Calcium is also essential to nerve and muscle activity, while phosphorus is a key constituent of the chemical energy carriers (e.g. *ATP) and the nucleic acids. Sulphur is needed primarily for amino acid synthesis (in plants and microorganisms). The trace elements may serve as *cofactors or as constituents of complex molecules, e.g. iron in haem and cobalt in vitamin B_{12}.

essential fatty acids *Fatty acids that must normally be present in the diet of certain animals, including man. Essential fatty acids, which include *linoleic and *linolenic acids, all possess double bonds at the same two positions along their hydrocarbon chain and so can act as precursors of *prostaglandins. Deficiency of essential fatty acids can cause dermatosis, weight loss, irregular oestrus, etc. An adult human requires 2–10 g linoleic acid or its equivalent per day.

essential oil A natural oil with a distinctive scent secreted by the glands of certain aromatic plants.

*Terpenes are the main constituents. Essential oils are extracted from plants by steam distillation, extraction with cold neutral fats or solvents (e.g. alcohol), or pressing and used in perfumes, flavourings, and medicine. Examples are citrus oils, flower oils (e.g. rose, jasmine), and oil of cloves.

esterification A reaction of an alcohol with an acid to produce an ester and water; e.g.

$$CH_3OH + C_6H_5COOH \rightleftharpoons CH_3OOCC_6H_5 + H_2O$$

The reaction is an equilibrium and is slow under normal conditions, but can be speeded up by addition of a strong acid catalyst. The ester can often be distilled off so that the reaction can proceed to completion. The reverse reaction is ester hydrolysis or *saponification. *See also* labelling.

esters Organic compounds formed by reaction between alcohols and acids (see illustration). Esters formed from carboxylic acids have the general formula RCOOR′. Examples are ethyl ethanoate, $CH_3COOC_2H_5$, and methyl propanoate, $C_2H_5COOCH_3$. Esters containing simple hydrocarbon groups are volatile fragrant substances used as flavourings in the food industry. Triesters, molecules containing three ester groups, occur in nature as oils and fats. *See also* glycerides.

etaerio A cluster of fruits formed from the unfused carpels of a single flower. For example, the anemone has an etaerio of *achenes, larkspur an etaerio of *follicles, and blackberry an etaerio of *drupes.

ethanal (acetaldehyde) A colourless highly flammable liquid aldehyde, CH_3CHO; r.d. 0.78; m.p. −124.6°C; b.p. 20.8°C. It is made from ethene by the *Wacker process and used as

methanol + ethanoic acid ⇌ methyl ethanoate + water

Ester formation

247

a starting material for making many organic compounds. The compound polymerizes if dilute acid is added to give *ethanal trimer* (or *paraldehyde*), which contains a six-membered ring of alternating carbon and oxygen atoms with a hydrogen atom and a methyl group attached to each carbon atom. It is used as a drug for inducing sleep. Addition of dilute acid below 0°C gives *ethanal tetramer* (or *metaldehyde*), which has a similar structure to the trimer but with an eight-membered ring. It is used as a solid fuel in portable stoves and in slug pellets.

ethanamide (acetamide) A colourless solid crystallizing in the form of long white crystals with a characteristic smell of mice, CH_3CONH_2; r.d. 1.159; m.p. 82.3°C; b.p. 221.25°C. It is made by the dehydration of ammonium ethanoate or by the action of ammonia on ethanoyl chloride, ethanoic anhydride, or ethyl ethanoate.

ethane A colourless flammable gaseous hydrocarbon, C_2H_6; m.p. −183°C; b.p. −89°C. It is the second member of the *alkane series of hydrocarbons and occurs in natural gas.

ethanedioic acid *See* oxalic acid.

ethane-1,2-diol (ethylene glycol; glycol) A colourless viscous hygroscopic liquid, CH_2OHCH_2OH; m.p. −13°C; b.p. 197°C. It is made by hydrolysis of epoxyethane (from ethene) and used as an antifreeze and a raw material for making *polyesters (e.g. Terylene).

ethanoate (acetate) A salt or ester of ethanoic acid (acetic acid).

ethanoic acid (acetic acid) A clear viscous liquid or glassy solid *carboxylic acid, CH_3COOH, with a characteristically sharp odour of vinegar; r.d. 1.049; m.p. 16.7°C; b.p. 118.5°C. The pure compound is called *glacial ethanoic acid*. It is manufactured by the oxidation of ethanol or by the

oxidation of butane in the presence of dissolved manganese(II) or cobalt(II) ethanoates at 200°C, and is used in making ethanoic anhydride for producing cellulose ethanoates. It is also used in making ethenyl ethanoate (for polyvinylacetate). The compound is formed by the fermentation of alcohol and is present in vinegar, which is made by fermenting beer or wine. 'Vinegar' made from ethanoic acid with added colouring matter is called 'nonbrewed condiment'. In living organisms it combines with *coenzyme A to form acetyl coenzyme A, which plays a crucial role in energy metabolism.

ethanol (ethyl alcohol) A colourless water-soluble *alcohol, C_2H_5OH; r.d. 0.61 (0°C); m.p. −169°C; b.p. −102°C. It is the active principle in intoxicating drinks, in which it is produced by fermentation of sugar using yeast

$$C_6H_{12}O_6 \rightarrow 2C_2H_5OH + 2CO_2$$

The ethanol produced kills the yeast and fermentation alone cannot produce ethanol solutions containing more than 15% ethanol by volume. Distillation can produce a constant-boiling mixture containing 95.6% ethanol and 4.4% water. Pure ethanol (*absolute alcohol*) is made by removing this water by means of drying agents.

The main industrial use of ethanol is as a solvent although at one time it was a major starting point for making other chemicals. For this it was produced by fermentation of molasses. Now ethene has replaced ethanol as a raw material and industrial ethanol is made by hydrolysis of ethene.

ethanoyl chloride (acetyl chloride) A colourless liquid acyl chloride (*see* acyl halides), CH_3COCl, with a pungent smell; r.d. 1.104; m.p. −112.15°C; b.p. 55°C. It is made by reacting ethanoic acid with a halogenating agent such as phosphorus(III) chloride, phosphorus(V) chloride, or sulphur dichloride oxide and is used to introduce ethanoyl groups into organic compounds containing –OH, $-NH_2$, and –SH groups. *See* acylation.

ethanoyl group (acetyl group) The organic group CH_3CO-.

ethene (ethylene) A colourless flammable gaseous hydrocarbon, C_2H_4; m.p. $-169°C$; b.p. $-102°C$. It is the first member of the *alkene series of hydrocarbons. It is made by cracking hydrocarbons from petroleum and is now a major raw material for making other organic chemicals (e.g. ethanal, ethanol, ethane-1,2-diol). It can be polymerized to *polyethene. It occurs naturally in plants, in which it acts as a *growth substance promoting the ripening of fruits.

ethenyl ethanoate (vinyl acetate) An unsaturated organic ester, $CH_2:CHOOCCH_3$; r.d. 0.9; m.p. $-100°C$; b.p. 73°C. It is made by catalytic reaction of ethanoic acid and ethene and used to make polyvinylacetate.

ether 1. (aether) A hypothetical medium once believed to be necessary to support the propagation of electromagnetic radiation. It is now regarded as unnecessary and in modern theory electromagnetic radiation can be propagated through empty space. The existence of the ether was first called into question as a result of the *Michelson–Morley experiment. **2.** *See* ethoxyethane; ethers.

ethers Organic compounds containing the group –O– in their molecules. Examples are dimethyl ether, CH_3OCH_3, and diethyl ether, $C_2H_5OC_2H_5$ (*see* ethoxyethane). They are volatile highly flammable compounds made by dehydrating alcohols using sulphuric acid.

ethology The study of the biology of *animal behaviour. Central to the ethologist's approach is the principle that animal behaviour (like physical characteristics) is subject to evolution through natural selection. Ethologists therefore seek to explain how the behaviour of an animal in its natural environment may contribute to the survival of the maximum number of its relatives and offspring. This involves recognizing the stimuli that are important in nature (*see* sign stimulus) and how innate predispositions interact with *learning in the development of behaviour (*see* instinct). Studies of this sort were pioneered by Konrad Lorenz (1903–89) and Niko Tinbergen (1907–88).

ethoxyethane (diethyl ether; ether) A colourless flammable volatile *ether, $C_2H_5OC_2H_5$; r.d. 0.71; m.p. $-116°C$; b.p. 34.5°C. It can be made by *Williamson's synthesis. It is an anaesthetic and useful organic solvent.

ethyl acetate *See* ethyl ethanoate.

ethyl alcohol *See* ethanol.

ethylamine A colourless flammable volatile liquid, $C_2H_5NH_2$; r.d. 0.69; m.p. $-81°C$; b.p. 16.6°C. It is a primary amine made by reacting chloroethane with ammonia and used in making dyes.

ethylbenzene A colourless flammable liquid, $C_6H_5C_2H_5$; r.d. 0.8; m.p. $-95°C$; b.p. 136°C. It is made from ethene and ethybenzene by a *Friedel–Crafts reaction and is used in making phenylethene (for polystyrene).

ethyl bromide *See* bromoethane.

ethylene *See* ethene.

ethylene glycol *See* ethane-1,2-diol.

ethylene oxide *See* epoxyethane.

ethyl ethanoate (ethyl acetate) A colourless flammable liquid ester, $C_2H_5OOCCH_3$; r.d. 0.69; m.p. $-81°C$; b.p. 16.6°C. It is used as a solvent and in flavourings and perfumery.

ethyl group The organic group C_2H_5-.

ethyl iodide *See* iodoethane.

ethyne **(acetylene)** A colourless unstable gas, C_2H_2, with a characteristic sweet odour; r.d. 0.618; m.p. $-83.25°C$; b.p. $-79.85°C$. It is the simplest member of the *alkyne series of unsaturated hydrocarbons, and is prepared by the action of water on calcium dicarbide or by adding alcoholic potassium hydroxide to 1,2-dibromoethane. It can be manufactured by heating methane to 1500°C in the presence of a catalyst. It is used in oxyacetylene welding and in the manufacture of ethanal and ethanoic acid. Ethyne can be polymerized easily at high temperatures to give a range of products. The inorganic salt-like dicarbides contain the ion C_2^{2-}, although ethyne itself is a neutral compound (i.e. not a protonic acid).

etiolation The abnormal form of growth observed when plants grow in darkness or severely reduced light. Such plants characteristically have blanched leaves and shoots, excessively long shoots, and reduced leaves and root systems.

eubacteria The true *bacteria, i.e. those typically having simple unbranched cells, rigid cell walls, and flagella for movement.

eucaryote *See* eukaryote.

euchromatin *See* chromatin.

Euclidean geometry The form of geometry set forth in the 13 volumes, called the *Elements*, by the Greek mathematician Euclid (*fl* c. 300 BC). It is based on a series of axioms and applies to plane figures. *Compare* non-Euclidean geometry.

eudiometer An apparatus for measuring changes in volume of gases during chemical reactions. A simple example is a graduated glass tube sealed at one end and inverted in mercury. Wires passing into the tube allow the gas mixture to be sparked to initiate the reaction between gases in the tube.

eugenics The study of methods of improving the quality of human populations by the application of genetic principles. *Positive eugenics* would seek to do this by selective breeding programmes. *Negative eugenics* aims to eliminate harmful genes (e.g. those causing haemophilia and colour blindness) by counselling any prospective parents who are likely to be *carriers.

Euglenophyta A division containing *algae (including *Euglena*) that in certain respects resemble animals more than they do plants. For example, they lack a cell wall and some forms are colourless and thus ingest food since they cannot photosynthesize. Members of the Euglenophyta are unicellular and move by means of flagella; they usually inhabit fresh water. These algae are often classified by zoologists into the class Flagellata of the *Protozoa; alternatively they are classified as protists (*see* Protista).

eukaryote **(eucaryote)** An organism consisting of cells in which the genetic material is contained within a distinct nucleus. All organisms except bacteria are eukaryotes. *See* cell. *Compare* prokaryote.

europium Symbol Eu. A soft silvery metallic element belonging to the *lanthanoids; a.n. 63; r.a.m. 151.96; r.d. 5.245 (20°C); m.p. 822°C; b.p. 1529°C. It occurs in small quantities in bastanite and monazite. Two stable isotopes occur naturally: europium-151 and europium-153, both of which are neutron absorbers. Experimental europium alloys have been tried for nuclear-reactor parts but until recently the metal has not been available in sufficient quantities. It is widely used in the form of the oxide in phosphors for television screens. It was discovered by Sir William Crookes in 1889.

Eustachian tube The tube that connects the *middle ear to the back of the throat (pharynx) in vertebrates. It is normally closed, but during swallowing it opens to allow air into the middle ear, which equalizes the pressure on each side of the *tympanum (eardrum). It was named after the Italian anatomist Bartolomeo Eustachio (?1520–74).

eutectic mixture A solid solution consisting of two or more substances and having the lowest freezing point of any possible mixture of these components. The minimum freezing point for a set of components is called the *eutectic point*. Low-melting-point alloys are usually eutectic mixtures.

Eutheria (Placentalia) A subclass of mammals in which the embryos are retained in a uterus in the mother's body and nourished by a *placenta. The young are thus fully protected during their embryonic development and kept at a constant temperature. Placental mammals evolved during the Cretaceous period (about 100 million years ago). Modern placentals are a highly diverse group that occupy all types of habitat in all parts of the world. They include the orders *Artiodactyla, *Carnivora, *Cetacea, *Chiroptera, *Insectivora, *Perissodactyla, *Primates, *Proboscidea, and *Rodentia. *Compare* Metatheria; Prototheria.

eutrophic Describing a body of water (e.g. a lake) with an abundant supply of nutrients and a high rate of formation of organic matter by photosynthesis. Pollution of a lake by sewage or fertilizers renders it eutrophic (a process called *eutrophication*). This stimulates excessive growth of algae; the death and subsequent bacterial decomposition of these depletes the oxygen content of the lake, resulting in the death of the lake's fish and other animals. *Compare* oligotrophic.

evaporation The change of state of a liquid into a vapour at a temperature below the boiling point of the liquid. Evaporation occurs at the surface of a liquid, some of those molecules with the highest kinetic energies escaping into the gas phase. The result is a fall in the average kinetic energy of the molecules of the liquid and consequently a fall in its temperature.

even–even nucleus An atomic nucleus containing an even number of protons and an even number of neutrons.

even–odd nucleus An atomic nucleus containing an even number of protons and an odd number of neutrons.

event horizon *See* black hole.

evergreen (Describing) a plant that bears leaves throughout the year, each leaf being shed independently of the others after two or three years. The leaves of evergreens are often reduced or adapted in some way to prevent excessive water loss; examples are the needles of conifers and the leathery waxy leaves of holly. *Compare* deciduous.

evocation The ability of experimental stimuli (e.g. chemicals or tissue implants) to cause unspecialized embryonic tissue to develop into specialized tissue.

evolute The locus of the centres of curvature of all the points on a given curve (called the *involute*).

evolution The gradual process by which the present diversity of plant and animal life arose from the earliest and most primitive organisms, which is believed to have been continuing for at least the past 3000 million years. Until the middle of the 18th century it was generally believed that each species was divinely created and fixed in its form throughout its existence (*see* special creation). Lamarck was the first biologist to publish a theory to explain how one species could have evolved into another (*see* Lamarckism), but it was not until the publication of Darwin's *On the Origin* *of Species* in 1859 that special creation was seriously challenged. Unlike Lamarck, Darwin proposed a feasible mechanism for evolution and backed it up with evidence from the fossil record and studies of comparative anatomy and embryology (*see* Darwinism; natural selection). The modern version of Darwinism, which incorporates discoveries in genetics made since Darwin's time, probably remains the most acceptable theory of species evolution. More controversial, however, and still to be firmly clarified, are the relationships and evolution of groups above the species level.

exa- Symbol E. A prefix used in the metric system to denote 10^{18} times. For example, 10^{18} metres = 1 exametre (Em).

excess electron An electron in a *semiconductor that is not required in the bonding system of the crystal lattice and has been donated by an impurity atom. It is available for conduction (*excess conduction*).

exchange force 1. A force resulting from the continued interchange of particles in a manner that bonds their hosts together. Examples are the covalent bond involving electrons, and the strong interaction (*see* fundamental interactions) in which mesons are exchanged between nucleons or gluons are exchanged between quarks (*see* elementary particles). **2.** *See* magnetism.

excitation 1. A process in which a nucleus, electron, atom, ion, or molecule acquires energy that raises it to a quantum state (*excited state*) higher than that of its *ground state. The difference between the energy in the ground state and that in the excited state is called the *excitation energy*. *See* collective excitation; energy level; quasiparticle. **2.** The process of applying current to the winding of an electromagnet, as in an electric motor. **3.** The process of applying a signal to the base of a transistor or the control electrode of a thermionic valve.

exciton An electron–hole pair in a crystal that is bound in a manner analogous to the electron and proton of a hydrogen atom. It behaves like an atomic excitation that passes from one atom to another and may be long-lived. Exciton behaviour in *semiconductors is important.

exclusion principle *See* Pauli exclusion principle.

excretion The elimination by an organism of the waste products that arise as a result of metabolic activity. These products include water, carbon dioxide, and nitrogenous compounds. In plants and simple animals waste products are excreted by simple diffusion from the body, but higher animals have specialized organs and organ systems devoted to this function. Examples of excretory organs in vertebrates are the lungs (for carbon dioxide and water), and the *kidneys (for nitrogenous compounds (urea) and water). In addition, mammals excrete small amounts of urea, salts, and water from the skin in sweat.

exitance Symbol M. The radiant or luminous flux emitted per unit area of a surface. The *radiant exitance* (M_e) is measured in watts per square metre (W m^{-2}), while the *luminous exitance* (M_v) is measured in lumens per square metre (lm m^{-2}). Exitance was formerly called *emittance*.

exocarp *See* pericarp.

exocrine gland A gland that discharges its secretion into a body cavity (such as the gut) or onto the body surface. Examples are the *sebaceous and *sweat glands, the *mammary glands, and part of the pancreas. Exocrine glands are formed in the embryo from the invagination of epithelial cells. Their secretions pass initially into a cavity (an *alveolus* or *acinus*) and then out through a duct or duct network, along which the secretion may become modified by exchange with the blood across the duct epithelium.

exodermis *See* hypodermis.

exoergic Denoting a nuclear process that gives out energy. *Compare* endoergic.

exogamy The fusion of reproductive cells from distantly related or unrelated organisms, i.e. *outbreeding. *Compare* endogamy.

exogenous Describing substances, stimuli, etc., that originate outside an organism. For example, vitamins that cannot be synthesized by an animal are said to be supplied exogenously in the diet. *Compare* endogenous.

exon A nucleotide sequence in a gene that codes for part or all of the gene product or for some control function. In eukaryotes, exons are separated by noncoding sequences called *introns.

exonuclease An enzyme that catalyses the cleavage of nucleotides from the end of a nucleic acid molecule. *Compare* endonuclease.

exoskeleton A rigid external covering for the body in certain animals, such as the hard chitinous cuticle of arthropods. An exoskeleton protects and supports the body and provides points of attachment for muscles. The cuticle of arthropods must be shed at intervals to allow growth to occur (*see* ecdysis). Other examples of exoskeletons are the shells of molluscs and the bony plates of tortoises and armadillos. *Compare* endoskeleton.

exosphere *See* earth's atmosphere.

exothermic Denoting a chemical reaction that releases heat into its surroundings. *Compare* endothermic.

exotoxin *See* toxin.

expansion The writing of a function or quantity as a *series of terms. The series may be finite or infinite. *See* binomial theorem; Taylor series.

expansion of the universe The hypothesis, based on the evidence of the *redshift, that the distance between the galaxies is continuously increasing. The original theory, which was proposed in 1929 by Edwin Hubble (1889–1953), assumes that the galaxies are flying apart as a consequence of the big bang with which the universe originated. Several variants have since been proposed. *See also* big-bang theory; Hubble constant.

expansivity (thermal expansion) 1. *Linear expansivity* is the fractional increase in length of a specimen of a solid, per unit rise in temperature. If a specimen increases in length from l_1 to l_2 when its temperature is raised $\theta°$, then the expansivity (α) is given by:
$$l_2 = l_1(1 + \alpha\theta).$$
This relationship assumes that α is independent of temperature. This is not, in general, the case and a more accurate relationship is:
$$l_2 = l_1(1 + a\theta + b\theta^2 + c\theta^3 \ldots),$$
where a, b, and c are constants. **2.** *Superficial expansivity* is the fractional increase in area of a solid surface caused by unit rise in temperature, i.e.
$$A_2 = A_1(1 + \beta\theta),$$
where β is the superficial expansivity. To a good approximation $\beta = 2\alpha$. **3.** *Volume expansivity* is the fractional increase in volume of a solid, liquid, or gas per unit rise in temperature, i.e.
$$V_2 = V_1(1 + \gamma\theta),$$
where γ is the cubic expansivity and $\gamma = 3\alpha$. For liquids, the expansivity observed directly is called the *apparent expansivity* as the container will also have expanded with the rise in temperature. The *absolute expansivity* is the apparent expansivity plus the volume expansivity of the container. For the expansion of gases, *see* Charles' Law.

Graph showing exponential growth of the human population

explantation The removal of cells, tissues, or organs of animals and plants for observation of their growth and development in appropriate culture media. *See also* tissue culture; organ culture.

explosives Substances that undergo rapid chemical reaction evolving heat and causing a sudden increase in pressure. The volume of gas produced by an explosive is great compared with the volume of the original substance. Examples of explosives are gunpowder, cellulose nitrate, and TNT.

exponent A number or symbol that indicates the power to which another number or expression is raised. For example, $(x + y)^n$ indicates that the expression $(x + y)$ is raised to the nth power; n is the exponent. Any number or expression in which the exponent is zero equals 1, i.e. $x^0 = 1$.

exponential A function that varies as the power of another quantity. If $y = a^x$, y varies exponentially with x. The function e^x, also written as $\exp(x)$, is called the *exponential function* (*see* e). It is equal to the sum of the *exponential series*, i.e.

$$e^x = 1 + x + x^2/2! + x^3/3! \ldots + x^n/n!$$

exponential growth A form of population growth in which the rate of growth is related to the number of individuals present. Increase is slow when numbers are low but rises sharply as numbers increase. If population number is plotted against time on a graph a characteristic J-shaped curve results (see graph). In animal and plant populations, such factors as overcrowding, lack of nutrients, and disease limit population increase beyond a certain point and the J-shaped exponential curve tails off giving an S-shaped (sigmoid) curve.

exposure meter A photocell that operates a meter to indicate the correct exposure for a specified film in photography. It enables the correct shutter speed and aperture to be chosen for any photographic circumstances. Some cameras have a built-in exposure meter that automatically sets the aperture according to the light available and the chosen shutter speed.

extender An inert substance added to a product (paint, rubber, washing powder, etc.) to dilute it or to modify its physical properties.

extensometer Any device for measuring the extension of a specimen of a material under longitudinal stress. A common method is to make the specimen form part of a capacitor, the capacitance of which will change with a change in the specimen's dimensions.

extinction coefficient A measure of the extent by which the intensity of a beam of light is reduced by passing through a distance d of a solution having a molar concentration c of the dissolved substance. If the intensity of the light is reduced from I_1 to I_2, the extinction coefficient is $[\log(I_1/I_2)]/cd$.

extracellular Located or occurring outside the cell.

extraction 1. The process of obtaining a metal from its ore. **2.** The separation of a component from a mixture by selective solubility. *See* partition.

Structure of the vertebrate eye

extraembryonic membranes (embryonic membranes) The tissues produced by an animal *embryo for protection and nutrition but otherwise taking no part in its development. The three membranes, which are called *fetal membranes* in man, are the *chorion, *amnion, and *allantois.

extraordinary ray *See* double refraction.

extremely high frequency (EHF) A radio frequency between 30 000 megahertz and 300 gigahertz.

extrinsic semiconductor *See* semiconductor.

eye The organ of sight. The most primitive eyes are the *eyespots of some unicellular organisms. More advanced eyes are the *ocelli and *compound eyes of arthropods (e.g. insects). The cephalopod molluscs (e.g. the octopus and squid) and vertebrates possess the most highly developed eyes (see illustration). These normally occur in pairs, are nearly spherical, and filled with fluid. Light is refracted by the *cornea through the pupil in the *iris and onto the *lens, which focuses images onto the retina. These images are received by light-sensitive cells in the retina (*see* cone; rod), which transmit impulses to the brain via the optic nerve.

eyepiece (ocular) The lens or sys-

tem of lenses in an optical instrument that is nearest to the eye. It usually produces a magnified image of the previous image formed by the instrument.

eyespot (stigma) 1. (in botany) A structure found in some free-swimming unicellular algae and in plant reproductive cells that contains orange or red pigments (carotenoids) and is sensitive to light. It enables the cell to move in relation to a light source (*see* phototaxis). **2.** (in zoology) A spot of pigment found in some lower animals, e.g. jellyfish.

eye tooth A *canine tooth in the upper jaw.

F

F_1 (first filial generation) The first generation of offspring resulting from an arranged cross between selected parents in breeding experiments. *See* monohybrid.

F_2 (second filial generation) The second generation of offspring in breeding experiments, obtained by crosses between individuals of the *F_1 generation. *See* monohybrid.

Fabry–Pérot interferometer A type of *interferometer in which monochromatic light is passed through a pair of parallel half-silvered glass plates producing circular interference fringes. One of the glass plates is adjustable, enabling the separation of the plates to be varied. The wavelength of the light can be determined by observing the fringes while adjusting the separation. This type of instrument is used in spectroscopy.

face-centred cubic (f.c.c.) *See* cubic crystal.

facilitated diffusion The transport of molecules across the outer membrane of a living cell by a process that involves a specific carrier located within the cell membrane but does not require expenditure of energy by the cell. The carrier combines with a molecule on one side of the membrane, moves through the membrane, and releases the molecule on the other side. *Compare* active transport.

factorial The product of a given number and all the whole numbers below it. It is usually writen $n!$, e.g. factorial $4 = 4! = 4 \times 3 \times 2 \times 1 = 24$. Factorial 0 is defined as 1.

Factor VIII (antihaemophilic factor) One of the blood *clotting factors. Factor VIII is a soluble protein that stimulates the activation of Factor X by Factor IXa, which in turn converts *prothrombin to thrombin, thus causing the fibrin matrix of a blood clot to form. *Haemophilia is due to a deficiency or defect of Factor VIII and is treated by administration of blood plasma or plasma concentrate containing the factor. Factor VIII can now be obtained from genetically engineered cell cultures. Such preparations avoid the risk of contamination with viruses, notably HIV (the AIDS virus).

FAD (flavin adenine dinucleotide) A *coenzyme important in various biochemical reactions. It comprises a phosphorylated vitamin B_2 (riboflavin) molecule linked to the nucleotide adenine monophosphate (AMP). FAD is usually tightly bound to the enzyme forming a *flavoprotein*. It functions as a hydrogen acceptor in dehydrogenation reactions, being reduced to $FADH_2$. This in turn is oxidized to FAD by the *electron transport chain, thereby generating ATP (two molecules of ATP per molecule of $FADH_2$).

faeces Waste material that is eliminated from the alimentary canal through the *anus. Faeces consist of the indigestible residue of food that remains after the processes of digestion and absorption of nutrients and water have taken place, together with bacteria and dead cells shed from the gut lining.

Fahrenheit scale A temperature scale in which (by modern definition) the temperature of boiling water is taken as 212 degrees and the temperature of melting ice as 32 degrees. It was invented in 1714 by the German scientist G. D. Fahrenheit (1686 –1736), who set the zero at the lowest temperature he knew how to obtain in the laboratory (by mixing ice and common salt) and took his own body temperature as 96°F. The scale is no longer in scientific use. To convert to the *Celsius scale the formula is $C = 5(F - 32)/9$.

Fajans' rules Rules indicating the extent to which an ionic bond has covalent character caused by polarization of the ions. Covalent character is more likely if:

(1) the charge of the ions is high;

(2) the positive ion is small or the negative ion is large;

(3) the positive ion has an outer electron configuration that is not a noble-gas configuration.

The rules were introduced by the Polish–American chemist Kasimir Fajans (1887–1975).

fallopian tube (oviduct) The tube that carries egg cells from the *ovary to the womb in mammals. The eggs are carried by the action of muscles and cilia. It was named after the Italian anatomist Gabriel Fallopius (1523 –62).

fall-out 1. (*or* **radioactive fall-out**) Radioactive particles deposited from the atmosphere either from a nuclear explosion or from a nuclear accident. *Local fall-out*, within 250 km of an explosion, falls within a few hours of the explosion. *Tropospheric fall-out* consists of fine particles deposited all round the earth in the approximate latitude of the explosion within about one week. *Stratospheric fall-out* may fall anywhere on earth over a period of years. The most dangerous radioactive isotopes in fall-out are the fission fragments iodine–131 and strontium–90. Both can be taken up by grazing animals and passed on to human populations in milk, milk products, and meat. Iodine–131 accumulates in the thyroid gland and strontium–90 accumulates in bones. **2.** (*or* **chemical fall-out**) Hazardous chemicals discharged into and subsequently released from the atmosphere, especially by factory chimneys.

false fruit *See* pseudocarp.

family A category used in the *classification of organisms that consists of one or several similar or closely related genera. Similar families are grouped into an order. Family names end in *-aceae* or *-ae* in botany (e.g. Cactaceae) and *-idae* in zoology (e.g. Equidae). The names are usually derived from a type genus (*Cactus* and *Equus* in the examples above) that is characteristic of the whole family (*see* type specimen). In botany, families are sometimes called *natural orders*.

farad Symbol F. The SI unit of capacitance, being the capacitance of a capacitor that, if charged with one coulomb, has a potential difference of one volt between its plates. $1 \text{ F} = 1 \text{ C V}^{-1}$. The farad itself is too large for most applications; the practical unit is the microfarad (10^{-6} F). The unit is named after Michael Faraday (1791–1867).

Faraday cage An earthed screen made of metal wire that surrounds an electric device in order to shield it from external electrical fields.

Faraday constant Symbol *F*. The electric charge carried by one mole of electrons or singly ionized ions, i.e. the product of the *Avogadro constant and the charge on an electron (disregarding sign). It has the value $9.648 5309(29) \times 10^4$ coulombs per mole. This number of coulombs is sometimes treated as a unit of electric charge called the *faraday*.

Faraday effect The rotation of the plane of polarization of electromagnetic radiation on passing through an isotropic medium exposed to a magnetic field. The angle of rotation is proportional to Bl, where l is the length of the path of the radiation in the medium and B is the magnetic flux density.

Faraday's laws Two laws describing electrolysis:
(1) The amount of chemical change during electrolysis is proportional to the charge passed.
(2) The charge required to deposit or liberate a mass m is given by $Q = Fmz/M$, where F is the Faraday constant, z the charge of the ion, and M the relative ionic mass.
These are the modern forms of the laws. Originally, they were stated by Faraday in a different form:
(1) The amount of chemical change produced is proportional to the quantity of electricity passed.
(2) The amount of chemical change produced in different substances by a fixed quantity of electricity is proportional to the electrochemical equivalent of the substance.

Faraday's laws of electromagnetic induction (1) An e.m.f. is induced in a conductor when the magnetic field surrounding it changes. (2) The magnitude of the e.m.f. is proportional to the rate of change of the field. (3) The sense of the induced e.m.f.

depends on the direction of the rate of change of the field.

fascia A sheet of fibrous connective tissue occurring beneath the skin and also enveloping glands, vessels, nerves, and forming tendon sheaths.

fascicle *See* vascular bundle.

fast green A green dye used in optical microscopy that stains cellulose, cytoplasm, collagen, and mucus green. It is frequently used to stain plant tissues, with *safranin as a counterstain. Unlike *light green*, a similar dye, it does not fade easily.

fast neutron A neutron resulting from nuclear fission that has an energy in excess of 0.1 MeV (1.6 × 10^{-14} J), having lost little of its energy by collision. In some contexts *fast fission* is defined as fission brought about by fast neutrons, i.e. neutrons having energies in excess of 1.5 MeV (2.4 × 10^{-13} J), the fission threshold of uranium-238. *See also* nuclear reactor; slow neutron.

fast reactor *See* nuclear reactor.

fat A mixture of lipids, chiefly *triglycerides, that is solid at normal body temperatures. Fats occur widely in plants and animals as a means of storing food energy, having twice the calorific value of carbohydrates. In mammals, fat is deposited in a layer beneath the skin (subcutaneous fat) and deep within the body as a specialized *adipose tissue (*see also* brown fat). The insulating properties of fat are also important, especially in animals lacking fur and those inhabiting cold climates (c.g. seals and whales).

Fats derived from plants and fish generally have a greater proportion of unsaturated *fatty acids than those from mammals. Their melting points thus tend to be lower, causing a softer consistency at room temperatures. Highly unsaturated fats are liquid at room temperatures and are therefore more properly called *oils.

fat body 1. An abdominal organ in amphibians attached to the anterior of each kidney. It contains a reserve of fat that nourishes the gonads during the winter hibernation in readiness for the spring breeding season. **2.** A mass of fatty tissue spreading throughout the body cavity of insects in which fats, proteins, and glycogen are stored as a reserve for hibernation or pupation.

fathom A unit used to describe a depth of water. It is equal to 6 feet (1.83 metres).

fatigue *See* metal fatigue.

fatty acid An organic compound consisting of a hydrocarbon chain and a terminal carboxyl group (*see* carboxylic acids). Chain length ranges from one hydrogen atom (methanoic, or formic, acid, HCOOH) to nearly 30 carbon atoms. Ethanoic (acetic), propanoic (propionic), and butanoic (butyric) acids are important in metabolism. Long-chain fatty acids (more than 8–10 carbon atoms) most commonly occur as constituents of certain lipids, notably glycerides, phospholipids, sterols, and waxes, in which they are esterified with alcohols. These long-chain fatty acids generally have an even number of carbon atoms; unbranched chains predominate over branched chains. They may be saturated (e.g. *palmitic (hexadecanoic) acid and *stearic (octadecanoic) acid) or unsaturated, with one double bond (e.g. *oleic (cis-octodec-9-enoic) acid) or two or more double bonds, in which case they are called *polyunsaturated fatty acids* (e.g. *linoleic acid and *linolenic acid). *See also* essential fatty acids.

The physical properties of fatty acids are determined by chain length, degree of unsaturation, and chain branching. Short-chain acids are pungent liquids, soluble in water. As chain length increases, melting points are raised and water-solubility

Types of fault

decreases. Unsaturation and chain branching tend to lower melting points.

fault A fracture in the earth's crust along which there has been displacement of rock on one side relative to the other. The displacement ranges from a few centimetres to a few kilometres and may occur in a horizontal, oblique, or vertical direction. The extent of vertical displacement of the strata is the *throw*; the horizontal displacement is the *heave*. The side of the fault on which the strata have moved relatively downward is known as the downthrow side; the other is the upthrow side.

f-block elements The block of elements in the *periodic table consisting of the lanthanoid series (from cerium to lutetium) and the actinoid series (from thorium to lawrencium). They are characterized by having two *s*-electrons in their outer shell (n) and *f*-electrons in their inner ($n-1$) shell.

f.c.c. Face-centred cubic. *See* cubic crystal.

feathers The body covering of birds, formed as outgrowths of the epidermis and composed of the protein *keratin. Feathers provide heat insulation, they give the body its streamlined shape, and those of the wings and tail are important in flight. Basically a feather consists of a *quill*, which is embedded in the skin attached to a feather follicle and is continuous with the shaft (*rachis*) of the feather, which carries the *barbs. This basic structure is modified depending on the type of feather (*see* contour feathers; down feathers; filoplumes).

fecundity The fertility of an organism (in higher animals, generally of the female of the species). Normally all organisms, assuming they reach reproductive age, are sufficiently fecund to replace themselves several times over. Darwin noted this, together with the fact that population numbers nevertheless tended to remain fairly constant: these observations led him to formulate his theory of evolution by *natural selection.

feedback The use of part of the output of a system to control its performance. In *positive feedback*, the output is used to enhance the input; an example is an electronic oscillator, or the howl produced by a loudspeaker that is placed too close to a microphone in the same circuit. A small random noise picked up by the microphone is amplified and reproduced by the loudspeaker. The microphone now picks it up again; it is further amplified, and fed from the speaker to microphone once again. This continues until the system is overloaded. In *negative feedback*, the output is used to reduce the input. In electronic amplifiers, stability is achieved, and distortion reduced, by using a system in which the input is decremented in proportion as the output increases. A similar negative feedback is used in *governors that reduce the fuel supply to an engine as its speed increases.

Fehling's test

Many biological processes rely on negative feedback. As the population of a species expands, so its food supply per individual is diminished; the result is that the population then begins to fall. Many biochemical processes are controlled by feedback *inhibition.

Fehling's test A chemical test to detect reducing sugars and aldehydes in solution, devised by the German chemist H. C. von Fehling (1812–85). *Fehling's solution* consists of Fehlings A (copper(II) sulphate solution) and Fehling's B (alkaline 2,3-dihydroxybutanedioate (sodium tartrate) solution), equal amounts of which are added to the test solution. After boiling, a positive result is indicated by the formation of a brick-red precipitate of copper(I) oxide. Methanal, being a strong reducing agent, also produces copper metal; ketones do not react.

feldspars A group of silicate minerals, the most abundant minerals in the earth's crust. They have a structure in which $(Si,Al)O_4$ tetrahedra are linked together with potassium, sodium, and calcium and very occasionally barium ions occupying the large spaces in the framework. The chemical composition of feldspars may be expressed as combinations of the four components:
anorthite (An), $CaAl_2Si_2O_8$;
albite (Ab), $NaAlSi_3O_8$;
orthoclase (Or), $KAlSi_3O_8$;
celsian (Ce), $BaAl_2Si_2O_8$.
The feldspars are subdivided into two groups: the *alkali feldspars* (including microcline, orthoclase, and sanidine), in which potassium is dominant with a smaller proportion of sodium and negligible calcium; and the *plagioclase feldspars*, which vary in composition in a series that ranges from pure sodium feldspar (albite) through to pure calcium feldspar (anorthite) with negligible potassium. Feldspars form

colourless, white, or pink crystals with a hardness of 6 on the Mohs' scale.

feldspathoids A group of alkali aluminosilicate minerals that are similar in chemical composition to the *feldspars but are relatively deficient in silica and richer in alkalis. The structure consists of a framework of $(Si,Al)O_4$ tetrahedra with aluminium and silicon atoms at their centres. The feldspathoids occur chiefly with feldspars but do not coexist with free quartz (SiO_2) as they react with silica to yield feldspars. The chief varieties of feldspathoids are:
nepheline, $KNa_3(AlSiO_4)_4$;
leucite, $KAlSi_2O_6$;
analcime, $NaAlSi_2O_6.H_2O$;
cancrinite, $Na_8(AlSiO_4)_6(HCO_3)_2$;
and the sodalite subgroup comprising
sodalite, $3(NaAlSiO_4).NaCl$;
nosean, $3(NaAlSiO_4).Na_2SO_4$;
haüyne, $3(NaAlSiO_4).CaSO_4$;
lazurite $(Na,Ca)_8(Al,Si)_{12}O_{24}(S,SO_4)$
(*see* lapis lazuli).

femto- Symbol f. A prefix used in the metric system to denote 10^{-15}. For example, 10^{-15} second = 1 femtosecond (fs).

femur 1. The thigh bone of terrestrial vertebrates. It articulates at one end with the pelvic girdle at the hip joint and at the other (via two *condyles) with the *tibia. **2.** The third segment of an insect's leg, attached to the *trochanter. *See also* coxa.

fenestra Either of the two delicate membranes between the *middle ear and the *inner ear. The lower membrane is the *fenestra ovalis (fenestra vestibuli*, or *oval window*). The stapes (the third *ear ossicle) transmits to the fenestra ovalis the vibrations it has received (via the other two ear ossicles) from the *tympanum (eardrum). Vibrations in the fenestra ovalis are transmitted to the *cochlea in the inner ear. The lower membrane, the *fenestra rotunda (fenestra cochleae*, or *round window*), vibrates in response to changes in pressure in the fluid (perilymph) surrounding the cochlea.

Fermat's principle The path taken by a ray of light between any two points in a system is always the path that takes the least time. This principle leads to the law of the rectilinear propagation of light and the laws of reflection and refraction. It was discovered by the French mathematician, Pierre de Fermat (1601–65).

fermentation A form of *anaerobic respiration occurring in certain microorganisms, e.g. yeasts. It comprises a series of biochemical reactions by which pyruvate (the end product of *glycolysis) is converted to ethanol and carbon dioxide. Fermentation is the basis of the baking, wine, and beer industries.

fermi A unit of length formerly used in nuclear physics. It is equal to 10^{-15} metre. In SI units this is equal to 1 femtometre (fm). It was named after the Italian-born US physicist Enrico Fermi (1901–54).

Fermi–Dirac statistics See quantum statistics.

Fermi level The energy in a solid at which the average number of particles per quantum state is $\frac{1}{2}$; i.e. one half of the quantum states are occupied. The Fermi level in conductors lies in the conduction band (see energy bands), in insulators it lies in the valence band, and in semiconductors it falls in the gap between the conduction band and the valence band. At absolute zero all the electrons would occupy energy levels up to the Fermi level and no higher levels would be occupied.

fermion An *elementary particle (or bound state of an elementary particle, e.g. an atomic nucleus or an atom) with half-integral spin; i.e. a particle that conforms to Fermi–Dirac statistics (see quantum statistics). Compare boson.

fermium Symbol Fm. A radioactive metallic transuranic element belonging to the *actinoids; a.n. 100; mass number of the most stable isotope 257 (half-life 10 days). Ten isotopes are known. The element was first identified by A. Ghiorso and associates in debris from the first hydrogen-bomb explosion in 1952.

ferns See Filicinae.

Ferrel's law A law proposed by the US meteorologist William Ferrel (1817–91). It states that a body moving across the surface of the earth will tend to be deflected to the right in the northern hemisphere and to the left in the southern hemisphere as a result of the earth's rotation.

ferric compounds Compounds of iron in its +3 oxidation state; e.g. ferric chloride is iron(III) chloride, $FeCl_3$.

ferricyanide A compound containing the complex ion $[Fe(CN)_6]^{3-}$, i.e. the hexacyanoferrate(III) ion.

ferrimagnetism See magnetism.

ferrite 1. A member of a class of mixed oxides $MO.Fe_2O_3$, where M is a metal such as cobalt, manganese, nickel, or zinc. The ferrites are ceramic materials that show either ferrimagnetism or ferromagnetism, but are not electrical conductors. For this reason they are used in high-frequency circuits as magnetic cores. **2.** See steel.

ferroalloys Alloys of iron with other elements made by smelting mixtures of iron ore and the metal ore; e.g. ferrochromium, ferrovanadium, ferromanganese, ferrosilicon, etc. They are used in making alloy *steels.

ferrocene An orange-red crystalline solid, $Fe(C_5H_5)_2$; m.p. 173°C. It can be made by adding the ionic compound $Na^+C_5H_5^-$ (cyclopentadienyl sodium, made from sodium and cyclopentadiene) to iron(III) chloride. In ferrocene, the two rings are parallel, with the iron ion sandwiched

ferrocyanide

Ferrocene

between them (hence the name *sandwich compound*: see formula). The bonding is between pi orbitals on the rings and *d*-orbitals on the Fe^{2+} ion. The compound can undergo electrophilic substitution on the C_5H_5 rings (they have some aromatic character). It can also be oxidized to the blue ion $(C_5H_5)_2Fe^+$. Ferrocene is the first of a class of similar complexes called *metallocenes*. Its systematic name is *di-π-cyclopentadienyl iron(II)*.

ferrocyanide A compound containing the complex ion $[Fe(CN)_6]^{4-}$, i.e. the hexacyanoferrate(II) ion.

ferroelectric materials Ceramic dielectrics, such as Rochelle salt and barium titanate, that have a domain structure making them analogous to ferromagnetic materials (*see* magnetism). They exhibit hysteresis and usually the *piezoelectric effect.

ferromagnetism *See* magnetism.

ferrosoferric oxide *See* tri-iron tetroxide.

ferrous compounds Compounds of iron in its +2 oxidation state; e.g. ferrous chloride is iron(II) chloride, $FeCl_2$.

fertile material A nuclide that can absorb a neutron to form a *fissile material. Uranium–238, for example, absorbs a neutron to form uranium–239, which decays to plutonium–239. This is the type of conversion that occurs in a breeder reactor (*see* nuclear reactor).

fertilization (syngamy) The fusion of the nuclei of the male and female gametes (reproductive cells) during the process of sexual reproduction to form a *zygote*. As each gamete contains only half the correct number of chromosomes, fertilization and zygote formation results in a cell with the full complement of chromosomes, half of which are derived from each of the parents. In animals the process involves fusion of the nuclei of a spermatozoan and an ovum. In most aquatic animals (e.g. fish) this takes place in the surrounding water, into which the gametes are shed. Among most terrestrial animals (e.g. insects, many mammals) fertilization occurs in the body of the female, into which the sperms are introduced. In flowering plants, after *pollination, the male gamete (pollen) produces a *pollen tube, which grows down into the female reproductive organ (carpel) to enable a pollen nucleus to fuse with the egg nucleus.

In *self-fertilization* the male and female gametes are derived from the same individual; in *cross-fertilization* they are derived from different individuals. Among plants, self-fertilization (also called *autogamy*) is common in many cultivated species, e.g. wheat and oats. In cross-fertilization (also called *allogamy*) the pollen comes either from another flower of the same plant or from a different plant (*see also* incompatibility).

FET *See* transistor.

fetal membranes *See* extraembryonic membranes.

fetus (foetus) The embryo of a mammal, especially a human, when development has reached a stage at which the main features of the adult form are recognizable. In humans the embryo from eight weeks to birth is called a fetus.

Feulgen's test A histochemical test in which the distribution of DNA in the chromosomes of dividing cell nuclei can be observed. It was devised by the German chemist R.

Feulgen (1884–1955). A tissue section is first treated with dilute hydrochloric acid to remove the purine bases of the DNA, thus exposing the aldehyde groups of the sugar deoxyribose. The section is then immersed in *Schiff's reagent, which combines with the aldehyde groups to form a magenta-coloured compound.

Feynman diagram *See* quantum electrodynamics.

fibre 1. An elongated plant cell whose walls are extensively (usually completely) thickened with lignin. Fibres are found in the vascular tissue, usually in the xylem, where they provide structural support. The term is often used loosely to mean any kind of xylem element. The fibres of many species, e.g. flax, are of commercial importance. **2.** Any of various threadlike structures in the animal body, such as a muscle fibre, a nerve fibre, or a collagen fibre. **3.** (*or* **dietary fibre**) The part of food that cannot be digested and absorbed to produce energy. Dietary fibre falls into four groups: *cellulose*, *hemicelluloses*, *lignins*, and *pectins*. Highly refined foods, such as sucrose, contain no dietary fibre. Foods with a high fibre content include wholemeal cereals and flour, root vegetables, nuts, and fruit. Dietary fibre is considered by some to be helpful in the prevention of many of the diseases of Western civilization, such as diverticulosis, constipation, appendicitis, obesity, and diabetes mellitus.

fibre optics *See* optical fibres.

fibrin The insoluble protein that forms fibres at the site of an injury and is the foundation of a blood clot. *See* blood clotting.

fibrinogen The protein dissolved in the blood plasma that, when suitably activated, is converted to insoluble *fibrin fibres. *See* blood clotting.

fibrinolysis The breakdown of the protein *fibrin by the enzyme *plasmin*

(or *fibrinolysin*), which occurs when blood clots are removed from the circulation.

fibroblast A cell that secretes fibres in the intercellular substance of *connective tissue. The cells are long, flat, and star-shaped and lie close to collagen fibres.

fibula The smaller and outer of the two bones between the knee and the ankle in terrestrial vertebrates. *Compare* tibia.

field A region in which a body experiences a *force as the result of the presence of some other body or bodies. A field is thus a method of representing the way in which bodies are able to influence each other. For example, a body that has mass is surrounded by a region in which another body that has mass experiences a force tending to draw the two bodies together. This is the gravitational field (*see* Newton's law of gravitation). The other three *fundamental interactions can also be represented by means of fields of force. However in the case of the *magnetic field and *electric field that together create the electromagnetic interaction, the force can vary in direction according to the character of the field. For example, in the field surrounding a negatively charged body, a positively charged body will experience a force of attraction, while another negatively charged body is repelled.

The strength of any field can be described as the ratio of the force experienced by a small appropriate specimen to the relevant property of that specimen, e.g. force/mass for the gravitational field. *See also* quantum field theory.

field coil The coil in an electrical machine that produces the magnetic field.

field-effect transistor (FET) *See* transistor.

field-emission microscope

field-emission microscope A type
of electron microscope in which a
high negative voltage is applied to a
metal tip placed in an evacuated vessel some distance from a glass screen
with a fluorescent coating. The tip
produces electrons by *field emission*,
i.e. the emission of electrons from an
unheated sharp metal part as a result
of a high electric field. The emitted
electrons form an enlarged pattern on
the fluorescent screen, related to the
individual exposed planes of atoms.
As the resolution of the instrument is
limited by the vibrations of the metal
atoms, it is helpful to cool the tip in
liquid helium. Although the individual
atoms forming the point are not displayed, individual adsorbed atoms of
other substances can be, and their
activity is observable.

**field-ionization microscope (field-
ion microscope)** A type of electron
microscope that is similar in principle
to the *field-emission microscope,
except that a high positive voltage is
applied to the metal tip, which is surrounded by low-pressure gas (usually
helium) rather than a vacuum. The
image is formed in this case by *field
ionization*: ionization at the surface of
an unheated solid as a result of a
strong electric field creating positive
ions by electron transfer from surrounding atoms or molecules. The
image is formed by ions striking the
fluorescent screen. Individual atoms
on the surface of the tip can be
resolved and, in certain cases,
adsorbed atoms may be detected.

field lens The lens in the compound
eye-piece of an optical instrument
that is furthest from the eye. Its function is to increase the field of view by
refracting towards the main eye lens
rays that would otherwise miss it.

field magnet The magnet that provides the magnetic field in an electrical machine. In some small dynamos
and motors it is a permanent magnet

but in most machines it is an electromagnet.

filament 1. (in zoology) A long slender hairlike structure, such as any of
the *barbs of a bird's feather. **2.** (in
botany) The stalk of the *stamen in a
flower. It bears the anther and consists mainly of conducting tissue.

file A collection of data stored in a
computer. It may consist of program
instructions or numerical, textual, or
graphical information. It usually consists of a set of similar or related
records.

Filicinae A class of mainly terrestrial
vascular plants – the ferns – belonging to the *Pteropsida. Ferns are perennial plants bearing large conspicuous leaves (fronds) usually arising
from either a rhizome or a short erect
stem. Bracken is a common example.
Only the tree ferns have stems that
reach an appreciable height. There is
a characteristic uncurling of the
young leaves as they expand into the
adult form. Spores are borne on the
underside of specialized leaves
(sporophylls).

filler A solid inert material added to
a synthetic resin or rubber, either to
change its physical properties or
simply to dilute it for economy.

film badge A lapel badge containing
masked photographic film worn by
personnel who could be exposed to
ionizing radiation. The film is
developed to indicate the extent that
the wearer has been exposed to harmful radiation.

filoplumes Minute hairlike *feathers
consisting of a shaft (*rachis*) bearing a
few unattached barbs. They are found
between the contour feathers.

filter 1. (in chemistry) A device for
separating solid particles from a liquid or gas. The simplest laboratory
filter for liquids is a funnel in which
a cone of paper (*filter paper*) is
placed. Special containers with a
porous base of sintered glass are also

used. *See also* Gooch crucible. **2.** (in physics) A device placed in the path of a beam of radiation to alter its frequency distribution. For example, a plane pigmented piece of glass may be placed over a camera lens to alter the relative intensity of the component wavelengths of the beam entering the camera.

filter feeding A method of feeding in which tiny food particles are ingested from the surrounding water. It is used by many aquatic invertebrates. *See also* ciliary feeding.

filter pump A simple laboratory vacuum pump in which air is removed from a system by a jet of water forced through a narrow nozzle. The lowest pressure possible is the vapour pressure of water.

filtrate The clear liquid obtained by filtration.

filtration The process of separating solid particles using a filter. In *vacuum filtration*, the liquid is drawn through the filter by a vacuum pump. *Ultrafiltration* is filtration under pressure; in the *nephrons of the vertebrate kidney blood is subjected to ultrafiltration to remove excess water and waste products that make up the urine.

finder A small low-powered astronomical telescope, with a wide field of view, that is fixed to a large astronomical telescope so that the large telescope can be pointed in the correct direction to observe a particular celestial body.

fine chemicals Chemicals produced industrially in relatively small quantities and with a high purity; e.g. dyes and drugs.

fineness of gold A measure of the purity of a gold alloy, defined as the parts of gold in 1000 parts of the alloy by mass. Gold with a fineness of 750 contains 75% gold, i.e. 18 *carat gold.

fine structure Closely spaced optical spectral lines arising from *transitions between energy levels that are split by the vibrational or rotational motion of a molecule or by electron spin. They are visible only at high resolution. *Hyperfine structure*, visible only at very high resolution, results from the influence of the atomic nucleus on the allowed energy levels of the atom.

finite series *See* series.

fins The locomotory organs of aquatic vertebrates. In fish there are typically one or more *dorsal* and *ventral fins* (sometimes continuous), whose function is balance; a *caudal fin* around the tail, which is the main propulsive organ; and two paired fins: the *pectoral fins* attached to the pectoral (shoulder) girdle and the *pelvic fins* attached to the pelvic (hip) girdle, which are used in steering. These paired fins are homologous with the limbs of tetrapods. Fins are strengthened by a number of flexible fin rays, which may be cartilaginous, bony and jointed, horny, or fibrous and jointed.

firedamp Methane formed in coal mines.

first-order reaction *See* order.

Fischer–Tropsch process An industrial method of making hydrocarbon fuels from carbon monoxide and hydrogen. The process was invented in 1933 and used by Germany in World War II to produce motor fuel. Hydrogen and carbon monoxide are mixed in the ratio 2:1 (water gas was used with added hydrogen) and passed at 200°C over a nickel or cobalt catalyst. The resulting hydrocarbon mixture can be separated into a higher-boiling fraction for Diesel engines and a lower-boiling gasoline fraction. The gasoline fraction contains a high proportion of straight-chain hydrocarbons and has to be reformed for use in motor fuel. Alcohols, aldehydes, and ketones are also present. The process is also used in the manufacture of SNG from coal. It is named after the German chemist Franz Fischer (1852–1932) and the Czech Hans Tropsch (1839–1935).

fish *See* Pisces.

fissile material A nuclide of an element that undergoes nuclear fission, either spontaneously or when irradiated by neutrons. Fissile nuclides, such as uranium–235 and plutonium–239, are used in *nuclear reactors and nuclear weapons. *Compare* fertile material.

fission 1. (in biology) A type of asexual reproduction occurring in some unicellular organisms, e.g. diatoms, protozoans, and bacteria, in which the parent cell divides to form two (*binary fission*) or more (*multiple fission*) similar daughter cells. **2.** (in physics) *See* nuclear fission.

fission products *See* nuclear fission.

fission-track dating A method of estimating the age of glass and other mineral objects by observing the tracks made in them by the fission fragments of the uranium nuclei that they contain. By irradiating the objects with neutrons to induce fission and comparing the density and number of the tracks before and after irradiation it is possible to estimate the time that has elapsed since the object solidified.

fitness (in evolution) The condition of an organism that is well adapted to its environment, as measured by its ability to reproduce itself. *See also* inclusive fitness.

Fittig reaction *See* Wurtz reaction.

Fitzgerald contraction *See* Lorentz–Fitzgerald contraction.

fixation 1. The first stage in the preparation of a specimen for microscopical examination, in which the tissue is killed and preserved in as natural a state as possible by immersion in a chemical *fixative*. The fixa-tive prevents the distortion of cell components by denaturing its constituent protein. Some commonly used fixatives are formaldehyde, ethanol, and Bouin's fluid (for light microscopy), and osmium tetroxide and gluteraldehyde (for electron microscopy). Fixation may also be brought about by heat. **2.** *See* nitrogen fixation.

fixed action pattern *See* instinct.

fixed point A temperature that can be accurately reproduced to enable it to be used as the basis of a *temperature scale.

fixed star One of very many heavenly bodies that does not appear to alter its position on the *celestial sphere. They were so called to distinguish them from the planets, which were once known as *wandering stars*. The discovery of the *proper motion of stars in the 18th century established that stars are not fixed in the sky although, because of their immense distances from the solar system, they may appear to be so.

Fizeau's method A method of measuring the speed of light, invented by the French physicist Armand Fizeau (1819–96) in 1849. A cogwheel rotating at high speed enables a series of flashes to be transmitted to a distant mirror. The light reflected back to the cogwheel is observed and the speed of light calculated from the rates of rotation of the wheel required to produce an eclipse of the returning light.

flaccid (in botany) Describing plant tissue that has become soft and less rigid than normal because the cytoplasm within its cells has shrunk and contracted away from the cell walls through loss of water (*see* plasmolysis).

flagellum A relatively long (up to 150 μm) fine whiplike structure present on the surface of certain cells, particularly motile reproductive cells (e.g. spermatozoa), bacteria, and certain protozoans. Flagella occur singly or in small groups. Beating of flagella produces movement of the cell, but some flagella (e.g. in *Hydra*) beat to cause movement of the surrounding fluid. *Compare* cilium.

flame A hot luminous mixture of gases undergoing combustion. The chemical reactions in a flame are mainly free-radical chain reactions and the light comes from fluorescence of excited molecules or ions or from incandescence of small solid particles (e.g. carbon).

flame test A simple test for metals, in which a small amount of the sample (usually moistened with hydrochloric acid) is placed on the end of a platinum wire and held in a Bunsen flame. Certain metals can be detected by the colour produced: barium (green), calcium (brick red), lithium (crimson), potassium (lilac), sodium (yellow), strontium (red).

flash photolysis A technique for studying free-radical reactions in gases. The apparatus used typically consists of a long glass or quartz tube holding the gas, with a lamp outside the tube suitable for producing an intense flash of light. This dissociates molecules in the sample creating free radicals, which can be detected spectroscopically by a beam of light passed down the axis of the tube. It is possible to focus the spectrometer on an absorption line for a particular product and measure its change in intensity with time using an oscilloscope. In this way the kinetics of very fast free-radical gas reactions can be studied.

flash point The temperature at which the vapour above a volatile liquid forms a combustible mixture with air. At the flash point the application of a naked flame gives a momentary flash rather than sustained combustion, for which the temperature is too low.

flatworms *See* Platyhelminthes.

flavin adenine dinucleotide *See* FAD.

flavonoid One of a group of naturally occurring phenolic compounds many of which are plant pigments. They include the *anthocyanins, flavonols,* and *flavones.* Patterns of flavonoid distribution have been used in taxonomic studies of plant species.

flavoprotein *See* FAD.

flavour *See* elementary particles.

fleas *See* Siphonaptera.

Fleming's rules Rules to assist in remembering the relative directions of the field, current, and force in electrical machines. The left hand refers to motors, the right hand to generators. If the forefinger, second finger, and thumb of the left hand are extended at right angles to each other, the forefinger indicates the direction of the field, the second finger the direction of the current, and the thumb the direction of the force. If the right hand is used the digits indicate these directions in a generator. The mnemonic was invented by Sir John Ambrose Fleming (1849–1945).

flies *See* Diptera.

flint (chert) Very hard dense nodules of microcrystalline quartz and chalcedony found in chalk and limestone.

flip-flop (bistable circuit) An electronic circuit that has two stable states. It is switched from one stable state to the other by means of a triggering pulse. They are extensively used as *logic circuits in computers.

flocculation The process of aggregating into larger clumps.

flocculent Aggregated in woolly masses; used to describe precipitates.

floppy disk A flexible plastic disk with a magnetic coating encased in a stiff envelope. It is used to store information in a small computer system. *See* magnetic disk.

flower The structure in angiosperms (flowering plants) that bears the

Section through a monocarpellary flower at the time of pollination

organs for sexual reproduction. Flowers are very variable in form, ranging from the small green insignificant wind-pollinated flowers of many grasses to spectacular brightly coloured insect-pollinated flowers. Flowers are often grouped together into *inflorescences, some of which (e.g. that of dandelion) are so compacted as to resemble a single flower. Typically flowers consist of a receptacle that bears sepals, petals, stamens, and carpels (see illustration). The flower parts are adapted to bring about pollination and fertilization resulting in the formation of seeds and fruits. The sepals are usually green and leaflike and protect the flower bud. The petals of insect-pollinated flowers are adapted in many ingenious ways to attract insects and, in some instances, other animals. For example, some flowers are adapted to attract short-tongued insects by having an open shallow *corolla tube and nectar situated in an exposed position. Flowers adapted for pollination by long-tongued insects have a long corolla tube of fused petals with nectar in a concealed position. The tongue of the insect brushes against the anthers and stigma before reaching the nectar. Wind-pollinated flowers, in contrast, are inconspicuous. The anthers dangle outside the corolla and the stigmas have a feathery surface to catch the pollen grains. Some species are adapted for self-pollination and have small flowers, no nectar, and stamens and carpels that mature simultaneously.

flowering plants *See* Angiospermae.

fluidics The use of jets of fluid in pipes to perform many of the control functions usually performed by electronic devices. Being about one million times slower than electronic devices, fluidic systems are useful where delay lines are required. They are also less sensitive to high temperatures, strong magnetic fields, and ionizing radiation than electronic devices.

fluidization A technique used in some industrial processes in which solid particles suspended in a stream of gas are treated as if they were in the liquid state. Fluidization is useful for transporting powders, such as coal dust. *Fluidized beds*, in which solid particles are suspended in an upward stream, are extensively used in the chemical industry, particularly in catalytic reactions where the powdered catalyst has a high surface area.

fluid mechanics The study of fluids at rest and in motion. *Fluid statics* is concerned with the pressures and forces exerted on liquids and gases at rest. *Hydrostatics is specifically concerned with the behaviour of liquids at rest. In *fluid dynamics* the forces exerted on fluids, and the motion that results from these forces, are examined. It can be divided into *hydrodynamics: the motion of liquids (not only water); and aerodynamics: the motion of gases.

Fluid dynamics is an important science used to solve many of the problems arising in aeronautical, chemical, mechanical, and civil engineering. It also enables many natural phenomena, such as the flight of birds, the swimming of fish, and the development of weather conditions, to be studied scientifically.

flukes *See* Trematoda.

fluorescein A yellowish-red dye that produces yellow solutions with a green fluorescence. It is used in tracing water flow and as an *absorption indicator.

fluorescence *See* luminescence.

fluorescent light *See* electric lighting.

fluoridation The process of adding very small amounts of fluorine salts (e.g. sodium fluoride, NaF) to drinking water to prevent tooth decay. The fluoride becomes incorporated into the substance of the growing teeth and reduces the incidence of caries.

fluoride *See* halide.

fluorination A chemical reaction in which a fluorine atom is introduced into a molecule. *See* halogenation.

fluorine Symbol F. A poisonous pale yellow gaseous element belonging to group VII of the periodic table (the *halogens); a.n. 9; r.a.m. 18.9984; d. 1.7 g dm^{-3}; m.p. $-219.62°C$; b.p. $-188.1°C$. The main mineral sources are *fluorite (CaF_2) and *cryolite (Na_3AlF). The element is obtained by electrolysis of a molten mixture of potassium fluoride and hydrogen fluoride. It is used in the synthesis of organic fluorine compounds. Chemically, it is the most reactive and electronegative of all elements. It is a highly dangerous compound, causing severe chemical burns on contact with the skin. The element was identified by Scheele in 1771 and first isolated by Moissan in 1886.

fluorite (fluorspar) A mineral form of calcium fluoride, CaF_2, crystallizing in the cubic system. It is variable in colour; the most common fluorites are green and purple (blue john), but other forms are white, yellow, or brown. Fluorite is used chiefly as a flux material in the smelting of iron and steel; it is also used as a source of fluorine and hydrofluoric acid and in the ceramic and optical-glass industries.

fluorite structure *See* calcium fluoride structure.

fluorocarbons Compounds obtained by replacing the hydrogen atoms of hydrocarbons by fluorine atoms. Their inertness and high stability to temperature make them suitable for a variety of uses as oils, polymers, etc. *See also* chlorofluorocarbon; halon.

flux 1. A substance applied to the surfaces of metals to be soldered to inhibit oxidation. 2. A substance used in the smelting of metals to assist in the removal of impurities as slag. 3. The number of particles flowing per unit area of cross section in a beam of particles. 4. *See* luminous flux. 5. *See* magnetic flux. 6. *See* electric flux.

flux density 1. *See* magnetic flux density. 2. *See* electric flux density.

monoclinal fold   anticlinal fold   synclinal fold

direction of pressure   direction of pressure

overturned fold   recumbent fold

Types of fold

fluxional molecule A molecule that undergoes alternate very rapid rearrangements of its atoms and thus only has a specific structure for a very short period of time. For example, the molecule ClF_3 has a T-shape at low temperatures ($-60°C$); at room temperature the fluorine atoms change position very rapidly and appear to have identical positions.

fluxmeter An instrument used to measure *magnetic flux. It is used in conjunction with a coil (the *search coil*) and resembles a moving-coil galvanometer except that there are no restoring springs. A change in the magnetic flux induces a momentary current in the search coil and in the coil of the meter, which turns in proportion and stays in the deflected position. This type of instrument has been largely superseded by a type using the Hall probe (*see* Hall effect).

FM (frequency modulation) *See* modulation.

f-number *See* aperture.

foam A dispersion of bubbles in a liquid. Foams can be stabilized by *surfactants. Solid foams (e.g. expanded polystyrene or foam rubber) are made by foaming the liquid and allowing it to set. *See also* colloid.

focal length The distance between the *optical centre of a lens or pole of a spherical mirror and its *principal focus.

focal point *See* focus.

focal ratio *See* aperture.

focus 1. (in optics) Any point in an optical system through or towards which rays of light are converged. It is sometimes called the *focal point* and sometimes loosely used to mean *principal focus or (particularly by photographers) *focal length. 2. (in mathematics) *See* conic; ellipse.

foetus *See* fetus.

folacin *See* folic acid.

fold A wavelike form in layered sedimentary rock strata that results from deformational processes in the earth's crust. Basin-shaped folds in which the beds of rock dip towards each other are known as *synclines*; those in which the beds of rock are folded into an arch shape are known as *anticlines*. More complex folds result where the rock strata are subjected to intense horizontal pressures.

folic acid (folacin) A vitamin of the *vitamin B complex. In its active form, tetrahydrofolic acid, it is a *coenzyme in various reactions involved in the metabolism of amino acids, purines, and pyrimidines. It is synthesized by intestinal bacteria and is widespread in food, especially green leafy vegetables. Deficiency causes poor growth and nutritional anaemia.

follicle 1. (in animal anatomy) Any enclosing cluster of cells that protects and nourishes a cell or structure within. For example, follicles in the *ovary contain developing egg cells, while *hair follicles envelop the roots of hairs. **2.** (in botany) A dry fruit that, when ripe, splits along one side to release its seeds. It is formed from a single carpel containing one or more seeds. Follicles do not occur singly but are grouped to form clusters (*etaerios*). Examples include larkspur, columbine, and monk's hood.

follicle-stimulating hormone (FSH) A hormone, secreted by the anterior pituitary gland in mammals, that stimulates, in female mammals, ripening of specialized structures in the ovary (*Graafian follicles) that produce ova and, in males, the formation of sperm in the testis. It is a major constituent of fertility drugs, used to treat failure of ovulation and decreased sperm production. *See also* gonadotrophin.

food chain The transfer of energy from green plants (the primary producers) through a sequence of organisms in which each eats the one below it in the chain and is eaten by the one above. Thus plants are eaten by herbivores, which are then eaten by carnivores. These may in turn be eaten by different carnivores. The position an organism occupies in a food chain is known as its *trophic level. In practice, many animals feed at several different trophic levels, resulting in a more complex set of feeding relationships known as a *food web*. *See* bioenergetics; consumer; producer; pyramid of numbers.

fool's gold *See* pyrite.

foot The unit of length in *f.p.s units. It is equal to one-third of a yard and is now therefore defined as 0.3048 metre. Several units based on the foot were formerly used in sci-

ence, including the units of work, the *foot-pound-force* and the *foot-poundal*, and the illumination units, the *foot-candle* and the *foot-lambert*. These have all been replaced by SI units.

foramen An aperture in an animal part or organ, especially one in a bone or cartilage. For example, the *foramen magnum* is the opening at the base of the skull through which the *spinal cord passes.

forbidden band *See* energy bands.

force Symbol *F*. The agency that tends to change the momentum of a massive body, defined as being proportional to the rate of increase of momentum. For a body of mass *m* travelling at a velocity *v*, the momentum is *mv*. In any coherent system of units the force is therefore given by $F = d(mv)/dt$. If the mass is constant $F = m dv/dt = ma$, where *a* is the acceleration (*see* Newton's laws of motion). The SI unit of force is the newton. Forces occur always in equal and opposite action–reaction pairs between bodies, though it is often convenient to think of one body being in a force *field.

forced convection *See* convection.

force ratio (mechanical advantage) The ratio of the output force (load) of a machine to the input force (effort).

forebrain (prosencephalon) One of the three sections of the brain of a vertebrate embryo. The forebrain develops to form the *cerebrum, *hypothalamus, and *thalamus in the adult. *Compare* hindbrain; midbrain.

foregut 1. The anterior region of the alimentary canal of vertebrates, up to the anterior part of the duodenum. **2.** The anterior part of the alimentary canal of arthropods. *See also* hindgut; midgut.

form 1. A category used in the *classification of organisms into which different types of a variety may be placed. **2.** Any distinct variant within a species. Seasonal variants, e.g. the tawny brown (summer) and blue-white (winter) forms of the blue hare, may be called forms, as may the different types that constitute a *polymorphism.

formaldehyde *See* methanal.

formalin A colourless solution of methanal (formaldehyde) in water with methanol as a stabilizer; r.d. 1.075–1.085. When kept at temperatures below 25°C a white polymer of methanal separates out. It is used as a disinfectant and preservative for biological specimens.

formate *See* methanoate.

formic acid *See* methanoic acid.

formula 1. (in chemistry) A way of representing a chemical compound using symbols for the atoms present. Subscripts are used for the numbers of atoms. The *molecular formula* simply gives the types and numbers of atoms present. For example, the molecular formula of ethanoic acid is $C_2H_4O_2$. The *empirical formula* gives the atoms in their simplest ratio; for ethanoic acid it is CH_2O. The *structural formula* gives an indication of the way the atoms are arranged. Commonly, this is done by dividing the formula into groups; ethanoic acid can be written $CH_3.CO.OH$ (or more usually simply CH_3COOH). Structural formulae can also show the arrangement of atoms or groups in space. **2.** (in mathematics and physics) A rule or law expressed in algebraic symbols.

formula weight The relative molecular mass of a compound as calculated from its molecular formula.

formyl group The group $HCO-$.

Fortin barometer *See* barometer.

fossil The remains or traces of any organism that lived in the geological past. In general only the hard parts of organisms become fossilized (e.g. bones, teeth, shells, and wood) but under certain circumstances the entire organism is preserved. For example, virtually unaltered fossils of extinct mammals, such as the woolly mammoth and woolly rhinoceros, have been found preserved in ice in the Arctic. Small organisms or parts of organisms (e.g. insects, leaves, flowers) have been preserved in *amber.

In the majority of fossils the organism has been turned to stone – a process known as *petrification*. This may take one of three forms. In *permineralization*, solutions originating underground fill the microscopic cavities in the organism. Minerals in these solutions (e.g. silica or calcite) may actually replace the original material of the organism so that even microscopic structures may be preserved; this process is known as *replacement* (or *mineralization*). A third form of petrification – *carbonization* (or *distillation*) – occurs in certain soft tissues that are composed chiefly of compounds of carbon, hydrogen, and oxygen (e.g. cellulose). After the organism has been buried, and in the absence of oxygen, carbon dioxide and water are liberated until only free carbon remains. This forms a black carbon film in the rock outlining the original organism. *Moulds* are formed when the original fossil is dissolved away leaving a mould of its outline in the solid rock. The deposition of mineral matter from underground solutions in a mould forms a *cast*. Palaeontologists often produce casts from moulds using such substances as dental wax. Moulds of thin organisms (e.g. leaves) are commonly known as *imprints*. *Trace fossils* are the fossilized remnants of the evidence of animal life, such as tracks, trails, footprints, burrows, and *coprolites* (fossilized faeces).

The ideal conditions for the formation of fossils occur in areas of rapid sedimentation, especially those parts of the seabed that lie below the zone of wave disturbance. *See also* chemical fossil; index fossil; microfossil.

fossil fuel Coal, oil, and natural gas, the fuels used by man as a source of energy. They are formed from the remains of living organisms and all have a high carbon or hydrogen content. Their value as fuels relies on the exothermic oxidation of carbon to form carbon dioxide ($C + O_2 \rightarrow CO_2$) and the oxidation of hydrogen to form water ($H_2 + \frac{1}{2}O_2 \rightarrow H_2O$).

fossil hominid *See* hominid.

Foucault pendulum A simple pendulum in which a heavy bob attached to a long wire is free to swing in any direction. As a result of the earth's rotation, the plane of the pendulum's swing slowly turns (at the poles of the earth it makes one complete revolution in 24 hours). It was devised by the French physicist Jean Bernard Léon Foucault (1819–68) in 1851, when it was used to demonstrate the earth's rotation.

Fourier series An expansion of a periodic function as a series of trigonometric functions. Thus,

$$f(x) = a_0 + (a_1\cos x + b_1\sin x) + (a_2\cos 2x + b_2\sin 2x) + \ldots,$$

where a_0, a_1, b_1, b_2, etc., are constants, called *Fourier coefficients*. The series was first formulated by J. B. J. Fourier (1768–1830) and is used in harmonic analysis (*Fourier analysis*) to determine the harmonic components of a complex wave.

fourth dimension *See* space–time.

fovea (fovea centralis) A shallow depression in the *retina of the eye, opposite the lens, that is present in some vertebrates. This area contains a large concentration of *cones with only a thin layer of overlying nerves. It is therefore specialized for the perception of colour and sharp intense images. The clarity is enhanced when light is focused on the foveae of both retinas simultaneously. *See* binocular vision.

f.p.s. units The British system of units based on the foot, pound, and second. It has now been replaced for all scientific purposes by SI units.

fractal A curve or surface generated by a process involving successive subdivision. For example, a *snowflake curve* can be produced by starting with an equilateral triangle and dividing each side into three segments. The middle segments are then replaced by two equal segments, which would form the sides of a smaller equilateral triangle. This gives a 12-sided star-shaped figure. The next stage is to subdivide each of the sides of this figure in the same way, and so on. The result is a developing figure that resembles a snowflake. In the limit, this figure has 'fractional dimension' – i.e. a dimension between that of a line (1) and a surface (2); the dimension of the snowflake curve is 1.26. The study of this type of 'self-similar' figure is used in certain branches of physics – for example, crystal growth. Fractals are also important in *chaos theory.

fraction *See* fractional distillation.

fractional crystallization A method of separating a mixture of soluble solids by dissolving them in a suitable hot solvent and then lowering the temperature slowly. The least soluble component will crystallize out first, leaving the other components in solution. By controlling the temperature, it is sometimes possible to remove each component in turn.

fractional distillation (fractionation) The separation of a mixture of liquids by distillation. Effective separation can be achieved by using a long vertical column (*fractionating column*) attached to the distillation vessel and filled with glass beads. Vapour from the liquid rises up the column until it condenses and runs back into the vessel. The rising vapour in the column flows over the descending liquid, and eventually a steady state is reached in which there is a decreasing temperature gradient up the column. The vapour in the column has more volatile components towards the top and less volatile components at the bottom. Various *fractions* of the mixture can be drawn off at points on the column. Industrially, fractional distillation is performed in large towers containing many perforated trays. It is used extensively in petroleum refining.

fractionating column See fractional distillation.

fractionation See fractional distillation.

fragmentation A method of asexual reproduction, occurring in some invertebrate animals, in which parts of the organism break off and subsequently differentiate and develop into new individuals. It occurs especially in certain coelenterates and annelids. In some, regeneration may occur before separation, producing chains of individuals budding from the parent.

frame of reference A set of axes, taken as being for practical purposes at rest, that enables the position of a point (or body) in space to be defined at any instant of time. In a four-dimensional continuum (*see* space–time) a frame of reference consists of a set of four coordinate axes, three spatial and one of time.

francium Symbol Fr. A radioactive element belonging to *group I of the periodic table; a.n. 87; r.d. 2.4; m.p. $27 \pm 1°C$; b.p. $677 \pm 1°C$. The element is found in uranium and thorium ores. All 22 known isotopes are radioactive, the most stable being francium –223. The existence of francium was confirmed by Marguerite Perey in 1939.

Frasch process A method of obtaining sulphur from underground deposits using a tube consisting of three concentric pipes. Superheated steam is passed down the outer pipe to melt the sulphur, which is forced up through the middle pipe by com-pressed air fed through the inner tube. The steam in the outer casing keeps the sulphur molten in the pipe.

fraternal twins (dizygotic twins) Two individuals that result from a single pregnancy, each having developed from a separate fertilized egg. The two egg cells contain different combinations of *alleles and so do the two sperm that fertilize them. The twins therefore have no more genetic similarity than brothers or sisters from single births. *Compare* identical twins.

Fraunhofer diffraction A form of *diffraction in which the light source and the receiving screen are in effect at infinite distances from the diffracting object, so that the wave fronts can be treated as planar rather than spherical. In practice it involves parallel beams of light. It can be regarded as an extreme case of *Fresnel diffraction but is of more practical use in explaining single and multiple slit patterns. It was studied by the German optician Joseph von Fraunhofer (1787–1826).

Fraunhofer lines Dark lines in the solar spectrum that result from the absorption by elements in the solar chromosphere of some of the wavelengths of the visible radiation emitted by the hot interior of the sun.

free electron See electron.

free energy A measure of a system's ability to do work. The *Gibbs free energy* (or *Gibbs function*), G, is defined by $G = H - TS$, where G is the energy liberated or absorbed in a reversible process at constant pressure and constant temperature (T), H is the *enthalpy and S the *entropy of the system. Changes in Gibbs free energy, ΔG, are useful in indicating the conditions under which a chemical reaction will occur. If ΔG is positive the reaction will only occur if energy is supplied to force it away from the equilibrium position (i.e. when $\Delta G = 0$). If ΔG is negative the reaction will proceed spontaneously to equilibrium.

The *Helmholtz free energy* (or *Helmholtz function*), *F*, is defined by $F = U - TS$, where *U* is the *internal energy. For a reversible isothermal process, ΔF represents the useful work available.

free fall Motion resulting from a gravitational field that is unimpeded by a medium that would provide a frictional retarding force or buoyancy. In the earth's gravitational field, free fall takes place at a constant acceleration, known as the *acceleration of free fall.

free radical An atom or group of atoms with an unpaired valence electron. Free radicals can be produced by photolysis or pyrolysis in which a bond is broken without forming ions (*see* homolytic fission). Because of their free valency, most free radicals are extremely reactive. *See also* chain reaction.

free space A region in which there is no matter and no electromagnetic or gravitational fields. It has a temperature of absolute zero, unit refractive index, and the speed of light is its maximum value. The electric constant (*see* permittivity) and the magnetic constant (*see* permeability) are defined for free space.

freeze drying A process used in dehydrating food, blood plasma, and other heat-sensitive substances. The product is deep-frozen and the ice trapped in it is removed by reducing the pressure and causing it to sublime. The water vapour is then removed, leaving an undamaged dry product.

freezing mixture A mixture of components that produces a low temperature. For example, a mixture of ice and sodium chloride gives a temperature of $-20°C$.

Frenkel defect *See* defect.

freon *See* chlorofluorocarbon.

frequency Symbol *f*. The rate of repetition of a regular event. The number of cycles of a wave, or some other oscillation or vibration, per second is expressed in *hertz (cycles per second). The frequency (*f*) of a wave motion is given by $f = c/\lambda$, where *c* is the velocity of propagation and λ is the wavelength. The frequency associated with a quantum of electromagnetic energy is given by $f = E/h$, where *E* is the quantum's energy and *h* is the Planck constant.

frequency modulation (FM) *See* modulation; radio.

fresnel A unit of frequency equal to 10^{12} hertz. In SI units this is equal to 1 terahertz (THz). It was named after the French physicist A. J. Fresnel (1788–1827).

Fresnel diffraction A form of *diffraction in which the light source or the receiving screen, or both, are at finite distances from the diffracting object, so that the wavefronts are not plane, as in *Fraunhofer diffraction. It was studied by A. J. Fresnel (1788–1827).

Fresnel lens A lens with one face cut into a series of steps. It enables a relatively light and robust lens of short focal length though poor optical quality to be used in projectors (as condenser lenses), searchlights, spotlights, car headlights, and lighthouses. Such lenses have been made fine enough to serve as magnifiers for reading small print.

friction The force that resists the motion of one surface relative to another with which it is in contact.

Friction

Friedel–Crafts reaction

benzene   chloromethane   methylbenzene (toluene)

Friedel–Crafts methylation

benzene   acetyl chloride   phenyl methyl ketone

Friedel–Crafts acetylation

For a body resting on a horizontal surface there is a normal contact force, R, between the body and surface, acting perpendicularly to the surface. If a horizontal force B is applied to the body with the intention of moving it to the right, there will be an equal horizontal friction force, F, to the left, resisting the motion (see illustration). If B is increased until the body just moves, the value of F will also increase until it reaches the *limiting frictional force* (F_L), which is the maximum value of F. F_L is then equal to $\mu_s R$, where μ_s is the *coefficient of static friction*, the value of which depends on the nature of the surfaces. Once the body is moving with constant velocity, the value of F falls to a value F_k, which is equal to $\mu_k R$, where μ_k is the *coefficient of kinetic friction*. Both μ_s and μ_k are independent of the surface area of the body unless this is very small and μ_k is almost independent of the relative velocity of the body and surface.

The cause of friction is that surfaces, however smooth they may look to the eye, on the microscopic scale have many humps and crests. Therefore the actual area of contact is very small indeed, and the consequent very high pressure leads to local pressure welding of the surfaces. During motion the welds are broken and remade continually. *See also* rolling friction.

Friedel–Crafts reaction A type of reaction in which an alkyl group (from a haloalkane) or an acyl group (from an acyl halide) is substituted on a benzene ring (see illustration). The product is an alkylbenzene (for alkyl substitution) or an alkyl aryl ketone (for acyl substitution). The reactions occur at high temperature (about 100°C) with an aluminium chloride catalyst. The catalyst acts as an electron acceptor for a lone pair on the halide atom. This polarizes the haloalkane or acyl halide, producing a positive charge on the alkyl or acyl group. The mechanism is then electrophilic substitution. Alcohols and alkenes can also undergo Friedel–Crafts reactions. The reaction is named after the French chemist Charles Friedel (1832–99) and the US chemist James M. Craft (1839–1917).

frogs *See* Amphibia.

frond *See* megaphyll.

front (in meteorology) The sloping interface between two *air masses of different temperature and humidity. Where the air masses come together the warm air, being lighter, rises and slopes over the colder air. Distinctive weather phenomena are associated with fronts. At a *warm front* (the boundary zone at the front of the warm sector of a *depression), warm air is overtaking cold air and rising above it with a slope of about 1:150. The weather phenomena associated with the warm front include cirrus clouds followed by cirrostratus, altostratus, then nimbostratus clouds, as the warm front passes. At a *cold front*, which is usually found to the rear of a depression, warm air is forced to rise by an advancing undercutting wedge of cold air. This is accompanied by a sudden veering of the wind, a fall in temperature, and heavy rainfall. An *occluded front* (or *occlusion*) occurs when a cold front overtakes a warm front in an atmospheric depression.

froth flotation A method of separating mixtures of solids, used industrially for separating ores from the unwanted gangue. The mixture is ground to a powder and water and a frothing agent added. Air is blown through the water. With a suitable frothing agent, the bubbles adhere only to particles of ore and carry them to the surface, leaving the gangue particles at the bottom.

fructification *See* sporophore.

fructose (fruit sugar; laevulose) A simple sugar, $C_6H_{12}O_6$, stereoisomeric with glucose. (Although natural fructose is the D-form, it is in fact laevorotatory.) Fructose occurs in green plants, fruits, and honey and tastes sweeter than sucrose (cane sugar), of which it is a constituent. Derivatives of fructose are important in the energy metabolism of living organisms. Some polysaccharide derivatives

(fructans) are carbohydrate energy stores in certain plants.

fruit The structure formed from the ovary of a flower, usually after the ovules have been fertilized. It consists of the *fruit wall* (*see* pericarp) enclosing the seed(s). Other parts of the flower, such as the receptacle, may develop and contribute to the structure, resulting in a *false fruit* (*see* pseudocarp). The fruit may retain the seeds and be dispersed whole (an *indehiscent fruit*), or it may open (dehisce) to release the seeds (a *dehiscent fruit*). Fruits are divided into two main groups depending on whether the ovary wall remains dry or becomes fleshy (*succulent*). Succulent fruits are generally dispersed by animals and dry fruits by wind, water, or by some mechanical means. See illustration. *See also* composite fruit.

fruit sugar *See* fructose.

frustum A solid figure produced when two parallel planes cut a larger solid or when one plane parallel to the base cuts it.

FSH *See* follicle-stimulating hormone.

fucoxanthin The major *carotenoid pigment present, with chlorophyll, in the brown algae (*see* Phaeophyta).

fuel A substance that is oxidized or otherwise changed in a furnace or heat engine to release useful heat or energy. For this purpose wood, vegetable oil, and animal products have largely been replaced by *fossil fuels since the 18th century.
The limited supply of fossil fuels and the expense of extracting them from the earth has encouraged the development of nuclear fuels to produce electricity (*see* nuclear energy).

fuel cell A cell in which the chemical energy of a fuel is converted directly into electrical energy. The simplest fuel cell is one in which hydrogen is oxidized to form water over porous

sycamore

ash

samaras

parachute of hairs

dandelion

thistle

fruit and seed

cypselas

pores through which seeds are dispersed

dispersed seeds

capsule (poppy)

pod (laburnum)

hooked bracts

fruit

achenes in axils of hooked bracts (burdock)

Dry fruits: seeds dispersed by wind, water, or other mechanical means

epicarp

stony endocarp

epicarp

seed

fleshy mesocarp

berry (grape)

seed

fleshy mesocarp

drupe (cherry)

achenes

receptacle

strawberry

seed

receptacle

pericarp (core)

apple (pome)

Succulent edible fruits: seeds dispersed by animals

pseudocarps (false fruits)

Different types of fruit and methods of seed dispersal

sintered nickel electrodes. A supply of gaseous hydrogen is fed to a compartment containing the porous cathode and a supply of oxygen is fed to a compartment containing the porous anode; the electrodes are separated by a third compartment containing a hot alkaline electrolyte, such as potassium hydroxide. The electrodes are porous to enable the gases to react with the electrolyte, with the nickel in the electrodes acting as a catalyst. At the cathode the hydrogen reacts with the hydroxide ions in the electrolyte to form water, with the release of two electrons per hydrogen molecule:

$$H_2 + 2OH^- \rightarrow 2H_2O + 2e^-$$

At the anode, the oxygen reacts with the water, taking up electrons, to form hydroxide ions:

$$\tfrac{1}{2}O_2 + H_2O + 2e^- \rightarrow 2OH^-$$

The electrons flow from the cathode to the anode through an external circuit as an electric current. The device is a more efficient converter of electric energy than a heat engine, but it is bulky and requires a continuous supply of gaseous fuels. Their use to power electric vehicles is being actively explored.

fuel element *See* nuclear reactor.

fugacity Symbol f. A thermodynamic function used in place of partial pressure in reactions involving real gases and mixtures. For a component of a mixture, $d(\ln f) = d\mu/RT$, where μ is the chemical potential. It has the same units as pressure and the fugacity of a gas is equal to the pressure if the gas is ideal. The fugacity of a liquid or solid is the fugacity of the vapour with which it is in equilibrium. The ratio of the fugacity to the fugacity in some standard state is the *activity. For a gas, the standard state is chosen to be the state at which the fugacity is 1. The activity then equals the fugacity.

fuller's earth A naturally occurring clay material (chiefly montmorillonite) that has the property of decolorizing oil and grease. In the past raw wool was cleaned of grease and whitened by kneading it in water with fuller's earth; a process known as *fulling*. Fuller's earth is now widely used to decolorize fats and oils and also as an insecticide carrier and drilling mud. The largest deposits occur in the USA, UK, and Japan.

full-wave rectifier *See* rectifier.

fulminate *See* cyanic acid.

fulminic acid *See* cyanic acid.

fumaric acid *See* butenedioic acid.

function Any operation or procedure that relates one variable to one or more other variables. If y is a function of x, written $y = f(x)$, a change in x produces a change in y, and if x is known, y can be determined. x is known as the *independent variable* and y is the *dependent variable*.

functional group The group of atoms responsible for the characteristic reactions of a compound. The functional group is $-OH$ for alcohols, $-CHO$ for aldehydes, $-COOH$ for carboxylic acids, etc.

fundamental *See* harmonic.

fundamental constants (universal constants) Those parameters that do not change throughout the universe. The charge on an electron, the speed of light in free space, the Planck constant, the gravitational constant, the electric constant, and the magnetic constant are all thought to be examples.

fundamental interactions The four different types of interaction that can occur between bodies. These interactions can take place even when the bodies are not in physical contact and together they account for all the observed forces that occur in the universe. While the unification of these four types of interaction into one model, theory, or set of equations has long been the aim of physicists, this has not yet been achieved, although progress has been made in the unification of the electromagnetic and weak interactions. *See also* elementary particles; gauge theory; standard model; unified-field theory.

The *gravitational interaction*, some 10^{40} times weaker than the electromagnetic interaction, is the weakest of all. The force that it generates acts between all bodies that have mass and the force is always attractive. The interaction can be visualized in terms of a classical *field of force in which the strength of the force falls off with the square of the distance between the interacting bodies (*see* Newton's law of gravitation). The hypothetical gravitational quantum, the *graviton*, is also a useful concept in some contexts. On the atomic scale the gravitational force is negligibly weak, but on the cosmological scale, where masses are enormous, it is immensely important in holding the components of the universe together. Because gravitational interactions are long-ranged, there is a well-defined macroscopic theory in general relativity. At present, there is no satisfactory quantum theory of gravitational interaction. It is possible that *superstring theory may give a consistent quantum theory of gravity as well as unifying gravity with the other fundamental interactions.

The *weak interaction*, some 10^{10} times weaker than the electromagnetic interaction, occurs between *leptons and in the decay of hadrons. It is responsible for the *beta decay of particles and nuclei. In the current model, the weak interaction is visualized as a force mediated by the exchange of virtual particles, called intermediate vector bosons. The weak interactions are described by *electroweak theory, which unifies them with the electromagnetic interactions.

The *electromagnetic interaction* is responsible for the forces that control atomic structure, chemical reactions, and all electromagnetic phenomena. It accounts for the forces between charged particles, but unlike the gravitational interaction, can be either attractive or repulsive. Some neutral particles decay by electromagnetic interaction. The interaction is either visualized as a classical field of force (*see* Coulomb's law) or as an exchange of virtual *photons. As with gravitational interactions, the fact that electromagnetic interactions are long-ranged means that they have a well-defined classical theory given by *Maxwell's equations. The quantum theory of electromagnetic interactions is described by *quantum electrodynamics, which is a simple form of gauge theory.

The *strong interaction*, some 10^2 times stronger than the electromagnetic interaction, functions only between *hadrons and is responsible for the force between nucleons that gives the atomic nucleus its great stability. It operates at very short range inside the nucleus (as little as 10^{-15} metre) and is visualized as an exchange of virtual mesons. The strong interactions are described by a gauge theory called *quantum chromodynamics.

fundamental units A set of independently defined *units of measurement that forms the basis of a system of units. Such a set requires three mechanical units (usually of length, mass, and time) and, in some systems, one electrical unit; it has also been found convenient to treat certain other quantities as fundamental, even though they are not strictly independent. In the metric system the centimetre–gram–second (c.g.s.) system was replaced by the metre–kilogram–second (m.k.s.) system; the latter has now been adapted to provide the basis for *SI units. In British Imperial units the foot–pound–second (f.p.s.) system was formerly used.

fungi A group of organisms formerly regarded as simple plants lacking chlorophyll but now usually classified in a separate kingdom, Mycota or Fungi. They can either exist as single cells or make up a multicellular body called a *mycelium*, which consists of filaments known as *hyphae. Most fungal cells are multinucleate and have cell walls composed chiefly of *chitin. Fungi exist primarily in damp situations on land and, because of the absence of chlorophyll, are either parasites or saprophytes on other organisms. The principal criteria used in classification are the nature of the spores produced and the presence or absence of cross walls within the hyphae (*see* ascomycetes; basidiomycetes; imperfect fungi; phycomycetes). *See also* lichens; slime moulds.

Fungi Imperfecti *See* imperfect fungi.

funicle The stalk that attaches an ovule to the placenta in the ovary of a flowering plant. It contains a strand of conducting tissue leading from the placenta into the chalaza.

furan A colourless liquid compound, C_4H_4O; r.d. 0.94; m.p. $-86°C$; b.p. 31.4°C. It has a five-membered ring consisting of four CH_2 groups and one oxygen atom.

furanose A *sugar having a five-membered ring containing four carbon atoms and one oxygen atom.

fuse A length of wire made of a metal alloy of low melting point that is designed to melt at a specified current loading in order to protect an electrical device or circuit from overloading. The wire is often enclosed in a small glass or ceramic cartridge with metal ends.

fused ring *See* ring.

fusible alloys Alloys that melt at low temperature (around 100°C). They have a number of uses, including constant-temperature baths, pipe bending, and automatic sprinklers to provide a spray of water to prevent fires from spreading. Fusible alloys are usually *eutectic mixtures of bismuth, lead, tin, and cadmium. Wood's metal and Lipowitz's alloy are examples of alloys that melt at about 70°C.

fusion 1. Melting. **2.** *See* nuclear fusion.

fusion reactor *See* thermonuclear reactor.

G

gadolinium Symbol Gd. A soft silvery metallic element belonging to the *lanthanoids; a.n. 64; r.a.m. 157.25; r.d. 7.901 (20°C); m.p. 1312°C; b.p. 3273°C. It occurs in gadolinite, xenotime, monazite, and residues from uranium ores. There are seven stable natural isotopes and eleven artificial isotopes are known. Two of the natural isotopes, gadolinium-155 and gadolinium-157, are the best neutron absorbers of all the elements. The metal has found limited applications in nuclear technology and in ferromagnetic alloys (with cobalt, copper, iron, and cerium). Gadolinium compounds are used in electronic components. The element was discovered by J. C. G. Marignac in 1880.

gain *See* amplifier.

galactose A simple sugar, $C_6H_{12}O_6$, stereoisomeric with glucose, that occurs naturally as one of the products of the enzymic digestion of milk sugar (lactose) and as a constituent of gum arabic.

galaxy A vast collection of stars, dust, and gas held together by the gravitational attraction between its components. Galaxies are usually classified as elliptical, spiral, or irregular in shape. *Elliptical galaxies* appear like ellipsoidal clouds of stars, with very little internal structure apart from (in some cases) a denser nucleus. *Spiral galaxies* are flat disc-shaped collections of stars with prominent spiral arms. *Irregular galaxies* have no apparent structure or shape.

The sun belongs to a spiral galaxy known as the *Galaxy* (with a capital G) or the *Milky Way System* There are some 10^{11} stars in the system, which is about 30 000 parsecs across with a maximum thickness at the centre of about 4000 parsecs. The sun is about 10 000 parsecs from the centre of the Galaxy.

The galaxies are separated from each other by enormous distances, the nearest large galaxy to our own (the Andromeda galaxy) being about 6.7 $\times$ 10^5 parsecs away.

galaxy cluster A group of *galaxies containing many hundreds of members extending over a radius of up to a few megaparsecs (there also exist small groups of galaxies, such as the *Local Group, with a few tens of

galena

members). The richest and most regular clusters, such as the *Coma cluster*, with thousands of members, are gravitationally bound systems; it is not certain whether other less regular and less concentrated clusters are also bound. As well as galaxies, the clusters contain hot *intracluster gas*, at temperatures between 10^7 and 10^8 K; this can be detected by its X-ray emission. On a scale larger than clusters there are also *superclusters*, with extents of the order of a hundred megaparsecs, containing about a hundred galaxies. It is not known whether superclusters are gravitationally bound. *See also* missing mass.

galena A mineral form of lead(II) sulphide, PbS, crystallizing in the cubic system; the chief ore of lead. It usually occurs as grey metallic cubes, frequently in association with silver, arsenic, copper, zinc, and antimony. Important deposits occur in Australia (at Broken Hill), Germany, the USA (especially in Missouri, Kansas, and Oklahoma), and the UK.

Galilean telescope *See* telescope.

Galilean transformations A set of equations for transforming the position and motion parameters from a frame of reference with origin at O and coordinates (x,y,z) to a frame with origin at O' and coordinates at (x',y',z'). They are:

$$x' = x - vt$$
$$y' = y$$
$$z' = z$$
$$t' = t$$

The equations conform to Newtonian mechanics. *Compare* Lorentz transformations.

gall bladder A small pouch attached to the *bile duct, present in most vertebrates. *Bile, produced in the *liver, is stored in the gall bladder and released when food (especially fatty substances) enters the duodenum.

gallium Symbol Ga. A soft silvery metallic element belonging to group IIIA of the periodic table; a.n. 31; r.a.m. 69.72; r.d. 5.90 (20°C); m.p. 29.78°C; b.p. 2403°C. It occurs in zinc blende, bauxite, and kaolin, from which it can be extracted by fractional electrolysis. It also occurs in gallite, $CuGaS_2$, to an extent of 1%; although bauxite only contains 0.01% this is the only commercial source. The two stable isotopes are gallium–69 and gallium–71; there are eight radioactive isotopes, all with short half-lives. The metal has only a few minor uses (e.g. as an activator in luminous paints), but gallium arsenide is extensively used as a semiconductor in many applications. Gallium corrodes most other metals because it rapidly diffuses into their lattices. Most gallium(I) and some gallium(II) compounds are unstable. The element was first identified by François Lecoq de Boisbaudran in 1875.

gallon 1. (*or* **Imperial gallon**) The volume occupied by exactly ten pounds of distilled water of density 0.998 859 gram per millilitre in air of density 0.001 217 gram per millilitre. 1 gallon = 4.546 09 litres (cubic decimetres). **2.** A unit of volume in the US Customary system equal to 0.832 68 Imperial gallon, i.e. 3.785 44 litres.

galvanic cell *See* voltaic cell.

galvanized iron Iron or steel that has been coated with a layer of zinc to protect it from corrosion. Corrugated mild-steel sheets for roofing and mild-steel sheets for dustbins, etc., are usually galvanized by dipping them in molten zinc. The formation of a brittle zinc–iron alloy is prevented by the addition of small quantities of aluminium or magnesium. Wire is often galvanized by a cold electrolytic process as no alloy forms in this process. Galvanizing is an effective method of protecting steel because even if the surface is scratched, the zinc still protects the underlying metal. *See* sacrificial protection.

galvanometer An instrument for detecting and measuring small electric currents. In the moving-coil instrument a pivoted coil of fine insulated copper wire surrounds a fixed soft-iron core between the poles of a permanent magnet. The interaction between the field of the permanent magnet and the sides of the coil, produced when a current flows through it, causes a torque on the coil. The moving coil carries either a pointer or a mirror that deflects a light beam when it moves; the extent of the deflection is a measure of the strength of the current. The galvanometer can be converted into an *ammeter or a *voltmeter. Digital electronic instruments are increasingly replacing the moving-coil type. *See also* ballistic galvanometer.

gametangium A plant sex organ that produces gametes. The term is usually restricted to the sex organs of algae and fungi.

gamete A reproductive cell that fuses with another gamete to form a zygote. Examples of gametes are ova and spermatozoa. Gametes are *haploid, i.e. they contain half the normal (diploid) number of chromosomes; thus when two fuse, the diploid number is restored (*see* fertilization). Gametes are formed by *meiosis. *See also* sexual reproduction.

gametophyte The generation in the life cycle of a plant that bears the gamete-producing sex organs. The gametophyte is *haploid. It is the dominant phase in the life cycle of mosses and other bryophytes, the *sporophyte generation depending on it either partially or completely. In pteridophytes it is the *prothallus. In seed plants it is very much reduced. For example, in angiosperms the pollen grain is the male gametophyte and the embryo sac is the female gametophyte. *See also* alternation of generations.

gamma globulin *See* globulin.

gamma-iron *See* iron.

gamma radiation Electromagnetic radiation emitted by excited atomic nuclei during the process of passing to a lower excitation state. Gamma radiation ranges in energy from about 10^{-15} to 10^{-10} joule (10 keV to 10 MeV) corresponding to a wavelength range of about 10^{-10} to 10^{-14} metre. A common source of gamma radiation is cobalt-60, the decay process of which is:
$$^{60}_{27}\text{Co} \xrightarrow{\beta} {}^{60}_{28}\text{Ni} \xrightarrow{\gamma} {}^{60}_{28}\text{Ni}$$
The de-excitation of nickel-60 is accompanied by the emission of gamma-ray photons having energies 1.17 MeV and 1.33 MeV.

ganglion A mass of nervous tissue containing many *cell bodies and *synapses, usually enclosed in a connective-tissue sheath. In vertebrates most ganglia occur outside the central nervous system; exceptions are the *basal ganglia in the brain. In invertebrates ganglia occur along the nerve cords and the most anterior pair (*cerebral ganglia*) are analogous to the vertebrate brain; invertebrate ganglia constitute a part of the central nervous system.

gangue Rock and other waste material present in an ore.

ganoid scale *See* scales.

garnet Any of a group of silicate minerals that conform to the general formula $A_3B_2(SiO_4)_3$. The elements representing A may include magnesium, calcium, manganese, and iron(II); those representing B may include aluminium, iron(III), chromium, or titanium. Six varieties of garnet are generally recognized:
pyrope, $Mg_3Al_2Si_3O_{12}$;
almandine, $Fe_3^{2+}Al_2Si_3O_{12}$;
spessartite, $Mn_3Al_2Si_3O_{12}$;
grossularite, $Ca_3Al_2Si_3O_{12}$;
andradite, $Ca_3(Fe^{3+},Ti)_2Si_3O_{12}$;
uvarovite, $Ca_3Cr_2Si_3O_{12}$.
Varieties of garnet are used as gemstones and abrasives.

gas A state of matter in which the matter concerned occupies the whole of its container irrespective of its quantity. In an *ideal gas, which obeys the *gas laws exactly, the molecules themselves would have a negligible volume and negligible forces between them, and collisions between molecules would be perfectly elastic. In practice, however, the behaviour of real gases deviates from the gas laws because their molecules occupy a finite volume, there are small forces between molecules, and in polyatomic gases collisions are to a certain extent inelastic (*see* equation of state).

gas chromatography A technique for separating or analysing mixtures of gases by *chromatography. The apparatus consists of a very long tube containing the stationary phase. This may be a solid, such as kieselguhr (*gas–solid chromatography*, or *GSC*), or a nonvolatile liquid, such as a hydrocarbon oil coated on a solid support (*gas–liquid chromatography*, or *GLC*). The sample is often a volatile liquid mixture, which is vaporized and swept through the column by a carrier gas (e.g. hydrogen). The components of the mixture pass through the column at different rates and are detected as they leave, either by measuring the thermal conductivity of the gas or by a flame detector.

Gas chromatography is usually used for analysis; components can be identified by the time they take to pass through the column. It is sometimes also used for separating mixtures.

Gas chromatography is often used to separate a mixture into its components, which are then directly injected into a mass spectrometer. This technique is known as *gas chromatography–mass spectroscopy* or *GCMS*.

gas constant (universal molar gas constant) Symbol R. The constant that appears in the *universal gas equa-tion* (*see* gas laws). It has the value $8.314\,510(70)\ \mathrm{J\,K^{-1}\,mol^{-1}}$.

gas-cooled reactor See nuclear reactor.

gas equation See gas laws.

gas laws Laws relating the temperature, pressure, and volume of an *ideal gas. *Boyle's law states that the pressure (p) of a specimen is inversely proportional to the volume (V) at constant temperature ($pV = $ constant). The modern equivalent of *Charles' law states that the volume is directly proportional to the thermodynamic temperature (T) at constant pressure ($V/T = $ constant); originally this law stated the constant expansivity of a gas kept at constant pressure. The pressure law states that the pressure is directly proportional to the thermodynamic temperature for a specimen kept at constant volume. The three laws can be combined in the *universal gas equation*, $pV = nRT$, where n is the amount of gas in the specimen and R is the *gas constant. The gas laws were first established experimentally for real gases, although they are obeyed by real gases to only a limited extent; they are obeyed best at high temperatures and low pressures. *See also* equation of state.

gasoline See petroleum.

gas thermometer A device for measuring temperature in which the working fluid is a gas. It provides the most accurate method of measuring temperatures in the range 2.5 to 1337 K. Using a fixed mass of gas a *constant-volume thermometer* measures the pressure of a fixed volume of gas at relevant temperatures, usually by means of a mercury *manometer and a *barometer.

gastric juice An acidic mixture of inorganic salts, hydrochloric acid, and digestive enzymes (e.g. *pepsin) produced by glands in the stomach lining.

gastric mill (proventriculus) A type of *gizzard occurring in many crustaceans. It is situated in the anterior region of the stomach and consists of a set of bones (ossicles) and muscles that grind food particles. The food particles are then filtered by bristles in the posterior section of the stomach.

gastrin A hormone, produced in the stomach, that controls the release of gastric juice. The secretion of gastrin is stimulated by the presence of food in the stomach. It is one of the hormones that integrates and controls digestive processes (*see also* secretin).

Gastropoda A class of molluscs that includes the snails, whelks, limpets, land and sea slugs, and conches. Molluscs have a well-developed head with tentacles, a large flattened foot, and a coiled twisted shell. They occupy marine, freshwater, and terrestrial habitats; in the terrestrial and some freshwater gastropods the *mantle cavity acts as a lung instead of enclosing gills.

gastrula The stage in the development of an animal embryo that succeeds the *blastula. It begins with the production of the *germ layers and the embryo becomes converted to a cup-shaped structure containing a cavity (the *archenteron).

gas turbine An internal-combustion engine in which the products of combustion of a fuel burnt in compressed air are expanded through a turbine. Atmospheric air is compressed by a rotary compressor driven by the turbine, fed into a combustion chamber, and mixed with the fuel (kerosene, natural gas, etc.); the expanding gases drive the turbine and power is taken from the unit by means of rotation of the turbine shaft (as in locomotives) or thrust from a jet (as in aircraft).

gate 1. An electronic circuit with a single output and one or more inputs; the output is a function of the input or inputs. In the *transmission gate* the output waveform is a replica of a selected input during a specific interval. In the *switching gate* a constant output is obtained for a specified combination of inputs. These gates are the basic components of digital computers. *See* logic circuits. **2.** The electrode in a field-effect *transistor that controls the current through the channel.

Gattermann reaction A variation of the *Sandmeyer reaction for preparing chloro- or bromoarenes by reaction of the diazonium compound. In the Gattermann reaction the aromatic amine is added to sodium nitrite and the halogen acid (10°C), then fresh copper powder (e.g. from Zn + $CuSO_4$) is added and the solution warmed. The diazonium salt then forms the haloarene, e.g.

$$C_6H_5N_2{}^+Cl^- \rightarrow C_6H_5Cl + N_2$$

The copper acts as a catalyst. The reaction is easier to perform than the Sandmeyer reaction and takes place at lower temperature, but generally gives lower yields. It was discovered in 1890 by the German chemist Ludwig Gattermann (1860–1920).

gauche conformation *See* conformation.

gauge theory Any of a number of *quantum field theories put forward to explain fundamental interactions. A gauge theory involves a symmetry *group for the fields and potentials (the *gauge group*). In the case of electrodynamics, the group is Abelian whereas the gauge theories for strong and weak interactions use non-Abelian groups. Non-Abelian gauge theories are known as *Yang–Mills theories*. This difference explains why *quantum electrodynamics is a much simpler theory than *quantum chromodynamics, which describes the strong interactions, and *electroweak theory, which is the unified theory of the weak and electromagnetic interactions. In the case of quantum gravity, the gauge group is even more complicated than the gauge groups for either the strong or weak interactions.

In gauge theories the interactions between particles can be explained by the exchange of particles (intermediate vector bosons, or *gauge bosons*), such as gluons, photons, and W and Z bosons. The *graviton has been postulated as a particle exchanged in the gravitational interaction.

gauss Symbol G. The c.g.s. unit of magnetic flux density. It is equal to 10^{-4} tesla.

Gaussian units A system of units for electric and magnetic quantities based upon c.g.s. electrostatic and electromagnetic units. Although replaced by *SI units in most branches of science, they are, like Heaviside–Lorentz units, still used in relativity theory and in particle physics. In Gaussian units, the electric and magnetic constants are both equal to unity.

Gauss' law The total electric flux normal to a closed surface in an electric field is proportional to the algebraic sum of the electric charges within the surface. A similar law applies to surfaces drawn in a magnetic field and the law can be generalized for any vector field through a closed surface. It was first stated by Karl Gauss (1777–1855).

gaussmeter A *magnetometer, especially one calibrated in gauss.

Gay-Lussac's law 1. When gases combine chemically the volumes of the reactants and the volume of the product, if it is gaseous, bear simple relationships to each other when measured under the same conditions of temperature and pressure. The law was first stated in 1808 by J. L. Gay-Lussac (1778–1850) and led to *Avogadro's law. **2.** *See* Charles' law.

gaylussite A mineral consisting of a hydrated mixed carbonate of sodium and calcium, $Na_2CO_3.CaCO_3.5H_2O$.

GCMS *See* gas chromatography.

Gegenschein (German: counterglow) A faint elliptical patch of light visible on a moonless night on the ecliptic at a point 180° from the position of the sun. It is caused by the reflection of sunlight by meteoric particles (*see also* zodiacal light).

Geiger counter (Geiger–Müller counter) A device used to detect and measure *ionizing radiation. It consists of a tube containing a low-pressure gas (usually a mixture of methane with argon or neon) and a cylindrical hollow cathode through the centre of which runs a fine-wire anode. A potential difference of about 1000 volts is maintained between the electrodes. An ionizing particle or photon passing through a window into the tube will cause an ion to be produced and the high p.d. will accelerate it towards its appropriate electrode, causing an avalanche of further ionizations by collision. The consequent current pulses can be counted in electronic circuits or simply amplified to work a small loudspeaker in the instrument. It was first devised in 1908 by Hans Geiger (1882–1947). Geiger and W. Müller produced an improved design in 1928.

Geissler tube An early form of gas-discharge tube designed to demonstrate the luminous effects of an electric discharge passing through a low-pressure gas between two electrodes. Modified forms are used in spectroscopy as a source of light. It was invented in 1858 by Heinrich Geissler (1814–79).

gel A lyophilic *colloid that has coagulated to a rigid or jelly-like solid. In a gel, the disperse medium has formed a loosely-held network of linked molecules through the dispersion medium. Examples of gels are silica gel and gelatin.

gelatin(e) A colourless or pale yellow water-soluble protein obtained by boiling collagen with water and evaporating the solution. It swells when water is added and dissolves in hot water to form a solution that sets to a gel on cooling. It is used in photographic emulsions and adhesives, in bacteriology for preparing culture media, in pharmacy for preparing capsules and suppositories, and in jellies and other foodstuffs.

gel filtration A type of column *chromatography in which a mixture of liquids is passed down a column containing a gel. Small molecules in the mixture can enter pores in the gel and move slowly down the column; large molecules, which cannot enter the pores, move more quickly. Thus, mixtures of molecules can be separated on the basis of their size. The technique is used particularly for separating proteins but it can also be applied to other polymers and to cell nuclei, viruses, etc.

gelignite A high explosive made from nitroglycerin, cellulose nitrate, sodium nitrate, and wood pulp.

gem Designating molecules in which two functional groups are attached to the same atom in a molecule. For example, the compound 1,1-dichloroethane (CH_3CHCl_2) is a gem dihalide and can be named gem-dichloroethane. Compare vicinal.

gemmation A type of *vegetative propagation in which small clumps of undifferentiated cells (gemmae) develop on the surface of a plant. These are shed and dispersed to other areas, where they grow to produce new individuals. Gemmation is found only in certain lower plants, such as mosses and liverworts.

gene A unit of heredity composed of DNA. In classical genetics (see Mendelism; Mendel's laws) a gene is visualized as a discrete particle, forming part of a *chromosome, that determines a particular characteristic. It can exist in different forms called *alleles, which determine which aspect of the characteristic is shown (e.g.

tallness or shortness for the characteristic of height).

A gene may be defined as the shortest length of chromosome that cannot be broken by recombination (see meiosis) or that can undergo *mutation, both units being recognized as a single base pair in the nucleotide sequence that makes up the DNA. The gene as a unit of function may be defined as the sequence of nucleotides concerned with the synthesis of a single polypeptide chain, corresponding to a particular sequence of the *genetic code. One or more of these structural genes, coding for protein, may be associated with other genes controlling their expression (see operon).

gene cloning The production of exact copies (clones) of a particular gene or genes using genetic engineering techniques. The DNA containing the target gene(s) is split into fragments using *restriction enzymes. These fragments are then inserted into cloning *vectors, such as bacterial plasmids or bacteriophages, which transfer the recombinant DNA to suitable host cells, such as the bacterium E. coli. Alternatively, *complementary DNA is inserted into the vectors, or 'naked' DNA fragments can be taken up directly by a host bacterium from its medium (this is less efficient than vector transfer).

Inside the host cell the recombinant DNA undergoes replication; thus, a bacterial host will give rise to a colony of cells containing the cloned target gene. Various screening methods may be used to identify such colonies, enabling them to be selected and cultured. Gene cloning facilitates *gene sequencing; it also enables large quantities of a desired protein product to be produced: human insulin, for example, is now produced by bacteria containing the cloned insulin gene.

gene library A collection of cloned DNA fragments representing the entire genetic material of an organism. This facilitates screening and isolation of any particular gene. Gene libraries are created by fractionating the genomic DNA into fragments using *restriction enzymes and/or physical methods. These fragments are cloned (see gene cloning) and the host cells containing the recombinant fragments are centrifuged and frozen; alternatively, the phage *vectors are maintained in culture. Individual genes in the library are identified using specific *gene probes with the *Southern blotting technique or, via their protein products, using *Western blotting. Gene libraries are thus repositories of raw material for use in genetic engineering. A human gene library may contain some 500 000 separate clones.

gene pool All the *genes and their different alleles that are present in a population of a plant or animal species. See also population genetics.

gene probe A single-stranded DNA or RNA fragment used in genetic engineering to search for a particular gene or other DNA sequence. The probe has a base sequence complementary to the target sequence and will thus attach to it by *base pairing. By labelling the probe with a radioactive isotope it can be identified on subsequent separation and purification. Probes of varying lengths, up to about 100 nucleotides, can be constructed in the laboratory. They are used in the *Southern blotting technique to identify particular DNA fragments, for instance in conjunction with *restriction mapping to diagnose gene abnormalities or to map certain sequences.

general theory of relativity See relativity.

generation time 1. (in physics) The average time that elapses between the creation of a neutron by fission in a nuclear reactor and a fission produced by that neutron. **2.** (in biology) The interval between the beginnings of consecutive cell divisions. It may be as short as 20 minutes in bacteria. See also interphase.

generative nucleus One of the two male gametes in the *pollen tube of angiosperms.

generator Any machine that converts mechanical power into electrical power. Electromagnetic generators are the main source of electricity and may be driven by steam turbines, water turbines, internal-combustion engines, windmills, or by some moving part of any other machine. In power stations, generators produce alternating current and are often called *alternators*.

gene sequencing The process of elucidating the nucleotide sequence of a gene. Two techniques are used. The Maxam–Gilbert method (named after Allan Maxam and Walter Gilbert) involves cleaving the DNA with a *restriction enzyme and labelling each of the resulting fragments with ^{32}P-phosphate at one end. The fragments are subjected to four different sets of reactions, each set specifically cleaving DNA at a particular base or bases. The cleaved fragments are separated by electrophoresis according to their chain length and identified by autoradiography. The base (nucleotide) sequence is deduced from the position of bands in each of the four lanes in the gel.

The Sanger method (named after Frederick Sanger), also called the dideoxy method, involves synthesizing a new DNA strand using as template single-stranded DNA from the gene being sequenced. Synthesis of the new strand can be stopped at any of the four bases by adding the corresponding dideoxy (dd) derivative of the deoxyribonucleoside phosphates; for example, by adding ddATP the synthesis terminates at an adenosine;

by adding ddGTP it terminates at a guanosine, etc. As in the first method, the fragments, which comprise radiolabelled nucleotides, are finally subjected to electrophoresis and autoradiography. Gene sequencing is now employed on a major scale, for example in determining the nucleotide sequence of the entire human genome.

gene splicing A stage in the processing of messenger *RNA (*see* transcription), occurring only in eukaryote cells, in which noncoding *introns are removed from the primary mRNA transcript and the coding *exons are spliced together to form the functional mRNA molecule. Splicing is catalysed by a complex of small RNA molecules and proteins called a *spliceosome*. In some organisms, self-splicing of mRNA is known to occur.

gene therapy The application of genetic engineering techniques to alter or replace defective genes. The techniques are still at the experimental stage but the eventual aim is to prevent such genetic diseases as sickle cell anaemia and thalassaemia.

genetic code The means by which genetic information in *DNA controls the manufacture of specific proteins by the cell. The code takes the form of a series of triplets (*codons*) of bases in DNA, from which is transcribed a complementary sequence of codons in messenger *RNA (*see* transcription). The sequence of these codons determines the sequence of amino acids during *protein synthesis. There are 64 possible codes from the combinations of the four bases present in DNA and messenger RNA and 20 amino acids present in body proteins: some of the amino acids are coded by more than one codon, and some codons have other functions.

genetic engineering (recombinant DNA technology) The techniques involved in altering the characters of an organism by inserting genes from another organism into its DNA. This altered DNA (known as *recombinant DNA*) is usually produced by *gene cloning. Genetic engineering has many applications, ranging from the commercial production of hormones, vaccines, etc., to the creation of *transgenic animals and crop plants in agriculture. *See also* gene library; gene probe.

genetic fingerprinting A technique in which an individual's DNA is analysed to reveal the pattern of repetition of particular nucleotide sequences (the marker sequences) throughout the genome. This pattern is claimed to be unique to the individual concerned, and the technique is therefore used for identification purposes in forensic science and paternity disputes, and in veterinary science. Sufficient DNA can be obtained from very small samples of body tissue, such as blood or semen. *Restriction enzymes are used to cleave the DNA and the marker sequences are revealed by specific *gene probes and *Southern blotting.

genetic mapping *See* chromosome map; restriction mapping.

genetics The branch of biology concerned with the study of heredity and variation. *Classical genetics* is based on the work of Gregor Mendel (*see* Mendelism). During the 20th century genetics has expanded to overlap with the fields of ecology and animal behaviour (*see* behavioural genetics; population genetics), and important advances in biochemistry and microbiology have led to clarification of the chemical nature of *genes and the ways in which they can replicate and be transmitted. *See also* genetic engineering.

gene tracking A method for determining the inheritance of a particular gene in a family. It is used in the diagnosis of genetic diseases, such as cystic fibrosis and Huntington's chorea. *Restriction fragment length polymorphisms (RFLPs) situated in or near the locus of interest are identified using *gene probes, and suitable marker RFLPs selected. These can then be traced through members of the family and used to detect the presence or absence of the disease locus prenatally in future at-risk pregnancies.

genome All the genes contained in a single set of chromosomes, i.e. in a *haploid nucleus. Each parent, through its reproductive cells, contributes its genome to its offspring.

genotype The genetic composition of an organism, i.e. the combination of *alleles it possesses. *Compare* phenotype.

genus (*pl.* **genera**) A category used in the *classification of organisms that consists of a number of similar or closely related species. The common name of an organism (especially a plant) is sometimes similar or identical to that of the genus, e.g. *Lilium* (lily), *Crocus*, *Antirrhinum*. Similar genera are grouped into families. *See also* binomial nomenclature.

geocentric universe A view of the universe in which the earth is regarded as being at its centre. Galileo finally established that the earth revolves round the sun (not the other way round, as the church believed); during the 20th century it has become clear from advances in observational astronomy that the earth is no more than one of nine planets orbiting the sun, which is one of countless millions of similar stars, many of which undoubtedly possess planetary bodies on which life could have evolved.

geochemistry The scientific study of the chemical composition of the earth. It includes the study of the abundance of the earth's elements and their isotopes and the distribution of the elements in environments of the earth (lithosphere, atmosphere, biosphere, and hydrosphere).

geochronology *See* varve dating.

geodesic (geodesic line) The shortest distance between two points on a curved surface.

geodesy The science concerned with surveying and mapping the earth's surface to determine, for example, its exact size, shape, and gravitational field. The information supplied by geodesy in the form of locations, distances, directions, elevations, and gravity information is of use in civil engineering, navigation, geophysics, and geography.

geography The study of the features of the earth's surface, their distribution and interaction, and the interaction of man with them. It is divided between the physical and social sciences. *Human geography* includes economic geography, political geography, historical geography, and urban geography; *physical geography* includes geomorphology, biogeography, climatology, pedology, and hydrology. The methods used by geographers include mapmaking (cartography), remote sensing techniques (e.g. aerial photography and satellite imagery), surveying techniques, and statistical analysis.

geological time scale A time scale that covers the earth's history from its origin, estimated to be about 4600 million years ago, to the present. The chronology is divided into a hierarchy of time intervals: eras, periods, epochs, ages, and chrons (see Appendix).

geology The study of the origin, structure, and composition of the earth. It is commonly subdivided into *historical geology*, which includes stratigraphy, palaeontology, and geochronology; and *physical geology*, which includes geomorphology, geophysics, geochemistry, mineralogy, petrology, crystallography, and economic geology.

geomagnetism The science concerned with the earth's magnetic field. If a bar magnet is suspended at any point on the earth's surface so that it can move freely in all planes, the north-seeking end of the magnet (N-pole) will point in a broadly northerly direction. The angle (D) between the horizontal direction in which it points and the geographic meridian at that point is called the *magnetic declination*. This is taken to be positive to the east of geographic north and negative to the west. The needle will not, however, be horizontal except on the *magnetic equator*. In all other positions it will make an angle (*I*) with the horizontal, called the *inclination* (or *magnetic dip*). At the *magnetic poles* $I = 90°$ ($+90°$ at the N-pole, $-90°$ at the S-pole) and the needle will be vertical. The positions of the poles, which vary with time, were in the 1970s approximately 76.1°N, 100°W (N) and 65.8°S, 139°E (S). The vector intensity F of the geomagnetic field is specified by I, D, and F, where F is the local magnetic intensity of the field measured in gauss or tesla (1 gauss = 10^{-4} tesla). F, I, and D, together with the horizontal and vertical components of F, and its north and east components, are called the *magnetic elements*. The value of F varies from about 0.2 gauss to 0.6 gauss, in general being higher in the region of the poles than at the equator, but values vary irregularly over the earth's surface with no correlation with surface features. There is also a slow unpredictable change in the local values of the magnetic elements called the *secular magnetic variation*. For example, in London between 1576 and 1800 D changed from $+11°$ to $-24°$ and I varied between 74° and 67°. The study of *palaeomagnetism has extended knowledge of the secular magnetic variation into the geological past and it is clear that the direction of the geomagnetic field has reversed several times. The source of the field and the cause of the variations are not known with any certainty but the source is believed to be associated with *dynamo action in the earth's liquid core.

geometrical isomerism *See* isomerism.

geometrical optics *See* optics.

geometric average *See* average.

geometric series A series of numbers or terms in which the ratio of any term to the subsequent term is constant. For example, 1, 4, 16, 64, 256, ... has a *common ratio* of 4. In general, a geometric series can be written:

$$a + ar + ar^2 \ldots + ar^{n-1}$$

and the sum of *n* terms is:

$$a(r^n - 1)/(r - 1).$$

geometrized units A system of units, used principally in general relativity, in which all quantities that have dimensions involving length, mass, and time are given dimensions of a power of length only. This is equivalent to setting the gravitational constant and the speed of light both equal to unity. *See also* Gaussian units; Heaviside–Lorentz units; natural units; Planck units.

geomorphology The study of the origin and development of landforms, excluding the major forms of the earth's surface (e.g. mountain chains and ocean basins).

geophysics The branch of science in which the principles of mathematics and physics are applied to the study of the earth's crust and interior. It includes the study of earthquake waves, geomagnetism, gravitational fields, and electrical conductivity using precise quantitative principles. In applied geophysics the techniques are applied to the discovery and location of economic minerals (e.g. petroleum). Meteorology and physical oceanography can also be considered as geophysical sciences.

geostationary orbit *See* synchronous orbit.

geosynchronous orbit *See* synchronous orbit.

geothermal energy Heat within the earth's interior that is a potential source of energy. Volcanoes, geysers, hot springs, and fumaroles are all sources of geothermal energy. The main areas of the world in which these energy sources are used to generate power include Larderello (Italy), Wairakei (New Zealand), Geysers (California, USA), and Reykjavik (Iceland). High-temperature porous rock also occurs in the top few kilometres of the earth's crust. Thermal energy from these reservoirs can be tapped by drilling into them and extracting their thermal energy by conduction to a fluid. The hot fluid can then be used for direct heating or to raise steam to drive a turbogenerator.

geotropism The growth of plant organs in response to gravity. A main root is positively geotropic and a main stem negatively geotropic, growing downwards and upwards respectively, irrespective of the positions in which they are placed. For example, if a stem is placed in a horizontal position it will still grow upwards. Geotropism is thought to be controlled by *auxin, a plant growth substance. *See* tropism.

geraniol An alcohol, $C_9H_{15}CH_2OH$, present in a number of essential oils.

germanium Symbol Ge. A lustrous hard metalloid element belonging to group IV of the periodic table; a.n. 32; r.a.m. 72.59; r.d. 5.36; m.p. 937°C; b.p. 2830°C. It is found in zinc sulphide and in certain other sulphide ores, and is mainly obtained as a by-product of zinc smelting. It is also present in some coal (up to 1.6%). Small amounts are used in specialized alloys but the main use depends on its semiconductor properties. Chemically, it forms compounds in the +2 and +4 oxidation states, the germanium(IV) compounds being the more stable. The element also forms a large number of organometallic compounds. Predicted in 1871 by Mendeleev (eka-silicon), it was discovered by Winkler in 1886.

German silver (nickel silver) An alloy of copper, zinc, and nickel, often in the proportions 5:2:2. It resembles silver in appearance and is used in cheap jewellery and cutlery and as a base for silver-plated wire. *See also* electrum.

germ cell Any cell in the series of cells (the *germ line*) that eventually produces *gametes. In mammals the germinal epithelium of the ovaries and testes contain the germ cells.

germination 1. The initial stages in the growth of a seed to form a seedling. The embryonic shoot (plumule) and embryonic root (radicle) emerge and grow upwards and downwards respectively. Food reserves for germination come from *endosperm tissue within the seed and/or from the seed leaves (cotyledons). *See also* epigeal; hypogeal. **2.** The first signs of growth of spores and pollen grains.

germ layers The layers of cells in an animal embryo at the *gastrula stage, from which are derived the various organs of the animal's body. There are two or three germ layers: an outer layer (*see* ectoderm), an inner layer (*see* endoderm), and in most animal groups a middle layer (*see* mesoderm).

germ plasm *See* Weismannism.

gestation The period in animals bearing live young (especially mammals) from the fertilization of the egg to birth of the young (parturition). In man gestation is known as *pregnancy* and takes about nine months (40 weeks).

getter A substance with a strong affinity for specified other substances. A getter may be used to remove the last traces of a gas when achieving a high vacuum. It may also be used to remove impurities from semiconductors.

GeV Gigaelectronvolt, i.e. 10^9 eV. In the USA this is often written BeV, billion-electronvolt.

giant fibre A nerve fibre with a very large diameter, found in many types of invertebrate (e.g. earthworms and squids). Its function is to allow extremely rapid transmission of nervous impulses and hence rapid escape movements in emergencies.

giant star A very large star that is highly luminous. Lying above the main sequence on a *Hertzsprung–Russell diagram, giant stars represent a late stage in *stellar evolution. *See also* red giant; supergiant.

gibberellin Any of a group of plant *growth substances chemically related to gibberellic acid. Gibberellins promote shoot elongation in certain plants, often overcoming genetic dwarfism. They are also involved in the release of buds from dormancy and in promoting seed germination.

gibbous *See* phases of the moon.

Gibbs free energy (Gibbs function) *See* free energy. It is named after the US chemist J. W. Gibbs (1839–1903)

gibbsite A mineral form of hydrated *aluminium hydroxide ($Al(OH)_3$). It is named after the US mineralogist George Gibbs (d. 1833).

giga- Symbol G. A prefix used in the metric system to denote one thousand million times. For example, 10^9 joules = 1 gigajoule (GJ).

gilbert Symbol Gb. The c.g.s. unit of *magnetomotive force equal to $10/4\pi$ (= 0.795 77) ampere-turn. It is named after the English physician William Gilbert (1544–1603).

gill 1. (in zoology) A respiratory organ used by aquatic animals to obtain oxygen from the surrounding water. A gill consists essentially of a membrane or outgrowth from the body, with a large surface area and a plentiful blood supply, through which diffusion of oxygen and carbon dioxide between the water and blood occurs. Fishes have *internal gills*, formed as outgrowths from the pharynx wall and contained within *gill slits. Water entering the mouth is pumped out through these slits and over the gills. The gills of most aquatic invertebrates and amphibian larvae are *external gills*, which project from the body so that water passes over them as the animal moves. **2.** (in botany) One of the ridges of tissue that radiate from the centre of the underside of the cap of mushrooms. The spores are produced on these gills.

gill bar A cartilaginous support for the tissue between the gill slits in lower chordates, such as amphioxus.

gill slit An opening leading from the pharynx to the exterior in aquatic vertebrates and amphioxus. In amphioxus they function in *filter feeding. In fish they contain the *gills and are usually in the form of a series of long slits. They are absent in adult tetrapod vertebrates (except for some amphibians) but their presence in some form in the embryos of all vertebrates is a characteristic of the phylum *Chordata.

gimbal A type of mount for an instrument (such as a *gyroscope or compass) in which the instrument is free to rotate about two perpendicular axes.

gingiva (gum) The part of the epithelial tissue lining the mouth that covers the jaw bones. It is continuous with the sockets surrounding the roots of the teeth.

Giorgi units *See* m.k.s. units.

gizzard A muscular compartment of the alimentary canal of many animals that is specialized for breaking up food. In birds the gizzard lies between the *proventriculus and the duodenum and contains small stones and grit, which assist in breaking up the food when the gizzard contracts. *See also* gastric mill.

glacial ethanoic (*or* **acetic**) **acid** *See* ethanoic acid.

gland A group of cells or a single cell in animals or plants that is specialized to secrete a specific substance. In animals there are two types of glands, both of which synthesize their secretions. *Endocrine glands discharge their products directly into the blood vessels; *exocrine glands secrete through a duct or network of ducts into a body cavity or onto the body surface. Secretory cells are characterized by having droplets (*vesicles*) containing their products. *See also* secretion.

In plants glands are specialized to secrete certain substances produced by the plant. The secretions may be retained within a single cell, secreted into a special cavity or duct, or secreted to the outside. Examples are the water glands (*hydathodes) of certain leaves, nectaries (*see* nectar), and the digestive glands of certain carnivorous plants.

Glashow–Weinberg–Salam model (GWS model) *See* electroweak theory.

glass Any noncrystalline solid; i.e. a solid in which the atoms are random and have no long-range ordered pattern. Glasses are often regarded as supercooled liquids. Characteristically they have no definite melting point, but soften over a range of temperatures.

The common glass used in windows, bottles, etc., is *soda glass*, which is made by heating a mixture of lime (calcium oxide), soda (sodium carbonate), and sand (silicon(IV) oxide). It is a form of calcium silicate. Borosilicate glasses (e.g. *Pyrex*) are made by incorporating some boron oxide, so

that silicon atoms are replaced by boron atoms. They are tougher than soda glass and more resistant to temperature changes, hence their use in cooking utensils and laboratory apparatus. Glasses for special purposes (e.g. optical glass) have other elements added (e.g. barium, lead).

glass electrode A type of *half cell having a glass bulb containing an acidic solution of fixed pH, into which dips a platinum wire. The glass bulb is thin enough for hydrogen ions to diffuse through. If the bulb is placed in a solution containing hydrogen ions, the electrode potential depends on the hydrogen-ion concentration. Glass electrodes are used in pH measurement.

glass fibres Melted glass drawn into thin fibres some 0.005 mm–0.01 mm in diameter. The fibres may be spun into threads and woven into fabrics, which are then impregnated with resins to give a material that is both strong and corrosion resistant for use in car bodies and boat building.

glauberite A mineral consisting of a mixed sulphate of sodium and calcium, $Na_2SO_4.CaSO_4$.

Glauber's salt *Sodium sulphate decahydrate, $Na_2SO_4.10H_2O$, used as a laxative. It is named after J. R. Glauber (1604–68).

GLC (gas–liquid chromatography) *See* gas chromatography.

glenoid cavity The socket-shaped cavity in the *scapula (shoulder blade) that holds the head of the *humerus in a ball-and-socket joint.

global warming *See* greenhouse effect.

globular cluster *See* star cluster.

globulin Any of a group of globular proteins that are generally insoluble in water and present in blood, eggs, milk, and as a reserve protein in seeds. Blood serum globulins comprise four types: α_1-, α_2-, and β-globulins, which serve as carrier proteins; and γ-globulins (*gamma globulins*), which include the *immunoglobulins responsible for immune responses.

glomerulus A tangled mass of blood capillaries enclosed by the cup-shaped end (*Bowman's capsule*) of a kidney tubule (*see* nephron). Waste products and water pass from these capillaries into the Bowman's capsule and down the nephron.

glottis The opening from the pharynx to the trachea (windpipe). In mammals it also serves as the space for the *vocal cords. *See also* epiglottis, larynx.

glove box A metal box that has gloves fitted to ports in its walls. It is used to manipulate mildly radioactive materials and in laboratory techniques in which an inert, sterile, dry, or dust-free atmosphere has to be maintained.

glow discharge An electrical discharge that passes through a gas at low pressure and causes the gas to become luminous. The glow is produced by the decay of excited atoms and molecules.

glucagon A hormone, secreted by the *islets of Langerhans in the pancreas, that increases the concentration of glucose in the blood by stimulating the metabolic breakdown of glycogen. It thus antagonizes the effects of *insulin (*see* antagonism).

glucocorticoid *See* corticosteroid.

gluconeogenesis *See* glycolysis.

gluconic acid An optically active hydroxycarboxylic acid,
$$CH_2(OH)(CHOH)_4COOH.$$
The carboxylic acid corresponding to the aldose sugar glucose, it can be made by the action of certain moulds.

glucosamine *See* amino sugar.

glucose (dextrose; grape sugar) A white crystalline sugar, $C_6H_{12}O_6$, occurring widely in nature. Like other *monosaccharides, glucose is optically active: most naturally occurring glucose is dextrorotatory. Glucose and its derivatives are crucially important in the energy metabolism of living organisms. It is a major energy source, being transported around the body in blood, lymph, and cerebrospinal fluid to the cells, where energy is released in the process of *glycolysis. Glucose is present in the sap of plants, in fruits, and in honey and is also a constituent of many polysaccharides, most notably of starch and cellulose. These yield glucose when broken down, for example by enzymes during digestion.

glueball A hypothetical state consisting of two or more gluons, having no overall colour. There is, as yet, no conclusive evidence for the existence of glueballs.

gluino *See* supersymmetry.

gluon *See* elementary particles.

glutamic acid *See* amino acid.

glutamine *See* amino acid.

glutathione A *peptide comprising the amino acids glutamic acid, cysteine, and glycine. It occurs widely in plants, animals, and microorganisms, serving chiefly as an antioxidant. Reduced glutathione reacts with potentially harmful oxidizing agents and is itself oxidized. This is important in ensuring the proper functioning of proteins, haemoglobin, membrane lipids, etc. Glutathione is also involved in amino acid transport across cell membranes.

gluten A mixture of two proteins, gliadin and glutenin, occurring in the endosperm of wheat grain. Their amino acid composition varies but glutamic acid (33%) and proline (12%) predominate. The composition of wheat glutens determines the 'strength' of the flour and whether or not it is suitable for biscuit or bread making. Sensitivity of the lining of the intestine to gluten occurs in *coe-*

The principal stages of glycolysis

liac disease, a condition that must be treated by a gluten-free diet.

glyceride (acylglycerol) A fatty-acid ester of glycerol. Esterification can occur at one, two, or all three hydroxyl groups of the glycerol molecule producing mono-, di-, and triglycerides respectively. *Triglycerides are the major constituent of fats and oils found in living organisms. Alternatively, one of the hydroxyl groups may be esterified with a phosphate group forming a phosphoglyceride (*see* phospholipid) or to a sugar forming a *glycolipid*.

glycerine *See* glycerol.

glycerol (glycerine; propane-1,2,3,-triol) A trihydric alcohol, $HOCH_2CH(OH)CH_2OH$. Glycerol is a colourless sweet-tasting viscous liquid, miscible with water but insoluble in ether. It is widely distributed in all living organisms as a constituent of the *glycerides, which yield glycerol when hydrolysed.

glycine *See* amino acid.

glycogen (animal starch) A *polysaccharide consisting of a highly branched polymer of glucose occurring in animal tissues, especially in liver and muscle cells. It is the major store of carbohydrate energy in animal cells and is present as granular clusters of minute particles.

glycol *See* ethane-1,2-diol.

glycolipid *See* glyceride.

glycolysis (Embden–Meyerhof pathway) The series of biochemical reactions in which glucose is broken down to pyruvate with the release of usable energy in the form of *ATP (see illustration). One molecule of glucose undergoes two phosphorylation reactions and is then split to form two triose-phosphate molecules. Each of these is converted to pyruvate. The net energy yield is two ATP molecules per glucose molecule. In *aerobic respiration pyruvate then enters the *Krebs cycle. Alternatively, when oxygen is in short supply or absent, the pyruvate is converted to various products by *anaerobic respiration. Other simple sugars, e.g. fructose and galactose, and glycerol (from fats) enter the glycolysis pathway at intermediate stages. Many of the glycolytic reactions are reversible and are used to reform glucose from pyruvate in the process of *gluconeogenesis*.

Gnathostomata A subphylum consisting of all vertebrates that possess jaws. It includes the fishes, amphibians, reptiles, birds, and mammals. *Compare* Agnatha.

gneiss A coarse-grained rock that is characterized by compositional banding of metamorphic origin. It consists chiefly of irregular granular bands of quartz and feldspar alternating with thin undulating bands of micas and amphiboles. Gneisses are formed during high-grade regional metamorphism; those derived from sedimentary origins are known as *paragneisses* and those from igneous origins as *orthogneisses*.

gnotobiotic Designating germ-free conditions, especially those in which experimental animals are inoculated with known strains of microorganisms.

goethite A yellow-brown mineral, FeO.OH, crystallizing in the orthorhombic system. It is formed as a result of the oxidation and hydration of iron minerals or as a direct precipitate from marine or fresh water (e.g. in swamps and bogs). Most *limonite is composed largely of cryptocrystalline goethite. Goethite is mined as an ore of iron.

gold Symbol Au. A soft yellow malleable metallic *transition element; a.n. 79; r.a.m. 196.967; r.d. 19.32; m.p. 1064.43°C; b.p. 2807±2°C. It is found as the free metal in gravel or in quartz veins, and is also present in some lead and copper sulphide ores. It also occurs combined with silver in the telluride sylvanite, $(Ag,Au)Te_2$. It is used in jewellery, dentistry, and electronic devices. Chemically, it is unreactive, being unaffected by oxygen. It reacts with chlorine at 200°C to form gold(III) chloride. It forms a number of complexes with gold in the +1 and +3 oxidation states.

Goldschmidt process A method of extracting metals by reducing the oxide with aluminium powder, e.g.

$$Cr_2O_3 + 2Al \rightarrow 2Cr + Al_2O_3$$

The reaction can also be used to produce molten iron (*see* thermite). It was discovered by the German chemist Hans Goldschmidt (1861–1923).

Goldstone's theorem The theorem in relativistic quantum field theory that if there is an exact continuous symmetry of the *Hamiltonian or *Langrangian defining the system, and this is not a symmetry of the *vacuum state (i.e. there is *broken symmetry), then there must be at least one spin-zero massless particle called a *Goldstone boson*. In the quantum theory of many-body systems Goldstone bosons are *collective excitations. An important exception to Goldstone's theorem is provided in *gauge theories with the Higgs mechanism, whereby the Goldstone bosons gain mass and become *Higgs bosons.

Golgi apparatus An assembly of vesicles and folded membranes within the cytoplasm of plant and animal *cells that stores and transports secretory products (such as enzymes and hormones) and plays a role in formation of a cell wall (when this is present). It is named after its discoverer, the Italian cytologist Camillo Golgi (1843–1926). *See also* endoplasmic reticulum.

gonad Any of the usually paired organs in animals that produce reproductive cells (gametes). The most important gonads are the male *testis, which produces spermatozoa, and the female *ovary, which produces ova (egg cells). The gonads also produce hormones that control secondary sexual characteristics.

gonadotrophin (gonadotrophic hormone) Any of several hormones, secreted by the mammalian anterior *pituitary gland, that stimulate reproductive activity of the testes or ovaries (the gonads). Gonadotrophins include *follicle-stimulating hormone, *luteinizing hormone, and a hormone

Gooch crucible

produced by the placenta (*chorionic gonadotrophin*) that maintains the *corpus luteum and whose presence in urine in large amounts is an indication of pregnancy.

Gooch crucible A porcelain dish with a perforated base over which a layer of asbestos is placed, used for filtration in gravimetric analysis.

governor A device that maintains a motor or engine at a constant speed despite variations in the load, using the principle of negative feedback. A common method uses a set of flying balls that reduce the fuel intake as the speed increases. The balls, attached by flexible steel strips to a collar capable of moving vertically up and down a rotating shaft, move outwards as the speed increases. The collar rises as the balls fly out and is coupled to a lever that controls the fuel intake.

Graafian follicle The fluid-filled cavity that surrounds and protects the developing egg cell in the ovary of a mammal. After the release of the ovum it develops into a *corpus luteum. It is named after the Dutch anatomist Reinier de Graaf (1641–73).

grad *See* gradient operator.

gradient 1. The slope of a line. In Cartesian coordinates, a straight line $y = mx + c$, has a gradient m. For a curve, $y = f(x)$, the gradient at a point is the derivative dy/dx at that point, i.e. the slope of the tangent to the curve at that point. **2.** *See* gradient operator.

gradient operator (grad) The *operator

$$\nabla = i\,\partial/\partial x + j\,\partial/\partial y + k\,\partial/\partial z,$$

where i, j, and k are unit vectors in the x, y, and z directions. Given a scalar function f and a unit vector n, the *scalar product $n.\nabla$f is the rate of change of f in the direction of n. *See also* curl; divergence.

graft An isolated portion of living tissue that is joined to another tissue, either in the same or a different organism, the consequent growth resulting in fusion of the tissues. (The word is also used for the process of joining the tissues.)

Grafting of plant tissues is a horticultural practice used to propagate plants, especially certain bushes and fruit trees, artificially. A shoot or bud of the desired variety (the *scion*) is grafted onto a rootstock of either a common or a wild related species (the *stock*). The scion retains its desirable characteristics (e.g. flower form or fruit yield) and supplies the stock with food made by photosynthesis. The stock supplies the scion with water and mineral salts and affects only the size and vigour of the scion.

Animal and human grafts are used to replace faulty or damaged parts of the body. An *autograft* is taken from one part of the body and transferred to another part of the same individual, e.g. a skin graft used for severe burns. A *homograft* (*allograft*) is taken from one individual (the *donor*) and implanted in another (the *recipient*), the process being known as *transplantation*, e.g. a heart or kidney transplant. In such cases the graft may be regarded by the body as foreign (a state of *incompatibility*): an *immune response follows and the graft is rejected (*see also* histocompatibility).

graft copolymer *See* polymer.

graft hybrid A type of plant *chimaera that may be produced when a part of one plant (the scion) is grafted onto another plant of a different genetic constitution (the stock). Shoots growing from the point of union of the graft contain tissues from both the stock and the scion.

Graham's law The rates at which gases diffuse is inversely proportional to the square roots of their densities. This principle is made use of in the diffusion method of separating isotopes. The law was formulated in 1829 by Thomas Graham (1805–69).

gram Symbol g. One thousandth of a kilogram. The gram is the fundamental unit of mass in *c.g.s. units and was formerly used in such units as the *gram-atom*, *gram-molecule*, and *gram-equivalent*, which have now been replaced by the *mole.

Gram's stain A staining method used to differentiate bacteria. The bacterial sample is smeared on a microscope slide, stained with a violet dye, treated with acetone-alcohol (a decolourizer), and finally counter stained with a red dye. *Gram-positive* bacteria retain the first dye, appearing blue-black under the microscope. In *Gram-negative* bacteria, the acetone-alcohol washes out the violet dye and the counterstain is taken up, the cells appearing red. It is named after the Danish bacteriologist H. C. J. Gram (1853–1938), who first described the technique (since modified) in 1884.

grand unified theory (GUT) A theory that attempts to combine the strong, weak, and electromagnetic interactions into a single *gauge theory with a single symmetry group. There are a number of different theories, most of which postulate that the interactions merge at high energies into a single interaction (the *standard model emerges from the GUT as a result of *broken symmetry). The energy above which the interactions are the same is around 10^{15} GeV, which is much higher than those obtainable with existing accelerators. One prediction of GUTs is the occurrence of *proton decay. Some also predict that the neutrino has nonzero mass. There is no conclusive evidence for either at the moment. *See also* superstring theory.

granum (*pl.* **grana**) A stack of platelike bodies (thylakoids), many of which are found in plant *chloroplasts. Grana bear the light-receptive pigment chlorophyll and

contain the enzymes responsible for the light reactions of photosynthesis.

grape sugar *See* glucose.

graph A diagram that illustrates the relationship between two variables. It usually consists of two perpendicular axes, calibrated in the units of the variables and crossing at a point called the *origin*. Points are plotted in the spaces between the axes and the points are joined to form a curve. *See also* Cartesian coordinates; polar coordinates.

graphite *See* carbon.

graphite-modulated reactor *See* nuclear reactor.

graptolites A group of extinct marine colonial animals that were common in the Palaeozoic era. Graptolites are generally regarded as being related to the *Coelenterata. They had chitinous outer skeletons in the form of simple or branched stems, the individual polyps occupying minute cups (*thecae*) along these stems. Fossils of these skeletons are found in Palaeozoic rocks of all continents; they are particularly abundant in Ordovician and Silurian rock strata, for which they are used as *index fossils. At the end of the Silurian many graptolites became extinct but a few groups continued into the early Carboniferous.

graticule (in optics) A network of fine wires or a scale in the eyepiece of a telescope or microscope or on the stage of a microscope for measuring purposes.

grating *See* diffraction grating.

gravimetric analysis A type of quantitative analysis that depends on weighing. For instance, the amount of silver in a solution of silver salts could be measured by adding excess hydrochloric acid to precipitate silver chloride, filtering the precipitate, washing, drying, and weighing.

gravitation *See* Newton's law of gravitation.

gravitational constant Symbol G. The constant that appears in *Newton's law of gravitation; it has the value $6.672\,59(85) \times 10^{-11}$ N m^2 kg^{-2}. G is usually regarded as a universal constant although, in some models of the universe, it is proposed that it decreases with time as the universe expands.

gravitational field The region of space surrounding a body that has the property of *mass. In this region any other body that has mass will experience a force of attraction. The ratio of the force to the mass of the second body is the *gravitational field strength*.

gravitational interaction See fundamental interactions.

gravitational mass See mass.

gravitational shift See redshift.

gravitational waves 1. (in physics) Waves propagated through a *gravitational field. The prediction that an accelerating mass will radiate gravitational waves (and lose energy) comes from the general theory of *relativity. Many attempts have been made to detect waves from space directly using large metal detectors. The theory suggests that a pulse of gravitational radiation (as from a supernova explosion or *black hole) causes the detector to vibrate, and the disturbance is detected by a transducer. The interaction is very weak and extreme care is required to avoid external disturbances and the effects of thermal noise in the detecting system. So far, no accepted direct observations have been made. However, indirect evidence of gravitational waves has come from observations of a pulsar in a binary system with another star. 2. (in oceanography) Water surface waves transmitted primarily because of the weight of the water in the crests, which causes them to collapse. Ocean waves are of this type.

graviton A hypothetical particle or quantum of energy exchanged in a gravitational interaction (see fundamental interactions). Such a particle has not been observed but is postulated to make the gravitational interaction consistent with quantum mechanics. It would be expected to travel at the speed of light and have zero rest mass and charge, and spin 2.

gravity The phenomenon associated with the gravitational force acting on any object that has mass and is situated within the earth's *gravitational field. The weight of a body (see mass) is equal to the force of gravity acting on the body. According to Newton's second law of motion $F = ma$, where F is the force producing an acceleration a on a body of mass m. The weight of a body is therefore equal to the product of its mass and the acceleration due to gravity (g), which is now called the *acceleration of free fall. By combining the second law of motion with *Newton's law of gravitation ($F = GM_1M_2/d^2$) it follows that: $g = GM/d^2$, where G is the *gravitational constant, M is the mass of the earth, and d is the distance of the body from the centre of the earth. For a body on the earth's surface $g = 9.806\,65$ m s^{-2}.

A force of gravity also exists on other planets, moons, etc., but because it depends on the mass of the planet and its diameter, the strength of the force is not the same as it is on earth. If F_e is the force acting on a given mass on earth, the force F_p acting on the same mass on another planet will be given by:

$$F_p = F_e d_e^2 M_p / M_e d_p^2,$$

where M_p and d_p are the mass and diameter of the planet, respectively. Substituting values of M_p and d_p for the moon shows that the force of gravity on the moon is only 1/6 the value on earth.

gray Symbol Gy. The derived SI unit of absorbed *dose of ionizing radiation (*see* radiation units). It is named after the British radiobiologist L. H. Gray (1905–65).

great circle Any circle on a sphere formed by a plane that passes through the centre of the sphere. The equator and the meridians of longitude are all great circles on the earth's surface.

green algae *See* Chlorophyta.

greenhouse effect An effect occurring in the atmosphere because of the presence of certain gases (*greenhouse gases*) that absorb infrared radiation. Light and ultraviolet radiation from the sun are able to penetrate the atmosphere and warm the earth's surface. This energy is re-radiated as infrared radiation, which, because of its longer wavelength, is absorbed by such substances as carbon dioxide. The overall effect is that the average temperature of the earth and its atmosphere is increasing (so-called *global warming*). The effect is similar to that occurring in a greenhouse, where light and long-wavelength ultraviolet radiation can pass through the glass into the greenhouse but the infrared radiation is absorbed by the glass and part of it is re-radiated into the greenhouse.

The greenhouse effect is seen as a major environmental hazard. Average increases in temperature could change weather patterns and agricultural output. It might also lead to melting of the polar ice caps and a corresponding rise in sea level. Carbon dioxide, from coal-fired power stations and car exhausts, is the main greenhouse gas. Other contributory pollutants are nitrogen oxides, ozone, methane, and *chlorofluorocarbons. *See also* pollution.

greenhouse gas *See* greenhouse effect.

greenockite A mineral form of cadmium sulphide, CdS.

green vitriol *See* iron(II) sulphate.

Gregorian telescope *See* telescope.

grey matter Part of the tissue that makes up the central nervous system of vertebrates. It is brown-grey in colour, consisting largely of nerve *cell bodies, *synapses, and *dendrites. The grey matter is the site of coordination between nerves of the central nervous system. *Compare* white matter.

grid 1. (in electricity) The system of overhead wires or underground cables by which electrical power is distributed from power stations to users. The grid is at a high voltage, up to 750 kV in some countries. **2.** (in electronics) *See* control grid. **3.** (in cartography) A network of horizontal and vertical lines on a map that provide a means of locating a specific point.

Grignard reagents A class of organometallic compounds of magnesium, with the general formula RMgX, where R is an organic group and X a halogen atom (e.g. CH_3MgCl, C_2H_5MgBr, etc.). They actually have the structure $R_2Mg.MgCl_2$, and can be made by reacting a haloalkane with magnesium in ether; they are rarely isolated but are extensively used in organic synthesis, when they are made in one reaction mixture. Grignard reagents have a number of reactions that make them useful in organic synthesis. With methanal they give a primary alcohol
$$CH_3MgCl + HCHO \rightarrow CH_3CH_2OH$$
Other aldehydes give a secondary alcohol
$$CH_3CHO + CH_3MgCl \rightarrow$$
$$(CH_3)_2CHOH$$
With alcohols, hydrocarbons are formed
$$CH_3MgCl + C_2H_5OH \rightarrow C_2H_5CH_3$$
Water also gives a hydrocarbon
$$CH_3MgCl + H_2O \rightarrow CH_4$$
The compounds are named after their discoverer, the French chemist F. A. V. Grignard (1871–1935).

grooming The actions of an animal of rearranging fur or feathers and cleaning the body surface by biting, scratching, licking, etc., which is important for removing parasites and spreading essential oils over the body surface. In many mammals, especially primates, grooming between individuals (*allogrooming*) has an important role in maintaining social cohesion.

ground state The lowest stable energy state of a system, such as a molecule, atom, or nucleus. *See* energy level.

ground tissues All the plant tissues formed by the *apical meristems except the epidermis and vascular tissue. The principal ground tissues are the *cortex, *pith, and primary *medullary rays, and they consist chiefly of *parenchyma. *Collenchyma* is a form of ground tissue less frequently observed. It consists of cells with limited amounts of additional cellulose thickening in the walls and is most commonly found in the stem cortex.

ground wave A radio wave that travels in approximately a straight line between points on the earth's surface. For transmission over longer distances sky waves have to be involved. *See* radio transmission.

group 1. (in physics) A set of elements A, B, C, etc., for which there exists a law of composition, referred to as 'multiplication'. Any two elements can be combined to give a 'product' AB.
(1) Every product of two elements is an element of the set.
(2) The operation is associative, i.e. $A(BC) = (AB)C$.
(3) The set has an element I, called the *identity element*, such that $IA = AI = A$ for all A in the set.
(4) Each element of the set has an *inverse* A^{-1} belonging to the set such that $AA^{-1} = A^{-1}A = I$.
Although the law of combination is called 'multiplication' this does not necessarily have its usual meaning. For example, the set of integers forms a group if the law of composition is addition.
Two elements A, B of a group *commute* if $AB = BA$. If all the elements of a group commute with each other the group is said to be *Abelian*. If this is not the case the group is said to be *non-Abelian*. The distinction between Abelian and non-Abelian groups is of fundamental importance in *gauge theories.
The interest of group theory in physics is in analysing symmetry. *Discrete groups* have a finite number of elements, such as the symmetries involved in rotations and reflections of molecules, which give rise to *point groups*. *Continuous groups* have an infinite number of elements where the elements are continuous. An example of a continuous group is the set of rotations about a fixed axis. The *rotation group* thus formed underlies the *quantum theory of *angular momentum, which has many applications to *atoms and *nuclei. More abstract and more general continuous groups describe fundamental interactions by gauge theories.
2. (in chemistry) *See* periodic table.

group 0 elements *See* noble gases.

group I elements A group of elements in the *periodic table, divided into two subgroups: group IA (the main group, *see* alkali metals) and group IB. Group IB consists of the *coinage metals, copper, silver, and gold, which are usually classified with the *transition elements.

group II elements A group of elements in the *periodic table, divided into two subgroups: group IIA (the main group, *see* alkaline-earth metals) and group IIB. Group IIB consists of the three metals zinc (Zn), cadmium (Cd), and mercury (Hg), which have two *s*-electrons outside filled *d*-subshells. Moreover, none of their compounds have unfilled *d*-levels, and the metals are regarded as nontransition elements. They are sometimes called the *zinc group*. Zinc and cadmium are relatively electropositive metals, forming compounds containing divalent ions Zn^{2+} or Cd^{2+}. Mercury is more unreactive and also unusual in forming mercury(I) compounds, which contain the ion Hg_2^{2+}.

group III elements A group of elements in the *periodic table, divided into subgroup IIIB (the main group) and subgroup IIIA. The IIIA subgroup consists of scandium (Sc), yttrium (Y), and lanthanum (La), which are generally classified with the *lanthanoids, and actinium (Ac), generally classified with the *actinoids. The term *group III elements* usually refers to the main-group elements: boron (B), aluminium (Al), gallium (Ga), indium (In), and thallium (Tl), which all have outer electronic configurations ns^2np^1 with no partly filled inner levels. They are the first members of the p-block. The group differs from the alkali metals and alkaline-earth metals in displaying a considerable variation in properties as the group is descended.

Boron has a small atomic radius and a relatively high ionization energy. In consequence its chemistry is largely covalent and it is generally classed as a metalloid. It forms a large number of volatile hydrides, some of which have the uncommon bonding characteristic of *electron-deficient compounds. It also forms a weakly acidic oxide. In some ways, boron resembles silicon (*see* diagonal relationship).

As the group is descended, atomic radii increase and ionization energies are all lower than for boron. There is an increase in polar interactions and the formation of distinct M^{3+} ions. This increase in metallic character is clearly illustrated by the increasing basic character of the hydroxides: boron hydroxide is acidic, aluminium and gallium hydroxides are ampho-

teric, indium hydroxide is basic, and thallium forms only the oxide. As the elements of group III have a vacant p-orbital they display many electron-acceptor properties. For example, many boron compounds form adducts with donors such as ammonia and organic amines (acting as Lewis acids). A large number of complexes of the type $[BF_4]^-$, $[AlCl_4]^-$, $[InCl_4]^-$, $[TlI_4]^-$ are known and the heavier members can expand their coordination numbers to six as in $[AlF_6]^{3-}$ and $[TlCl_6]^{3-}$. This acceptor property is also seen in bridged dimers of the type Al_2Cl_6. Another feature of group III is the increasing stability of the monovalent state down the group. The electron configuration ns^2np^1 suggests that only one electron could be lost or shared in forming compounds. In fact, for the lighter members of the group the energy required to promote an electron from the s-subshell to a vacant p-subshell is small. It is more than compensated for by the resulting energy gain in forming three bonds rather than one. This energy gain is less important for the heavier members of the group. Thus, aluminium forms compounds of the type AlCl in the gas phase at high temperatures. Gallium similarly forms such compounds and gallium(I) oxide (Ga_2O) can be isolated. Indium has a number of known indium(I) compounds (e.g. InCl, In_2O, $In^I[In^{III}Cl_6]$). Thallium has stable monovalent compounds. In aqueous solution, thallium(I) compounds are more stable than the corresponding thallium(III) compounds. *See* inert-pair effect.

group IV elements A group of elements in the *periodic table, divided into subgroup IVB (the main group) and subgroup IVA. The IVA subgroup consists of titanium (Ti), zirconium (Zr), and hafnium (Hf), which are generally classified with the *transition elements. The term *group IV elements* usually refers to the main-group elements: carbon (C), silicon (Si), germanium (Ge), tin (Sn), and lead (Pb), which all have outer electronic configurations ns^2np^2 with no partly filled inner levels.

The main valency of the elements is 4, and the members of the group show a variation from nonmetallic to metallic behaviour in moving down the group. Thus, carbon is a nonmetal and forms an acidic oxide (CO_2) and a neutral oxide. Carbon compounds are mostly covalent. One allotrope (diamond) is an insulator, although graphite is a fairly good conductor. Silicon and germanium are metalloids, having semiconductor properties. Tin is a metal, but does have a nonmetallic allotrope (grey tin). Lead is definitely a metal. Another feature of the group is the tendency to form divalent compounds as the size of the atom increases. Thus carbon has only the highly reactive carbenes. Silicon forms analogous silylenes. Germanium has an unstable hydroxide ($Ge(OH)_2$), a sulphide (GeS), and halides. The sulphide and halides disproportionate to germanium and the germanium(IV) compound. Tin has a number of tin(II) compounds, which are moderately reducing, being oxidized to the tin(IV) compound. Lead has a stable lead(II) state. See inert-pair effect.

In general, the reactivity of the elements increases down the group from carbon to lead. All react with oxygen on heating. The first four form the dioxide; lead forms the monoxide (i.e. lead(II) oxide, PbO). Similarly, all will react with chlorine to form the tetrachloride (in the case of the first four) or the dichloride (for lead). Carbon is the only one capable of reacting directly with hydrogen. The hydrides all exist from the stable methane (CH_4) to the unstable plumbane (PbH_4).

group V elements A group of elements in the *periodic table, divided into subgroup VB (the main group) and subgroup VA. The VA subgroup consists of vanadium (V), niobium (Nb), and tantalum (Ta), which are generally classified with the *transition elements. The term *group V elements* usually refers to the main-group elements: nitrogen (N), phosphorus (P), arsenic (As), antimony (Sb), and bismuth (Bi), which all have outer electronic configurations ns^2np^3 with no partly filled inner levels.

The lighter elements (N and P) are nonmetals; the heavier elements are metalloids. The lighter elements are electronegative in character and have fairly large ionization energies. Nitrogen has a valency of 3 and tends to form covalent compounds. The other elements have available d-sublevels and can promote an s-electron into one of these to form compounds with the V oxidation state. Thus, they have two oxides P_2O_3, P_2O_5, Sb_2O_3, Sb_2O_5, etc. In the case of bismuth, the pentoxide Bi_2O_5 is difficult to prepare and unstable – an example of the increasing stability of the III oxidation state in going from phosphorus to bismuth. The oxides also show how there is increasing metallic (electropositive) character down the group. Nitrogen and phosphorus have oxides that are either neutral (N_2O, NO) or acidic. Bismuth trioxide (Bi_2O_3) is basic. Bismuth is the only member of the group that forms a well-characterized positive ion Bi^{3+}.

group VI elements A group of elements in the *periodic table, divided into group VIB (the main group) and group VIA. The VIA subgroup consists of chromium (Cr), molybdenum (Mo), and tungsten (W), which are generally classified with the *transition elements. The term *group VI elements* usually refers to the main-group elements: oxygen (O), sulphur (S), selenium (Se), tellurium (Te), and polonium (Po), which all have outer electronic configurations ns^2np^4 with no partly filled inner levels. They are also called the *chalcogens*.

The configurations are just two electrons short of the configuration of a noble gas and the elements are characteristically electronegative and almost entirely nonmetallic. Ionization energies are high, (O 1314 to Po 813 kJ mol^{-1}) and monatomic cations are not known. Polyatomic cations do exist, e.g. O_2^+, S_8^{2+}, Se_8^{2+}, Te_4^{2+}. Electronegativity decreases down the group but the nearest approach to metallic character is the occurrence of 'metallic' allotropes of selenium, tellurium, and polonium along with some metalloid properties, in particular, marked photoconductivity. The elements of group VI combine with a wide range of other elements and the bonding is largely covalent. The elements all form hydrides of the type XH_2. Apart from water, these materials are all toxic foul-smelling gases; they show decreasing thermal stability with increasing relative atomic mass of X. The hydrides dissolve in water to give very weak acids (acidity increases down the group). Oxygen forms the additional hydride H_2O_2 (hydrogen peroxide), but sulphur forms a range of sulphanes, such as H_2S_2, H_2S_4, H_2S_6.

Oxygen forms the fluorides O_2F_2 and OF_2, both powerful fluorinating agents; sulphur forms analogous fluorides along with some higher fluorides, S_2S_2, SF_2, SF_4, SF_6, S_2F_{10}. Selenium and tellurium form only the higher fluorides MF_4 and MF_6; this is in contrast to the formation of lower valence states by heavier elements observed in groups III, IV, and V. The chlorides are limited to M_2Cl_2 and MCl_4; the bromides are similar except that sulphur only forms S_2Br_2. All metallic elements form oxides and sulphides and many form selenides.

group VII elements A group of elements in the *periodic table divided into two subgroups: group VIIA and group VIIB. Group VIIB is the main group (*see* halogens) and group VIIA consists of the elements manganese (Mn), technetium (Tc), and rhenium (Re), which are usually classified with the *transition elements.

group VIII elements A group of nine elements in the *periodic table consisting of the *platinum metals – cobalt (Co), rhodium (Rh), iridium (Ir), nickel (Ni), palladium (Pd), and platinum (Pt) – together with iron (Fe), ruthenium (Ru), and osmium (Os). These metals are now classified with the *transition elements.

growth An increase in the dry weight or volume of an organism through cell division and cell enlargement. Growth may continue throughout the life of the organism, as occurs in woody plants, or it may cease at maturity, as in man and other mammals. *See also* allometric growth; exponential growth.

growth factor A *polypeptide that is produced by cells and stimulates them to proliferate. It is thought that some are involved in the abnormal regulation of growth seen in cancer when produced in excessive amounts. An example is *platelet-derived growth factor* (*PDGF*), normally produced by the platelets to repair injured tissue.

growth hormone (GH; somatotrophin) A hormone, secreted by the mammalian pituitary gland, that stimulates protein synthesis and growth of the long bones in the legs and arms. Production of growth hormone is greatest during early life. In man, overproduction results in gigantism and underproduction in dwarfism. Bovine somatotrophin is used to increase milk and meat production in cattle.

growth ring (annual ring) Any of the rings that can be seen in a cross-section of a woody stem (e.g. a tree trunk). It represents the *xylem formed in one year as a result of fluctuating activity of the vascular

cortex cork cork cambium

phloem

vascular cambium

pith

medullary ray

autumn wood
spring wood } growth ring year 3

autumn wood
spring wood } growth ring year 2

autumn wood
spring wood } growth ring year 1

Transverse section through a three-year-old woody stem to show the growth rings

*cambium. In temperate climates pale soft *spring wood*, characterized by large xylem vessels, is formed in spring and early summer. Growth slows down in late summer and a darker dense *autumn wood* with smaller xylem vessels is formed (see illustration). The age of a tree can be determined by counting the rings. Under certain circumstances two or more growth rings may form in one year, giving rise to false annual rings.

growth substance (phytohormone; plant hormone) Any of a number of organic chemicals that are synthesized by plants and regulate growth and development. They are usually made in a particular region, such as the shoot tip, and transported to other regions, where they take effect. *See* abscisic acid; auxin; cytokinin; ethene (ethylene); gibberellin.

GSC (gas−solid chromatography) *See* gas chromatography.

guanidine A crystalline basic compound $HN:C(NH_2)_2$, related to urea.

guanine A *purine derivative. It is one of the major component bases of *nucleotides and the nucleic acids *DNA and *RNA.

guanosine A nucleoside consisting of one guanine molecule linked to a D-ribose sugar molecule. The derived nucleotides, guanosine mono-, di-, and triphosphate (GMP, GDP, and GTP, respectively), participate in various metabolic reactions.

guard cell *See* stoma.

gum 1. Any of a variety of substances obtained from plants. Typically they are insoluble in organic solvents but form gelatinous or sticky solutions with water. Gum resins are mixtures of gums and natural resins. Gums are produced by the young xylem vessels of some plants (mainly trees) in response to wounding or pruning. The exudate hardens when it reaches the plant surface and thus provides a temporary protective seal while the cells below divide to form a permanent repair. Excessive gum formation is a symptom of some plant diseases. *See also* mucilage. **2.** *See* gingiva.

guncotton *See* cellulose nitrate.

306

gun metal A type of bronze usually containing 88–90% copper, 8–10% tin, and 2–4% zinc. Formerly used for cannons, it is still used for bearings and other parts that require high resistance to wear and corrosion.

gunpowder An explosive consisting of a mixture of potassium nitrate, sulphur, and charcoal.

gut *See* alimentary canal.

GUT *See* grand unified theory.

guttation *See* hydathode.

GWS model Glashow–Weinberg–Salam model. *See* electroweak theory.

Gymnospermae A subdivision of the *Spermatophyta or a class of the *Pteropsida containing the conifers (*see* Coniferales) and related species, such as cycads (*see* Cycadales) and ginkgos, as well as the extinct seed ferns (*see* Pteridospermales) Gymnosperms have an extensive fossil record going back to the late Devonian. The gametes are carried in male and female *cones, fertilization usually being achieved by wind-borne pollen. The ovules and the seeds into which they develop are borne unprotected (rather than enclosed in a carpel, as are those of the *Angiospermae). (The term gymnosperm means naked seed.) Internal tissue and cell structure of these species is not as advanced as in the angiosperms.

gynoecium (gynaecium) The female sex organs (*carpels) of a flower. *Compare* androecium.

gypsum A monoclinic mineral form of hydrated *calcium sulphate, $CaSO_4.2H_2O$. It occurs in five varieties: *rock gypsum*, which is often red stained and granular; *gypsite*, an impure earthy form occurring as a surface deposit; *alabaster*, a pure fine-grained translucent form; *satin spar*, which is fibrous and silky; and *selenite*, which occurs as transparent crystals in muds and clays. It is used in the building industry and in the manufacture of cement, rubber, paper, and plaster of Paris.

gyrocompass A *gyroscope that is driven continuously so that it can be used as a nonmagnetic compass. When the earth rotates the gyroscope experiences no torque if its spin axis is parallel to the earth's axis; if these axes are not parallel, however, the gyroscope experiences a sequence of restoring torques that tend to make it align itself with the earth's axis. The gyrocompass is therefore an accurate north-seeking device that is uninfluenced by metallic or magnetic objects and it is also more consistent than the magnetic compass. It is therefore widely used on ships, aircraft, missiles, etc.

gyromagnetic ratio Symbol γ. The ratio of the angular momentum of an atomic system to its magnetic moment. The inverse of the gyromagnetic ratio is called the *magneto-mechanical ratio*.

gyroscope A disc with a heavy rim mounted in a double *gimbal so that its axis can adopt any orientation in space. When the disc is set spinning the whole contrivance has two useful properties: (1) Gyroscopic inertia, i.e. the direction of the axis of spin resists change so that if the gimbals are turned the spinning disc maintains the same orientation in space. This property forms the basis of the *gyrocompass and other navigational devices. (2) Precession, i.e. when a gyroscope is subjected to a torque that tends to alter the direction of its axis, the gyroscope turns about an axis at right angles both to the axis about which the torque was applied and to its main axis of spin. This is a consequence of the need to conserve *angular momentum.

In the gyrostabilizer for stabilizing a ship, aircraft, or platform, three gyroscopes are kept spinning about mutually perpendicular axes so that any torque tending to alter the orientation of the whole device affects one of the gyroscopes and thereby activates a servomechanism that restores the original orientation.

H

Haber process An industrial process for producing ammonia by reaction of nitrogen with hydrogen:

$$N_2 + 3H_2 \rightleftharpoons 2NH_3$$

The reaction is reversible and exothermic, so that a high yield of ammonia is favoured by low temperature (*see* Le Chatelier's principle). However, the rate of reaction would be too slow for equilibrium to be reached at normal temperatures, so an optimum temperature of about 450°C is used, with a catalyst of iron containing potassium and aluminium oxide promoters. The higher the pressure the greater the yield, although there are technical difficulties in using very high pressures. A pressure of about 250 atmospheres is commonly employed.

The process is of immense importance for the fixation of nitrogen for fertilizers. It was developed in 1908 by the German chemist Fritz Haber (1886–1934) and was developed for industrial use by Carl Bosch (1874–1940), hence the alternative name *Haber–Bosch process*. The nitrogen is obtained from liquid air. Formerly, the hydrogen was from *water gas and the water-gas shift reaction (the *Bosch process*) but now the raw material (called *synthesis gas*) is obtained by steam *reforming natural gas.

habit *See* crystal habit.

habitat The place in which an organism lives, which is characterized by its physical features or by the dominant plant types. Freshwater habitats,

for example, include streams, ponds, rivers, and lakes.

habituation 1. A simple type of learning consisting of a gradual waning in the response of an animal to a continuous or repeated stimulus that is not associated with *reinforcement. **2.** The condition of being psychologically, but not physically, dependent on a drug, with a desire to continue its use but not to increase the dosage.

hadron Any of a class of subatomic particles that interact by the strong interaction (*see* fundamental interactions). The class includes protons, neutrons, and pions. Hadrons are believed to have an internal structure and to consist of quarks; they are therefore not truly elementary. Hadrons are either *baryons, which decay into protons and are believed to consist of three quarks, or *mesons, which decay into *leptons and photons or into proton pairs and are believed to consist of a quark and an antiquark. *See* elementary particles.

haem (heme) An iron-containing molecule that binds with proteins as a *cofactor or *prosthetic group to form the *haemoproteins*. These are *haemoglobin, *myoglobin, and the *cytochromes. Essentially, haem comprises a *porphyrin with its four nitrogen atoms holding the iron(II) atom as a chelate. This iron can reversibly bind oxygen (as in haemoglobin and myoglobin) or (as in the cytochromes) conduct electrons by conversion between iron(II) and iron(III) species.

haemagglutination *See* agglutination.

haematite A mineral form of iron(III) oxide, Fe_2O_3. It is the most important ore of iron and usually occurs in two main forms: as a massive red kidney-shaped ore (*kidney ore*) and as grey to black metallic crystals known as *specular iron ore*. Haematite is the major red colouring agent in rocks; the largest deposits are of sedimentary origin. In industry haematite is also used as a polishing agent (jeweller's rouge) and in paints.

haematoxylin A compound used in its oxidized form (*haematein*) as a blue dye in optical microscopy, particularly for staining smears and sections of animal tissue. It stains nuclei blue and is frequently used with *eosin as a counterstain for cytoplasm. Haematoxylin requires a mordant, such as iron alum, which links the dye to the tissue. Different types of haematoxylin can be made up depending on the mordant used, the method of oxidation, and the pH. Examples are *Delafield's haematoxylin* and *Ehrlich's haematoxylin*.

haemocyanin (hemocyanin) Any of a group of copper-containing respiratory proteins found in solution in the blood of certain arthropods and molluscs. Haemocyanins contain two copper atoms that reversibly bind oxygen, changing between the colourless deoxygenated form (CuI) and the blue oxygenated form (CuII). In some species, haemocyanin molecules form giant polymers with molecular weights of several million.

haemoglobin One of a group of globular proteins occurring widely in animals as oxygen carriers in blood. Vertebrate haemoglobin comprises two pairs of polypeptide chains (forming the *globin* protein) with each chain folded to provide a binding site for a *haem group. Each of the four haem groups binds one oxygen molecule to form *oxyhaemoglobin*. Dissociation occurs in oxygen-depleted tissues: oxygen is released and haemoglobin is reformed. The haem groups also bind other inorganic molecules, including carbon monoxide (to form *carboxyhaemoglobin*). Carbon monoxide binds more strongly than oxygen and competes with it (hence its toxicity). In vertebrates, haemoglobin is contained in the red blood cells (*erythrocytes).

haemolysis The breakdown of red blood cells. It may be due to the action of disease-causing microorganisms, poisons, antibodies in mismatched blood transfusions, or certain allergic reactions. It produces anaemia.

haemophilia A hereditary sex-linked disease (*see* sex linkage) in which there is a deficiency or defect of *Factor VIII, causing the blood to clot very slowly. There may be prolonged bleeding following injury, and in severe cases spontaneous bleeding into the joints and muscles. The disorder is due to a defective recessive allele of the Factor VIII gene, which is located on the X chromosome. Female carriers of the defective allele are unaffected, whereas all males who inherit a defective allele exhibit the disease.

hafnium Symbol Hf. A silvery lustrous metallic *transition element; a.n. 72; r.a.m. 178.49; r.d. 13.3; m.p. 2230±20°C; b.p. 4602°C. The element is found with zirconium and is extracted by formation of the chloride and reduction by the Kroll process. It is used in tungsten alloys in filaments and electrodes and as a neutron absorber. The metal forms a passive oxide layer in air. Most of its compounds are hafnium(IV) complexes; less stable hafnium(III) complexes also exist. The element was first reported by Urbain in 1911, and its existence was finally established by D. Coster and G. C. de Hevesey in 1923.

hahnium *See* transactinide elements.

hair 1. A multicellular threadlike structure, consisting of many dead keratinized cells, that is produced by the epidermis in mammalian *skin. The section of a hair below the skin surface (the *root*) is contained within a *hair follicle, the base of which produces the hair cells. Hair assists in maintaining body temperature by reducing heat loss from the skin.

Bristles and whiskers are specialized types of hair. **2.** Any of various threadlike structures on plants, such as a *trichome.

hair follicle A narrow tubular depression in mammalian skin containing the root of a *hair. It is lined with epidermal cells and extends down through the epidermis and dermis to its base in the subcutaneous tissue. The ducts of *sebaceous glands empty into hair follicles.

half cell An electrode in contact with a solution of ions, forming part of a *cell. Various types of half cell exist, the simplest consisting of a metal electrode immersed in a solution of metal ions. Gas half cells have a gold or platinum plate in a solution with gas bubbled over the metal plate. The commonest is the *hydrogen half cell. Half cells can also be formed by a metal in contact with an insoluble salt or oxide and a solution. The *calomel half cell is an example of this. Half cells are commonly referred to as *electrodes*.

half-life *See* decay.

half-thickness The thickness of a specified material that reduces the intensity of a beam of radiation to half its original value.

half-wave plate *See* retardation plate.

half-wave rectifier *See* rectifier.

half-width The width of a spectral line measured at half its height. In some contexts, the term is used for half the width of the line measured at half its height.

halide A compound of a halogen with another element or group. The halides of typical metals are ionic (e.g. sodium fluoride, Na^+F^-). Metals can also form halides in which the bonding is largely covalent (e.g. aluminium chloride, $AlCl_3$). Organic compounds are also sometimes referred to as halides; e.g. the alkyl halides (*see* haloalkanes) and the

*acyl halides. Halides are named *fluorides*, *chlorides*, *bromides*, or *iodides*.

halite (rock salt) Naturally occurring *sodium chloride (common salt, NaCl), crystallizing in the cubic system. It is chiefly colourless or white (sometimes blue) when pure but the presence of impurities may colour it grey, pink, red, or brown. Halite often occurs in association with anhydrite and gypsum.

Hall effect The production of an e.m.f. within a conductor or semiconductor through which a current is flowing when there is a strong transverse magnetic field. The potential difference develops at right angles to both the current and the field. It is caused by the deflection of charge carriers by the field and was first discovered by Edwin Hall (1855–1938). The strength of the electric field E_H produced is given by the relationship $E_H = R_H jB$, where j is the current density, B is the magnetic flux density, and R_H is a constant called the *Hall coefficient*. The value of R_H can be shown to be $1/ne$, where n is the number of charge carriers per unit volume and e is the electronic charge. The effect is used to investigate the nature of charge carriers in metals and semiconductors, in the *Hall probe* for the measurement of magnetic fields, and in magnetically operated switching devices. *See also* quantum Hall effect.

Halley's comet A bright *comet with a period of 76 years. Its last visit was in 1986. The comet moves around the sun in the opposite direction to the planets. Its orbit was first calculated in 1705 by Edmund Halley (1656–1742), after whom it is named.

Hall–Heroult cell An electrolytic cell used industrially for the extraction of aluminium from bauxite. The bauxite is first purified by dissolving it in sodium hydroxide and filtering off insoluble constituents. Aluminium hydroxide is then precipitated (by adding CO_2) and this is decomposed by heating to obtain pure Al_2O_3. In the Hall–Heroult cell, the oxide is mixed with cryolite (to lower its melting point) and the molten mixture electrolysed using graphite anodes. The cathode is the lining of the cell, also of graphite. The electrolyte is kept in a molten state (about 850°C) by the current. Molten aluminium collects at the bottom of the cell and can be tapped off. Oxygen forms at the anode, and gradually oxidizes it away. The cell is named after the US chemist Charles Martin Hall (1863–1914), who discovered the process in 1886, and the French chemist Paul Heroult (1863–1914), who discovered it independently in the same year.

hallux The innermost digit on the hindlimb of a tetrapod vertebrate. In man it is the big toe and contains two phalanges. The hallux is absent in some mammals and in many birds it is directed backwards as an adaptation to perching. *Compare* pollex.

halo A luminous ring that sometimes can be observed around the sun or the moon. It is caused by diffraction of their light by particles in the earth's atmosphere; the radius of the ring is inversely proportional to the predominant particle radius.

haloalkanes (alkyl halides) Organic compounds in which one or more hydrogen atoms of an alkane have been substituted by halogen atoms. Examples are chloromethane, CH_3Cl, dibromoethane, CH_2BrCH_2Br, etc. Haloalkanes can be formed by direct reaction between alkanes and halogens using ultraviolet radiation. They are usually made by reaction of an alcohol with a halogenating agent.

haloform reaction A reaction for producing haloforms from methyl ketones. An example is the production of chloroform from propanone using sodium chlorate(I) (or bleaching powder):

$$CH_3COCH_3 + 3NaOCl \rightarrow$$
$$CH_3COCl_3 + 3NaOH$$

The substituted ketone then reacts to give chloroform (trichloromethane):

$$CH_3COCCl_3 + NaOH \rightarrow$$
$$NaOCOCH_3 + CHCl_3$$

The reaction can also be used for making carboxylic acids, since $RCOCH_3$ gives the product $NaOCOR$. It is particularly useful for aromatic acids as the starting ketone can be made by a Friedel–Crafts acylation.

The reaction of methyl ketones with sodium iodate(I) gives iodoform (triiodomethane), which is a yellow solid with a characteristic smell. This reaction is used in the *iodoform test* to identify methyl ketones. It also gives a positive result with a secondary alcohol of the formula $RCH(OH)CH_3$ (which is first oxidized to a methylketone) or with ethanol (oxidized to ethanal, which also undergoes the reaction).

haloforms The four compounds with formula CHX_3, where X is a halogen atom. They are *chloroform* ($CHCl_3$), and, by analogy, *fluoroform* (CHF_3), *bromoform* ($CHBr_3$), and *iodoform* (CHI_3). The systematic names are trichloromethane, trifluoromethane, etc.

halogenating agent *See* halogenation.

halogenation A chemical reaction in which a halogen atom is introduced into a compound. Halogenations are described as *chlorination*, *fluorination*, *bromination*, etc., according to the halogen involved. Halogenation reactions may take place by direct reaction with the halogen. This occurs with alkanes, where the reaction involves free radicals and requires high temperature, ultraviolet radiation, or a chemical initiator; e.g.

$$C_2H_6 + Br_2 \rightarrow C_2H_5Br + HBr$$

The halogenation of aromatic compounds can be effected by electrophilic substitution using an aluminium chloride catalyst:

$$C_6H_6 + Cl_2 \rightarrow C_6H_5Cl + HCl$$

Halogenation can also be carried out using compounds, such as phosphorus halides (e.g. PCl_3) or sulphur dihalide oxides (e.g. $SOCl_2$), which react with $-OH$ groups. Such compounds are called *halogenating agents*. Addition reactions are also referred to as halogenations; e.g.

$$C_2H_4 + Br_2 \rightarrow CH_2BrCH_2Br$$

halogens (group VII elements) A group of elements in the *periodic table (group VIIB): fluorine (F), chlorine (Cl), bromine (Br), iodine (I), and astatine (At). All have a characteristic electron configuration of noble gases but with outer ns^2np^5 electrons. The outer shell is thus one electron short of a noble-gas configuration. Consequently, the halogens are typical nonmetals; they have high electronegativities – high electron affinities and high ionization energies. They form compounds by gaining an electron to complete the stable configuration; i.e. they are good oxidizing agents. Alternatively, they share their outer electrons to form covalent compounds, with single bonds.

All are reactive elements with the reactivity decreasing down the group. The electron affinity decreases down the group and other properties also show a change from fluorine to astatine. Thus, the melting and boiling points increase; at $20\,°C$, fluorine and chlorine are gases, bromine a liquid, and iodine and astatine are solids. All exist as diatomic molecules.

The name 'halogen' comes from the Greek 'salt-producer', and the elements react with metals to form ionic halide salts. They also combine with nonmetals, the activity decreasing down the group: fluorine reacts with all nonmetals except nitrogen and the noble gases helium, neon, and argon; iodine does not react with any noble gas, nor with carbon, nitrogen, oxy-

gen, or sulphur. The elements fluorine to iodine all react with hydrogen to give the acid, with the activity being greatest for fluorine, which reacts explosively. Chlorine and hydrogen react slowly at room temperature in the dark (sunlight causes a free-radical chain reaction). Bromine and hydrogen react if heated in the presence of a catalyst. Iodine and hydrogen react only slowly and the reaction is not complete. There is a decrease in oxidizing ability down the group from fluorine to iodine. As a consequence, each halogen will displace any halogen below it from a solution of its salt, for example:

$$Cl_2 + 2Br^- \rightarrow Br_2 + 2Cl^-$$

The halogens also form a wide variety of organic compounds in which the halogen atom is linked to carbon. In general, the aryl compounds are more stable than the alkyl compounds and there is decreasing resistance to chemical attack down the group from the fluoride to the iodide.

Fluorine has only a valency of 1, although the other halogens can have higher oxidation states using their vacant d-electron levels. There is also evidence for increasing metallic behaviour down the group. Chlorine and bromine form compounds with oxygen in which the halogen atom is assigned a positive oxidation state. Only iodine, however, forms positive ions, as in $I^+NO_3^-$.

halon A compound obtained by replacing the hydrogen atoms of a hydrocarbon by bromine along with other halogen atoms, for instance halon 1211 is bromochlorodifluoromethane (CF_2BrCl) and halon 1301 is bromotrifluoromethane (CF_3Br). Halons are very stable and unreactive and are widely used in fire extinguishers. There is concern that they are being broken down in the atmosphere to bromine, which reacts with ozone, leading to depletion of the *ozone layer, and their use is being curtailed. Although more *chloro-fluorocarbons are present in the atmosphere, halons are between three and ten times more destructive of ozone.

halophyte A plant that can tolerate a high concentration of salt in the soil. Such conditions occur in salt marshes and mudflats. Examples of halophytes are the thrift (*Armeria*), sea lavender (*Limonium*), and rice grass (*Spartina*).

Hamiltonian Symbol *H*. A function used to express the energy of a system in terms of its momentum and positional coordinates. In simple cases this is the sum of its kinetic and potential energies. In *Hamiltonian equations*, the usual equations used in mechanics (based on forces) are replaced by equations expressed in terms of momenta. This method of formulating mechanics (*Hamiltonian mechanics*) was first introduced by Sir William Rowan Hamilton (1805–65).

haploid Describing a nucleus, cell, or organism with a single set of unpaired chromosomes. The haploid number is designated as *n*. Reproductive cells, formed as a result of *meiosis, are haploid. Fusion of two such cells (*see* fertilization) restores the normal (*diploid) number.

haptotropism *See* thigmotropism.

hardening of oils The process of converting unsaturated esters of *fatty acids into (more solid) saturated esters by hydrogenation using a nickel catalyst. It is used in the manufacture of margarine from vegetable oils.

hard ferromagnetic materials *See* soft iron.

hardness of water The presence in water of dissolved calcium or magnesium ions, which form a scum with soap and prevent the formation of a lather. The main cause of hard water is dissolved calcium hydrogen-carbonate ($Ca(HCO_3)_2$), which is formed in limestone or chalk regions by the action of dissolved carbon dioxide on calcium carbonate. This type is known as *temporary hardness* because it is removed by boiling:

$$Ca(HCO_3)_2(aq) \rightarrow CaCO_3(s) + H_2O(l) + CO_2(g)$$

The precipitated calcium carbonate is the 'fur' (or 'scale') formed in kettles, boilers, pipes, etc. In some areas, hardness also results from dissolved calcium sulphate ($CaSO_4$) or calcium fluoride (CaF_2). These compounds can not be removed by boiling (*permanent hardness*).

Hard water is a considerable problem in washing and in reducing the efficiency of boilers, heating systems, and certain industrial processes. Various methods of *water softening* are used. In public supplies, the temporary hardness can be removed by adding lime (calcium hydroxide), which precipitates calcium carbonate

$$Ca(OH)_2(aq) + Ca(HCO_3)_2(aq) \rightarrow 2CaCO_3(s) + 2H_2O(l)$$

This is known as the *Clark process* (or as '*clarking*'). It does not remove permanent hardness. Both temporary and permanent hardness can be treated by precipitating calcium carbonate by added sodium carbonate – hence its use as washing soda and in bath salts. Calcium (and other) ions can also be removed from water by ion-exchange using zeolites (e.g. *Permutit*). This method is used in small domestic water-softeners. Another technique is not to remove the Ca^{2+} ions but to complex them and prevent them reacting further. For domestic use polyphosphates (containing the ion $P_6O_{18}^{6-}$, e.g. *Calgon*) are added. Other sequestering agents are also used for industrial water. *See also* sequestration.

hard radiation Ionizing radiation of high penetrating power, usually gamma rays or short-wavelength X-rays. *Compare* soft radiation.

hardware *See* computer.

hardwood *See* wood.

harmonic

1 st harmonic (fundamental)

2 nd harmonic (1 st overtone)

3 rd harmonic (2 nd overtone)

Harmonics

harmonic An oscillation having a frequency that is a simple multiple of a *fundamental* sinusoidal oscillation. The fundamental frequency of a sinusoidal oscillation is usually called the *first harmonic*. The *second harmonic* has a frequency twice that of the fundamental and so on (see illustration). A taut string or column of air, as in a violin or organ, will sound upper harmonics at the same time as the fundamental sounds. This is because the string or column of air divides itself into sections, each section then vibrating as if it were a whole. The upper harmonics are also called *overtones*, but the second harmonic is the first overtone, and so on. Musicians, however, often regard harmonic and overtone as synonymous, not counting the fundamental as a harmonic.

harmonic motion *See* simple harmonic motion.

harmonic series (harmonic progression) A series or progression in which the reciprocals of the terms have a constant difference between them, e.g. $1 + 1/2 + 1/3 + 1/4 \dots + 1/n$.

Harvard classification *See* spectral class.

haustorium A specialized structure of certain parasitic plants that penetrates the cells of the host plant to absorb nutrients. In parasitic fungi haustoria are formed from enlarged hyphae and in parasitic flowering plants, such as the dodder (*Cuscuta*), they are outgrowths of the stem.

Haversian canals Narrow tubes within compact *bone containing blood vessels and nerves. They generally run parallel to the bone surface. Each canal surrounded by a series of rings of bone (*lamellae*) is known as a *Haversian system*. Haversian systems are joined to each other by bone material.

Hawking process Emission of particles by a *black hole as a result of qantum-mechanical effects. The process was first suggested by Stephen Hawking (1942–). The gravitational field of the black hole causes production of particle–antiparticle pairs in the vicinity of the event horizon (the process is analogous to that of pair production). One member of each pair (either the particle or the antiparticle) falls into the black hole, while the other escapes. To an external observer it appears that the black hole is emitting radiation (*Hawking radiation*). The energy of the particles that fall in is negative and exactly balances the (positive) energy of the escaping particles. This negative energy reduces the mass of the black hole and the net result of the process is that the emitted particle flux appears to carry off the black-hole mass. It can be shown that the black hole radiates like a *black body, with the energy distribution of the particles obeying *Planck's radiation law for a temperature that is inversely proportional to the mass of the hole. For a black hole of the mass of the sun, this temperature is only about 10^{-7} K, so the energy loss is negligible. However, for a 'mini' black hole, such as might have been formed in the early universe, with a mass of order 10^{12} kg (and a radius of order 10^{-15} m), the temperature would be of order 10^{11} K and the hole would radi-

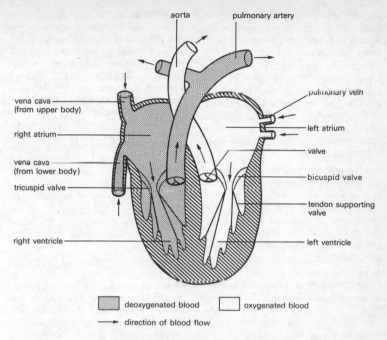

aorta   pulmonary artery

vena cava
(from upper body)

right atrium

vena cava
(from lower body)

tricuspid valve

right ventricle

pulmonary vein

left atrium

valve

bicuspid valve

tendon supporting
valve

left ventricle

▨ deoxygenated blood   ☐ oxygenated blood

➤ direction of blood flow

Structure of the mammalian heart

ate copiously (at a rate of about 6 ×
10^9 W) a flux of gamma rays, neutrinos, and electron–positron pairs.
(The observed levels of cosmic
gamma rays put strong constraints on
the number of such 'mini' black
holes, suggesting that there are too
few of them to solve the *missing-
mass problem.)

h.c.p. Hexagonal close packing. *See*
close packing.

health physics The branch of medical physics concerned with the protection of medical, scientific, and industrial workers from the hazards of ionizing radiation and other dangers
associated with atomic physics. Establishing the maximum permissible
*dose of radiation, the disposal of
radioactive waste, and the shielding
of dangerous equipment are the principal activities in this field.

heart A hollow muscular organ that,
by means of regular contractions,
pumps blood through the circulatory
system. Mammals have a four-
chambered heart consisting of two
atria and two ventricles; the right and
left sides are completely separate
from each other so there is no mixing
of oxygenated and deoxygenated
blood (see illustration). Oxygenated
blood from the pulmonary veins
enters the heart through the left
atrium, passes to the left ventricle,
and leaves the heart through the
*aorta. Deoxygenated blood from the
*venae cavae enters the right atrium
and is pumped through the right ventricle to the pulmonary artery, which
conveys it to the lungs for oxygena-
tion. The mitral and bicuspid valves
ensure that there is no backflow of
blood. The contractions of the heart

are initiated and controlled by the *pacemaker; in an average adult human the heart contracts about 70 times per minute.

The hearts of other vertebrates are similar except in the number of atria and ventricles (there may be one or two) and in the degree of separation of oxygenated and deoxygenated blood. Invertebrates, however, show great variation in the form and functioning of the heart.

heartwood (duramen) The wood at the centre of a tree trunk or branch. It consists of dead *xylem cells heavily thickened with lignin and provides structural support. Many heartwood cells contain oils, gums, and resins, which darken the wood. *Compare* sapwood.

heat The process of energy transfer from one body or system to another as a result of a difference in temperature. The energy in the body or system before or after transfer is also sometimes referred to as heat.

A body in equilibrium with its surroundings contains energy (the kinetic and potential energies of its atoms and molecules) but this is called *internal energy*, U, rather than heat. When such a body changes its temperature or phase there is a change in internal energy, ΔU, which (according to the first law of *thermodynamics) is given by $\Delta U = Q - W$, where Q is the heat absorbed by the body from the surroundings and W is the work done simultaneously on the surroundings. To use the word 'heat' for both U and Q is clearly confusing. Note also that certain physical quantities are described as *heat of atomization, *heat of combustion, etc. What is usually used is a standard molar *enthalpy change for the process under consideration. The units are kJ mol^{-1}; a negative value indicates that energy is liberated. *See also* heat capacity; heat transfer; latent heat.

heat balance 1. A balance sheet showing all the heat inputs to a system (such as a chemical process, furnace, etc.) and all the heat outputs. **2.** The equilibrium that exists on the average between the radiation received from the sun by the earth and its atmosphere and that reradiated or reflected by the earth and the atmosphere. In general, the regions of the earth nearer the equator than about 35°N or S receive more energy from the sun than they are able to reradiate, whereas those regions polewards of 35°N or S receive less energy than they lose. The excess of heat received by the low latitudes is carried to the higher latitudes by atmospheric and oceanic circulations.

heat capacity (thermal capacity) The ratio of the heat supplied to an object or specimen to its consequent rise in temperature. The *specific heat capacity* is the ratio of the heat supplied to unit mass of a substance to its consequent rise in temperature. The *molar heat capacity* is the ratio of the heat supplied to unit amount of a substance to its consequent rise in temperature. In practice, heat capacity (C) is measured in joules per kelvin, specific heat capacity (c) in J K^{-1} kg^{-1}, and molar heat capacity (C_m) in J K^{-1} mol^{-1}. For a gas, the values of c and C_m are commonly given either at *constant volume*, when only its *internal energy is increased, or at *constant pressure*, which requires a greater input of heat as the gas is allowed to expand and do work against the surroundings. The symbols for the specific and molar heat capacities at constant volume are c_v and C_v, respectively; those for the specific and molar heat capacities at constant pressure are c_p and C_p.

heat death of the universe The condition of the universe when *entropy is maximized and all large-scale samples of matter are at a uniform temperature. In this condition no energy is available for doing work and the universe is finally unwound. The condition was predicted by Rudolph Clausius (1822–88), the German physicist, who introduced the concept of entropy. Clausius's dictum that "the energy of the universe is constant, its entropy tends to a maximum" is a statement of the first two laws of thermodynamics. These laws apply in this sense only to closed systems and, for the predicted heat death to occur, the universe must be a closed system.

heat engine A device for converting heat into work. The heat is derived from the combustion of a fuel. In an *internal-combustion engine the fuel is burnt inside the engine, whereas in a *steam engine or steam turbine, examples of external-combustion engines, the fuel is used to raise steam outside the engine and then some of the steam's internal energy is used to do work inside the engine. Engines usually work on cycles of operation, the most efficient of which would be the *Carnot cycle. This cannot be realized in practice, but the *Rankine cycle is approximated by some engines.

heat exchanger A device for transferring heat from one fluid to another without permitting the two fluids to contact each other. A simple industrial heat exchanger consists of a bundle of parallel tubes, through which one fluid flows, enclosed in a container, through which the other fluid flows in the opposite direction (a *counter-current heat exchanger*).

heat of atomization The energy required to dissociate one mole of a given substance into atoms.

heat of combustion The energy liberated when one mole of a given substance is completely oxidized.

heat of crystallization The energy liberated when one mole of a given

substance crystallizes from a saturated solution of the same substance.

heat of dissociation The energy absorbed when one mole of a given substance is dissociated into its constituent elements.

heat of formation The energy liberated or absorbed when one mole of a compound is formed from its constituent elements.

heat of neutralization The energy liberated in neutralizing one mole of an acid or base.

heat of reaction The energy liberated or absorbed as a result of the complete chemical reaction of molar amounts of the reactants.

heat of solution The energy liberated or absorbed when one mole of a given substance is completely dissolved in a large volume of solvent (strictly, to infinite dilution).

heat pump A device for transferring heat from a low temperature source to a high temperature region by doing work. It is essentially a refrigerator with a different emphasis. The working fluid is at one stage a vapour, which is compressed by a pump adiabatically so that its temperature rises. It is then passed to the radiator, where heat is given out to the surroundings (the space to be heated) and the fluid condenses to a liquid. It is then expanded into an evaporator where it takes up heat from its surroundings and becomes a vapour again. The cycle is completed by returning the vapour to the compressor. Heat pumps are sometimes adapted to be used as dual-purpose space-heating-in-winter and air-conditioning-in-summer devices.

heat radiation (radiant heat) Energy in the form of electromagnetic waves emitted by a solid, liquid, or gas as a result of its temperature. It can be transmitted through space; if there is a material medium this is not warmed by the radiation except to the extent that it is absorbed. Although it covers the whole electromagnetic spectrum, the highest proportion of this radiation lies in the infrared portion of the spectrum at normal temperatures. *See* black body.

heat shield A specially prepared surface that prevents a spacecraft or capsule from overheating as it re-enters the earth's atmosphere. The surface is coated with a plastic impregnated with quartz fibres, which is heated by friction with air molecules as the craft enters the atmosphere, causing the outer layer to vaporize. In this way about 80% of the energy is reradiated and the craft is safeguarded from an excessive rise in temperature.

heat transfer The transfer of energy from one body or system to another as a result of a difference in temperature. The heat is transferred by *conduction (*see also* conductivity), *convection, and radiation (*see* heat radiation).

Heaviside–Kennelly layer *See* earth's atmosphere.

Heaviside–Lorentz units A system of units for electric and magnetic quantities based upon c.g.s. electrostatic and electromagnetic units. They are the rationalized forms of *Gaussian units and, like the latter, are widely used in particle physics and relativity in preference to the *SI units now employed for general purposes in physics.

heavy-fermion system A substance in which the electrons have a very high effective mass; i.e. they act as if they had masses several hundred times the normal mass of the electron. An example of a heavy-fermion system is the cerium–copper–silicon compound $CeCuSi_2$. The electrons with high effective mass are f-electrons in narrow energy bands associated with strong many-body effects. Substances containing such electrons have unusual thermodynamic, magnetic,

and superconducting properties, which are still not completely understood. The *superconductivity of such materials has a more complicated mechanism than that for metals described by the BCS theory, since the Cooper pairs are formed from *quasiparticles with very high effective masses rather than from electrons.

heavy hydrogen *See* deuterium.

heavy metal A metal with a high relative atomic mass. The term is usually applied to common transition metals, such as copper, lead, and zinc. These metals are a cause of environmental *pollution from a number of sources, including lead in petrol, industrial effluents, and leaching of metal ions from the soil into lakes and rivers by acid rain.

heavy spar A mineral form of *barium sulphate, $BaSO_4$.

heavy water (deuterium oxide) Water in which hydrogen atoms, ^1H, are replaced by the heavier isotope deuterium, ^2H (symbol D). It is a colourless liquid, which forms hexagonal crystals on freezing. Its physical properties differ from those of 'normal' water; r.d. 1.105; m.p. 3.8°C; b.p. 101.4°C. Deuterium oxide, D_2O, occurs to a small extent (about 0.003% by weight) in natural water, from which it can be separated by fractional distillation or by electrolysis. It is useful in the nuclear industry because of its ability to reduce the energies of fast neutrons to thermal energies (*see* moderator) and because its absorption cross-section is lower than that of hydrogen and consequently it does not appreciably reduce the neutron flux. In the laboratory it is used for *labelling other molecules for studies of reaction mechanisms. Water also contains the compound HDO.

hecto- Symbol h. A prefix used in the metric system to denote 100 times. For example, 100 coulombs = 1 hectocoulomb (hC).

Heisenberg uncertainty principle *See* uncertainty principle.

helicate A type of inorganic molecule containing a double helix of bipyridyl-derived molecules formed around a chain of up to five copper(I) ions. *See also* supramolecular chemistry.

heliocentric universe A view of the universe in which the sun is taken to be at its centre. The model was first proposed by Aristarchus of Samos (310–230 BC) but dropped in favour of the *geocentric system proposed by Ptolemy (*c.* 90–168 AD). Copernicus (1473–1543) revived an essentially heliocentric view, which was upheld by Galileo (1564–1642) against strong opposition from the church on the grounds that if the earth was not at the centre of the universe man's position in it was diminished. In the modern view the sun is at the centre of the *solar system, but the solar system is one of an enormous number of stars in the Galaxy, which is itself one of an enormous number of *galaxies.

heliotropism *See* phototropism.

helium Symbol He. A colourless odourless gaseous nonmetallic element belonging to group 0 of the periodic table (*see* noble gases); a.n. 2; r.a.m. 4.0026; d. 0.178 $g\,dm^{-3}$; m.p. $-272.2°C$ (at 20 atm.); b.p. $-268.93°C$. The element has the lowest boiling point of all substances and can be solidified only under pressure. Natural helium is mostly helium–4, with a small amount of helium–3. There are also two short-lived radioactive isotopes: helium–5 and –6. It occurs in ores of uranium and thorium and in some natural-gas deposits. It has a variety of uses, including the provision of inert atmospheres for welding and semiconductor manufacture, as a refrigerant for superconductors, and as a diluent in breathing apparatus. It is also used in filling balloons. Chemically it is totally inert and has no known compounds. It was discovered in the solar spectrum in 1868 by Lockyer.

Helmholtz free energy *See* free energy.

heme *See* haem.

hemiacetals *See* acetals.

hemihydrate A crystalline hydrate containing two molecules of compound per molecule of water (e.g. $2CaSO_4.H_2O$).

hemiketals *See* ketals.

Hemiptera An order of insects comprising the true bugs. Hemipterans typically have oval flattened bodies with two pairs of wings, which are folded back across the abdomen at rest. The forewings are hardened, either at their bases only (in the suborder Heteroptera) or uniformly (in the suborder Homoptera). The mouthparts are modified for piercing and sucking, with long slender stylets forming a double tube. Many bugs feed on plant sap and are serious agricultural pests, including aphids, leaf-hoppers, scale insects, mealy bugs, etc. Others are carnivorous, and the order contains many aquatic species, such as the water boatmen, which have legs adapted for swimming and the exchange of respiratory gases.

henry Symbol H. The *SI unit of inductance equal to the inductance of a closed circuit in which an e.m.f. of one volt is produced when the electric current in the circuit varies uniformly at a rate of one ampere per second. It is named after Joseph Henry (1797 –1878), a US physicist.

Henry's law At a constant temperature the mass of gas dissolved in a liquid at equilibrium is proportional to the partial pressure of the gas. The law, discovered in 1801 by the British chemist and physician William Henry (1775–1836), is a special case of the

319

partition law. It applies only to gases that do not react with the solvent.

heparin A mucopolysaccharide with *anticoagulant properties, occurring in vertebrate tissues, especially the lungs and blood vessels.

Hepaticae A class of plants within the *Bryophyta comprising the liverworts, which occur in moist situations (including fresh water) and as epiphytes on other plants. The plant body may be a thallus, growing closely pressed to the ground (thallose liverworts, e.g. *Pellia*), or it may bear many leaflike lobes (leafy liverworts). It gives rise to leafless stalks bearing capsules. Spores formed in the capsules are released and grow to produce new plants.

hepatic portal system The vein (*hepatic portal vein*) or veins that transport blood containing the absorbed products of digestion from the intestine directly to the liver.

heptahydrate A crystalline hydrate that has seven molecules of water per molecule of compound.

heptane A liquid straight-chain alkane obtained from petroleum, C_7H_{16}; r.d. 0.684; m.p. $-90.6°C$; b.p. $98.4°C$. In standardizing *octane numbers, heptane is given a value zero.

heptaoxodiphosphoric(V) acid *See* phosphoric(V) acid.

heptavalent (septivalent) Having a valency of seven.

herbaceous Describing a plant that contains little permanent woody tissue. The aerial parts of the plant die back after the growing season. In *annuals the whole plant dies; in *biennials and herbaceous *perennials the plant has organs (e.g. bulbs or corms) that are modified to survive beneath the soil in unfavourable conditions.

herbivore An animal that eats vegetation, especially any of the plant-eating mammals, such as ungulates (cows, horses, etc.). Herbivores are characterized by having teeth adapted for grinding plants and alimentary canals specialized for digesting cellulose (*see* caecum).

hermaphrodite (bisexual) 1. An animal, such as the earthworm, that has both male and female reproductive organs. **2.** A plant whose flowers contain both stamens and carpels. This is the usual arrangement in most plants. *Compare* monoecious; dioecious.

herpesvirus One of a group of complex DNA-containing viruses causing infections in man and most other vertebrates that tend to recur. The group includes *herpes simplex*, the agent of cold sores; *herpes varicella/zoster*, the virus causing chickenpox and shingles; *Epstein–Barr (EB) virus*, the causal agent of glandular fever and also implicated in the cancer Burkitt's lymphoma; and the *cytomegalovirus.

hertz Symbol Hz. The *SI unit of frequency equal to one cycle per second. It is named after Heinrich Hertz (1857–94), a German physicist.

Hertzsprung–Russell diagram (H–R diagram) A graphical representation of the absolute magnitude of stars (usually along the y-axis) plotted against the spectral class or colour index (x-axis): see illustration. The y-axis then represents the energy output of the star and the x-axis its surface temperature. The majority of stars on such a diagram fall on a band running from the top left to the bottom right of the graph. These are called *main-sequence stars* (the sun falls into this class). The few stars falling in the lower left portion are called *white dwarfs. The *giants fall in a cluster above the main sequence and the *supergiants are above them. The diagram, which was first devised in 1911 by Ejnar Hertzsprung (1873–1969) and in 1913 by H. N. Russell (1897–

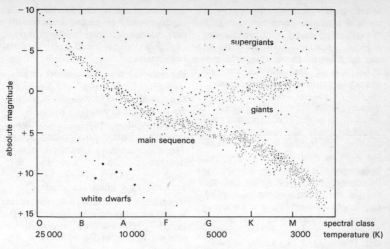

Hertzsprung–Russell diagram

1957), forms the basis of the theory of *stellar evolution.

hesperidium *See* berry.

Hess's law If reactants can be converted into products by a series of reactions, the sum of the heats of these reactions (with due regard to their sign) is equal to the heat of reaction for direct conversion from reactants to products. More generally, the overall energy change in going from reactants to products does not depend on the route taken. The law can be used to obtain thermodynamic data that cannot be measured directly. For example, the heat of formation of ethane can be found by considering the reactions:

$$2C(s) + 3H_2(g) + 3\frac{1}{2}O_2(g) \rightarrow$$
$$2CO_2(g) + 3H_2O(l)$$

The heat of this reaction is $2\Delta H_C + 3\Delta H_H$, where ΔH_C and ΔH_H are the heats of combustion of carbon and hydrogen respectively, which can be measured. By Hess' law, this is equal to the sum of the energies for two stages:

$$2C(s) + 3H_2(g) \rightarrow C_2H_6(g)$$

(the heat of formation of ethane, ΔH_f) and

$$C_2H_6(g) + 3\frac{1}{2}O_2 \rightarrow 2CO_2(g) + 3H_2O(l)$$

(the heat of combustion of ethane, ΔH_E). As ΔH_E can be measured and as

$$\Delta H_f + \Delta H_E = 2\Delta H_c + 3\Delta H_H$$

ΔH_f can be found. Another example is the use of the *Born–Haber cycle to obtain lattice energies. The law was first put forward in 1840 by the Russian chemist Germain Henri Hess (1802–50). It is sometimes called the *law of constant heat summation* and is a consequence of the law of conservation of energy.

hetero atom An odd atom in the ring of a heterocyclic compound. For instance, nitrogen is the hetero atom in pyridine.

heterochromatin *See* chromatin.

heterocyclic *See* cyclic.

heterodont Describing animals that possess teeth of more than one type (i.e. *incisors, *canine teeth, *premolars, and *molars), each with a particular function. Most mammals are heterodont. *Compare* homodont.

heterodyne Denoting a device or method of radio reception in which *beats are produced by superimposing a locally generated radio wave on an incoming wave. In the *superheterodyne receiver the intermediate frequency is amplified and demodulated. In the *heterodyne wavemeter*, a variable-frequency local oscillator is adjusted to give a predetermined beat frequency with the incoming wave, enabling the frequency of the incoming wave to be determined.

heterogametic sex The sex that is determined by possession of two dissimilar *sex chromosomes (i.e. XY). In humans and many other mammals this is the male sex. The heterogametic sex produces reproductive cells (gametes) of two kinds, half containing an X chromosome and half a Y chromosome. *Compare* homogametic sex.

heterogeneous Relating to two or more phases, e.g. a heterogeneous *catalyst. *Compare* homogeneous.

heterolytic fission The breaking of a bond in a compound in which the two fragments are oppositely charged ions. For example, HCl $\rightarrow$ H$^+$ + Cl$^-$. *Compare* homolytic fission.

heteropolar bond *See* chemical bond.

heteropoly compound *See* cluster compound.

heteropolymer *See* polymer.

heterosis *See* hybrid vigour.

heterotrophism A type of nutrition in which energy is derived from the intake and digestion of organic substances, normally plant or animal tissues. The breakdown products of digestion are used to synthesize the organic materials required by the organism. All animals obtain their food this way: they are *heterotrophs*. *Compare* autotrophism.

heterozygous Describing an organism that has two different *alleles controlling a particular feature. The aspect of the feature displayed by the organism will be that determined by the *dominant allele. Heterozygous organisms do not breed true. *Compare* homozygous.

heuristic Denoting a method of solving a problem for which no *algorithm exists. It involves trial and error, as in *iteration.

Heusler alloys Ferromagnetic alloys containing no ferromagnetic elements. The original alloys contained copper, manganese, and tin and were first made by Conrad Heusler (19th-century mining engineer).

hexadecane (cetane) A colourless liquid straight-chain alkane hydrocarbon, $C_{16}H_{34}$, used in standardizing *cetane ratings of Diesel fuel.

hexadecanoate *See* palmitate.

hexadecanoic acid *See* palmitic acid.

hexagonal close packing *See* close packing.

hexagonal crystal *See* crystal system.

hexanedioate (adipate) A salt or ester of hexanedioic acid.

hexanedioic acid (adipic acid) A carboxylic acid, $(CH_2)_4(COOH)_2$; r.d. 1.366; m.p. 149°C; b.p. 265°C (100 mmHg). It is used in the manufacture of *nylon 6,6. *See also* polymerization.

hexanoate (caproate) A salt or ester of hexanoic acid.

hexanoic acid (caproic acid) A liquid fatty acid, $CH_3(CH_2)_4COOH$; r.d. 0.93; m.p. -3.4°C; b.p. 205°C. Glycerides of the acid occur naturally in cow and goat milk and in some vegetable oils.

hexose A *monosaccharide that has six carbon atoms in its molecules.

hexyl group *or* **radical** The organic group

$$CH_3CH_2CH_2CH_2CH_2CH_2-,$$

derived from hexane.

hibernation A sleeplike state in which some animals pass the winter months as a way of surviving food scarcity and cold weather. Various physiological changes occur, such as lowering of the body temperature and slowing of the pulse rate and other vital processes, and the animal lives on its reserve of body fat. Animals that hibernate include bats, hedgehogs, and many fish, amphibians, and reptiles. *See also* dormancy. *Compare* aestivation.

hidden matter *See* missing mass.

Higgs boson A spin-zero particle with a nonzero mass, predicted by Peter Higgs (1929–) to exist in certain *gauge theories, in particular in *electroweak theory (the Weinberg–Salam model). The Higgs boson has not yet been found but it is thought likely that it will be found by larger *accelerators in the next few years, especially since other associated features of the theory, including W and Z bosons, have been found.

high frequency (HF) A radio frequency in the range 3–30 megahertz; i.e. having a wavelength in the range 10–100 metres.

high-performance liquid chromatography (HPLC) A sensitive technique for separating or analysing mixtures, in which the sample is forced through the chromatography column under pressure.

high-speed steel A steel that will remain hard at dull red heat and can therefore be used in cutting tools for high-speed lathes. It usually contains 12–22% tungsten, up to 5% chromium, and 0.4–0.7% carbon. It may also contain small amounts of vanadium, molybdenum, and other metals.

high-temperature superconductivity *See* superconductivity.

high tension (HT) A high potential difference, usually one of several hundred volts or more. However, batteries supplying the anode circuits of radio devices using valves are usually called *high-tension batteries*, even when they supply a lower potential difference.

hilum A scar on the seed coat of a plant marking the point at which the seed was attached to the fruit wall by the *funicle. It is a feature that distinguishes seeds from fruits.

hindbrain (rhombencephalon) One of the three sections of the brain of a vertebrate embryo. It develops to form the *cerebellum, *pons, and *medulla oblongata, which control and coordinate fundamental physiological processes (including respiration and circulation of blood). *Compare* forebrain; midbrain.

hindgut 1. The posterior part of the alimentary canal of vertebrates, comprising the posterior section of the colon 2. The posterior section of the alimentary canal of arthropods. *See also* foregut; midgut.

hip girdle *See* pelvic girdle.

hippocampus A part of the vertebrate brain consisting of two ridges, one over each of the two lateral *ventricles. It is highly developed in advanced mammals (primates and whales) and its function appears to be related to the expression of responses that generate emotion (such as fear and anger).

Hirudinea A class of freshwater and terrestrial annelid worms that comprises the leeches. They have suckers at both anterior and posterior ends but no bristles. Some are blood-sucking parasites of vertebrates and invertebrates but the majority are predators.

histamine A substance that is released during allergic reactions, e.g. hay fever. Formed from the amino acid histidine, histamine can occur in various tissues but is concentrated in connective tissue. It causes dilation and increased permeability of small blood vessels, which results in such symptoms as localized swelling, itching, sneezing, and runny eyes and nose. The effects of histamine can be countered by the administration of *antihistamine drugs.

histidine *See* amino acid.

histochemistry The study of the distribution of the chemical constituents of tissues by means of their chemical reactions. It utilizes such techniques as *staining, light and electron microscopy, *autoradiography, and *chromatography.

histocompatibility The degree to which tissue from one organism will be tolerated by the immune system of another organism. The body recognizes a tissue *graft or invading organism as foreign because of the *histocompatibility antigens* present on its surface. Each individual has a unique set of such antigens, consequently a foreign tissue graft will be rejected by the recipient's immune system. In mice and humans the set of antigens is determined by the *major histocompatibility complex (MHC) – a complex cluster of genes. This encodes two distinct classes of histocompatibility antigens, both of which are glycoproteins. Class I antigens determine graft rejection and help cytotoxic *T cells recognize virus-infected cells. They are found on the surface of most cells in the body. Class II antigens act as receptors for presenting foreign antigen to helper T cells. Their distribution is restricted to certain cell types of the immune system, including macrophages, B cells, and activated T cells.

histocompatibility antigen *See* histocompatibility; HLA system.

histology The microscopic study of the tissues of living organisms. The study of cells, a specialized branch of histology, is known as *cytology.

histone Any of a group of water-soluble proteins found in association with the *DNA of plant and animal chromosomes. They contain a large proportion of the basic (positively charged) amino acids lysine, arginine, and histidine. They are believed to be involved in the condensation and coiling of chromosomes during cell division and have also been implicated in nonspecific suppression of gene activity. Histones do not occur in vertebrate sperm cells (*see* protamine) or in bacteria.

HIV (human immunodeficiency virus) The *retrovirus that causes *AIDS in humans. It has a specific affinity for the helper *T cells of its host. The membrane envelope glycoproteins encasing the virus show great variability in their amino-acid sequences, hence the difficulty of preparing an effective AIDS vaccine.

HLA system (human leucocyte antigen system) A series of four gene loci (*A, B, C,* and *D*) in humans that code for a group of antigens, present on the surface of cell membranes, that are important in determining the acceptance or rejection by the body of a tissue or organ transplant (*see* graft). These antigens are one group of the so-called *histocompatibility antigens* (*see* histocompatibility). Two individuals with identical HLA types are said to be *histocompatible*. Successful transplantation requires a minimum number of HLA differences between the donor's and recipient's tissues. *See also* major histocompatibility complex.

hole A vacant electron position in the lattice structure of a solid that behaves like a mobile positive *charge carrier with a negative *rest energy. *See* semiconductor.

holmium Symbol Ho. A soft silvery metallic element belonging to the *lanthanoids; a.n. 67; r.a.m. 164.93; r.d. 8.795 (20°C); m.p. 1472°C; b.p. 2700°C. It occurs in apatite, xenotime, and some other rare-earth minerals. There is one natural isotope, holmium–165; eighteen artificial isotopes have been produced. There are no uses for the element, which was discovered by P. T. Cleve and J. L. Soret in 1879.

Holocene (Recent) The most recent geological epoch of the *Quaternary period, comprising roughly the past 10 000 years since the end of the *Pleistocene up to the present. It follows the final glacial of the Pleistocene and thus is sometimes known as the *Postglacial* epoch. Some geologists consider the Holocene to be an interglacial phase of the Pleistocene that will be followed by another glacial.

holocrine secretion *See* secretion.

holoenzyme A complex comprising an enzyme molecule and its *cofactor. Only in this state is an enzyme catalytically active. *Compare* apoenzyme.

holography A method of recording and displaying a three-dimensional image of an object, usually using *coherent radiation from a *laser and photographic plates (see illustration). The light from a laser is divided so that some of it (the reference beam) falls directly on a photographic plate. The other part illuminates the object, which reflects it back onto the photographic plate. The two beams form interference patterns on the plate, which when developed is called the

Holography

hologram. To reproduce the image of the object, the hologram is illuminated by coherent light, ideally the original reference beam. The hologram produces two sets of diffracted waves; one set forms a virtual image coinciding with the original object position and the other forms a real image on the other side of the plate. Both are three-dimensional. The method was invented by Dennis Gabor (1900–79) in 1948. More recent techniques can produce holograms visible in white light.

holophytic Describing organisms that feed like plants, i.e. that are photoautotrophic. *See* autotrophism.

holotype *See* type specimen.

holozoic Describing organisms that feed like animals, i.e. that are heterotrophic. *See* heterotrophism.

homeostasis The regulation by an organism of the chemical composition of its blood and body fluids and other aspects of its *internal environment so that physiological processes can proceed at optimum rates. Examples of homeostatic regulation are the maintenance of the *acid–base balance and body temperature (*see* homoiothermy; poikilothermy).

hominid Any member of the primate family Hominidae, which includes man and his fossil ancestors (*fossil hominids*) in the genus *Homo.

Homo The genus of primates that includes modern man (*H. sapiens*, the only living representative) and various extinct species. The oldest *Homo* fossils are those of *H. habilis*, found in the Olduvai Gorge in Tanzania in 1960 and estimated to be about 1.75 million years old. The find was significant as it had been thought that man emerged in Asia rather than Africa. *H. habilis* appears to have been 1–1.5 m tall and had more manlike features and a larger brain than *Australopithecus*. However, the direct ancestor of modern man is thought to be *H. erectus*, which includes Java man and Peking man and appeared about 1.5 million years ago. Fossils of *H. erectus*, which are sometimes called *Pithecanthropus* (ape man), are similar to present-day man except that there was a prominent ridge above the eyes and no forehead or chin. They had crude stone tools and used fire. *H. sapiens* appeared about 100 000 years ago as *Neanderthal man and subsequently (about 35 000 years ago) as *Cromagnon man.

homocyclic *See* cyclic.

homodont Describing animals whose teeth are all of the same type. Most vertebrates except mammals are homodont. *Compare* heterodont.

homogametic sex The sex that is determined by possession of two similar *sex chromosomes (i.e. XX). In humans and many other mammals this is the female sex. All the reproductive cells (gametes) produced by the homogametic sex have the same kind of sex chromosome (i.e. an X chromosome). *Compare* heterogametic sex.

homogamy The condition in a flower in which the male and female reproductive organs mature at the same time, thereby allowing self-fertilization. *Compare* dichogamy.

homogeneous Relating to only one phase, e.g. a homogeneous mixture, a homogeneous *catalyst. *Compare* hetereogeneous.

homoiothermy (warm-bloodedness) The maintenance by an animal of its internal body temperature at a relatively constant value by using metabolic processes to counteract fluctuations in the temperature of the environment. Homoiothermy occurs in birds and mammals. The heat produced by their tissue metabolism and the heat lost to the environment are balanced by various means to keep body temperature constant: 36–38°C in mammals and 38–40°C in birds. The *hypothalamus in the brain monitors blood temperature and controls thermoregulation by both nervous and hormonal means. This produces both short-term responses, such as shivering or sweating, and long-term adjustments to metabolism according to seasonal changes in climate (acclimatization). Warm-blooded animals generally possess insulating feathers or fur. Their relatively high internal temperature permits fast action of muscles and nerves and enables them to lead highly active lives even in cold climates. However, in certain animals, homoiothermy is abandoned during periods of *hibernation. *Compare* poikilothermy.

homologous Describing features of organisms that have the same evolutionary origin but have developed different functions. For example the wings of a bat, the flippers of a dolphin, and the arms of a man are homologous organs, having evolved from the paired pectoral fins of a fish ancestor. *Compare* analogous.

homologous chromosomes Chromosomes having the same structural features. In *diploid nuclei, pairs of homologous chromosomes can be identified at the start of meiosis. One member of each pair comes from the female parent and the other from the male. Homologous chromosomes have the same pattern of genes along the chromosome but the nature of the genes may differ (*see* allele).

homologous series A series of related chemical compounds that have the same functional group(s) but differ in formula by a fixed group of atoms. For instance, the simple carboxylic acids: methanoic (HCOOH), ethanoic (CH_3COOH), propanoic (C_2H_5COOH), etc., form a homologous series in which each member differs from the next by CH_2. Successive members of such a series are called *homologues*.

homolytic fission The breaking of a bond in a compound in which the fragments are uncharged free radicals. For example, $Cl_2 \rightarrow Cl\cdot + Cl\cdot$. *Compare* heterolytic fission.

homoplastic Denoting similarity resulting from *convergent evolution. *Compare* patristic.

homopolar bond *See* chemical bond.

homopolymer *See* polymer.

homozygous Describing an organism having two identical *alleles controlling a particular feature (these may be either dominant or recessive). Homozygous organisms breed true when crossed with genetically identical organisms. *Compare* heterozygous.

Hooke's law The *stress applied to any solid is proportional to the *strain it produces within the elastic limit for that solid. The ratio of longitudinal stress to strain is equal to the Young modulus of elasticity (*see* elastic modulus). The law was first stated by the English scientist Robert Hooke (1635–1703) in the form "Ut tensio, sic vis."

hormone 1. A substance that is manufactured and secreted in very small quantities into the bloodstream by an *endocrine gland or a specialized nerve cell (*see* neurohormone) and regulates the growth or functioning of a specific tissue or organ in a distant part of the body. For example, the hormone *insulin controls the rate and manner in which glucose is used by the body. Other hormones include the *sex hormones, *corticosteroids, *adrenaline, *thyroxine, and *growth hormone. **2.** A plant *growth substance.

hornblende Any of a group of common rock-forming minerals of the amphibole group with the generalized formula:
$(Ca,Na)_2(Mg,Fe,Al)_5(Al,Si)_8O_{22}(OH,F)_2$
Hornblendes consist mainly of calcium, iron, and magnesium silicate.

horsepower (hp) An imperial unit of power originally defined as 550 foot-pound force per second; it is equal to 745.7 watts.

horsetails *See* Sphenopsida.

host An organism whose body provides nourishment and shelter for another. *See* parasitism.

hot-wire instrument An electrical measuring instrument (basically an ammeter) in which the current to be measured is passed through a thin wire and causes its temperature to rise. The temperature rise, which is proportional to the square of the current, is measured by the expansion of the wire. Such instruments can be used for either direct current or alternating current.

HPLC *See* high-performance liquid chromatography.

Hubble constant The rate at which the velocity of recession of the galaxies increases with distance as determined by the *redshift. The value is not agreed upon but current measurements indicate that it lies between 49 and 95 km s^{-1} per megaparsec. The reciprocal of the Hubble constant, the *Hubble time*, is a measure of the age of the universe, assuming that the expansion rate has remained constant. If, as seems likely, the expansion rate has slowed down with time as a result of gravitational attraction between the galaxies, the Hubble time gives an upper limit to the age of the universe of some 2×10^{10} years. The constant is named after the US astronomer Edwin Hubble (1889–1953).

Hückel rule *See* aromatic compound.

hue *See* colour.

human immunodeficiency virus *See* HIV.

humerus The long bone of the upper arm of tetrapod vertebrates. It articulates with the *scapula (shoulder blade) at the *glenoid cavity and with the *ulna and *radius (via a *condyle) at the elbow.

humidity The concentration of water vapour in the atmosphere. The *absolute humidity* is the mass of water vapour per unit volume of air, usually expressed in $kg\,m^{-3}$. A useful measure is the *relative humidity*, the ratio, expressed as a percentage, of the moisture in the air to the moisture it would contain if it were saturated at the same temperature and pressure. The *specific humidity* is also sometimes used: this is the mass of water vapour in the atmosphere per unit mass of air.

humus The dark-coloured amorphous colloidal material that constitutes the organic component of soil. It is formed by the decomposition of plant and animal remains and excrement and has a complex and variable chemical composition. Being a colloid, it can hold water and therefore improves the water-retaining properties of soil; it also enhances soil fertility and workability. Acidic humus (*mor*) is found in regions of coniferous forest, where the decay is brought about mainly by fungi. Alkaline humus (*mull*) is typically found in grassland and deciduous forest: it supports an abundance of microorganisms and small animals (e.g. earthworms).

hurricane 1. A tropical cyclone with surface wind speeds in excess of 64 knots that occurs in the North Atlantic Ocean, Caribbean Sea, and Gulf of Mexico. (Tropical cyclones that occur in the North Pacific Ocean are known as *typhoons*.) Hurricanes cause widespread damage over land areas and are a considerable hazard to shipping. In order for a hurricane to form, an extensive area of ocean (at least 5° of latitude from the equator) must have a surface temperature in excess of 27°C. At the earth's surface a hurricane appears as a roughly circular vortex with the *eye* at the centre, in which pressure is very low. Winds spiral around this core at great speed in an anticlockwise direction in the northern hemisphere. A hurricane is usually accompanied by deep clouds and torrential rain over a wide area. **2.** *See* Beaufort wind scale.

Huygens' construction (Huygens' principle) Every point on a wavefront may itself be regarded as a source of secondary waves. Thus, if the position of a wavefront at any instant is known, a simple construction enables its position to be drawn at any subsequent time. The construction was first used by Christian Huygens (1629–95).

hybrid The offspring of a mating in which the parents differ in at least one characteristic. The term is usually used of offspring of widely different parents, e.g. different varieties or species. Hybrids between different animal species are usually sterile, as is the mule (a cross between a horse and a donkey). *See also* hybrid vigour.

hybrid orbital *See* orbital.

hybrid vigour (heterosis) The increased vigour displayed by the offspring from a cross between genetically different parents. Hybrids from crosses between different crop varieties (F_1 hybrids) are often stronger and produce better yields than the original varieties. Mules, the offspring of mares crossed with donkeys, have greater strength and resistance to disease and a longer lifespan than either parent.

hydathode A pore found in the *epidermis of the leaves of certain plants. Like *stomata, hydathodes are surrounded by two crescent-shaped cells but these, unlike guard cells, do not regulate the size of the aperture. Hydathodes are used by the plant to secrete water under conditions in which *transpiration is inhibited; for example, when the atmosphere is very humid. This process of water loss is called *guttation*.

Hydraulic press

hydracid *See* binary acid.

hydrate A substance formed by combination of a compound with water. *See* water of crystallization.

hydrated alumina *See* aluminium hydroxide.

hydrated aluminium hydroxide *See* aluminium hydroxide.

hydration *See* solvation.

hydraulic press A device in which a force (F_1) applied to a small piston (A_1) creates a pressure (p), which is transmitted through a fluid to a larger piston (A_2), where it gives rise to a larger force (F_2): see illustration. This depends on Pascal's principle that the pressure applied anywhere in an enclosed fluid is transmitted equally in all directions. The principle of the hydraulic press is widely used in jacks, vehicle brakes, presses, and earth-moving machinery, usually with oil as the working fluid.

hydraulics The study of water or other fluids at rest or in motion, particularly with respect to their engineering uses. The study is based on the principles of *hydrostatics and *hydrodynamics.

hydrazine A colourless liquid or white crystalline solid, N_2H_4; r.d. 1.01 (liquid); m.p. 1.4°C; b.p. 113.5°C. It is very soluble in water and soluble in ethanol. Hydrazine is prepared by the *Raschig synthesis* in which ammonia reacts with sodium(I) chlorate (sodium hypochlorite) to give NH_2Cl, which then undergoes further reaction with ammonia to give N_2H_4. Industrial production must be carefully controlled to avoid a side reaction leading to NH_4Cl. The compound is a weak base giving rise to two series of salts, those based on $N_2H_5^+$, which are stable in water (sometimes written in the form $N_2H_4.HCl$ rather than $N_2H_5^+Cl^-$), and a less stable and extensively hydrolysed series based on $N_2H_6^{2+}$. Hydrazine is a powerful reducing agent and reacts violently with many oxidizing agents, hence its use as a rocket propellant.

hydrazoic acid *See* hydrogen azide.

hydrazones Organic compounds containing the group $=C:NNH_2$, formed by condensation of substituted hydrazines with with aldehydes and ketones (see illustration). *Phenylhydrazones* contain the group $=C:NNHC_6H_5$.

hydrobromic acid *See* hydrogen bromide.

hydrocarbons Chemical compounds that contain only carbon and hydrogen. A vast number of different hydrocarbon compounds exist, the main types being the *alkanes, *alkenes, *alkynes, and *arenes.

hydrochloric acid *See* hydrogen chloride.

hydrochloride *See* amine salts.

Formation of a hydrazone from a ketone. The same reaction occurs with an aldehyde (R′=H). If R″=C_6H_5, the product is phenylhydrazone

hydrocortisone

hydrocortisone *See* cortisol.

hydrocyanic acid *See* hydrogen cyanide.

hydrodynamics The study of the motion of incompressible fluids and the interaction of such fluids with their boundaries.

hydroelectric power Electric power generated by a flow of water. A natural waterfall provides a source of energy, in the form of falling water, which can be used to drive a water *turbine. This turbine can be coupled to a generator to provide electrical energy. Hydroelectric generators can be arranged to work in reverse so that during periods of low power demand current can be fed to the generator, which acts as a motor. This motor drives the turbine, which then acts as a pump. The pump then raises water to an elevated reservoir so that it can be used to provide extra power at peak-load periods.

hydrofluoric acid *See* hydrogen fluoride.

hydrogen Symbol H. A colourless odourless gaseous chemical element; a.n. 1; r.a.m. 1.008; d. 0.0899 g dm^{-3}; m.p. $-259.14°C$; b.p. $-252.87°C$. It is the lightest element and the most abundant in the universe. It is present in water and in all organic compounds. There are three isotopes: naturally occurring hydrogen consists of the two stable isotopes hydrogen-1 (99.985%) and *deuterium. The radioactive *tritium is made artificially. The gas is diatomic and has two forms: *orthohydrogen*, in which the nuclear spins are parallel, and *parahydrogen*, in which they are antiparallel. At normal temperatures the gas is 25% parahydrogen. In the liquid it is 99.8% parahydrogen. The main source of hydrogen is steam *reforming of natural gas. It can also be made by the Bosch process (*see* Haber process) and by electrolysis of water. The main use is in the Haber process for making ammonia. Hydro-

gen is also used in various other industrial processes, such as the reduction of oxide ores, the refining of petroleum, the production of hydrocarbons from coal, and the hydrogenation of vegetable oils. Considerable interest has also been shown in its potential use in a 'hydrogen fuel economy' in which primary energy sources not based on fossil fuels (e.g. nuclear, solar, or geothermal energy) are used to produce electricity, which is employed in electrolysing water. The hydrogen formed is stored as liquid hydrogen or as metal *hydrides. Chemically, hydrogen reacts with most elements. It was discovered by Henry Cavendish in 1776.

hydrogenation 1. A chemical reaction with hydrogen; in particular, an addition reaction in which hydrogen adds to an unsaturated compound. Nickel is a good catalyst for such reactions. **2.** The process of converting coal to oil by making the carbon in the coal combine with hydrogen to form hydrocarbons. *See* Fischer–Tropsch process; Bergius process.

hydrogen azide (hydrazoic acid; azoimide) A colourless liquid, HN$_3$; r.d. 1.09; m.p. $-80°C$; b.p. 37°C. It is highly toxic and a powerful reductant, which explodes in the presence of oxygen and other oxidizing agents. It may be prepared by the reaction of sodium amide and sodium nitrate at 175°C followed by distillation of a mixture of the resulting sodium azide and a dilute acid. *See also* azides.

hydrogen bomb *See* nuclear weapons.

hydrogen bond A type of electrostatic interaction between electronegative atoms (fluorine, nitrogen, or oxygen) in one molecule and hydrogen atoms bound to electronegative atoms in another molecule. It is a strong dipole–dipole attraction caused by the electron-withdrawing properties of the electronegative atom. Thus, in the water molecule the oxygen atom attracts the electrons in the O–H bonds. The hydrogen atom has no inner shells of electrons to shield the nucleus, and there is an electrostatic interaction between the hydrogen proton and a lone pair of electrons on an oxygen atom in a neighbouring molecule. Each oxygen atom has two lone pairs and can make hydrogen bonds to two different hydrogen atoms. The strengths of hydrogen bonds are about one tenth of the strengths of normal covalent bonds. Hydrogen bonding does, however, have significant effects on physical properties. Thus it accounts for the unusual properties of *water and for the relatively high boiling points of H_2O, HF, and NH_3 (compared with H_2S, HCl, and PH_3). It is also of great importance in living organisms. Hydrogen bonding occurs between bases in the chains of DNA (*see* base pairing). It also occurs between the C=O and N–H groups in proteins, and is responsible for maintaining the secondary structure.

hydrogen bromide A colourless gas, HBr; m.p. −86°C; b.p. −66.4°C. It can be made by direct combination of the elements using a platinum catalyst. It is a strong acid dissociating extensively in solution (*hydrobromic acid*).

hydrogencarbonate (bicarbonate) A salt of *carbonic acid in which one hydrogen atom has been replaced; it thus contains the hydrogencarbonate ion HCO_3^-.

hydrogen chloride A colourless fuming gas, HCl; m.p. −114°C; b.p. −85°C. It can be prepared in the laboratory by heating sodium chloride with concentrated sulphuric acid (hence the former name *spirits of salt*). Industrially it is made directly from the elements at high temperature and used in the manufacture of PVC and other chloro compounds. It

is a strong acid and dissociates fully in solution (*hydrochloric acid*).

hydrogen cyanide (hydrocyanic acid; prussic acid) A colourless liquid or gas, HCN, with a characteristic odour of almonds; r.d. 0.699 (liquid at 22°C); m.p. −14°C; b.p. 26°C. It is an extremely poisonous substance formed by the action of acids on metal cyanides. Industrially, it is made by catalytic oxidation of ammonia and methane with air and is used in producing acrylate plastics. Hydrogen cyanide is a weak acid (K_a = 2.1 × 10⁻⁹ mol dm⁻³). With organic carbonyl compounds it forms *cyanohydrins.

hydrogen electrode *See* hydrogen half cell.

hydrogen fluoride A colourless liquid, HF; r.d. 0.99; m.p. −83°C; b.p. 19.5°C. It can be made by the action of sulphuric acid on calcium fluoride. The compound is an extremely corrosive fluorinating agent, which attacks glass. It is unlike the other hydrogen halides in being a liquid (a result of *hydrogen-bond formation). It is also a weaker acid than the others because the small size of the fluorine atom means that the H–F bond is shorter and stronger. Solutions of hydrogen fluoride in water are known as *hydrofluoric acid*.

hydrogen half cell (hydrogen electrode) A type of *half cell in which a metal foil is immersed in a solution of hydrogen ions and hydrogen gas is bubbled over the foil. The standard hydrogen electrode, used in measuring standard *electrode potentials, uses a platinum foil with a 1.0 M solution of hydrogen ions, the gas at 1 atmosphere pressure, and a temperature of 25°C. It is written Pt(s)|H_2(g), H⁺(aq), the effective reaction being
$$H_2 \rightarrow 2H^+ + 2e.$$

hydrogen iodide A colourless gas, HI; m.p. −51°C; b.p. −36°C. It can be made by direct combination of the elements using a platinum catalyst. It is a strong acid dissociating extensively in solution (*hydroiodic acid*). It is also a reducing agent.

hydrogen ion *See* acids; pH.

hydrogen peroxide A colourless or pale blue viscous unstable liquid, H_2O_2; r.d. 1.44; m.p. $-0.89°C$; b.p. 151.4°C. As with water, there is considerable hydrogen bonding in the liquid, which has a high dielectric constant. It can be made in the laboratory by adding dilute acid to barium peroxide at 0°C. Large quantities are made commercially by electrolysis of $KHSO_4.H_2SO_4$ solutions. Another industrial process involves catalytic oxidation (using nickel, palladium, or platinum with an anthraquinone) of hydrogen and water in the presence of oxygen. Hydrogen peroxide readily decomposes in light or in the presence of metal ions to give water and oxygen. It is usually supplied in solutions designated by volume strength. For example, 20-volume hydrogen peroxide would yield 20 volumes of oxygen per volume of solution. Although the *peroxides are formally salts of H_2O_2, the compound is essentially neutral. Thus, the acidity constant of the ionization

$$H_2O_2 + H_2O \leftrightharpoons H_3O^+ + HO_2^-$$

is 1.5×10^{-12} mol dm^{-3}. It is a strong oxidizing agent, hence its use as a mild antiseptic and as a bleaching agent for cloth, hair, etc. It has also been used as an oxidant in rocket fuels.

hydrogen spectrum The atomic spectrum of hydrogen is characterized by lines corresponding to radiation quanta of sharply defined energy. A graph of the frequencies at which these lines occur against the ordinal number that characterizes their position in the series of lines, produces a smooth curve indicating that they obey a formal law. In 1885 J. J. Balmer (1825–98) discovered the law having the form:

$$1/\lambda = R(1/n_1^2 + 1/n_2^2)$$

This law gives the so-called *Balmer series* of lines in the visible spectrum in which $n_1 = 2$ and $n_2 = 3,4,5 \ldots$, λ is the wavelength associated with the lines, and R is the *Rydberg constant.

In the *Lyman series*, discovered by Theodore Lyman (1874–1954), $n_1 = 1$ and the lines fall in the ultraviolet. The Lyman series is the strongest feature of the solar spectrum as observed by rockets and satellites above the earth's atmosphere. In the *Paschen series*, discovered by F. Paschen (1865–1947), $n_1 = 3$ and the lines occur in the far infrared. The *Brackett series* also occurs in the far infrared, with $n_1 = 4$.

hydrogensulphate (bisulphate) A salt containing the ion HSO_4^- or an ester of the type $RHSO_4$, where R is an organic group. It was formerly called *hydrosulphate*.

hydrogen sulphide (sulphuretted hydrogen) A gas, H_2S, with an odour of rotten eggs; r.d. 1.54 (liquid); m.p. $-85.5°C$; b.p. $-60.7°C$. It is soluble in water and ethanol and may be prepared by the action of mineral acids on metal sulphides, typically hydrochloric acid on iron(II) sulphide (*see* Kipp's apparatus). Solutions in water (known as *hydrosulphuric acid*) contain the anions HS^- and minute traces of S^{2-} and are weakly acidic. Acid salts (those containing the HS^- ion) are known as *hydrogensulphides* (formerly *hydrosulphides*). In acid solution hydrogen sulphide is a mild reducing agent. Hydrogen sulphide has an important role in traditional qualitative chemical analysis, where it precipitates metals with insoluble sulphides (in acid solution: Cu, Pb, Hg, Cd, Bi, As, Sb, Sn). The formation of a black precipitate with alkaline solutions of lead salts may be used as a test for hydrogen sulphide but the characteristic smell is usually sufficient. Hydrogen sulphide is

The hydrological cycle

exceedingly poisonous (more toxic than hydrogen cyanide).

The compound burns in air with a blue flame to form sulphur(IV) oxide (SO_2); solutions of hydrogen sulphide exposed to the air undergo oxidation but in this case only to elemental sulphur. North Sea gas contains some hydrogen sulphide (from S-proteins in plants) as do volcanic emissions.

hydrogensulphite (bisulphite) A salt containing the ion $^-HSO_3$ or an ester of the type $RHSO_3$, where R is an organic group.

hydrography The study of the waters of the earth's surface, including the oceans, seas, lakes, and rivers. It involves the measurement of these features and the presentation of the information on hydrographic charts.

hydroiodic acid *See* hydrogen iodide.

hydrological cycle (water cycle) The circulation of water between the atmosphere, land, and oceans on the earth. Water evaporates from the oceans and other water bodies on earth to form water vapour in the atmosphere. This may condense to form clouds and be returned to the earth's surface as precipitation (e.g. rainfall, hail, and snow). Some of this precipitation is returned to the atmosphere directly through evaporation or transpiration by plants; some flows off the land surface as overland flow, eventually to be returned to the oceans via rivers; and some infiltrates the ground to flow underground forming groundwater storage.

hydrology The scientific study of terrestrial water, in particular inland water before its discharge into the oceans or evaporation into the atmosphere. It includes the study of the occurrence and movement of water and ice on or under the earth's surface. The science has many important applications, for example in flood control, irrigation, domestic and industrial uses, and hydroelectric power.

hydrolysis A chemical reaction of a compound with water. For instance, salts of weak acids or bases hydrolyse in aqueous solution, as in

$$Na^{+ -}OOCCH_3 + H_2O \rightleftharpoons Na^+ + OH^- + CH_3COOH$$

The reverse reaction of *esterification is another example. *See also* solvolysis.

hydromagnesite A mineral form of basic *magnesium carbonate, $3MgCO_3 \cdot Mg(OH)_2 \cdot 3H_2O$.

hydrometer An instrument for measuring the density or relative density of liquids. It usually consists of a glass tube with a long bulb at one end. The bulb is weighted so that the device floats vertically in the liquid, the relative density being read off its calibrated stem by the depth of immersion.

hydrophilic Having an affinity for water. *See* lyophilic.

hydrophily A rare form of pollination in which pollen is carried to a flower by water. It occurs by one of two methods. In Canadian pondweed (*Elodea canadensis*) the male flowers break off and float downstream until they contact the female flowers. In *Zostera*, a marine species, the filamentous pollen grains are themselves carried in the water. *Compare* anemophily; entomophily.

hydrophobic Lacking affinity for water. *See* lyophobic.

hydrophyte Any plant that lives either in very wet soil or completely or partially submerged in water. Structural modifications of hydrophytes include the reduction of mechanical and supporting tissues and vascular tissue, the absence or reduction of a root system, and specialized leaves that may be either floating or finely divided, with little or no cuticle. Examples of hydrophytes are waterlilies and certain pondweeds. *Compare* mesophyte; xerophyte.

hydroponics A commercial technique for growing certain crop plants in culture solutions rather than in soil. The roots are immersed in an aerated solution containing the correct proportions of essential mineral salts. The technique is based on various water culture methods used in the laboratory to assess the effects of the absence of certain mineral elements on plant growth.

hydroquinone *See* benzene-1,4-diol.

hydrosol A sol in which the continuous phase is water. *See* colloid.

hydrosphere The water on the surface of the earth. Some 74% of the earth's surface is covered with water, 97% (or some 10^{21} kilograms) of which is in the oceans. Icecaps and glaciers contain about 3×10^{19} kg, rivers about 10^{15} kg, lakes and inland seas about 2×10^{17} kg, and groundwater (down to 4000 metres) about 8×10^{19} kg. Water in the atmosphere contains only about 10^{16} kg.

hydrostatics The study of liquids at rest, with special reference to storage tanks, dams, bulkheads, and hydraulic machinery.

hydrosulphate *See* hydrogensulphate.

hydrosulphide *See* hydrogen sulphide.

hydrosulphuric acid *See* hydrogen sulphide.

hydrotropism The growth of a plant part in response to water. Roots, for example, grow towards water in the soil. *See* tropism.

hydroxide A metallic compound containing the ion OH^- (*hydroxide ion*) or containing the group $-OH$ (hydroxyl group) bound to a metal atom. Hydroxides of typical metals are basic; those of *metalloids are amphoteric.

hydroxycerussite *See* lead(II) carbonate hydroxide.

hydroxyl group The group $-OH$ in a chemical compound.

2-hydroxypropanoic acid *See* lactic acid.

5-hydroxytryptamine *See* serotonin.

hygrometer An instrument for measuring *humidity in the atmosphere. The mechanical type uses an organic material, such as human hair, which expands and contracts with changes in atmospheric humidity. The expansion and contraction is used to operate a needle. In the electric type, the change in resistance of a hygroscopic substance is used as an indication of humidity. In *dew-point hygrometers* a polished surface is reduced in temperature until water vapour from the atmosphere forms on it. The temperature of this dew point enables the relative humidity of the atmosphere to be calculated. In the *wet-and-dry bulb hygrometer*, two thermometers are mounted side by side, the bulb of one being surrounded by moistened muslin. The thermometer with the wet bulb will register a lower temperature than that with a dry bulb owing to the cooling effect of the evaporating water. The temperature difference enables the relative humidity to be calculated. Only the dew-point hygrometer can be operated as an absolute instrument; all the others must ultimately be calibrated against this.

hygroscopic Describing a substance that can take up water from the atmosphere. *See also* deliquescence.

Hymenoptera An order of insects that includes the ants, bees, wasps, ichneumon flies, and sawflies. Hymenopterans generally have a narrow waist between thorax and abdomen. The smaller hindwings are interlocked with the larger forewings by a row of tiny hooks on the leading edges of the hindwings. Some species are wingless. The mouthparts are typically adapted for biting, although some advanced forms (e.g. bees) possess a tubelike proboscis for sucking liquid food, such as nectar. The long slender *ovipositor can serve for sawing, piercing, or stinging. Metamorphosis occurs via a pupal stage to the adult form. *Parthenogenesis is common in the group.

Ants and some bees and wasps live in colonies, often comprising numerous individuals divided into *castes and organized into a coordinated and complex society. The colony of the honeybee (*Apis mellifera*), for example, consists of workers (sterile females), drones (fertile males), and usually a single fertile female – the queen. The sole concern of the queen is egg laying. She determines the gender of the egg by either withholding or releasing stored sperm. Unfertilized eggs become males; fertilized eggs become females. The workers fulfil a variety of tasks, including nursing the developing larvae, building the wax cells (combs) of the hive, guarding the colony, and foraging for nectar and pollen. The single function of the larger drones is to mate with the young queen on her nuptial flight.

hyoid arch The second of seven bony V-shaped arches that support the gills of primitive vertebrates. In advanced vertebrates, part of the hyoid arch has evolved to form the stapes (one of the *ear ossicles). The rest of it forms the *hyoid bone*, which supports the tongue.

hyper- A prefix denoting over, above, high; e.g. hypersonic, hyperpolarization, hypertonic.

hyperbola A *conic with eccentricity $e > 1$. It has two branches (see graph). For a hyperbola centred at the origin, the *transverse axis* runs along the x-axis between the vertices and has length $2a$. The *conjugate axis* runs along the y-axis and has length $2b$. There are two *foci* on the x-axis at $(ae, 0)$ and $(-ae, 0)$. The *latus rectum*, the chords through the foci perpendicular to the transverse axis, have length $2b^2/a$. The equation of the hyperbola is:
$$x^2/a^2 - y^2/b^2 = 1,$$
and the asymptotes are $y = \pm\, bx/a$.

A hyperbola

Hypermetropia

hyperbolic functions A set of functions; *sinh*, *cosh*, and *tanh*, that have similar properties to *trigonometric functions but are related to the hyperbola in the manner that trigonometric functions are related to the circle. The hyperbolic sine (sinh) of the angle x is defined by:

$$\sinh x = \tfrac{1}{2}(e^x - e^{-x}).$$

Similarly,

$$\cosh x = \tfrac{1}{2}(e^x + e^{-x})$$
$$\tanh x = (e^x - e^{-x})/(e^x + e^{-x})$$

Hyperbolic secant (sech), cosecant (cosech), and cotangent (coth) are the reciprocals of cosh, sinh, and tanh, respectively.

hypercharge A quantized property of *baryons (*see* elementary particles) that provides a formal method of accounting for the nonoccurrence of certain expected decays by means of the strong interaction (*see* fundamental interactions). Hypercharge is in some respects analogous to electric charge but it is not conserved in weak interactions. Nucleons have a hypercharge of $+1$, and the *pion has a value of 0. Quarks would be expected to have fractional hypercharges.

hyperfine structure *See* fine structure.

hypermetropia (hyperopia) Long-sightedness. A vision defect in which the lens of the eye is unable to accommodate sufficiently to throw the image of near objects onto the retina. It is caused usually by shortness of the eyeball rather than any fault in the lens system. The subject requires spectacles with converging lenses to bring the image from behind the retina back on to its surface (see illustration).

hyperon A shortlived *elementary particle; it is classified as a *baryon and has a nonzero *strangeness.

hyperplasia Increase in the size of a tissue or organ due to an increase in the number of its component cells. *Compare* hypertrophy.

hypersonic Denoting a velocity in excess of Mach 5 (*see* Mach number). *Hypersonic flight* is flight at hypersonic speeds in the earth's atmosphere.

hypertension *See* blood pressure.

hypertonic solution A solution that has a higher osmotic pressure than some other solution. *Compare* hypotonic solution.

hypertrophy An increase in the size of a tissue or organ due to an increase in the size of its component cells. Hypertrophy often occurs in response to an increased workload in an organ, which may result from malfunction or disease. *Compare* hyperplasia.

hypha A delicate filament in fungi many of which may form either a loose network (*mycelium) or a tightly packed interwoven mass of *pseudoparenchyma, as in the fruiting body of mushrooms. Hyphae may be branched or unbranched and may or may not possess cross walls. The cell wall consists of either fungal cellulose or a nitrogenous compound called *chitin. The cell wall is lined with cytoplasm, which often contains oil globules and glycogen, and there is a central vacuole. The hyphae produce enzymes that in parasitic fungi digest the host tissue, and in saprophytic fungi digest dead organic matter.

hypo- A prefix denoting under, below, low; e.g. hypogyny, hyponasty, hypotonic.

hypochlorite *See* chlorates.

hypochlorous acid *See* chloric(I) acid.

hypocotyl The region of the stem beneath the stalks of the seed leaves (*cotyledons) and directly above the young root of an embryo plant. It grows rapidly in seedlings showing *epigeal germination and lifts the cotyledons above the soil surface. In this region (the *transition zone*) the arrangement of vascular bundles in the root changes to that of the stem. *Compare* epicotyl.

hypodermis (exodermis) The outermost layer of cells in the plant *cortex, lying immediately below the epidermis. These cells are sometimes modified to give additional structural support or to store food materials or water. After the loss of the *piliferous layer of the root the hypodermis takes over the protective functions of the epidermis.

hypogeal 1. Describing seed germination in which the seed leaves (cotyledons) remain below ground. Examples of hypogeal germination are seen in oak and runner bean. *Compare* epigeal. **2.** Describing fruiting bodies that develop underground, such as truffles and peanuts.

hypophosphorus acid *See* phosphinic acid.

hypophysis *See* pituitary gland.

hyposulphite *See* sulphinate.

hyposulphurous acid *See* sulphinic acid.

hypotenuse The longest side of a right-angle triangle.

hypothalamus Part of the vertebrate brain that is derived from the *fore-brain and located on the ventral surface below the *thalamus and the *cerebrum. The hypothalamus regulates a wide variety of physiological processes, including maintenance of body temperature, water balance, sleeping, and feeding, via both the *autonomic nervous system (which it controls) and the *neuroendocrine system. Its endocrine functions are largely mediated by the *pituitary gland. The pituitary responds to releasing hormones produced by the hypothalamus, which in this way indirectly controls hormone production in other glands.

hypotonic solution A solution that has a lower osmotic pressure than some other solution. *Compare* hypertonic solution.

hypsometer A device for calibrating thermometers at the boiling point of water. As the boiling point depends on the atmospheric pressure, which in turn depends on the height above sea level, the apparatus can be used to measure height above sea level.

hysteresis A phenomenon in which two physical quantities are related in a manner that depends on whether one is increasing or decreasing in relation to the other.

The repeated measurement of *stress against *strain, with the stress first increasing and then decreasing, will produce for some specimens a graph

Hysteresis

that has the shape of a closed loop. This is known as a *hysteresis cycle*. The most familiar hysteresis cycle, however, is produced by plotting the magnetic flux density (B) within a ferromagnetic material against the applied magnetic field strength (H).

If the material is initially unmagnetized at O it will reach saturation at P as H is increased. As the field is reduced and again increased the loop PQRSTP is formed (see graph). The area of this loop is proportional to the energy loss (*hysteresis loss*) occurring during the cycle. The value of B equal to OQ is called the *remanance* (or retentivity) and is the magnetic flux density remaining in the material after the saturating field has been reduced to zero. This is a measure of the tendency of the magnetic domain patterns (*see* magnetism) to remain distorted even after the distorting field has been removed. The value of H equal to OR is called the *coercive force* (or coercivity) and is the field strength required to reduce the remaining flux density to zero. It is a measure of the difficulty of restoring the symmetry of the domain patterns.

I

IAA (indoleacetic acid) *See* auxin.

ice *See* water.

ice age A period in the earth's history during which ice advanced towards the equator and a general lowering of temperatures occurred. The last major ice age, that of the Pleistocene period (sometimes known as the *Ice Age*), ended about 10 000 years ago. At least four major ice advances (glacials) occurred during the Pleistocene period; these were separated by interglacials during which the ice retreated and temperatures rose. At present it is not known if the earth is between ice ages or is in an interglacial. It has been established that ice ages also occurred during the Precambrian (over 500 million years ago) and during the Permo-Carboniferous (about 250 million years ago).

ice point The temperature at which there is equilibrium between ice and water at standard atmospheric pressure (i.e. the freezing or melting point under standard conditions). It was used as a fixed point (0°) on the Celsius scale, but the kelvin and the International Practical Temperature Scale are based on the *triple point of water.

iconoscope A form of television camera tube (*see* camera) in which the beam of light from the scene is focused on to a thin mica plate. One side of the plate is faced with a thin metallic electrode, the other side being coated with a mosaic of small globules of a photoemissive material. The light beam falling on the mosaic causes photoemission of electrons, creating a pattern of positive charges in what is effectively an array of tiny capacitors. A high-velocity electron beam scans the mosaic, discharging each capacitor in turn through the metallic electrode. The resulting current is fed to amplification circuits, the current from a particular section of the mosaic depending on the illumination it has received. In this way the optical information in the light beam is converted into an electrical signal.

ICSH *See* luteinizing hormone.

ideal crystal A single crystal with a perfectly regular lattice that contains no impurities, imperfections, or other defects.

ideal gas (perfect gas) A hypothetical gas that obeys the *gas laws exactly. An ideal gas would consist of molecules that occupy negligible space and have negligible forces between them. All collisions made between molecules and the walls of the container or between molecules and other gas molecules would be perfectly elastic, because the molecules would have no means of storing energy except as translational kinetic energy.

ideal solution *See* Raoult's law.

identical twins (monozygotic twins) Two individuals that develop from a single fertilized egg cell by its division into two genetically identical parts. Each part eventually gives rise to a separate individual and these twins are therefore identical in every respect. *Compare* fraternal twins.

identity Symbol $\equiv$. A statement of equality that applies for all values of the unknown quantity. For example, $5y \equiv 2y + 3y$.

idiogram *See* karyogram.

IE Ionization energy. *See* ionization potential.

igneous rocks A group of rocks formed from the crystallization of magma (molten silicate liquid). Igneous rocks form one of the three major rock categories (*see also* metamorphic rocks; sedimentary rocks). According to the depth at which the magma solidifies, igneous rocks may be classified as plutonic, hypabyssal, or volcanic. Plutonic rocks solidify slowly at great depth, typically forming large intrusive masses (e.g. batholiths and stocks), and generally have the coarsest texture. Examples of plutonic rocks are granite, syenite, diorite, and gabbro. Volcanic (extrusive) rocks are formed from magma that has been poured out onto the earth's surface; these rocks (e.g. basalt, andesite) are characteristically fine-grained or glassy as a result of their rapid cooling. Hypabyssal rocks (e.g. diorite), which cool at shallower depths than the plutonic, are intermediate in character and medium-grained. They commonly occur in the form of small intrusions, such as

dykes and sills. Igneous rocks may also be classified chemically according to their silica content as acidic (over 66% silica), intermediate (55–66%), basic (45–55%), or ultrabasic (under 45%).

ignition temperature 1. The temperature to which a substance must be heated before it will burn in air. **2.** The temperature to which a *plasma has to be raised in order that nuclear fusion will occur.

ileum The portion of the mammalian *small intestine that follows the *jejunum and precedes the *large intestine. It is a site of digestion and absorption. The internal lining of the ileum bears numerous small outgrowths (*see* villus), which increase its absorptive surface area.

ilium The largest of the three bones that make up each half of the *pelvic girdle. The ilium bears a flattened wing of bone that is attached by ligaments to the sacrum (*see* sacral vertebrae). *See also* ischium; pubis.

illuminance (illumination) Symbol *E*. The energy in the form of visible radiation reaching a surface per unit area in unit time; i.e. the luminous flux per unit time. It is measured in *lux (lumens per square metre).

image A representation of a physical object formed by a lens, mirror, or other optical instrument. If the rays of light actually pass through the image, it is called a *real image*. If a screen is placed in the plane of a real image it will generally become visible. If the image is seen at a point from which the rays appear to come to the observer, but do not actually do so, the image is called a *virtual image*. No image will be formed on a screen placed at this point. Images may be *upright* or *inverted* and they may be *magnified* or *diminished*.

image converter An electronic device in which an image formed by invisible radiation (usually gamma rays, X-rays, ultraviolet, or infrared) is converted into a visible image. Commonly the invisible radiation is focused on to a photocathode, which emits electrons when it is illuminated. These electrons fall on a fluorescent anode screen, after acceleration and focusing by a system of electron lenses. The fluorescent screen produces a visible image. The device is used in fluoroscopes, infrared telescopes, ultraviolet microscopes, and other devices.

imaginary number A number that is a multiple of $\sqrt{-1}$, which is denoted by i; for example $\sqrt{-3} = i\sqrt{3}$. *See also* complex number.

imago The adult sexually mature stage in the life cycle of an insect after metamorphosis.

imbibition The uptake of water by substances that do not dissolve in water, so that the process results in swelling of the substance. Imbibition is a property of many biological substances, including cellulose (and other constituents of plant cell walls), starch, and some proteins. It occurs in dry seeds before they germinate and – together with osmosis – is responsible for the uptake of water by growing plant cells.

imides Organic compounds containing the group –CO.NH.CO.– (the *imido group*).

imido group *See* imides.

imines Compounds containing the group –NH– (the *imino group*) joined to two other groups; i.e. secondary *amines.

imino group *See* imines.

immersion objective An optical microscope objective in which the front surface of the lens is immersed in a liquid on the cover slip of the microscope specimen slide. Cedarwood oil (for an *oil-immersion lens*) or sugar solution is frequently used. It has the same refractive index as the glass of the cover slip, so that the

object is effectively immersed in it. The presence of the liquid increases the effective aperture of the objective, thus increasing the resolution.

immune response The reaction of the body to foreign or potentially dangerous substances (*antigens), particularly disease-producing microorganisms. The response involves the production by specialized white blood cells (*lymphocytes) of proteins known as *antibodies, which react with the antigens to render them harmless. The antibody–antigen reaction is highly specific. *See also* anaphylaxis; immunity.

immunity The state of relative insusceptibility of an animal to infection by disease-producing organisms or to the harmful effects of their poisons (toxins). Immunity depends on the presence in the blood of *antibodies and white blood cells (*lymphocytes), which produce an *immune response. *Inherited (natural) immunity* is that with which an individual is born. *Acquired immunity* is of two types, active and passive. *Active immunity* arises when the body produces antibodies against an invading foreign substance (*antigen), either through infection or *immunization; this type of immunity may be *humoral*, in which B lymphocytes produce free antibodies that circulate in the bloodstream, or *cell-mediated*, caused by the action of T lymphocytes (*see* T cell). *Passive immunity* is induced by injection of serum taken from an individual already immune to a particular antigen. Active immunity tends to be long-lasting; passive immunity is short-lived. *See also* autoimmunity.

immunization The production of *immunity in an individual by artificial means. *Active immunization (vaccination)* involves the introduction, either orally or by injection (*inoculation*), of specially treated bacteria, viruses, or their toxins to stimulate the production of *antibodies (*see* vaccine). *Passive immunization* is induced by the injection of preformed antibodies.

immunoassay Any of various techniques that measure the amount of a particular substance by virtue of its binding antigenically to a specific antibody. In *solid-phase immunoassay* the specific antibody is attached to a solid supporting medium, such as a PVC sheet. The sample is added and any test antigens will bind to the antibody. A second antibody, specific for a different site on the antigen, is added. This carries a radioactive or fluorescent label, enabling its concentration, and thus that of the test antigen, to be determined by comparison with known standards. Variations on this technique include *ELISA and *Western blotting.

immunoglobulin One of a group of proteins (*globulins) in the body that act as *antibodies. They are produced by specialized white blood cells (*lymphocytes) and are present in blood serum and other body fluids. There are several classes (e.g. IgE, IgG, and IgM) having different functions.

impedance Symbol Z. The quantity that measures the opposition of a circuit to the passage of a current and therefore determines the amplitude of the current. In a d.c. circuit this is the resistance (R) alone. In an a.c. circuit, however, the *reactance (X) also has to be taken into account, according to the equation: $Z^2 = R^2 + X^2$, where Z is the impedance. The *complex impedance* is given by $Z = R + iX$, where $i = \sqrt{-1}$. The real part of the complex impedance, the resistance, represents the loss of power according to *Joule's law. The ratio of the imaginary part, the reactance, to the real part is an indication of the difference in phase between the voltage and the current.

imperfect fungi A group of fungi in which sexual reproduction is absent.

Otherwise they are generally similar to *ascomycetes and are usually regarded as ascomycetes that have lost the ability to produce ascospores. They may be placed in an artificial subdivision, Fungi Imperfecti or Deuteromycotina. Examples of imperfect fungi are the *Penicillium* moulds.

Imperial units The British system of units based on the pound and the yard. The former f.p.s. system was used in engineering and was loosely based on Imperial units; for all scientific purposes *SI units are now used. Imperial units are also being replaced for general purposes by metric units.

implantation (nidation) (in embryology) The embedding of a fertilized mammalian egg into the wall of the uterus (womb) where it will continue its development. After fertilization in the fallopian tube the egg passes into the womb in the form of a ball of cells (*blastocyst*). Its outer cells destroy cells of the uterine wall, forming a cavity into which the blastocyst sinks.

implosion An inward collapse of a vessel, especially as a result of evacuation.

imprinting A specialized form of *learning in which young animals, during a critical period in their early development, learn to recognize and approach some large moving object nearby. In nature this is usually the mother, though simple models or individuals of a different species (including man) may suffice. Imprinting was first described by Konrad Lorenz (1903–89), working with young ducks and geese.

impulse 1. (in physiology) (*or* **nerve impulse**) The signal that travels along the length of a *nerve fibre and is the means by which information is transmitted through the nervous system. It is marked by the flow of ions across the membrane of the *axon caused by changes in the permeability of the membrane, producing a reduction in potential difference that can be detected as the *action potential. The strength of the impulse produced in any nerve fibre is constant (*see* all-or-none response). **2.** (in physics) Symbol J. The product of a force F and the time t for which it acts. If the force is variable, the impulse is the integral of $F\mathrm{d}t$ from t_0 to t_1. The impulse of a force acting for a given time interval is equal to the change in momentum produced over that interval, i.e. $J = m(v_1 - v_0)$, assuming that the mass (m) remains constant while the velocity changes from v_0 to v_1.

inbreeding Mating between closely related individuals, the extreme condition being self-fertilization, which occurs in many plants and some primitive animals. A population of inbreeding individuals generally shows less variation than an *outbreeding population. Continued inbreeding among a normally outbreeding population leads to *inbreeding depression* (the opposite of *hybrid vigour) and an increased incidence of harmful characteristics. For example, in humans, certain mental and other defects tend to occur more often in families with a history of cousin marriages.

incandescence The emission of light by a substance as a result of raising it to a high temperature. An *incandescent lamp* is one in which light is emitted by an electrically heated filament. *See* electric lighting.

incisor A sharp flattened chisel-shaped tooth in mammals that is adapted for biting food and – in rodents – for gnawing. In humans there are normally two pairs of incisors (central and lateral) in each jaw.

inclination 1. *See* geomagnetism. **2.** The angle between the orbital plane of a planet, satellite, or comet and the plane of the earth's *ecliptic.

inclusive fitness The quality that organisms attempt (unconsciously) to maximize as the result of natural selection acting on genes that are influential in controlling their behaviour and physiology. It includes the individual's own *reproductive success* (usually taken as the number of its offspring that survive to adulthood) and also the effects of the individual's actions on the reproductive success of its relatives, because relatives have a higher probability of sharing some identical genes with the individual than do other members of the population. When interactions between relatives are likely to occur (which happens during the lives of many animals and plants) *kin selection will operate.

incompatibility 1. The condition that exists when foreign grafts or blood transfusions evoke a marked *immune response and are rejected. **2.** The phenomenon in which pollen from one flower fails to fertilize other flowers on the same plant (*self-incompatibility*) or on other genetically similar plants. This genetically determined mechanism prevents self-fertilization (breeding between likes) and promotes cross-fertilization (breeding between individuals with different genetic compositions). *See also* allogamy; fertilization; pollination.

incomplete dominance The condition that arises when neither *allele controlling a characteristic is dominant and the aspect displayed by the organism results from the partial influence of both alleles. For example, a snapdragon plant with alleles for red and for white flowers produces pink flowers. *Compare* co-dominance.

incus (anvil) The middle of the three *ear ossicles of the mammalian middle ear.

indefinite inflorescence *See* racemose inflorescence.

indefinite integral *See* integration.

indehiscent Describing a fruit or fruiting body that does not open to release its seeds or spores when ripe. Instead, release occurs when the fruit wall decays or, if eaten by an animal, is digested. *Compare* dehiscence.

indene A colourless flammable hydrocarbon, C_9H_8; r.d. 1.01; m.p. $-35°C$; b.p. $182°C$. Indene is an aromatic hydrocarbon with a five-membered ring fused to a benzene ring. It is present in coal tar and is used as a solvent and raw material for making other organic compounds.

Independent assortment The separation of the alleles of one gene into the reproductive cells (gametes) independently of the way in which the alleles of other genes have segregated. By this process all possible combinations of alleles should occur equally frequently in the gametes. In practice this does not happen because alleles situated on the same chromosome tend to be inherited together. However, if the allele pairs *Aa* and *Bb* are on different chromosomes, the combinations *AB*, *Ab*, *aB*, and *ab* will normally be equally likely to occur in the gametes. *See* meiosis; Mendel's laws.

Indeterminacy *See* uncertainty principle.

index fossil (zone fossil) An animal *fossil of a group that existed continuously during a particular span of geological time and can therefore be used to date the rock in which it is found. Index fossils are found chiefly in sedimentary rocks. They are an essential tool in stratigraphy for comparing the geological ages of sedimentary rock formations. Examples are *ammonites and *graptolites.

indicator A substance used to show the presence of a chemical substance or ion by its colour. *Acid–base indicators* are compounds, such as phenolphthalein and methyl orange, that change colour on going from acidic to basic solutions. They are usually weak acids in which the un-ionized form HA has a different colour from the negative ion A^-. In solution the indicator dissociates slightly

$$HA \leftrightarrows H^+ + A^-$$

In acid solution the concentration of H^+ is high, and the indicator is largely undissociated HA; in alkaline solutions the equilibrium is displaced to the right and A^- is formed. Useful acid–base indicators show a sharp colour change over a range of about 2 pH units. In titration, the point at which the reaction is complete is the *equivalence point* (i.e. the point at which equivalent quantities of acid and base are added). The *end point* is the point at which the indicator just changes colour. For accuracy, the two must be the same. During a titration the pH changes sharply close to the equivalence point, and the indicator used must change colour over the same range.

Other types of indicator can be used for other reactions. Starch, for example, is used in iodine titrations because of the deep blue complex it forms. *Oxidation–reduction indicators* are substances that show a reversible colour change between oxidized and reduced forms. *See also* absorption indicator.

indicator species A plant or animal species that is very sensitive to a particular environmental factor, so that its presence (or absence) in an area can provide information about the levels of that factor. For example, some lichens are very sensitive to the concentration of sulphur dioxide (a major pollutant) in the atmosphere. Examination of the lichens present in an area can provide a good indication of the prevailing levels of sulphur dioxide.

indigenous Describing a species that occurs naturally in a certain area, as distinct from one introduced by man; native.

indigo A blue vat dye, $C_{16}H_{10}N_2O_2$. It occurs as the glucoside *indican* in the leaves of plants of the genus *Indigofera*, from which it was formerly extracted. It is now made synthetically.

indium Symbol In. A soft silvery element belonging to group IIIA of the periodic table; a.n. 49; r.a.m. 114.82; r.d. 7.31 (20°C); m.p. 156.6°C; b.p. 2080±2°C. It occurs in zinc blende and some iron ores and is obtained from zinc flue dust in total quantities of about 40 tonnes per annum. Naturally occurring indium consists of 4.23% indium–113 (stable) and 95.77% indium–115 (half-life 6 × 10^{14} years). There are a further five short-lived radioisotopes. The uses of the metal are small – some special-purpose electroplates and some special fusible alloys. Several semiconductor compounds are used, such as InAs, InP, and InSb. With only three electrons in its valency shell, indium is an electron acceptor; it forms stable indium(I), indium(II), and indium(III) compounds. The element was discovered in 1863 by Reich and Richter.

indoleacetic acid (IAA) *See* auxin.

inductance The property of an electric circuit or component that causes an e.m.f. to be generated in it as a result of a change in the current flowing through the circuit (*self inductance*) or of a change in the current flowing through a neighbouring circuit with which it is magnetically linked (*mutual inductance*). In both cases the changing current is associated with a changing magnetic field, the linkage with which in turn induces the e.m.f. In the case of self inductance, L, the e.m.f., E, generated is given by $E = -L.dI/dt$, where I is the instantaneous current and the minus sign indicates that the e.m.f. induced is in opposition to the change of current. In the case of mutual inductance, M, the e.m.f., E_1, induced in one circuit

is given by $E_1 = -M.dI_2/dt$, where I_2 is the instantaneous current in the other circuit.

induction 1. (in embryology) The ability of natural stimuli to cause unspecialized embryonic tissue to develop into specialized tissue. **2.** (in obstetrics) The initiation of childbirth by artificial means; for example, by injection of the hormone *oxytocin. **3.** (in physics) A change in the state of a body produced by a field. *See* electromagnetic induction; inductance.

induction coil A type of *transformer used to produce a high-voltage alternating current or pulses of high-voltage current from a low-voltage direct-current source. The induction coil is widely used in spark-ignition *internal-combustion engines to produce the spark in the sparking plugs. In such an engine the battery is connected to the primary winding of the coil through a circuit-breaking device driven by the engine and the e.m.f. generated in the secondary winding of the coil is led to the sparking plugs through the distributor. The primary coil consists of relatively few turns, whereas the secondary consists of many turns of fine wire.

induction heating The heating of an electrically conducting material by *eddy currents induced by a varying electromagnetic field. Induction heating may be an undesirable effect leading to power loss in transformers and other electrical devices. It is, however, useful for melting, forging, rolling, and heat-treating metals, as well as for welding, brazing, and soldering. The material to be heated is inserted into a coil through which an alternating current flows and acts as the short-circuited secondary of a *transformer. Eddy currents induced in the material within the coil cause the temperature of the material to rise.

induction motor *See* electric motor.

inductive effect The effect of a group or atom of a compound in pulling electrons towards itself or in pushing them away. Inductive effects can be used to explain some aspects of organic reactions. For instance, electron-withdrawing groups, such as $-NO_2$, $-CN$, $-CHO$, $-COOH$, and the halogens substituted on a benzene ring, reduce the electron density on the ring and decrease its susceptibility to further (electrophilic) substitution. Electron-releasing groups, such as $-OH$, $-NH_2$, $-OCH_3$, and $-CH_3$, have the opposite effect.

indusium The kidney-shaped covering of the *sorus of certain ferns that protects the developing sporangia. It withers when the sorus ripens to expose the sporangia.

industrial melanism The increase of melanic (dark) forms of an animal in areas darkened by industrial pollution. The example most often quoted is that of the peppered moth (*Biston betularia*), melanic forms of which markedly increased in the industrial north of England during the 19th century. Experiments have shown that the dark forms increase in polluted regions because they are less easily seen by birds against a dark background; conversely the paler forms survive better in unpolluted areas.

inelastic collision A collision in which some of the kinetic energy of the colliding bodies is converted into internal energy in one body so that kinetic energy is not conserved. In collisions of macroscopic bodies some kinetic energy is turned into vibrational energy of the atoms, causing a heating effect. Collisions between molecules of a gas or liquid may also be inelastic as they cause changes in vibrational and rotational *energy levels. In nuclear physics, an inelastic collision is one in which the incoming particle causes the nucleus it strikes to become excited or to break up.

inert gases *See* noble gases.

inertia The property of matter that causes it to resist any change in its motion. Thus, a body at rest remains at rest unless it is acted upon by an external force and a body in motion continues to move at constant speed in a straight line unless acted upon by an external force. This is a statement of Newton's first law of motion. The *mass of a body is a measure of its inertia. *See* Mach's principle; inertial frame.

inertial frame A *frame of reference in which bodies move in straight lines with constant speeds unless acted upon by external forces, i.e. a frame of reference in which free bodies are not accelerated. Newton's laws of motion are valid in an inertial system but not in a system that is itself accelerated with respect to such a frame.

inertial mass *See* mass.

inert-pair effect An effect seen especially in groups III and IV of the periodic table, in which the heavier elements in the group tend to form compounds with a valency two lower than the expected group valency. It is used to account for the existence of thallium(I) compounds in group III and lead(II) in group IV. In forming compounds, elements in these groups promote an electron from a filled *s*-level state to an empty *p*-level. The energy required for this is more than compensated for by the extra energy gain in forming two more bonds. For the heavier elements, the bond strengths or lattice energies in the compounds are lower than those of the lighter elements. Consequently the energy compensation is less important and the lower valence states become favoured.

infection The invasion of any living organism by disease-causing microorganisms (*see* pathogen), which proceed to establish themselves, multiply, and produce various symptoms in their host. Pathogens may invade via a wound or (in animals) through the mucous membranes lining the alimentary, respiratory, and reproductive tracts, and may be transmitted by an infected individual, a *carrier, or an insect *vector. Symptoms in animals appear after an initial symptomless *incubation period* and typically consist of localized *inflammation, often with pain and fever. Infections are combatted by the body's natural defences (*see* immune response). Treatment with drugs (*see* antibiotic; antiseptic) is effective against most bacterial, fungal, and protozoan infections but not against viral infections. *See also* immunization.

inferior Describing a structure that is positioned below or lower than another structure in the body. For example, in flowering plants the ovary is described as inferior when it is located below the other organs of the flower (*see* epigyny). *Compare* superior.

infinite series *See* series.

infinitesimal Vanishingly small but not zero. Infinitesimal changes are notionally made in the *calculus, which is sometimes called the *infinitesimal calculus*.

infinity Symbol ∞. A quantity having a value that is greater than any assignable value. Minus infinity, −∞, is a quantity having a value that is less than any assignable value.

inflammation The defence reaction of tissue to injury, infection, or irritation by chemicals or physical agents. Cells in the affected tissue release a chemical (*histamine) that causes localized dilatation of blood vessels so that fluid leaks out and blood flow is increased. This leads to swelling, redness, heat, and often pain. White blood cells, particularly *phagocytes, enter the tissue and an *immune response is stimulated. A gradual healing process usually follows.

inflation *See* early universe.

inflationary universe *See* early universe.

inflection A point on a curve at which the tangent changes from rotation in one direction to rotation in the opposite direction. If the curve y = f(x) has a stationary point dy/dx = 0, there is either a maximum, minimum, or inflection at this point. If d^2y/dx^2 = 0, the stationary point is a point of inflection.

inflorescence A particular arrangement of flowers on a single main stalk of a plant. There are many different types of inflorescence, which are classified into two main groups depending on whether the tip of the flower axis goes on producing new flower buds during growth (*see* racemose inflorescence) or loses this ability (*see* cymose inflorescence).

infrared astronomy The study of radiation from space in the infrared region of the spectrum (*see* infrared radiation). Some infrared radiation is absorbed by water and carbon dioxide molecules in the atmosphere but there are several narrow atmospheric *windows in the near-infrared (1.15 – 1.3 μm, 1.5 – 1.75 μm, 2 – 2.4 μm, 3.4 – 4.2 μm, 4.6 – 4.8 μm, 8 – 13 μm, and 16 – 18 μm). Longer wavelength observations must be made from balloons, rockets, or satellites. Infrared sources are either thermal, i.e. emitted by gas or dust atoms or molecules in the temperature range 100 – 3000 K, or electronic, i.e. emitted by high-energy electrons interacting with magnetic fields as in *synchrotron radiation. Detectors are either modified reflecting *telescopes or solid-state photon detectors, usually incorporating *photovoltaic devices.

infrared radiation (IR) Electromagnetic radiation with wavelengths longer than that of red light but shorter than radiowaves, i.e. radiation in the wavelength range 0.7 micrometre to 1 millimetre. It was discovered in 1800 by William Herschel (1738 – 1822) in the sun's spectrum. The natural vibrational frequencies of atoms and molecules and the rotational frequencies of some gaseous molecules fall in the infrared region of the electromagnetic spectrum. The infrared absorption spectrum of a molecule is highly characteristic of it and the spectrum can therefore be used for molecular identification. Glass is opaque to infrared radiation of wavelength greater than 2 micrometres and other materials, such as germanium, quartz, and polyethene, have to be used to make lenses and prisms. Photographic film can be made sensitive to infrared up to about 1.2 μm.

infrared spectroscopy (IR spectroscopy) A technique for chemical analysis and the determination of structure. It is based on the principles that molecular vibrations occur in the infrared region of the electromagnetic spectrum and functional groups have characteristic absorption frequencies. The frequencies of most interest range from 2.5 to 16 μm; however, in IR spectroscopy it is common to use the reciprocal of the wavelength, and thus this range becomes 4000–625 cm⁻¹. Examples of typical vibrations are centred on 2900 cm⁻¹ for C–H stretching in alkanes, 1600 cm⁻¹ for N–H stretching in amino groups, and 2200 cm⁻¹ for C≡C stretching in alkynes. In an IR spectrometer there is a source of IR light, covering the whole frequency range of the instrument, which is split into two beams of equal intensity. One beam is passed through the sample and the other is used as a reference against which the first is then compared. The spectrum is usually obtained as a chart showing absorption peaks, plotted against wavelength or frequency. The sample can be a gas, liquid, or solid.

infrasound Soundlike waves with frequencies below the audible limit of about 20 hertz.

inhibition 1. (in chemistry) A reduction in the rate of a catalysed reaction by substances called *inhibitors*. In biochemical reactions, in which the catalysts are *enzymes, if the inhibitor molecules resemble the substrate molecules they may bind to the active site of the enzyme, so preventing normal enzymatic activity. Alternatively they may form a complex with the substrate–enzyme intermediate or irreversibly destroy the enzyme configuration and active-site properties. The toxic effects of many substances are produced in this way. Inhibition by reaction products (*feedback inhibition*) is important in the control of enzyme activity. **2.** (in physiology) The prevention or reduction of the activity of effectors (such as muscles) by means of certain nerve impulses. Inhibitory activity often provides a balance to stimulation of a process; for example, the impulse to stimulate contraction of a voluntary muscle may be accompanied by an inhibitory impulse to prevent contraction of its antagonist.

initial One of a group of cells (or, in lower plants, a single cell) that divides to produce the cells of a plant tissue or organ. The cells of the apical meristem, cambium, and cork cambium are initials.

innate behaviour An inherited pattern of behaviour that appears in a similar form in all normally reared individuals of the same sex and species. *See* instinct.

inner Describing a chemical compound formed by reaction of one part of a molecule with another part of the same molecule. Thus, a lactam is an inner amide; a lactone is an inner ester.

inner ear The structure in vertebrates, surrounded by the temporal bone of the skull, that contains the organs of balance and hearing. It consists of soft hollow sensory structures (the *membranous labyrinth*), containing fluid (*endolymph*), surrounded by fluid (*perilymph*), and encased in a bony cavity (the *bony labyrinth*). It consists of two chambers, the *sacculus* and *utriculus*, that bear the *cochlea and *semicircular canals respectively.

inner transition series *See* transition elements.

innervation The supply of nerve fibres to and from an organ.

innominate artery A short artery that branches from the aorta to divide into the *subclavian artery (the main artery to the arm) and the right *carotid artery (which supplies blood to the head).

innominate bone One of the two bones that form each half of the *pelvic girdle in adult vertebrates. This bone is formed by the fusion of the *ilium, *ischium, and *pubis.

inorganic chemistry The branch of chemistry concerned with compounds of elements other than carbon. Certain simple carbon compounds, such as CO, CO_2, CS_2, and carbonates and cyanides, are usually treated in inorganic chemistry.

inositol A cyclic alcohol, $C_6H_{12}O_6$, that is a constituent of certain cell phosphoglycerides. It is sometimes classified as a member of the vitamin B complex but it can be synthesized by many animals and it is not regarded as an essential nutrient in humans. *Phosphatidyl inositol*, a constituent of cell membranes, is a precursor of the intracellular messenger molecules, *inositol 1,4,5-trisphosphate* (*IP₃*) and *diacylglycerol*; these are produced in response to the binding of a substance, e.g. serotonin, to receptors on the cell surface. They mediate such cellular events as smooth muscle contraction, adrenaline secretion, and histamine secretion.

Insecta (Hexapoda) A class of arthropods comprising some 700 000 known species, but many more are thought to exist. They are distributed worldwide in nearly all terrestrial habitats. Ranging in length from 0.5 to over 300 mm, an insect's body consists of a head, a thorax of three segments and usually bearing three pairs of legs and one or two pairs of wings, and an abdomen of eleven segments. The head possesses a pair of sensory *antennae and a pair of large *compound eyes, between which are three simple eyes (*ocelli). The *mouthparts are variously adapted for either chewing or sucking, enabling insects to feed on a wide range of plant and animal material. Insects owe much of their success to having a highly waterproof *cuticle (to resist desiccation) and wings – outgrowths of the body wall that confer the greater mobility of flight. Breathing occurs through a network of tubes (*see* trachea).

Most insect species have separate sexes and undergo sexual reproduction. In some, this may alternate with asexual *parthenogenesis and in a few, males are unknown and reproduction is entirely asexual. The newly hatched young grow by undergoing a series of moults. In the more primitive groups, e.g. *Dermaptera, *Dictyoptera, and *Hemiptera, the young (called a *nymph) resembles the adult. More advanced groups, e.g. *Coleoptera, *Diptera, *Lepidoptera, and *Hymenoptera, undergo *metamorphosis, in which the young (called a *larva) is transformed into a quiescent *pupa from which the fully formed adult emerges. Insects are of vital importance in many ecosystems and many are of economic significance to man – as animal or plant pests or disease vectors or beneficially as crop pollinators or producers of silk, honey, etc.

Insectivora An order of small, mainly nocturnal, mammals that includes the hedgehogs, moles, and shrews. They have long snouts covered with stiff tactile hairs and their teeth are specialized for seizing and crushing insects and other small prey. The insectivores have changed very little since they evolved in the Cretaceous period, 130 million years ago.

insectivore An animal that eats insects, especially a mammal of the order Insectivora (hedgehogs, shrews, etc.).

insectivorous plant *See* carnivorous plant.

insertion sequence *See* transposon.

insolation (from *in*coming *sol*ar radi*ation*) The solar radiation that is received at the earth's surface per unit area. It is related to the *solar constant, the duration of daylight, the altitude of the sun, and the latitude of the receiving surface. It is measured in MJ m^{-2}.

instantaneous value The value of any varying quantity at a specified instant.

instar A stage in the larval development of an insect between two moults (ecdyses). There are usually a number of larval instars before the pupal stage and metamorphosis.

instinct An innate tendency to behave in a particular way, which does not depend critically on particular learning experiences for its development and therefore is seen in a similar form in all normally reared individuals of the same sex and species. Much instinctive behaviour takes the form of *fixed action patterns*. These are movements that – once started – are performed in a stereotyped way unaffected by external stimuli. For example, a frog's prey-catching tongue flick is performed in the same way whether or not anything is caught. Some complex instinctive behaviour, however, requires some learning by the animal before it is perfected. Birdsong, for example, consists of an innate component that is modified and made more

complex by the influence of other birds, the habitat, etc.

insulator A substance that is a poor conductor of heat and electricity. Both properties usually occur as a consequence of a lack of mobile electrons. *See* energy bands.

insulin A protein hormone, secreted by the *islets of Langerhans in the pancreas, that promotes the uptake of glucose by body cells and thereby controls its concentration in the blood. Insulin was the first protein whose amino-acid sequence was fully determined (in 1955). Underproduction of insulin results in the accumulation of large amounts of glucose in the blood and its subsequent excretion in the urine. This condition, known as *diabetes mellitus*, can be treated successfully by insulin injections.

integer Any one of the positive or negative whole numbers.

integral calculus *See* calculus.

integrand *See* integration.

integrated circuit A miniature electronic circuit produced within a single crystal of a *semiconductor, such as silicon. They range from simple logic circuits, little more than 1 mm square, to large-scale circuits measuring up to 8 mm square and containing a million or so transistors (active components) and resistors or capacitors (passive components). They are widely used in memory circuits, microcomputers, pocket calculators, and electronic watches on account of their low cost and bulk, reliability, and high speed. They are made by introducing impurities into specific regions of the semiconductor crystal by a variety of techniques.

integration 1. (in physiology) The coordination within the brain of separate but related nervous processes. For example, sensory information from the inner ear and the eye are both necessary for the sense of balance. These stimuli must be integrated by the brain not only with each other but also with various motor nerves, which coordinate the muscles that control posture. **2.** (in mathematics) The process of continuously summing changes in a function $f(x)$. It is the basis of the integral *calculus and the opposite process to *differentiation. The function to be integrated is called the *integrand* and the result of integration on the integrand is called the *integral*. For example, the integration of $f(x)$ is written $\int f(x)dx$, the differential dx being added to indicate that $f(x)$ must be integrated with respect to x. To complete the integration, a *constant of integration*, C, must be added where no interval over which the integration takes place is given. This is called an *indefinite integral*. If the interval is specified, e.g.

$$\int_{-r}^{r} f(x)dx,$$

no constant of integration is required and the result is a *definite integral*. This means that $f(x)$ is to be integrated between the values $x = +r$ and $x = -r$.

integument 1. The outermost body layer of an animal, characteristically comprising a layer of living cells – the *epidermis – together with a superficial protective coat, which may be a secreted hardened *cuticle, as in arthropods, or dead keratinized cells, as in vertebrates (*see* skin). **2.** The outer protective covering of a plant *ovule. It is perforated by a small pore, the *micropyle. Usually two integuments are present in angiosperms and one in gymnosperms. After fertilization the integuments form the *testa of the seed.

intensity 1. The rate at which radiant energy is transferred per unit area. *See* radiant intensity. **2.** The rate at which sound energy is transferred as measured relative to some reference value. *See* decibel. **3.** Mag-

netic intensity. *See* magnetic field. **4.** Electric intensity. *See* electric field. **5.** *See* luminous intensity.

interaction An effect involving a number of bodies, particles, or systems as a result of which some physical or chemical change takes place to one or more of them. *See also* fundamental interaction.

intercalation compound A type of compound in which atoms, ions, or molecules are trapped between layers in a crystal lattice. There is no formal chemical bonding between the host crystal and the trapped molecules (*see also* clathrate). Such compounds are formed by *lamellar solids and are often nonstoichiometric; examples are graphitic oxide (graphite–oxygen) and the mineral *muscovite.

intercellular (in biology) Located or occurring between cells. *Compare* intracellular.

intercostal muscles The muscles located between the *ribs, surrounding the lungs. They play an essential role in breathing (*see* respiratory movement).

interference The interaction of two or more wave motions affecting the same part of a medium so that the instantaneous disturbances in the resultant wave are the vector sum of the instantaneous disturbances in the interfering waves.

The phenomenon was first described by Thomas Young (1773–1829) in 1801 in light waves; it provided strong evidence for the wave theory of light. In the apparatus known as *Young's slits*, light is passed from a small source through a slit in a screen and the light emerging from this slit is used to illuminate two adjacent slits on a second screen. By allowing the light from these two slits to fall on a third screen, a series of parallel interference fringes is formed. Where the maximum values of the two waves from the slits coincide a bright fringe occurs (*constructive interference*) and where the maxima of one wave coincide with the minima of the other dark fringes are produced (*destructive interference*). *Newton's rings are also an interference effect. Because *lasers produce *coherent radiation they are also used to produce interference effects, one application of their use being *holography. *See also* interferometer.

interferometer An instrument designed to produce optical *interference fringes for measuring wavelengths, testing flat surfaces, measuring small distances, etc. *See also* echelon; Fabry–Pérot interferometer; Michelson–Morley experiment. In astronomy, radio interferometers are one of the two basic types of *radio telescopes.

interferon Any of a number of proteins (*see* cytokine) that increase the resistance of a cell to attack by viruses by unmasking genes that synthesize antiviral proteins. In humans, three groups of interferons have been discovered: α-interferons from white blood cells; β-interferons from connective tissue fibroblasts; and γ-interferons from lymphocytes (*see* interleukin). Interferons are also produced by lymphocyte *killer cells, which attack altered tissue cells, such as cancer cells. This converts other normal lymphocytes to killer cells and effects other changes in the immune system. Interferon is therefore of great potential use in treating both acute viral infections and cancer and attempts are underway to produce it on a large scale by means of *gene cloning in bacterial hosts.

interleukin Any of several *cytokines that act specifically as mediators between leucocytes. *Interleukin-1* (*IL-1*) is secreted by antigen-activated macrophages; it induces activated helper *T cells to secrete *interleukin-2* (*IL-2*). IL-2 stimulates the production of other cytokines, including B-cell growth factor, B-cell differentiation factor, *colony-stimulating factor, and γ-*interferon. *Interleukin-3* is involved in regulating mast-cell proliferation, and *interleukin-4* is thought to be a B-cell growth factor.

intermediate bond *See* chemical bond.

intermediate frequency *See* heterodyne; superheterodyne.

intermediate neutron A *neutron with kinetic energy in the range 10^2 – 10^5 electronvolts (1.6×10^{-17} – 1.6×10^{-14} joule).

intermediate vector boson *See* W boson; Z boson.

intermetallic compound A compound consisting of two or more metallic elements present in definite proportions in an alloy.

intermolecular forces Weak forces occurring between molecules. *See* van der Waals' forces; hydrogen bond.

internal-combustion engine A *heat engine in which fuel is burned in combustion chambers within the engine rather than in a separate furnace (as with the steam engine). The first working engine was the four-stroke *Otto engine* produced in 1876 by Nikolaus Otto (1832–91). In this type of engine a piston descends in a cylinder, drawing in a charge of fuel and air through an inlet valve; after reaching the bottom of its stroke the piston rises in the cylinder with the valves closed and compresses the charge; at or near the top of its stroke the charge is ignited by a spark and the resulting increase in pressure from the explosion forces the piston down again; on the subsequent upstroke the exhaust valve opens and the burnt gases are pushed out of the combustion chamber. The cycle is then repeated. Otto's engine used gas as a fuel; however, the invention of the carburettor and the development of the oil industry at the end of the 19th century enabled the Otto engine to become the source of power for the emerging motor car. A variation of the Otto four-stroke engine is the two-stroke engine that has no complicated valve system, the explosive charge entering and leaving the cylinder through ports in the cylinder that are covered and uncovered by the moving piston.

An alternative to the Otto engine, especially for heavy vehicles where weight is not a problem, is the compression-ignition *Diesel engine* invented by Rudolf Diesel (1858–1913) in about 1896. In this type of engine there are no sparking plugs; instead air is compressed in the cylinder, causing its temperature to rise to about 550°C. Oil is then sprayed into the combustion chamber and ignites on contact with the hot air. While the spark-ignition petrol engine typically works on a *compression ratio of 8 or 9 to 1, the Diesel engine has to have a compression ratio of 15 to 20 to 1. This requires a much heavier, and therefore more expensive, engine. *See also* Wankel engine; gas turbine.

internal conversion A process in which an excited atomic nucleus (*see* excitation) decays to the *ground state and the energy released is transferred by electromagnetic coupling to one of the bound electrons of that atom rather than being released as a photon. The coupling is usually with an electron in the K-, L-, or M-shell of the atom, and this *conversion electron* is ejected from the atom with a kinetic energy equal to the difference between the nuclear transition energy and the binding energy of the electron. The resulting ion is itself in an excited state and usually subsequently emits an *Auger electron or an X-ray photon.

internal energy Symbol U. The total of the kinetic energies of the atoms and molecules of which a system consists and the potential energies associated with their mutual interactions. It does not include the kinetic and potential energies of the system as a whole nor their nuclear energies or other intra-atomic energies. The value of the absolute internal energy of a system in any particular state cannot be measured; the significant quantity is the change in internal energy, ΔU. For a closed system (i.e. one that is not being replenished from outside its boundaries) the change in internal energy is equal to the heat absorbed by the system (Q) from its surroundings, less the work done (W) by the system on its surroundings, i.e. $\Delta U = Q - W$. *See also* energy; heat; thermodynamics.

internal environment The conditions that prevail within the body of an organism, particularly with respect to the composition of the intercellular fluid. Selective absorption of materials across cell membranes plays a large part in controlling the internal environment of both animals and plants. Animals in addition can regulate their body fluids by the action of hormones and the nervous system. *See* homeostasis.

internal resistance The resistance within a source of electric current, such as a cell or generator. It can be calculated as the difference between the e.m.f. (E) and the potential difference (V) between the terminals divided by the current being supplied (I), i.e. $r = (E - V)/I$, where r is the internal resistance.

international candle A former unit of *luminous intensity. It has now been replaced by the *candela, to which it is approximately equal.

international date line An imaginary line on the earth's surface that joins the north and south poles and approximately follows the 180° meridian through the Pacific Ocean. This line has been agreed internationally to mark the beginning and end of a day. A traveller moving towards the east, against the sun's apparent movement, gains 1 hour for every 15°

of longitude; westward he loses time at the same rate. In crossing the dateline therefore he is deemed to compensate for this by losing or gaining (respectively) one day. The 180° meridian was chosen as the date line by the International Meridian Conference in 1884.

International Practical Temperature Scale See temperature scales.

internode 1. (in botany) The part of a plant stem between two *nodes. **2.** (in neurology) The myelinated region of a nerve fibre between two nodes of Ranvier. See myelin sheath.

interphase The period following the completion of *cell division, when the nucleus is not dividing. During this period changes in both the nucleus and the cytoplasm result in the complete development of the daughter cells.

interplanetary space The space between the sun and the planets within the *solar system. The interplanetary matter that occupies this region of space mostly originates in the sun, from which it flows as the *solar wind. The solar wind consists primarily of protons emerging from the sun at a rate of about 10⁹ kilograms per second. At the earth's distance from the sun the particle density has fallen to a few particles per cm³. Apart from this very tenuous gas, there are dust particles in interplanetary space, largely believed to originate in the belt of asteroids. Particles weighing about 1 g produce visible meteors in the earth's atmosphere; micrometeorites as small as 1 nanogram can be detected by their impact on spacecraft.

intersex An organism displaying characteristics that are intermediate between those of the typical male and typical female of its species. For example, a human intersex may have testes that fail to develop, so that although he is technically a man he has the external appearance of a

woman. Intersexes may be produced in various ways; for example, by malfunctioning of the sex hormones. See also hermaphrodite.

interstellar space The space between the stars. The interstellar matter that occupies this space constitutes several percent of the Galaxy's total mass and it is from this matter that new stars are formed. The matter is primarily hydrogen, in which a number of other molecules and radicals have been detected, together with small solid dust grains. On average the density of matter in interstellar space is about 10⁶ hydrogen atoms per cubic metre, but the gas is not uniformly distributed, being clumped into interstellar clouds of various sizes and densities.

interstitial See defect.

interstitial-cell-stimulating hormone See luteinizing hormone.

interstitial compound A compound in which ions or atoms of a nonmetal occupy interstitial positions in a metal lattice. Such compounds often have metallic properties. Examples are found in the *carbides, *borides, and *silicides.

intestine The portion of the *alimentary canal posterior to the stomach. Its major functions are the final digestion of food matter from the stomach, the absorption of soluble food matter, the absorption of water, and the production of *faeces. See large intestine; small intestine.

intracellular (in biology) Located or occurring within cells. Compare intercellular.

intrinsic semiconductor See semiconductor.

intron (intervening sequence) A nucleotide sequence in a gene that does not code for the gene product (compare exon). Introns, which are confined to eukaryotes, are transcribed into messenger *RNA but are subsequently removed from the tran-

inulin

script before translation (*see* gene splicing). Their functional significance is still subject to controversy.

inulin A polysaccharide, made up from fructose molecules, that is stored as a food reserve in the roots or tubers of many plants, such as the dahlia.

Invar A trade name for an alloy of iron (63.8%), nickel (36%), and carbon (0.2%) that has a very low *expansivity over a a restricted temperature range. It is used in watches and other instruments to reduce their sensitivity to changes in temperature.

inverse Compton effect The gain in energy of low-energy photons when they are scattered by free electrons of much higher energy. As a consequence, the electrons lose energy. *See also* Compton effect.

inverse functions If $y = f(x)$ and a function can be found so that $x = g(y)$, then $g(y)$ is said to be the inverse function of $f(x)$. If y is a trigonometrical function of the angle x, say $y = \sin x$, then x is the *inverse trigonometrical function* of y, written $x = \arcsin y$ or $\sin^{-1}y$. Similarly, the other trigonometrical functions form the inverse trigonometrical functions $\cos^{-1}y$, $\tan^{-1}y$, $\cot^{-1}y$, $\sec^{-1}y$, and $\csc^{-1}y$. *Inverse hyperbolic functions* are also formed in this way, e.g. arcsinh$y$ or $\sinh^{-1}y$, $\cosh^{-1}y$, and $\tanh^{-1}y$.

inverse-square law A law in which the magnitude of a physical quantity is proportional to the reciprocal of the square of the distance from the source of that property. *Newton's law of gravitation and *Coulomb's law are both examples.

inversion A chemical reaction involving a change from one optically active configuration to the opposite configuration. The Walden inversion is an example. *See* nucleophilic substitution.

inversion layer *See* transistor.

inversion temperature *See* Joule–Kelvin effect.

in vitro Describing biological processes that are made to occur outside the living body, in laboratory apparatus (literally 'in glass', i.e. in a test tube). In *in vitro* fertilization, mature egg cells are removed from the ovary of a woman unable to conceive normally and fertilized externally; the resultant *blastocyst is implanted into her uterus. *Compare in vivo.*

in vivo Describing biological processes as they are observed to occur in their natural environment, i.e. within living organisms. *Compare in vitro.*

involucre A protective structure in some flowering plants and bryophytes. In flowering plants it consists of a ring of *bracts arising beneath the flower cluster of those species with a *capitulum (i.e. members of the dandelion family) or an *umbel (i.e. members of the carrot family). In bryophytes the involucre is a projection of tissue from the thallus that arches over the developing *archegonium.

involuntary muscle (smooth muscle) Muscle whose activity is not under the control of the will; it is supplied by the *autonomic nervous system. Involuntary muscle comprises long spindle-shaped cells without striations. These cells occur singly, in groups, or as sheets in the skin, around hair follicles, and in the digestive tract, respiratory tract, urinogenital tract, and the circulatory system. The cells contract slowly in spontaneous rhythms or when stretched; they may show sustained contraction (tonus) for long periods without fatigue. *Compare* voluntary muscle.

involute *See* evolute.

involution 1. A decrease in the size of an organ or the body. It may be associated with functional decline, as

354

occurs in the ageing process, or follow enlargement, as when the uterus returns to its normal size after pregnancy. **2.** The turning or rolling inwards of cells that occurs during the development of some vertebrate embryos.

iodic acid Any of various oxoacids of iodine, such as iodic(V) acid and iodic(VII) acid. When used without an oxidation state specified, the term usually refers to iodic(V) acid (HIO₃).

iodic(V) acid A colourless or very pale yellow solid, HIO₃; r.d. 4.63; decomposes at 110°C. It is soluble in water but insoluble in pure ethanol and other organic solvents. The compound is obtained by oxidizing iodine with concentrated nitric acid, hydrogen peroxide, or ozone. It is a strong acid and a powerful oxidizing agent.

iodic(VII) acid (periodic acid) A hygroscopic white solid, H₅IO₆, which decomposes at 138°C and is very soluble in water, ethanol, and ethoxyethane. Iodic(VII) acid may be prepared by electrolytic oxidation of concentrated solutions of iodic(V) acid at low temperatures. It is a weak acid but a strong oxidizing agent.

iodide *See* halide.

iodine Symbol I. A dark violet nonmetallic element belonging to group VII of the periodic table (*see* halogens); a.n. 53; r.a.m. 126.9045; r.d. 4.94; m.p. 113.5°C; b.p. 183.45°C. The element is insoluble in water but soluble in ethanol and other organic solvents. When heated it gives a violet vapour that sublimes. Iodine is required as a trace element (*see* essential element) by living organisms; in animals it is concentrated in the thyroid gland as a constituent of thyroid hormones. The element is present in sea water and was formerly extracted from seaweed. It is now obtained from oil-well brines (displacement by chlorine). There is one stable isotope, iodine–127, and fourteen radioactive

isotopes. It is used in medicine as a mild antiseptic (dissolved in ethanol as *tincture of iodine*), and in the manufacture of iodine compounds. Chemically, it is less reactive than the other halogens and the most electropositive (metallic) halogen. It was discovered in 1812 by Courtois.

iodine(V) oxide (iodine pentoxide) A white solid, I₂O₅; r.d. 4.79; decomposes at 300°C. It dissolves in water to give iodic(V) acid and also acts as an oxidizing agent.

iodine value A measure of the amount of unsaturation in a fat or vegetable oil (i.e. the number of double bonds). It is obtained by finding the percentage of iodine by weight absorbed by the sample in a given time under standard conditions.

iodoethane (ethyl iodide) A colourless liquid *haloalkane*, C₂H₅I; r.d. 1.9; m.p. –108°C; b.p. 72°C. It is made by reacting ethanol with a mixture of iodine and red phosphorus.

iodoform *See* tri-iodomethane.

iodoform test *See* haloform reaction.

iodomethane (methyl iodide) A colourless liquid haloalkane, CH₃I; r.d. 2.28; m.p. –66.45°C; b.p. 42.4°C. It can be made by reacting methanol with a mixture of iodine and red phosphorus.

ion An atom or group of atoms that has either lost one or more electrons, making it positively charged (a cation), or gained one or more electrons, making it negatively charged (an anion). *See also* ionization.

ion engine A type of jet-propulsion engine that may become important for propelling or controlling spacecraft. It consists of a unit producing a beam of ions, which are accelerated by an electric or electromagnetic field. Reaction forces from the high-speed ions causes propulsion in much the same way as that caused by exhaust gas of a rocket. However, a separate beam of electrons or ions of opposite polarity to the propelling beam must also be ejected from the engine to enable recombination to take place behind the vehicle (to avoid the vehicle becoming charged). Ion engines provide high *specific impulse and therefore low propellant consumption. The three main components of an ion engine are the power generator, the propellant feed, and the thruster. The power generator may be a nuclear reactor or a solar-energy collector. If it is the former, a gas turbine is coupled to the reactor and the turbine drives an electric generator. A solar-energy unit provides electricity direct. The propellant chosen needs to have an ion of medium mass (low mass for high specific impulse, high mass for high thrust) and a low first *ionization potential. Caesium and mercury are materials currently envisaged as suitable propellants. The thruster consists of an ionizer to produce the ions, an accelerator to provide and shape the accelerating field, and a neutralizer (usually an electron emitter) to neutralize the fast-moving ion beam after ejection.

ion exchange The exchange of ions of the same charge between a solution (usually aqueous) and a solid in contact with it. The process occurs widely in nature, especially in the absorption and retention of water-soluble fertilizers by soil. For example, if a potassium salt is dissolved in water and applied to soil, potassium ions are absorbed by the soil and sodium and calcium ions are released from it.

The soil, in this case, is acting as an ion exchanger. Synthetic *ion-exchange resins* consist of various copolymers having a cross-linked three-dimensional structure to which ionic groups have been attached. An *anionic resin* has negative ions built into its structure and therefore exchanges positive ions. A *cationic resin* has positive ions built in and exchanges negative ions.

Ion-exchange resins, which are used in sugar refining to remove salts, are synthetic organic polymers containing side groups that can be ionized. In anion exchange, the side groups are ionized basic groups, such as $-NH_3^+$ to which anions X^- are attached. The exchange reaction is one in which different anions in the solution displace the X^- from the solid. Similarly, cation exchange occurs with resins that have ionized acidic side groups such as $-COO^-$ or $-SO_2O^-$, with positive ions M^+ attached.

Ion exchange also occurs with inorganic polymers such as *zeolites, in which positive ions are held at sites in the silicate lattice. These are used for water-softening, in which Ca^{2+} ions in solution displace Na^+ ions in the zeolite. The zeolite can be regenerated with sodium chloride solution. *Ion-exchange membranes* are used as separators in electrolytic cells to remove salts from sea water (*see also* desalination) and in producing deionized water.

ionic bond *See* chemical bond.

ionic crystal *See* crystal.

ionic product The product of the concentrations of ions present in a given solution taking the stoichiometry into account. For a sodium chloride solution the ionic product is $[Na^+][Cl^-]$; for a calcium chloride solution it is $[Ca^{2+}][Cl^-]^2$. In pure water, there is an equilibrium with a small amount of self-ionization:

$$H_2O \leftrightharpoons H^+ + OH^-$$

The equilibrium constant of this dissociation is given by

$$K_W = [H^+][OH^-]$$

since the concentration $[H_2O]$ can be taken as constant. K_W is referred to as the ionic product of water. It has the value 10^{-14} $mol^2\,dm^{-6}$ at $25°C$. In pure water (i.e. no added acid or added alkali) $[H^+] = [OH^-] = 10^{-7}$ $mol\,dm^{-3}$. *See also* solubility product; pH scale.

ionic radius A value assigned to the radius of an ion in a crystalline solid, based on the assumption that the ions are spherical with a definite size. X-ray diffraction can be used to measure the internuclear distance in crystalline solids. For example, in NaF the Na – F distance is 0.231 nm, and this is assumed to be the sum of the Na^+ and F^- radii. By making certain assumptions about the shielding effect that the inner electrons have on the outer electrons, it is possible to assign individual values to the ionic radii – Na^+ 0.096 nm; F^- 0.135 nm. In general, negative ions have larger ionic radii than positive ions. The larger the negative charge, the larger the ion; the larger the positive charge, the smaller the ion.

ionic strength Symbol I. A function expressing the effect of the charge of the ions in a solution, equal to the sum of the molality of each type of ion present multiplied by the square of its charge. $I = \Sigma m_i z_i^2$.

ion implantation The technique of implanting ions in the lattice of a semiconductor crystal in order to modify its electronic properties. It is used as an alternative to diffusion, or in conjunction with it, in the manufacture of integrated circuits and solid-state components.

ionization The process of producing *ions. Certain molecules (*see* *electrolytes) ionize in solution; for example, *acids ionize when dissolved in water (*see also* solvation):
$$HCl \rightarrow H^+ + Cl^-$$
Electron transfer also causes ionization in certain reactions; for example, sodium and chlorine react by the transfer of a valence electron from the sodium atom to the chlorine atom to form the ions that constitute a sodium chloride crystal:
$$Na + Cl \rightarrow Na^+Cl^-$$
Ions may also be formed when an atom or molecule loses one or more electrons as a result of energy gained in a collision with another particle or a quantum of radiation (*see* photoionization). This may occur as a result of the impact of *ionizing radiation or of *thermal ionization and the reaction takes the form
$$A \rightarrow A^+ + e$$
Alternatively, ions can be formed by electron capture, i.e.
$$A + e \rightarrow A^-$$

ionization chamber An instrument for detecting *ionizing radiation. It consists of two electrodes contained in a gas-filled chamber with a potential difference maintained between them. Ionizing radiation entering the chamber ionizes gas atoms, creating electrons and positive ions. The electric field between the electrodes drives the electrons to the anode and the positive ions to the cathode. This current is, in suitable conditions, proportional to the intensity of the radiation. *See also* Geiger counter.

ionization energy (IE) *See* ionization potential.

ionization gauge A vacuum gauge consisting of a three-electrode system inserted into the container in which the pressure is to be measured. Electrons from the cathode are attracted to the grid, which is positively biased. Some pass through the grid but do not reach the anode, as it is maintained at a negative potential. Some of these electrons do, however, collide with gas molecules, ionizing them and converting them to positive ions. These ions are attracted to the anode; the resulting anode current can be used as a measure of the number of gas molecules present. Pressure as low as 10^{-6} pascal can be measured in this way.

ionization potential (IP) Symbol I. The minimum energy required to remove an electron from a specified atom or molecule to such a distance that there is no electrostatic interaction between ion and electron. Originally defined as the minimum potential through which an electron would have to fall to ionize an atom, the ionization potential was measured in volts. It is now, however, defined as the energy to effect an ionization and is conveniently measured in electronvolts (although this is not an SI unit) or joules per mole. The synonymous term *ionization energy* (*IE*) is often used.

The energy to remove the least strongly bound electron is the *first ionization potential*. Second, third, and higher ionization potentials can also be measured, although there is some ambiguity in terminology. Thus, in chemistry the second ionization potential is often taken to be the minimum energy required to remove an electron from the singly charged ion; the second IP of lithium would be the energy for the process

$$Li^+ \rightarrow Li^{2+} + e$$

In physics, the second ionization potential is the energy required to remove an electron from the next to highest energy level in the neutral atom or molecule; e.g.

$$Li \rightarrow Li^{*+} + e,$$

where Li^{*+} is an excited singly charged ion produced by removing an electron from the K-shell.

ionizing radiation Radiation of sufficiently high energy to cause *ionization in the medium through which it passes. It may consist of a stream of high-energy particles (e.g. electrons, protons, alpha-particles) or short-wavelength electromagnetic radiation (ultraviolet, X-rays, gamma-rays). This type of radiation can cause extensive damage to the molecular structure of a substance either as a result of the direct transfer of energy to its atoms or molecules or as a result of the *secondary electrons released by ionization. In biological tissue the effect of ionizing radiation can be very serious, usually as a consequence of the ejection of an electron from a water molecule and the oxidizing or reduc-

ing effects of the resulting highly reactive species:

$$H_2O \rightarrow e^- + H_2O^* + H_2O^+ \rightarrow$$
$$\cdot OH + H_3O^+ + \cdot H,$$

where the dot before a radical indicates an unpaired electron and an * denotes an excited species.

ion-microprobe analysis A technique for analysing the surface composition of solids. The sample is bombarded with a narrow beam (as small as 2 μm diameter) of high-energy ions. Ions ejected from the surface by sputtering are detected by mass spectrometry. The technique allows quantitative analysis of both chemical and isotopic composition for concentrations as low as a few parts per million.

ionosphere *See* earth's atmosphere; radio transmission.

ionospheric wave *See* radio transmission.

ion pair A pair of oppositely charged ions produced as a result of a single ionization; e.g.

$$HCl \rightarrow H^+ + Cl^-.$$

Sometimes a positive ion and an electron are referred to as an ion pair, as in

$$A \rightarrow A^+ + e^-.$$

ion pump A type of *vacuum pump that can reduce the pressure in a container to about 1 nanopascal by passing a beam of electrons through the residual gas. The gas is ionized and the positive ions formed are attracted to a cathode within the container where they remain trapped. The pump is only useful at very low pressures, i.e. below about 1 micropascal. The pump has a limited capacity because the absorbed ions eventually saturate the surface of the cathode. A more effective pump can be made by simultaneously *sputtering a film of metal, so that fresh surface is continuously produced. The device is then known as a *sputter-ion pump*.

IP *See* ionization potential.

IR *See* infrared radiation.

iridium Symbol Ir. A silvery metallic *transition element (*see also* platinum metals); a.n. 77; r.a.m. 192.20; r.d. 22.42; m.p. 2410°C; b.p. 4130°C. It occurs with platinum and is mainly used in alloys with platinum and osmium. The element forms a range of iridium(III) and iridium(IV) complexes. It was discovered in 1804 by Tennant.

iridium anomaly The occurrence of unusually high concentrations of the relatively scarce metal iridium at the boundaries of certain geological strata. Two such layers have been discovered, one at the end of the Cretaceous, 65 million years ago, and the second at the end of the Eocene, 34 million years ago. One theory to account for these suggests that on each occasion a huge iridium-containing meteorite may have collided with the earth, producing a cloud of dust that settled out to form an iridium-rich layer. The environmental consequences of such an impact, notably in causing a general warming of the earth by the *greenhouse effect, may have led to the extinction of the dinosaurs at the end of the Cretaceous and the extinction of many radiolarians at the end of the Eocene.

iris 1. (in anatomy) The pigmented ring of muscular tissue, lying between the cornea and the lens, in the eyes of vertebrates and some cephalopod molluscs. It has a central hole (the *pupil*) through which light enters the eye and it contains both circular and radial muscles. Reflex contraction of the former occurs in bright light to reduce the diameter of the pupil; contraction of the radial muscles in dim light increases the pupil diameter and therefore the amount of light entering the eye. Colour is determined by the amount of the pigment melanin in the iris. Blue eyes result from relatively little melanin; grey and brown eyes from increasingly larger amounts. **2.** (in physics) *See* diaphragm.

iron Symbol Fe. A silvery malleable and ductile metallic *transition element; a.n. 26; r.a.m. 55.847; r.d. 7.87; m.p. 1535°C; b.p. 2750°C. The main sources are the ores *haematite (Fe_2O_3), *magnetite (Fe_3O_4), limonite ($FeO(OH)_nH_2O$), ilmenite ($FeTiO_3$), siderite ($FeCO_3$), and pyrite (FeS_2). The metal is smelted in a *blast furnace to give impure *pig iron, which is further processed to give *cast iron, *wrought iron, and various types of *steel. The pure element has three crystal forms: *alpha-iron*, stable below 906°C with a body-centred-cubic structure; *gamma-iron*, stable between 906°C and 1403°C with a nonmagnetic face-centred-cubic structure; and *delta-iron*, which is the body-centred-cubic form above 1403°C. Alpha-iron is ferromagnetic up to its Curie point (768°C). The element has nine isotopes (mass numbers 52–60), and is the fourth most abundant in the earth's crust. It is required as a trace element (*see* essential element) by living organisms. Iron is quite reactive, being oxidized by moist air, displacing hydrogen from dilute acids, and combining with nonmetallic elements. It forms ionic salts and numerous complexes with the metal in the +2 or +3 oxidation states. Iron(VI) also exists in the ferrate ion FeO_4^{2-}, and the element also forms complexes in which its oxidation number is zero (e.g. $Fe(CO)_5$).

iron(II) chloride A green-yellow deliquescent compound, $FeCl_2$; hexagonal; r.d. 3.16; m.p. 670°C. It also exists in hydrated forms: $FeCl_2.2H_2O$ (green monoclinic; r.d. 2.36) and $FeCl_2.4H_2O$ (blue-green monoclinic deliquescent; r.d. 1.93). Anhydrous iron(II) chloride can be made by passing a stream of dry hydrogen chloride over the heated metal; the hydrated forms can be made using dilute hydrochloric acid or by recrystallizing with water. It is converted into iron(III) chloride by the action of chlorine.

iron(III) chloride A black-brown solid, $FeCl_3$; hexagonal; r.d. 2.9; m.p. 306°C; decomposes at 315°C. It also exists as the hexahydrate $FeCl_3.6H_2O$, a brown-yellow deliquescent crystalline substance (m.p. 37°C; b.p. 280–285°C). Iron(III) chloride is prepared by passing dry chlorine over iron wire or steel wool. The reaction proceeds with incandescence when started and iron(III) chloride sublimes as almost black iridescent scales. The compound is rapidly hydrolysed in moist air. In solution it is partly hydrolysed; hydrolysis can be suppressed by the addition of hydrochloric acid. The compound dissolves in many organic solvents, forming solutions of low electrical conductivity: in ethanol, ethoxyethane, and pyridine the molecular weight corresponds to $FeCl_3$ but is higher in other solvents corresponding to Fe_2Cl_6. The vapour is also dimerized. In many ways the compound resembles aluminium chloride, which it may replace in Friedel–Crafts reactions.

iron(II) oxide A black solid, FeO; cubic; r.d. 5.7; m.p. 369°C. It can be obtained by heating iron(II) oxalate; the carbon monoxide formed produces a reducing atmosphere thus preventing oxidation to iron(III) oxide. The compound has the sodium chloride structure, indicating its ionic nature, but the crystal lattice is deficient in iron(II) ions and it is nonstoichiometric. Iron(II) oxide dissolves readily in dilute acids.

iron(III) oxide A red-brown to black insoluble solid, Fe_2O_3; trigonal; r.d. 5.24; m.p. 1565°C. There is also a hydrated form, $Fe_2O_3xH_2O$, which is a red-brown powder; r.d. 2.44–3.60. (*See* rusting.)

Iron(III) oxide occurs naturally as *haematite and can be prepared by heating iron(III) hydroxide or iron(II)

sulphate. It is readily reduced on heating with carbon or in a stream of carbon monoxide, hydrogen, or coal gas:

$$Fe_2O_3 + 3C \rightarrow 2Fe + 3CO$$

iron pyrites *See* pyrite.

iron(II) sulphate An off-white solid, $FeSO_4.H_2O$; monoclinic; r.d. 2.970. There is also a heptahydrate form, $FeSO_4.7H_2O$; blue-green monoclinic; r.d. 1.898; m.p. 64°C. The heptahydrate is the best known iron(II) salt and is sometimes called *green vitriol* or *copperas*. It is obtained by the action of dilute sulphuric acid on iron in a reducing atmosphere. The anhydrous compound is very hygroscopic. It decomposes at red heat to give iron(III) oxide, sulphur trioxide, and sulphur dioxide. A solution of iron(II) sulphate is gradually oxidized on exposure to air, a basic iron(III) sulphate being deposited.

iron(III) sulphate A yellow hygroscopic compound, $Fe_2(SO_4)_3$; rhombic; r.d. 3.097; decomposes above 480°C. It is obtained by heating an aqueous acidified solution of iron(II) sulphate with hydrogen peroxide:

$$2FeSO_4 + H_2SO_4 + H_2O_2 \rightarrow$$
$$Fe_2(SO_4)_3 + 2H_2O$$

On crystallizing, the hydrate $Fe_2(SO_4)_3.9H_2O$ is formed. The acid sulphate $Fe_2(SO_4)_3.H_2SO_4.8H_2O$ is deposited from solutions containing a sufficient excess of sulphuric acid.

irradiance Symbol E. The *radiant flux per unit area reaching a surface; in SI units it is measured in watts per square metre (W m⁻²). Irradiance refers to electromagnetic radiation of all kinds, whereas *illuminance refers only to visible radiation.

irradiation Exposure to any form of radiation; often exposure to *ionizing radiation is implied.

irrational number A number that cannot be expressed as the ratio of two integers. An irrational number may be a *surd, such as $\sqrt{2}$ or $\sqrt{3}$, which can be expressed to any desired degree of accuracy but cannot be assigned an exact value. Alternatively, it may be a *transcendental, such as π or e. *Compare* rational number.

irreversible process *See* reversible process.

irreversible reaction *See* chemical reaction.

irritability One of the fundamental properties of all organisms: the capacity to detect, interpret, and respond to changes in the environment (e.g. the stimuli of light, touch, chemicals, etc.). Multicellular animals have specialized *sense organs and *effector organs for this purpose; in unicellular animals and plants, which lack a nervous system, the reception of, and response to, a stimulus occur in the same cell.

IR spectroscopy *See* infrared spectroscopy.

ischium The most posterior of the three bones that make up each half of the *pelvic girdle. *See also* ilium; pubis.

isentropic process Any process that takes place without a change of *entropy. The quantity of heat transferred, δQ, in a reversible process is proportional to the change in entropy, δS, i.e. $\delta Q = T\delta S$, where T is the thermodynamic temperature. Therefore, a reversible *adiabatic process is isentropic, i.e. when $\delta Q = 0$, δS also equals 0.

islets of Langerhans Small groups of cells in the pancreas that function as an endocrine gland. The α cells secrete the hormone *glucagon and the β cells secrete *insulin. The islets are named after their discoverer, the German anatomist and microscopist Paul Langerhans (1847–88).

iso- Prefix denoting that a compound is an *isomer, e.g. isopentane $(CH_3CH(CH_3)C_2H_5$, 2-methylbutane) is an isomer of pentane.

isobar 1. A line on a map or chart that joins points or places that have the same atmospheric pressure. **2.** A curve on a graph representing readings taken at constant pressure. **3.** One of two or more nuclides that have the same number of nucleons but different *atomic numbers. Radium–88, actinium–89, and thorium–90 are isobars as each has a *nucleon number of 228.

isobaric spin *See* isotopic spin.

isocline A line on a map or chart joining points or places of equal magnetic dip (*see* geomagnetism)

isocyanate *See* cyanic acid.

isocyanic acid *See* cyanic acid.

isocyanide *See* isonitrile.

isocyanide test A test for primary amines by reaction with an alcoholic solution of potassium hydroxide and trichloromethane.

$$RNH_2 + 3KOH + CHCl_3 \rightarrow$$
$$RNC + 3KCl + 3H_2O$$

The isocyanide RNC is recognized by its unpleasant smell. This reaction of primary amines is called the *carbylamine reaction*.

isodiaphere One of two or more nuclides in which the difference between the number of neutrons and the number of protons is the same. A nuclide and its product after losing an *alpha particle are isodiapheres.

isodynamic line A line on a map or chart joining points or places at which the total strengths of the earth's magnetic field are equal (*see* geomagnetism).

isoelectronic Describing compounds that have the same numbers of valence electrons. For example, nitrogen (N_2) and carbon monoxide (CO) are isoelectronic molecules.

isogamy Sexual reproduction involving the production and fusion of gametes that are similar in size and structure. It occurs only in some lower animals and plants, e.g. certain protozoans and algae. *Compare* anisogamy.

isogonal line A line on a map or chart joining points or places of equal magnetic declination (*see* geomagnetism).

isolating mechanism Any of the factors that prevent interbreeding (and therefore exchange of genes) between all the members of a species. Probably the most important isolating mechanism is geographical isolation, when different populations of the same species occupy different habitats and therefore never come into contact with each other. Ultimately this may lead to the populations becoming separate species by means of *adaptive radiation.

isoleucine *See* amino acid.

isomerism 1. (in chemistry) The existence of chemical compounds (*isomers*) that have the same molecular formulae but different molecular structures or different arrangements of atoms in space. In *structural isomerism* the molecules have different molecular structures: i.e. they may be different types of compound or they may simply differ in the position of the functional group in the molecule. Structural isomers generally have different physical and chemical properties. In *stereoisomerism*, the isomers have the same formula and functional groups, but differ in the arrangement of groups in space. Optical isomerism is one form of this (*see* optical activity). Another type is *cis*–*trans isomerism* (formerly *geometrical isomerism*), in which the isomers have different positions of groups with respect to a double bond or central atom (see illustration). 2. (in physics) The existence of atomic nuclei that have the same atomic number and the same mass number but different energy states.

isomers *See* isomerism.

isometric 1. (in technical drawing) Denoting a projection in which the three axes are equally inclined to the surface of the drawing and lines are drawn to scale. 2. (in crystallography) Denoting a system in which the axes are perpendicular to each other, as in cubic crystals. 3. (in physics) Denoting a line on a graph illustrating the way in which temperature and pressure are interrelated at constant volume.

isomorphism The existence of two or more substances (*isomorphs*) that have the same crystal structure, so that they are able to form *solid solutions.

isonitrile (isocyanide; carbylamine) An organic compound containing the group –NC, in which the bonding is to the nitrogen atom.

iso-octane *See* octane; octane number.

isopoly compound *See* cluster compound.

isoprene A colourless liquid diene, $CH_2:C(CH_3)CH:CH_2$. The systematic name is *2-methylbuta-1,3-diene*. It is the structural unit in *terpenes and natural *rubber, and is used in making synthetic rubbers.

isospin *See* isotopic spin.

isostasy The theoretical equilibrium that tends to exist in the earth's crust. If this equilibrium is disturbed, for example as a result of erosion or deposition, compensatory movements in the earth's crust occur: areas of deposition sink, whereas areas of erosion rise. Continental ice sheets have been an important cause of isostatic movements of the earth's crust. The growth of ice sheets and the resulting additional mass of the ice is compensated by the downward deflection of the lithospheric plate and the displacement of asthenospheric material (*see* asthenosphere) beneath it. When the ice melts, the displaced material flows back causing the overlying lithospheric plate to rise.

1-chloropropane

2-chloropropane

structural isomers in which the functional group has different positions

methoxymethane

ethanol

structural isomers in which the functional groups are different

trans-but-2-ene

cis-but-2-ene

cis–trans isomers in which the groups are distributed on a double bond

cis–trans isomers in a square-planar complex

keto form   enol form

keto–enol tautomerism

Isomerism

isotactic polymer *See* polymer.

isotherm 1. A line on a map or chart joining points or places of equal temperature. **2.** A curve on a graph representing readings taken at constant temperature (e.g. the relationship between the pressure and volume of a gas at constant temperature).

isothermal process Any process that takes place at constant temperature. In such a process heat is, if necessary, supplied or removed from the system at just the right rate to maintain constant temperature. *Compare* adiabatic process.

isotone One of two or more nuclides that contain the same number of neutrons but different numbers of protons. The naturally occurring isotones, for example, strontium–88 and yttrium–89 (both with 50 neutrons), give an indication of the stability of certain nuclear configurations.

isotonic Describing solutions that have the same osmotic pressure.

isotope One of two or more atoms of the same element that have the same number of protons in their nucleus but different numbers of neutrons. Hydrogen (1 proton, no neutrons), deuterium (1 proton, 1 neutron), and tritium (1 proton, 2 neutrons) are isotopes of hydrogen. Most elements in nature consist of a mixture of isotopes. *See* isotope separation.

isotope separation The separation of the *isotopes of an element from each other on the basis of slight differences in their physical properties. For laboratory quantities the most suitable device is often the mass spectrometer. On a larger scale the methods used include gaseous diffusion (widely used for separating isotopes of uranium in the form of the gas uranium hexafluoride), distillation (formerly used to produce heavy water), electrolysis (requiring cheap electrical power), thermal diffusion (formerly used to separate uranium isotopes, but now considered uneconomic), centrifuging (a method in which there is renewed interest), and laser methods (involving the excitation of one isotope and its subsequent separation by electromagnetic means).

isotopic number (neutron excess) The difference between the number of neutrons in an isotope and the number of protons.

isotopic spin (isospin; isobaric spin) A quantum number applied to hadrons (*see* elementary particles) to distinguish between members of a set of particles that differ in their electromagnetic properties but are otherwise apparently identical. For example if electromagnetic interactions and weak interactions are ignored, the proton cannot be distinguished from the neutron in their strong interactions: isotopic spin was introduced to make a distinction between them. The use of the word 'spin' implies only an analogy to angular momentum, to which isotopic spin has a formal resemblance.

isotropic Denoting a medium whose physical properties are independent of direction. *Compare* anisotropic.

iteration The process of successive approximations used as a technique for solving a mathematical problem. The technique can be used manually but is widely used by computers.

J

jade A hard semiprecious stone consisting either of jadeite or nephrite. *Jadeite*, the most valued of the two, is a sodium aluminium pyroxene, $NaAlSi_2O_6$. It is prized for its intense translucent green colour but white, green and white, brown, and orange varieties also occur. The only important source of jadeite is in the Mogaung region of upper Burma. *Nephrite* is one of the amphibole group of rock-forming minerals. It occurs in a variety of colours, including green, yellow, white, and black. Important sources include the Soviet Union, New Zealand, Alaska, China, and W USA.

jadeite *See* jade.

Jahn–Teller effect If a likely structure of a nonlinear molecule or ion would have degenerate orbitals (i.e. two molecular orbitals with the same energy levels) the actual structure of the molecule or ion is distorted so as to split the energy levels ('raise' the degeneracy). The effect is observed in inorganic complexes. For example, the ion $[Cu(H_2O)_6]^{2+}$ is octahedral and the six ligands might be expected to occupy equidistant positions at the corners of a regular octahedron. In fact, the octahedron is distorted, with four ligands in a square and two opposite ligands further away. If the 'original' structure has a centre of symmetry, the distorted structure must also have a centre of symmetry. The effect was predicted theoretically by H. A. Jahn and Edward Teller in 1937.

jasper An impure variety of *chalcedony. It is associated with iron ores and as a result contains iron oxide impurities that give the mineral its characteristic red or reddish-brown colour. Jasper is used as a gemstone.

jejunum The portion of the mammalian *small intestine that follows the *duodenum and precedes the *ileum. The surface area of the lining of the jejunum is greatly increased by numerous small outgrowths (*see* villus). This facilitates the absorption of digested material, which is the prime function of the jejunum.

jellyfish *See* Coelenterata.

jet A variety of *coal that can be cut and polished and is used for jewellery, ornaments, etc.

jet propulsion (reaction propulsion) The propulsion of a body by means of a force produced by discharging a fluid in the form of a jet. The backward-moving jet of fluid reacts on the body in which it was produced, in accordance with Newton's third law of motion, to create a reactive force that drives the body forward. Jet propulsion occurs in nature, the squid using a form of it to propel itself through water. Although jet-propelled boats and cars have been developed, the main use of jet propulsion is in aircraft and spacecraft. Jet propulsion is the only known method of propulsion in space. In the atmosphere, jet propulsion becomes more efficient at higher altitudes, as efficiency is inversely proportional to the density of the medium through which a body is flying. The three principal means of providing jet propulsion are the turbojet, the ramjet, and the rocket. The *turbojet* is an air-breathing *heat engine based on the *gas turbine, used to power jet aircraft. The *ramjet* is also an air-breathing engine, but compression of the oxidant is achieved by the forward motion of the device through the atmosphere. This enables the compressor and turbine of the gas turbine to be dispensed with and the remaining system consists simply of an inlet diffuser, a combustion chamber in which fuel is burnt, and a jet nozzle through which the products of combustion are discharged. Used in guided missiles, the ramjet must be accelerated to its operating velocity before it can fly (*see also* pulse jet). These two forms of jet propulsion, being air-breathing engines, can only be used in the earth's atmosphere. The *rocket, however, carries its own oxidant and can

a hinge joint (the elbow)

a ball-and-socket joint (the hip)

Types of freely movable joints

the vertebrae of the spinal column; and (3) *freely movable* or *synovial joints*, e.g. those that occur between the limb bones. Synovial joints include the *ball-and-socket joints* (between the limbs and the hip and shoulder girdles), which allow movement in all directions; and the *hinge joints* (e.g. at the knee and elbow), which allow movement in one plane only (see illustration). A synovial joint is bound by ligaments and lined with *synovial membrane.

Joly's steam calorimeter An apparatus invented by John Joly (1857–1933) to measure the specific heat capacity of a gas at constant volume. Two equal spherical containers are suspended from the opposite ends of a balance arm. One sphere is evacuated and the other contains the sample gas. The whole apparatus is enclosed in a steam bath, the specific heat capacity of the sample gas being calculated from the difference between the masses of the water that condenses on each sphere.

Josephson effects Electrical effects observed when two superconducting materials (at low temperature) are separated by a thin layer of insulating material (typically a layer of oxide less than 10^{-8} m thick). If normal metallic conductors are separated by such a barrier it is possible for a small current to flow between the conductors by the *tunnel effect. If the materials are superconductors (see superconductivity), several unusual phenomena occur:

(1) A supercurrent can flow through the barrier; i.e. it has zero resistance.

(2) If this current exceeds a critical value, this conductivity is lost; the barrier then only passes the 'normal' low tunnelling current and a voltage develops across the junction.

(3) If a magnetic field is applied below the critical current value, the current density changes regularly with distance across the junction. The net

therefore be used in space. *See also* ion engine.

jet stream A narrow wind current that occurs in the earth's atmosphere above the lower troposphere (*see* earth's atmosphere). The wind flows towards the east at speeds of between 60 kilometres per hour (summer) and 125 km/h (winter).

jeweller's rouge Red powdered haematite, iron(III) oxide, Fe_2O_3. It is a mild abrasive used in metal cleaners and polishes.

joint The point of contact between two (or more) bones, together with the tissues that surround it. Joints fall into three classes that differ in the degree of freedom of movement they allow: (1) *immovable joints*, e.g. the *sutures between the bones that form the cranium; (2) *slightly movable joints*, e.g. the *symphyses between current through the barrier depends on the magnetic field applied. As the field is increased the net current increases from zero to a maximum, decreases to zero, increases again to a (lower) maximum, decreases, and so on. If the field exceeds a critical value the superconductivity in the barrier vanishes and a potential difference develops across the junction.

(4) If a potential difference is applied across the junction, a high-frequency alternating current flows through the junction. The frequency of this current depends on the size of the potential difference.

A junction of this type is called a *Josephson junction*; two or more junctions joined by superconducting paths form a *Josephson interferometer*. Such junctions can be used in measuring fundamental constants, in defining a voltage standard, and in the highly accurate measurement of magnetic fields. An important potential use is in logic components in high-speed computers. Josephson junctions can switch states very quickly (as low as 6 picoseconds). Moreover they have very low power consumption and can be packed closely without generating too much heat. It is possible that computers based on such devices could operate 50 times faster than the best existing machines. The effects are named after B. D. Josephson (1940–), who predicted them theoretically in 1962.

joule Symbol J. The *SI unit of work and energy equal to the work done when the point of application of a force of one newton moves, in the direction of the force, a distance of one metre. 1 joule = 10^7 ergs = 0.2388 calorie. It is named after James Prescott Joule (1818–89).

Joule heating The production of heat in a conductor as a result of the passage of an electric current through the conductor. The quantity of heat

produced is given by *Joule's law (def. 1).

Joule's laws 1. The heat (Q) produced when an electric current (I) flows through a resistance (R) for a time (t) is given by $Q = I^2Rt$. **2.** The *internal energy of a given mass of gas is independent of its volume and pressure, being a function of temperature alone. This law applies only to *ideal gases (for which it provides a definition of thermodynamic temperature) as in a real gas intermolecular forces would cause changes in the internal energy should a change of volume occur. *See also* Joule–Thomson effect.

Joule–Thomson effect (Joule–Kelvin effect) The change in temperature that occurs when a gas expands through a porous plug into a region of lower pressure. For most real gases the temperature falls under these circumstances as the gas has to do internal work in overcoming the intermolecular forces to enable the expansion to take place. This is a deviation from *Joule's law (def. 2). There is usually also a deviation from *Boyle's law, which can cause either a rise or a fall in temperature since any increase in the product of pressure and volume is a measure of external work done. At a given pressure, there is a particular temperature, called the *inversion temperature* of the gas, at which the rise in temperature from the Boyle's law deviation is balanced by the fall from the Joule's law deviation. There is then no temperature change. Above the inversion temperature the gas is heated by expansion, below it, it is cooled. The effect was discovered by James Joule working in collaboration with William Thomson (later Lord Kelvin; 1824–1907).

Jovian Relating to the planet Jupiter.

JUGFET *See* transistor.

jugular vein A paired vein in the neck of mammals that returns blood from the head to the heart. It joins the subclavian vein at the base of the neck.

junction detector (solid-state detector) A sensitive detector of *ionizing radiation in which the output is a current pulse proportional to the energy falling in or near the depletion region of a reverse-biased *semiconductor junction. The first types were made by evaporating a thin layer of gold on to a polished wafer of n-type germanium; however, gold–silicon devices can be operated at room temperature and these have superseded the germanium type, which have to be operated at the temperature of liquid nitrogen to reduce noise. When the gold–silicon junction is reverse-biased a depletion region, devoid of charge carriers (electrons and holes), forms in the silicon. Incoming ionizing radiation falling in this depletion region creates pairs of electrons and holes, which both have to be collected in order to give an output pulse proportional to the energy of the detected particle.

Junction detectors are used in medicine and biology as well as in space systems.

junction transistor *See* transistor.

Jupiter The largest planet in the *solar system and the fifth in order from the sun (778 × 10⁶ km distant). Its equatorial diameter is 142 700 km and its *sidereal period is 11.86 years. Viewed through an optical telescope it appears as an elliptical disc crossed by a number of bands. With a *radio telescope thermal radiation can be detected from the Jovian stratosphere indicating a temperature between 100 and 130 K. Microwave nonthermal emission can also be detected from the Jovian *Van Allen belts. The planet has 16 known satellites, the largest of which, Io, has a diameter of 3242 km. A planetary ring of rocks was discovered in 1979 by the US Galileo probe.

Jurassic The second geological period of the Mesozoic era. It followed the Triassic, which ended about 190 million years ago, and extended until the beginning of the Cretaceous period, about 139 million years ago. It was named in 1829 by A. Brongniart after the Jura Mountains on the borders of France and Switzerland. Jurassic rocks include clays and limestones in which fossil flora and fauna are abundant. Plants included ferns, cycads, ginkgos, rushes, and conifers. Important invertebrates included *ammonites (on which the Jurassic is zoned), corals, brachiopods, bivalves, and echinoids. Reptiles dominated the vertebrates and the first flying reptiles – the pterosaurs – appeared. The first primitive bird, *Archaeopteryx*, also made its appearance.

juvenile hormone A hormone secreted by insects from a pair of endocrine glands (*corpora allata*) close to the brain. It inhibits metamorphosis and maintains the presence of larval features.

K

kainite A naturally occurring double salt of magnesium sulphate and potassium chloride, $MgSO_4.KCl.3H_2O$.

Kainozoic *See* Cenozoic.

kalinite A mineral form of *aluminium potassium sulphate ($Al_2(SO_4)_3$. $K_2SO_4.24H_2O$).

kallidin *See* kinin.

Kaluza–Klein theory A type of *unified-field theory that postulates a generalization of the general theory of relativity to higher than four space–time dimensions. In five space–time dimensions this gives general relativity and electromagnetic interactions. In higher space–time dimensions Kaluza–Klein theories give general relativity and more general *gauge theories. A combination of Kaluza–Klein theory and *supersymmetry gives rise to *supergravity, which needs eleven space–time dimensions. In these theories it is proposed that the higher dimensions are 'rolled up' to become microscopically small (a process known as *spontaneous compactification*) with four macroscopic space–time dimensions remaining.

kaolin (china clay) A soft white clay that is composed chiefly of the mineral kaolinite (*see* clay minerals). It is formed during the weathering and hydrothermal alteration of other clays or feldspar. Kaolin is mined in the UK, France, Czechoslovakia, and USA. Besides its vital importance in the ceramics industry it is also used extensively as a filler in the manufacture of rubber, paper, paint, and textiles and as a constituent of medicines.

kaon A K-meson. *See* meson.

karyogram (idiogram) A diagram representing the characteristic features of the *chromosomes of a species.

karyokinesis The division of a cell nucleus. *See* meiosis; mitosis.

karyotype The number and structure of the *chromosomes in the nucleus of a cell. The karyotype is identical in all the *diploid cells of an organism.

Kater's pendulum A complex *pendulum designed by Henry Kater (1777–1835) to measure the acceleration of free fall. It consists of a metal bar with knife edges attached near the ends and two weights that can slide between the knife edges. The bar is pivoted from each knife edge in turn and the positions of the weights are adjusted so that the period of the pendulum is the same with both pivots. The period is then given by the formula for a simple pendulum, which enables g to be calculated.

katharometer An instrument for comparing the thermal conductivities of two gases by comparing the rate of loss of heat from two heating coils surrounded by the gases. The instrument can be used to detect the presence of a small amount of an impurity in air and is also used as a detector in gas chromatography.

keel (carina) The projection of bone from the sternum (breastbone) of a bird or bat, to which the powerful flight muscles are attached. The sterna of flightless birds (e.g. ostrich and emu) lack keels.

keeper A piece of soft iron used to bridge the poles of a permanent magnet when it is not in use. It reduces the leakage field and thus preserves the magnetization.

Kekulé structure A proposed structure of *benzene in which the molecule has a hexagonal ring of carbon atoms linked by alternating double and single bonds. Kekulé structures contribute to the resonance hybrid of benzene. The structure was suggested in 1865 by Friedrich August Kekulé (1829–69).

kelp Any large brown seaweed (*see* Phaeophyta) or its ash, used as a source of iodine.

kelvin Symbol K. The *SI unit of thermodynamic *temperature equal to the fraction 1/273.16 of the thermodynamic temperature of the *triple point of water. The magnitude of the kelvin is equal to that of the degree celsius (centigrade), but a temperature expressed in degrees celsius is numerically equal to the temperature in kelvins less 273.15 (i.e. $°C = K - 273.15$). The *absolute zero of temperature has a temperature of 0 K ($-273.15°C$). The former name *degree kelvin* (symbol °K) became obsolete by international agreement in 1967. The unit is named after Lord Kelvin (1824–1907).

Kelvin effect *See* Thomson effect.

Kepler's laws Three laws of planetary motion formulated by Johannes Kepler (1571–1630) in about 1610 on the basis of observations made by Tycho Brahe (1546–1601). They state that: (1) the orbits of the planets are elliptical with the sun at one *focus of the ellipse; (2) each planet revolves around the sun so that an imaginary line (the *radius vector*) connecting the planet to the sun sweeps out equal areas in equal times; (3) the ratio of the square of each planet's *sidereal period to the cube of its distance from the sun is a constant for all the planets.

keratin Any of a group of fibrous *proteins occurring in hair, feathers, hooves, and horns. Keratins have coiled polypeptide chains that combine to form supercoils of several polypeptides linked by disulphide bonds between adjacent cysteine amino acids. Aggregates of these supercoils form microfibrils, which are embedded in a protein matrix. This produces a strong but elastic structure.

keratinization (cornification) The process in which the cytoplasm of the outermost cells of the mammalian *epidermis is replaced by *keratin. Keratinization occurs in the *stratum corneum, feathers, hair, claws, nails, hooves, and horns.

kerosine *See* petroleum.

Kerr effect The ability of certain substances to refract differently light waves whose vibrations are in two directions (*see* double refraction) when the substance is placed in an electric field. The effect, discovered in 1875 by John Kerr (1824–1907), is caused by the fact that certain molecules have electric *dipoles, which tend to be orientated by the applied field; the normal random motions of the molecules tends to destroy this orientation and the balance is struck by the relative magnitudes of the field

strength, the temperature, and the magnitudes of the dipole moments.

The Kerr effect is observed in a *Kerr cell*, which consists of a glass cell containing the liquid or gaseous substance; two capacitor plates are inserted into the cell and light is passed through it at right angles to the electric field. There are two principal indexes of refraction: n_o (the ordinary index) and n_e (the extraordinary index). The difference in the velocity of propagation in the cell causes a phase difference, δ, between the two waves formed from a beam of monochromatic light, wavelength λ, such that

$$\delta = (n_o - n_e)x/\lambda,$$

where x is the length of the light path in the cell. Kerr also showed empirically that the ratio

$$(n_o - n_e)\lambda = BE^2,$$

where E is the field strength and B is a constant, called the *Kerr constant*, which is characteristic of the substance and approximately inversely proportional to the thermodynamic temperature.

The *Kerr shutter* consists of a Kerr cell filled with a liquid, such as nitrobenzene, placed between two crossed polarizers; the electric field is arranged to be perpendicular to the axis of the light beam and at 45° to the axis of the polarizers. In the absence of a field there is no optical path through the device. When the field is switched on the nitrobenzene becomes doubly refracting and a path opens between the crossed polarizers.

ketals Organic compounds, similar to *acetals, formed by addition of an alcohol to a ketone. If one molecule of ketone (RR′CO) reacts with one molecule of alcohol R″OH, then a *hemiketal* is formed. The rings of ketose sugars are hemiketals. Further reaction produces a full ketal (RR′C(OR″)$_2$).

keto–enol tautomerism A form of tautomerism in which a compound containing a $-CH_2-CO-$ group (the *keto form* of the molecule) is in equilibrium with one containing the $-CH=C(OH)-$ group (the *enol*). It occurs by migration of a hydrogen atom between a carbon atom and the oxygen on an adjacent carbon. *See* isomerism.

keto form *See* keto-enol tautomerism.

ketohexose *See* monosaccharide.

ketone body Any of three compounds, acetoacetic acid (3-oxobutanoic acid, CH_3COCH_2COOH), β-hydroxybutyric acid (3-hydroxybutanoic acid, $CH_3CH(OH)CH_2COOH$), and acetone (propanone, CH_3COCH_3), all of which are produced by the liver as a result of the metabolism of body fat deposits. Ketone bodies are normally used as energy sources by peripheral tissues. However, if carbohydrate supply is limited (e.g. during starvation or in diabetics), the blood level of ketone bodies rises and they may be present in urine, giving it a characteristic 'pear drops' odour. This condition is called *ketosis*.

ketones Organic compounds that contain the group $-CO-$ linked to two hydrocarbon groups. The *ketone group* is a carbonyl group with two single bonds to other carbon atoms. In systematic chemical nomenclature, ketone names end with the suffix *-one*. Examples are propanone (acetone), CH_3COCH_3, and butanone (methyl ethyl ketone), $CH_3COC_2H_5$. Ketones can be made by oxidizing secondary alcohols to convert the $C-OH$ group to $C=O$. Certain ketones form addition compounds with sodium hydrogensulphate(IV) (sodium hydrogensulphite). They also form addition compounds with hydrogen cyanide to give *cyanohydrins* and with alcohols to give *ketals*. They undergo condensation reactions to yield *oximes*, *hydrazones*, phenylhydrazones, and *semicarbazones*. These are reactions that they share with aldehydes. Unlike aldehydes, they do not affect Fehling's solution or Tollen's reagent and do not easily oxidize. Strong oxidizing agents produce a mixture of carboxylic acids; butanone, for example, gives ethanoic and propanoic acids.

ketopentose *See* monosaccharide.

ketose *See* monosaccharide.

The kidneys of a mammal

kidney The main organ of *excretion* of vertebrates, through which nitrogenous waste material (usually in the form of *urine*) is eliminated from the body. In mammals there is a pair of kidneys situated in the abdomen (see illustration). They are made up of tubular units called *nephrons, through which nitrogenous waste is filtered from the blood, with the formation of urine. The nephrons drain into a basin-like cavity in the kidney (the *renal pelvis*), which leads to the *ureter and *bladder.

kieselguhr A soft fine-grained deposit consisting of the siliceous skeletal remains of diatoms, formed in lakes and ponds. Kieselguhr is used as an absorbent, filtering material, filler, and insulator.

kieserite A mineral form of *magnesium sulphate monohydrate, $MgSO_4.H_2O$.

killer cell Either of two types of *lymphocyte that destroy infected or cancerous host cells. *Natural killer cells* are distinct from both T cells and B cells. They can act without needing the presence on the surface of the target cell of foreign antigen combined with host histocompatibility proteins. This contrasts with the other type of killer cells – cytotoxic *T cells – which do require such antigen presentation by the target cell. Killer cells destroy their target cells by releasing proteins (perforins) that create pores through the target cell membrane, thus causing lysis and cell death.

kilo- Symbol k. A prefix used in the metric system to denote 1000 times. For example, 1000 volts = 1 kilovolt (kV).

kilogram Symbol kg. The *SI unit of mass defined as a mass equal to that of the international platinum–iridium prototype kept by the International Bureau of Weights and Measures at Sèvres, near Paris.

kiloton weapon A nuclear weapon with an explosive power equivalent to one thousand tons of TNT. *Compare* megaton weapon.

kilowatt-hour Symbol kWh. The commercial unit of electrical energy. It is equivalent to a power consumption of 1000 watts for 1 hour. 1 kWh = 3.6×10^6 joules.

kimberlite A rare igneous rock that often contains diamonds. It occurs as narrow pipe intrusions but is often altered and fragmented. It consists of olivine and phlogopite mica, usually with calcite, serpentine, and other minerals. The chief occurrences of kimberlite are in South Africa, especially at Kimberley (after which the rock is named), and in the Yakutia area of Siberia.

kinematic equation *See* equation of motion.

kinematics The branch of mechanics concerned with the motions of objects without being concerned with the forces that cause the motion. In this latter respect it differs from *dynamics, which is concerned with the forces that affect motion. *See also* equation of motion.

kinematic viscosity Symbol v. The ratio of the *viscosity of a liquid to its density. The SI unit is $m^2 s^{-1}$.

kinesis The movement of a cell or organism in response to a stimulus in which the rate of movement depends on the intensity (rather than the direction) of the stimulus. For example, a woodlouse moves slowly in a damp atmosphere and quickly in a dry one.

kinetic effect A chemical effect that depends on reaction rate rather than on thermodynamics. For example, diamond is thermodynamically less stable than graphite; its apparent stability depends on the vanishingly slow rate at which it is converted. *Overvoltage in electrolytic cells is another example of a kinetic effect. *Kinetic isotope effects* are changes in reaction rates produced by isotope substitution. For example, if the slow step in a chemical reaction is the breaking of a C–H bond, the rate for the deuterated compound would be slightly lower because of the lower vibrational frequency of the C–D bond. Such effects are used in investigating the mechanisms of chemical reactions.

kinetic energy *See* energy.

kinetics The branch of physical chemistry concerned with measuring and studying the rates of chemical reactions. The main aim of chemical kinetics is to determine the mechanism of reactions by studying the rate under different conditions (temperature, pressure, etc.).

kinetic theory A theory, largely the work of Count Rumford (1753–1814), James Joule (1818–89), and James Clerk Maxwell (1831–79), that explains the physical properties of matter in terms of the motions of its constituent particles. In a gas, for example, the pressure is due to the incessant impacts of the gas molecules on the walls of the container. If it is assumed that the molecules occupy negligible space, exert negligible forces on each other except during collisions, are perfectly elastic, and make only brief collisions with each other, it can be shown that the pressure p exerted by one mole of gas containing n molecules each of mass m in a container of volume V, will be given by:

$$p = nm\bar{c}^2/3V,$$

where $\bar{c}^2$ is the mean square speed of the molecules. As according to the *gas laws for one mole of gas: $pV = RT$, where T is the thermodynamic temperature, and R is the molar *gas constant, it follows that:

$$RT = nm\bar{c}^2/3$$

Thus, the thermodynamic temperature of a gas is proportional to the mean square speed of its molecules. As the average kinetic *energy of translation of the molecules is $m\bar{c}^2/2$, the temperature is given by:

$$T = (m\bar{c}^2/2)(2n/3R)$$

The number of molecules in one mole of any gas is the *Avogadro constant, N_A; therefore in this equation $n = N_A$. The ratio R/N_A is a constant called the *Boltzmann constant (k). The average kinetic energy of translation of the molecules of one mole of any gas is therefore $3kT/2$. For monatomic gases this is proportional to the *internal energy (U) of the gas, i.e.

$$U = N_A 3kT/2$$

and as $k = R/N_A$

$$U = 3RT/2$$

For diatomic and polyatomic gases the rotational and vibrational energies also have to be taken into account (*see* degrees of freedom).

In liquids, according to the kinetic theory, the atoms and molecules still move around at random, the temperature being proportional to their average kinetic energy. However, they are sufficiently close to each other for the attractive forces between molecules to be important. A molecule that approaches the surface will experience a resultant force tending to keep it within the liquid. It is, therefore, only some of the fastest moving molecules that escape; as a result the average kinetic energy of those that fail to escape is reduced. In this way evaporation from the surface of a liquid causes its temperature to fall.

In a crystalline solid the atoms, ions, and molecules are able only to vibrate about the fixed positions of a *crystal lattice; the attractive forces are so strong at this range that no free movement is possible.

kinetochore Two parallel brushlike filaments situated in the *centromere of a chromosome. Kinetochores attach chromatids to the *spindle fibres during nuclear division.

kingdom The highest category into which organisms are classified. Traditionally two kingdoms have been recognized, the Plantae (*see* plant) and the Animalia (*see* animal). More recently a number of other kingdoms have been suggested. For example, the fungi are now considered to be sufficiently distinct from green plants to merit being placed in a separate kingdom, Mycota or Fungi. The differences between *prokaryotes and *eukaryotes have led some to recognize the former as a separate kingdom, Prokaryota. *See also* Protista.

kinin 1. One of a group of peptides, occurring in blood, that are involved in inflammation. Kinins are formed by the enzymatic splitting of blood plasma globulins (*kininogens*) at the site of inflammation. Kinins so far identified include *bradykinin* and *kallidin*. They cause local increases in the permeability of small blood vessels. 2. *See* cytokinin.

kinomere *See* centromere.

kin selection Natural selection of genes that tend to cause the individuals bearing them to be altruistic to close relatives. These relatives therefore have a higher probability of bearing identical copies of those same genes than do other members of the population. Thus kin selection for a gene that tends to cause an animal to share food with a close relative will result in the gene being spread through the population because it (unconsciously) benefits itself. The more closely two animals are related, the higher the probability that they share some identical genes and therefore the more closely their interests coincide. Parental care is a special case of kin selection. *See* inclusive fitness.

Kipp's apparatus A laboratory apparatus for making a gas by the reaction of a solid with a liquid (e.g. the reaction of hydrochloric acid with iron sulphide to give hydrogen sulphide). It consists of three interconnected glass globes arranged vertically, with the solid in the middle globe. The upper and lower globes are connected by a tube and contain the liquid. The middle globe has a tube with a tap for drawing off gas. When the tap is closed, pressure of gas forces the liquid down in the bottom reservoir and up into the top, and reaction does not occur. When the tap is opened, the release in pressure allows the liquid to rise into the middle globe, where it reacts with the solid. It is named after Petrus Kipp (1808–64).

Kirchhoff's law of radiation A law stating that the emissivity of a body is equal to its absorptance at the same temperature.

Kirchhoff's laws Two laws relating to electric circuits, first formulated by G. R. Kirchhoff (1824–87). (a) The current law states that the algebraic sum of the currents flowing through all the wires in a network that meet at a point is zero. (b) The voltage law states that the algebraic sum of the e.m.f.s within any closed circuit is equal to the sum of the products of the currents and the resistances in the various portions of the circuit.

Kjeldahl's method A method for measuring the percentage of nitrogen in an organic compound. The compound is boiled with concentrated sulphuric acid and copper(II) sulphate catalyst to convert any nitrogen to ammonium sulphate. Alkali is added and the mixture heated to distil off ammonia. This is passed into a standard acid solution, and the amount of ammonia can then be found by estimating the amount of unreacted acid by titration. The amount of nitrogen in the original specimen can then be calculated. The method was developed by the Danish chemist Johan Kjeldahl (1849–1900).

klystron An electron tube that generates or amplifies microwaves by *velocity modulation*. Several types are used; in the simple two-cavity klystron a beam of high-energy electrons from an electron gun is passed through a *resonant cavity, where it interacts with high-frequency radio waves. This microwave energy modulates the velocities of the electrons in the beam, which then enters a drift space where the faster electrons overtake the slower ones to form bunches. The bunched beam now has an alternating component, which is transferred to an output cavity and thence to an output waveguide.

knocking The metallic sound produced by a spark-ignition petrol engine under certain conditions. It is caused by rapid combustion of the unburnt explosive mixture in the combustion chambers ahead of the flame front. As the flame travels from the sparking plug towards the piston it compresses and heats the unburnt gases ahead of it. If the flame front moves fast enough, normal combustion occurs and the explosive mixture is ignited progressively by the flame. If it moves too slowly, ignition of the last part of the unburnt gas can occur very rapidly before the flame reaches it, producing a shock wave that travels back and forth across the combustion chamber. The result is overheating, possible damage to the plugs, an undesirable noise, and loss of power (probably due to preignition caused by overheated plugs). Knocking can be avoided by an engine design that increases turbulence in the combustion chamber and thereby increases flame speed. It also can be avoided by reducing the compression ratio, but this involves loss of efficiency. The most effective method is to use high-octane fuel (*see* octane number), which has a longer self-ignition delay than low-octane fuels. This can be achieved by the addition of lead(IV) tetraethyl to the fuel, although lead-free petrol is now preferred to petrol containing lead tetraethyl owing to environmental changes arising from lead in the atmosphere.

Knudsen flow *See* molecular flow.

Kohlrausch's law If a salt is dissolved in water, the conductivity of the (dilute) solution is the sum of two values – one depending on the positive ions and the other on the negative ions. The law, which depends on the independent migration of ions, was deduced experimentally by the German chemist Friedrich Kohlrausch (1840–1910).

Kolbe's method A method of making alkanes by electrolysing a solution of a carboxylic acid salt. For a salt Na^+RCOO^-, the carboxylate ions lose electrons at the cathode to give radicals:

$$RCOO^- - e \rightarrow RCOO\cdot$$

These decompose to give alkyl radicals

$$RCOO\cdot \rightarrow R\cdot + CO_2$$

Two alkyl radicals couple to give an alkane

$$R\cdot + R\cdot \rightarrow RR$$

The method can only be used for hydrocarbons with an even number of carbon atoms, although mixtures of two salts can be electrolysed to give a mixture of three products. The method was discovered by the German chemist Herman Kolbe (1818–84), who electrolysed pentanoic acid (C_4H_9COOH) in 1849 and obtained a hydrocarbon, which he assumed was the substance 'butyl' C_4H_9 (actually octane, C_8H_{18}).

Kovar A trade name for an alloy of iron, cobalt, and nickel with an *expansivity similar to that of glass. It is therefore used in making glass-to-metal seals, especially in circumstances in which a temperature variation can be expected.

Krebs cycle (citric acid cycle; tricarboxylic acid cycle; TCA cycle) A cyclical series of biochemical reactions that is fundamental to the metabolism of aerobic organisms, i.e. animals, plants, and many microorganisms (see illustration). The enzymes of the Krebs cycle are located in the *mitochondria and are in close association with the components of the *electron transport chain. The two-carbon acetyl coenzyme A (acetyl CoA) reacts with the four-carbon oxaloacetate to form the six-carbon citric acid. In a series of seven reactions, this is reconverted to oxaloacetate and produces two molecules of carbon dioxide. Most importantly, the cycle generates one molecule of guanosine triphosphate (GTP – equivalent to 1 ATP) and reduces three molecules of the coenzyme *NAD to NADH and one molecule of the coenzyme *FAD to $FADH_2$. NADH and $FADH_2$ are

Kroll process

Krebs cycle

then oxidized by the electron transport chain to generate three and two molecules of ATP respectively. This gives a net yield of 12 molecules of ATP per molecule of acetyl CoA.

Acetyl CoA can be derived from carbohydrates (via *glycolysis), fats, or certain amino acids. (Other amino acids may enter the cycle at different stages.) Thus the Krebs cycle is the central 'crossroads' in the complex system of metabolic pathways and is involved not only in degradation and energy production but also in the synthesis of biomolecules. It is named after its principal discoverer, Sir Hans Adolf Krebs (1900–81).

Kroll process A process for producing certain metals by reducing the chloride with magnesium metal, e.g.

$$TiCl_4 + 2Mg \rightarrow Ti + 2MgCl_2$$

krypton Symbol Kr. A colourless gaseous element belonging to group 0 (the *noble gases) of the periodic table; a.n. 36; r.a.m. 83.80; d. 3.73 $g\,m^{-3}$; m.p. $-156.6°C$; b.p. $-152.3°C$.

Krypton occurs in the air (0.0001% by volume) from which it can be extracted by fractional distillation of liquid air. Usually, the element is not isolated but is used with other inert gases in fluorescent lamps, etc. The element has five natural isotopes (mass numbers 78, 80, 82, 83, 84) and there are five radioactive isotopes (76, 77, 79, 81, 85). Krypton –85 (half-life 10.76 years) is produced in fission reactors and it has been suggested that an equilibrium amount will eventually occur in the atmosphere. The element is practically inert and forms very few compounds (certain fluorides, such as KrF_2, have been reported).

Kundt's tube An apparatus designed by August Kundt (1839–94) in 1866 to measure the speed of sound in various fluids. It consists of a closed glass tube into which a dry powder (such as lycopodium) has been sprinkled. The source of sound in the original device was a metal rod clamped at its centre with a piston at one end, which is inserted into the tube. When the rod is stroked, sound waves generated by the piston enter the tube. If the position of the piston in the tube is adjusted so that the gas column is a whole number of half wavelengths long, the dust will be disturbed by the resulting *stationary waves forming a series of striations, enabling distances between *nodes to be measured. The vibrating rod can be replaced by a small loudspeaker fed by an oscillator.

Kupfer nickel A naturally occurring form of nickel arsenide, NiAs; an important ore of nickel.

kurchatovium *See* transactinide elements.

L

labelling The process of replacing a stable atom in a compound with a radioisotope of the same element to enable its path through a biological or mechanical system to be traced by the radiation it emits. In some cases a different stable isotope is used and the path is detected by means of a mass spectrometer. A compound containing either a radioactive or stable isotope is called a *labelled compound*. If a hydrogen atom in each molecule of the compound has been replaced by a tritium atom, the compound is called a *tritiated compound*. A radioactive labelled compound will behave chemically and physically in the same way as an otherwise identical stable compound, and its presence can easily be detected using a *Geiger counter. This process of *radioactive tracing* is widely used in chemistry, biology, medicine, and engineering. For example, it can be used to follow the course of the reaction of a car-

boxylic acid with an alcohol to give an ester, e.g.

$$CH_3COOH + C_2H_5OH \rightarrow$$
$$C_2H_5COOCH_3 + H_2O$$

To determine whether the non-carbonyl oxygen in the ester comes from the acid or the alcohol, the reaction is performed with the labelled compound $CH_3CO^{18}OH$, in which the oxygen in the hydroxyl group of the acid has been 'labelled' by using the ^{18}O isotope. It is then found that the water product is $H_2^{18}O$; i.e. the oxygen in the ester comes from the alcohol, not the acid.

labile Describing a chemical compound in which certain atoms or groups can easily be replaced by other atoms or groups. The term is applied to coordination complexes in which ligands can easily be replaced by other ligands in an equilibrium reaction.

labium The lower lip in the *mouthparts of an insect, which is used in feeding and is formed by the fusion of a pair of appendages (the second *maxillae).

labrum The upper lip in the *mouthparts of an insect. It is formed from a plate of cuticle hinged to the head above the mouth and is used in feeding.

labyrinth The system of cavities and tubes that comprises the *inner ear of vertebrates. It consists of a system of membranous structures (*membranous labyrinth*) housed in a similar shaped bony cavity (*bony labyrinth*).

***lac* operon** The *operon that regulates lactose metabolism in the bacterium *Escherichia coli*. Its form was first postulated in 1961 by François Jacob (1920–) and Jacques Monod (1910–76) to explain control of β-galactosidase synthesis, and it is used as a model for the structure of all other operons.

lacrimal gland The tear gland, present in the eyelids of some amino acid → γ lactam (γ butyrolactam) ⇌ γ lactim

Lactam formation

vertebrates. The fluid (tears) produced by this gland cleanses and lubricates the exposed surface of the eye; it drains into the nose through the lacrimal duct.

lactams Organic compounds containing a ring of atoms in which the group –NH.CO.– forms part of the ring. Lactams can be formed by reaction of an –NH₂ group in one part of a molecule with a –COOH group in the other to give a cyclic amide (see illustration). They can exist in an alternative tautomeric form, the *lactim* form, in which the hydrogen atom on the nitrogen has migrated to the oxygen of the carbonyl to give –N=C(OH)–. The pyrimidine base uracil is an example of a lactam.

lactate A salt or ester of lactic acid (i.e. a 2-hydroxypropanoate).

lactation The discharge of milk from the *mammary glands. This generally only occurs after birth of the young and is stimulated by the sucking action of the infants. Lactation is under the control of hormones, notably *prolactin and *oxytocin.

lacteal A minute blind-ended lymph vessel that occurs in each *villus of the small intestine. Digested fats are absorbed into the lacteals (*see* chyle) and transported to the bloodstream through the *thoracic duct.

lactic acid (2-hydroxypropanoic acid) A clear odourless hygroscopic syrupy liquid, CH₃CH(OH)COOH, with a sour taste; r.d. 1.206; m.p. 18°C; b.p. 122°C. It is prepared by the hydrolysis of ethanal cyanohydrin or the oxidation of propan-1,2-diol

using dilute nitric acid. Lactic acid is manufactured by the fermentation of lactose (from milk) and used in the dyeing and tanning industries. It is an alpha hydroxy *carboxylic acid. *See also* optical activity.

Lactic acid is produced from pyruvic acid in active muscle tissue when oxygen is limited and subsequently removed for conversion to glucose by the liver. During strenuous exercise it may build up in the muscles, causing cramplike pains. It is also produced by fermentation in certain bacteria and is characteristic of sour milk.

lactims *See* lactams.

lactogenic hormone *See* prolactin.

lactones Organic compounds containing a ring of atoms in which the group –CO.O– forms part of the ring. Lactones can be formed (or regarded as formed) by reaction of an –OH group in one part of a molecule with a –COOH group in the other to give a cyclic ester (see illustration). This type of reaction occurs with γ-hydroxy carboxylic acids such as CH₂(OH)CH₂CH₂COOH (in which

hydroxy carboxylic acid → γ lactone (γ butyrolactone)

Lactone formation

the hydroxyl group is on the third carbon from the carboxyl group). The resulting γ-lactone has a five-membered ring. Similarly, δ-lactones have six-membered rings. β-lactones, with a four-membered ring, are not produced directly from β-hydroxy acids, but can be synthesized by other means.

lactose (milk sugar) A sugar comprising one glucose molecule linked to a galactose molecule. Lactose is manufactured by the mammary gland and occurs only in milk. For example, cows' milk contains about 4.7% lactose. It is less sweet than sucrose (cane sugar).

lacuna A gap or cavity in the tissues of an organism; for example, the hollow centre of certain plant stems or any of the small cavities in bone in which the bone-forming cells are found.

Ladenburg benzene An (erroneous) structure for *benzene proposed by Albert Ladenburg (1842–1911), in which the six carbon atoms were arranged at the corners of a triangular prism and linked by single bonds to each other and to the six hydrogen atoms.

laevorotatory Designating a chemical compound that rotates the plane of plane-polarized light to the left (anticlockwise for someone facing the oncoming radiation). *See* optical activity.

laevulose *See* fructose.

lag 1. The time delay between a specified maximum value of one wave or vibration and the maximum value of another wave or vibration. A *lagging current* is an alternating current that reaches its maximum at a later instant in the cycle than the e.m.f. producing it. **2.** This time delay expressed as an angle.

Lagrangian Symbol *L*. A function used to define a dynamical system in terms of functions of coordinates, velocities, and times given by:

$$L = T - V$$

where *T* is the kinetic energy of the system and *V* is the potential energy of the system. The Lagrangian formulation of dynamics has the advantage that it does not deal with many vector quantities, such as forces and accelerations, but only with two scalar functions, *T* and *V*. This leads to great simplifications. *Lagrangian dynamics* was formulated by J. L. Lagrange (1736–1813).

LAH Lithium aluminium hydride; *see* lithium tetrahydroaluminate(III).

lake A pigment made by combining an organic dyestuff with an inorganic compound (usually an oxide, hydroxide, or salt). Absorption of the organic compound on the inorganic substrate yields a coloured complex, as in the combination of a dyestuff with a *mordant. Lakes are used in paints and printing inks.

Lamarckism One of the earliest superficially plausible theories of *evolution, proposed by the French biologist Jean-Baptiste de Lamarck (1744–1829) in 1809. He suggested that changes in an individual are acquired during its lifetime, chiefly by increased use or disuse of organs in response to "a need that continues to make itself felt", and that these changes are inherited by its offspring. Thus the long neck and limbs of a giraffe are explained as having evolved by the animal stretching its neck to browse on the foliage of trees. This so-called inheritance of acquired characteristics has never unquestionably been demonstrated to occur and the theory was largely displaced by *Darwinism. Lamarckism is also incompatible with the *Central Dogma of molecular biology. *See also* Lysenkoism.

lambda point Symbol λ. The temperature of 2.186 K below which helium–4 becomes a superfluid. The name derives from the shape of the curve of specific heat capacity against temperature, which is shaped like a Greek letter lambda (λ) at this point. *See* superfluidity.

lambert A former unit of *luminance equal to the luminance of a uniformly diffusing surface that emits or reflects one lumen per square centimetre. It is approximately equal to 3.18×10^3 Cd m^{-2}. It is named after Johann H. Lambert (1728–77).

Lambert's laws (1) The *illuminance of a surface illuminated by light falling on it perpendicularly from a point source is inversely proportional to the square of the distance between the surface and the source. (2) If the rays make an angle θ with the normal to the surface, the illuminance is proportional to $\cos\theta$. (3) (Also called *Bouguer's law*) The *luminous intensity (I) of light (or other electromagnetic radiation) decreases exponentially with the distance d that it enters an absorbing medium, i.e.

$$I = I_0 \exp(-\alpha d)$$

where I_0 is the intensity of the radiation that enters the medium and α is its *linear absorption coefficient*. These laws were first stated (for light) by Johann H. Lambert (1728–77).

Lamb shift A small energy difference between two levels ($^2S_{1/2}$ and $^2P_{1/2}$) in the *hydrogen spectrum. The shift results from the quantum interaction between the atomic electron and the electromagnetic radiation. It was first explained by Willis Eugene Lamb (1913–).

lamella 1. (in botany) **a.** Any of the paired folds of membranes seen between the *grana in a plant chloroplast. **b.** Any of the spore-bearing gills on the underside of the cap of many mushrooms and toadstools. *See also* middle lamella. **2.** (in zoology) Any of various thin layers of membranes, especially any of the thin layers of tissue of which compact bone is formed.

lamellar solids Solid substances in which the crystal structure has distinct layers (i.e. has a layer lattice). The *micas are an example of this type of compound. *Intercalation compounds* are lamellar compounds formed by interposition of atoms, ions, etc., between the layers of an existing element or compound. For example, graphite is a lamellar solid. With strong oxidizing agents (e.g. a mixture of concentrated sulphuric and nitric acids) it forms a nonstoichiometric 'graphitic oxide', which is an intercalation compound having oxygen atoms between the layers of carbon atoms.

Lamellibranchia A class of aquatic molluscs (the bivalves) that include the oysters, mussels, and clams. They are characterized by a laterally flattened body and a shell consisting of two hinged shells (i.e. a bivalved shell). The enlarged gills are covered with cilia and have the additional function of filtering microscopic food particles from the water flowing over them. Bivalves live on the sea bed or lake bottom and are sedentary, so the head and foot are reduced.

lamina 1. The thin and usually flattened blade of a leaf, in which photosynthesis and transpiration occurs. The bulk of the lamina is made up of *mesophyll cells interspersed by a network of veins (*vascular bundles). The mesophyll is enclosed by a protective epidermis that produces a waxy cuticle. **2.** The leaflike part of the thallus of certain algae, notably kelps. *See also* stipe.

laminar flow *Streamline flow of a fluid in which the fluid moves in layers without fluctuations or turbulence so that successive particles passing the same point have the same velocity. It occurs at low *Reynolds numbers, i.e. low velocities, high viscosities, low densities or small dimensions. The flow of lubricating oil in bearings is normally laminar because of the thinness of the lubricant layer.

laminated core A core for a transformer or other electrical machine in which the ferromagnetic alloy is made into thin sheets (laminations), which are oxidized or varnished to provide a relatively high resistance between them. This has the effect of reducing *eddy currents, which occur when alternating currents are used.

lamp black A finely divided (microcrystalline) form of carbon made by burning organic compounds in insufficient oxygen. It is used as a black pigment and filler.

lanolin An emulsion of purified wool fat in water, containing cholesterol and certain terpene alcohols and esters. It is used in cosmetics.

lansfordite A mineral form of *magnesium carbonate pentahydrate, $MgCO_3.5H_2O$.

lanthanides *See* lanthanoids.

lanthanoid contraction *See* lanthanoids.

lanthanoids (lanthanides; lanthanons; rare-earth elements) A series of elements in the *periodic table, generally considered to range in proton number from cerium (58) to lutetium (71) inclusive. The lanthanoids all have two outer s-electrons (a $6s^2$ configuration), follow lanthanum, and are classified together because an increasing proton number corresponds to increase in number of $4f$ electrons. In fact, the $4f$ and $5d$ levels are close in energy and the filling is not smooth. The outer electron configurations are as follows:

57 lanthanum (La) $5d^16s^2$
58 cerium (Ce) $4f5d^16s^2$ (or $4f^26s^2$)
59 praseodymium (Pr) $4f^36s^2$
60 neodymium (Nd) $4f^46s^2$
61 promethium (Pm) $4f^56s^2$
62 samarium (Sm) $4f^66s^2$
63 europium (Eu) $4f^76s^2$
64 gadolinium (Gd) $4f^75d^16s^2$
65 terbium (Tb) $4f^96s^2$

66 dysprosium (Dy) $4f^{10}6s^2$
67 holmium (Ho) $4f^{11}6s^2$
68 erbium (Er) $4f^{12}6s^2$
69 thulium (Tm) $4f^{13}6s^2$
70 ytterbium (Yb) $4f^{14}6s^2$
71 lutetium (Lu) $4f^{14}5d^16s^2$

Note that lanthanum itself does not have a $4f$ electron but it is generally classified with the lanthanoids because of its chemical similarities, as are yttrium (Yt) and scandium (Sc). Scandium, yttrium, and lanthanum are d-block elements; the lanthanoids and *actinoids make up the f-block.

The lanthanoids are sometimes simply called the *rare earths*, although strictly the 'earths' are their oxides. Nor are they particularly rare: they occur widely, usually together. All are silvery very reactive metals. The f-electrons do not penetrate to the outer part of the atom and there is no f-orbital participation in bonding (unlike the d-orbitals of the main *transition elements) and the elements form very few coordination compounds. The main compounds contain M^{3+} ions. Cerium also has the highly oxidizing Ce^{4+} state and europium and ytterbium have a M^{2+} state.

The $4f$ orbitals in the atoms are not very effective in shielding the outer electrons from the nuclear charge. In going across the series the increasing nuclear charge causes a contraction in the radius of the M^{3+} ion – from 0.1061 nm in lanthanum to 0.0848 nm in lutetium. This effect, the *lanthanoid contraction*, accounts for the similarity between the transition elements zirconium and hafnium.

lanthanons *See* lanthanoids.

lanthanum Symbol La. A silvery metallic element belonging to group IIIB of the periodic table and often considered to be one of the *lanthanoids; a.n. 57; r.a.m. 138.91; r.d. 6.146 (20°C); m.p. 918°C; b.p. 3464°C. Its principal ore is bastnasite, from which it is separated by an ion-exchange process. There are two natural isotopes, lanthanum–139 (stable) and lanthanum–138 (half-life 10^{10}–10^{15} years). The metal, being pyrophoric, is used in alloys for lighter flints and the oxide is used in some optical glasses. The largest use of lanthanum, however, is as a catalyst in cracking crude oil. Its chemistry resembles that of the lanthanoids. The element was discovered by C. G. Mosander in 1839.

lapis lazuli A blue rock that is widely used as a semiprecious stone and for ornamental purposes. It is composed chiefly of the deep blue mineral *lazurite* embedded in a matrix of white calcite and usually also contains small specks of pyrite. It occurs in only a few places in crystalline limestones as a contact metamorphic mineral. The chief source is Afghanistan; lapis lazuli also occurs near Lake Baikal in Siberia and in Chile. It was formerly used to make the artists' pigment ultramarine.

Laplace equation The partial differential equation:
$$\partial^2 u/\partial x^2 + \partial^2 u/\partial y^2 + \partial^2 u/\partial z^2 = 0$$
It may also be written in the form $\nabla^2 u = 0$, where ∇^2 is called the *Laplace operator*. It was formulated by the French mathematician P. S. Laplace (1749–1827).

large intestine The portion of the alimentary canal of vertebrates between the *small intestine and the *anus. It consists of the *caecum, *colon, and *rectum and its principal function is the absorption of water and formation of faeces.

Larmor precession A precession of the motion of charged particles in a magnetic field. It was first deduced in 1897 by Sir Joseph Larmor (1857–1942). Applied to the orbital motion of an electron around the nucleus of an atom in a magnetic field of flux density B, the frequency of precession is given by $eB/4\pi m v\mu$, where e and m are the electronic charge and mass respectively, μ is the permeability,

and v is the velocity of the electron. This is known as the *Larmor frequency*.

larva The juvenile stage in the life cycle of most invertebrates, amphibians, and fish, which hatches from the egg, is unlike the adult in form, and is usually incapable of sexual reproduction (*see* paedogenesis). It develops into the adult by undergoing *metamorphosis. Larvae can feed themselves and are otherwise self-supporting. Examples are the tadpoles of frogs, the caterpillars of butterflies, and the ciliated planktonic larvae of many marine animals. *Compare* nymph.

larynx The anterior portion of the *trachea (windpipe) of tetrapod vertebrates, which in amphibians, reptiles, and mammals contains the *vocal cords. Movement of the cartilage in the walls of the larynx (by means of the laryngeal muscles) alters the tension of the vocal cords. This changes the pitch of the sound emitted by the vocal cords when they vibrate. The final voiced sound is modified by resonance within the oral and nasal cavities.

laser (*l*ight *a*mplification by *s*timulated *e*mission of *r*adiation) A light amplifier usually used to produce monochromatic coherent radiation in the infrared, visible, and ultraviolet regions of the *electromagnetic spectrum. Lasers that operate in the X-ray region of the spectrum are also being developed.

Nonlaser light sources emit radiation in all directions as a result of the spontaneous emission of photons by thermally excited solids (filament lamps) or electronically excited atoms, ions, or molecules (fluorescent lamps, etc.). The emission accompanies the spontaneous return of the excited species to the *ground state and occurs randomly, i.e. the radiation is not coherent. In a laser, the atoms, ions, or molecules are first 'pumped' to an excited state and then stimulated to emit photons by collision of a photon of the same energy. This is called *stimulated emission*. In order to use it, it is first necessary to create a condition in the amplifying medium, called *population inversion*, in which the majority of the relevant entities are excited. Random emission from one entity can then trigger coherent emission from the others that it passes. In this way amplification is achieved.

The laser amplifier is converted to an oscillator by enclosing the amplifying medium within a resonator. Radiation then introduced along the axis of the resonator is reflected back and forth along its path by a mirror at one end and by a partially transmitting mirror at the other end. Between the mirrors the waves are amplified by stimulated emission. The radiation emerges through the semitransparent mirror at one end as a powerful coherent monochromatic parallel beam of light. The emitted beam is uniquely parallel because waves that do not bounce back and forth between the mirrors quickly escape through the sides of the oscillating medium without amplification.

Some lasers are solid, others are liquid or gas devices. Population inversion can be achieved by *optical pumping* with flashlights or with other lasers. It can also be achieved by such methods as chemical reactions, discharges in gases, and recombination emission in semiconducting materials (*see* recombination process).

Lasers have found many uses since their invention in 1960, including laser welding, surgery, *holography, printing, optical communications, and the reading of digital information. In chemistry, their main use has been in the study of photochemical reactions and in the spectroscopic investigation of molecules.

latent heat Symbol L. The quantity of heat absorbed or released when a substance changes its physical phase at constant temperature (e.g. from solid to liquid at the melting point or from liquid to gas at the boiling point). For example, the latent heat of vaporization is the energy a substance absorbs from its surroundings in order to overcome the attractive forces between its molecules as it changes from a liquid to a gas and in order to do work against the external atmosphere as it expands. In thermodynamic terms the latent heat is the *enthalpy of evaporation (ΔH), i.e. $L = \Delta H = \Delta U + p\Delta V$, where ΔU is the change in the internal energy, p is the pressure, and ΔV is the change in volume.

The *specific latent heat* (symbol l) is the heat absorbed or released per unit mass of a substance in the course of its isothermal change of phase. The *molar latent heat* is the heat absorbed or released per unit amount of substance during an isothermal change of state.

lateral inversion (perversion) The type of reversal that occurs with an image formed by a plane mirror. A person with a mole on his left cheek sees an image in a plane mirror of a person with a mole on his right cheek. Since, however, that is (correctly) to the observer's left, the real reversal is of front and back; the image is 'turned through' itself to face the object – hence the alternative name.

lateral velocity The component of a celestial body's velocity that is at 90° to its *line-of-sight velocity.

laterite A layer of deposits composed largely of hydroxides of iron and aluminium formed from the weathering of rocks in humid tropical and subtropical climates. Laterites range from soft earthy materials to hard dense rocks. It hardens on exposure to the atmosphere and is used as a building material. In the weathering process silica, alkalis, and alkaline earths are

latex

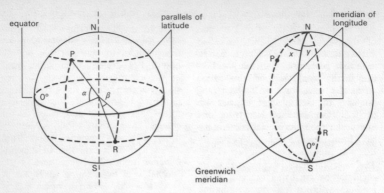

The latitude of P is given by the angle α. In this case it would be $\alpha°$ N. The latitude of R is $\beta°$ S.

The longitude of P is given by the angle x. In this case it would be $x°$ W. R has a longitude of $y°$ E.

Latitude and longitude

removed leaving behind concentrations of iron and aluminium oxides; this process is known as *laterization*.

latex A milky fluid of mixed composition found in some herbaceous plants and trees. Its function is not clear but it may assist in protecting wounds (*compare* gum) and it may be involved in the nutrition of the plant. The latex of some species, notably rubber trees, is collected for commercial purposes.

latitude and longitude 1. (in geography) Imaginary lines on the earth's surface, enabling any point to be defined in terms of two angles subtended at its centre (see illustration). *Parallels of latitude* are circles drawn round the earth parallel to the equator; their diameters diminish as they approach the poles. These parallels are specified by the angle subtended at the centre of the earth by the arc formed between a point on the parallel and the equator. All points on the equator therefore have a latitude of 0°, while the north pole has a latitude of 90°N and the south pole of 90°S. Parallels of latitude 1° apart are separated on the earth's surface by about 100 km.

Meridians of longitude are half *great circles passing through both poles; they cross parallels of latitude at right angles. In 1884 the meridian through Greenwich, near London, was selected as the prime meridian and designated as 0°. Other meridians are defined by the angle between the plane of the meridian and the plane of the prime meridian specifying whether it is E or W of the prime meridian. At the equator meridians 1° apart are separated by about 112 km.

2. (in astronomy) The *celestial latitude* of a star, or other celestial body, is its angular distance north (taken as positive) or south (taken as negative) of the ecliptic measured along the great circle through the body and the poles of the ecliptic. The *celestial longitude* is the angular distance from the vernal equinox measured eastwards along the ecliptic to the intersection of the body's circle of longitude; it is measured in the same direction as the sun's apparent annual motion.

lattice The regular arrangement of atoms, ions, or molecules in a crystalline solid. *See* crystal lattice.

lattice energy A measure of the stability of a *crystal lattice, given by the energy that would be released per mole if atoms, ions, or molecules of the crystal were brought together from infinite distances apart to form the lattice. *See* Born–Haber cycle.

lattice vibrations The periodic vibrations of the atoms, ions, or molecules in a *crystal lattice about their mean positions. On heating, the amplitude of the vibrations increases until they are so energetic that the lattice breaks down. The temperature at which this happens is the melting point of the solid and the substance becomes a liquid. On cooling, the amplitude of the vibrations diminishes. At *absolute zero a residual vibration persists, associated with the *zero-point energy of the substance. The increase in the electrical resistance of a conductor is due to increased scattering of the free conduction electrons by the vibrating lattice particles.

latus rectum *See* ellipse; hyperbola; parabola.

laughing gas *See* dinitrogen oxide.

launch vehicle A rocket used to launch a satellite, spaceprobe, space station, etc. Multistage rockets are usually used, the empty tanks and engine of the first two stages being jettisoned before the desired orbit is reached. The *launch window* is the time interval during which the vehicle must be launched to achieve the orbit.

lauric acid *See* dodecanoic acid.

law of chemical equilibrium *See* equilibrium constant.

law of conservation of energy *See* conservation law.

law of conservation of mass *See* conservation law.

law of constant composition *See* chemical combination.

law of definite proportions *See* chemical combination.

law of mass action *See* mass action.

law of multiple proportions *See* chemical combination.

law of octaves (Newlands' law) An attempt at classifying elements made by J. A. R. Newlands (1837–98) in 1863. He arranged 56 elements in order of increasing atomic mass in groups of eight, pointing out that each element resembled the element eight places from it in the list. He drew an analogy with the notes of a musical scale. *Newlands' octaves* were groups of similar elements distinguished in this way: e.g. oxygen and sulphur; nitrogen and phosphorus; and fluorine, chlorine, bromine, and iodine. In some cases it was necessary to put two elements in the same position. The proposal was rejected at the time. *See* periodic table.

law of reciprocal proportions *See* chemical combination.

lawrencium Symbol Lr. A radioactive metallic transuranic element belonging to the *actinoids; a.n. 103; mass number of the first discovered isotope 257 (half-life 8 seconds). A number of very short-lived isotopes have now been synthesized. The element was identified by A. Ghiorso and associates in 1961. The alternative name *unniltrium* has been proposed.

laws of chemical combination *See* chemical combination.

Lawson criterion A condition for the release of energy from a *thermonuclear reactor first laid down by J. D. Lawson. It is usually stated as the minimum value for the product of the density (n_G) of the fusion-fuel particles and the *containment time (τ) for energy breakeven, i.e. it is a measure of the density of the reacting particles required and the time for which they need to react in order to produce more energy than was used in raising the temperature of the reacting particles to the *ignition temperature. For a 50:50 mixture of deuterium and tritium at the ignition temperature, the value of $n_G\tau$ is between 10^{14} and 10^{15} cm^{-3} s.

layer lattice A crystal structure in which the atoms are chemically bonded in plane layers, with relatively weak forces between atoms in adjacent layers. Graphite and micas are examples of substances having layer lattices (i.e. they are *lamellar solids).

lazurite See lapis lazuli.

LCP See liquid-crystal polymer.

LD$_{50}$ Lethal dose 50, or median lethal dose: the amount of a pharmacological or toxic substance (such as ionizing radiation) that causes death in 50% of a group of experimental animals. For each LD$_{50}$ the species and weight of the animal and the route of administration of the substance is specified. LD$_{50}$s are used both in toxicology and in the *bioassay of therapeutic compounds.

L-dopa See dopa.

L–D process See basic-oxygen process.

leaching Extraction of soluble components of a solid mixture by percolating a solvent through it.

lead Symbol Pb. A heavy dull grey soft ductile metallic element belonging to *group IV of the periodic table; a.n. 82; r.a.m. 207.19; r.d. 11.35; m.p. 327.5°C; b.p. 1740°C. The main ore is the sulphide galena (PbS); other minor sources include anglesite (PbSO$_4$), cerussite (PbCO$_3$), and litharge (PbO). The metal is extracted by roasting the ore to give the oxide, followed by reduction with carbon. Silver is also recovered from the ores. Lead has a variety of uses including building construction, lead-plate accumulators, bullets, and shot, and is a constituent of such alloys as solder, pewter, bearing metals, type metals, and fusible alloys. Chemically, it forms compounds with the +2 and +4 oxidation states, the lead(II) state being the more stable.

lead(II) acetate See lead(II) ethanoate.

lead–acid accumulator An accumulator in which the electrodes are made of lead and the electrolyte consists of dilute sulphuric acid. The electrodes are usually cast from a lead alloy containing 7–12% of antimony (to give increased hardness and corrosion resistance) and a small amount of tin (for better casting properties). The electrodes are coated with a paste of lead(II) oxide (PbO) and finely divided lead; after insertion into the electrolyte a 'forming' current is passed through the cell to convert the PbO on the negative plate into a sponge of finely divided lead. On the positive plate the PbO is converted to lead(IV) oxide (PbO$_2$). The equation for the overall reaction during discharge is:
$$PbO_2 + 2H_2SO_4 + Pb \rightarrow$$
$$2PbSO_4 + 2H_2O$$
The reaction is reversed during charging. Each cell gives an e.m.f. of about 2 volts and in motor vehicles a 12-volt battery of six cells is usually used. The lead–acid battery produces 80–120 kJ per kilogram. *Compare* nickel–iron accumulator.

lead(II) carbonate A white solid, PbCO$_3$, insoluble in water; rhombic; r.d. 6.6. It occurs as the mineral *cerussite, which is isomorphous with aragonite and may be prepared in the laboratory by the addition of cold ammonium carbonate solution to a cold solution of a lead(II) salt (acetate or nitrate). It decomposes at 315°C to lead(II) oxide and carbon dioxide.

lead(II) carbonate hydroxide (white lead; basic lead carbonate) A powder, 2PbCO$_3$.Pb(OH)$_2$, insoluble in water, slightly soluble in aqueous carbonate solutions; r.d. 6.14; decomposes at 400°C. Lead(II) carbonate hydroxide occurs as the mineral *hydroxycerussite* (of variable composition). It was previously manufactured from lead in processes using spent tanning bark or horse manure, which released carbon dioxide. It is currently made by electrolysis of mixed solutions (e.g. ammonium nitrate, nitric acid, sulphuric acid, and acetic acid) using lead anodes. For the highest grade product the lead must be exceptionally pure (known in the trade as 'corroding lead') as small amounts of metallic impurity impart grey or pink discolorations. The material was used widely in paints, both for art work and for commerce, but it has the disadvantage of reacting with hydrogen sulphide in industrial atmospheres and producing black lead sulphide. The poisonous nature of lead compounds has also contributed to the declining importance of this material.

lead-chamber process An obsolete method of making sulphuric acid by the catalytic oxidation of sulphur dioxide with air using a potassium nitrate catalyst in water. The process was carried out in lead containers (which was expensive) and only produced dilute acid. It was replaced in 1876 by the *contact process.

lead dioxide *See* lead(IV) oxide.

lead equivalent A measure of the absorbing power of a radiation screen, expressed as the thickness of a lead screen in millimetres that would afford the same protection as the material being considered.

lead(II) ethanoate (lead(II) acetate) A white crystalline solid, $Pb(CH_3COO)_2$, soluble in water and slightly soluble in ethanol. It exists as the anhydrous compound (r.d. 3.25; m.p. 280°C), as a trihydrate, $Pb(CH_3COO)_2.3H_2O$ (monoclinic; r.d. 2.55; loses water at 75°C), and as a decahydrate, $Pb(CH_3COO)_2.10H_2O$

(rhombic; r.d. 1.69). The common form is the trihydrate. Its chief interest stems from the fact that it is soluble in water and it also forms a variety of complexes in solution. It was once known as *sugar of lead* because of its sweet taste.

lead(IV) ethanoate (lead tetra-acetate) A colourless solid, $Pb(CH_3COO)_4$, which decomposes in water and is soluble in pure ethanoic acid; monoclinic; r.d. 2.228; m.p. 175°C. It may be prepared by dissolving dilead(II) lead(IV) oxide in warm ethanoic acid. In solution it behaves essentially as a covalent compound (no measurable conductivity) in contrast to the lead(II) salt, which is a weak electrolyte.

lead(IV) hydride *See* plumbane.

lead monoxide *See* lead(II) oxide.

lead(II) oxide (lead monoxide) A solid yellow compound, PbO, which is insoluble in water; m.p. 886°C. It exists in two crystalline forms: *litharge* (tetrahedral; r.d. 9.53) and *massicot* (rhombic; r.d. 8.0). It can be prepared by heating the nitrate, and is manufactured by heating molten lead in air. If the temperature used is lower than the melting point of the oxide, the product is massicot; above this, litharge is formed. Variations in the temperature and in the rate of cooling give rise to crystal vacancies and red, orange, and brown forms of litharge can be produced. The oxide is amphoteric, dissolving in acids to give lead(II) salts and in alkalis to give *plumbates.

lead(IV) oxide (lead dioxide) A dark brown or black solid with a rutile lattice, PbO_2, which is insoluble in water and slightly soluble in concentrated sulphuric and nitric acids; r.d. 9.375; decomposes at 290°C. Lead(IV) oxide may be prepared by the oxidation of lead(II) oxide by heating with alkaline chlorates or nitrates, or by anodic oxidation of lead(II) solutions. It is an oxidizing agent and readily reverts to the lead(II) oxidation state, as illustrated by its conversion to Pb_3O_4 and PbO on heating. It reacts with hydrochloric acid to evolve chlorine. Lead(IV) oxide has been used in the manufacture of safety matches and was widely used until the mid-1970s as an adsorbent for sulphur dioxide in pollution monitoring.

lead(II) sulphate A white crystalline solid, $PbSO_4$, which is virtually insoluble in water and soluble in solutions of ammonium salts; r.d. 6.2; m.p. 1170°C. It occurs as the mineral *anglesite*; it may be prepared in the laboratory by adding any solution containing sulphate ions to solutions of lead(II) ethanoate. The material known as *basic lead(II) sulphate* may be made by shaking together lead(II) sulphate and lead(II) hydroxide in water. This material has been used in white paint in preference to lead(II) carbonate hydroxide, as it is not so susceptible to discoloration through reaction with hydrogen sulphide. The toxicity of lead compounds has led to a decline in the use of these compounds.

lead(II) sulphide A black crystalline solid, PbS, which is insoluble in water; r.d. 7.5; m.p. 1114°C. It occurs naturally as the metallic-looking mineral *galena (the principal ore of lead). It may be prepared in the laboratory by the reaction of hydrogen sulphide with soluble lead(II) salts. Lead(II) sulphide has been used as an electrical rectifier.

lead tetra-acetate *See* lead(IV) ethanoate.

lead(IV) tetraethyl (tetraethyl lead) A colourless liquid, $Pb(C_2H_5)_4$, insoluble in water, soluble in benzene, ethanol, ether, and petroleum; r.d. 1.659; m.p. −137°C; b.p. 200°C. It may be prepared by the reaction of hydrogen and ethene with lead but a more convenient laboratory and industrial method is the reaction of a

sodium–lead alloy with chloroethane. A more recent industrial process is the electrolysis of ethylmagnesium chloride (the Grignard reagent) using a lead anode and slowly running additional chloroethane onto the cathode. Lead tetraethyl is used in fuel for internal-combustion engines (along with 1,2-dibromoethane) to increase the *octane number and reduce knocking. The use of lead(IV) tetraethyl in petrol results in the emission of hazardous lead compounds into the atmosphere. Pressure from environmental groups has encouraged a reduction in the use of lead(IV) tetraethyl and an increasing use of lead-free petrol.

leaf A flattened structure that develops from a superficial group of tissues, the leaf buttress, on the side of the stem apex. Each leaf has a lateral bud in its axil. Leaves are arranged in a definite pattern (*see* phyllotaxis) and usually show limited growth. Each consists of a broad flat *lamina (leaf blade) and a leaf base, which attaches the leaf to the stem; a leaf stalk (*petiole*) may also be present. The leaves of bryophytes are simple appendages, which are not homologous with the leaves of vascular plants as they develop on the gametophyte generation.

Leaves show considerable variation in size, shape, arrangement of veins, type of attachment to the stem, and texture. They may be *simple* or divided into *leaflets*, i.e. *compound* (see illustration). Types of leaf include: *cotyledons (seed leaves); *scale leaves*, which lack chlorophyll and develop on rhizomes or protect the inner leaves of a bud; *foliage leaves*, which are the main organs for photosynthesis and transpiration; and *bracts and *floral leaves*, such as sepals, petals, stamens, and carpels, which are specialized for reproduction.

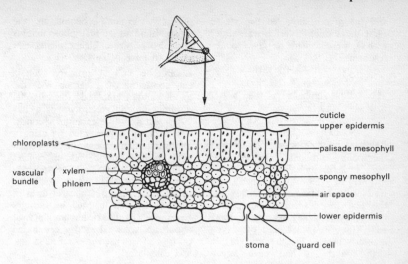

chloroplasts

vascular bundle { xylem phloem

cuticle
upper epidermis
palisade mesophyll
spongy mesophyll
air space
lower epidermis

stoma   guard cell

Transverse section through a leaf blade

leaflets

lamina (leaf blade)

petiole

petiole

Simple leaves   Compound leaves

Leaves may be modified for special purposes. For example the leaf bases of bulbs are swollen with food to survive the winter. In some plants leaves are reduced to spines for protection and their photosynthetic function is carried out by another organ, such as a *cladode.

leaf buttress *See* primordium.

learning A process by which an animal's response to a particular situation may be permanently altered, usually in a beneficial way, as a result of its experience. Learning allows an animal to respond more flexibly to the situations it encounters: learning

abilities in different species vary widely and are adapted to the species' environment. Numerous different categories of learning have been proposed, but there is no general agreement on how many different processes are involved. These categories include *habituation, associative learning (through *conditioning), and insight learning (which may include *imprinting).

Leblanc process An obsolete process for manufacturing sodium carbonate. The raw materials were sodium chloride, sulphuric acid, coke, and limestone (calcium carbonate), and the process involved two stages. First the sodium chloride was heated with sulphuric acid to give sodium sulphate:

$$2NaCl(s) + H_2SO_4(l) \rightarrow$$
$$Na_2SO_4(s) + 2HCl(g)$$

The sodium sulphate was then heated with coke and limestone:

$$Na_2SO_4 + 2C + CaCO_3 \rightarrow$$
$$Na_2CO_3 + CaS + 2CO_2$$

Calcium sulphide was a by-product, the sodium carbonate being extracted by crystallization. The process, invented in 1783 by the French chemist Nicolas Leblanc (1742–1806), was the first for producing sodium carbonate synthetically (earlier methods were from wood ash and other vegetable sources). By the end of the 19th century it had been largely replaced by the *Solvay process.

lechatelierite A mineral form of *silicon(IV) oxide, SiO_2.

Le Chatelier's principle If a system is in equilibrium, any change imposed on the system tends to shift the equilibrium to nullify the effect of the applied change. The principle, which is a consequence of the law of conservation of energy, was first stated in 1888 by Henri Le Chatelier (1850–1936). It is applied to chemical equilibria. For example, in the gas reaction

$$2SO_2 + O_2 \leftrightharpoons 2SO_3$$

an increase in pressure on the reaction mixture displaces the equilibrium to the right, since this reduces the total number of molecules present and thus decreases the pressure. The standard enthalpy change for the forward reaction is negative (i.e. the reaction is exothermic). Thus, an increase in temperature displaces the equilibrium to the left since this tends to reduce the temperature. The *equilibrium constant thus falls with increasing temperature.

lecithin Any of a group of *phospholipids containing the amino alcohol *choline esterified to the phosphate group. Lecithins are the most abundant animal phospholipids and also occur in higher plants, but rarely in microorganisms.

Leclanché cell A primary *voltaic cell consisting of a carbon rod (the anode) and a zinc rod (the cathode) dipping into an electrolyte of a 10–20% solution of ammonium chloride. *Polarization is prevented by using a mixture of manganese dioxide mixed with crushed carbon, held in contact with the anode by means of a porous bag or pot; this reacts with the hydrogen produced. This wet form of the cell, devised in 1867 by Georges Leclanché (1839–82), has an e.m.f. of about 1.5 volts. The *dry cell based on it is widely used in torches, radios, and calculators.

LED See light-emitting diode.

leeches See Hirudinea.

LEED Low-energy electron diffraction. See electron diffraction.

legume (pod) A dry fruit formed from a single carpel and containing one or more seeds, which are shed when mature. It is the characteristic fruit of the Leguminosae (pea family). It splits, often explosively, along both sides and the two halves of the fruit move apart to expose the seeds. A special form of the legume is the *lomentum.

Leishman's stain A neutral stain for blood smears devised by the British surgeon W. B. Leishman (1865–1926). It consists of a mixture of *eosin (an acidic stain), and *methylene blue (a basic stain) in alcohol and is usually diluted and buffered before use. It stains the different components of blood in a range of shades between red and blue. The similar *Wright's stain* is favoured by American workers.

lens 1. (in physics) A curved, ground, and polished piece of glass, moulded plastic, or other transparent material used for the refraction of light. A

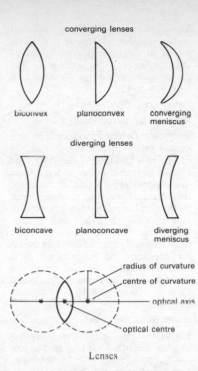

converging lenses

biconvex   planoconvex   converging meniscus

diverging lenses

biconcave   planoconcave   diverging meniscus

radius of curvature

centre of curvature

optical axis

optical centre

Lenses

converging lens is one that brings the rays of a parallel beam of light to a real *principal focus. They include biconvex, planoconvex, and converging meniscus lenses. *Diverging lenses* cause the rays of a parallel beam to diverge as if from a virtual principal focus; these include the biconcave, planoconcave, and diverging meniscus lenses. See illustrations.

The *centre of curvature* of a lens face is the centre of the sphere of which the surface of the lens is a part. The *optical axis* is the line joining the two centres of curvature of a lens or, in the case of a lens with one plane surface, the line through one centre of curvature that is normal to the plane surface. The *optical centre* of a lens is the point within a lens on the optical axis through which any rays entering the lens pass without deviation. The

distance between the optical centre and the principal focus of a lens is called the *focal length* (*f*). The distance (*v*) between the lens and the image it forms is related to the distance (*u*) between the lens and the object by:

$$1/v + 1/u = 1/f,$$

provided that the *real-is-positive convention is used. This takes distances to real objects, images, and foci as positive; those to virtual objects, images, and foci as negative. The equation does not always apply if the alternative New Cartesian convention (*see* sign convention) is used.

2. (in anatomy) A transparent biconvex structure in the eyes or analogous organs of many animals, responsible for directing light onto light-sensitive cells. In vertebrates it is a flexible structure centred behind the iris and attached to the *ciliary body. In terrestrial species its main function is to focus images onto the retina. To focus on near objects, the circular muscles in the ciliary body contract and the lens becomes more convex; contraction of the radial muscles in the ciliary body flattens the lens for focusing on distant objects (*see also* accommodation).

lenticel Any of the raised pores in the stems of woody plants that allow gas exchange between the atmosphere and the internal tissues. The pore is formed by the *cork cambium, which, at certain points, produces a loose bulky form of cork that pushes through the outer tissues to create the lenticel.

Lenz's law An induced electric current always flows in such a direction that it opposes the change producing it. This law, first stated by Heinrich Lenz (1804–65) in 1835, is essentially a form of the law of conservation of energy.

Lepidoptera An order of insects comprising the butterflies and moths, found mainly in tropical regions.

Adults possess two pairs of membranous wings, often brightly coloured and usually coupled together. The wings, body, and legs are covered with minute scales. Adult mouthparts are generally modified to form a long proboscis for sucking nectar, fruit juices, etc. Butterflies are typically small-bodied, active during daylight, and rest with their wings folded vertically; moths have larger bodies, are nocturnal, and rest with their wings in various positions. The larvae (caterpillars) have a prominent head and a segmented wormlike body, most segments bearing a pair of legs. They chew leaves and stems, sometimes causing considerable damage to crop plants. The larvae undergo metamorphosis via a *pupa (chrysalis) to the adult form. In some groups, the pupa is enclosed in a cocoon of silk derived from silk glands (modified salivary glands); others use leaves, etc. to build a cocoon.

lepton Any of a class of *elementary particles that consists of the *electron, muon, tau particle, and three types of *neutrino (one associated with each of the other types of lepton). For each lepton there is an equivalent antiparticle. The antileptons have a charge opposite that of the leptons; the antineutrinos, like the neutrinos, have no charge. The electron, muon, and tau particle all have a charge of -1. These three particles differ from each other only in mass: the muon is 200 times more massive than the electron and the tau particle is 3500 times more massive than the electron. Leptons interact by the electromagnetic interaction and the weak interaction (*see* fundamental interactions).

leptotene The beginning of the first prophase of *meiosis, when the chromatids can be seen and *pairing begins.

Leslie's cube A metal box in the shape of a cube in which each of the four vertical sides have different surface finishes. When hot water is placed in the cube, the emissivity of the finishes can be compared. The device was first used by Sir John Leslie (1766–1832).

lethal dose 50 *See* LD$_{50}$.

leucine *See* amino acid.

leucocyte (white blood cell) A colourless cell with a nucleus, found in blood and lymph. Leucocytes are formed in lymph nodes and red bone marrow and are capable of amoeboid movement. They can produce *antibodies and move through the walls of vessels to migrate to the sites of injuries, where they surround and isolate dead tissue, foreign bodies, and bacteria. There are two major types: those without granules in the cytoplasm (*lymphocytes and *monocytes) and those with granular cytoplasm (*granulocytes*).

leuco form *See* dyes.

leucoplast Any of various organelles in plant cells that contain no pigment and are therefore colourless. Leucoplasts are usually found in tissues not normally exposed to light and frequently contain reserves of starch, protein, or oil. *See* plastid. *Compare* chromoplast.

level An instrument used in *surveying to determine heights. It usually consists of a telescope and attached spirit level mounted on a tripod. The level is set up between a point of known height and a point for which the height is required. Before use it is adjusted until the line of sight is exactly horizontal. Sightings are then made onto a graduated levelling staff at the two points. The difference in elevation between the two points can then be calculated from the readings taken at these points.

lever A simple machine consisting of a rigid bar pivoted about a fulcrum. The mechanical advantage or *force ratio of a lever (the ratio of load to

first-order lever

effort

load

second-order lever

load

effort

effort

third-order lever

load

Levers

effort) is equal to the ratio of the perpendicular distance of the line of action of the effort from the fulcrum to the perpendicular distance of the line of action of the load from the fulcrum. In a first-order lever the fulcrum comes between load and effort. In a second-order lever the load comes between the fulcrum and the effort. In a third-order lever the effort comes between the fulcrum and the load. See illustrations.

Lewis acid and base *See* acid.

Leyden jar An early form of *capacitor consisting of a glass jar with a layer of metal foil on the outside and a similar layer on the inside. Contact to the inner foil is by means of a loose chain hanging inside the jar. It was invented in the Dutch town of Leyden in about 1745.

LF *See* low frequency.

LH *See* luteinizing hormone.

libration The phenomenon that enables 59% of the moon's surface to be observed from earth over a 30-year period, in spite of its *synchronous rotation. *Physical libration* arises from slight variations in the rotation of the moon on its axis, caused by minor distortions in its physical shape. *Geometric librations* are apparent oscillations arising from the fact that the moon is observed from slightly different directions at different times. The geometric *libration in longitude* results from the nonuniform orbital motion of the moon. The geometric *libration in latitude* arises because the moon's axis of rotation is not perpendicular to its orbital plane; it enables more of the lunar polar regions to be observed.

lice *See* Mallophaga (bird lice); Siphunculata (sucking lice).

lichens A group of organisms that are symbiotic associations (*see* symbiosis) between a fungus (usually one of the *ascomycetes) and a green alga or a blue-green alga. The fungus usually makes up most of the plant body and the cells of the alga are distributed within it. The alga photosynthesizes and passes most of its food to the fungus and the fungus protects the algal cells. The lichen reproduces by means of fungal spores, which must find a suitable alga on germination. Lichens are slow growing but can live in regions that are too cold or exposed for other plants. They may form a flattened crust or be erect and branching. Many grow as *epiphytes, especially on tree trunks. Some species are very sensitive to air pollution and have been used as *indicator species. Formerly placed together in the taxon Lichenes, lichens are now variously classified according to their fungal constituents.

Liebig condenser A laboratory condenser having a straight glass tube surrounded by a coaxial glass jacket through which cooling water is passed. It is named after the German organic chemist Justus von Liebig (1803–73).

life cycle The complete sequence of events undergone by organisms of a particular species from the fusion of gametes in one generation to the same stage in the following generation. In most animals gametes are formed by *meiosis of germ cells in the reproductive organs of the parents. The zygote, formed by the fusion of two gametes, eventually develops into an organism essentially similar to the parents. In plants, however, the products of meiosis are spores, which develop into plants (the *gametophyte generation) often very different in form from the spore-forming (*sporophyte) generation. The sporophyte generation is restored when gametes, formed by the gametophyte generation, fuse. See alternation of generations.

ligament A resilient but flexible band of tissue (chiefly *collagen) that holds two or more bones together at a movable *joint. Ligaments restrain the movement of bones at a joint and are therefore important in preventing dislocation.

ligand An ion or molecule that donates a pair of electrons to a metal atom or ion in forming a coordination *complex. Molecules that function as ligands are acting as Lewis bases (see acid). For example, in the complex hexaquocopper(II) ion $[Cu(H_2O)_6]^{2+}$ six water molecules coordinate to a central Cu^{2+} ion. In the tetrachloroplatinate(II) ion $[PtCl_4]^{2-}$ four Cl^- ions are coordinated to a central Pt^{2+} ion. A feature of such ligands is that they have lone pairs of electrons, which they donate to empty metal orbitals. A certain class of ligands also have empty p- or d-orbitals in addition to their lone pair of electrons and can produce complexes in which the metal has low oxidation state. A double bond is formed between the metal and the ligand: a sigma bond by donation of the lone pair from ligand to metal, and a pi bond by back donation of electrons on the metal to empty d-orbitals on the ligand. Carbon monoxide is the most important such ligand, forming metal carbonyls (e.g. $Ni(CO)_4$).

The examples given above are examples of monodentate ligands (literally: 'having one tooth'), in which there is only one point on each ligand at which coordination can occur. Some ligands are polydentate; i.e. they have two or more possible coordination points. For instance, 1,2-diaminoethane, $H_2NC_2H_4NH_2$, is a bidentate ligand, having two coordination points. Certain polydentate ligands can form *chelates.

ligand-field theory A modification of *crystal-field theory in which the overlap of orbitals is taken into account. See also complex.

ligase Any of a class of enzymes that catalyse the formation of covalent bonds using the energy released by the cleavage of ATP. Ligases are important in the synthesis and repair of many biological molecules, including DNA, and are used in genetic engineering to insert foreign DNA into cloning *vectors.

light The form of *electromagnetic radiation to which the human eye is sensitive and on which our visual awareness of the universe and its contents relies (see colour).

The finite velocity of light was suspected by many early experimenters in optics, but it was not established until 1676 when Ole Rømer (1644–1710) measured it. Sir Isaac Newton (1642–1727) investigated the optical *spectrum and used existing knowledge to establish a primarily corpuscular theory of light, in which it was regarded as a stream of particles that set up disturbances in the 'aether' of space. His successors adopted the corpuscles but ignored the wavelike disturbances until Thomas Young (1773–1829) rediscovered the *interference of light in 1801 and showed that a wave theory was essential to

ohmic contact

p-type layer of gallium arsenide phosphide

n-type layer of gallium arsenide phosphide

n-type gallium arsenide substrate

ohmic contact

Light emitting diode

interpret this type of phenomenon. This view was accepted for most of the 19th century and it enabled James Clerk Maxwell (1831–79) to show that light forms part of the *electromagnetic spectrum. He believed that waves of electromagnetic radiation required a special medium to travel through, and revived the name 'luminiferous ether' for such a medium. The *Michelson–Morley experiment in 1887 showed that, if the medium existed, it could not be detected; it is now generally accepted that the ether is an unnecessary hypothesis. In 1905 Albert Einstein (1879–1955) showed that the *photoelectric effect could only be explained on the assumption that light consists of a stream of discrete *photons of electromagnetic energy. This renewed conflict between the corpuscular and wave theories has gradually been resolved by the evolution of the *quantum theory and *wave mechanics. While it is not easy to construct a model that has both wave and particle characteristics, it is accepted, according to Bohr's theory of *complementarity, that in some experiments light will appear wavelike, while in others it will appear to be corpuscular. During the course of the evolution of wave mechanics it has also become evident that electrons and other elementary particles have dual wave and particle properties.

light bulb *See* electric lighting.

light-emitting diode (LED) A rectifying *semiconductor device that converts electrical energy into light or infrared radiation in the range 550 nm (green light) to 1300 nm (infrared radiation). The most commonly used LED (see illustration) emits red light and consists of gallium arsenide–phosphide on a gallium arsenide substrate, light being emitted at a p–n junction, when electrons and holes recombine (*see* recombination process). LEDs are extensively used for displaying letters and numbers in digital instruments in which a self-luminous display is required.

light green *See* fast green.

lightning A high-energy luminous electrical discharge that passes between a charged cloud and a point on the surface of the earth, between two charged clouds, or between oppositely charged layers of the same cloud. In general, the upper parts of clouds are positively charged and the lower parts are negatively charged; the reasons for this separation of charge are complex. Lightning usually occurs in the form of a downward step leader followed by an intensely luminous return stroke, which can produce instantaneous temperatures as high as 30 000°C. In the typical step leader a surge of electrons descends in approximately 50-metre steps with about 50-microsecond pauses between steps. When this leader reaches the earth a surge of charge returns up the preionized path taken by the leader. Cloud-to-cloud strokes also involve a leader and return stroke. The average current in a lightning stroke is about 10 000 amperes, but maximum currents in the return stroke can reach 20 000 A.

light reactions *See* photosynthesis.

light year A unit of distance used in astronomy; the distance travelled by light in a vacuum during one year. It is equal to 9.4650×10^{15} metres or 5.8785×10^{12} miles.

lignin A complex organic polymer that is deposited within the cellulose of plant cell walls during secondary thickening. Lignification makes the walls woody and therefore rigid.

lignite *See* coal.

ligule **1.** A membranous scalelike outgrowth from the leaves of certain flowering plants. Many grasses have a ligule at the base of the leaf blade. **2.** A small membranous structure that develops on the upper surface of a young leaf base in certain pteridophytes, for example *Selaginella*. It withers as the plant matures. **3.** A strap-shaped extension from the corolla tube in certain florets of a *capitulum. They are termed *ligulate* (or *ray*) *florets*.

lime *See* calcium oxide.

limestone A sedimentary rock that is composed largely of carbonate minerals, especially carbonates of calcium and magnesium. *Calcite and *aragonite are the chief minerals; *dolomite is also present in the dolomitic limestones. There are many varieties of limestones but most are deposited in

shallow water. *Organic limestones* (e.g. *chalk) are formed from the calcareous skeletons of organisms; *precipitated limestones* include oolite, which is composed of ooliths – spherical bodies formed by the precipitation of carbonate around a nucleus; and *clastic limestones* are derived from fragments of pre-existing calcareous rocks.

limewater A saturated solution of *calcium hydroxide in water. When carbon dioxide gas is bubbled through limewater, a 'milky' precipitate of calcium carbonate is formed:

$$Ca(OH)_2(aq) + CO_2(g) \rightarrow$$
$$CaCO_3(s) + H_2O(l)$$

If the carbon dioxide continues to be bubbled through, the calcium carbonate eventually redissolves to form a clear solution of calcium hydrogencarbonate:

$$CaCO_3(s) + CO_2(g) + H_2O(g) \rightarrow$$
$$Ca(HCO_3)_2(aq)$$

If cold limewater is used the original calcium carbonate precipitated has a calcite structure; hot limewater yields an aragonite structure.

liming The application of lime (calcium hydroxide) to soils to increase levels of calcium and decrease acidity.

limit The value that a function approaches as the independent variable approaches a specified value.

limit cycle *See* attractor.

limiting friction The friction force that just balances a moving force applied to a solid body resting on a solid surface when the body fails to move. If the moving force exceeds the limiting friction, the body will begin to move.

limnology The study of the physical and biological characteristics of lakes and other bodies of fresh water.

limonite A generic term for a group of hydrous iron oxides, mostly amorphous. *Goethite and *haematite are important constituents, together with colloidal silica, clays, and manganese oxides. Limonite is formed by direct precipitation from marine or fresh water in shallow seas, lagoons, and bogs (thus it is often called *bog iron ore*) and by oxidation of iron-rich minerals. It is used as an ore of iron and as a pigment.

linac *See* linear accelerator.

Linde process A process for the *liquefaction of gases by the Joule–Thomson effect. In this process, devised by Carl von Linde (1842–1934) for liquefying air, the air is freed of carbon dioxide and water and compressed to 150 atmospheres. The compressed gas is passed through a copper coil to an expansion nozzle within a Dewar flask. The emerging air is cooled by the Joule–Thomson effect as it expands and then passes back within a second copper coil that surrounds the first coil. Thus the expanded gas cools the incoming gas in a process that is said to be *regenerative*. Eventually the air is reduced to its *critical temperature and, at the pressure of 150 atmospheres (well above its critical pressure), liquefies. The process is also used for other gases, especially hydrogen and helium. Hydrogen has first to be cooled below its inversion temperature (*see* Joule–Thomson effect) using liquid air; helium has first to be cooled below its inversion temperature using liquid hydrogen.

linear absorption coefficient *See* Lambert's laws.

linear accelerator (linac) A type of particle *accelerator in which charged particles are accelerated in a straight line, either by a steady electric field or by means of radio-frequency electric fields. In the latter variety the passage of the particles is synchronized with the phase of the accelerating field by means of either travelling-wave acceleration or standing-wave acceleration. In the first of these the particles are carried forward by the electric component of the field,

the phase velocity of which is equal to the particle velocity. In the standing-wave type, forward and backward waves are superimposed so that the particles are alternately repelled by cylindrical electrodes through which they have passed and attracted by the cylindrical electrodes they are approaching. At room temperatures megawatt radio-frequency powers are required but in recent superconducting linacs only a few watts are needed.

linear energy transfer (LET) The energy transferred per unit path length by a moving high-energy charged particle (such as an electron or a proton) to the atoms and molecules along its path. It is of particular importance when the particles pass through living tissue as the LET modifies the effect of a specific dose of radiation. LET is proportional to the square of the charge on the particle and increases as the velocity of the particle decreases.

linear equation An equation between two variables that gives a straight line when plotted on a graph. It has the general form $y = mx + c$, where m is the gradient of the line and c is the intercept of the line on the y-axis (in Cartesian coordinates).

linear expansivity *See* expansivity.

linear molecule A molecule in which the atoms are in a straight line, as in carbon dioxide, $O=C=O$.

linear momentum *See* momentum.

linear motor A form of induction motor in which the stator and armature are linear and parallel, rather than cylindrical and coaxial. In some experimental trains the magnetic force between the primary winding in the vehicle and the secondary winding on the ground support the vehicle on an air cushion thus eliminating track friction. However, because of the high cost of the installation and the low efficiency the device has not yet found commercial application.

line defect *See* defect.

line-of-sight velocity (radial velocity) The component of a celestial body's velocity along the line of sight of the observer. It is usually given in relation to the sun to avoid complications arising from the earth's orbital motion. Line-of-sight velocity is normally calculated from the *Doppler effect on the body's spectrum, a *redshift indicating a receding body (taken as a positive velocity) and a blueshift indicating an approaching body (taken as negative).

line printer An output device used with computers. It prints a line of characters at a time, typically operating at 200 to 3000 lines per minute.

lines of force Imaginary lines in a *field of force that enable the direction and strength of the field to be visualized. They are used primarily in electric and magnetic fields; in electric fields they are sometimes called *tubes of force*, to express their characteristic of being perpendicular to a conducting surface. The tangent to a line of force at any point gives the direction of the field at that point and the number of lines per unit area perpendicular to the force represents the *intensity of the field.

line spectrum *See* spectrum.

linkage The association of a number of characteristics that are inherited together because the genes responsible for them are situated on the same chromosome. Linkage can be broken by *crossing over during mitosis, when sections of chromosomes are exchanged and new combinations of genes are produced. *See also* sex linkage.

linkage map *See* chromosome map.

Linnaean system *See* binomial nomenclature.

Linnz–Donnewitz process *See* basic-oxygen process.

linoleic acid A liquid polyunsaturated *fatty acid with two double bonds, $CH_3(CH_2)_4CH:CHCH_2CH:CH-(CH_2)_7COOH$. Linoleic acid is abundant in many plant fats and oils, e.g. linseed oil, groundnut oil, and soyabean oil. It is an *essential fatty acid.

linolenic acid A liquid polyunsaturated *fatty acid with three double bonds in its structure: $CH_3CH_2CH:CHCH_2CH:CHCH_2CH:CH(CH_2)_7-COOH$. Linolenic acid occurs in certain plant oils, e.g. linseed and soyabean oil, and in algae. It is one of the *essential fatty acids.

linseed oil A pale yellow oil pressed from flax seed. It contains a mixture of glycerides of fatty acids, including linoleic acid and linolenic acid. It is a *drying oil, used in oil paints, varnishes, linoleum, etc.

lipase An enzyme secreted by the pancreas and the glands of the small intestine of vertebrates that catalyses the breakdown of fats into fatty acids and glycerol.

lipid Any of a diverse group of organic compounds, occurring in living organisms, that are insoluble in water but soluble in organic solvents, such as chloroform, benzene, etc. Lipids are broadly classified into two categories: *complex lipids*, which are esters of long-chain fatty acids and include the *glycerides (which constitute the *fats and *oils of animals and plants), glycolipids, *phospholipids, and *waxes; and *simple lipids*, which do not contain fatty acids and include the *steroids and *terpenes.

Lipids have a variety of functions in living organisms. Fats and oils are a convenient and concentrated means of storing food energy in plants and animals. Phospholipids and *sterols, such as cholesterol, are major components of cell membranes. Waxes provide vital waterproofing for body surfaces. Terpenes include vitamins A, E, and K, and phytol (a component of chlorophyll) and occur in essential oils, such as menthol and camphor. Steroids include the adrenal hormones, sex hormones, and bile acids. Lipids can combine with proteins to form *lipoproteins, e.g. in cell membranes. In bacterial cell walls, lipids may associate with polysaccharides to form *lipopolysaccharides*.

lipoic acid A vitamin of the *vitamin B complex. It is one of the *coenzymes involved in the decarboxylation of pyruvate by the enzyme pyruvate dehydrogenase. This reaction has to take place before carbohydrates can enter the *Krebs cycle during aerobic respiration. Good sources of lipoic acid include liver and yeast.

lipolysis The breakdown of storage lipids in living organisms. Most long-term energy reserves are in the form of triglycerides in fats and oils. When these are needed, e.g. during starvation, lipase enzymes convert the triglycerides into glycerol and the component fatty acids. These are then transported to tissues and oxidized to provide energy.

lipoprotein One of a group of compounds consisting of a lipid combined with a protein. Lipoproteins are the main structural materials of the membranes of cells and cell organelles. They also occur in blood and lymph, being the form in which lipids are transported in these media. *Cholesterol is transported in the bloodstream in the form of *low-density lipoproteins (LDLs)* and is removed by means of LDL receptors in cell membranes; the LDLs are bound to the receptors, which are then taken into the cells. Lack of LDL receptors, occurring as a genetic defect in some individuals, is believed to be a cause of high levels of cholesterol in the blood, predisposing to atherosclerosis.

liposome A microscopic spherical membrane-enclosed vesicle or sac (20 – 30 nm in diameter) made artificially in the laboratory by the addition of an aqueous solution to a phospholipid gel. The membrane resembles a cell membrane and the whole vesicle is similar to a cell organelle. Liposomes can be incorporated into living cells and are used to transport relatively toxic drugs into diseased cells, where they can exert their maximum effects. For example, liposomes containing the drug methotrexate, used in the treatment of cancer, can be injected into the patient's blood. The cancerous organ is heated to a temperature higher than body temperature, so that when the liposome passes through its blood vessels, the membrane melts and the drug is released. The study of the behaviour of liposome membranes is used in research into membrane function, particularly to observe the behaviour of membranes during anaesthesia with respect to permeability changes.

lipotrophin Either of two peptide hormones produced in the anterior pituitary gland that trigger the mobilization of fat deposits and the transfer of lipid components to the bloodstream. β-lipotrophin is formed by cleavage of the precursor pro-opiocortin, and is itself cleaved to form γ-lipotrophin and β-endorphin.

liquation The separation of mixtures of solids by heating to a temperature at which lower-melting components liquefy.

liquefaction of gases The conversion of a gaseous substance into a liquid. This is usually achieved by one of four methods or by a combination of two of them:
(1) by vapour compression, provided that the substance is below its *critical temperature;
(2) by refrigeration at constant pressure, typically by cooling it with a colder fluid in a countercurrent heat exchanger;
(3) by making it perform work adiabatically against the atmosphere in a reversible cycle;

(4) by the *Joule–Thomson effect (*see also* Linde process).

Large quantities of liquefied gases are now used commercially, especially *liquefied petroleum gas and liquefied natural gas.

liquefied petroleum gas (LPG) Various petroleum gases, principally propane and butane, stored as a liquid under pressure. It is used as an engine fuel and has the advantage of causing very little cylinder-head deposits.

Liquefied natural gas (LNG) is a similar product and consists mainly of methane. However, it cannot be liquefied simply by pressure as it has a low critical temperature of 190 K and must therefore be cooled to below this temperature before it will liquefy. Once liquefied it has to be stored in well-insulated containers. It provides a convenient form in which to ship natural gas in bulk from oil wells to users. It is also used as an engine fuel.

liquid A phase of matter between that of a crystalline solid and a *gas. In a liquid, the large-scale three-dimensional atomic (or ionic or molecular) regularity of the solid is absent but, on the other hand, so is the total disorganization of the gas. Although liquids have been studied for many years there is still no comprehensive theory of the liquid state. It is clear, however, from diffraction studies that there is a short-range structural regularity extending over several molecular diameters. These bundles of ordered atoms, molecules, or ions move about in relation to each other, enabling liquids to have almost fixed volumes, which adopt the shape of their containers.

liquid crystal A substance that flows like a liquid but has some order in its arrangement of molecules. *Nematic crystals* have long molecules all aligned in the same direction, but otherwise randomly arranged. *Choles-teric* and *smectic* liquid crystals also have aligned molecules, which are arranged in distinct layers. In cholesteric crystals, the axes of the molecules are parallel to the plane of the layers; in smectic crystals they are perpendicular.

liquid-crystal display A digital display unit used in watches, calculators, etc. It provides a source of clearly displayed digits for a very low power consumption. In the display unit a thin film of *liquid crystal is sandwiched between two transparent electrodes (glass with a thin metal or oxide coating). In the commonly used field-effect display, twisted nematic crystals are used. The nematic liquid crystal cell is placed between two crossed polarizers. Polarized light entering the cell follows the twist of the nematic liquid crystal, is rotated through 90°, and can therefore pass through the second polarizer. When an electric field is applied the molecular alignment in the liquid crystal is altered, the polarization of the entering light is unchanged, and no light is therefore transmitted. In these circumstances, a mirror placed behind the second polarizer will cause the display to appear black. One of the electrodes, shaped in the form of a digit, will then provide a black digit when the voltage is applied.

liquid-crystal polymer A polymer with a liquid-crystal structure, this being the most thermodynamically stable. Liquid-crystal polymers contain long rigid chains and combine strength with lightness. They are, however, difficult to produce commercially.

liquid-drop model A model of the atomic nucleus in which the nucleons are regarded as being analogous to the molecules in a liquid, the interactions between which maintain the droplet shape by surface tension. The model has been useful in the theory of nuclear fission.

***l*-isomer** *See* optical activity.

L-isomer *See* absolute configuration.

Lissajous figures A curve in one plane traced by a point moving under the influence of two independent harmonic motions. In the common case the harmonic motions are simple, perpendicular to each other, and have a simple frequency ratio. They can be displayed by applying sinusoidal alternating potentials to the X- and Y-inputs of a *cathode-ray oscilloscope.

Listeria A genus of rod-shaped aerobic motile Gram-positive bacteria. Only one species, *L. monocytogenes*, causes disease (*listeriosis*). It is resistant to physical and chemical treatments and can occur as a contaminant in certain foods, in faeces, etc. Listeriosis can take various forms, depending on the site of infection: localization in the central nervous system causes meningoencephalitis, while uterine infection can result in abortion or congenital handicap in the fetus.

litharge *See* lead(II) oxide.

lithia *See* lithium oxide.

lithium Symbol Li. A soft silvery metal, the first member of *group I of the periodic table; a.n. 3; r.a.m. 6.939; r.d. 0.534; m.p. 180.54°C; b.p. 1340°C. It is a rare element found in spodumene (LiAlSi$_2$O$_6$), petalite (LiAlSi$_4$O$_{10}$), the mica lepidolite, and certain brines. It is usually extracted by treatment with sulphuric acid to give the sulphate, which is converted to the chloride. This is mixed with a small amount of potassium chloride, melted, and electrolysed. The stable isotopes are lithium–6 and lithium–7. Lithium–5 and lithium–8 are short-lived radioisotopes. The metal is used to remove oxygen in metallurgy and as a constituent of some Al and Mg alloys. It is also used in batteries and is a potential tritium source for fusion research. Lithium salts are used in psychomedicine. The element reacts with oxygen and water; on heating it also reacts with nitrogen and hydrogen. Its chemistry differs somewhat from that of the other group I elements because of the small size of the Li$^+$ ion.

lithium aluminium hydride *See* lithium tetrahydroaluminate(III).

lithium carbonate A white solid, Li$_2$CO$_3$; r.d. 2.11; m.p. 735°C; decomposes above 1200°C. It is produced commercially by treating the ore with sulphuric acid at 250°C and leaching the product to give a solution of lithium sulphate. The carbonate is then obtained by precipitation with sodium carbonate solution. Lithium carbonate is used in the prevention and treatment of manic-depressive disorders. It is also used industrially in ceramic glazes.

lithium deuteride *See* lithium hydride.

lithium hydride A white solid, LiH; cubic; r.d. 0.82; m.p. 686°C; decomposes at about 850°C. It is produced by direct combination of the elements at temperatures above 500°C. The bonding in lithium hydride is believed to be largely ionic; i.e. Li$^+$H$^-$ as supported by the fact that hydrogen is released from the anode on electrolysis of the molten salt. The compound reacts violently and exothermically with water to yield hydrogen and lithium hydroxide. It is used as a reducing agent to prepare other hydrides and the ^{2}H isotopic compound, *lithium deuteride*, is particularly valuable for deuterating a range of organic compounds. Lithium hydride has also been used as a shielding material for thermal neutrons.

lithium hydrogencarbonate A compound, LiHCO$_3$, formed by the reaction of carbon dioxide with aqueous lithium carbonate and known only in solution. It has found medicinal uses similar to those of lithium carbonate and is sometimes included in proprietary mineral waters.

lithium hydroxide A white crystalline solid, LiOH, soluble in water, slightly soluble in ethanol and insoluble in ether. It is known as the monohydrate (monoclinic; r.d. 1.51) and in the anhydrous form (tetragonal, r.d. 1.46; m.p. 450°C; decomposes at 924°C). The compound is made by reacting lime with lithium salts or lithium ores. Lithium hydroxide is basic but has a closer resemblance to group II hydroxides than to the other group I hydroxides (an example of the first member of a periodic group having atypical properties).

lithium oxide (lithia) A white crystalline compound, Li_2O; cubic; r.d. 2.01; m.p. 1700°C. It can be obtained from a number of lithium ores; the main uses are in lubricating greases, ceramics, glass and refractories, alkaline storage batteries, and as a flux in brazing and welding.

lithium sulphate A white or colourless crystalline material, Li_2SO_4, soluble in water and insoluble in ethanol. It forms a monohydrate (monoclinic; r.d. 1.88) and an anhydrous form, which exists in α- (monoclinic), β- (hexagonal) and γ- (cubic) forms; r.d. 2.23. The compound is prepared by the reaction of the hydroxide or carbonate with sulphuric acid. It is not isomorphous with other group I sulphates and does not form alums.

lithium tetrahydroaluminate(III) (lithium aluminium hydride; LAH) A white or light grey powder, $LiAlH_4$; r.d. 0.917; decomposes at 125°C. It is prepared by the reaction of excess lithium hydride with aluminium chloride. The compound is soluble in ethoxyethane, reacts violently with water to release hydrogen, and is widely used as a powerful reducing agent in organic chemistry. It should always be treated as a serious fire risk in storage.

lithosphere The earth's crust (*see* earth). Sometimes the lithosphere is also understood to include the mantle and sometimes the mantle and the core.

litmus A water-soluble dye extracted from certain lichens. It turns red under acid conditions and blue under alkaline conditions, the colour change occurring over the pH range 4.5–8.3 (at 25°C). It is not suitable for titrations because of the wide range over which the colour changes, but is used as a rough *indicator of acidity or alkalinity, both in solution and as litmus paper (absorbent paper soaked in litmus solution).

litre Symbol l or L. A unit of volume in the metric system regarded as a special name for the cubic decimetre. It was formerly defined as the volume of 1 kilogram of pure water at 4°C at standard pressure, which is equivalent to $1.000\ 028\ dm^3$.

littoral Designating or occurring in the marginal shallow-water zone of a sea or lake, especially (in the sea) between high and low tide lines. In this zone enough light penetrates to the bottom to support rooted aquatic plants. *Compare* profundal; sublittoral.

liver A large lobed organ in the abdomen of vertebrates that plays an essential role in many metabolic processes by regulating the composition and concentration of nutrients and toxic materials in the blood. It receives the products of digestion dissolved in the blood via the *hepatic portal vein and its most important functions are to convert excess glucose to the storage product *glycogen, which serves as a food reserve; to break down excess amino acids to ammonia, which is converted to *urea and excreted via the kidneys; and to store and break down fats (*see* lipolysis). Other functions of the liver are (1) the production of *bile; (2) the breakdown (detoxification) of poisonous substances in the blood; (3) the removal of damaged red blood cells; (4) the synthesis of vitamin A and the blood-clotting substances prothrombin and fibrinogen; and (5) the storage of iron.

liverworts *See* Hepaticae.

living fossil Any organism whose closest relatives are extinct and that was once itself thought to be extinct. An example is the coelacanth, a primitive fish that was common in the Devonian era, the first recent living specimen of which was discovered in 1938.

lixiviation The separation of mixtures by dissolving soluble constituents in water.

lizards *See* Squamata.

Lloyd's mirror An optical arrangement for producing interference fringes. A slit is illuminated by monochromatic light and placed close to a plane mirror. Interference occurs between direct light from the slit and light reflected from the mirror. It was first used by Humphrey Lloyd (1800–81) in 1834.

loaded concrete Concrete containing elements (such as iron or lead) with a high mass number; it is used in making the radiation shield around nuclear reactors.

Local Group The group of *galaxies of which our own Galaxy is a member. It consists of some 25 galaxies, the most massive of which are the Galaxy and the Andromeda galaxy.

localized bond A *chemical bond in which the electrons forming the bond remain between (or close to) the linked atoms. *Compare* delocalization.

local oscillator An *oscillator in a *heterodyne or *superheterodyne radio receiver. It supplies the radio-frequency signal that beats with the incoming signal to produce the intermediate frequency.

loculus A small cavity in a plant or animal body. In plants the loculus of the ovary is the cavity containing the ovules and the loculi of the anther contain the developing pollen grains.

locus 1. (in mathematics) A set of points whose location is specified by an equation. For example, if a point moves so that the sum of its distances from two fixed points is constant, the locus of the point is an *ellipse. **2.** (in genetics) The position of a gene on a chromosome. The two alleles of a gene occupy the same locus on *homologous chromosomes.

lodestone *See* magnetite.

logarithm The power to which a number, called the *base*, has to be raised to give another number. Any number y can be written in the form $y = x^n$. n is then the logarithm to the base x of y, i.e. $n = \log_x y$. If the base is 10, the logarithms are called *common logarithms*. *Natural* (or *Napierian*) *logarithms* are to the base $e = 2.718\,28\ldots$, written $\log_e y$ or $\ln y$. Logarithms were formerly used to facilitate calculations, before the advent of electronic calculators.

A logarithm contains two parts, an integer and a decimal. The integer is called the *characteristic*, and the decimal is called the *mantissa*. For example, the logarithm to the base 10 of 210 is 2.3222, where 2 is the characteristic and 0.3222 is the mantissa.

logarithmic scale 1. A scale of measurement in which an increase or decrease of one unit represents a tenfold increase or decrease in the quantity measured. Decibels and pH measurements are common examples of logarithmic scales of measurement. **2.** A scale on the axis of a graph in which an increase of one unit represents a tenfold increase in the variable quantity. If a curve $y = x^n$ is plotted on graph paper with logarithmic scales on both axes, the result is a straight line of slope n, i.e. $\log y$

logarithmic series

$= n\log x$, which enables n to be determined.

logarithmic series The expansion of a logarithmic function, such as $\log_e(1 + x)$, i.e. $x - x^2/2 + x^3/3 - \ldots + (-1)^n x^n/n$, or $\log_e(1 - x)$, i.e. $-x - x^2/2 - x^3/3 \ldots - x^n/n$.

logic circuits The basic switching circuits or *gates used in digital computers and other digital electronic devices. The output signal, using a *binary notation, is controlled by the logic circuit in accordance with the input system. The three basic logic circuits are the *and*, *or*, and *not* circuits. The *and* circuit gives a binary 1 output if a binary 1 is present on each input circuit; otherwise the output is a binary 0. The *or* circuit gives a binary 1 output if a binary 1 is present on at least one input circuit; otherwise the output is binary 0. The *not* circuit inverts the input signal, giving a binary 1 output for a binary 0 input or a 0 output for a 1 input. Often these basic logic circuits are used in combination, e.g. a *nand* circuit consists of *not* + *and* circuits. In terms of electronic equipment, logic circuits are now almost exclusively embodied into *integrated circuits. They formerly made use of discrete transistors, although they were originally based on thermionic valves.

lomentum A type of dry dehiscent fruit formed from a single carpel but divided into one-seeded compartments by constrictions between the seeds. *Legumes (e.g. those of *Acacia*) and *siliquas (e.g. those of wild radish) can be divided in this way.

lone pair A pair of electrons having opposite spin in an orbital of an atom. For instance, in ammonia the nitrogen atom has five electrons, three of which are used in forming single bonds with hydrogen atoms. The other two occupy a filled atomic orbital and constitute a lone pair (see

Lone pair of electrons in ammonia

illustration). The orbital containing these electrons is equivalent to a single bond (sigma orbital) in spatial orientation, accounting for the pyramidal shape of the molecule. In the water molecule, there are two lone pairs on the oxygen atom. In considering the shapes of molecules, repulsions between bonds and lone pairs can be taken into account: lone pair–lone pair → lone pair–bond → bond–bond.

long-day plant A plant in which flowering can be induced or enhanced by long days, usually of more than 12 hours of daylight. Examples are spinach and spring barley. *See* photoperiodism. *Compare* day-neutral plant; short-day plant.

longitude *See* latitude and longitude.

longitudinal wave *See* wave.

long period *See* periodic table.

long-sightedness *See* hypermetropia.

Lorentz–Fitzgerald contraction (Fitzgerald contraction) The contraction of a moving body in the direction of its motion. It was proposed independently by H. A. Lorentz (1853–1928) and G. F. Fitzgerald (1851–1901) in 1892 to account for the null result of the *Michelson–Morley experiment. The contraction was given a theoretical background in Einstein's special theory of *relativity. In this theory, an object of length l_0 at rest in one *frame of reference will appear, to an observer in another frame moving with relative velocity v with respect to the first, to have length $l_0\sqrt{(1 - v^2/c^2)}$, where c is the speed of light. The original hypothesis regarded this contraction as a real one accompanying the absolute motion of the body. The contraction is in any case negligible unless v is of the same order as c.

Lorentz transformations A set of equations for transforming the position and motion parameters from a frame of reference with origin at O and coordinates (x,y,z) to a frame moving relative to it with origin at O' and coordinates (x',y',z'). They replace the *Galilean transformations used in *Newtonian mechanics and are used in relativistic mechanics. They are:
$$x' = \beta(x - vt)$$
$$y' = y$$
$$z' = z$$
$$t' = \beta(t - vx/c^2),$$
where v is the relative velocity of separation of O and O', c the speed of light, and $\beta = 1/\sqrt{(1 - v^2/c^2)}$. The above equations apply for constant v in the xx' direction with O and O' coinciding at $t = t' = 0$.

Loschmidt's constant (Loschmidt number) Symbol N_L. The number of particles per unit volume of an *ideal gas at STP. It has the value $2.686\,763(23) \times 10^{75}$ m^{-3} and was first worked out by Joseph Loschmidt (1821–95).

loudness The physiological perception of sound intensity. As the ear responds differently to different frequencies, for a given intensity loudness is dependent on frequency. Sounds with frequencies between 1000 hertz and 5000 Hz are louder than sounds of the same intensity at higher or lower frequencies. Duration is also a factor in loudness, long bursts of sound being louder than short bursts. Loudness increases up to a duration of about 0.2 second; above this limit loudness does not increase with duration.

Relative loudness is usually measured on the assumption of proportionality to the logarithm of the intensity (for a given frequency), i.e. proportionality to the relative intensity on the *decibel scale. A subjective judgment is made of the relative intensity above threshold that a note of 1000 Hz must have to match the specimen sound; the loudness of this, in *phons, is then equal to that relative intensity in decibels.

loudspeaker A transducer for converting an electrical signal into an acoustic signal. Usually it is important to preserve as many characteristics of the electrical waveform as possible. The device must be capable of reproducing frequencies in the range 150–8000 hertz for speech and 20–20 000 Hz for music.

The most common loudspeaker consists of a moving-coil device. In this a cone-shaped diaphragm is attached to a coil of wire and made to vibrate in accordance with the electrical signal by the interaction between the current passing through the coil and a steady magnetic field from a permanent magnet surrounding it.

lowering of vapour pressure A reduction in the saturated vapour pressure of a pure liquid when a solute is introduced. If the solute is a solid of low vapour pressure, the decrease in vapour pressure of the liquid is proportional to the concentration of particles of solute; i.e. to the number of dissolved molecules or ions per unit volume. To a first approximation, it does not depend on the nature of the particles. *See* colligative property; Raoult's law.

low frequency (LF) A radio frequency in the range 30–300 kilohertz; i.e. having a wavelength in the range 1–10 kilometre.

Lowry–Brønsted theory *See* acid.

L-series *See* absolute configuration.

lubrication The use of a substance to prevent contact between solid surfaces in relative motion in order to reduce friction, wear, overheating, and rusting. Liquid hydrocarbons (oils), either derived from petroleum or made synthetically, are the most widely used lubricants as they are relatively inexpensive, are good coolants, provide the appropriate range of viscosities, and are thermally stable. Additives include polymeric substances that maintain the desired viscosity as the temperature increases, antioxidants that prevent the formation of a sludge, and alkaline-earth phenates that neutralize acids and reduce wear. At high temperatures, solid lubricants, such as graphite or molybdenum disulphide, are often used. Semifluid lubricants (greases) are used to provide a seal against moisture and dirt and to remain attached to vertical surfaces. They are made by adding gelling agents, such as metallic soaps, to liquid lubricants.

Recent technology has made increasing use of gases as lubricants, usually in air bearings. Their very low viscosities minimize energy losses at the bearings but necessitate some system for pumping the gas continuously to the bearings. The principle is that of the hovercraft.

lumbar vertebrae The *vertebrae in the region of the lower back. They occur below the *thoracic vertebrae and above the *sacral vertebrae. In mammals they bear processes for the attachment of back muscles.

lumen 1. The space enclosed by a vessel, duct, or other tubular or saclike organ. The central cavity of blood vessels and of the digestive tract are examples. **2.** Symbol lm. The SI unit of *luminous flux equal to the flux emitted by a uniform point source of 1 candela in a solid angle of 1 steradian.

luminance (photometric brightness) Symbol L. The *luminous intensity of any surface in a given direction per unit projected area of the surface, viewed from that direction. It is given by the equation $L = dI/(dA\cos\theta)$, where I is the luminous intensity and θ is the angle between the line of sight and the normal to the surface area A being considered. It is measured in candela per square metre.

luminescence The emission of light by a substance for any reason other than a rise in its temperature. In general, atoms of substances emit *photons of electromagnetic energy when they return to the *ground state after having been in an excited state (*see* excitation). The causes of the excitation are various. If the exciting cause is a photon, the process is called *photoluminescence*; if it is an electron it is called *electroluminescence*. *Chemiluminescence* is luminescence resulting from a chemical reaction (such as the slow oxidation of phosphorus); *bioluminescence is the luminescence produced by a living organism (such as a firefly). If the luminescence persists significantly after the exciting cause is removed it is called *phosphorescence*; if it does not it is called *fluorescence*. This distinction is arbitrary since there must always be some delay; in some definitions a persistence of more than 10 nanoseconds (10^{-8} s) is treated as phosphorescence.

luminosity 1. *Luminous intensity in a particular direction; the apparent brightness of an image. **2.** The brightness of a star defined as the total energy radiated in unit time. It is related to the surface area (A) and the *effective temperature* (T_e; the temperature of a black body having the same radius as the star and radiating the same amount of energy per unit area in one second) by a form of *Stefan's law, i.e.

$$L = A\sigma T_e^4$$

where σ is the Stefan constant and L is the luminosity.

luminous exitance *See* exitance.

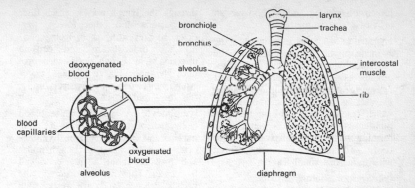

The lungs and air passages of a mammal (right lung cut open to show internal structure)

luminous flux Symbol Φ_v. A measure of the rate of flow of light, i.e. the radiant flux in the wavelength range 380–760 nanometres, corrected for the dependence on wavelength of the sensitivity of the human eye. It is measured by reference to emission from a standard source, usually in lumens.

luminous intensity Symbol I_v. A measure of the light-emitting ability of a light source, either generally or in a particular direction. It is measured in candelas.

lunar eclipse *See* eclipse.

lunation *See* synodic month.

lung The *respiratory organ of air-breathing vertebrates. A pair of lungs is situated in the thorax, within the ribcage. Each consists essentially of a thin moist membrane that is folded to increase its surface area. Exchange of oxygen and carbon dioxide takes place between blood capillaries on one side of the membrane and air on the other. The lung is supplied with air through a *bronchus. In mammals and reptiles the membrane of the lung takes the form of numerous sacs (*see* alveolus) that are connected to the bronchus via *bronchioles (see illustration). The lungs themselves contain no muscular tissue and are

ventilated by *respiratory movements, the mechanisms of which vary with the species.

lungfish *See* Dipnoi.

luteinizing hormone (LH; interstitial-cell-stimulating hormone; ICSH) A hormone, secreted by the anterior pituitary gland in mammals, that stimulates in males the production of sex hormones (*androgens) by secretory cells in the testes (known as *interstitial cells*) and in females ovulation, *progesterone synthesis, and *corpus luteum formation.

luteotrophic hormone *See* prolactin.

lutetium Symbol Lu. A silvery metallic element belonging to the *lanthanoids; a.n. 71; r.a.m. 174.97; r.d. 9.842 (20°C); m.p. 1663°C; b.p. 3402°C. Lutetium is the least abundant of the elements and the little quantities that are available have been obtained by processing other metals. There are two natural isotopes, lutetium–175 (stable) and lutetium–176 (half-life 2.2×10^{10} years). The element is used as a catalyst. It was first identified by G. Urban in 1907.

lux Symbol lx. The SI unit of *illuminance equal to the illumination produced by a *luminous flux of 1 lumen distributed uniformly over an area of 1 square metre.

Lycopsida A subdivision of the *Tracheophyta or a class of the *Pteridophyta containing the club-mosses (genus *Lycopodium*) and related genera (including *Selaginella*) as well as numerous extinct forms, which reached their peak in the Carboniferous period with giant coal-forming tree species. Lycopsids have roots and their stems are covered with numerous small leaves.

lye *See* potassium hydroxide.

Lyman series *See* hydrogen spectrum.

lymph The colourless liquid found within the *lymphatic system, into which it drains from the spaces between the cells. Lymph (called *tissue fluid* in the intercellular spaces) resembles *blood plasma, consisting mostly of water with dissolved salts and proteins. Fats are found in suspension and their presence varies wth food intake. The lymph eventually enters the bloodstream near the heart.

lymphatic system The network of vessels that conveys *lymph from the tissue fluids to the bloodstream. Tiny lymph capillaries drain into larger tubular vessels that converge to form the right lymphatic duct and the *thoracic duct, which connect with the venous blood supply to the heart. Associated with the lymphatic vessels at intervals along the system are the *lymph nodes. The lymph capillary walls are very permeable, so lymph bathing the body's tissues can drain away molecules that are too large to pass through blood capillary walls. Lymph is pumped by cycles of contraction and relaxation of the lymphatic vessels and also by the action of adjoining muscles.

lymph node A mass of *lymphoid tissue, many of which occur at intervals along the *lymphatic system. Lymph in the lymphatic vessels flows through the lymph nodes, which filter out bacteria and other foreign particles, so preventing them from entering the bloodstream and causing infection. The lymph nodes also produce *lymphocytes. In humans, major lymph nodes occur in the neck, under the arms, and in the groin.

lymphocyte A type of white blood cell (*leucocyte) that has a large nucleus and little cytoplasm. Lymphocytes are formed in the *lymph nodes and provide about a quarter of all leucocytes. They are important in the body's defence and are responsible for immune reactions as the presence of *antigens stimulates them to produce *antibodies. There are two populations of lymphocytes: *B lymphocytes*, which produce circulating antibodies and are responsible for humoral *immunity; and *T lymphocytes* (*see* T cell), which are responsible for cell-mediated immunity. *See also* killer cell.

lymphoid tissue The type of tissue found in the *lymph nodes, *tonsils, *spleen, and *thymus. It is responsible for producing lymphocytes and antibodies and therefore contributes to the body's defence against infection.

lymphokine Any soluble factor secreted by lymphocytes that mediates the activities of lymphoid cells. Examples include *interleukin-2, B-cell growth factor (BCGF), B-cell differentiation factor (BCDF), and macrophage migration inhibition factor (MIF). *See also* cytokine.

lyophilic Having an affinity for a solvent ('solvent-loving'; if the solvent is water the term *hydrophilic* is used). *See* colloid.

lyophobic Lacking any affinity for a solvent ('solvent-hating'; if the solvent is water the term *hydrophobic* is used). *See* colloid.

Lysenkoism The official Soviet science policy governing the work of geneticists in the USSR from about 1940 to 1960. It was named after its chief promoter, the agriculturalist Trofim Lysenko (1898–1976). Lysenkoism dismissed all the advances that had been made in classical genetics, denying the existence of genes, and held that the variability of organisms was produced solely by environmental changes. There was also a return to a belief in the inheritance of acquired characteristics (*see* Lamarckism). This state of affairs continued, despite overwhelming conflicting evidence from Western scientists, because it provided support for communist theory.

lysigeny The localized disruption of plant cells to form a cavity (surrounded by remnants of the broken cells) in which secretions accumulate. Examples are the oil cavities in the leaves of citrus trees. *Compare* schizogeny.

lysine *See* amino acid.

lysis The destruction of a living cell. This may be effected by *lysosomes or *lymphocytes, either as part of the normal metabolic process (as when cells are damaged or worn out) or as a reaction against invading cells (e.g. bacteria). *Bacteriophages eventually cause lysis of their host cells.

lysosome A membrane-bound sac within the cytoplasm of animal *cells that contains enzymes responsible for the digestion of material in food vacuoles, the dissolution of foreign particles entering the cell (such as bacteria) and, on the death of the cell, the breakdown of all cell structures. A similar structure is present in plant cells.

lysozyme An antibacterial enzyme widely distributed in body fluids and secretions, including tears and saliva. It disrupts the polysaccharide components of bacterial cell walls, leaving them susceptible to destruction.

M

machine A device capable of making the performance of mechanical work easier, usually by overcoming a force of resistance (the load) at one point by the application of a more convenient force (the effort) at some other point. In physics, the six so-called *simple machines* are the lever, wedge, inclined plane, screw, pulley, and wheel and axle.

Mach number The ratio of the relative speeds of a fluid and a rigid body to the speed of sound in that fluid under the same conditions of temperature and pressure. If the Mach number exceeds 1 the fluid or body is moving at a *supersonic speed*. If the Mach number exceeds 5 it is said to be *hypersonic*. The number is named after Ernst Mach (1838–1916).

Mach's principle The *inertia of any particular piece of matter is attributable to the interaction between that piece of matter and the rest of the universe. A body in isolation would have zero inertia. This principle was stated by Ernst Mach in the 1870s and was made use of by Einstein in his general theory of *relativity.

Maclaurin's series *See* Taylor series.

macromolecular crystal A crystalline solid in which the atoms are all linked together by covalent bonds. Carbon (in diamond), boron nitride, and silicon carbide are examples of substances that have macromolecular crystals. In effect, the crystal is a large molecule (hence the alternative description *giant-molecular*), which accounts for the hardness and high melting point of such materials.

macromolecule A very large molecule. Natural and synthetic polymers have macromolecules, as do such substances as haemoglobin. *See also* colloids.

macronutrient

macronutrient A nutrient required by an organism in relatively large amounts. Macronutrients are typically used as sources of energy or structural components and include water, carbon dioxide, carbohydrates, proteins, lipids, and certain minerals (*see also* essential element). *Compare* micronutrient.

macrophage (histiocyte; clasmatocyte) A large phagocytic cell (*see* phagocyte) normally found in tissues that produce blood cells. Macrophages ingest bacteria and cell debris. They develop from *monocytes and assume different shapes depending on their position. *Reticuloendothelial cells* are similar but are found only lining blood vessels and do not move around the body.

macrophyll *See* megaphyll.

macula 1. One of a number of patches of sensory cells in the *utriculus and *sacculus of the inner ear that are involved in maintaining balance and may also assist in assessing the degree of acceleration and deceleration when the body is moving. 2. An area of the *retina of the vertebrate eye with increased visual acuity. Maculae occur in some animals that lack *foveae and often surround foveae in those animals that possess them.

Magellanic clouds Two small galaxies situated close to the Milky Way that are only visible from the southern hemisphere. They were first recorded by Ferdinand Magellan (1480–1521) in 1519.

magic numbers Numbers of neutrons or protons that occur in atomic nuclei to produce very stable structures. The magic numbers for both protons and neutrons are 2, 8, 20, 28, 50, and 82. For neutrons 126 and 184 are also magic numbers and for protons 114 is a magic number. The relationship between stability and magic numbers led to a nuclear *shell model in analogy to the electron shell model of the atom.

magma Hot molten material that originates within the earth's crust or mantle and when cooled and solidified forms igneous rock. Most magmas are composed largely of silicates with suspended crystals and dissolved gases. Magma is extruded as *lava onto the surface of the earth as a result of volcanic activity; magma that cools and solidifies within the earth's crust may form either plutonic (at great depths) or hypabyssal (at intermediate depths) rocks.

Magnadur A trade name for a ceramic material used to make permanent magnets. It consists of sintered iron oxide and barium oxide.

Magnalium A trade name for an aluminium-based alloy of high reflectivity for light and ultraviolet radiation that contains 1–2% of copper and between 5% and 30% of magnesium. Strong and light, these alloys also sometimes contain other elements, such as tin, lead, and nickel.

magnesia *See* magnesium oxide.

magnesite A white, colourless, or grey mineral form of *magnesium carbonate, $MgCO_3$, crystallizing in the trigonal system. It is formed as a replacement mineral of magnesium-rich rocks when carbon dioxide is available. Magnesite is mined both as an ore for magnesium and as a source of magnesium carbonate. It occurs in Austria, USA, Greece, Norway, India, Australia, and South Africa.

magnesium Symbol Mg. A silvery metallic element belonging to group II of the periodic table (*see* alkaline-earth metals); a.n. 12; r.a.m. 24.312; r.d. 1.74; m.p. 651°C; b.p. 1107°C. The element is found in a number of minerals, including magnesite ($MgCO_3$), dolomite ($MgCO_3.CaCO_3$), and carnallite ($MgCl_2.KCl.6H_2O$). It is also present in sea water, and it is an *essential element for living organisms. Extraction is by electrolysis of the fused chloride. The element is used in a number of light alloys (e.g. for aircraft). Chemically, it is very reactive. In air it forms a protective oxide coating but when ignited it burns with an intense white flame. It also reacts with the halogens, sulphur, and nitrogen. Magnesium was first isolated by Bussy in 1828.

magnesium bicarbonate *See* magnesium hydrogencarbonate.

magnesium carbonate A white compound, $MgCO_3$, existing in anhydrous and hydrated forms. The anhydrous material (trigonal; r.d. 2.96) is found in the mineral *magnesite. There is also a trihydrate, $MgCO_3.3H_2O$ (rhombic; r.d. 1.85), which occurs naturally as *nesquehonite*, and a pentahydrate, $MgCO_3.5H_2O$ (monoclinic; r.d. 1.73), which occurs as *lansfordite*. Magnesium carbonate also occurs in the mixed salt *dolomite ($CaCO_3.MgCO_3$) and as *basic magnesium carbonate* in the two minerals *artinite* ($MgCO_3.Mg(OH)_2.3H_2O$) and *hydromagnesite*
($3MgCO_3.Mg(OH)_2.3H_2O$).
The anhydrous salt can be formed by heating magnesium oxide in a stream of carbon dioxide:
$$MgO(s) + CO_2(g) \rightarrow MgCO_3(s)$$
Above 350°C, the reverse reaction predominates and the carbonate decomposes. Magnesium carbonate is used in making magnesium oxide and is a drying agent (e.g. in table salt). It is also used as a medical antacid and laxative (the basic carbonate is used) and is a component of certain inks and glasses.

magnesium chloride A white solid compound, $MgCl_2$. The anhydrous salt (hexagonal; r.d. 2.32; m.p. 714°C; b.p. 1412°C) can be prepared by the direct combination of dry chlorine with magnesium:
$$Mg(s) + Cl_2(g) \rightarrow MgCl_2(s)$$

The compound also occurs naturally as a constituent of carnallite ($KCl.MgCl_2$). It is a deliquescent compound that commonly forms the hexahydrate, $MgCl_2.6H_2O$ (monoclinic; r.d. 1.57). When heated, this hydrolyses to give magnesium oxide and hydrogen chloride gas. The fused chloride is electrolysed to produce magnesium and it is also used for fireproofing wood, in magnesia cements and artificial leather, and as a laxative.

magnesium hydrogencarbonate (magnesium bicarbonate) A compound, $Mg(HCO_3)_2$, that is stable only in solution. It is formed by the action of carbon dioxide on a suspension of magnesium carbonate in water:
$$MgCO_3(s) + CO_2(g) + H_2O(l) \rightarrow$$
$$Mg(HCO_3)_2(aq)$$
On heating, this process is reversed. Magnesium hydrogencarbonate is one of the compounds responsible for temporary *hardness in water.

magnesium hydroxide A white solid compound, $Mg(OH)_2$; trigonal; r.d. 2.36; decomposes at 350°C. Magnesium hydroxide occurs naturally as the mineral *brucite* and can be prepared by reacting magnesium sulphate or chloride with sodium hydroxide solution. It is used in the refining of sugar and in the processing of uranium. Medicinally it is important as an antacid (*milk of magnesia*) and as a laxative.

magnesium oxide (magnesia) A white compound, MgO; cubic; r.d. 3.58; m.p. 2800°C. It occurs naturally as the mineral *periclase* and is prepared commercially by thermally decomposing the mineral *magnesite:
$$MgCO_3(s) \rightarrow MgO(s) + CO_2(g)$$
It has a wide range of uses, including reflective coatings on optical instruments and aircraft windscreens and in semiconductors. Its high melting point makes it useful as a refractory lining in metal and glass furnaces.

magnesium peroxide A white solid, MgO_2. It decomposes at 100°C to release oxygen and also releases oxygen on reaction with water:

$$2MgO_2(s) + 2H_2O \rightarrow 2Mg(OH)_2 + O_2$$

The compound is prepared by reacting sodium peroxide with magnesium sulphate solution and is used as a bleach for cotton and silk.

magnesium sulphate A white soluble compound, $MgSO_4$, existing as the anhydrous compound (rhombic; r.d. 2.66; decomposes at 1124°C) and in hydrated crystalline forms. The monohydrate $MgSO_4.H_2O$ (monoclinic; r.d. 2.45) occurs naturally as the mineral *kieserite*. The commonest hydrate is the heptahydrate, $MgSO_4.7H_2O$ (rhombic; r.d. 1.68), which is called *Epsom salt(s)*, and occurs naturally as the mineral *epsomite*. This is a white powder with a bitter saline taste, which loses $6H_2O$ at 150°C and $7H_2O$ at 200°C. It is used in sizing and fireproofing cotton and silk, in tanning leather, and in the manufacture of fertilizers, explosives, and matches. In medicine, it is used as a laxative. It is also used in veterinary medicine for treatment of local inflammations and infected wounds.

magnet A piece of magnetic material (*see* magnetism) that has been magnetized and is therefore surrounded by a *magnetic field. A magnet, often in the shape of a bar or horseshoe, that retains appreciable magnetization indefinitely (provided it is not heated, beaten, or exposed to extraneous magnetic fields) is called a *permanent magnet*. See also electromagnet.

magnetic bottle A nonuniform *magnetic field used to contain the plasma in a *thermonuclear reactor. At the temperature of a thermonuclear reaction (10^8 K) any known substance would vaporize and the plasma has therefore to be contained in such a way that it does not come into contact with a material surface. The magnetic bottle provides a means of achieving this, by deflecting away from its boundaries the moving charged particles that make up the plasma.

magnetic bubble memory A form of computer memory in which a small magnetized region of a substance is used to store information. Bubble memories consist of materials, such as magnetic garnets, that are easily magnetized in one direction but hard to magnetize in the perpendicular direction. A thin film of these materials deposited on a nonmagnetic substrate constitutes a bubble-memory chip. When a magnetic field is applied to such a chip, by placing it between two permanent magnets, cylindrical domains (called magnetic bubbles) are formed. These bubbles constitute a magnetic region of one polarity surrounded by a magnetic region of the opposite polarity. Information is represented as the presence or absence of a bubble at a specified storage location and is retrieved by means of a rotating magnetic field. Typically a chip measures 15 mm², or 25 mm² enclosed in two permanent magnets and two rotating field coils; each chip can store up to one million bits.

magnetic circuit A closed path containing a *magnetic flux. The path is clearly delimited only if it consists mainly or wholly of ferromagnetic or other good magnetic materials; examples include transformer cores and iron parts in electrical machines. The design of these parts can often be assisted by analogy with electrical circuits, treating the *magnetomotive force as the analogue of e.m.f., the magnetic flux as current, and the *reluctance as resistance. There is, however, no actual flow around a magnetic circuit.

magnetic compass *See* compass.

magnetic constant *See* permeability.

magnetic declination *See* geomagnetism.

magnetic dip *See* geomagnetism.

magnetic disk A smooth aluminium disk, usually 35.6 cm in diameter, both surfaces of which are coated with magnetic iron oxide. The disks are used as a recording medium in computers, up to ten such disks being mounted in a disk pack. Data is recorded in concentric tracks on both surfaces with up to 236 tracks per centimetre. The disks rotate at 3600 revolutions per minute, information being put onto the disk and removed from it by a record-playback head. *See also* floppy disk.

magnetic domain *See* magnetism.

magnetic elements *See* geomagnetism.

magnetic equator *See* equator; geomagnetism.

magnetic field A *field of force that exists around a magnetic body (*see* magnetism) or a current-carrying conductor. Within a magnetic field a magnetic dipole may experience a torque and a moving charge may experience a force. The strength and direction of the field can be given in terms of the *magnetic flux density* (or *magnetic induction*), symbol B; it can also be given in terms of the *magnetic field strength* (*magnetizing force* or *magnetic intensity*), symbol H.

The magnetic flux density is a vector quantity and is the *magnetic flux per unit area of a magnetic field at right angles to the magnetic force. It can be defined in terms of the effects the field has, for example by $B = F/qv\sin\theta$, where F is the force a moving charge q would experience if it was travelling at a velocity v in a direction making an angle θ with that of the field. The *SI unit is the tesla. The magnetic field strength is also a vector quantity and is related to B by: $H = B/\mu$, where μ is the *permeability of the medium. The SI unit of

field strength is the ampere per metre $(A\ m^{-1})$.

magnetic field strength *See* magnetic field.

magnetic flux Symbol Φ. A measure of quantity of magnetism, taking account of the strength and the extent of a *magnetic field. The flux $d\Phi$ through an element of area dA perpendicular to B is given by $d\Phi = BdA$. The *SI unit of magnetic flux is the weber.

magnetic flux density *See* magnetic field.

magnetic force The attractive or repulsive force exerted on a *magnetic pole or a moving electric charge in a *magnetic field.

magnetic induction *See* magnetic field.

magnetic intensity *See* magnetic field.

magnetic meridian *See* meridian.

magnetic mirror A device used to contain *plasma in thermonuclear experimental devices. It consists of a region of high magnetic field strength at the end of a containment tube. Ions entering the region reverse their motion and return to the plasma from which they have emerged. *See also* magnetic bottle.

magnetic moment The ratio between the maximum torque (T_{max}) exerted on a magnet, current-carrying coil, or moving charge situated in a *magnetic field and the strength of that field. It is thus a measure of the strength of a magnet or current-carrying coil. In the Sommerfeld approach this quantity (also called *electromagnetic moment* or *magnetic area moment*) is T_{max}/B. In the Kennelly approach the quantity (also called *magnetic dipole moment*) is T_{max}/H.

In the case of a magnet placed in a magnetic field of field strength H, the maximum torque T_{max} occurs when the axis of the magnet is perpendicular to the field. In the case of a coil of N turns and area A carrying a current I, the magnetic moment can be shown to be $m = T/B = NIA$ or $m = T/H = \mu NIA$. Magnetic moments are measured in *SI units in A m².

An orbital electron has an orbital magnetic moment IA, where I is the equivalent current as the electron moves round its orbit. It is given by $I = q\omega/2\pi$, where q is the electronic charge and ω is its angular velocity. The orbital magnetic moment is therefore $IA = q\omega A/2\pi$, where A is the orbital area. If the electron is spinning there is also a spin magnetic moment (*see* spin); atomic nuclei also have magnetic moments (*see* nuclear moment).

magnetic monopole A hypothetical magnetic entity consisting of an isolated elementary north or south pole. It has been postulated as a source of a *magnetic field by analogy with the way in which an electrically charged particle produces an electric field. Numerous ingenious experiments have been designed to detect the monopole but so far none has produced an unequivocal result. Magnetic monopoles are predicted to exist in certain *gauge theories with *Higgs bosons. In particular, some *grand unified theories predict very heavy monopoles (with mass of order 10^{16} GeV). Magnetic monopoles are also predicted to exist in *Kaluza–Klein theories and *superstring theory.

magnetic permeability *See* permeability.

magnetic poles 1. *See* geomagnetism. **2.** The regions of a *magnet from which the magnetic forces appear to originate. A magnetized bar has a pole at each end; if it is freely suspended in the earth's magnetic field (*see* geomagnetism) it will rotate so that one end points approximately towards the earth's geographical north pole. This end is called the north-seeking end or the north pole of the magnet. The other end is accordingly

called the south-seeking end or south pole. In the obsolete theory associated with the *c.g.s. electromagnetic system of units, a *unit magnetic pole* was treated as one of a pair, which repelled each other with a force of 1 dyne when separated by 1 cm in space.

magnetic potential *See* magnetomotive force.

magnetic quantum number *See* atom.

magnetic resonance imaging (MRI) *See* nuclear magnetic resonance.

magnetic susceptibility *See* susceptibility.

magnetic tape A plastic tape coated with a ferromagnetic iron oxide powder. The tape is used for recording data in tape recorders and computers. To record, the tape is passed over a recording head containing a gap in a magnetic circuit whose magnetization is modulated by the information to be recorded; the information is imprinted on the tape in the form of the direction of magnetization of the individual particles of iron oxide. The particles themselves are not rotated by the magnetizing field: it is their directions of magnetization that are orientated in accordance with the information. In audio-frequency recorders a high-frequency bias (in the range 75–100 kHz) is used to reduce distortion by facilitating the re-orientation. The playback procedure is the reverse of recording; the tape containing its orientation of tiny magnets is fed over the gap of the same (now the playback) head, in whose coil corresponding e.m.f.s are generated by induction.

magnetic variation (secular magnetic variation) *See* geomagnetism.

magnetism A group of phenomena associated with *magnetic fields. Whenever an electric current flows a magnetic field is produced; as the orbital motion and the *spin of atomic electrons are equivalent to tiny current loops, individual atoms create magnetic fields around them, when their orbital electrons have a net *magnetic moment as a result of their angular momentum. The magnetic moment of an atom is the vector sum of the magnetic moments of the orbital motions and the spins of all the electrons in the atom. The macroscopic magnetic properties of a substance arise from the magnetic moments of its component atoms and molecules. Different materials have different characteristics in an applied magnetic field; there are four main types of magnetic behaviour:

(a) In *diamagnetism* the magnetization is in the opposite direction to that of the applied field, i.e. the *susceptibility is negative. Although all substances are diamagnetic, it is a weak form of magnetism and may be masked by other, stronger, forms. It results from changes induced in the orbits of electrons in the atoms of a substance by the applied field, the direction of the change (in accordance with *Lenz's law) opposing the applied flux. There is thus a weak negative susceptibility and a relative permeability that is slightly less than one.

(b) In *paramagnetism* the atoms or molecules of the substance have net orbital or spin magnetic moments that are capable of being aligned in the direction of the applied field. They therefore have a positive (but small) susceptibility and a relative permeability slightly in excess of one. Paramagnetism occurs in all atoms and molecules with unpaired electrons; e.g. free atoms, free radicals, and compounds of transition metals containing ions with unfilled electron shells. It also occurs in metals as a result of the magnetic moments associated with the spins of the conducting electrons.

(c) In *ferromagnetic* substances, within a certain temperature range, there are net atomic magnetic moments, which line up in such a way that magnetization persists after the removal of the applied field. Below a certain temperature, called the *Curie point (or Curie temperature) an increasing magnetic field applied to a ferromagnetic substance will cause increasing magnetization to a high value, called the *saturation magnetization*. This is because a ferromagnetic substance consists of small (1–0.1 mm across) magnetized regions called *domains*. The total magnetic moment of a sample of the substance is the vector sum of the magnetic moments of the component domains. Within each domain the individual atomic magnetic moments are spontaneously aligned by *exchange forces*, related to whether or not the atomic electron spins are parallel or antiparallel. However, in an unmagnetized piece of ferromagnetic material the magnetic moments of the domains themselves are not aligned; when an external field is applied those domains that are aligned with the field increase in size at the expense of the others. In a very strong field all the domains are lined up in the direction of the field and provide the high observed magnetization. Iron, nickel, cobalt, and their alloys are ferromagnetic. Above the Curie point, ferromagnetic materials become paramagnetic.

(d) Some metals, alloys, and transition-element salts exhibit another form of magnetism called *antiferromagnetism*. This occurs below a certain temperature, called the *Néel temperature, when an ordered array of atomic magnetic moments spontaneously forms in which alternate moments have opposite directions. There is therefore no net resultant magnetic moment in the absence of an applied field. In manganese fluoride, for example, this antiparallel arrangement occurs below a Néel

magnetite

temperature of 72 K. Below this temperature the spontaneous ordering opposes the normal tendency of the magnetic moments to align with the applied field. Above the Néel temperature the substance is paramagnetic.

A special form of antiferromagnetism is *ferrimagnetism*, a type of magnetism exhibited by the *ferrites. In these materials the magnetic moments of adjacent ions are antiparallel and of unequal strength, or the number of magnetic moments in one direction is greater than those in the opposite direction. By suitable choice of rare-earth ions in the ferrite lattices it is possible to design ferrimagnetic substances with specific magnetizations for use in electronic components. *See also* geomagnetism.

magnetite A black mineral form of iron oxide crystallizing in the cubic system. It is a mixed iron(II)-iron(III) oxide, Fe_3O_4, and is one of the major ores of iron. It is strongly magnetic and some varieties, known as *lodestone*, are natural magnets; these were used as compasses in the ancient world. Magnetite is widely distributed and occurs as an accessory mineral in almost all igneous and metamorphic rocks. The largest deposits of the mineral occur in N Sweden.

magneto An alternating-current generator used as a high-tension source in the ignition systems of petrol engines in which there are no batteries, e.g. in some tractor, marine, and aviation engines. Most modern magnetos consist of a permanent-magnet rotor revolving within a primary (low-voltage) winding around which a secondary winding is placed in which to induce the high voltage needed to produce the spark across the points of the plugs. Magnetos are geared to the engine shaft, the speed depending on the number of poles of the magneto and the number of engine cylinders. A make-and-break

device is incorporated in the primary winding; when the primary current stops the change of flux within the secondary induces in it a large e.m.f.

magnetobremsstrahlung *See* synchrotron radiation.

magnetocaloric effect A reversible change of temperature resulting from a change in the magnetization of a ferromagnetic or paramagnetic substance (*see* magnetism). The change in temperature ΔT, accompanying an adiabatic change of magnetic field ΔH, is:

$$\Delta T/\Delta H = -T/C_H(\partial M/\partial T)_H$$

C_H is the specific heat capacity per unit volume at constant H and M is the magnetization.

magnetochemistry The branch of physical chemistry concerned with measuring and investigating the magnetic properties of compounds. It is used particularly for studying transition-metal complexes, many of which are paramagnetic because they have unpaired electrons. Measurement of the magnetic susceptibility allows the magnetic moment of the metal atom to be calculated, and this gives information about the bonding in the complex.

magnetohydrodynamics (MHD) The study of the interactions between a conducting fluid and a *magnetic field. MHD is important in the study of controlled thermonuclear reactions in which the conducting fluid is a *plasma confined by a magnetic field. Other important applications include the *magnetohydrodynamic power generator* (see illustration). In the open-cycle MHD generator a fossil fuel, burnt in oxygen or preheated compressed air, is seeded with an element of low *ionization potential (such as potassium or caesium). This element is thermally ionized at the combustion temperature (usually over 2500 K) producing sufficient free electrons (e.g. $K \rightarrow K^+ + e$) to provide adequate electrical conductivity. The

magnetic field

load

gas flow

positive electrodes

current

negative electrodes

gas flow

Magnetohydrodynamic generator

interaction between the moving conducting fluid and the strong applied magnetic field across it generates an e.m.f. on the Faraday principle, except that the solid conductor of the conventional generator is replaced by a fluid conductor. The power output per unit fluid volume (W) is given by $W = k\sigma v^2 B^2$, where σ is the conductivity of the fluid, v is its velocity, B is the magnetic flux density, and K is a constant. Devices of this kind are in use in some power stations, where they are suitable for helping to meet high short-term demands and have the ability of increasing the thermal efficiency of a steam-turbine generator from about 40% to 50%. In experimental closed-cycle systems the fluid is continuously recirculated through a compressor; the fluid consists of a heated and seeded noble gas or a liquid metal.

magnetomechanical ratio *See* gyromagnetic ratio.

magnetometer An instrument for measuring the magnitude, and sometimes the direction, of a magnetic field. *Absolute magnetometers* measure the field without reference to a standard magnetic instrument. The most

widely used are the *vibration magnetometer*, the *deflection galvanometer*, and the more modern *nuclear magnetometer*. The vibration instrument was devised by Gauss in 1832 and depends on the rate of oscillation of a small bar magnet suspended in a horizontal plane. The same magnet is then used as a fixed deflector to deflect a second similarly suspended magnet. The deflection galvanometer uses a Helmholtz coil system of known dimensions with a small magnet suspended at its centre. The deflected magnet comes to rest at a position controlled by the earth's magnetic field, the coil's magnetic field, and the angle through which the coil must be turned to keep the magnet and the coil in alignment. The sensitive nuclear magnetometers are based on measuring the audio-frequency voltage induced in a coil by the precessing protons in a sample of water. Various *relative magnetometers* are also in use, especially for measuring the earth's magnetic field and in calibrating other equipment.

magnetomotive force (m.m.f.) The analogue of *electromotive force in a *magnetic circuit. Mathematically, it

solar wind

shock front

magnetopause

magnetosphere

to sun

lines of force

Van Allen belt

tail

0   100 000 km

Magnetosphere and magnetopause

is the circular integral of $H\cos\theta$ ds, where $H\cos\theta$ is the component of the *magnetic field strength in the direction of a path of length ds. The m.m.f. is measured in *SI units in ampere-turns. It was formerly called the *magnetic potential*.

magneton A unit for measuring *magnetic moments of nuclear, atomic, or molecular magnets. The *Bohr magneton* μ_B has the value of the classical magnetic moment of an electron, given by

$$\mu_B = eh/4\pi m_e = 9.274 \times 10^{-24} \text{ A m}^2,$$

where e and m_e are the charge and mass of the electron and h is the Planck constant. The *nuclear magneton*, μ_N is obtained by replacing the mass of the electron by the mass of the proton and is therefore given by

$$\mu_N = \mu_B m_e/m_p = 5.05 \times 10^{-27} \text{ A m}^2.$$

magneto-optical effects Effects resulting from the influence of a *magnetic field upon matter that is in the process of emitting or absorbing light. Examples are the *Faraday effect and the *Zeeman effect.

magnetosphere A comet-shaped region surrounding the earth and other magnetic planets in which the charged particles of the *solar wind are controlled by the planet's magnetic field rather than the sun's magnetic field. It extends for some 60 000 km on the side facing the sun but on the opposite side it extends to a much greater extent. The boundary of the magnetosphere is known as the *magnetopause* (see illustration). The magnetosphere of the earth includes the *Van Allen belts.

magnetostriction The change in length of a ferromagnetic material (*see* magnetism) when it is magnetized. It results from changes in the boundaries of the domains. A ferromagnetic rod exposed to an alternating field will vibrate along its length. This appears to be a major source of transformer hum, which can be removed by using a magnetic steel containing 6.5% silicon. Magnetostriction of a nickel transducer is used to generate and receive ultrasonic waves.

magnetron A microwave generator in which electrons, generated by a heated cathode, move under the combined force of an electric field and a magnetic field. The cathode consists of a central hollow cylinder, the outer surface of which carries the barium and strontium oxide electron emitters. The anode is also a cylinder, arranged concentrically around the cathode, and it contains a series of quarter-wavelength *resonant cavities arranged around its inner surface. The electric field is applied radially between anode and cathode, the magnetic field is coaxial with the cathode. The whole device is maintained in a vacuum enclosure. The magnetron is extensively used as a generator for radar installations and can produce microsecond pulses of up to 10 MW.

magnification A measure of the extent to which an optical system enlarges or reduces an image. The *linear magnification*, *m*, is the ratio of the image height to the object height. If this ratio is greater than one the system is enlarging, if it is less than one, it is reducing. The *angular magnification*, *M* or γ, is the ratio of the angles formed by the final image and the object (when viewed directly, in the most favourable position available) at the eye. This is also sometimes called the *magnifying power* of an optical system.

magnifying power *See* magnification.

magnitude A measure of the relative brightness of a star or other celestial object. The *apparent magnitude* depends on the star's *luminosity, its distance, and the absorption of light between the object and the earth. In 1856 the astronomer N. R. Pogson devised a scale in which a difference of five magnitudes corresponds to a brightness ratio of 100 to 1. Two stars that differ by one magnitude therefore have a brightness ratio of $(100)^{0.2}:1 = 2.512$, known as the *Pogson ratio*. This scale is now universally adopted. Apparent magnitudes are not a measure of luminosity, which is defined in terms of the *absolute magnitude*. This is the apparent magnitude of a body if it was situ-

ated at a standard distance of 10 parsecs.

magnon *See* spin wave.

Magnox A group of magnesium alloys used to enclose uranium fuel elements in *nuclear reactors. They usually contain some aluminium as well as other elements, such as beryllium.

mainframe computer *See* computer.

main-sequence stars *See* Hertzsprung–Russell diagram.

major histocompatibility complex (MHC) A large gene cluster that encodes various components of the immune system, including the *histocompatibility antigens and components of the *complement system. In humans the MHC is located on chromosome 6 and includes the *HLA system. Other vertebrate species have similar MHC regions. Certain MHC genes can have many variant alleles; this produces an enormous diversity of antigens in a population, each individual possessing a unique set.

majority carrier *See* semiconductor.

Maksutov telescope *See* telescope.

malachite A secondary mineral form of copper carbonate–hydroxide, $CuCO_3.Cu(OH)_2$. It is bright green and crystallizes in the monoclinic system but usually occurs as aggregates of fibres or in massive form. It is generally found with *azurite in association with the more important copper ores and is itself mined as an ore of copper (e.g. in Zaïre). It is also used as an ornamental stone and as a gemstone.

maleic acid *See* butenedioic acid.

malic acid (2-hydroxybutanedioic acid) A white crystalline solid, $HOOCCH(OH)CH_2COOH$. L-malic acid occurs in living organisms as an intermediate metabolite in the *Krebs cycle and also (in certain plants) in photosynthesis. It is found especially in the juice of unripe fruits, e.g. green apples.

malleus (hammer) The first of the three *ear ossicles of the mammalian *middle ear.

Mallophaga An order of wingless insects comprising the bird lice. Bird lice are minute with dorsoventrally flattened ovoid bodies, reduced eyes, and biting mouthparts. They are ectoparasites of birds, feeding on particles of dead skin, feather fragments, and sometimes blood. The eggs hatch to form nymphs resembling the adults.

malonic acid *See* propanedioic acid.

Malpighian body (Malpighian corpuscle) The part of a *nephron in the kidney that consists of its cup-shaped end together with the *glomerulus that it encloses. It is named after its discoverer, the Italian anatomist M. Malpighi (1628–94).

Malpighian layer (stratum germinativum) The innermost layer of the *epidermis of mammalian *skin, separated from the underlying dermis by a fibrous *basement membrane*. It is only in this layer of the epidermis that active cell division (*mitosis) occurs. As the cells produced by these divisions age and mature, they migrate upwards through the layers of the epidermis to replace the cells being continuously worn away at the surface.

maltose (malt sugar) A sugar consisting of two linked glucose molecules that results from the action of the enzyme *amylase on starch. Maltose occurs in high concentrations in germinating seeds; malt, used in the manufacture of beer and malt whisky, is produced by allowing barley seeds to germinate and then slowly drying them.

malt sugar *See* maltose.

Mammalia A class of vertebrates containing some 4250 species. Mammals are warm-blooded animals (*see* homoiothermy), typically having sweat glands whose secretion cools the skin and an insulating body covering of hair. All mammals have *mammary glands, which secrete milk to nourish the young. Mammalian teeth are differentiated into incisors, canines, premolars, and molars and the middle ear contains three sound-conducting *ear ossicles. The four-chambered heart enables complete separation of oxygenated and deoxygenated blood and a muscular *diaphragm takes part in breathing movements, both of which ensure that the tissues are well supplied with oxygen. This, together with well-developed sense organs and brain, have enabled mammals to pursue an active life and to colonize a wide variety of habitats.

Mammals evolved from carnivorous reptiles in the Triassic period about 225 million years ago. There are three subclasses: the primitive egg-laying *Prototheria (monotremes), the pouched *Metatheria (marsupials), and the *Eutheria (placental mammals), which includes most of the living orders.

mammary glands The milk-producing organs (possibly modified sweat glands) of female mammals, which provide food for the young (*see* milk; colostrum). Their number (2 to 20) and position (on the chest or abdomen) vary according to the species. In most mammals the gland openings project as a nipple or teat. Nipples have a number of milk-duct openings; teats have one duct leading from a storage cavity.

mandible 1. One of a pair of horny *mouthparts in insects, crustaceans, centipedes, and millipedes. The mandibles lie in front of the weaker *maxillae and their lateral movements assist in biting and crushing the food. **2.** The lower jaw of vertebrates. **3.** Either of the two parts of a bird's beak.

manganate(VI) A salt containing the ion MnO_4^{2-}. Manganate(VI) ions are dark green; they are produced by manganate(VII) ions in basic solution.

manganate(VII) (permanganate) A salt containing the ion MnO_4^-. Manganate(VII) ions are dark purple and strong oxidizing agents.

manganese Symbol Mn. A grey brittle metallic *transition element, a.n. 25; r.a.m. 54.94; r.d. 7.4; m.p. 1244°C; b.p. 2040°C. The main sources are pyrolusite (MnO_2) and rhodochrosite ($MnCO_3$). The metal can be extracted by reduction of the oxide using magnesium (*Kroll process) or aluminium (*Goldschmidt process). Often the ore is mixed with iron ore and reduced in an electric furnace to produce ferromanganese for use in alloy steels. The element is quite electropositive, reacting with water and dilute acids to give hydrogen. It combines with oxygen, nitrogen, and other metals when heated. Salts of manganese contain the element in the +2 and +3 oxidation states. Manganese(II) salts are the more stable. It also forms compounds in higher oxidation states, such as manganese(IV) oxide and manganate(VI) and manganate(VII) salts. The element was discovered in 1774 by Scheele.

manganese(IV) oxide (manganese dioxide) A black oxide made by heating manganese(II) nitrate. The compound also occurs naturally as pyrolusite. It is a strong oxidizing agent, used as a depolarizing agent in voltaic cells.

manganic compounds Compounds of manganese in its +3 oxidation state; e.g. manganic oxide is manganese(III) oxide, Mn_2O_3.

manganin A copper alloy containing 13–18% of manganese and 1–4% of nickel. It has a high electrical resistance, which is relatively insensitive to temperature changes. It is therefore suitable for use in resistance wire.

manganous compounds Compounds of manganese in its +2 oxidation state; e.g. manganous oxide is manganese(II) oxide, MnO.

mannitol A polyhydric alcohol, $CH_2OH(CHOH)_4CH_2OH$, derived from mannose or fructose. It is the main soluble sugar in fungi and lichens and an important carbohydrate reserve in brown algae. Mannitol is used as a sweetener in certain foodstuffs.

mannose A *monosaccharide hexose, $C_6H_{12}O_6$, stereoisomeric with glucose, that occurs naturally only in polymerized forms called *mannans*. These are found in plants, fungi, and bacteria, serving as food energy stores.

manometer A device for measuring pressure differences, usually by the difference in height of two liquid columns. The simplest type is the U-tube manometer, which consists of a glass tube bent into the shape of a U. If a pressure to be measured is fed to one side of the U-tube and the other is open to the atmosphere, the difference in level of the liquid in the two limbs gives a measure of the unknown pressure.

mantissa *See* logarithm.

mantle 1. (in zoology) The fold of skin covering the dorsal surface of the body of molluscs, which extends into lateral flaps that protect the gills in the *mantle cavity* (the space between the body and mantle). The outer surface of the mantle secretes the shell (in species that have shells). **2.** (in geology) *See* earth.

map projections The methods used to represent the spherical surface of the earth on a plane surface. The circles of latitude and longitude are represented by a network or graticule of lines. Directions, areas, distances, and shape can never all be recreated accurately and the map projection chosen for a particular area will thus depend on the purpose for which the map is to be used and on the part of the world represented. There are three main groups of projections:

(1) *Cylindrical projections* are obtained by projecting the globe onto a cylinder that intersects the earth. If the axis of the cylinder is parallel to the axis of the earth the meridians and parallels will appear as straight lines. The three basic types of cylindrical projection are the simple cylindrical, the equal-area, and Mercator's projections.

(2) *Conic* (*conical*) *projections* result from the projection of the meridians and parallels onto a cone. In conic projections the axis of the cone is usually parallel to the earth's axis; as a result the meridians appear as radiating straight lines and the parallels are shown as concentric arcs. Scale is correct only along the standard parallel (i.e. the parallel along which the cone intersects the globe).

(3) *Azimuthal* (*zenithal*) *projections* are constructed as if a plane is placed at a tangent to the earth's surface and the portion of the earth covered is transferred onto the plane. On this projection all great circles that pass through the centre of the projection appear as straight lines; all points have their true compass bearings. Examples of azimuthal projections include the polar azimuthal and Lambert's azimuthal.

marble A metamorphic rock composed of recrystallized *calcite or *dolomite. Pure marbles are white but such impurities as silica or clay minerals result in variations of colour. Marble is extensively used for building purposes and ornamental use; the pure white marble from Carrara in Italy is especially prized by sculptors. The term is applied commercially to any limestone or dolomite that can be cut and polished.

marker gene 1. A gene used to identify a particular bacterial colony or bacteriophage plaque. Such genes are incorporated into cloning *vectors to enable the isolation and replication of colonies containing a desired vector. Typically, marker genes confer resistance to specific antibiotics or produce colour changes. **2.** A gene or short sequence of DNA that acts as a tag for another, closely linked, gene. Such markers are used in mapping the order of genes along chromosomes and in following the inheritance of particular genes: genes closely linked to the marker will generally be inherited with it. Markers must be readily identifiable in the phenotype, for instance by controlling an easily observable feature (such as eye colour). *Restriction fragment length polymorphisms (RFLPs) are now widely used as markers, for example in tracking the inheritance of genes associated with diseases (*see* gene tracking).

Markovnikoff's rule When an acid HA adds to an alkene, a mixture of products can be formed if the alkene is not symmetrical. For instance, the reaction between $C_2H_5CH{:}CH_2$ and HCl can give $C_2H_5CH_2CH_2Cl$ or $C_2H_5CHClCH_3$. In general, a mixture of products occurs in which one predominates over the other. In 1870, W. Markovnikoff proposed the rule that the main product would be the one in which the hydrogen atom adds to the carbon having the larger number of hydrogen atoms (the latter product above). This occurs when the mechanism is *electrophilic addition, in which the first step is addition of H^+. The electron-releasing effect of the alkyl group (C_2H_5) distorts the electron-distribution in the double bond, making the carbon atom furthest from the alkyl group negative. This is the atom attacked by H^+ giving the carbonium ion $C_2H_5C^+HCH_3$, which further reacts with the negative ion Cl^-.

In some circumstances *anti-Markovnikoff* behaviour occurs, in which the opposite effect is found. This happens when the mechanism involves free radicals and is common in addition of hydrogen bromide when peroxides are present.

Mars A planet having its orbit between the earth and Jupiter. Its mean distance from the sun is 227.94 × 10^6 km and its mean diameter 6762 km; it has a *sidereal period of 686.98 days. In 1976 two Viking Landers reached the surface of Mars revealing a thin atmosphere (less than 1% of the pressure at the earth's surface) of carbon dioxide with about 5% of nitrogen, argon, oxygen, and water vapour. The surface is covered with a thin layer of dust and frost and the surface material was revealed to be an iron-rich clay. Temperatures ranged between 150 K and 250 K. Mars has two small satellites, Phobos (approximately 20 × 23 × 28 km) and Deimos (10 × 12 × 16 km), neither of which is sufficiently massive to have contracted to a sphere.

marsh gas Methane formed by rotting vegetation in marshes.

Marsh's test A chemical test for arsenic in which hydrochloric acid and zinc are added to the sample, arsine being produced by the nascent hydrogen generated. Gas from the sample is led through a heated glass tube and, if arsine is present, it decomposes to give a brown deposit of arsenic metal. The arsenic is distinguished from antimony (which gives a similar result) by the fact that antimony does not dissolve in sodium chlorate(I) (hypochlorite). The test was devised in 1836 by the British chemist James Marsh.

marsupials See Metatheria.

martensite A solid solution of carbon in alpha-iron (see iron) formed when *steel is cooled too rapidly for pearlite to form from austenite. It is responsible for the hardness of quenched steel.

mascagnite A mineral form of *ammonium sulphate, $(NH_4)_2SO_4$.

mascon A gravitational anomaly on the surface of the moon resulting from a concentration of mass below the lunar surface. They occur in circular lunar maria and were caused either by the mare basalt as it flooded the basins or by uplift of high-density mantle material when the basins were formed.

maser (microwave amplification by stimulated emission of radiation) A device for amplifying or generating *microwaves by means of stimulated emission (see laser). As oscillators, masers are used in *atomic clocks, while they are used as amplifiers in *radio astronomy, being especially suitable for amplifying feeble signals from space.

In the ammonia gas maser (devised in 1954) a molecular beam of ammonia passes through a small orifice into a vacuum chamber, where it is subjected to a nonuniform electric field. This field deflects ground-state ammonia molecules, shaped like a pyramid with the three hydrogen atoms forming the plane of the base and the single nitrogen atom forming the apex. The ground-state molecule has a dipole moment on account of its lack of symmetry and it is for this reason that it suffers deflection. Excited molecules, in which the nitrogen atom vibrates back and forth through the plane of the hydrogen atoms, have no resultant dipole moment and are not deflected. The beam, now consisting predominately of excited molecules, is passed to a resonant cavity fed with the microwave radiation corresponding to the energy difference between the excited and the ground states. This causes stimulated emission as the excited molecules fall to the ground state and the input microwave radiation is amplified coherently. This arrangement can also be made to

mass

oscillate and in this form is the basis of the *ammonia clock.

In the more versatile *solid-state maser* a magnetic field is applied to the electrons of paramagnetic (*see* magnetism) atoms or molecules. The energy of these electrons is quantized into two levels, depending on whether or not their spins are parallel to the magnetic field. The situation in which there are more parallel magnetic moments than antiparallel can be reversed by sudden changes in the magnetic field. This electron-spin resonance in paramagnetic materials allows amplification over broader bandwidths than gas masers.

mass A measure of a body's *inertia, i.e. its resistance to acceleration. According to Newton's laws of motion, if two unequal masses, m_1 and m_2, are allowed to collide, in the absence of any other forces both will experience the same force of collision. If the two bodies acquire accelerations a_1 and a_2 as a result of the collision, then $m_1a_1 = m_2a_2$. This equation enables two masses to be compared. If one of the masses is regarded as a standard of mass, the mass of all other masses can be measured in terms of this standard. The body used for this purpose is a 1-kg cylinder of platinum–iridium alloy, called the international standard of mass. Mass defined in this way is called the *inertial mass* of the body. Mass can also be defined in terms of the gravitational force it produces. Thus, according to Newton's law of gravitation, $m_g = Fd^2/MG$, where M is the mass of a standard body situated a distance d from the body of mass m_g; F is the gravitational force between them and G is the *gravitational constant. The mass defined in this way is the *gravitational mass*. In the 19th century Roland Eötvös (1848 –1919) showed experimentally that gravitational and inertial mass are indistinguishable, i.e. $m_i = m_g$.

Although mass is formally defined in terms of its inertia, it is usually measured by gravitation. The *weight* (*W*) of a body is the force by which a body is gravitationally attracted to the earth corrected for the effect of rotation and equals the product of the mass of the body and the *acceleration of free fall (*g*), i.e. $W = mg$. In the general language, weight and mass are often used synonymously; however, for scientific purposes they are different. Mass is measured in kilograms; weight, being a force, is measured in newtons. Weight, moreover, depends on where it is measured, because the value of *g* varies at different localities on the earth's surface. Mass, on the other hand, is constant wherever it is measured, subject to the special theory of *relativity. According to this theory, announced by Albert Einstein in 1905, the mass of a body is a measure of its total energy content. Thus, if the energy of a body increases, for example by an increase in kinetic energy or temperature, then its mass will increase. According to this law an increase in energy ΔE is accompanied by an increase in mass Δm, according to the *mass–energy equation* $\Delta m = \Delta E/c^2$, where *c* is the speed of light. Thus, if 1 kg of water is raised in temperature by 100 K, its internal energy will increase by 4 × 10^{-12} kg. This is, of course, a negligible increase and the mass–energy equation is only significant for extremely high energies. For example, the mass of an electron is increased sevenfold if it moves relative to the observer at 99% of the speed of light.

mass action The law of mass action states that the rate at which a chemical reaction takes place at a given temperature is proportional to the product of the *active masses* of the reactants. The active mass of a reactant is taken to be its molar concentration. For example, for a reaction

424

$$A + B \rightarrow C$$

the rate is given by

$$R = k[A][B]$$

where k is the *rate constant. The principle was introduced by C. M. Guldberg and P. Waage in 1863. It is strictly correct only for ideal gases. In real cases *activities can be used.

mass concentration *See* concentration.

mass decrement *See* mass defect.

mass defect 1. The difference between the rest mass of an atomic nucleus and the sum of the rest masses of its individual nucleons in the unbound state. It is thus the mass equivalent of the *binding energy on the basis of the mass–energy equation (*see* mass; relativity). 2. (*or* **mass decrement**) The difference between the rest mass of a radioactive nucleus before decay and the total rest mass of the decay products.

mass–energy equation *See* mass; relativity.

massicot *See* lead(II) oxide.

mass number *See* nucleon number.

mass spectroscopy A technique used to determine relative atomic masses and the relative abundance of isotopes, and for chemical analysis and the study of ion reactions. In a *mass spectrometer* a sample (usually gaseous) is ionized and the positive ions produced are accelerated into a high-vacuum region containing electric and magnetic fields. These fields deflect and focus the ions onto a detector. The fields can be varied in a controlled way so that ions of different types can impinge on the detector. A *mass spectrum* is thus obtained consisting of a series of peaks of variable intensity to which mass/charge (m/e) values can be assigned. The original ions are usually produced by electron impact, although ion impact, photoionization, and field ionization are also used. For organic molecules, the mass spectrum consists of a series of peaks, one corresponding to the parent ion and the others to fragment ions produced by the ionization process. Different molecules can be identified by their characteristic pattern of lines. Analysis of mixtures can be done by gas chromatography–mass spectroscopy (*see* gas chromatography).

mass spectrum *See* spectrum.

mast cells Large amoeboid cells that are found in vertebrate *connective tissue. They may be connected with the deposition of fat. In certain situations they produce *histamine, responsible for allergy symptoms. They also produce *heparin, an anticoagulant, which may prevent fibrin from forming clots in the intercellular spaces.

mastoid process An outgrowth from the temporal bone of the skull containing air cavities that communicate with the cavity of the middle ear. In man it is a route through which infection may spread from the middle ear.

masurium A former name for *technetium.

matrix (*pl.* **matrices**) 1. (in chemistry) A continuous solid phase in which particles (atoms, ions, etc.) are embedded. Unstable species, such as free radicals, can be trapped in an unreactive substrate, such as solid argon, and studied by spectroscopy. The species under investigation are separated by the matrix, hence the term *matrix isolation* for this technique. 2. (in geology) The fine-grained material of rock in which the coarser-grained material is embedded. 3. (in mathematics) A set of quantities in a rectangular array, used in certain mathematical operations. The array is usually enclosed in large parentheses or in square brackets. 4. (in histology) The component of tissues (e.g. bone and cartilage) in which the cells of the tissue are embedded.

maxilla

maxilla 1. One of a pair of
*mouthparts in insects, crustaceans,
centipedes, and millipedes. They lie
behind the *mandibles and their lateral movements assist in feeding.
Crustaceans have two pairs of maxillae but in insects the second pair are
fused together forming the *labium.
2. One of a pair of large tooth-
bearing bones in the upper jaw of
vertebrates. In mammals they carry
all the upper teeth except the incisors.

**maximum and minimum thermom-
eter** A thermometer designed to
record both the maximum and minimum temperatures that have occurred
over a given time period. It usually
consists of a graduated capillary tube
at the base of which is a bulb containing ethanol. The capillary contains a thin thread of mercury with a
steel index at each end. As the temperature rises the index is pushed up
the tube, where it remains in position
to show the maximum temperature
reached; as the temperature falls the
lower index is pushed down the tube
and similarly remains in position at
the lowest temperature. The indexes
are reset by means of a permanent
magnet.

maximum permissible dose *See*
dose.

maxwell A unit of magnetic flux in
the *c.g.s. system, equal to the flux
through 1 square centimetre perpendicular to a magnetic field of 1 gauss.
1 maxwell is equal to 10^{-8} weber. It
is named after James Clerk Maxwell
(1831–79).

Maxwell–Boltzmann distribution
A law describing the distribution of
speeds among the molecules of a gas.
In a system consisting of N molecules
that are independent of each other
except that they exchange energy on
collision, it is clearly impossible to
say what velocity any particular molecule will have. However, statistical
statements regarding certain functions
of the molecules were worked out by

James Clerk Maxwell (1831–79) and
Ludwig Boltzmann (1844–1906). One
form of their law states that $n =
N\exp(-E/RT)$, where n is the number
of molecules with energy in excess of
E, T is the thermodynamic temperature, and R is the *gas constant.

McLeod gauge

Maxwell's equations A set of differential equations describing the space
and time dependence of the electromagnetic field and forming the basis
of classical electrodynamics. In *SI
units the equations are:

(1) $\mathrm{div}D = \rho$
(2) $\mathrm{curl}E = -\partial B/\partial t$
(3) $\mathrm{div}B = 0$
(4) $\mathrm{curl}H = \partial D/\partial t + J$

where D is the electric displacement,
E is the electric field strength, B is
the magnetic flux density, H is the
magnetic field strength, ρ is the volume charge density, and J is the electric current density. Note that in relativity and particle physics it is common to use *Gaussian or *Heaviside–

426

Lorentz units, in which case Maxwell's equations include 4π and the speed of light c. Maxwell's equations have the following interpretation. Equation (1) represents *Coulomb's law; equation (2) represents *Faraday's laws of electromagnetic induction; equation (3) represents the absence of *magnetic monopoles; equation (4) represents a generalization of *Ampère's law.

McLeod gauge A vacuum pressure gauge in which a relatively large volume of a low-pressure gas is compressed to a small volume in a glass apparatus (see illustration). The volume is reduced to an extent that causes the pressure to rise sufficiently to support a column of fluid high enough to read. This simple device, which relies on *Boyle's law, is suitable for measuring pressures in the range 10^3 to 10^{-1} pascal.

mean See average.

mean free path The average distance travelled between collisions by the molecules in a gas, the electrons in a metallic crystal, the neutrons in a moderator, etc. According to the *kinetic theory the mean free path between elastic collisions of gas molecules of diameter d (assuming they are rigid spheres) is $1/\sqrt{2}n\pi d^2$, where n is the number of molecules per unit volume in the gas. As n is proportional to the pressure of the gas, the mean free path is inversely proportional to the pressure.

mean free time The average time that elapses between the collisions of the molecules in a gas, the electrons in a crystal, the neutrons in a moderator, etc. See mean free path.

mean life See decay.

mean solar day See day.

meatus A small canal or passage in the body. An example is the *external auditory meatus* of the *outer ear in mammals, which connects the exterior opening to the eardrum.

mechanical advantage See force ratio.

mechanical equivalent of heat Symbol J. The ratio of a unit of mechanical energy to the equivalent unit of thermal energy, when a system of units is used in which they differ. J has the value 4.1868×10^7 ergs per calorie. The concept loses its usefulness in *SI units in which all forms of energy are expressed in joules and J therefore has a value of 1.

mechanics The study of the interactions between matter and the forces acting on it. *Statics is broadly concerned with the action of forces when no change of momentum is concerned, while *dynamics deals with cases in which there is a change of momentum. *Kinematics is the study of the motion of bodies without reference to the forces affecting the motion. These classical sciences are concerned with macroscopic bodies in the solid state, while *fluid mechanics is the science of the interaction between forces and fluids.

mechanism (in chemistry) The way in which a particular chemical reaction occurs, described in terms of the steps involved. For example, the hydrolysis of an alkyl chloride proceeds by the S_N1 mechanism (see nucleophilic substitution).

mechanoreceptor A *receptor that responds to such mechanical stimuli as touch, sound, and pressure. The skin is rich in mechanoreceptors.

median 1. The middle number or value in a series of numbers or values. **2.** A straight line in a triangle that joins the vertex to the mid-point of the base.

median eye (pineal eye) An eyelike structure, with a lens and retina, found on the top of the head of some lizards, *Sphenodon*, and the Cyclostomata (lampreys) as well as in many fossil vertebrates. Its function is unknown.

median lethal dose *See* LD$_{50}$.

mediastinum 1. A membrane in the midline of the *thorax of mammals that separates the lungs. **2.** The space between the two lungs, which is occupied by the heart and oesophagus.

medium frequency (MF) A radio frequency in the range 0.3–3 megahertz; i.e. having a wavelength in the range 100–1000 metres.

medulla 1. (in zoology) The central tissue of various organs, including the adrenal glands (*adrenal medulla*) and kidneys (*renal medulla*). **2.** (in botany) *See* pith.

medulla oblongata Part of the vertebrate *brainstem, derived from the *hindbrain, that is continuous with the spinal cord. Its function is to regulate the reflex responses controlling respiration, heart beat, blood pressure, and other involuntary processes. It gives rise to many of the *cranial nerves.

medullary ray (ray) Any of the vertical plates of *parenchyma cells running radially through the cylinder of vascular tissue in the stems and roots of plants. Each may be one to many cells in width. *Primary medullary rays* occur in young plants and in those not showing secondary thickening; they pass from the cortex through to the pith. *Secondary medullary rays* are produced by the vascular *cambium and terminate in xylem and phloem tissues. Medullary rays store and transport food materials.

medullated nerve fibre A nerve fibre that is characterized by a *myelin sheath, which insulates the axon.

medusa The free-swimming stage in the life cycle of the *Coelenterata. Medusae are umbrella-shaped, with tentacles round the edge and the mouth in the centre underneath. They swim by pulsations of the body and reproduce sexually. In the Hydrozoa (e.g. *Hydra*) they alternate in the life cycle with *polyps, from which they are produced by budding. In the Scyphozoa, which includes all the common jellyfish, the medusa is the dominant form and the polyp is reduced or absent.

mega- 1. Symbol M. A prefix used in the metric system to denote one million times. For example, 10^6 volts = 1 megavolt (MV). **2.** A prefix denoting large size; e.g. meganucleus, megasporangium.

megaphyll A type of foliage leaf in ferns and seed plants that has branched or parallel vascular bundles running through the lamina. The megaphylls of ferns are large pinnate leaves called *fronds*. A megaphyll was formerly called a *macrophyll*. *Compare* microphyll.

megaspore *See* sporophyll.

megasporophyll *See* sporophyll.

megaton weapon A nuclear weapon with an explosive power equivalent to one million tons of TNT. *Compare* kiloton weapon.

meiosis (reduction division) A type of cell division that gives rise to four reproductive cells (gametes) each with half the chromosome number of the parent cell. Two consecutive divisions occur (see illustration). In the first, *homologous chromosomes become paired and may exchange genetic material (*see* crossing over) before moving away from each other into separate daughter nuclei. This is the actual reduction division because each of the two nuclei so formed contains only half of the original chromosomes. The daughter nuclei then divide by mitosis and four *haploid cells are produced. *See also* prophase; metaphase; anaphase; telophase.

Meissner effect The falling off of the magnetic flux within a superconducting metal when it is cooled to a temperature below the critical temperature in a magnetic field. It was

1st Prophase: Leptotene

centriole centromere

1st Prophase: Pachytene

1st Prophase: Diplotene

chiasmata

1st Prophase: Diakinesis

nuclear membrane chromatids

1st Metaphase

pole equator

1st Anaphase

1st Telophase

developing cell membrane

2nd Metaphase

pole spindle fibres

2nd Anaphase

2nd Telophase

The stages of meiosis in a cell containing two pairs of homologous chromosomes

discovered by Walther Meissner in 1933 when he observed that the earth's magnetic field was expelled from the interior of tin crystals below 3.72 K, indicating that as *superconductivity appeared the material became perfectly diamagnetic. *See* magnetism.

melamine A white crystalline compound, $C_3N_6H_6$. Melamine is a cyclic compound having a six-membered ring of alternating C and N atoms, with three NH_2 groups. It can be copolymerized with methanal to give thermosetting *melamine resins*, which are used particularly for laminated coatings.

melanin Any of a group of polymers, derived from the amino acid tyrosine, that cause pigmentation of eyes, skin, and hair in vertebrates. Melanins are produced by specialized epidermal cells called *melanophores* (or *melanocytes*); their dispersion in these cells is controlled by *melanocyte-stimulating hormone* and *melatonin. Certain invertebrates, fungi, and microorganisms also produce melanin pigments. The 'ink' of the octopus and squid is a notable example. Hereditary *albinism is caused by the absence of the enzyme tyrosinase, which is necessary for melanin production.

melanism Black coloration of the body caused by overproduction of the pigment melanin, often as a reaction to the environment. There are several species of melanic moths in industrially polluted areas (*see* industrial melanism) and the panther is a melanic form of leopard.

melatonin A hormone derived from *serotonin and secreted by the pineal gland and retinas of vertebrates. Melatonin secretion by the pineal is linked to the dark–light cycle of the organism's environment, being greatest at night and lowest by day. The hormone is involved in regulating certain diurnal and seasonal changes in the body, such as the reproductive cycle in seasonally breeding animals. Melatonin also controls pigmentation changes; it triggers aggregation of the pigment *melanin into melanophores in the skin, causing the skin to turn pale.

melting point (m.p.) The temperature at which a solid changes into a liquid. A pure substance under standard conditions of pressure (usually 1 atmosphere) has a single reproducible melting point. If heat is gradually and uniformly supplied to a solid the consequent rise in temperature stops at the melting point until the fusion process is complete.

membrane A sheetlike tissue that covers, connects, or lines biological cells and their organelles, organs, and other structures. Membranes consist of polar lipids (predominantly the phospholipids lecithin and cephalin) and protein: the lipids form a double layer (bilayer) with protein molecules suspended within it. Membranes are permeable to water and fat-soluble substances but impermeable to polar substances (e.g. sugars). Movement of ions across a membrane occurs by carrier-mediated processes, including *active transport and *facilitated diffusion. Large molecules may be taken up by *phagocytosis. *See also* cell membrane; osmosis.

membrane bone (dermal bone) *Bone formed directly in connective tissue, rather than by replacing cartilage (*compare* cartilage bone). Some face bones, skull bones, and part of the clavicle are membrane bones. Small areas of membrane become jelly-like and attract calcium salts. Bone-forming cells break down these areas forming a bone lattice, which eventually fills in.

membranous labyrinth The soft tubular sensory structures that form the *inner ear of vertebrates and are housed within the bony labyrinth.

Mendeleev's law *See* periodic law.

mendelevium Symbol Md. A radioactive metallic transuranic element belonging to the *actinoids; a.n. 101; mass number of the first discovered nuclide 256 (half-life 1.3 hours). Several short-lived isotopes have now been synthesized. The element was first identified by A. Ghiorso, G. T. Seaborg, and associates in 1955. The alternative name *unnilunium* has been proposed.

Mendelism The theory of heredity that forms the basis of classical *genetics, proposed by Gregor Mendel (1822–84) in 1866 and formulated in two laws (*see* Mendel's laws; particulate inheritance). Mendel suggested that individual characteristics were determined by inherited 'factors', and when improved microscopes revealed details of cell structure the behaviour of Mendel's factors could be related to the behaviour of chromosomes during *meiosis.

Mendel's laws Two laws summarizing Gregor Mendel's theory of inheritance (*see also* Mendelism). The *Law of Segregation* states that each hereditary characteristic is controlled by two 'factors' (now called *alleles), which segregate (separate) and pass into separate germ (reproductive) cells. The *Law of Independent Assortment* states that pairs of 'factors' segregate independently of each other when germ cells are formed (*see also* independent assortment). These laws are the foundation of genetics.

Mendius reaction A reaction in which an organic nitrile is reduced by nascent hydrogen (e.g. from sodium in ethanol) to a primary amine:

$$RCN + 2H_2 \rightarrow RCH_2NH_2$$

meninges The three membranes that surround the brain and spinal cord of vertebrates: the *pia mater, the *arachnoid membrane, and the outer *dura mater. The pia and arachnoid are separated by the *subarachnoid space*, which contains *cerebrospinal fluid.

meniscus 1. A concave or convex upper surface that forms on a liquid in a tube as a result of *surface tension. **2.** See concave.

menstrual cycle The approximately monthly cycle of events associated with *ovulation that replaces the *oestrous cycle in most primates (including humans). The lining of the uterus becomes progressively thicker with more blood vessels in preparation for the *implantation of a fertilized egg cell (blastocyst). Ovulation occurs during the middle of the cycle (the fertile period). If fertilization does not occur the uterine lining breaks down and is discharged from the body (*menstruation*). In women the fertile period is 11–15 days after the end of the last menstruation.

menstruation See menstrual cycle.

menthol A white crystalline terpene alcohol, $C_{10}H_{19}OH$; r.d. 0.89; m.p. 42°C; b.p. 212°C. It has a minty taste and is found in certain essential oils (e.g. peppermint) and used as a flavouring.

mercaptans See thiols.

mercapto group See thiols.

mercuric compounds Compounds of mercury in its +2 oxidation state; e.g. mercuric chloride is mercury(II) chloride, $HgCl_2$.

mercurous compounds Compounds of mercury in its +1 oxidation state; e.g. mercury(I) chloride is mercurous chloride, HgCl.

mercury Symbol Hg. A heavy silvery liquid metallic element belonging to the *zinc group; a.n. 80; r.a.m. 200.59; r.d. 13.55; m.p. −38.87°C; b.p. 356.58°C. The main ore is the sulphide cinnabar (HgS), which can be decomposed to the elements. Mercury is used in thermometers, barometers, and other scientific apparatus, and in dental amalgams. The element is less reactive than zinc and cadmium and will not displace hydrogen from acids. It is also unusual in forming mercury(I) compounds containing the Hg_2^{2+} ion, as well as mercury(II) compounds containing Hg^{2+} ions. It also forms a number of complexes and organomercury compounds (e.g. the *Grignard reagents).

Mercury The smallest of the major planets in the *solar system and the nearest to the sun (57.91×10^6 km distant). Its equatorial diameter is 4878 km and its *sidereal period 87.969 days. The Mariner 10 spacecraft passed within 1000 km of Mercury in 1975 and revealed surface features similar to the moon, with impact craters dominating a bleak landscape that has been bombarded for some thousand million years by meteoric debris. There is no atmosphere and a maximum surface temperature of 700 K at the equator at perihelion, which rapidly cools to 110 K at night.

mercury cell A primary *voltaic cell consisting of a zinc anode and a cathode of mercury(II) oxide (HgO) mixed with graphite. The electrolyte is potassium hydroxide (KOH) saturated with zinc oxide, the overall reaction being:

$$Zn + HgO \rightarrow ZnO + Hg$$

The e.m.f. is 1.35 volts and the cell will deliver about 0.3 ampere-hour per cm^3.

mercury(I) chloride A white salt, Hg_2Cl_2; r.d. 7.0; m.p. 302°C; b.p. 384°C. It is made by heating mercury(II) chloride with mercury and is used in calomel cells (so called because the salt was formerly called *calomel*) and as a fungicide.

mercury(II) chloride A white salt, $HgCl_2$; r.d. 5.4; m.p. 276°C; b.p. 303°C. It is made by reacting mercury with chlorine and used in making other mercury compounds.

mercury(II) oxide A yellow or red oxide of mercury, HgO. The red form is made by heating mercury in oxygen at 350°C; the yellow form, which differs from the red in particle size, is precipitated when sodium hydroxide solution is added to a solution of mercury(II) nitrate. Both forms decompose to the elements at high temperature. The black precipitate formed when sodium hydroxide is added to mercury(I) nitrate solution is sometimes referred to as mercury(I) oxide (Hg_2O) but is probably a mixture of HgO and free mercury.

mercury(II) sulphide A red or black compound, HgS, occurring naturally as the minerals cinnabar (red) and metacinnabar (black). It can be obtained as a black precipitate by bubbling hydrogen sulphide through a solution of mercury(II) nitrate. The red form is obtained by sublimation. The compound is also called *vermilion* (used as a pigment).

mercury-vapour lamp A type of discharge tube in which a glow discharge takes place in mercury vapour. The discharge takes place in a transparent tube of fused silica or quartz into the ends of which molybdenum and tungsten electrodes are sealed; this tube contains argon and a small amount of pure mercury. A small arc is struck between a starter electrode and one of the main electrodes causing local ionization of some argon atoms. The ionized atoms diffuse through the tube causing the main discharge to strike; the heat from this vaporizes the mercury droplets, which become ionized current carriers. Radiation is confined to four visible wavelengths in the visible spectrum and several strong ultraviolet lines. The light is bluish but can be changed by the use of *phosphors on an outer tube. The outer tube is also usually used to filter out excessive ultraviolet radiation. The lamp is widely used for street lighting on account of its low cost and great reliability and as a source of ultraviolet radiation.

mericarp *See* schizocarp.

meridian 1. *See* latitude and longitude. 2. **(magnetic meridian)** An imaginary great circle on the earth's surface that passes through the north and south magnetic poles. A compass needle on the earth's surface influenced only by the earth's magnetic field (*see* geomagnetism) comes to rest along a magnetic meridian. 3, **(celestial meridian)** A great circle of the *celestial sphere that passes through the zenith and the celestial poles. It meets the horizon at the north and south points.

meristem A plant tissue consisting of actively dividing cells that give rise to cells that differentiate into new tissues of the plant. The most important meristems are those occurring at the tip of the shoot and root (*see* apical meristem) and the lateral meristems in the older parts of the plant (*see* cambium; cork cambium).

merocrine secretion *See* secretion.

mesencephalon *See* midbrain.

mesentery A thin sheet of tissue, bounded on each side by *peritoneum, that supports the gut and other organs in the body cavities of animals. Vertebrates have a well-developed dorsal mesentery that anchors the stomach and intestine and contains blood vessels and nerves supplying the gut. The reproductive organs and their ducts are also supported by mesenteries.

mesocarp *See* pericarp.

mesoderm The layer of cells in the *gastrula that lies between the *ectoderm and *endoderm. It develops into the muscles, circulatory system, and sex organs and in vertebrates also into the excretory system and skeleton. *See also* germ layers.

mesoglea The gelatinous noncellular layer between the endoderm and ectoderm in the body wall of coelenterates. It may be thin, as in *Hydra*, or tough and fibrous, as in the larger jellyfish and sea anemones. It often contains cells that have migrated from the two body layers but these do not form tissues and organs and the mesoglea is not homologous with the mesoderm of *triploblastic animals.

meso-isomer *See* optical activity.

mesomerism A former name for *resonance in molecules.

meson Any of a class of *elementary particles that are a subclass of the *hadrons. According to current quark theory mesons consist of quark–antiquark pairs. They exist with positive, negative, and zero charges, but when charged the charge has the same magnitude as that of the electron. They include the *kaon*, *pion*, and *psi* particles. Mesons are believed to participate in the forces that hold nucleons together in the nucleus. The muon, originally called a mu-meson, was thought to be a meson but is now recognized as a *lepton.

meson-catalysed fusion *See* nuclear fusion.

mesophyll The internal tissue of a leaf blade (lamina), consisting of *parenchyma cells. There are two distinct forms. *Palisade mesophyll* lies just beneath the upper epidermis and consists of cells elongated at right angles to the leaf surface. They contain a large number of *chloroplasts

and their principal function is photosynthesis. *Spongy mesophyll* occupies most of the remainder of the lamina. It consists of spherical loosely arranged cells containing fewer chloroplasts than the palisade mesophyll. Between these cells are air spaces leading to the *stomata.

mesophyte Any plant adapted to grow in soil that is well supplied with water and mineral salts. Such plants wilt easily when exposed to drought conditions as they are not adapted to conserve water. The majority of flowering plants are mesophytes. *Compare* hydrophyte; xerophyte.

mesothelium A single layer of thin platelike cells covering the surface of the inside of the abdominal cavity and thorax and surrounding the heart, forming part of the *peritoneum and *pleura (*see* serous membrane). It is derived from the *mesoderm. *Compare* endothelium; epithelium.

Mesozoic The geological era that extended from the end of the *Palaeozoic era, about 225 million years ago, to the beginning of the *Cenozoic era, about 65 million years ago. It comprises the *Triassic, *Jurassic, and *Cretaceous periods. The Mesozoic era is often known as the *Age of Reptiles* as these animals, which included the dinosaurs, pterosaurs, and ichthyosaurs, became the dominant lifeform; most became extinct before the end of the era.

messenger RNA *See* RNA.

meta- **1.** Prefix designating a benzene compound in which two substituents are in the 1,3 positions on the benzene ring. The abbreviation *m*- is used; for example, *m*-xylene is 1,3-dimethylbenzene. *Compare* ortho-; para-. **2.** Prefix designating a lower oxo acid, e.g. metaphosphoric acid. *Compare* ortho-.

metabolism The sum of the chemical reactions that occur within living organisms. The various compounds that take part in or are formed by these reactions are called *metabolites*. In animals many metabolites are obtained by the digestion of food, whereas in plants only the basic starting materials (carbon dioxide, water, and minerals) are externally derived. The synthesis (*anabolism) and breakdown (*catabolism) of most compounds occurs by a number of reaction steps, the reaction sequence being termed a *metabolic pathway*. Some pathways (e.g. *glycolysis) are linear; others (e.g. the *Krebs cycle) are cyclic. The changes at each step in a pathway are usually small and are promoted by efficient biological catalysts – the enzymes. In this way the amounts of energy required or released at any given stage are minimal, which helps in maintaining a constant *internal environment. Various *feedback mechanisms exist to govern metabolic rates.

metaboric acid *See* boric acid.

metacarpal One of the bones in the *metacarpus.

metacarpus The hand (or corresponding part of the forelimb) in terrestrial vertebrates, consisting of a number of rod-shaped bones (*metacarpals*) that articulate with the bones of the wrist (*see* carpus) and those of the fingers (*see* phalanges). The number of metacarpals varies between species: in the basic *pentadactyl limb there are five, but this number is reduced in many species.

metal Any of a class of chemical elements that are typically lustrous solids that are good conductors of heat and electricity. Not all metals have all these properties (e.g. mercury is a liquid). In chemistry, metals fall into two distinct types. Those of the *s*-and *p*-blocks (e.g. sodium and aluminium) are generally soft silvery reactive elements. They tend to form positive ions and so are described as electropositive. This is contrasted with

typical nonmetallic behaviour of forming negative ions. The *transition elements (e.g. iron and copper) are harder substances and generally less reactive. They form coordination complexes. All metals have oxides that are basic.

metaldehyde A solid compound, $C_4O_4H_4(CH_3)_4$, formed by polymerization of ethanal (acetaldehyde) in dilute acid solutions below $0°C$. The compound, a tetramer of ethanal, is used in slug pellets and as a fuel for portable stoves.

metal fatigue A cumulative effect causing a metal to fail after repeated applications of *stress, none of which exceeds the ultimate *tensile strength. The *fatigue strength* (or *fatigue limit*) is the stress that will cause failure after a specified number (usually 10^7) of cycles. The number of cycles required to produce failure decreases as the level of stress or strain increases. Other factors, such as corrosion, also reduce the fatigue life.

metallic bond A chemical bond of the type holding together the atoms in a solid metal or alloy. In such solids, the atoms are considered to be ionized, with the positive ions occupying lattice positions. The valence electrons are able to move freely (or almost freely) through the lattice, forming an 'electron gas'. The bonding force is electrostatic attraction between the positive metal ions and the electrons. The existence of free electrons accounts for the good electrical and thermal conductivities of metals. *See also* energy bands.

metallic crystal A crystalline solid in which the atoms are held together by *metallic bonds. Metallic crystals are found in some *interstitial compounds as well as in metals and alloys.

metallized dye *See* dyes.

metallocene A type of organometallic complex in which one or more aromatic rings (e.g. $C_5H_5^-$ or C_6H_6) coordinate to a metal ion or atom by the pi electrons of the ring. *Ferrocene was the first such compound to be discovered.

metallography The microscopic study of the structure of metals and their alloys. Both optical *microscopes and *electron microscopes are used in this work.

metalloid (semimetal) Any of a class of chemical elements intermediate in properties between metals and nonmetals. The classification is not clear cut, but typical metalloids are boron, silicon, germanium, arsenic, and tellurium. They are electrical semiconductors and their oxides are amphoteric.

metallurgy The branch of applied science concerned with the production of metals from their ores, the purification of metals, the manufacture of alloys, and the use and performance of metals in engineering practice. *Process metallurgy* is concerned with the extraction and production of metals, while *physical metallurgy* concerns the mechanical behaviour of metals.

metameric segmentation (metamerism; segmentation) The division of an animal's body (except at the head region – *see* cephalization) into a number of compartments (*segments* or *metameres*) each containing the same organs. Metameric segmentation is most strongly marked in annelid worms (e.g. earthworms), in which the muscles, blood vessels, nerves, etc. are repeated in each segment. In these animals the segmentation is obvious both externally and internally. It also occurs internally in arthropods and in the embryonic development of all vertebrates, in which it is confined to parts of the muscular, skeletal, and nervous systems and does not show externally.

metamict state The amorphous state of a substance that has lost its crystalline structure as a result of the radioactivity of uranium or thorium. *Metamict minerals* are minerals whose structure has been disrupted by this process. The metamictization is caused by alpha-particles and the recoil nuclei from radioactive disintegration.

metamorphic rocks One of the three major rock categories (*see also* igneous rocks, sedimentary rocks). Metamorphic rock is formed when pre-existing rock is subjected to either chemical or physical alteration by heat, pressure, or chemically active fluids. It involves three main processes of formation. *Contact metamorphism* results from the intrusion of a mass of molten rock into sedimentary rock. Heat from the intrusion spreads into the surrounding sediments causing mineralogical changes to take place. *Regional metamorphism* is developed over large areas and is associated with mountain building. Sediments collect in large depressions, known as geosynclines, in the earth's crust. As successive layers accumulate the lower layers subside into the crust and are subjected to increasing heat and pressure, causing the rocks to be metamorphosed. Eventually these rocks may also be uplifted and folded to form mountain chains. *Dislocation metamorphism* results from the more localized mechanical shearing and crushing of rocks, for example along fault planes.

Metamorphic rocks are characteristically resistant and tend to form upland areas. Examples of metamorphic rocks include marble (metamorphosed limestone) and slate (metamorphosed shale).

metamorphosis The rapid transformation from the larval to the adult form that occurs in the life cycle of many invertebrates and amphibians. Examples are the changes from a tadpole to an adult frog and from a pupa to an adult insect. Metamorphosis often involves considerable destruction of larval tissues by lysosomes, and in both insects and amphibians it is controlled by hormones.

metaphase The second stage of cell division, during which the membrane around the nucleus breaks down, the *spindle forms, and centromeres attach the chromosomes to the equator of the spindle. In the first metaphase of *meiosis pairs of chromosomes (bivalents) are attached, while in *mitosis and the second metaphase of meiosis, individual chromosomes are attached.

metaphosphoric acid *See* phosphoric(VI) acid.

metaplasia The transformation of a tissue into a different type. This is an abnormal process; for example, metaplasia of the epithelium of the bronchi may be an early sign of cancer.

metaplumbate *See* plumbate.

metastable state A condition of a system in which it has a precarious stability that can easily be disturbed. It is unlike a state of stable equilibrium in that a minor disturbance will cause a system in a metastable state to fall to a lower energy level. A book lying on a table is in a state of stable equilibrium; a thin book standing on edge is in metastable equilibrium. Supercooled water is also in a metastable state. It is liquid below 0°C; a grain of dust or ice introduced into it will cause it to freeze. An excited state of an atom or nucleus that has an appreciable lifetime is also metastable.

metastannate *See* stannate.

metatarsal One of the bones in the *metatarsus.

metatarsus The foot (or corresponding part of the hindlimb) in terrestrial vertebrates, consisting of a number of rod-shaped bones (*metatarsals*) that articulate with the bones of the ankle (*see* tarsus) and those of the toes (*see* phalanges). The number of metatar-

sals varies between species: in the basic *pentadactyl limb there are five, but this number is reduced in some species.

Metatheria A subclass of mammals containing the marsupials. The female bears an abdominal pouch (*marsupium*) into which the newly born young, which are in a very immature state, move to complete their development. They obtain nourishment from the mother's mammary teats. Modern marsupials are restricted to Australasia (where they include the kangaroos, koala bears, phalangers, and bandicoots) and America (the opossums). Marsupials evolved during the late Cretaceous period, 80 million years ago. In Australia, where the marsupials have been isolated for millions of years, they show the greatest diversity of form, having undergone *adaptive radiation to many of the niches occupied by placental mammals elsewhere. *Compare* Eutheria; Prototheria.

metathesis *See* double decomposition.

Metazoa The subkingdom of animals in which the body is composed of many cells grouped into tissues and coordinated by a nervous system. It comprises all vertebrates and invertebrates except the *Protozoa and *Porifera (sponges).

meteor A streak of light observable in the sky when a particle of matter enters the earth's atmosphere and becomes incandescent as a result of friction with atmospheric atoms and molecules. These particles of matter are known collectively as *meteoroids*. Meteoroids that survive their passage through the atmosphere and strike the earth's surface are known as *meteorites*. Only some 2500 meteorites are known, excluding the *micrometeorites* (bodies less than 1 mm in diameter). Meteorites consist mainly of silicate materials (stony meteorites) or iron (iron meteorites). It is estimated that the earth collects over 10^8 kg of meteoritic material every year, mostly in the form of micrometeorites. Micrometeorites survive atmospheric friction because their small size enables them to radiate away the heat generated by friction before they vaporize.

meteorite *See* meteor.

meteorology The study of the physical phenomena and processes taking place in the atmosphere and its interactions with the ground surface. This knowledge is applied to weather forecasting. The chief branches of meteorology are *dynamical meteorology, *micrometeorology, and *synoptic meteorology.

methacrylate A salt or ester of methacrylic acid (2-methylpropenoic acid).

methacrylate resins *Acrylic resins obtained by polymerizing 2-methylpropenoic acid or its esters.

methacrylic acid *See* 2-methylpropenoic acid.

methanal (formaldehyde) A colourless gas, HCHO; r.d. 0.815 (at $-20°C$); m.p. $-92°C$; b.p. $-19°C$. It is the simplest *aldehyde, made by the catalytic oxidation of methanol ($500°C$; silver catalyst) by air. It forms two polymers: *methanal trimer and polymethanal. *See also* formalin.

methanal trimer A cyclic trimer of methanal, $C_3O_3H_6$, obtained by distillation of an acidic solution of methanal. It has a six-membered ring of alternating $-O-$ and $-CH_2-$ groups.

methane A colourless odourless gas, CH_4; m.p. $-182.5°C$; b.p. $-161°C$. Methane is the simplest hydrocarbon, being the first member of the *alkane series. It is the main constituent of natural gas ($\sim99\%$) and as such is an important raw material for producing other organic compounds. It can be

converted into methanol by catalytic oxidation.

methanide *See* carbide.

methanoate (formate) A salt or ester of methanoic acid.

methanoic acid (formic acid) A colourless pungent liquid, HCOOH; r.d. 1.2; m.p. $8°C$; b.p. $101°C$. It can be made by the action of concentrated sulphuric acid on the sodium salt (sodium methanoate), and occurs naturally in ants and stinging nettles. Methanoic acid is the simplest of the *carboxylic acids.

methanol (methyl alcohol) A colourless liquid, CH_3OH; r.d. 0.79; m.p. $-98°C$; b.p. $64°C$. It is made by catalytic oxidation of methane (from natural gas) using air. Methanol is used as a solvent (*see* methylated spirits) and as a raw material for making methanal (mainly for urea—formaldehyde resins). It was formerly made by the dry distillation of wood (hence the name *wood alcohol*).

methionine *See* amino acid.

method of mixtures A method of determining the specific heat capacities of liquids or a liquid and a solid by mixing known masses of the substances at different temperatures and measuring the final temperature of the mixture.

methoxy group The organic group CH_3O-.

methyl acetate *See* methyl ethanoate.

methyl alcohol *See* methanol.

methylamine A colourless flammable gas, CH_3NH_2; m.p. $-92°C$; b.p. $-7°C$. It can be made by a catalytic reaction between methanol and ammonia and is used in the manufacture of other organic chemicals.

methylated spirits A mixture consisting mainly of ethanol with added methanol ($\sim9.5\%$), pyridine ($\sim0.5\%$), and blue dye. The additives are included to make the ethanol undrinkable so that it can be sold without excise duty for use as a solvent and a fuel (for small spirit stoves).

methylation A chemical reaction in which a methyl group (CH_3-) is introduced in a molecule. A particular example is the replacement of a hydrogen atom by a methyl group, as in a *Friedel–Crafts reaction.

methylbenzene (toluene) A colourless liquid, $CH_3C_6H_5$; r.d. 0.9; m.p. $-94°C$; b.p. $111°C$. Methylbenzene is derived from benzene by replacement of a hydrogen atom by a methyl group. It can be obtained from coal tar or made from methylcyclohexane (extracted from crude oil) by catalytic dehydrogenation. Its main uses are as a solvent and as a raw material for producing TNT.

methyl bromide *See* bromomethane.

2-methylbuta-1,3-diene *See* isoprene.

methyl chloride *See* chloromethane.

methylene The highly reactive *carbene, $:CH_2$. The divalent CH_2 group in a compound is the *methylene group*.

methylene blue A blue dye used in optical microscopy to stain nuclei of animal tissues. It is also suitable as a vital stain and a bacterial stain.

methyl ethanoate (methyl acetate) A colourless volatile fragrant liquid, CH_3COOCH_3; r.d. 0.92; m.p. $-98°C$; b.p. $54°C$. A typical *ester, it can be made from methanol and methanoic acid and is used mainly as a solvent.

methyl ethyl ketone *See* butanone.

methyl group *or* **radical** The organic group CH_3-.

methyl methacrylate An ester of methacrylic acid (2-methylpropenoic acid), $CH_2:C(CH_3)COOCH_3$, used in making *methacrylate resins.

methyl orange An organic dye used as an acid–base *indicator. It changes from red below pH 3.1 to yellow above pH 4.4 (at 25°C) and is used for titrations involving weak bases.

methylphenols (cresols) Organic compounds having a methyl group and a hydroxyl group bound directly to a benzene ring. There are three isomeric methylphenols with the formula $CH_3C_6H_4OH$, differing in the relative positions of the methyl and hydroxyl groups. A mixture of the three can be obtained by distilling coal tar and is used as a germicide and antiseptic.

2-methylpropenoic acid (methacrylic acid) A white crystalline unsaturated soluble carboxylic acid, $CH_2:C(CH_3)COOH$, used in making *methacrylate resins.

methyl red An organic dye similar in structure and use to methyl orange. It changes from red below pH 4.4 to yellow above pH 6.0 (at 25°C).

metre Symbol m. The SI unit of length, being the length of the path travelled by light in vacuum during a time interval of $1/(2.99\,792\,458 \times 10^8)$ second. This definition, adopted by the General Conference on Weights and Measures in October, 1983, replaced the 1967 definition based on the krypton lamp, i.e. $1\,650\,763.73$ wavelengths in a vacuum of the radiation corresponding to the transition between the levels $2p^{10}$ and $5d^5$ of the nuclide krypton–86. This definition (in 1958) replaced the older definition of a metre based on a platinum–iridium bar of standard length. When the *metric system was introduced in 1791 in France, the metre was intended to be one ten-millionth of the earth's meridian quadrant passing through Paris. However, the original geodetic surveys proved the impractibility of such a standard and the original platinum metre bar, the *mètre des archives*, was constructed in 1793.

metre bridge *See* Wheatstone bridge.

metric system A decimal system of units originally devised by a committee of the French Academy, which included J. L. Lagrange and P. S. Laplace, in 1791. It was based on the *metre, the gram defined in terms of the mass of a cubic centimetre of water, and the second. This centimetre-gram-second system (*see* c.g.s. units) later gave way for scientific work to the metre-kilogram-second system (*see* m.k.s. units) on which *SI units are based.

metric ton (tonne) A unit of mass equal to 1000 kg or 2204.61 lb. 1 tonne = 0.9842 ton.

metrology The scientific study of measurement, especially the definition and standardization of the units of measurement used in science.

MHC *See* major histocompatibility complex.

MHD *See* magnetohydrodynamics.

mho A reciprocal ohm, the former name of the unit of electrical *conductance now known as the siemens.

mica Any of a group of silicate minerals with a layered structure. Micas are composed of linked SiO_4 tetrahedra with cations and hydroxyl groupings between the layers. The general formula of the micas is $X_2Y_{4-6}Z_8O_{20}(OH,F)_4$, where X = K,Na,Ca; Y = Al,Mg,Fe,Li; and Z = Si,Al. The three main mica minerals are:

*muscovite, $K_2Al_4(Si_6Al_2O_{20})(OH,F)_4$;
*biotite, $K_2(Mg,Fe^{2+})_{6-4}(Fe^{3+},Al,Ti)_{0-2}$ $(Si_{6-5}Al_{2-3}O_{20})(OH,F)_4$;
lepidolite, $K_2(Li,Al)_{5-6}(Si_{6-7}Al_{2-1}O_{20})$-$(OH,F)_4$.
Micas have perfect basal cleavage and the thin cleavage flakes are flexible and elastic. Flakes of mica are used as electrical insulators and as the dielectric in capacitors.

micelle An aggregate of molecules in a *colloid. For example, when soap or other *detergents dissolve in water they do so as micelles – small clusters of molecules in which the nonpolar hydrocarbon groups are in the centre and the hydrophilic polar groups are on the outside solvated by the water molecules. Phospholipids in aqueous solution also form micelles. The products of fat digestion are dispersed into micelles by the action of bile salts, which facilitates their absorption in the small intestine.

Michelson–Morley experiment An experiment, conducted in 1887 by Albert Michelson (1852–1931) and Edward Morley (1838–1923), that attempted to measure the velocity of the earth through the *ether. Using a modified *Michelson interferometer* (see illustration) they expected to observe a shift in the interference fringes formed when the instrument was rotated through 90°, showing that the speed of light measured in the direction of the earth's rotation, or orbital motion, is not identical to its speed at right angles to this direction. No shift was observed. An explanation was finally provided by the *Lorentz–Fitzgerald contraction, which provided an important step in the formulation of Einstein's special theory of *relativity and the abandonment of the ether concept.

Michelson interferometer

micro- 1. A prefix denoting very small size; e.g. microgamete, micronucleus. 2. Symbol μ. A prefix used in the metric system to denote one millionth. For example, 10^{-6} metre = 1 micrometre (μm).

microbalance A sensitive *balance capable of weighing masses of the order 10^{-6} to 10^{-9} kg.

microbiology The scientific study of microorganisms (e.g. bacteria, viruses, and fungi). Originally this was directed towards their effects (e.g. in causing disease and decay), but during the 20th century the emphasis has shifted to their physiology, biochemistry, and genetics. Microbes are now recognized as important vehicles for the study of biochemical and genetic processes common to all living organisms, and their rapid growth enables their laboratory culture in large numbers for studies in genetics.

microdissection (micromanipulation) A technique used for the dissection of living cells under the high power of an optical microscope. It utilizes minute mechanically manipulated instruments, such as needles, scalpels, and pipettes. For example, the instruments may be used to remove a single nucleus from one cell and to implant it in another.

microelectronics The techniques of designing and making electronic circuits of very small size. As a result of these techniques a single *silicon chip measuring less than a centimetre in either direction can contain many thousands of transistors and may constitute the central processing unit of a microcomputer. In addition to an enormous drop in size, compared to an equivalent valve-operated device, these microelectronic circuits are some 100 000 times more reliable than their thermionic predecessors.

microfossil A *fossil that is so small that it can only be studied under a microscope. Microfossils include bacteria, diatoms, and Protozoa and parts of organisms, such as plant pollen and skeletal fragments. Microfossils are important in the correlation of rocks where only small samples are available. The study of microfossils, particularly pollen, is known as *palynology.

micromanipulation *See* microdissection.

micrometeorite *See* meteor.

micrometeorology The branch of meteorology concerned with small-scale processes at work within the lowest layers of the atmosphere, including the interaction of the atmosphere with the ground surface. Examples of the processes studied include mountain and valley winds and land and sea breezes.

micrometer A gauge for measuring small diameters, thicknesses, etc., accurately. It consists of a G-shaped device in which the gap between the measuring faces is adjusted by means of an accurately calibrated screw, the end of which forms one of the measuring faces.

micron The former name for the *SI unit now called the micrometre, i.e. 10^{-6} m.

micronutrient A nutrient required by organisms in relatively small quantities. Micronutrients are typically found in cofactors and coenzymes. They include *vitamins and trace elements (*see* essential element). *Compare* macronutrient.

microorganism (microbe) Any organism that can be observed only with the aid of a microscope. Microorganisms include bacteria, viruses, rickettsiae, protozoans, and some algae and fungi. *See* microbiology.

microphone A *transducer in which sound waves are converted into corresponding variations in an electrical signal for amplification, transmission to a distant point, or recording. Various types of device are used. In the

eye

eyepiece

first real image

objective

object

final virtual image

Compound microscope

dynamic microphone the sound waves impinge on a conductor of low mass supported in a magnetic field and cause it to oscillate at the frequency of the sound waves. These movements induce an e.m.f. in the conductor that is proportional to its velocity. The moving conductor consists of a metal ribbon, a wire, or a coil of wire. In the *moving-iron microphone*, sound waves cause a light armature to oscillate so that it varies the reluctance of a magnetic circuit. In a coil surrounding this path the varying reluctance is experienced as a variation in the magnetic flux within it, which induces a corresponding e.m.f. In the *carbon microphone*, widely used in telephones, a diaphragm constitutes a movable electrode in contact with carbon granules, which are also in contact with a fixed electrode. The movement of the diaphragm, in response to the

sound waves, varies the resistance of the path through the granules to the fixed electrode. *See also* capacitor microphone; crystal microphone.

microphyll A type of foliage leaf seen in certain pteridophytes (e.g. *Selaginella, Lycopodium*) that has a single unbranched midrib. Such leaves are generally no more than a few millimetres long. *Compare* megaphyll.

microprocessor *See* computer.

micropyle A small opening in the surface of a plant ovule through which the pollen tube passes prior to fertilization. It results from the incomplete covering of the nucellus by the integuments. It remains as an opening in the testa of most seeds through which water is absorbed.

microscope A device for forming a magnified image of a small object. The *simple microscope* consists of a biconvex magnifying glass or an equivalent system of lenses, either hand-held or in a simple frame. The *compound microscope* (see illustration) uses two lenses or systems of lenses, the second magnifying the real image formed by the first. The lenses are usually mounted at the opposite ends of a tube that has mechanical controls to move it in relation to the object. An optical condenser and mirror, often with a separate light source, provide illumination of the object. The widely used *binocular microscope* consists of two separate instruments fastened together so that one eye looks through one while the other eye looks through the other. This gives stereoscopic vision and reduces eye strain. *See also* atomic force microscope; electron microscope; field-emission microscope; field-ionization microscope; phase-contrast microscope; scanning tunnelling microscope; ultraviolet microscope.

microspore *See* sporophyll.

microsporophyll *See* sporophyll.

microtome A machine used to cut thin sections (3–5 μm thick) of plant or animal tissue for microscopical observation. There are various designs of microtome, each basically consisting of a steel knife, a block for supporting the specimen, and a device for moving the specimen towards the knife. The specimen is usually supported by being embedded in wax; if a *freezing microtome* is used, the specimen is frozen. An *ultramicrotome* is used to cut much thinner sections (20–100 nm thick) for electron microscopy. The biological material is embedded in plastic or resin, sectioned with a glass or diamond knife, and the cut sections are allowed to float on the surface of water in an adjacent water bath.

microtubule A minute protein filament in living cells that occurs singly or in pairs, triplets, or bundles. Microtubules help certain cells (e.g. nerve cells) to maintain their shape; they also occur in *cilia, *flagella, and the *centrioles and form the *spindle during nuclear division.

microvillus One of a number of minute finger-like projections on the free surfaces of epithelial cells. Microvilli are covered with cell membrane and their cytoplasm is continuous with the main cell cytoplasm. Their purpose is probably to increase the absorptive or secretory surface area of the cell, and they are abundant on the villi of the intestine.

microwave background radiation A cosmic background of radiation in the frequency range 3×10^{11} hertz to 3×10^8 hertz discovered in 1965. Believed to have emanated from the primordial fireball of the big bang with which the universe is thought to have originated (*see* big-bang theory), the radiation has an energy density in intergalactic space of some 4×10^{-14} J m^{-3}.

microwave optics The study of the behaviour of microwaves by analogy with the behaviour of light waves. On the large scale microwaves are propagated in straight lines and, like light waves, they undergo reflection, refraction, diffraction, and polarization.

microwaves Electromagnetic waves with wavelengths in the range 10^{-3} to 0.03 m.

microwave spectroscopy A sensitive technique for chemical analysis and the determination of molecular structure (bond lengths, bond angles, and dipole moments), and also relative atomic masses. It is based on the principle that microwave radiation (*see* microwaves) causes changes in the rotational energy levels of molecules and absorption consequently occurs at characteristic frequencies. In a microwave spectrometer a microwave source, usually a klystron valve, produces a beam that is passed through a gaseous sample. The beam then impinges on the detector, usually a crystal detector, and the signal (wavelength against intensity) is displayed, either as a printed plot or on an oscilloscope. As microwaves are readily absorbed by air the instrument must be evacuated.

midbrain (mesencephalon) One of the three sections of the brain of a vertebrate embryo. Unlike the *forebrain and the *hindbrain, the midbrain does not undergo further subdivision to form additional zones. In mammals it becomes part of the *brainstem, but in amphibians, reptiles, and birds the roof of the midbrain becomes enlarged as the *tectum*, a dominant centre for integration, and may include a pair of *optic lobes*.

middle ear (tympanic cavity) The air-filled cavity within the skull of vertebrates that lies between the *outer ear and the *inner ear. It is linked to the pharynx (and therefore to outside air) via the *Eustachian tube and in mammals contains the three *ear ossicles, which transmit auditory vibrations from the outer ear (via the *tympanum) to the inner ear (via the *fenestra ovalis).

middle lamella A thin layer of material, consisting mainly of pectins, that binds together the walls of adjacent plant cells.

midgut 1. The middle section of the alimentary canal of vertebrates, which is concerned with digestion and absorption. It comprises most of the small intestine. **2.** The middle section of the alimentary canal of arthropods. *See also* foregut; hindgut.

migration 1. (in chemistry) The movement of a group, atom, or double bond from one part of a molecule to another. **2.** (in physics) The movement of ions under the influence of an electric field. **3.** (in biology) The seasonal movement of complete populations of animals to a more favourable environment. It is usually a response to lower temperatures resulting in a reduced food supply, and is often triggered by a change in day length (*see* photoperiodism). Migration is common in mammals (e.g. porpoises), fish (e.g. eels and salmon), and some insects but is most marked in birds. The Arctic tern, for example, migrates annually from its breeding ground in the Arctic circle to the Antarctic – a distance of some 1/600 km. Migrating animals possess considerable powers of orientation; birds seem to possess a compass sense, using the sun, pole stars, and (in cloud) the earth's magnetic lines of force as reference points.

milk The fluid secreted by the *mammary glands of mammals. It provides a balanced and highly nutritious food for offspring. Cows' milk comprises about 87% water, 3.6% lipids (triglycerides, phospholipids, cholesterol, etc.), 3.3% protein (largely casein), 4.7% lactose (milk sugar), and, in much smaller amounts, vitamins (especially vitamin A and many B vitamins) and minerals (notably calcium, phosphorus, sodium, potassium,

magnesium, and chlorine). Composition varies among species; human milk contains less protein and more lactose.

milk of magnesia *See* magnesium hydroxide.

milk sugar *See* lactose.

milk teeth *See* deciduous teeth.

Milky Way *See* galaxy.

milli- Symbol m. A prefix used in the metric system to denote one thousandth. For example, 0.001 volt = 1 millivolt (mV).

millibar *See* bar.

millipedes *See* Myriapoda.

mimicry The resemblance of one animal to another, which has evolved as a means of protection. In one form of mimicry the markings of certain harmless insects closely resemble the *warning coloration of another insect (the *model*). Predators that have learnt to avoid the model will also avoid good mimics of it. This phenomenon is often found among butterflies. A second form of mimicry involves the mutual resemblance of a group of animals, all harmful, such as the wasp, bee, and hornet, so that a predator, having experienced one, will subsequently avoid them all.

mineral A naturally occurring substance that has a characteristic chemical composition and, in general, a crystalline structure. The term is also often applied generally to organic substances that are obtained by mining (e.g. coal, petroleum, and natural gas) but strictly speaking these are not minerals, being complex mixtures without definite chemical formulas. Rocks are composed of mixtures of minerals. Minerals may be identified by the properties of their crystal system, hardness (measured on the Mohs' scale), relative density, lustre, colour, cleavage, and fracture. Many names of minerals end in *-ite*.

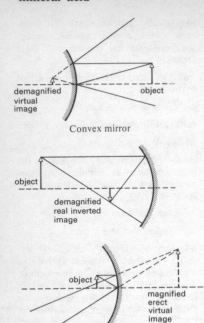

Convex mirror

Concave mirrors

mineral acid A common inorganic acid, such as hydrochloric acid, sulphuric acid, or nitric acid.

mineralocorticoid *See* corticosteroid.

mineralogy The branch of geology concerned with the study of *minerals.

minority carrier *See* semiconductor.

minor planets *See* asteroids.

minute 1. One sixtieth of an hour. **2.** One sixtieth of a degree (angle).

mirabilite A mineral form of *sodium sulphate, $Na_2SO_4.10H_2O$.

mirage An optical phenomenon that occurs as a result of the bending of light rays through layers of air having very large temperature gradients. An *inferior mirage* occurs when the ground surface is strongly heated and the air near the ground is much warmer that the air above. Light rays from the sky are strongly refracted upwards near the surface giving the appearance of a pool of water. A *superior mirage* occurs if the air close to the ground surface is much colder than the air above. Light is bent downwards from the object towards the viewer so that it appears to be elevated or floating in the air.

mirror A surface that reflects most of the light falling on it. A *plane mirror* is a flat surface that produces an erect virtual *image of a real object, in which front and back are reversed. *Spherical mirrors* are formed from the surfaces of spheres and form images of real objects in much the same way as lenses. A convex mirror forms erect virtual images. They are commonly used as rear-view mirrors in road vehicles, and give a diminished wide-angle image. A concave mirror can form either inverted real images or erect virtual images. (See illustrations.) Spherical mirrors obey the *lens equation (using the real-positive sign convention) and are subject to some *aberrations similar to those of lenses.

misch metal An alloy of cerium (50%), lanthanum (25%), neodymium (18%), praseodymium (5%), and other rare earths. It is used alloyed with iron (up to 30%) in lighter flints, and in small quantities to improve the malleability of iron. It is also added to copper alloys to make them harder, to aluminium alloys to make them stronger, to magnesium alloys to reduce creep, and to nickel alloys to reduce oxidation.

missing mass The mass of matter in the universe that cannot be observed by direct observations of its emitted or absorbed electromagnetic radiation. There are a number of astrophysical observations that suggest that the actual mass of the universe is much greater than that estimated by observations using optical telescopes, radiotelescopes, etc. It is thought that

Prophase

centriole — nuclear membrane — cytoplasm

chromosome — chromatids — centromere

(a)   (b)

Metaphase   Anaphase

pole of spindle — equator of spindle

pole of spindle — spindle fibres

Telophase

The stages of mitosis in a cell containing two pairs of homologous chromosomes

there is a considerable amount of *dark matter* (or *hidden matter*) causing this discrepancy. Various explanations have been put forward for the missing mass, including black holes, brown dwarfs, cosmic strings, axions, neutrinos, monopoles, and various exotic particles, such as *weakly interacting massive particles* (*WIMPS*).

mitochondrion A structure within the cytoplasm of plant and animal *cells that carry out aerobic respiration: it is the site of the *Krebs cycle and *electron transport chain, and therefore the cell's energy production. Mitochondria vary greatly in shape, size, and number but are typically oval or sausage-shaped and bounded by two membranes, the inner one being folded into finger-like projections (*cristae*). They are most numerous in cells with a high level of metabolic activity.

mitosis The division of a cell to form two daughter cells each having a nucleus containing the same number and kind of chromosomes as the mother cell. The changes during divisions are clearly visible with a light microscope. Each chromosome divides lengthwise into two *chromatids, which separate and form the chromosomes of the resulting daughter nuclei. The process is divided into four stages, *prophase, *metaphase, *anaphase, and *telophase, which merge into each other (see illustration). Mitotic divisions ensure that all the cells of an individual are genetically identical to each other and to the original fertilized egg.

mitral valve (biscuspid valve) A valve, consisting of two flaps, situated between the left atrium and the left ventricle of the heart of birds and mammals. When the left ventricle contracts, forcing blood into the aorta, the mitral valve closes the aperture to the left atrium, thereby preventing any backflow of blood. The valve reopens to allow blood to flow from the atrium into the ventricle. *Compare* tricuspid valve.

Mitscherlich's law (law of isomorphism) Substances that have the same crystal structure have similar chemical formulae. The law can be used to determine the formula of an unknown compound if it is isomorphous with a compound of known formula.

mixture A system of two or more distinct chemical substances. *Homogeneous mixtures* are those in which the atoms or molecules are interspersed, as in a mixture of gases or in a solution. *Heterogeneous mixtures* have distinguishable phases, e.g. a mixture of iron filings and sulphur. In a mixture there is no redistribution of valence electrons, and the components retain their individual chemical properties. Unlike compounds, mixtures can be separated by physical means (distillation, crystallization, etc.).

m.k.s. units A *metric system of units devised by A. Giorgi (and sometimes known as *Giorgi units*) in 1901. It is based on the metre, kilogram, and second and grew from the earlier *c.g.s. units. The electrical unit chosen to augment these three basic units was the ampere and the *permeability of space (magnetic constant) was taken as 10^{-7} H m^{-1}. To simplify electromagnetic calculations the magnetic constant was later changed to $4\pi \times 10^{-7}$ H m^{-1} to give the *rationalized MKSA system*. This system, with some modifications, formed the basis of *SI units, now used in most scientific work.

m.m.f. *See* magnetomotive force.

mmHg A unit of pressure equal to that exerted under standard gravity by a height of one millimetre of mercury, or 133.322 pascals.

mobility (of an ion) Symbol *u*. The terminal speed of an ion in an electric field divided by the field strength.

moderator A substance that slows down free neutrons in a *nuclear reactor, making them more likely to cause fissions of atoms of uranium–235 and less likely to be absorbed by atoms of uranium–238. Moderators are light elements, such as deuterium (in heavy water), graphite, and beryl-lium, to which neutrons can impart some of their kinetic energy on collision without being captured. Neutrons that have had their energies reduced in this way (to about 0.025 eV, equivalent to a speed of 2200 m s^{-1}) are said to have been *thermalized* or to have become *thermal neutrons*.

modern synthesis *See* neo-Darwinism.

modulation The process of superimposing the characteristics of a periodic signal onto a *carrier wave so that the information contained in the signal can be transmitted by the carrier wave. In *radio transmission, the simplest form of modulation is *amplitude modulation* (*AM*), in which the amplitude of the carrier is increased or diminished as the signal amplitude increases and diminishes. In *frequency modulation* (*FM*), the frequency of the carrier is increased or diminished as the signal amplitude increases and diminishes but the carrier amplitude

carrier

sine-wave signal

amplitude-modulated wave

phase-modulated wave

frequency-modulated wave

Modulation

remains constant. In *phase modulation*, the relative phase of the carrier is varied in accordance with the signal amplitude. (See illustrations.) Both frequency modulation and phase modulation are forms of *angle modulation*.

In *pulse modulation* the information is transmitted by controlling the amplitude, duration, position, or presence of a series of pulses. Morse code is a simple form of a pulse modulation.

modulus *See* absolute value.

modulus of elasticity *See* elastic modulus.

Moho (Mohorovičić discontinuity) A discontinuity within the *earth that marks the junction between the crust and the underlying mantle. Below the discontinuity earthquake seismic waves undergo a sudden increase in velocity, a feature that was first observed in 1909 by the Yugoslavian geophysicist Andrija Mohorovičić (1857–1936), after whom the discontinuity was named. The Moho lies at a depth of about 10–12 km below the oceans and about 33–35 km below the continents.

Mohs' scale A hardness scale in which a series of ten minerals are arranged in order, each mineral listed being scratched by and therefore softer than those below it. The minerals are: (1) talc; (2) gypsum; (3) calcite; (4) fluorite; (5) apatite; (6) orthoclase; (7) quartz; (8) topaz; (9) corundum; (10) diamond. As a rough guide a mineral with a value up to 2.5 on this scale can be scratched by a fingernail, up to 4 can be scratched by a coin, and up to 6 by a knife. The scale was devised by Friedrich Mohs (1773–1839).

molal concentration *See* concentration.

molality *See* concentration.

molar 1. (in physics and chemistry) Denoting that an extensive physical property is being expressed per *amount of substance, usually per mole. For example, the molar heat capacity of a compound is the heat capacity of that compound per unit amount of substance; in SI units it would be expressed in $J K^{-1} mol^{-1}$. **2.** (in chemistry) Having a concentration of one mole per dm^3. **3.** (in anatomy) A broad ridged tooth in the adult dentition of mammals, found at the back of the jaws behind the premolars. There are two or more molars on each side of both jaws; their surfaces are raised into ridges or *cusps for grinding food during chewing. In man the third (and most posterior) molar does not appear until young adulthood: these molars are known as *wisdom teeth*.

molar conductivity Symbol Λ. The conductivity of that volume of an electrolyte that contains one mole of solution between electrodes placed one metre apart.

molar heat capacity *See* heat capacity.

molarity *See* concentration.

molar latent heat *See* latent heat.

molar volume (molecular volume) The volume occupied by a substance per unit amount of substance.

mole Symbol mol. The SI unit of *amount of substance. It is equal to the amount of substance that contains as many elementary units as there are atoms in 0.012 kg of carbon–12. The elementary units may be atoms, molecules, ions, radicals, electrons, etc., and must be specified. 1 mole of a compound has a mass equal to its *relative molecular mass expressed in grams.

molecular beam A beam of atoms, ions, or molecules at low pressure, in which all the particles are travelling in the same direction and there are few collisions between them. They are formed by allowing a gas or vapour to pass through an aperture into an enclosure, which acts as a collimator by containing several additional apertures and vacuum pumps to remove any particles that do not pass through the apertures. Molecular beams are used in studies of surfaces and chemical reactions and in spectroscopy.

molecular biology The study of the structure and function of large molecules associated with living organisms, in particular proteins and the nucleic acids *DNA and *RNA. *Molecular genetics* is a specialized branch, concerned with the analysis of genes (*see* gene sequencing).

molecular distillation Distillation in high vacuum (about 0.1 pascal) with the condensing surface so close to the surface of the evaporating liquid that the molecules of the liquid travel to the condensing surface without collisions. This technique enables very much lower temperatures to be used than are used with distillation at atmospheric pressure and therefore heat-sensitive substances can be distilled. Oxidation of the distillate is also eliminated as there is no oxygen present.

molecular flow (Knudsen flow) The flow of a gas through a pipe in which the mean free path of gas molecules is large compared to the dimensions of the pipe. This occurs at low pressures; because most collisions are with the walls of the pipe rather than other gas molecules, the flow characteristics depend on the relative molecular mass of the gas rather than its viscosity. The effect was studied by M. H. C. Knudsen (1871–1949).

molecular formula *See* formula.

molecularity The number of molecules involved in forming the activated complex in a step of a chemical reaction. Reactions are said to be *unimolecular*, *bimolecular*, or *trimolecular* according to whether 1, 2, or 3 molecules are involved.

molecular orbital *See* orbital.

molecular sieve Porous crystalline substances, especially aluminosilicates (*see* zeolite), that can be dehydrated with little change in crystal structure. As they form regularly spaced cavities, they provide a high surface area for the adsorption of smaller molecules.

The general formula of these substances is $M_nO.Al_2O_3.xSiO_2.yH_2O$, where M is a metal ion and n is twice the reciprocal of its valency. Molecular sieves are used as drying agents and in the separation and purification of fluids. They can also be loaded with chemical substances, which remain separated from any reaction that is taking place around them, until they are released by heating or by displacement with a more strongly adsorbed substance. They can thus be used as cation exchange mediums and as catalysts and catalyst supports.

molecular volume *See* molar volume.

molecular weight *See* relative molecular mass.

molecule One of the fundamental units forming a chemical compound; the smallest part of a chemical compound that can take part in a chemical reaction. In most covalent compounds, molecules consist of groups of atoms held together by covalent or coordinate bonds. Covalent substances that form *macromolecular crystals have no discrete molecules (in a sense, the whole crystal is a molecule). Similarly, ionic compounds do not have single molecules, being collections of oppositely charged ions.

mole fraction Symbol X. A measure of the amount of a component in a mixture. The mole fraction of component A is given by $X_A = n_A/N$, where n_A is the amount of substance of A (for a given entity) and N is the total amount of substance of the mixture (for the same entity).

Moment of a force. For equilibrium $mgb + F_2(b+c) = F_1 a$, where mg is the weight of the beam acting through its centre of mass

Molisch's test *See* alpha-naphthol test.

Mollusca A phylum of soft-bodied invertebrates characterized by an unsegmented body differentiated into a *head*, a ventral muscular *foot* used in locomotion, and a dorsal *visceral hump* covered by a fold of skin – the *mantle – which secretes a protective shell in many species. Respiration is by means of gills or a lunglike organ and the feeding organ is a *radula. Molluscs occur in marine, freshwater, and terrestrial habitats and there are six classes, including the *Gastropoda (snails, slugs, limpets, etc.), *Lamellibranchia (bivalves, e.g. mussels, oysters), and *Cephalopoda (squids and octopuses).

molybdenum Symbol Mo. A silvery hard metallic *transition element; a.n. 42; r.a.m. 95.94; r.d. 10.22; m.p. 2610°C; b.p. 5560°C. It is found in molybdenite (MoS_2), the metal being extracted by roasting to give the oxide, followed by reduction with hydrogen. The element is used in alloy steels. Molybdenum(IV) sulphide (MoS_2) is used as a lubricant. Chemically, it is unreactive, being unaffected by most acids. It oxidizes at high temperatures and can be dissolved in molten alkali to give a range of molybdates and polymolybdates. Molybdenum was discovered in 1778 by Scheele.

moment of a force A measure of the turning effect produced by a force about an axis. The magnitude of the moment is the product of the force and the perpendicular distance from the axis to the line of action of the force. An object will be in rotational equilibrium if the algebraic sum of all the moments of the forces on it about any axis is zero. See illustration.

moment of inertia Symbol I. The moment of inertia of a massive body about an axis is the sum of all the products formed by multiplying the magnitude of each element of mass (δm) by the square of its distance (r) from the line, i.e. $I_m = \Sigma r^2 \delta m$. It is the analogue in rotational dynamics of mass in linear dynamics. The basic equation is $T = I\alpha$, where T is the torque causing angular acceleration α about the specified axis.

momentum The *linear momentum* (p) of a body is the product of its mass (m) and its velocity (v), i.e. $p = mv$. *See also* angular momentum.

monatomic molecule A 'molecule' consisting of only one atom (e.g. Ar or He), distinguished from diatomic and polyatomic molecules.

Mond process A method of obtaining pure nickel by heating the impure metal in a stream of carbon monoxide at 50–60°C. Volatile nickel carbonyl ($Ni(CO)_4$) is formed, and this can be decomposed at higher temperatures (180°C) to give pure nickel. The method was invented by the German–British chemist Ludwig Mond (1839–1909).

Monel metal An alloy of nickel (60–70%), copper (25–35%), and small quantities of iron, manganese, silicon, and carbon. It is used to make acid-resisting equipment in the chemical industry.

Monera In some systems of classification, the kingdom that includes all microorganisms with prokaryotic *cells. *See also* Protista.

monobasic acid An *acid that has only one acidic hydrogen atom in its molecules. Hydrochloric (HCl) and nitric (HNO_3) acids are common examples.

monochasium *See* cymose inflorescence.

monochromatic radiation Electromagnetic radiation, especially visible radiation, of only one frequency or wavelength. Completely monochromatic radiation cannot be produced, but *lasers produce radiation within a very narrow frequency band. *Compare* polychromatic radiation.

monochromator A device that provides *monochromatic radiation from a polychromatic source. In the case of visible radiation, for example, a prism can be used together with slits to select a small range of wavelengths.

monoclinic *See* crystal system.

monoclonal antibody A specific *antibody produced by one of numerous identical cells derived from a single parent cell. (The population of these cells comprises a *clone and each cell is said to be *monoclonal*.) The parent cell is obtained by the fusion of a normal antibody-producing cell (a lymphocyte) with a cell derived from a malignant tumour of *lymphoid tissue of a mouse. This hybrid cell then multiplies rapidly and yields large amounts of antibody. Monoclonal antibodies are used to identify a particular antigen within a mixture and can therefore be used for identifying blood groups; they also enable the production of highly specific, and therefore effective, *vaccines.

Monocotyledonae One of the two classes or subclasses of plants within the *Angiospermae, distinguished by having one seed leaf (*cotyledon) within the seed. The monocotyledons generally have parallel leaf veins, scattered vascular bundles within the stems, and flower parts in threes or multiples of three. Monocotyledon species include some crop plants (e.g. cereals, onions, fodder grasses), ornamentals (e.g. tulips, orchids, lilies), and a very limited number of trees (e.g. the palms). *Compare* Dicotyledonae.

monocyte The largest form of white blood cell (*leucocyte) in vertebrates. It has a kidney-shaped nucleus and is actively phagocytic, ingesting bacteria and cell debris (*see* phagocyte).

monoecious Describing plant species that have separate male and female flowers on the same plant. Examples of monoecious plants are maize and birch. *Compare* dioecious.

monohybrid The offspring of a cross between parents that differ in the alleles they possess for one particular gene, one parent having two dominant alleles and the other two recessives. All the offspring have one dominant and one recessive allele for that gene (i.e. they are hybrid at that one locus). Crossing between these offspring yields a characteristic 3:1 (monohybrid) ratio in the following generation of dominant:recessive phenotypes (see illustration).

monohydrate A crystalline compound having one molecule of water per molecule of compound.

monokine *See* cytokine.

monomer A molecule (or compound) that joins with others in forming a dimer, trimer, or polymer.

monophyletic Denoting any group of organisms that are assumed to have originated from the same ancestor, i.e. any family, class, etc., of a natural classification. Sometimes the term has a more limited meaning and designates only those groups that include *all* the descendants of a common ancestor. In this restricted sense the birds are considered monophyletic because they are the sole descendants of a group of arboreal Triassic reptiles but the modern reptiles are not, because their common amphibian ancestor also gave rise to the birds and mammals. Such groups as the reptiles are described as *paraphyletic*. *Compare* polyphyletic.

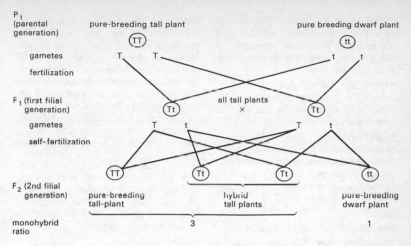

The inheritance of stem lengths in garden peas

monophyodont Describing a type of dentition that consists of a single set of teeth that last for the entire life-span of an animal. *Compare* diphyodont; polyphyodont.

monopodium The primary axis of growth in such plants as pine trees. It consists of a single main stem that continues to grow from the tip and gives rise to lateral branches. *Compare* sympodium.

monosaccharide (simple sugar) A carbohydrate that cannot be split into smaller units by the action of dilute acids. Monosaccharides are classified according to the number of carbon atoms they possess: *trioses* have three carbon atoms; *tetroses*, four; *pentoses*, five; *hexoses*, six; etc. Each of these is further divided into *aldoses* and *ketoses*, depending on whether the molecule contains an aldehyde group (–CHO) or a ketone group (–CO–). For example glucose, having six carbon atoms and an aldehyde group, is an *aldohexose* whereas fructose is a *ketohexose*. These aldehyde and ketone groups confer reducing properties on monosaccharides: they can be oxidized to yield sugar acids. They also react with phosphoric acid to produce phosphate esters (e.g. in *ATP), which are important in cell metabolism. Monosaccharides can exist as either straight-chain or ring-shaped molecules. They also exhibit *optical activity, giving rise to both dextrorotatory and laevorotatory forms.

monotremes *See* Prototheria.

monotropy *See* allotropy.

monovalent (univalent) Having a valency of one.

monozygotic twins *See* identical twins.

moon The earth's only natural satellite, which orbits the earth at a mean distance of 384 400 km. It has a diameter of 3476 km. It has no atmosphere or surface water. Its surface temperature varies between 80 K (night minimum) and 400 K (noon at the equator). It is the only celestial body outside the earth to have been reached by man (1969).

mordant A substance used in certain dyeing processes. Mordants are often inorganic oxides or salts, which are absorbed on the fabric. The dyestuff then forms a coloured complex with the mordant, the colour depending on the mordant used as well as the dyestuff. *See also* lake.

morphine An alkaloid present in opium. It is an analgesic and narcotic, used medically for the relief of severe pain.

morphogenesis The development, through growth and differentiation, of form and structure in an organism.

morphology The study of the form and structure of organisms, especially their external form. *Compare* anatomy.

mosaic gold *See* tin(IV) sulphide.

Moseley's law The frequencies of the lines in the *X-ray spectra of the elements are related to the atomic numbers of the elements. If the square roots of the frequencies of corresponding lines of a set of elements are plotted against the atomic numbers a straight line is obtained. The law was discovered by H. G. Moseley (1887–1915).

moss agate *See* agate.

Mössbauer effect An effect occurring when certain nuclides decay with emission of gamma radiation. For an isolated nucleus, the gamma radiation would usually have a spread of energies because the energy of the process is partitioned between the gamma-ray photon and the recoil energy of the nucleus. In 1957 R. L. Mössbauer (1929–) found that in certain solids, in which the emitting nucleus is held by strong forces in the lattice, the recoil energy is taken up by the whole lattice. Since this may typically contain $10^{10}-10^{20}$ atoms, the recoil energy is negligible and the energy of the emitted photon is sharply defined in a very narrow energy spread.

The effect is exploited in *Mössbauer spectroscopy* in which a gamma-ray source is mounted on a moving platform and a similar sample is mounted nearby. A detector measures gamma rays scattered by the sample. The source is moved slowly towards the sample at a varying speed, so as to continuously change the frequency of the emitted gamma radiation by the Doppler effect. A sharp decrease in the signal from the detector at a particular speed (i.e. frequency) indicates resonance absorption in the sample nuclei. The effect is used to investigate nuclear energy levels. In chemistry, Mössbauer spectroscopy can also give information about the bonding and structure of compounds because *chemical shifts* in the resonance energy are produced by the presence of surrounding atoms.

mosses *See* Musci.

moths *See* Lepidoptera.

motion A change in the position of a body or system with respect to time, as measured by a particular observer in a particular *frame of reference. Only relative motion can be measured; absolute motion is meaningless. *See also* equation of motion; Newton's laws of motion.

motivation The internal conditions responsible for temporary reversible changes in the responsiveness of an animal to external stimulation. Thus an animal that has been deprived of food will accept less palatable food than one that has not been deprived: the difference is attributed to a change in feeding motivation. Changes in responsiveness due to maturation, *learning, or injury are not usually readily reversible and are therefore not considered to be due to changes in motivation. Early attempts to describe motivation in terms of a number of separate 'drives' (e.g. food drive, sex drive) have not found general favour, partly because 'drives' interact with one another; for example, water deprivation often affects an animal's willingness to feed.

motor Any device for converting chemical energy or electrical energy into mechanical energy. *See* electric motor; internal-combustion engine; linear motor.

motor generator An electric motor mechanically coupled to an electric generator. The motor is driven by a supply of specified voltage, frequency, or number of phases and the generator provides an output in which one or more of these parameters is different to suit a particular purpose.

motor neurone A *neurone that transmits nerve impulses from the central nervous system to an effector organ (such as a muscle or gland) and thereby initiates a physiological response (e.g. muscle contraction).

moulting 1. The seasonal loss of hair, fur, or feathers that occurs in mammals and birds. **2.** The periodic loss of the integument of arthropods and reptiles. *See* ecdysis.

mouthparts Modified paired appendages on the head segments of arthropods, used for feeding. A typical insect has a *labium (lower lip), one pair each of *mandibles and *maxillae, and a *labrum (upper lip), although in many the mouthparts are modified to form piercing stylets or a sucking proboscis. Crustaceans, centipedes, and millipedes have one pair of mandibles and two pairs of maxillae used for cutting and holding the food. Crustaceans also have several pairs of *maxillipedes*.

moving-coil instrument A measuring instrument in which current or voltage is determined by the couple on a small coil pivoted between the poles of a magnet with curved poles, giving a radial magnetic field. When a current flows through the coil it turns against a return spring. If the angle through which it turns is α, the current I is given by $I = k\alpha/BAN$, where B is the magnetic flux density, A is the area of the coil and N is its number of turns; k is a constant depending on the strength of the return spring. The instrument is suit-

able for measuring d.c. but can be converted for a.c. by means of a rectifier network. It is usually made as a *galvanometer and converted to an ammeter or voltmeter by means of a *shunt or a *multiplier.

moving-iron instrument A measuring instrument in which current or voltage is determined by the force of attraction on a bar of soft iron pivoted within the magnetic field of a fixed coil or by the repulsion between the poles induced in two soft iron rods within the coil. As the deflection caused by the passage of a current through the coil does not depend on the direction of the current, moving-iron instruments can be used with either d.c. or a.c. without a rectifier. They are, however, less sensitive than *moving-coil instruments.

moving-iron microphone *See* microphone.

mucilage A gumlike substance frequently present in the cell walls of aquatic plants and in the seed coats of certain other species. Mucilages are hard when dry and slimy when wet. Like *gums they probably have a general protective function or serve to anchor the plant. Some organisms (e.g. certain bacteria) are completely covered with mucilage and in such cases it probably prevents water loss.

mucous membrane A layer of tissue comprising an epithelium supported on connective tissue. Within the epithelium are special *goblet cells*, which secrete *mucus onto the surface, and the epithelium often bears cilia. Mucous membranes line body cavities communicating with the exterior, including the alimentary and respiratory tracts. *Compare* serous membrane.

mucus The slimy substance secreted by goblet cells onto the surface of a *mucous membrane to protect and lubricate it and to trap bacteria, dust particles, etc. Mucus consists of water, the glycoprotein *mucin*, cells, and salts.

multicellular Describing tissues, organs, or organisms that are composed of a number of cells. *Compare* unicellular.

multicentre bond A bond formed between three, and sometimes more, atoms that contains only a single pair of electrons. The structure of *boranes can be explained by considering them to be *electron-deficient compounds containing multicentre bonds.

multifactorial inheritance The determination of a particular characteristic, e.g. height or skin colour, by many genes, each having a small effect individually. Characteristics controlled in this way show *continuous variation.

multimeter An electrical measuring instrument designed to measure potential differences or currents over a number of ranges. It also usually has an internal dry cell enabling resistances to be measured. Most multimeters are moving-coil instruments with a switch to enable series resistors or parallel resistors to be incorporated into the circuit.

multiple bond A bond between two atoms that contains more than one pair of electrons. Such bonds primarily involve sigma bonding with secondary contribution from pi bonding. *See* orbital.

multiple proportions *See* chemical combination.

multiplet 1. A spectral line formed by more than two (*see* doublet) closely spaced lines. **2.** A group of *elementary particles that are identical in all respects except that of electric charge.

multiplication factor Symbol *k*. The ratio of the average number of neutrons produced in a *nuclear reactor per unit time to the number of neutrons lost by absorption or leakage in

the same time. If $k = 1$, the reactor is said to be *critical*. If $k > 1$ it is *supercritical* and if $k < 1$ it is *subcritical*. *See also* critical reaction.

multiplier A fixed resistance used with a voltmeter, usually a *moving-coil instrument, to vary its range. Many voltmeters are provided with a series of multipliers from which the appropriate value can be selected. If the original instrument requires *i* amperes for full-scale deflection and the resistance of the moving coil is *r* ohms, the value *R* of the resistance of the multiplier required to give a full-scale deflection when a voltage *V* is applied across the terminals is given by $R = V/i - r$.

multivibrator An electronic *oscillator consisting of two active devices, usually transistors, interconnected in an electrical network. The purpose of the device is to generate a continuous square wave with which to store information in binary form in a logic circuit. This is achieved by applying a portion of the output voltage or current of each active device to the input of the other with the appropriate magnitude and polarity, so that the devices are conducting alternately for controllable periods.

Mumetal The original trade name for a ferromagnetic alloy, containing 78% nickel, 17% iron, and 5% copper, that had a high *permeability and a low *coercive force. More modern versions also contain chromium and molybdenum. These alloys are used in some transformer cores and for shielding various devices from external magnetic fields.

Muntz metal A form of *brass containing 60% copper, 39% zinc, and small amounts of lead and iron. Stronger than alpha-brass, it is used for hot forgings, brazing rods, and large nuts and bolts. It is named after G. F. Muntz (1794–1857).

muon *See* lepton; elementary particles.

Musci A class of plants within the *Bryophyta comprising the mosses, which are found in both damp (including freshwater) and drier situations. Mosses possess erect or prostrate leafy stems, which give rise to leafless stalks bearing capsules. Spores formed in the capsules are released and grow to produce new plants. A common moss is *Funaria*.

muscle A tissue consisting of sheets or bundles of cells (*muscle fibres*) that are capable of contracting, so producing movement or tension in the body. There are three types of muscle. *Voluntary muscle produces voluntary movement (e.g. at joints); *involuntary muscle mainly effects the movements of hollow organs (e.g. intestine and bladder); and *cardiac muscle occurs only in the heart.

muscovite (white mica; potash mica) A mineral consisting of potassium aluminosilicate, $K_2Al_4(Si_6Al_2)O_{20}(OH,F)_4$; one of the most important members of the *mica group of minerals. It is chemically complex and has a sheetlike crystal structure (*see* intercalation compound). It is usually silvery-grey in colour, sometimes tinted with green, brown, or pink. Muscovite is a common constituent of certain granites and pegmatites. It is also common in metamorphic and sedimentary rocks. It is widely used in industry, for example in the manufacture of electrical equipment and as a filler in roofing materials, wallpapers, and paint.

mustard gas A highly poisonous gas, $(ClCH_2CH_2)_2S$; dichlorodiethyl sulphide. It is made from ethene and disulphur dichloride (S_2Cl_2), and used as a war gas.

mutagen An agent that causes an increase in the number of mutants (*see* mutation) in a population. Mutagens operate either by causing changes in the DNA of the *genes, so interfering with the coding system, or by causing chromosome damage.

Various chemicals (e.g. *colchicine) and forms of radiation (e.g. X-rays) have been identified as mutagens.

mutarotation Change of optical activity with time as a result of spontaneous chemical reaction.

mutation A sudden random change in the genetic material of a cell that may cause it and all cells derived from it to differ in appearance or behaviour from the normal type. An organism affected by a mutation (especially one with visible effects) is described as a *mutant. Somatic mutations affect the nonreproductive cells and are therefore restricted to the tissues of a single organism but *germline mutations*, which occur in the reproductive cells or their precursors, may be transmitted to the organism's descendants and cause abnormal development.

Mutations occur naturally at a low rate but this may be increased by radiation and by some chemicals (*see* mutagen). Most (the *gene mutations*) consist of invisible changes in the DNA of the chromosomes, but some (the *chromosome mutations*) affect the appearance or the number of the chromosomes. An example of a chromosome mutation is that giving rise to *Down's syndrome.

The majority of mutations are harmful, but a very small proportion may increase an organism's *fitness; these spread through the population over successive generations by natural selection. Mutation is therefore essential for evolution, being the ultimate source of genetic variation.

mutual inductance *See* inductance.

mutualism An interaction between two species in which both species benefit. (The term *symbiosis is often used synonymously with mutualism.) A well-known example of mutualism is the association between termites and the specialized protozoans that inhabit their guts. The protozoans, unlike the termites, are able to digest the cellulose of the wood that the termites eat and release sugars that the termites absorb. The termites benefit by being able to use wood as a foodstuff, while the protozoans are supplied with food and a suitable environment.

mycelium A network of *hyphae that forms the body of a fungus. It consists of feeding hyphae together with reproductive hyphae, which produce *sporangia and *gametangia.

mycology The scientific study of *fungi.

Mycota *See* fungi.

myelin sheath (medullary sheath) The layer of fatty material that surrounds and electrically insulates the axons of most vertebrate and some invertebrate neurones. The myelin sheath enables a more rapid transmission of nerve impulses (at speeds up to 120 m s^{-1}). It consists of layers of membrane derived from *Schwann cells. The sheath is interrupted at intervals along the axon by *nodes of Ranvier*; myelinated sections of axon are called *internodes*.

myeloid tissue Tissue within red *bone marrow that produces the blood cells. It is found around the blood vessels and contains various cells that are precursors of the blood cells.

myoglobin A globular protein occurring widely in muscle tissue as an oxygen carrier. It comprises a single polypeptide chain and a *haem group, which reversibly binds a molecule of oxygen. This is only relinquished at relatively low external oxygen concentrations, e.g. during strenuous exercise when muscle oxygen demand outpaces supply from the blood. Myoglobin thus acts as an emergency oxygen store.

myopia Short-sightedness. It results from the lens of the eye refracting the parallel rays of light entering it to a focus in front of the retina generally because of an abnormally long eyeball. The condition is corrected by using diverging spectacle lenses to move the image back to the retina.

Myriapoda A group of terrestrial arthropods generally found in damp conditions, comprising centipedes (class Chilopoda), millipedes (class Diplopoda), pauropods (class Pauropoda), and symphylids (class Symphyla). Myriapods characteristically possess a distinct head with a single pair of antennae and a wormlike body with numerous similar segments, each bearing one or two pairs of legs. Young myriapods may hatch with fewer segments than the adult, gaining more with each successive moult. Centipedes are fast-moving predators with 15–177 body segments, relatively long antennae, and the first pair of appendages specialized as poison fangs; the remaining segments each bear a single pair of legs. Millipedes are slow moving and feed on decaying leaves. The antennae are short and the body consists of 20 to over 60 segments, each bearing two pairs of legs.

Myxomycota *See* slime moulds.

myxovirus One of a group of RNA-containing viruses associated with various diseases of man and other vertebrates. *Orthomyxoviruses* produce diseases of the respiratory tract, e.g. influenza; *paramyxoviruses* include the causal agents of mumps, measles, and fowl pest.

N

NAD (nicotinamide adenine dinucleotide) A *coenzyme, derived from the B vitamin *nicotinic acid, that participates in many biological dehydrogenation reactions. NAD is characteristically loosely bound to the enzymes concerned. It normally carries a positive charge and can accept one hydrogen atom and two electrons to become the reduced form, *NADH*. NADH is generated during the oxidation of food, especially by the reactions of the *Krebs cycle. It then gives up its two electrons (and single proton) to the *electron transport chain, thereby reverting to NAD$^+$ and generating three molecules of ATP per molecule of NADH.

NADP (*nicotinamide adenine dinucleotide phosphate*) differs from NAD only in possessing an additional phosphate group. It functions in the same way as NAD although anabolic reactions (*see* anabolism) generally use NADPH (reduced NADP) as a hydrogen donor rather than NADH. Enzymes tend to be specific for either NAD or NADP as coenzyme.

nadir The point opposite the *zenith on the *celestial sphere.

nand circuit *See* logic circuit.

nano- Symbol n. A prefix used in the metric system to denote 10^{-9}. For example, 10^{-9} metre = 1 nanometre (nm).

α or 1 position

β or 2 position

Naphthalene

naphthalene A white volatile solid, $C_{10}H_8$ (see formula); r.d. 1.14; m.p. 802°C; b.p. 218°C. Naphthalene is an aromatic hydrocarbon with an odour of mothballs and is obtained from crude oil. It is a raw material for making certain synthetic resins.

naphthols Two phenols derived from naphthalene with the formula $C_{10}H_7OH$, differing in the position of the –OH group. The most important is naphthalen-2-ol (β-naphthol), with the –OH in the 2-position. It is a white solid (r.d. 1.2; m.p. 122°C; b.p.

285°C) used in rubber as an antioxidant. Naphthalen-2-ol will couple with diazonium salts at the 1-position to form red *azo compounds, a reaction used in testing for primary amines (by making the diazonium salt and adding naphthalen-2-ol).

naphthyl group The group $C_{10}H_7-$ obtained by removing a hydrogen atom from naphthalene. There are two forms depending on whether the hydrogen is removed from the 1- or 2-position.

Napierian logarithm *See* logarithm.

narcotic Any drug that induces stupor and relieves pain, especially morphine and other *opiates. Such narcotics are addictive and cause dependence, and their medical use is strictly controlled.

nares (nostrils) The paired openings of the *nasal cavity in vertebrates. All vertebrates have *external nares*, which open to the exterior; in some species these are situated on a *nose. *Internal nares* (or *choanae*) are present only in air-breathing vertebrates (including lungfish) and open into the mouth cavity. In mammals they open posteriorly, beyond the secondary *palate.

nasal cavity The cavity in the head of a vertebrate that is lined by a membrane rich in sensitive olfactory receptors (*see* olfaction). It is connected to the exterior by external nostrils and (in air-breathing vertebrates) to the respiratory system by internal *nares.

nascent hydrogen A reactive form of hydrogen generated *in situ* in the reaction mixture (e.g. by the action of acid on zinc). Nascent hydrogen can reduce elements and compounds that do not readily react with 'normal' hydrogen. It was once thought that the hydrogen was present as atoms, but this is not the case. Probably hydrogen molecules are formed in an excited state and react before they revert to the ground state.

nastic movements Movements of plant organs in response to external stimuli that are independent of the direction of the stimuli. Examples are the opening of crocus and tulip flowers in response to a rise in temperature (*thermonasty*), the opening of evening primrose flowers at night (*photonasty*), and the folding up and drooping of leaves of the sensitive plant (*Mimosa pudica*) when lightly touched (*seismonasty*). Compare tropism. *See also* nyctinasty.

natron A mineral form of hydrated sodium carbonate, $Na_2CO_3.H_2O$.

Natta process *See* Ziegler process.

natural abundance *See* abundance.

natural frequency 1. The frequency of the free oscillation of a system. **2.** The frequency at which resonance occurs in an electrical circuit.

natural gas A naturally occurring mixture of gaseous hydrocarbons that is found in porous sedimentary rocks in the earth's crust, usually in association with *petroleum deposits. It consists chiefly of methane (about 85%), ethane (up to about 10%), propane (about 3%), and butane. Carbon dioxide, nitrogen, oxygen, hydrogen sulphide, and sometimes helium may also be present. Natural gas, like petroleum, originates in the decomposition of organic matter. It is widely used as a fuel and also to produce carbon black and some organic chemicals. Natural gas occurs on every continent, the major reserves occurring in the USSR, USA, Algeria, Canada, and the Middle East.

natural group A group of organisms of any taxonomic rank that are believed to be descended from a common ancestor (*see* monophyletic). For example, man and the apes are often regarded as a natural group, descended from fossil ancestors, the dryopithecines (or their near relatives). In an ideal natural classifica-tion all taxa should be natural groups. *See also* cladistics.

natural logarithm *See* logarithm.

natural order In the classification of plants, the former name for a *family.

natural selection The process that, according to *Darwinism, brings about the evolution of new species of animals and plants. Darwin noted that the size of any population tends to remain constant despite the fact that more offspring are produced than are needed to maintain it. He also saw that variations existed between individuals of the population and concluded that disease, competition, and other forces acting on the population eliminated those individuals less well adapted to their environment. The survivors would pass on any inheritable advantageous characteristics (i.e. characteristics with survival value) to their offspring and in time the composition of the population would change in adaptation to a changing environment. Over a long period of time this process could give rise to organisms so different from the original population that new species are formed. *See also* adaptive radiation. *Compare* punctuated equilibrium.

natural units A system of units, used principally in particle physics, in which all quantities that have dimensions involving length, mass, and time are given dimensions of a power of energy (usually expressed in electronvolts). This is equivalent to setting the rationalized *Planck constant and the speed of light both equal to unity. *See also* Gaussian units; geometrized units; Heaviside–Lorentz units; Planck units.

nature and nurture The combined effects of inherited factors (nature) and environmental factors (nurture) on the development of an organism. The genetic potential of an organism will only be realized under appropriate environmental conditions. *See also* phenotype.

nautical mile A measure of distance used at sea. In the UK it is defined as 6080 feet but the international definition is 1852 metres. 1 international nautical mile is therefore equivalent to 1.15078 land (statute) miles.

Neanderthal man A form of fossil man that appeared in Europe and Western Asia about 100 000 years ago. He was thought to be a different species (*Homo neanderthalensis*) from modern man but is now generally regarded as a subspecies of *H. sapiens*. The fossil remains indicate that Neanderthals were fairly short and had low brows but that the brain size was the same as modern man's. They were nomadic hunters and are thought to have had a language. *Cromagnon man began to replace Neanderthal man about 35 000 years ago. The name is derived from the site in the Neander valley, West Germany, where fossils were found in 1856.

near point The nearest point at which the human eye can focus an object. As the lens becomes harder with age, the extent to which accommodation can bring a near object into focus decreases. Therefore with advancing age the near point recedes – a condition known as *presbyopia.

nebula Originally a fixed, extended, and somewhat fuzzy white haze observed in the sky with a telescope. Many of these objects can now be resolved into clouds of individual stars and have been identified as *galaxies. They are still sometimes referred to as *extragalactic nebulae*. The *gaseous nebulae*, however, cannot be resolved into individual stars and consist, for the most part, of interstellar dust and gas. In some of these gaseous nebulae the gas atoms have been ionized by ultraviolet radiation from nearby stars and light is emitted as these ions interact with the free electrons in the gas. These are called *emission nebulae*. In the *dark nebulae*, there are no nearby stars and these objects are consequently dark; they can only be detected by what they obscure.

nectar A sugary liquid produced in plants by *nectaries*, regions of secretory cells on the receptacle or other parts of a flower. It attracts pollinating insects or other animals.

Néel temperature The temperature above which an antiferromagnetic substance becomes paramagnetic (*see* magnetism). The *susceptibility increases with temperature, reaching a maximum at the Néel temperature, after which it abruptly declines. The phenomenon was discovered around 1930 by L. E. F. Néel (1904–).

negative charge *See* charge.

negative feedback *See* feedback.

nekton *Pelagic organisms that actively swim through the water. Examples are fish, jellyfish, turtles, and whales. *Compare* plankton.

nematic crystal *See* liquid crystal.

nematoblast *See* thread cell.

Nematoda A phylum of invertebrates comprising the roundworms. They are characterized by a smooth narrow cylindrical unsegmented body tapered at both ends. They shed their tough outer cuticle four times during life to allow growth. The microscopic free-living forms are found in all parts of the world, where they play an important role in the destruction and recycling of organic matter. The many parasitic nematodes are much larger; they include the filaria (*Wuchereria*) and Guinea worm (*Dracunculus*), which cause serious diseases in man.

neo-Darwinism (modern synthesis) The current theory of the process of *evolution, formulated between about 1920 and 1950, that combines evidence from classical genetics with the Darwinian theory of evolution by *natural selection (*see* Darwinism). It makes use of modern knowledge of genes and chromosomes to explain the source of the genetic variation upon which selection works. This aspect was unexplained by traditional Darwinism.

neodymium Symbol Nd. A soft silvery metallic element belonging to the *lanthanoids; a.n. 60; r.a.m. 144.24; r.d. 7.004 (20°); m.p. 1016°C; b.p. 3068°C. It occurs in bastnasite and monazite, from which it is recovered by an ion-exchange process. There are seven naturally occurring isotopes, all of which are stable, except neodymium–144, which is slightly radioactive (half-life 10^{10}–10^{15} years). Seven artificial radioisotopes have been produced. The metal is used to colour glass violet-purple and to make it dichroic. It is also used in mischmetal (18% neodymium). It was discovered by C. A. von Welsbach in 1885.

neo-Lamarckism Any of the comparatively modern theories of evolution based on Lamarck's theory of the inheritance of acquired characteristics (*see* Lamarckism). These include the unfounded dogma of *Lysenkoism and the recent controversial experiments on the inheritance of acquired immunological tolerance in mice.

Neolithic The New Stone Age, beginning in the Middle East approximately 9000 BC and lasting until 6000 BC, during which man first developed agriculture. Grinding and polishing of stone tools was also practised.

neon Symbol Ne. A colourless gaseous element belonging to group I of the periodic table (the *noble gases); a.n. 10; r.a.m. 20.179; d. 0.9 $g\,dm^{-3}$; m.p. -248.67°C; b.p. -246.05°C. Neon occurs in air (0.0018% by volume) and is obtained by fractional distillation of liquid air. It is used in discharge tubes and neon lamps, in which it has a characteristic red glow. It forms hardly any compounds (neon fluorides have been reported). The element was discovered in 1898 by Sir William Ramsey and M. W. Travers.

neon lamp A small lamp consisting of a pair of electrodes, treated to emit electrons freely, sealed in a glass bulb containing neon gas at low pressure. When a minimum voltage of between 60 and 90 volts is applied across the electrodes, the kinetic energy of the electrons is sufficient to ionize the neon atoms around the cathode, causing the emission of a reddish light. With d.c. the glow is restricted to the cathode; with a.c. both electrodes act alternately as cathodes and a glow appears to emanate from both electrodes. The device consumes a very low power and is widely used as an indicator light showing that a circuit is live.

neoplasm Any new abnormal growth of cells, forming either a harmless (benign) tumour or a malignant one (*see* cancer).

neoprene A synthetic rubber made by polymerizing the compound 2-chlorobuta-1,2-diene. Neoprene is often used in place of natural rubber in applications requiring resistance to chemical attack.

neoteny The retention of the juvenile body form, or particular features of it, in a mature animal. For example, the axolotl, a salamander, retains the gills of the larva in the adult. Neoteny is thought to have been an important mechanism in the evolution of certain groups, such as man, who is believed to have developed from the juvenile form of apes.

neper A unit used to express a ratio of powers, currents, etc., used especially in telecommunications to denote the attenuation of an amplitude A_1 to an amplitude A_2 as N nepers, where

$$N = \ln(A_2/A_1).$$

1 neper = 8.686 decibels. The unit is named after John Napier (1550–

1617), the inventor of natural logarithms.

nephrite *See* jade.

distal convoluted tubule

arteriole

Bowman's capsule

glomerulus

proximal convoluted tubule

collecting duct

to ureter

loop of Henle

Structure of a single nephron

nephron The excretory unit of the vertebrate *kidney (see illustration). Many constituents of the blood are filtered from the *glomerulus into the Bowman's capsule at one end of the nephron. The glomerular filtrate passes along the length of the nephron and most of its water, plus some salts, glucose, and amino acids, are reabsorbed into the surrounding blood capillaries. The resulting concentrated solution of nitrogen-containing waste matter (*urea) plus inorganic salts passes into the collecting ducts of the nephrons and is discharged as *urine into the *ureter.

Neptune A planet in the *solar system, the eighth in order from the sun

(4496 × 10⁶ km distant). Its diameter is about 48 600 km and its *sidereal period is 164.8 years. It was discovered in 1846 by J. G. Galle on the basis of predictions made by J. C. Adams and U. J. J. Le Verrier, who had separately observed perturbations in the orbit of Uranus. Spectroscopic evidence suggests that there is an atmosphere predominantly composed of methane and hydrogen. The surface temperature appears to be about −200°C. Neptune has two main satellites; the inner one, Triton, has a diameter of about 4025 km, while the outer one, Nereid, is about one tenth of this size. Six other smaller satellites were discovered by the US Voyager II probe in 1989.

neptunium Symbol Np. A radioactive metallic transuranic element belonging to the *actinoids; a.n. 93; r.a.m. 237.0482. The most stable isotope, neptunium−237, has a half-life of 2.2 × 10⁶ years and is produced in small quantities as a by-product by nuclear reactors. Other isotopes have mass numbers 229−236 and 238−241. The only other relatively long-lived isotope is neptunium−236 (half-life 5 × 10³ years). The element was first produced by McMillan and Abelson in 1940.

neptunium series *See* radioactive series.

neritic zone The region of the sea over the continental shelf, which is less than 200 metres deep (approximately the maximum depth for organisms carrying out photosynthesis). *Compare* oceanic zone.

Nernst effect An effect in which a temperature gradient along an electric conductor or semiconductor placed in a perpendicular magnetic field, causes a potential difference to develop in the third perpendicular direction between opposite edges of the conductor. This effect, an analogue of the *Hall effect, was discovered in 1886 by Walter Nernst (1864−1941).

Nernst heat theorem A statement of the third law of *thermodynamics in a restricted form: if a chemical change takes place between pure crystalline solids at *absolute zero there is no change of entropy.

nerve A strand of tissue comprising many *nerve fibres plus supporting tissues, enclosed in a connective-tissue sheath. Nerves connect the central nervous system with the organs and tissues of the body. A nerve may carry only motor nerve fibres (*motor nerve*) or only sensory fibres (*sensory nerve*) or it may be mixed and carry both types (*mixed nerve*). Although the nerve fibres are in close proximity within the nerve, their physiological responses are independent of each other.

nerve cell *See* neurone.

nerve cord A large bundle of nerve fibres, running down the longitudinal axis of the body, that forms an important part of the *central nervous system. Most invertebrates have a pair of solid nerve cords, situated ventrally and bearing segmentally arranged *ganglia. All animals of the phylum *Chordata have a dorsal hollow nerve cord; in vertebrates this is the *spinal cord.

nerve fibre The *axon of a *neurone together with the tissues associated with it (such as a *myelin sheath). The length and diameter of nerve fibres are very variable, even within the same organism. *See also* giant fibre.

nerve impulse *See* impulse.

nerve net A network of nerve cells connected with each other by synapses or fusion. The nervous system of certain invertebrates (e.g. coelenterates and echinoderms) consists exclusively of a nerve net in the body wall.

nervous system The system of cells and tissues in multicellular animals by which information is conveyed between sensory cells and organs and effectors (such as muscles and glands). It consists of the *central nervous system (in vertebrates the *brain and *spinal cord; in invertebrates the *nerve cord and *ganglia) and the *peripheral nervous system. Its function is to receive, transmit, and interpret information and then to formulate appropriate responses for the effector organs. It also serves to coordinate responses that require more than one physiological process. Nervous tissue consists of *neurones, which convey the information in the form of *impulses, and supporting tissue.

nesquehonite A mineral form of *magnesium carbonate trihydrate, $MgCO_3.3H_2O$.

Nessler's reagent A solution of mercury(II) iodide (HgI_2) in potassium iodide and potassium hydroxide. It is used in testing for ammonia, with which it forms a brown coloration or precipitate.

Neumann's law The magnitude of an electromagnetically induced e.m.f. (E) is given by $E = -d\Phi/dt$, where Φ is the magnetic flux. This is a quantitative statement of *Faraday's second law of electromagnetic induction and is sometimes known as the Faraday–Neumann law.

neuroendocrine system Any of the systems of dual control of certain activities in the body of some higher animals by nervous and hormonal stimulation. For example, the posterior *pituitary gland and the medulla of the *adrenal gland receive direct nervous stimulation to secrete their hormones, whereas the anterior pituitary gland is stimulated by releasing hormones from the hypothalamus.

neurohormone Any hormone that is produced not by an endocrine gland but by a specialized nerve cell and is secreted from nerve endings into the bloodstream or directly to the tissue

Sensory neurone

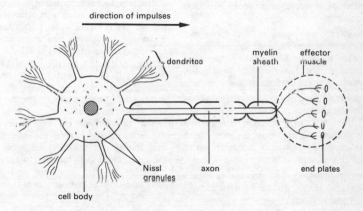

Motor neurone

or organ whose growth or function it controls. Examples of neurohormones are *noradrenaline, *vasopressin, and hormones associated with metamorphosis and moulting in insects (*see* ecdysone; juvenile hormone).

neurone (nerve cell) An elongated branched cell that is the fundamental unit of the *nervous system, being specialized for the conduction of *impulses. A neurone consists of a *cell body, containing the nucleus and *Nissl granules; *dendrites, which receive incoming impulses and

pass them towards the cell body; and an *axon, which conducts impulses away from the cell body, sometimes over long distances. Impulses are passed from one neurone to the next via *synapses. *Sensory neurones* transmit information from *receptors to the central nervous system. *Motor neurones* conduct information from the central nervous system to *effectors (e.g. muscles). See illustration.

neurone theory The hypothesis, now accepted, that the nervous system consists of nerve cells (*neurones), which are functionally linked at *synapses but are physically separate. This theory has superseded the idea that the cytoplasm of the cells of the nervous system is continuous.

neurotransmitter A chemical that mediates the transmission of a nerve impulse across a *synapse. Examples are *adrenaline and *noradrenaline (in adrenergic nerves) and *acetylcholine (in cholinergic nerves). The neurotransmitter is released at the tip of the *axon into the synaptic space. It diffuses across to the opposite membrane (the postsynaptic membrane), where it initiates the transmission of a nerve impulse in the next neurone.

neuter An organism that does not possess either male or female reproductive organs. Cultivated ornamental flowers that have neither pistils nor stamens are called neuters.

neutral Describing a compound or solution that is neither acidic nor basic. A neutral solution is one that contains equal numbers of both protonated and deprotonated forms of the solvent.

neutralization The process in which an acid reacts with a base to form a salt and water.

neutrino A *lepton (*see also* elementary particles) that exists in three forms, one in association with the electron, one with the muon, and one with the tau particle. Each form has its own antiparticle. The neutrino, which was postulated in 1931 to account for the 'missing' energy in *beta decay, was identified tentatively in 1953 and, more definitely, in 1956. Neutrinos have no charge, are thought to have zero rest mass, and travel at the speed of light. In some *grand unified theories they are predicted to have nonzero mass, but there is no conclusive evidence for this.

neutron A neutral hadron (*see* elementary particles) that is stable in the atomic nucleus but decays into a proton, an electron, and an antineutrino with a mean life of 12 minutes outside the nucleus. Its rest mass (symbol m_n) is slightly greater than that of the proton, being $1.674\,9286(10) \times 10^{-27}$ kg. Neutrons occur in all atomic nuclei except normal hydrogen. The neutron was first reported in 1932 by James Chadwick (1891–1974).

neutron diffraction The scattering of neutrons by atoms in solids, liquids, or gases. This process has given rise to a technique, analogous to *X-ray diffraction techniques, using a flux of thermal neutrons from a nuclear reactor to study solid-state phenomena. Thermal neutrons have average kinetic energies of about 0.025 eV (4×10^{-21} J) giving them an equivalent wavelength of about 0.1 nanometre, which is suitable for the study of interatomic interference. There are two types of interaction in the scattering of neutrons by atoms: one is the interaction between the neutrons and the atomic nucleus, the other is the interaction between the *magnetic moments of the neutrons and the spin and orbital magnetic moments of the atoms. The latter interaction has provided valuable information on antiferromagnetic and ferrimagnetic materials (*see* magnetism). Interaction with the atomic nucleus gives diffraction patterns that complement those from X-rays. X-rays, which interact with the extranuclear electrons, are not suitable for investigating light elements (e.g. hydrogen), whereas neutrons do give diffraction patterns from such atoms because they interact with nuclei.

neutron drip *See* neutron star.

neutron excess *See* isotopic number.

neutron number Symbol N. The number of neutrons in an atomic nucleus of a particular nuclide. It is equal to the difference between the *nucleon number and the *atomic number.

neutron star A compact stellar object that is supported against collapse under self-gravity by the *degeneracy pressure of the neutrons of which it is primarily composed. Neutron stars are believed to be formed as the end products of the evolution of stars of mass greater than a few (4–10) solar masses. The core of the evolved star collapses and (assuming that its mass is greater than the *Chandrasekhar limit for a *white dwarf), at the very high densities involved (about 10^{14} kg m^{-3}), electrons react with protons in atomic nuclei to produce neutrons. The neutron-rich nuclei thus formed release free neutrons in a process known as *neutron drip*. The density increases to about 10^{17} kg m^{-2}, at which most of the electrons and protons have been converted to a *degenerate gas of neutrons and the atomic nuclei have lost their separate identities. If the mass of the core exceeds the Chandrasekhar limit for a neutron star, then further collapse will occur, leading to the formation of a *black hole. *Pulsars are believed to be rapidly rotating magnetized neutron stars and many X-ray sources are thought to be neutron stars in binary systems with another star, from which material is drawn into an accretion disc. This material, heated to a very high temperature, emits radiation in the X-ray region.

neutron temperature A concept used to express the energies of neutrons that are in thermal equilibrium with their surroundings, assuming that they behave like a monatomic gas. The neutron temperature T, on the Kelvin scale, is given by $T = 2E/3k$, where E is average neutron energy and k the *Boltzmann constant.

Newlands' law *See* law of octaves.

Newtonian fluid

newton Symbol N. The *SI unit of force, being the force required to give a mass of one kilogram an acceleration of 1 m s^{-2}. It is named after Sir Isaac Newton (1642–1727).

Newtonian fluid A fluid in which the velocity gradient is directly proportional to the shear stress. If two flat plates of area A are separated by a layer of fluid of thickness d and move relative to each other at a velocity v, then the rate of shear is v/d and the shear stress is F/A, where F is the force applied to each (see illustration). For a Newtonian fluid $F/A = \mu v/d$, where μ is the constant of proportionality and is called the Newtonian *viscosity. Many liquids are Newtonian fluids over a wide range of temperatures and pressures. However, some are not; these are called *non-Newtonian fluids*. In such fluids there is a departure from the simple Newtonian relationships. For example, in some liquids the viscosity increases as the velocity gradient increases, i.e. the faster the liquid moves the more viscous it becomes. Such liquids are said to be *dilatant* and the phenomenon they exhibit is called *dilatancy*. It occurs in some pastes and suspensions. More common, however, is the opposite effect in which the viscosity depends not only on the velocity gradient but also on the time for which it has been applied. These liquids are said to exhibit *thixotropy*. The faster a *thixotropic liquid* moves the less viscous it becomes. This property is used in nondrip paints (which are more viscous on the brush than on the wall) and in lubricating oils (which become thinner when the parts they are lubricating start to move). Another exam-

Newtonian mechanics

ple is the non-Newtonian flow of macromolecules in solution or in polymer melts. In this case the shearing force F is not parallel to the shear planes and the linear relationship does not apply. In general, the many types of non-Newtonian fluid are somewhat complicated and no theory has been developed to accommodate them fully.

Newtonian mechanics The system of *mechanics that relies on *Newton's laws of motion. Newtonian mechanics is applicable to bodies moving at speeds relative to the observer that are small compared to the speed of light. Bodies moving at speeds comparable to the speed of light require an approach based on *relativistic mechanics, in which the mass of a body changes with its speed.

Newtonian telescope *See* telescope.

Newton's formula For a lens, the distances p and q between two conjugate points and their respective foci is given by $pq = f^2$, where f is the focal length of the lens.

Newton's law of cooling The rate at which a body loses heat is proportional to the difference in temperature between the body and the surroundings. It is an empirical law that is only true for substantial temperature differences if the heat loss is by forced convection or conduction.

Newton's law of gravitation There is a force of attraction between any two massive particles in the universe. For any two point masses m_1 and m_2, separated by a distance d, the force of attraction F is given by $F = m_1 m_2 G/d^2$, where G is the *gravitational constant. Real bodies having spherical symmetry act as point masses positioned at their centres of mass.

Newton's laws of motion The three laws of motion on which *Newtonian mechanics is based. (1) A body con-

tinues in a state of rest or uniform motion in a straight line unless it is acted upon by external forces. (2) The rate of change of momentum of a moving body is proportional to and in the same direction as the force acting on it, i.e. $F = \mathrm{d}(mv)/\mathrm{d}t$, where F is the applied force, v is the velocity of the body, and m its mass. If the mass remains constant, $F = m\mathrm{d}v/\mathrm{d}t$ or $F = ma$, where a is the acceleration. (3) If one body exerts a force on another, there is an equal and opposite force, called a *reaction*, exerted on the first body by the second.

Newton's rings 1. (in optics) *Interference fringes formed by placing a slightly convex lens on a flat glass plate. If monochromatic light is reflected by the two close surfaces into the observer's eye at a suitable angle, the point of contact of the lens is seen as a dark spot surrounded by a series of bright and dark rings. The radius of the nth dark ring is given by $r_n = \sqrt{nR\lambda}$, where λ is the wavelength and R is the radius of curvature of the lens. The phenomenon is used in the quality testing of lens surfaces. With white light, coloured rings are formed. **2.** (in photography) The irregular patterns produced by thin film interference between a projected transparency and its cover glass.

niacin *See* nicotinic acid.

niche *See* ecological niche.

Nichrome Trade name for a group of nickel–chromium alloys used for wire in heating elements as they possess good resistance to oxidation and have a high resistivity. Typical is Nichrome V containing 80% nickel and 19.5% chromium, the balance consisting of manganese, silicon, and carbon.

nickel Symbol Ni. A malleable ductile silvery metallic *transition element; a.n. 28; r.a.m. 58.70; r.d. 8.9; m.p. 1450°C; b.p. 2840°C. It is found in the minerals pentlandite (NiS), pyrrhoite ((Fe,Ni)S), and garnierite ((Ni,Mg)$_6$(OH)$_6$Si$_4$O$_{11}$.H$_2$O). Nickel is also present in certain iron meteorites (up to 20%). The metal is extracted by roasting the ore to give the oxide, followed by reduction with carbon monoxide and purification by the *Mond process. Alternatively electolysis is used. Nickel metal is used in special steels, in Invar, and, being ferromagnetic, in magnetic alloys, such as *Mumetal. It is also an effective catalyst, particularly for hydrogenation reactions (see also Raney nickel). The main compounds are formed with nickel in the +2 oxidation state; the +3 state also exists (e.g. the black oxide, Ni$_2$O$_3$). Nickel was discovered by A. F. Cronstedt in 1751.

nickel carbonyl A colourless volatile liquid, Ni(CO)$_4$; m.p. −25°C; b.p. 43°C. It is formed by direct combination of nickel metal with carbon monoxide at 50–60°C. The reaction is reversed at higher temperatures, and the reactions are the basis of the *Mond process for purifying nickel. The nickel in the compound has an oxidation state of zero, and the compound is a typical example of a complex with pi-bonding *ligands, in which filled d-orbitals on the nickel overlap with empty p-orbitals on the carbon.

nickelic compounds Compounds of nickel in its +3 oxidation state; e.g. nickelic oxide is nickel(III) oxide (Ni$_2$O$_3$).

nickel–iron accumulator (Edison cell; NIFE cell) A *secondary cell devised by Thomas Edison (1847–1931) having a positive plate of nickel oxide and a negative plate of iron both immersed in an electrolyte of potassium hydroxide. The reaction on discharge is

$$2NiOOH.H_2O + Fe \rightarrow 2Ni(OH)_2 + Fe(OH)_2,$$

the reverse occurring during charging. Each cell gives an e.m.f. of about 1.2 volts and produces about 100 kJ per kilogram during each discharge. Compare lead–acid accumulator.

nickelous compounds Compounds of nickel in its +2 oxidation state, e.g. nickelous oxide is nickel(II) oxide (NiO).

nickel(II) oxide A green powder, NiO; r.d. 6.6. It can be made by heating nickel(II) nitrate or carbonate with air excluded.

nickel(III) oxide (nickel peroxide; nickel sesquioxide) A black or grey powder, Ni$_2$O$_3$; r.d. 4.8. It is made by heating nickel(II) oxide in air and used in *nickel–iron accumulators.

nickel silver See German silver.

Nicol prism A device for producing *plane-polarized light. It consists of two pieces of calcite cut with a 68° angle and stuck together with Canada balsam. The extraordinary ray (see double refraction) passes through the prism while the ordinary ray suffers total internal reflection at the interface between the two crystals, as the refractive index of the calcite is 1.66 for the ordinary ray and that of the Canada balsam is 1.53. Modifications of the prism using different shapes and cements are used for special purposes. It was devised in 1828 by William Nicol (1768–1851).

nicotinamide adenine dinucleotide See NAD.

nicotine A colourless poisonous *alkaloid present in tobacco. It is used as an insecticide.

nicotinic acid (niacin) A vitamin of the *vitamin B complex. It can be manufactured by plants and animals from the amino acid tryptophan. The amide derivative, nicotinamide, is a component of the coenzymes *NAD and NADP. These take part in many metabolic reactions as hydrogen acceptors. Deficiency of nicotinic acid causes the disease pellagra in humans.

Apart from tryptophan-rich protein, good sources are liver and groundnut and sunflower meals.

nictitating membrane A clear membrane forming a third eyelid in amphibians, reptiles, birds, and some mammals (but not man). It can be drawn across the cornea independently of the other eyelids, thus clearing the eye surface and giving added protection without interrupting the continuity of vision.

nidation *See* implantation.

nielsbohrium *See* transactinide elements.

NIFE cell *See* nickel–iron accumulator.

niobium Symbol Nb. A soft ductile grey-blue metallic transition element; a.n. 41; r.a.m. 92.91; r.d. 8.57; m.p. 2468°C; b.p. 4742°C. It occurs in several minerals, including niobite ($Fe(NbO_3)_2$), and is extracted by several methods including reduction of the complex fluoride K_2NbF_7 using sodium. It is used in special steels and in welded joints (to increase strength). Niobium–zirconium alloys are used in superconductors. Chemically, the element combines with the halogens and oxidizes in air at 200°C. It forms a number of compounds and complexes with the metal in oxidation states 2, 3, or 5. The element was discovered by Charles Hatchett in 1801 and first isolated by Blomstrand in 1864. Formerly, it was called *columbium*.

Nissl granules (Nissl bodies) Particles seen within the cell bodies of *neurones. They are rich in RNA and stain strongly with basic dyes. They are named after F. Nissl (1860–1919), the German neurologist who discovered them.

nit A unit of *luminance equal to one *candela per square metre.

nitrate A salt or ester of nitric acid.

nitration A type of chemical reaction in which a nitro group ($-NO_2$) is added to or substituted in a molecule. Nitration can be carried out by a mixture of concentrated nitric and sulphuric acids. An example is electrophilic substitution of benzene (and benzene compounds), where the electrophile is the nitryl ion NO_2^+.

nitre (saltpetre) Commercial *potassium nitrate; the name was formerly applied to natural crustlike efflorescences, occurring in some arid regions.

nitre cake *See* sodium hydrogensulphate.

nitric acid A colourless corrosive poisonous liquid, HNO_3; r.d. 1.50; m.p. −42°C; b.p. 83°C. Nitric acid may be prepared in the laboratory by the distillation of a mixture of an alkali-metal nitrate and concentrated sulphuric acid. The industrial production is by the oxidation of ammonia to nitrogen monoxide, the oxidation of this to nitrogen dioxide, and the reaction of nitrogen dioxide with water to form nitric acid and nitrogen monoxide (which is recycled). The first reaction (NH_3 to NO) is catalysed by platinum or platinum/rhodium in the form of fine wire gauze. The oxidation of NO and the absorption of NO_2 to form the product are noncatalytic and proceed with high yields but both reactions are second-order and slow. Increases in pressure reduce the selectivity of the reaction and therefore rather large gas absorption towers are required. In practice the absorbing acid is refrigerated to around 2°C and a commercial 'concentrated nitric acid' at about 67% is produced.

Nitric acid is a strong acid (highly dissociated in aqueous solution) and dilute solutions behave much like other mineral acids. Concentrated nitric acid is a strong oxidizing agent. Most metals dissolve to form nitrates but with the evolution of nitrogen oxides. Concentrated nitric acid also reacts with several nonmetals to give the oxo acid or oxide. Nitric acid is generally stored in dark brown bottles because of the photolytic decomposition to dinitrogen tetroxide. *See also* nitration.

nitric oxide *See* nitrogen monoxide.

nitrides Compounds of nitrogen with a more electropositive element. Boron nitride is a covalent compound having macromolecular crystals. Certain electropositive elements, such as lithium, magnesium, and calcium, react directly with nitrogen to form ionic nitrides containing the N^{3-} ion. Transition elements form a range of interstitial nitrides (e.g. Mn_4N, W_2N), which can be produced by heating the metal in ammonia.

nitrification A chemical process in which nitrogen (mostly in the form of ammonia) in plant and animal wastes and dead remains is oxidized at first to nitrites and then to nitrates. These reactions are effected mainly by the nitrifying bacteria *Nitrosomonas* and *Nitrobacter* respectively. Unlike ammonia, nitrates are readily taken up by plant roots; nitrification is therefore a crucial part of the *nitrogen cycle. Nitrogen-containing compounds are often applied to soils deficient in this element, as fertilizer. *Compare* denitrification.

nitrile rubber A copolymer of buta-1,3-diene and propenonitrile. It is a commercially important synthetic rubber because of its resistance to oil and many solvents.

nitriles (cyanides) Organic compounds containing the group $-CN$ bound to an organic group. Nitriles are made by reaction between potassium cyanide and haloalkanes in alcoholic solution, e.g.

$$KCN + CH_3Cl \rightarrow CH_3CN + KCl$$

An alternative method is dehydration of amides

$$CH_3CONH_2 - H_2O \rightarrow CH_3CN$$

They can be hydrolysed to amides and carboxylic acids and can be reduced to amines.

nitrite A salt or ester of nitrous acid; the nitrite ion, NO_2^-, has a bond angle of $115°$.

nitrobenzene A yellow oily liquid, $C_6H_5NO_2$; r.d. 1.2; m.p. $6°C$; b.p. $211°C$. It is made by the *nitration of benzene using a mixture of nitric and sulphuric acids.

nitrocellulose *See* cellulose nitrate.

nitro compounds Organic compounds containing the group $-NO_2$ (the *nitro group*) bound to a carbon atom in a benzene ring. Nitro compounds are made by *nitration reactions. They can be reduced to aromatic amines (e.g. nitrobenzene can be reduced to phenylamine).

nitrogen Symbol N. A colourless gaseous element belonging to *group V of the periodic table; a.n. 7; r.a.m. 14.0067; d. 1.2506 $g\,dm^{-3}$; m.p. $-209.86°C$; b.p. $-195.8°C$. It occurs in air (about 78% by volume) and is an essential constituent of proteins and nucleic acids in living organisms (*see* nitrogen cycle). Nitrogen is obtained for industrial purposes by fractional distillation of liquid air. Pure nitrogen can be obtained in the laboratory by heating a metal azide. There are two natural isotopes: nitrogen 14 and nitrogen 15 (about 3%). The element is used in the *Haber process for making ammonia and is also used to provide an inert atmosphere in welding and metallurgy. The gas is diatomic and relatively inert – it reacts with hydrogen at high temperatures and with oxygen in electric discharges. It also forms *nitrides with certain metals. Nitrogen was discovered in 1772 by D. Rutherford.

nitrogen cycle One of the major cycles of chemical elements in the environment (see illustration).

Nitrates in the soil are taken up by plant roots and may then pass along *food chains into animals. Decomposing bacteria convert nitrogen-containing compounds (especially ammonia) in plant and animal wastes and dead

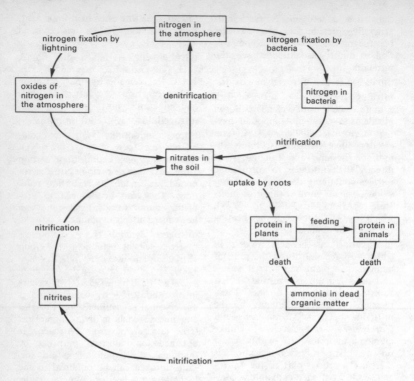

The nitrogen cycle

remains back into nitrates, which are released into the soil and can again be taken up by plants (*see* nitrification). Though nitrogen is essential to all forms of life, the huge amount present in the atmosphere is not directly available to most organisms (*compare* carbon cycle). It can, however, be assimilated by some specialized bacteria (*see* nitrogen fixation) and is thus made available to other organisms indirectly. Lightning flashes also make some nitrogen available to plants by causing the combination of atmospheric nitrogen and oxygen to form oxides of nitrogen, which enter the soil and form nitrates. Some nitrogen is returned from the soil to the atmosphere by denitrifying bacteria (*see* denitrification).

nitrogen dioxide *See* dinitrogen tetroxide.

nitrogen fixation A chemical process in which atmospheric nitrogen is assimilated into organic compounds in living organisms and hence into the *nitrogen cycle. The ability to fix nitrogen is limited to certain bacteria (e.g. *Azotobacter*) and blue-green algae (e.g. *Anabaena*). *Rhizobium* bacteria are able to fix nitrogen in association with cells in the roots of leguminous plants such as peas and beans, in which they form characteristic root nodules; cultivation of legumes is therefore one way of increasing soil nitrogen. Various chemical processes are used to fix atmospheric nitrogen in the manufacture of fertilizers. These include the *Birkeland–Eyde process, the cyanamide process (*see* calcium dicarbide), and the *Haber process.

nitrogen monoxide (nitric oxide) A colourless gas, NO; m.p. −163.6°C; b.p. −151.8°C. It is soluble in water, ethanol, and ether. In the liquid state nitrogen monoxide is blue in colour (r.d. 1.26). It is formed in many reactions involving the reduction of nitric acid, but more convenient reactions for the preparation of reasonably pure NO are reactions of sodium nitrite, sulphuric acid, and either sodium iodide or iron(II) sulphate. Nitrogen monoxide reacts readily with oxygen to give nitrogen dioxide and with the halogens to give the nitrosyl halides XNO (X = F,Cl,Br). It is oxidized to nitric acid by strong oxidizing agents and reduced to dinitrogen oxide by reducing agents. The molecule has one unpaired electron, which accounts for its paramagnetism and for the blue colour in the liquid state. This electron is relatively easily removed to give the *nitrosyl ion* NO⁺, which is the ion present in such compounds as NOClO₄, NOBF₄, NOFeCl₄, (NO)₂PtCl₆ and a ligand in complexes such as Co(CO)₃NO.

nitrogenous base A basic compound containing nitrogen. The term is used especially of organic ring compounds, such as adenine, guanine, cytosine, and thymine, which are constituents of nucleic acids. *See* amine salts.

nitroglycerine An explosive made by reacting 1,2,3-trihydroxypropane (glycerol) with a mixture of concentrated sulphuric and nitric acids. Despite its name and method of preparation, it is not a nitro compound, but an ester of nitric acid, CH₂(NO₃)CH(NO₃)CH₂(NO₃). It is used in dynamites.

nitro group *See* nitro compounds.

nitronium ion *See* nitryl ion.

nitrosyl ion The ion NO⁺. *See* nitrogen monoxide.

nitrous acid A weak acid, HNO₂, known only in solution and in the gas phase. It is prepared by the action of acids upon nitrites, preferably using a combination that removes the salt as an insoluble precipitate (e.g. Ba(NO₂)₂ and H₂SO₄). The solutions are unstable and decompose on heating to give nitric acid and nitrogen monoxide. Nitrous acid can function both as an oxidizing agent (forms NO) with I⁻ and Fe²⁺, or as a reducing agent (forms NO₃⁻) with, for example, Cu²⁺; the latter is most common. It is widely used (prepared *in situ*) for the preparation of diazonium compounds in organic chemistry. The full systematic name is *dioxonitric(III) acid*.

nitrous oxide *See* dinitrogen oxide.

nitryl ion (nitronium ion) The ion NO₂⁺, found in mixtures of nitric acid and sulphuric acid and solutions of nitrogen oxides in nitric acid. Nitryl salts, such as NO₂⁺ClO₄⁻, can be isolated but are extremely reactive. Nitryl ions generated *in situ* are used for *nitration in organic chemistry.

NMR *See* nuclear magnetic resonance.

nobelium Symbol No. A radioactive metallic transuranic element belonging to the *actinoids; a.n. 102; mass number of most stable element 254 (half-life 55 seconds). Seven isotopes are known. The element was first identified with certainty by A. Ghiorso and G. T. Seaborg in 1966. The alternative name *unnilbium* has been proposed.

noble gases (inert gases; rare gases; group 0 elements) A group of monatomic gaseous elements forming group 0 (sometimes called group VIII) of the *periodic table: helium (He), neon (Ne), argon (Ar), krypton (Kr), xenon (Xe), and radon (Rn).

The electron configuration of helium is $1s^2$. The configurations of the others terminate in ns^2np^6 and all inner shells are fully occupied. The elements thus represent the termination of a period and have closed-shell configuration and associated high ionization energies (He 2370 to Rn 1040 kJ mol^{-1}) and lack of chemical reactivity. Being monatomic the noble gases are spherically symmetrical and have very weak interatomic interactions and consequent low enthalpies of vaporization. The behaviour of the lighter members approaches that of an ideal gas at normal temperatures; with the heavier members increasing polarizability and dispersion forces lead to easier liquefaction under pressure. Four types of 'compound' have been described for the noble gases but of these only one can be correctly described as compounds in the normal sense. One type consists of such species as HHe^+, He_2^+, Ar_2^+, $HeLi^+$, which form under highly energetic conditions, such as those in arcs and sparks. They are short-lived and only detected spectroscopically. A second group of materials described as inert-gas–metal compounds do not have defined compositions and are simply noble gases adsorbed onto the surface of dispersed metal. The third type, previously described as 'hydrates' are in fact clathrate compounds with the noble gas molecule trapped in a water lattice. True compounds of the noble gases were first described in 1962 and several fluorides, oxyfluorides, fluoroplatinates, and fluoroantimonates of *xenon are known. A few krypton fluorides and a radon fluoride are also known although the short half-life of radon and its intense alpha activity restrict the availability of information. Apart from argon, the noble gases are present in the atmosphere at only trace levels. Helium may be found along with natural gas (up to 7%), arising from the radioactive decay of heavier elements (via alpha particles).

nodal points Two points on the axis of a system of lenses; if the incident ray passes through one, the emergent ray will pass through the other.

node 1. (in botany) The part of a plant stem from which one or more leaves arise. The nodes at the stem apex are very close together and remain so in species of monocotyledons that form bulbs. In older regions of the stem they are separated by areas of stem called *internodes*. 2. (in anatomy) A natural thickening or bulge in an organ or part of the body. Examples are the *sinoatrial node* that controls the heartbeat (*see* pacemaker) and the *lymph nodes. 3. (in physics) A point of minimum disturbance in a *stationary-wave system. 4. (in astronomy) Either of two points at which the orbit of a celestial body intersects a reference plane, usually the plane of the *ecliptic or the celestial equator (*see* celestial sphere).

node of Ranvier *See* myelin sheath.

noise 1. Any undesired sound. It is measured on a *decibel scale ranging from the threshold of hearing (0 dB) to the threshold of pain (130 dB). Between these limits a whisper registers about 20 dB, heavy urban traffic about 90 dB, and a heavy hammer on steel plate about 110 dB. A high noise level (industrial or from overamplified music, for example) can cause permanent hearing impairment. 2. Any unwanted disturbance within a useful frequency band in a communication channel.

nomad (in cytology) A cell that migrates or wanders from its site of formation. Certain types of *phagocytes are nomads.

nomogram A graph consisting of three lines, each with its own scale, each line representing the values of a variable over a specified range. A ruler laid between two points on two of the lines enables the value of the third variable to be read off the third line.

nonbenzenoid aromatics Aromatic compounds that have rings other than benzene rings. Examples are the cyclopentadienyl anion, $C_5H_5^-$, and the tropylium cation, $C_7H_7^+$.

non-Euclidean geometry A type of geometry that does not comply with the basic postulates of *Euclidean geometry, particularly a form of geometry that does not accept Euclid's postulate that only one straight line can be drawn through a point in space parallel to a given straight line. Several types of non-Euclidean geometry exist.

nonmetal Any of a class of chemical elements that are typically poor conductors of heat and electricity and that do not form positive ions. Nonmetals are electronegative elements, such as carbon, nitrogen, oxygen, phosphorus, sulphur, and the halogens. They form compounds that contain negative ions or covalent bonds. Their oxides are either neutral or acidic.

non-Newtonian fluid See Newtonian fluid.

nonpolar compound A compound that has covalent molecules with no permanent dipole moment. Examples of nonpolar compounds are methane and benzene.

nonpolar solvent See solvent.

nonrelativistic quantum theory See quantum theory.

nonstoichiometric compound (Berthollide compound) A chemical compound in which the elements do not combine in simple ratios. For example, rutile (titanium(IV) oxide) is often deficient in oxygen, having such a formula as $TiO_{1.8}$.

noradrenaline (norepinephrine) A hormone produced by the *adrenal glands and also secreted from nerve endings in the *sympathetic nervous system as a chemical transmitter of nerve impulses (see neurotransmitter). Many of its general actions are similar to those of *adrenaline, but it is more concerned with maintaining normal body activity than with preparing the body for emergencies.

Nordhausen sulphuric acid See disulphuric(VI) acid.

norepinephrine See noradrenaline.

normal 1. (in mathematics) A line drawn at right angles to a surface. **2.** (in chemistry) Having a concentration of one gram equivalent per dm^3.

normalizing The process of heating steel to above an appropriate critical temperature followed by cooling in still air. The process promotes the formation of a uniform internal structure and the elimination of internal stress.

Northern blotting See Southern blotting.

nose The protuberance on the face of some vertebrates that contains the nostrils (see nares) and part of the *nasal cavity. It therefore forms part of the olfactory system (see olfaction) and the external opening of the respiratory system.

nostrils See nares.

not circuit See logic circuit.

note 1. A musical sound of specified pitch. **2.** A representation of such a sound in a musical score. Such a representation has a specified duration as well as a specified pitch.

notochord An elastic skeletal rod lying lengthwise beneath the nerve cord and above the alimentary canal in the embryos or adults of all chordate animals (see Chordata). Its function is to strengthen and support the body and act as a protagonist for the muscles. It is found in both the adult and larva of amphioxus but in adult vertebrates it is largely replaced by the *vertebral column.

nova A star that, over a period of only a few days, becomes $10^3 - 10^4$ times brighter than it was. Some 10–15 such events occur in the Milky Way each year. Novae are believed to be close *binaries, one component of which is usually a *white dwarf and the other a *red giant. Matter is transferred from the red giant to the white dwarf, on whose surface it accumulates, eventually leading to a thermonuclear explosion. *See also* supernova.

N.T.P. *See* s.t.p.

n-type conductivity *See* semiconductor; transistor.

nucellus The tissue that makes up the greater part of the ovule of seed plants. It contains the *embryo sac and nutritive tissue. It is enclosed by the integuments except for a small gap, the *micropyle. In certain flowering plants it may persist after fertilization and provide nutrients for the embryo.

nuclear battery A single cell, or battery of cells, in which the energy of particles emitted from the atomic nucleus is converted internally into electrical energy. In the high-voltage type, a beta-emitter, such as strontium–90, krypton–85, or tritium, is sealed into a shielded glass vessel, the electrons being collected on an electrode that is separated from the emitter by a vacuum or by a solid dielectric. A typical cell delivers some 160 picoamperes at a voltage proportional to the load resistance. It can be used to maintain the voltage of a charged capacitor. Of greater use, especially in space technology, are the various types of low-voltage nuclear batteries. Typical is the gas-ionization device in which a beta-emitter ionizes a gas in an electric field. Each beta-particle produces about 200 ions, thus multiplying the current. The electric field is obtained by the contact potential difference between two electrodes, such as lead dioxide and magnesium. Such a cell, containing argon and tritium, gives about 1.6 nanoamperes at 1.5 volts. Other types use light from a phosphor receiving the beta-particles to operate photocells or heat from the nuclear reaction to operate a thermopile.

nuclear energy Energy obtained as a result of *nuclear fission or *nuclear fusion. The nuclear fission of one uranium atom yields about 3.2×10^{-11} joule, whereas the combustion of one carbon atom yields about 6.4×10^{-19} joule. Mass for mass, uranium yields about 2 500 000 times more energy by fission than carbon does by combustion. The nuclear fusion of deuterium to form helium releases about 400 times as much energy as the fission of uranium (on a mass basis).

nuclear fission A nuclear reaction in which a heavy nucleus (such as uranium) splits into two parts (*fission products*), which subsequently emit either two or three neutrons, releasing a quantity of energy equivalent to the difference between the rest mass of the neutrons and the fission products and that of the original nucleus. Fission may occur spontaneously or as a result of irradiation by neutrons. For example, the fission of a uranium–235 nucleus by a *slow neutron may proceed thus:

$$^{235}U + n \rightarrow {}^{148}La + {}^{85}Br + 3n$$

The energy released is approximately 3×10^{-11} J per ^{235}U nucleus. For 1 kg of ^{235}U this is equivalent to 20 000 megawatt-hours – the amount of energy produced by the combustion of 3×10^6 tonnes of coal. Nuclear fission is the process used in *nuclear reactors and atom bombs (*see* nuclear weapons).

nuclear force A strong attractive force (resulting from the strong interaction) between *nucleons in the atomic nucleus that holds the nucleus together. At close range (up to about 2×10^{-15} metre) these forces are some 100 times stronger than electromagnetic forces. *See* fundamental interactions.

nuclear fuel A substance that will sustain a fission chain reaction so that it can be used as a source of *nuclear energy. The *fissile isotopes are uranium–235, uranium–233, plutonium–241, and plutonium–239. The first occurs in nature as 1 part in 140 of natural uranium, the others have to be made artificially. ^{233}U is produced when thorium–232 captures a neutron and ^{239}Pu is produced by neutron capture in ^{238}U. ^{232}Th and ^{238}U are called *fertile isotopes.

nuclear fusion A type of *nuclear reaction in which atomic nuclei of low atomic number fuse to form a heavier nucleus with the release of large amounts of energy. In *nuclear fission reactions a neutron is used to break up a large nucleus, but in nuclear fusion the two reacting nuclei themselves have to be brought into collision. As both nuclei are positively charged there is a strong repulsive force between them, which can only be overcome if the reacting nuclei have very high kinetic energies. These high kinetic energies imply temperatures of the order of 10^8 K. As the kinetic energy required increases with the nuclear charge (i.e. atomic number), reactions involving low atomic-number nuclei are the easiest to produce. At these elevated temperatures, however, fusion reactions are self-sustaining; the reactants at these temperatures are in the form of a *plasma (i.e. nuclei and free electrons) with the nuclei possessing sufficient energy to overcome electrostatic repulsion forces. The fusion bomb (*see* nuclear weapons) and the stars generate energy in this way. It is hoped that the method will be harnessed in the *thermonuclear reactor as a source of energy for man's use.

Typical fusion reactions with the energy release in joules are:

$$D + D = T + p + 6.4 \times 10^{-13} J$$

$$T + D = {}^4He + n + 28.2 \times 10^{-13} J$$

$$^6Li + D = 2^4He + 35.8 \times 10^{-13} J$$

By comparison the formation of a water molecule from hydrogen and oxygen is accompanied by the release of 1.5×10^{-19} J.

A large amount of work is currently being done on *cold fusion*; i.e. fusion that can occur at lower temperatures than those necessary to overcome the electrostatic repulsion between nuclei. The most productive approach is *meson-catalysed fusion*, in which the deuterium atoms have their electrons replaced by negative muons to give 'muonic atoms' of deuterium. The muon is 207 times heavier than the electron, so the muonic deuterium atom is much smaller and is able to approach another deuterium atom more closely, allowing nuclear fusion to occur. The muon is released to form another muonic atom, and the process continues. The limiting factor is the short lifetime of the muon, which restricts the number of fusion reactions it can catalyse. *See also* plasma physics.

nuclear isomerism A condition in which atomic nuclei with the same number of neutrons and protons have different lifetimes. This occurs when nuclei exist in different unstable quantum states, from which they decay to lower excited states or to the ground state, with the emission of gamma-ray photons. If the lifetime of a particular excited state is unusually long it is said to be isomeric, although there is no fixed limit separating isomeric decays from normal decays.

nuclear magnetic resonance (NMR) The absorption of electromagnetic radiation at a suitable precise frequency by a nucleus with a nonzero magnetic moment in an external magnetic field. The phenomenon occurs if the nucleus has nonzero *spin, in which case it behaves as a small magnet. In an external magnetic field, the nucleus's magnetic moment vector precesses about the field direction but only certain orientations are allowed by quantum rules. Thus, for hydrogen (spin of ½) there are two possible states in the presence of a field, each with a slightly different energy. Nuclear magnetic resonance is the absorption of radiation at a photon energy equal to the difference between these levels, causing a transition from a lower to a higher energy state. For practical purposes, the difference in energy levels is small and the radiation is in the radiofrequency region of the electromagnetic spectrum. It depends on the field strength.

NMR can be used for the accurate determination of nuclear moments. It can also be used in a sensitive form of magnetometer to measure magnetic fields. In medicine, *magnetic resonance imaging (MRI)* has been developed, in which images of tissue are produced by magnetic-resonance techniques.

The main application of NMR is as a technique for chemical analysis and structure determination, known as *NMR spectroscopy*. It depends on the fact that the electrons in a molecule shield the nucleus to some extent from the field, causing different atoms to absorb at slightly different frequencies (or at slightly different fields for a fixed frequency). Such effects are known as *chemical shifts*. In an NMR spectrometer, the sample is subjected to a strong field, which can be varied in a controlled way over a small region. It is irradiated with radiation at a fixed frequency, and a detector monitors the field at the sample. As the field changes, absorption corresponding to transitions occurs at certain values, and this causes oscillations in the field, which induce a signal in the detector. The most common nucleus studied is ^{1}H. For instance, an NMR spectrum of ethanol (CH_3CH_2OH) has three peaks in

the ratio 3:2:1, corresponding to the three different hydrogen-atom environments. The peaks also have a fine structure caused by interaction between spins in the molecule. Other nuclei can also be used for NMR spectroscopy (e.g. ^{13}C, ^{14}N, ^{19}F) although these generally have lower magnetic moment and natural abundance than hydrogen. *See also* electron spin resonance.

nuclear moment A property of atomic nuclei in which lack of spherical symmetry of the nuclear charge gives rise to electric moments and the intrinsic spin and rotational motion of the component nucleons give rise to magnetic moments.

nuclear physics The physics of atomic nuclei and their interactions, with particular reference to the generation of *nuclear energy.

nuclear power Electric power or motive power generated by a *nuclear reactor.

nuclear reaction Any reaction in which there is a change to an atomic nucleus. This may be a natural spontaneous disintegration or an artificial bombardment of a nucleus with an energetic particle, as in a *nuclear reactor. Nuclear reactions are commonly represented by enclosing within a bracket the symbols for the incoming and outgoing particles; the initial nuclide is shown before the bracket and the final nuclide after it. For example, the reaction:

$$^{12}C + {}^2H \rightarrow {}^{13}N + {}^1n$$

is shown as $^{12}C(d,n)^{13}N$, where d is the symbol for a deuteron.

nuclear reactor A device in which a *nuclear fission *chain reaction is sustained and controlled in order to produce *nuclear energy, radioisotopes, or new nuclides. The fuels available for use in a fission reactor are uranium–235, uranium–233, and plutonium–239; only the first occurs in nature (as 1 part in 140 of natural uranium), the others have to be produced artificially (*see* nuclear fuel). When a uranium–235 nucleus is made to undergo fission by the impact of a neutron it breaks into two roughly equal fragments, which release either two or three very high-energy neutrons. These *fast neutrons* need to be slowed down to increase the probability that they will cause further fissions of ^{235}U nuclei and thus sustain the chain reaction. This slowing down process occurs naturally to a certain extent when the neutrons collide with other nuclei; unfortunately, however, the predominant uranium isotope, ^{238}U, absorbs fast neutrons to such an extent that in natural uranium the fission reaction is not self-sustaining. In order to create a controlled self-sustaining chain reaction it is necessary either to slow down the neutrons (using a *moderator* in a *thermal reactor*) to greatly reduce the number absorbed by ^{238}U, or to reduce the predominance of ^{238}U in natural uranium by enriching it with more ^{235}U than it normally contains. In a *fast reactor* the fuel used is enriched uranium and no moderator is employed.

In thermal reactors, neutrons are slowed down by collisions with light moderator atoms (such as graphite, deuterium, or beryllium); they are then in thermal equilibrium with the surrounding material and are known as *thermal neutrons*. In a *heterogeneous thermal reactor* the fuel and moderator are in separate solid and liquid phases (e.g. solid uranium fuel and a heavy water moderator). In the *homogeneous thermal reactor* the fuel and moderator are mixed together, for example in a solution, molten dispersion, slurry, or suspension.

In the reactor *core* the *fuel elements* encase the fuel; in a heterogeneous reactor the fuel elements may fit into a lattice that also contains the moderator. The progress of the reaction is controlled by *control rods*, which when lowered into the core absorb

neutrons and so slow down or stop the chain reaction. The heat produced by the nuclear reaction in the core is used to generate electricity by the same means as in a conventional power station, i.e. by raising steam to drive a steam turbine that turns a generator. The heat is transferred to the steam-raising boiler or heat-exchanger by the *coolant. Water is frequently used as the coolant; in the case of the *boiling-water reactor* (BWR) and the *pressurized-water reactor* (PWR) water is both coolant and moderator. In the BWR the primary coolant drives the turbine; in the PWR the primary coolant raises steam in a secondary circuit for driving the turbine. In the *gas-cooled reactor* the coolant is a gas, usually carbon dioxide with an outlet temperature of about 350°C, or 600°C in the case of the *advanced gas-cooled reactor* (AGR).

In fast reactors, in which there is no moderator, the temperature is higher and a liquid-metal coolant is used, usually liquid sodium. Some fast reactors are used as converters or breeders. A *converter reactor* is one that converts *fertile material (such as ^{238}U) into *fissile material (such as ^{239}Pu). A *breeder reactor* produces the same fissile material as it uses. For example, a *fast breeder reactor* using uranium enriched with ^{239}Pu as the fuel can produce more ^{239}Pu than it uses by converting ^{238}U to ^{239}Pu. *See also* thermonuclear reactor.

nuclear waste *See* radioactive waste.

nuclear weapons Weapons in which an explosion is caused by *nuclear fission, *nuclear fusion, or a combination of both. In the fission bomb (*atomic bomb* or *A-bomb*) two subcritical masses (*see* critical mass) of a *fissile material (uranium–235 or plutonium–239) are brought together by a chemical explosion to produce one supercritical mass. The resulting nuclear explosion is typically in the *kiloton range with temperatures of the order 10^8 K being reached. The fusion bomb (*thermonuclear weapon, hydrogen bomb,* or *H-bomb*) relies on a nuclear-fusion reaction, which becomes self-sustaining at a critical temperature of about 35×10^6 K. This is achieved by means of an inner fission bomb, which is surrounded by a hydrogenous material, such as heavy hydrogen (deuterium) or lithium deuteride. The *megaton explosion produced in such a thermonuclear reaction has not yet been used in war, the largest explosion so far produced being the 58 megaton weapon exploded in the Arctic by the USSR in 1961.

nucleic acid A complex organic compound in living cells that consists of a chain of *nucleotides. There are two types: *DNA (deoxyribonucleic acid) and *RNA (ribonucleic acid).

nucleoid (nuclear region) The part of a cell of a bacterium (i.e. a prokaryotic *cell) that contains the genetic material *DNA and therefore controls the activity of the cell. It corresponds to the nucleus of the more advanced eukaryotic cells but is not bounded by a membrane.

nucleolus A small dense round body within the nondividing *nucleus of plant and animal cells that consists of protein, DNA, and ribosomal *RNA. It plays an important role in *ribosome manufacture (and therefore protein synthesis).

nucleon A *proton or a *neutron.

nucleonics The technological aspects of *nuclear physics, including the design of nuclear reactors, devices to produce and detect radiation, and nuclear transport systems. It is also concerned with the technology of *radioactive waste disposal and with radioisotopes.

nucleon number (mass number) Symbol A. The number of *nucleons in an atomic nucleus of a particular nuclide.

nucleophile An ion or molecule that can donate electrons. Nucleophiles arc often oxidizing agents and Lewis bases. They are either negative ions (e.g. Cl^-) or molecules that have electron pairs (e.g. NH_3). In organic reactions they tend to attack positively charged parts of a molecule. *Compare* electrophile.

nucleophilic addition A type of addition reaction in which the first step is attachment of a *nucleophile to a positive (electron-deficient) part of the molecule. *Aldehydes and *ketones undergo reactions of this type because of polarization of the carbonyl group (carbon positive).

nucleophilic substitution A type of substitution reaction in which a *nucleophile displaces another group or atom from a compound. For example, in
$$CR_3Cl + OH^- \rightarrow CR_3OH + Cl^-$$
the nucleophile is the OH^- ion. There are two possible mechanisms of nucleophilic substitution. In S_N1 *reactions*, a positive carbonium ion is first formed:
$$CR_3Cl \rightarrow CR_3^+ + Cl^-$$
This then reacts with the nucleophile
$$CR_3^+ + OH^- \rightarrow CR_3OH$$
The CR_3^+ ion is planar and the OH^- ion can attack from either side. Consequently, if the original molecule is optically active (the three R groups are different) then a racemic mixture of products results.

The alternative mechanism, the S_N2 *reaction*, is a concerted reaction in which the nucleophile approaches from the side of the R groups as the other group (Cl in the example) leaves. In this case the configuration of the molecule is inverted. If the original molecule is optically active, the product has the opposite activity, an effect known as *Walden inversion*.

nucleoprotein Any compound present in cells of living organisms that consists of a nucleic acid (DNA or RNA) combined with a protein. Chromosomes consist of nucleoprotein (DNA and proteins, mostly histones), as do ribosomes (RNA and protein).

nucleoside An organic compound consisting of a nitrogen-containing *purine or *pyrimidine base linked to a sugar (ribose or deoxyribose). An example is *adenosine. *Compare* nucleotide.

nucleotide An organic compound consisting of a nitrogen-containing *purine or *pyrimidine base linked to a sugar (ribose or deoxyribose) and a phosphate group. *DNA and *RNA are made up of long chains of nucleotides (i.e. *polynucleotides). *Compare* nucleoside.

nucleus (of atom) The central core of an atom that contains most of its mass. It is positively charged and consists of one or more nucleons (protons or neutrons). The positive charge of the nucleus is determined by the number of protons it contains (*see* atomic number) and in the neutral atom this is balanced by an equal number of electrons, which move around the nucleus. The simplest nucleus is the hydrogen nucleus, consisting of one proton only. All other nuclei also contain one or more neutrons. The neutrons contribute to the atomic mass (*see* nucleon number) but not to the nuclear charge. The most massive nucleus that occurs in nature is uranium–238, containing 92 protons and 146 neutrons. The symbol used for this *nuclide is $^{238}_{92}U$, the upper figure being the nucleon number and the lower figure the atomic number. In all nuclei the nucleon number (A) is equal to the sum of the atomic number (Z) and the neutron number (N), i.e. $A = Z + N$.

nucleus (of cell) The large body embedded in the cytoplasm of all plant and animal *cells (but not bacterial cells) that contains the genetic material *DNA. The nucleus functions as the control centre of the cell. It is bounded by a double membrane (the *nuclear membrane* or *envelope*), which contains many pores for the exchange of material between the nucleus and cytoplasm, and contains a viscous sap (*nucleoplasm*). When the cell is not dividing a *nucleolus is present in the nucleus and the genetic material is dispersed in the sap as *chromatin; in dividing cells the chromatin is organized into *chromosomes and the nucleolus disappears. In certain protozoans there are two nuclei per cell, a *macronucleus* (or *meganucleus*) concerned with vegetative functions and a *micronucleus* with sexual reproduction.

nuclide A type of atom as characterized by its *atomic number and its *neutron number. An *isotope refers to a member of a series of different atoms that have the same atomic number but different neutron numbers (e.g. uranium–238 and uranium–235 are isotopes of uranium), whereas a nuclide refers only to a particular nuclear species (e.g. the nuclides uranium–235 and plutonium –239 are fissile). The term is also used for a type of nucleus.

null method A method of making a measurement in which the quantity to be measured is balanced by another similar reading by adjusting the instrument to read zero (*see* Wheatstone bridge).

numerical taxonomy *See* taxonomy.

nut A dry single-seeded fruit that develops from more than one carpel and does not shed its seed when ripe. The fruit wall is woody or leathery. Many nuts are enclosed in a hard or membranous cup-shaped structure, the *cupule. The term nut is often loosely used of any hard fruit. For example, the walnut and coconut are in fact *drupes and the Brazil nut is a seed.

nutation

nutation 1. (in astronomy) An irregular periodic oscillation of the earth's poles. It causes an irregularity of the precessional circle traced by the celestial poles and results from the varying distances and relative directions of the sun and the moon. **2.** (in botany) The spiral movement of a plant organ during growth, also known as *circumnutation*. It is seen in climbing plants and helps the plant find a suitable support to twine around. Examples are the coiling movements of the shoot tips of runner beans and of the tendrils of sweet peas.

nyctinasty (sleep movements) *Nastic movements of plant organs in response to the changes in light and temperature that occur between day and night (and vice versa). Examples are the opening and closing of many flowers and the folding together of the leaflets of clover and other plants at night.

nylon Any of various synthetic polyamide fibres having a protein-like structure formed by the condensation between an amino group of one molecule and a carboxylic acid group of another. There are three main nylon fibres, nylon 6, nylon 6,6, and nylon 6,10. Nylon 6, for example Enkalon and Celon, is formed by the self-condensation of 6-aminohexanoic acid. Nylon 6,6, for example Bri nylon, is made by polycondensation of hexanedioic acid (adipic acid) and 1,6-diaminohexane (hexamethylenediamine) having an average formula weight between 12 000 and 15 000. Nylon 6,10 is prepared by polymerizing decanedioic acid and 1,6-diaminohexane.

nymph The juvenile stage of certain insects, such as dragonflies, grasshoppers, and earwigs, which resembles the adult except that the wings and reproductive organs are undeveloped. There is no pupal stage, and the nymph develops directly into the adult. *Compare* larva.

O

objective The *lens or system of lenses nearest to the object being examined through an optical instrument.

occipital condyle A single or paired bony knob that protrudes from the occipital bone of the skull and articulates with the first cervical vertebra (the *atlas). In humans there is a pair of occipital condyles, one on each side of the *foramen magnum. Occipital condyles are absent in most fish, which cannot move their heads.

occluded front *See* front.

occlusion 1. The trapping of small pockets of liquid in a crystal during crystallization. **2.** The absorption of a gas by a solid such that atoms or molecules of the gas occupy interstitial positions in the solid lattice. Palladium, for example, can occlude hydrogen.

occultation The disappearance of a star, planet, or other celestial body behind the moon or another planet, while it is being observed. A solar *eclipse is a form of occultation.

oceanic zone The region of the open sea beyond the edge of the continental shelf, where the depth is greater than 200 metres. *Compare* neritic zone.

oceanography The study of the oceans. It includes the origin, structure, and form of the oceans, the nature of the seafloor and its sediments, the characteristics of the ocean waters (e.g. tides, salinity, and currents), and the types of flora and fauna living within the oceans. The effects of human intervention also form an important aspect of oceanography.

ocellus A simple eye occurring in insects and other invertebrates. It typically consists of light-sensitive cells and a single cuticular lens.

ochre A yellow or red mineral form of iron(III) oxide, Fe_2O_3, used as a pigment.

octadecanoate See stearate.

octadecanoic acid See stearic acid.

octadecenoic acid A straight-chain unsaturated fatty acid with the formula $C_{17}H_{33}COOH$. *Cis-octadec-9-enoic acid* (see oleic acid) has the formula $CH_3(CH_2)_7CH:CH(CH_2)_7COOH$. The glycerides of this acid are found in many natural fats and oils.

octahedral See complex.

octahydrate A crystalline hydrate having eight molecules of water per molecule of compound.

octane A straight-chain liquid *alkane, C_8H_{18}; r.d. 0.7; m.p. $-56.79°C$; b.p. $125.7°C$. It is present in petroleum. The compound is isomeric with 2,2,4-trimethylpentane, $(CH_3)_3CCH_2CH(CH_3)_2$, *iso-octane*). See octane number.

octane number A number that provides a measure of the ability of a fuel to resist *knocking when it is burnt in a spark-ignition engine. It is the percentage by volume of iso-octane (C_8H_{18}; 2,2,4-trimethylpentane) in a blend with normal heptane (C_7H_{16}) that matches the knocking behaviour of the fuel being tested in a single cylinder four-stroke engine of standard design. *Compare* cetane number.

octanoic acid (caprylic acid) A colourless liquid straight-chain saturated *carboxylic acid, $CH_3(CH_2)_6COOH$; b.p. $237°C$.

octavalent Having a valency of eight.

octave 1. The interval between two musical notes that have fundamental frequencies in the ratio 2:1; the word describes the interval in terms of the eight notes of the diatonic scale. **2.** *See* law of octaves.

octet A stable group of eight electrons in the outer shell of an atom (as in an atom of an inert gas).

ocular See eyepiece.

odd–even nucleus An atomic nucleus containing an odd number of protons and an even number of neutrons.

odd–odd nucleus An atomic nucleus containing an odd number of protons and an odd number of neutrons.

Odonata An order of insects containing the dragonflies and damselflies, most of which occur in tropical regions. Adult dragonflies have a pair of prominent *compound eyes, a compact thorax bearing two pairs of delicate membranous wings, and a long slender abdomen. They are strong fliers and prey on other insects, either in flight or at rest. The eggs are laid near or in water, and the newly hatched nymphs are aquatic and resemble the adults, with rudimentary wings. They breathe through gills and feed on small aquatic animals. The nymph leaves the water for its final moult into the terrestrial adult.

odontoblast A cell that is responsible for producing the *dentine of vertebrate teeth. Odontoblasts are found around the lining of the *pulp cavity and have processes that extend into the dentine.

oersted Symbol Oe. The unit of magnetic field strength in the *c.g.s. electromagnetic system. A field has a strength of one oersted if it exerts a force of one dyne on a unit magnetic pole placed in it. It is equivalent to $10^3/4\pi$ A m^{-1}. The unit was named after Hans Christian Oersted (1777–1851).

oesophagus The section of the *alimentary canal that lies between the *pharynx and the stomach. It is a muscular tube whose function is to transfer food to the stomach by means of wavelike contractions (*peristalsis) along its length.

oestrogen One of a group of female sex hormones, produced principally by the ovaries, that promote the onset of *secondary sexual characteristics (such as breast enlargement and development in women) and control the *oestrous cycle. *Oestradiol* is the most important. Oestrogens are secreted at particularly high levels during ovulation, stimulating the uterus to prepare for pregnancy. They are used in *oral contraceptives (with *progestogens) and as treatment for various disorders of the female reproductive organs. Small amounts of oestrogens are produced by the adrenal glands and testes.

oestrous cycle The cycle of reproductive activity shown by most sexually mature nonpregnant female mammals except most primates (*compare* menstrual cycle). There are four phases:
(1) *pro-oestrus* – *Graafian follicles develop in the ovary and secrete oestrogens;
(2) *oestrus* (*heat*) – ovulation normally occurs, the female is ready to mate and becomes sexually attractive to the male;
(3) *metoestrus* – *corpus luteum develops from ruptured follicle;
(4) *dioestrus* – *progesterone secreted by corpus luteum prepares uterus for implantation.
The length of the cycle depends on the species: larger mammals typically have a single annual cycle with a well-defined breeding season (they are described as *monoestrous*). The males have a similar cycle of sexual activity. Other species may have many cycles per year (i.e. they are *polyoestrous*) and the male may be sexually active all the time.

oestrus (heat) *See* oestrous cycle.

offset *See* runner.

ohm Symbol Ω. The derived *SI unit of electrical resistance, being the resistance between two points on a conductor when a constant potential difference of one volt, applied between these points, produces a current of one ampere in the conductor. The former *international ohm* (sometimes called the 'mercury ohm') was defined in terms of the resistance of a column of mercury. The unit is named after Georg Ohm (1787–1854).

ohmmeter Any direct-reading instrument for measuring the value of a resistance in ohms. The instrument commonly used is a *multimeter capable of measuring also both currents and voltages. To measure resistance a dry cell and resistor are switched in series with the moving coil *galvanometer and the unknown resistance is connected across the instrument's terminals. The value of the resistance is then read off an ohms scale. Such instruments are increasingly being replaced by electronic digital multimeters.

Ohm's law The ratio of the potential difference between the ends of a conductor to the current flowing through it is constant. This constant is the *resistance of the conductor, i.e. $V = IR$, where V is the potential difference in volts, I is the current in amperes, and R is the resistance in ohms. The law was discovered in 1827 by Georg Ohm. Most materials do not obey this simple linear law; those that do are said to be *ohmic* but remain so only if physical conditions, such as temperature, remain constant. Metals are the most accurately ohmic conductors.

oil Any of various viscous liquids that are generally immiscible with water. Natural plant and animal oils are either volatile mixtures of terpenes and simple esters (e.g. *essential oils) or are *glycerides of fatty acids. Mineral oils are mixtures of hydrocarbons (e.g. *petroleum).

oil-immersion lens *See* immersion objective.

oil of vitriol *See* sulphuric acid.

oil sand (tar sand; bituminous sand) A sandstone or porous carbonate rock that is impregnated with hydrocarbons. The largest deposit of oil sand occurs in Alberta, Canada (the Athabasca tar sands); there are also deposits in the Orinoco Basin of Venezuela, the USSR, USA, Madagascar, Albania, Trinidad, and Romania.

oil shale A fine-grained carbonaceous sedimentary rock from which oil can be extracted. The rock contains organic matter – *kerogen* – which decomposes to yield oil when heated. Deposits of oil shale occur on every continent, the largest known reserves occurring in Colorado, Utah, and Wyoming in the USA. Commercial production of oil from oil shale is generally considered to be uneconomic unless the price of petroleum rises above the recovery costs for oil from oil shale. However, threats of declining conventional oil resources have resulted in considerable interest and developments in recovery techniques.

Olbers' paradox If the universe is infinite, uniform, and unchanging the sky at night would be bright, as in whatever direction one looked one would eventually see a star. The number of stars would increase in proportion to the square of the distance from the earth; the intensity of light reaching the earth from a given star is inversely proportional to the square of the distance. Consequently, the whole sky should be about as bright as the sun. The paradox, that this is not the case, was stated by Heinrich Olbers (1758–1840) in 1826. (It had been discussed earlier, in 1744, by J. P. L. Chesaux.) The paradox is resolved by the fact that, according to the *big-bang theory, the universe is not infinite, not uniform, and not unchanging. For instance, light from the most distant galaxies displays an extreme *redshift and ceases to be visible.

oleate A salt or ester of *oleic acid.

olefines *See* alkenes.

oleic acid An unsaturated *fatty acid with one double bond, $CH_3(CH_2)_7$-$CH:CH(CH_2)_7COOH$; r.d. 0.9; m.p. 13°C. Oleic acid is one of the most abundant constituent fatty acids of animal and plant fats, occurring in butterfat, lard, tallow, groundnut oil, soya-bean oil, etc. Its systematic chemical name is *cis-octadec-9-enoic acid*.

oleum *See* disulphuric(VI) acid.

olfaction The sense of smell or the process of detecting smells. This is achieved by receptors in *olfactory organs* (such as the *nose) that are sensitive to air- or water-borne chemicals. Stimulation of these receptors results in the transmission of information to the brain via the *olfactory nerve*.

Oligocene The third geological epoch of the *Tertiary period. It began about 38 million years ago, following the Eocene epoch, and extended for about 12 million years to the beginning of the Miocene epoch. The epoch was characterized by the continued rise of mammals; the first pigs, rhinoceroses, and tapirs made their appearance.

Oligochaeta A class of hermaphrodite annelid worms that bear only a few bristles (*chaetae). Oligochaetes are very abundant in freshwater and terrestrial habitats. The most familiar members of the class are the earthworm (*Lumbricus*) and the freshwater bloodworm (*Tubifex*).

oligopeptide *See* peptide.

oligotrophic Describing a body of water (e.g. a lake) with a poor supply of nutrients and a low rate of formation of organic matter by photosynthesis. *Compare* eutrophic.

olivine An important group of rock-forming silicate minerals crystallizing in the orthorhombic system. Olivine conforms to the general formula $(Mg,Fe)_2SiO_4$ and comprises a complete series from pure magnesium silicate (forsterite, Mg_2SiO_4) to pure iron silicate (fayalite, Fe_2SiO_4). It is green, brown-green, or yellow-green in colour.

omnivore An animal that eats both animal and vegetable matter. Pigs, for example, are omnivorous. *Compare* carnivore; herbivore.

oncogenic Describing a chemical, organism, or environmental factor that causes the development of cancer. Some viruses are oncogenic to vertebrates, notably the *retroviruses (including the Rous sarcoma virus of chickens), and some are suspected of being oncogenic (e.g. some of the *adenoviruses and *papovaviruses). Many of these viruses contain genes (known as *oncogenes*) that are responsible for the transformation of a normal host cell into a cancerous cell. *See also* growth factor.

onium ion An ion formed by adding a proton to a neutral molecule, e.g. the hydroxonium ion (H_3O^+) or the ammonium ion (NH_4^+).

ontogeny The developmental course of an organism from the fertilized egg through to maturity. It has been suggested that "ontogeny recapitulates *phylogeny", i.e. the stages of development, especially of the embryo, reflect the evolutionary history of the organism. This idea is now discredited. *See* recapitulation.

oocyte *See* oogenesis.

oogamy Sexual reproduction involving the formation and subsequent fusion of a large, usually stationary, female gamete and a small motile male gamete. The female gamete may contain nourishment for the development of the embryo, which is often retained and protected by the parent organism.

oogenesis The production and growth of the ova (egg cells) in the animal ovary. Special cells (*oogonia*) within the ovary divide repeatedly by mitosis to produce large numbers of prospective egg cells (*oocytes*). When mature, these undergo meiosis, which halves the number of chromosomes. During the first meiotic division a *polar body* and a secondary oocyte are produced. At the second meiotic division the secondary oocyte produces an ovum and a second polar body. Oocytes may be present in the ovaries at birth and may represent the total number of eggs to be produced.

oosphere (ovum; egg cell) The nonmotile female gamete in plants. In angiosperms it is a cell in the *embryo sac of the ovule. In gymnosperms, pteridophytes, and bryophytes it is situated in an *archegonium. In some lower plants, such as *Fucus*, the oosphere is protected by an *oogonium* until it is shed into the water prior to fertilization. Many oospheres store food in the form of starch or oil droplets.

oospore A zygote that is produced as a result of *oogamy in certain algae and fungi. It contains food reserves, develops a protective outer covering, and enters a resting phase before germination. *Compare* zygospore.

opacity The extent to which a medium is opaque to electromagnetic radiation, especially to light. It is the reciprocal of the *transmittance. A medium that is opaque to X-rays and gamma rays is said to be *radio-opaque*.

opal A hydrous amorphous form of silica. Many varieties of opal occur, some being prized as gemstones. Common opal is usually milk white but the presence of impurities may colour it yellow, green, or red. Precious opals, which are used as gemstones, display the property of *opales-*

cence – a characteristic internal play of colours resulting from the interference of light rays within the stone. Black opal has a black background against which the colours are displayed. The chief sources of precious opals are Australia and Mexico. Geyserite is a variety deposited by geysers or hot springs. Another variety, diatomite, is made up of the skeletons of diatoms.

open chain *See* chain.

open cluster (galactic cluster) *See* star cluster.

open-hearth process A traditional method for manufacturing steel by heating together scrap, pig iron, hot metal, etc., in a refractory-lined shallow open furnace heated by burning producer gas in air.

opera glasses *See* binoculars.

operator A mathematical symbol indicating that a specified operation should be carried out. For example, the operator $\sqrt{}$ in $\sqrt{x}$ indicates that the square root of x should be taken; the operator d/dx in dy/dx indicates that y should be differentiated with respect to x, etc.

operculum 1. (in zoology) A lid or flap of skin covering an aperture, such as the gill slit cover of fish and larval amphibians and the horny calcareous operculum secreted by many gastropod molluscs, which closes the opening of the shell when the animal is inside. **2.** (in botany) The cone-shaped lid of the *capsule of mosses, which is forcibly detached to release the spores.

operon A functionally integrated genetic unit comprising *structural genes*, coding for protein, and adjacent loci controlling their expression – an *operator site* and a *promoter site*. *Transcription of the structural genes is prevented by binding of a *repressor* molecule to the operator site. Another molecule, the *inducer*, can bind to the repressor molecule, preventing it from

binding to the operator and thus allowing the promoter site to bind the enzyme RNA polymerase, thereby initiating transcription. The repressor molecule is encoded by a *regulator gene*, which may be close to or distant from the operon. Some operons also have an *attenuator site* preceding the first structural gene, where transcription may either stall or proceed according to the amount of end-product in the cell. *See also lac* operon.

opiate One of a group of drugs derived from *opium*, an extract of the poppy plant *Papaver somniferum* that depresses brain function (a *narcotic* action). Opiates include morphine and its synthetic derivatives, such as heroin and codeine. They are used in medicine chiefly to relieve pain, but the use of morphine and heroin is strictly controlled since they can cause drug dependence and tolerance.

opposition The moment at which a planet having its orbit outside that of the earth is in a line with the earth and the sun. When a planet is in opposition it can be observed during the night and is near to its closest point to the earth; it is therefore a favourable opportunity for observation.

optical activity The ability of certain substances to rotate the plane of plane-polarized light as it passes through a crystal, liquid, or solution. It occurs when the molecules of the substance are asymmetric, so that they can exist in two different structural forms each being a mirror image of the other. The two forms are *optical isomers* or *enantiomers*. The existence of such forms is also known as *enantiomorphism* (the mirror images being *enantiomorphs*). One form will rotate the light in one direction and the other will rotate it by an equal amount in the other. The two possible forms are described as *dextrorotatory or *laevorotatory according to the direction of rotation, and

D-form   L-form   *meso*-form

Isomers of tartaric acid

prefixes are used to designate the isomer, as in *d*-tartaric and *l*-tartaric acids (see formulae). An equimolar mixture of the two forms is not optically active. It is called a *racemic mixture* (or *racemate*) and designated by *dl*-. The alternative prefixes (+)-, (−)-, and (±)-, corresponding to *d*-, *l*-, and *dl*-, respectively, are preferred and increasingly used. In addition, certain molecules can have a *meso-isomer* in which one part of the molecule is a mirror image of the other. Such molecules are not optically active.

Molecules that show optical activity have no plane of symmetry. The commonest case of this is in organic compounds in which a carbon atom is linked to four different groups. An atom of this type is said to be a *chiral centre*. Asymmetric molecules showing optical activity can also occur in inorganic compounds. For example, an octahedral complex in which the central ion coordinates to eight different ligands would be optically active. Many naturally occurring compounds show optical isomerism and usually only one isomer occurs naturally. For instance, glucose is found in the dextrorotatory form. The other isomer, *l*- or (−)-glucose, can be synthesized in the laboratory, but cannot be synthesized by living

organisms. *See also* absolute configuration.

optical axis (principal axis; optic axis) The line passing through the *optical centre and the centre of curvature of a *lens or spherical *mirror.

optical centre The point at the geometrical centre of a *lens through which a ray of light entering the lens passes without deviation.

optical fibre A *waveguide through which light can be transmitted with very little leakage through the sidewalls. In the *step-index fibre* a pure glass core, with a diameter between 6 and 250 micrometres, is surrounded by a coaxial glass or plastic cladding of lower refractive index. The cladding is usually between 10 and 150 micrometres thick. The interface between core and cladding acts as a cylindrical mirror at which *total internal reflection of the transmitted light takes place. This structure enables a beam of light to travel through many kilometres of fibre. In the *graded-index fibre*, each layer of glass, from the fibre axis to its outer wall, has a slightly lower refractive index than the layer inside it. This arrangement also prevents light from escaping through the fibre walls by a combination of refraction and total internal reflection, and can be made to give the same transit time for rays at different angles.

Fibre-optic systems use optical fibres to transmit information, in the form of coded pulses or fragmented images (using bundles of fibres), from a source to a receiver. Over moderate distances they are used in telecommunications, for which purpose they are becoming competitive with electric cables. They are also used in medical instruments (*fibrescopes*) to examine internal body cavities, such as the stomach and bladder.

optical flat A flat glass disc having very accurately polished surfaces so that the deviation from perfect flatness does not exceed (usually) 50 nanometres. It is used to test the flatness of such plane surfaces as gauge anvils by means of the *interference patterns formed when parallel beams of light pass through the flat and are reflected by the surface being inspected.

Surfaces are said to be *optically flat* if the deviation from perfect flatness is smaller than the wavelength of light.

optical glass Glass used in the manufacture of lenses, prisms, and other optical parts. It must be homogeneous and free from bubbles and strain. Optical *crown glass* may contain potassium or barium in place of the sodium of ordinary crown glass and has a refractive index in the range 1.51 to 1.54. *Flint glass* contains lead oxide and has a refractive index between 1.58 and 1.72. Higher refractive indexes are obtained by adding lanthanoid oxides to glasses; these are now known as lanthanum crowns and flints.

optical isomers *See* optical activity.

optical microscope *See* microscope.

optical pumping *See* laser.

optical pyrometer *See* pyrometer.

optical rotary dispersion (ORD) The effect in which the amount of rotation of plane-polarized light by

an optically active compound depends on the wavelength. A graph of rotation against wavelength has a characteristic shape showing peaks or troughs.

optical rotation Rotation of plane-polarized light. *See* optical activity.

optical telescope *See* telescope.

optical temperature *See* radiation temperature.

optic axis 1. The direction in a doubly refracting crystal in which light is transmitted without double refraction. **2.** *See* optical axis.

optic nerve The second *cranial nerve: a paired sensory nerve that runs from each eye to the brain. It is responsible for conveying visual stimuli received by the rods and cones in the retina to the brain for interpretation.

optics The study of *light and the phenomena associated with its generation, transmission, and detection. In a broader sense, optics includes all the phenomena associated with infrared and ultraviolet radiation. *Geometrical optics* assumes that light travels in straight lines and is concerned with the laws controlling the reflection and refraction of rays of light. *Physical optics* deals with phenomena that depend on the wave nature of light, e.g. diffraction, interference, and polarization.

oral contraceptive Any preparation taken in the form of a pill to prevent conception. The most common form is the combination pill, which contains the hormones *oestrogen and a *progestogen. Both act to suppress ovulation, while the progestogen additionally causes changes in the viscosity of cervical mucus and alters the lining of the womb, both of which decrease the chances of fertilization should ovulation occur. The so-called 'minipill' contains only a progestogen.

orbit 1. (in astronomy) The path through space of one celestial body about another. For one small body moving in the gravitational field of another the orbit is a *conic section. Most such orbits are elliptical and most planetary orbits in the solar system are nearly circular. The shape and size of an elliptical orbit is specified by its *eccentricity, e, and the length of its semimajor axis, a. **2.** (in physics) The path of an electron as it travels round the nucleus of an atom. *See* orbital. **3.** (in anatomy) Either of the two sockets in the skull of vertebrates that house the eyeballs.

orbital A region in which an electron may be found in an atom or molecule. In the original *Bohr theory of the atom the electrons were assumed to move around the nucleus in circular orbits, but further advances in quantum mechanics led to the view that it is not possible to give a definite path for an electron. According to *wave mechanics, the electron has a certain probability of being in a given element of space. Thus for a hydrogen atom the electron can be anywhere from close to the nucleus to out in space but the maximum probability in spherical shells of equal thickness occurs in a spherical shell around the nucleus with a radius equal to the Bohr radius of the atom. The probabilities of finding an electron in different regions can be obtained by solving the Schrödinger wave equation to give the wave function ψ, and the probability of location per unit volume is then proportional to $|\psi|^2$. Thus the idea of electrons in fixed orbits has been replaced by that of a probability distribution around the nucleus – an *atomic orbital* (see illustration). Alternatively, the orbital can be thought of as an electric charge distribution (averaged over time). In representing orbitals it is convenient to take a surface enclosing the space in which the electron is likely to be found with a high probability.

The possible atomic orbitals correspond to subshells of the atom. Thus there is one s-orbital for each shell (orbital quantum number l = 0). This is spherical. There are three p-orbitals (corresponding to the three values of l) and five d-orbitals. The shapes of orbitals depend on the value of l. For instance, p-orbitals each have two lobes; most d-orbitals have four lobes.

In molecules, the valence electrons move under the influence of two nuclei (in a bond involving two atoms) and there are corresponding *molecular orbitals* for electrons (see illustration). It is convenient in considering these to regard them as formed by overlap of atomic orbitals. In a hydrogen molecule the s-orbitals on the two atoms overlap and form a molecular orbital between the two nuclei. This is an example of a *sigma orbital*. In a double bond, as in ethene, one bond is produced by overlap along the line of axes to form a sigma orbital. The other is produced by sideways overlap of the lobes of the p-orbitals (see illustration). The resulting molecular orbital has two parts, one on each side of the sigma orbital – this is a *pi orbital*. In fact, the combination of two atomic orbitals produces two molecular orbitals with different energies. The one of lower energy is the *bonding orbital*, holding the atoms together; the other is the *antibonding orbital*, which would tend to push the atoms apart. In the case of valence electrons, only the lower (bonding) orbital is filled.

In considering the formation of molecular orbitals it is often useful to think in terms of *hybrid* atomic orbitals. For instance, carbon has in its outer shell one s-orbital and three p-orbitals. In forming methane (or other tetrahedral molecules) these can be regarded as combining to give four equivalent sp^3 hybrid orbitals, each with a lobe directed to a corner of a

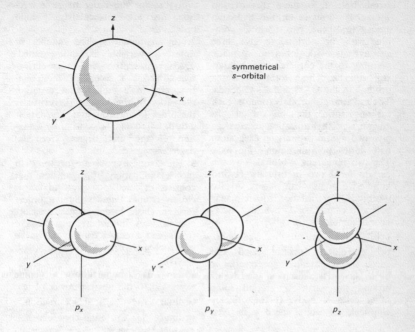

symmetrical
s−orbital

p_x p_y p_z

three equivalent p-orbitals, each having 2 lobes

Atomic orbitals

p-orbitals

pi orbital

hybrid sp_3 orbitals

sigma orbital

Molecular orbitals: formation of the double bond in ethene

Orbitals

tetrahedron. It is these that overlap with the s-orbitals on the hydrogen atoms. In ethene, two p-orbitals combine with the s-orbital to give three sp^2 hybrids with lobes in a plane pointing to the corners of an equilateral triangle. These form the sigma orbitals in the C–H and C–C bonds. The remaining p-orbitals (one on each carbon) form the pi orbital. In ethyne, sp^2 hybridization occurs to give two hybrid orbitals on each atom with lobes pointing along the axis. The two remaining p-orbitals on each carbon form two pi orbitals. Hybrid atomic orbitals can also involve d-orbitals. For instance, square-planar complexes use sp^2d hybrids; octahedral complexes use sp^3d^2.

orbital quantum number *See* atom.

orbital velocity (orbital speed) The speed of a satellite, spacecraft, or other body travelling in an *orbit around the earth or around some other celestial body. If the orbit is elliptical, the orbital speed, v, is given by:

$$v = \sqrt{[gR^2(2/r - 1/a)]},$$

where g is the accceleration of free fall, R is the radius of the orbited body, a is the semimajor axis of the orbit, and r is the distance between the orbiting body and the centre of mass of the system. If the orbit is circular, $r = a$ and $v = \sqrt{(gR^2/r)}$.

or circuit *See* logic circuits.

order 1. (in chemistry) In the expression for the rate of a chemical reaction, the sum of the powers of the concentrations is the overall order of the reaction. For instance, in a reaction

$$A + B \rightarrow C$$

the rate equation may have the form

$$R = k[A][B]^2$$

This reaction would be described as *first order* in A and *second order* in B. The overall order is three. The order of a reaction depends on the mechanism and it is possible for the rate to be independent of concentration (*zero order*) or for the order to be a fraction. *See also* molecularity; pseudo order.

2. (in mathematics) The number of times a variable is differentiated. dy/dx represents a first-order derivative, d^2y/dx^2 a second-order derivative, etc. In a *differential equation the order of the highest derivative is the order of the equation. $d^2y/dx^2 + 2dy/dx = 0$ is a second-order equation of the first degree. *See also* degree.

3. (in taxonomy) A category used in the *classification of organisms that consists of one or several similar or closely related families. Similar orders form a class. Order names typically end in *-ales* in botany, e.g. Rosales (roses and orchard fruits), and in *-a* in zoology, e.g. Carnivora (flesh eaters).

order of magnitude A value expressed to the nearest power of ten.

ordinary ray *See* double refraction.

ordinate *See* Cartesian coordinates.

Ordovician The second geological period of the Palaeozoic era, following the Cambrian and preceding the Silurian periods. It began about 500 million years ago and lasted for about 60 million years. The period was named by the British geologist Charles Lapworth (1842–1920) in 1879. *Graptolites, in deep-water deposits, are the dominant fossils. Other fossils include *trilobites, brachiopods, bryozoans, gastropods, bivalves, echinoids, crinoids, nautiloid cephalopods, and the first corals.

ore A naturally occurring mineral from which a metal can be extracted, usually on a commercial basis. The metal may be present in the ore as the native metal, but more commonly it occurs in a combined form as an oxide, sulphide, sulphate, silicate, etc.

ore dressing *See* beneficiation.

organ Any distinct part of an organism that is specialized to perform one or a number of functions. Examples are ears, eyes, lungs, and kidneys (in animals) and leaves, roots, and flowers (in plants). A given organ will contain many different *tissues.

organ culture The culture of complete living organs (*explants*) of animals and plants outside the body in a suitable culture medium. Animal organs must be small enough to allow the nutrients in the culture medium to penetrate all the cells. Whole plant roots and even root systems can be kept alive in such conditions for a considerable period of time. *See also* explantation.

organelle A minute structure within a plant or animal *cell that has a particular function. Examples of organelles are the nucleus, mitochondria, and lysosomes.

organic chemistry The branch of chemistry concerned with compounds of carbon.

organizer An area of an animal embryo that causes adjacent areas of the embryo to develop in a certain way. The *primary organizer* (blastopore lip or archenteron roof) causes the *gastrula to develop as a complete organism.

organo- Prefix used before the name of an element to indicate compounds of the elements containing organic groups (with the element bound to carbon atoms). For example, lead(IV) tetraethyl is an organolead compound.

organometallic compound A compound in which a metal atom or ion is bound to an organic group. Organometallic compounds may have single metal–carbon bonds, as in the aluminium alkyls (e.g. $Al(CH_3)_3$). In some cases, the bonding is to the pi electrons of a double bond, as in complexes formed between platinum and ethene, or to the pi electrons of a ring, as in ferrocene.

origin See Cartesian coordinates.

origin of life The process by which living organisms developed from inanimate matter, which is generally thought to have occurred on earth between 3500 and 4000 million years ago. It is supposed that the primordial atmosphere was like a chemical soup containing all the basic constituents of organic matter: ammonia, methane, hydrogen, and water vapour. These underwent a process of chemical evolution using energy from the sun and electric storms to combine into ever more complex molecules, such as amino acids, proteins, and vitamins. Eventually self-replicating nucleic acids, the basis of all life, could have developed. The very first organisms may have consisted of such molecules bounded by a simple membrane.

ornithine (orn) An *amino acid, $H_2N(CH_2)_3CH(NH_2)COOH$, that is not a constituent of proteins but is important in living organisms as an intermediate in the reactions of the *urea cycle and in arginine synthesis.

ornithine cycle See urea cycle.

orogenesis The process by which major mountain chains are formed. This includes the deformational processes of thrusting, folding, and faulting that result from the collision of two continents. Examples of mountains formed through orogenesis include the Alpine–Himalayan, Appalachian, and Cordilleran orogenic belts.

orpiment A natural yellow mineral form of arsenic(III) sulphide, As_2S_3. The name is also used for the synthetic compound, which is used as a pigment.

ortho- **1.** Prefix indicating that a benzene compound has two substituted groups in the 1,2 positions (i.e. on adjacent carbon atoms). The abbreviation *o*- is used; for example *o*-dichlorobenzene is 1,2-dichlorobenzene. *Compare* meta-; para-. **2.** Prefix formerly used to indicate the most hydrated form of an acid. For example, phosphoric(V) acid, H_3PO_4, was called orthophosphoric acid to distinguish it from the lower metaphosphoric acid, HPO_3, which is actually $(HPO_3)_n$.

orthoboric acid *See* boric acid.

orthoclase *See* feldspars.

orthogenesis An early theory of the nature of evolutionary change, which proposed that organisms evolve along particular paths predetermined by some factor in their genetic make-up. More recent understanding of selection pressure and other external forces that can be shown experimentally to affect the survival of organisms has proved the improbability of the theory.

orthohydrogen *See* hydrogen.

orthophosphoric acid *See* phosphoric(V) acid.

orthoplumbate *See* plumbate.

orthorhombic *See* crystal system.

orthosilicate *See* silicate.

orthostannate *See* stannate.

orthotropism The tendency for a *tropism (growth response of a plant) to be orientated directly in line with the stimulus concerned. An example is the vertical growth of main stems and roots in response to gravity (*orthogeotropism*). *Compare* plagiotropism.

oscillator An electronic device that produces an alternating output of known frequency. If the output voltage or current has the form of a sine wave with respect to time, the device is called a *sinusoidal* (or *harmonic*) *oscillator*. If the output voltage changes abruptly from one level to another (as in a *square wave or *sawtooth wave) it is called a *relaxation oscillator*. A harmonic oscillator consists of a frequency-determining circuit or device, such as a *resonant circuit, maintained in oscillation by a source of power that by positive feedback also makes up for the resistive losses. In some relaxation oscillators the circuit is arranged so that in each cycle energy is stored in a reactive element (a capacitor or inductor) and subsequently discharged over a different time interval. *See also* multivibrator.

oscilloscope *See* cathode-ray oscilloscope.

osmiridium A hard white naturally occurring alloy consisting principally of osmium (17–48%) and iridium (49%). It also contains small quantities of platinum, rhodium, and ruthenium. It is used for making small items subject to wear, e.g. electrical contacts or the tips of pen nibs.

osmium Symbol Os. A hard blue-white metallic *transition element; a.n. 76; r.a.m. 190.2; r.d. 22.57; m.p. 3045°C; b.p. 5027°C. It is found associated with platinum and is used in certain alloys with platinum and iridium (*see* osmiridium). Osmium forms a number of complexes in a range of oxidation states.

osmium(IV) oxide (osmium tetroxide) A yellow solid, OsO_4, made by heating osmium in air. It is used as an oxidizing agent in organic chemistry, as a catalyst, and as a fixative in electron microscopy.

osmometer *See* osmosis.

osmoregulation The control of the water content and the concentration of salts in the body of an animal. In freshwater animals osmoregulation must counteract the tendency for water to pass into the animal by *osmosis. Various methods have been developed to eliminate the excess, such as *contractile vacuoles in protozoans and *kidneys with well-developed glomeruli in freshwater fish. Marine vertebrates have the opposite problem: they prevent excessive water loss and enhance the excretion of salts by having kidneys with few glomeruli and short tubules. In terrestrial vertebrates the dangers of desiccation are reduced by the presence of long convoluted kidney tubules, which increase the reabsorption of water and salts.

osmosis The passage of a solvent through a *semipermeable membrane* separating two solutions of different concentrations. A semipermeable membrane is one through which the molecules of a solvent can pass but the molecules of most solutes cannot. There is a thermodynamic tendency for solutions separated by such a membrane to become equal in concentration, the water (or other solvent) flowing from the weaker to the stronger solution. Osmosis will stop when the two solutions reach equal concentration, and can also be stopped by applying a pressure to the liquid on the stronger-solution side of the membrane. The pressure required to stop the flow from a pure solvent into a solution is a characteristic of the solution, and is called the *osmotic pressure* (symbol Π). Osmotic pressure depends only on the concentration of particles in the solution, not on their nature (i.e. it is a *colligative property). For a solution of n moles in volume V at thermodynamic temperature T, the osmotic pressure is given by $\Pi V = nRT$, where R is the gas constant. Osmotic-pressure measurements are used in finding the relative molecular masses of compounds, particularly macromolecules. A device used to measure osmotic pressure is called an *osmometer*.

The distribution of water in living organisms is dependent to a large extent on osmosis, water entering the cells through their membranes. A cell membrane is not truly semipermeable as it allows the passage of certain solute molecules; it is described as *differentially permeable*. Animals have evolved various means to counteract the effects of osmosis (*see* osmoregulation); in plant cells, excessive osmo-

sis is prevented by the pressure exerted by the cell wall, which opposes the osmotic pressure. *See* water potential.

osmotic pressure *See* osmosis.

ossification The process of *bone formation. It is brought about by the action of special cells called *osteoblasts*, which deposit layers of bone in connective tissue. Some bones are formed directly in connective tissue (*see* membrane bone); others are formed by the replacement of cartilage (*see* cartilage bone).

Osteichthyes The class of vertebrates comprising the bony fishes – marine and freshwater fish with a bony skeleton. All have gills covered with a bony operculum, and a layer of thin overlapping bony *scales covers the entire body surface. Bony fish have a *swim bladder, which acts as a hydrostatic organ enabling the animal to remain suspended in the water at any depth. In some fish this bladder acts as a lung. *See also* Dipnoi; Teleostei. *Compare* Elasmobranchii.

Ostwald's dilution law An expression for the degree of dissociation of a weak electrolyte. For example, if a weak acid dissociates in water

$$HA \rightleftharpoons H^+ + A^-$$

the dissociation constant K_a is given by

$$K_a = \alpha^2 n/(1 - \alpha)V$$

where α is the degree of dissociation, n the initial amount of substance (before dissociation), and V the volume. If α is small compared with 1, then $\alpha^2 = KV/n$; i.e. the degree of dissociation is proportional to the square root of the dilution. The law was first put forward by W. Ostwald (1853–1932) to account for electrical conductivities of electrolyte solutions.

Otto engine *See* internal-combustion engine.

ounce 1. One sixteenth of a pound (avoirdupois), equal to 0.028 349 kg.

2. Eight drachms (Troy), equal to 0.031 103 kg. **3. (fluid ounce)** Eight fluid drachms, equal to 0.028 413 dm³.

outbreeding Mating between unrelated or distantly related individuals of a species. Outbreeding populations usually show more variation than *inbreeding ones and have a greater potential for adapting to environmental changes. Outbreeding increases the number of *heterozygous individuals, so that disadvantageous recessive characteristics tend to be masked by dominant alleles.

outer ear (external ear) The part of the ear external to the *tympanum (eardrum). It is present in mammals, birds, and some reptiles and consists of a tube (the *external auditory meatus*) that directs sound waves onto the tympanum. In mammals it may include an external *pinna, which extends beyond the skull.

ovary 1. The reproductive organ in female animals in which eggs (ova) are produced. In most vertebrates there are two ovaries (in some fish the ovaries fuse together to form a single structure and in birds the left ovary only is functional). As well as eggs, they produce steroid hormones (*see* oestrogen; progesterone). In mammals each ovary is situated close to the opening of a *fallopian tube (see illustration); it contains numerous follicles in which the eggs develop

fallopian tube — uterus — ovary — bladder — vagina

Human female reproductive system

and from which they are released in a regular cycle (*see* Graafian follicle; menstrual cycle; oogenesis; ovulation). **2.** The hollow base of the *carpel of a flower, containing one or more *ovules. After fertilization, the ovary wall develops into the fruit enclosing the seeds. In some species, the carpels are fused together to form a complex ovary.

overdamped *See* damping.

overpotential A potential that must be applied in an electrolytic cell in addition to the theoretical potential required to liberate a given substance at an electrode. The value depends on the electrode material and on the current density. It is a kinetic effect occurring because of the significant activation energy for electron transfer at the electrodes, and is particularly important for the liberation of such gases as hydrogen and oxygen. For example, in the electrolysis of a solution of zinc ions, hydrogen ($E^{\ominus} = 0.00$ V) would be expected to be liberated at the cathode in preference to zinc ($E^{\ominus} = -0.76$ V). In fact, the high overpotential of hydrogen on zinc (about 1 V under suitable conditions) means that zinc can be deposited instead.

overtones *See* harmonic.

oviduct The tube that conveys an animal egg cell from the ovary to other parts of the reproductive system or to the outside. Eggs are passed along the oviduct by the action of muscles and cilia. *See* fallopian tube.

oviparity Reproduction in which fertilized eggs are laid or spawned by the mother and hatch outside her body. It occurs in most animals except marsupial and placental mammals. *Compare* ovoviviparity; viviparity.

ovipositor An organ at the hind end of the abdomen of female insects through which eggs are laid. It consists of a pair of modified appendages and is often long and piercing, so that eggs can be laid in otherwise inaccessible places. The sting of bees and wasps is a modified ovipositor.

ovoviviparity Reproduction in which fertilized eggs develop and hatch in the oviduct of the mother. It occurs in many invertebrates and in some fish and reptiles (e.g. the viper). *Compare* oviparity; viviparity.

ovulation The release of an egg cell from the ovary, which in mammals is stimulated by *luteinizing hormone. The developing egg cell within its follicle migrates to the ovary surface; when mature, it is released from the follicle (which breaks open) into the body cavity, from where it passes into the oviduct. *See also* menstrual cycle.

ovule The part of the female reproductive organs of seed plants that consists of the *nucellus, *embryo sac, and *integuments. The ovules of gymnosperms are situated on ovuliferous scales of the female cones while those of angiosperms are enclosed in the carpel. After fertilization, the ovule becomes the seed.

ovuliferous scale One of a group of large woody specialized leaves that form the female *cone of conifers and related trees. It bears the ovules, which develop into seeds.

ovum (egg cell) **1.** (in zoology) The mature reproductive cell (*see* gamete) of female animals, which is produced by the ovary (*see* oogenesis). It is spherical, has a nucleus, is covered with a vitelline membrane, and is not mobile. **2.** (in botany) The *oosphere of plants.

oxalate A salt or ester of *oxalic acid.

oxalic acid (ethanedioic acid) A crystalline solid, $(COOH)_2$, that is slightly soluble in water. Oxalic acid is strongly acidic and very poisonous. It occurs in certain plants, e.g. sorrel and the leaf blades of rhubarb.

oxidant *See* oxidizing agent.

oxidation *See* oxidation–reduction.

oxidation number (oxidation state) *See* oxidation–reduction.

oxidation–reduction (redox) Originally, *oxidation* was simply regarded as a chemical reaction with oxygen. The reverse process – loss of oxygen – was called *reduction*. Reaction with hydrogen also came to be regarded as reduction. Later, a more general idea of oxidation and reduction was developed in which oxidation was loss of electrons and reduction was gain of electrons. This wider definition covered the original one. For example, in the reaction

$$4Na(s) + O_2(g) \rightarrow 2Na_2O(s)$$

the sodium atoms lose electrons to give Na^+ ions and are oxidized. At the same time, the oxygen atoms gain electrons and are reduced. These definitions of oxidation and reduction also apply to reactions that do not involve oxygen. For instance in

$$2Na(s) + Cl_2(g) \rightarrow 2NaCl(s)$$

the sodium is oxidized and the chlorine reduced. Oxidation and reduction also occurs at the electrodes in *cells. This definition of oxidation and reduction applies only to reactions in which electron transfer occurs – i.e. to reactions involving ions. It can be extended to reactions between covalent compounds by using the concept of *oxidation number* (or *state*). This is a measure of the electron control that an atom has in a compound compared to the atom in the pure element. An oxidation number consists of two parts:

(1) Its sign, which indicates whether the control has increased (negative) or decreased (positive).

(2) Its value, which gives the number of electrons over which control has changed.

The change of electron control may be complete (in ionic compounds) or partial (in covalent compounds). For example, in SO_2 the sulphur has an oxidation number $+4$, having gained partial control over 4 electrons compared to sulphur atoms in pure sulphur. The oxygen has an oxidation number -2, each oxygen having lost partial control over 2 electrons compared to oxygen atoms in gaseous oxygen. Oxidation is a reaction involving an increase in oxidation number and reduction involves a decrease. Thus in

$$2H_2 + O_2 \rightarrow 2H_2O$$

the hydrogen in water is $+1$ and the oxygen -2. The hydrogen is oxidized and the oxygen is reduced.

The oxidation number is used in naming inorganic compounds. Thus in H_2SO_4, sulphuric(VI) acid, the sulphur has an oxidation number of $+6$. Compounds that tend to undergo reduction readily are *oxidizing agents; those that undergo oxidation are *reducing agents.

oxidative phosphorylation A reaction occurring during the final stages of *aerobic respiration, in which ATP is formed from ADP and phosphate coupled to electron transport in the *electron transport chain. The reaction occurs in the mitochondria and takes place at three sites on the electron transport chain.

oxides Binary compounds formed between elements and oxygen. Oxides of nonmetals are covalent compounds having simple molecules (e.g. CO, CO_2, SO_2) or giant molecular lattices (e.g. SiO_2). They are typically acidic or neutral. Oxides of metals are ionic, containing the O^{2-} ion. They are generally basic or *amphoteric. Various other types of ionic oxide exist (*see* ozonides; peroxides; superoxides).

oxidizing acid An acid that can act as a strong oxidizing agent as well as an acid. Nitric acid is a common example. It is able to attack metals, such as copper, that are below hydrogen in the electromotive series, by oxidizing the metal:

$$2HNO_3 + Cu \rightarrow CuO + H_2O + 2NO_2$$

This is followed by reaction between the acid and the oxide:

$$2HNO_3 + CuO \rightarrow Cu(NO_3)_2 + H_2O$$

oxidizing agent (oxidant) A substance that brings about oxidation in other substances. It achieves this by being itself reduced. Oxidizing agents contain atoms with high oxidation numbers; that is the atoms have suffered electron loss. In oxidizing other substances these atoms gain electrons.

oximes Compounds containing the group C:NOH, formed by reaction of an aldehyde or ketone with hydroxylamine (H_2NOH) (see illustration). Ethanal (CH_3CHO), for example, forms the oxime $CH_3CH:NOH$.

oxo- Prefix indicating the presence of oxygen in a chemical compound.

oxo acid An *acid in which the acidic hydrogen atom(s) are bound to oxygen atoms. Sulphuric acid is an example: the two acidic hydrogens are on the $-OH$ groups bound to the sulphur. *Compare* binary acid.

oxonium ion An ion of the type R_3O^+, in which R indicates hydrogen or an organic group. The hydroxonium ion, H_3O^+, is formed when *acids dissociate in water.

oxo process An industrial process for making aldehydes by reaction between alkanes, carbon monoxide, and hydrogen (cobalt catalyst using high pressure and temperature).

oxyacetylene burner A welding or cutting torch that burns a mixture of oxygen and acetylene (ethyne) in a specially designed jet. The flame temperature of about 3300°C enables all ferrous metals to be welded. For cutting, the point at which the steel is to be cut is preheated with the oxyacetylene flame and a powerful jet of oxygen is then directed onto the steel. The oxygen reacts with the hot steel to form iron oxide and the heat of this reaction melts more iron, which is blown away by the force of the jet.

Formation of an oxime from a ketone
The same reaction occurs with an aldehyde (R′=H)

oxygen Symbol O. A colourless odourless gaseous element belonging to *group VI of the periodic table; a.n. 8; r.a.m. 15.9994; d. 1.429 g dm^{-3}; m.p. −214.4°C; b.p. −183°C. It is the most abundant element in the earth's crust (49.2% by weight) and is present in the atmosphere (28% by volume). Atmospheric oxygen is of vital importance for all organisms that carry out *aerobic respiration. For industrial purposes it is obtained by fractional distillation of liquid air. It is used in metallurgical processes, in high-temperature flames (e.g. for welding), and in breathing apparatus. The common form is diatomic (*dioxygen*, O$_2$); there is also a reactive allotrope *ozone (O$_3$). Chemically, oxygen reacts with most other elements forming *oxides. The element was discovered by Priestley in 1774.

oxygen debt The physiological state that exists in a normally aerobic animal when insufficient oxygen is available for metabolic requirements (e.g. during a period of strenuous physical activity). To meet the body's increased demand for energy, pyruvate is converted anaerobically (i.e. in the absence of oxygen) to lactic acid, which requires oxygen for its breakdown and accumulates in the tissues. When oxygen is available again lactic acid is oxidized in the liver, thus repaying the debt.

oxyhaemoglobin *See* haemoglobin.

oxytocin A hormone, secreted by the posterior pituitary gland in mammals, that stimulates the ejection of milk in lactating females and the contraction of the uterus during parturition. Like *vasopressin, oxytocin is synthesized in the hypothalamus.

ozone (trioxygen) A colourless gas, O$_3$, soluble in cold water and in alkalis; m.p. −192.7°C; b.p. −111.9°C. Liquid ozone is dark blue in colour and is diamagnetic (dioxygen, O$_2$, is paramagnetic). The gas is made by passing oxygen through a silent electric discharge and is usually used in mixtures with oxygen. It is produced in the stratosphere by the action of high-energy ultraviolet radiation on oxygen and its presence there acts as a screen for ultraviolet radiation (*see* ozone layer). Ozone is also one of the greenhouse gases (*see* greenhouse effect). It is a powerful oxidizing agent and is used to form ozonides by reaction with alkenes and subsequently by hydrolysis to carbonyl compounds.

ozone layer (ozonosphere) A layer of the *earth's atmosphere in which most of the atmosphere's ozone is concentrated. It occurs 15–50 km above the earth's surface and is virtually synonymous with the stratosphere. In this layer most of the sun's ultraviolet radiation is absorbed by the ozone molecules, causing a rise in the temperature of the stratosphere and preventing vertical mixing so that the stratosphere forms a stable layer. By absorbing most of the solar ultraviolet radiation the ozone layer protects living organisms on earth. The fact that the ozone layer is thinnest at the equator is believed to account for the high equatorial incidence of skin cancer as a result of exposure to unabsorbed solar ultraviolet radiation. In the 1980s it was found that depletion of the ozone layer was occurring over both the poles. This is thought to have been caused by a series of complex photochemical reactions involving nitrogen oxides produced from aircraft and, more seriously, *chlorofluorocarbons and *halons. *See also* pollution.

ozonides 1. A group of compounds formed by reaction of ozone with alkali metal hydroxides and formally containing the ion O_3^-. **2.** Unstable compounds formed by the addition of ozone to the $C=C$ double bond in alkenes. *See* ozonolysis.

ozonolysis A reaction of alkenes with ozone to form an ozonide. It was once used to investigate the structure of alkenes by hydrolysing the ozonide to give aldehydes or ketones. For instance

$$R_2C:CHR' \rightarrow R_2CO + R'CHO$$

These could be identified, and the structure of the original alkene determined.

P

P_1 (parental generation) The individuals that are selected to begin a breeding experiment, crosses between which yield the *F_1 generation.

pacemaker 1. (*or* **sinoatrial node**) A small mass of specialized muscle cells in the mammalian heart, found in the wall of the right atrium near the opening for the vena cava. The cells initiate and maintain the heart beat: by their rhythmic and spontaneous contractions they stimulate contractions in the rest of the heart muscle. The cells themselves are controlled by the autonomic nervous system, which determines the heart rate. Similar pacemakers occur in the hearts of other vertebrates. **2.** An electronic or nuclear battery-charged device that can be implanted surgically into the

chest to produce and maintain the heart beat. These devices are used when the heart's own pacemaker is defective or diseased.

pachytene The period in the first prophase of *meiosis when paired *homologous chromosomes are fully contracted and twisted around each other.

packing density 1. The number of devices (such as *logic circuits) or integrated circuits per unit area of a *silicon chip. **2.** The quantity of information stored in a specified space of a storage system associated with a computer, e.g. *bits per inch of magnetic tape.

packing fraction The algebraic difference between the relative atomic mass of an isotope and its mass number, divided by the mass number.

paedogenesis Reproduction by an animal that is still in the larval or pre-adult form. Paedogenesis is a form of *neoteny and is particularly marked in the axolotl, a larval form of the salamander, which retains its larval features owing to a thyroid deficiency but can breed, producing individuals like itself. If the thyroid hormone thyroxine is given, metamorphosis occurs.

pairing (synapsis) The close association between *homologous chromosomes that develops during the first prophase of *meiosis. The two chromosomes move together and an exact pairing of corresponding points along their lengths occurs as they lie side by side. The resulting structure is called a *bivalent*.

pair production The creation of an electron and a positron from a photon in a strong electric field, such as that surrounding an atomic nucleus. The electron and the positron each have a mass of about 9×10^{-31} kg, which is equivalent on the basis of the mass–energy equation ($E = mc^2$) to a total of 16×10^{-14} J. The frequency, ν, associated with a photon of this energy (according to $E = h\nu$) is 2.5×10^{20} Hz. Pair production thus requires photons of high quantum energy (Bremsstrahlung or gamma rays). Any excess energy is taken up as kinetic energy of the products.

palaeobotany The branch of *palaeontology concerned with the study of plants through geological time, as revealed by their *fossil remains (*see also* palynology). It overlaps with other aspects of plant study, including anatomy, ecology, evolution, and taxonomy.

Palaeocene The earliest geological epoch of the *Tertiary period. It began about 65 million years ago, following the Cretaceous period, and extended for about 11.5 million years to the beginning of the *Eocene (the Palaeocene is sometimes included in the Eocene). It was named by the palaeobotanist W. P. Schimper in 1874. A major floral and faunal discontinuity occurred between the end of the Cretaceous and the beginning of the Palaeocene: following the extinction of many reptiles the mammals became abundant on land. By the end of the epoch primates and rodents had evolved.

palaeoclimatology The study of climates of earlier geological periods. This is based largely on the study of sediments that were laid down during these periods and of fossils. The changes in the positions of the continents as a result of *continental drift and *plate tectonics complicate the study.

palaeoecology The study of the relationships of *fossil organisms to each other and to their environments. It involves the study both of the fossils and of the surrounding rocks in which they are found. Trace fossils may provide information on the behaviour of the organism.

Palaeolithic The Old Stone Age, lasting in Europe from about 2.5 million to 9000 years ago, during which man used primitive stone tools made by chipping stones and flints.

palaeomagnetism The study of magnetism in rocks, which provides information on variations in the direction and intensity of the earth's magnetic field with time. During the formation of an igneous or sedimentary rock containing magnetic minerals the polarity of the earth's magnetic field at that time becomes 'frozen' into the rock. Studies of this fossil magnetism in samples of rocks have enabled the former positions of magnetic poles at various geological times to be located. It has also revealed that periodic reversals in the geomagnetic field have taken place (i.e. the N-pole becomes the S-pole and vice versa). This information has been important in plate tectonics in establishing the movements of lithospheric plates over the earth's surface. The magnetic reversals provided crucial evidence for the sea-floor spreading hypothesis proposed in the early 1960s.

palaeontology The study of extinct organisms, including their structure, environment, evolution, and distribution, as revealed by their *fossil remains. Palaeontological work also makes important contributions to geology in revealing stratigraphic relationships between rock strata and determining the physical appearance and climate of the earth during past geological ages. *See also* palaeobotany; palaeoecology; palaeozoology.

Palaeozoic The first era of *Phanerozoic time. It follows the *Precambrian and is subdivided into the Lower Palaeozoic, comprising the *Cambrian, *Ordovician, and *Silurian periods, and the Upper Palaeozoic, comprising the *Devonian, *Carboniferous, and *Permian periods, It extended from about 570 million years ago to about 225 million years ago, when it was succeeded by the *Mesozoic era.

palaeozoology The branch of *palaeontology concerned with the study of animals throughout geological time, as revealed by their *fossil remains.

palate The roof of the mouth of vertebrates, which separates the *buccal and nasal cavities. In mammals it is divided into two zones, the bony *hard palate* and the *soft palate*, and completely separates the buccal cavity from the air passage to enable simultaneous eating and breathing.

palisade mesophyll *See* mesophyll.

palladium Symbol Pd. A soft white ductile *transition element (*see also* platinum metals); a.n. 46; r.a.m. 106.4; r.d. 12.26; m.p. $1551 \pm 1°C$; b.p. $3140 \pm 1°C$. It occurs in some copper and nickel ores and is used in jewellery and as a catalyst for hydrogenation reactions. Chemically, it does not react with oxygen at normal temperatures. It dissolves slowly in hydrochloric acid. Palladium is capable of occluding 900 times its own volume of hydrogen. It forms few simple salts, most compounds being complexes of palladium(II) with some palladium(IV). It was discovered by Woolaston in 1803.

pallium *See* cerebral cortex.

palmitate (hexadecanoate) A salt or ester of palmitic acid.

palmitic acid (hexadecanoic acid) A 16-carbon saturated fatty acid, $CH_3(CH_2)_{14}COOH$; r.d. 0.85; m.p. 63°C; b.p. 351°C. Glycerides of palmitic acid occur widely in plant and animal oils and fats.

palp An elongated sensory organ, usually near the mouth, in many invertebrates. Examples are the tactile head appendages of polychaete worms, the ciliated flap of tissue that produces feeding currents in bivalve molluscs, the distal part of the *mandibles of crustaceans, and the olfactory parts of the first and second *maxillae of some insects.

palynology (micropalaeontology) The study of fossil pollen and spores (*pollen analysis*) and various other *microfossils, such as coccoliths and dinoflagellates. Palynology is used in stratigraphy, palaeoclimatology, and archaeology. Pollen and spores are very resistant to decay and therefore their fossils are found in sedimentary rocks. They may be extracted by various methods, including boiling with potassium hydroxide solution, washing with strong oxidizing mixtures, and centrifuging repeatedly. Spores and pollen are classified according to shape, form of aperture, and both internal and external details of the exine (outer coat). They indicate the nature of the dominant flora, and therefore the climate and conditions of the period in which they lived.

pancreas A gland in vertebrates lying between the duodenum and the spleen. Under the influence of the hormone *secretin it secretes digestive enzymes (mainly trypsinogen (*see* trypsin), *amylase, and *lipase) into the duodenum via the pancreatic duct. It also contains groups of cells – the *islets of Langerhans – that function as an *endocrine gland, producing the hormones *insulin and *glucagon, which regulate blood sugar levels.

pancreozymin *See* cholecystokinin.

panicle A type of flowering shoot common in the grass family. The primary axis bears groups of *racemes and is itself racemose, as the youngest groups of flowers are at the top (e.g. oat). The term may be used loosely for any form of branched *racemose inflorescence; for example, the horse chestnut is a raceme of cymes. Both these arrangements are seen in the family Polygonaceae (docks and sorrels).

pantothenic acid A vitamin of the *vitamin B complex. It is a constituent of *coenzyme A, which performs a crucial role in the oxidation of fats, carbohydrates, and certain amino acids. Deficiency rarely occurs because the vitamin occurs in many foods, especially cereal grains, peas, egg yolk, liver, and yeast.

papain A protein-digesting enzyme (see protease) occurring in the fruit of the West Indian papaya tree (*Carica papaya*). It is used as a digestant and in the manufacture of meat tenderizers.

paper chromatography A technique for analysing mixtures by *chromatography, in which the stationary phase is absorbent paper. A spot of the mixture to be investigated is placed near one edge of the paper and the sheet is suspended vertically in a solvent, which rises through the paper by capillary action carrying the components with it. The components move at different rates, partly because they absorb to different extents on the cellulose and partly because of partition between the solvent and the moisture in the paper. The paper is removed and dried, and the different components form a line of spots along the paper. Colourless substances are detected by using ultraviolet radiation or by spraying with a substance that reacts to give a coloured spot (e.g. ninhydrin gives a blue coloration with amino acids). The components can be identified by the distance they move in a given time.

papilla Any cone-shaped protuberance projecting from the surface of an organ or organism. Papillae occur, for example, on the tongue and, in plants, on the surface of many petals.

papovavirus One of a group of *DNA-containing viruses that produce tumours in their hosts. *Papillomaviruses* produce non-malignant tumours (such as warts) in all vertebrates; *polyomaviruses* produce malignant tumours in only certain classes of vertebrates (not including man).

pappus A group of modified *sepals, often in the form of a ring of silky hairs. For example, when the fruit of the dandelion matures a pappus of hairs persists at the top of a thin stalk forming a parachute-like structure, which serves to disperse the fruit.

para- 1. Prefix designating a benzene compound in which two substituents are in the 1,4 positions, i.e. directly opposite each other, on the benzene ring. The abbreviation *p*- is used; for example, *p*-xylene is 1,4-dimethylbenzene. *Compare* ortho-; meta-. **2.** Prefix denoting the form of diatomic molecules in which the nuclei have opposite spins, e.g. parahydrogen. *Compare* ortho-.

parabola A *conic with eccentricity *e* = 1. It is the locus of a point that moves so that its distance from the *focus* is equal to its perpendicular distance from the *directrix*. A chord through the focus, perpendicular to the axis, is called the *latus rectum*. For a parabola with its vertex at the origin, lying symmetrically about the *x*-axis, the equation is $y^2 = 4ax$, where *a* is the distance from the vertex to the focus. The directrix is the line $x = -a$, and the latus rectum is 4*a*. See illustration.

A parabola

parabolic reflector (paraboloidal reflector) A reflector having a section that is a parabola. A concave parabolic reflector will reflect a parallel beam of radiation through its focus and, conversely, will produce a parallel beam if the source of the radiation is placed at its focus. Parabolic mirrors are used in reflecting optical *telescopes to collect the light and in some light sources that require a parallel beam of light. In radio telescopes a dish aerial may also consist of a parabolic reflector.

paraboloid A solid formed by rotating a parabola about its axis of symmetry.

paraffin See petroleum.

paraffins See alkanes.

paraffin wax See petroleum.

paraformaldehyde See methanal.

parahydrogen See hydrogen.

paraldehyde See ethanal.

parallax 1. An apparent displacement of a distant object (with respect to a more distant background) when viewed from two different positions. If such an object is viewed from two points at either end of a base line, the angle between the lines joining the object to the ends of the base line is the *angle of parallax*. If the base line is the distance between the two eyes of an observer the angle is called the *binocular parallax*. **2.** The angular displacement in the apparent position of a celestial body when observed from two different points. *Diurnal parallax* results from the earth's daily rotation, the celestial body being viewed from the surface of the earth rather than from its centre. *Annual parallax* is caused by the earth's motion round the sun, the celestial body being viewed from the earth rather than from the centre of the sun. *Secular parallax* is caused by the motion of the solar system relative to the fixed stars.

parallel circuits A circuit in which the circuit elements are connected so that the current divides between them. For resistors in parallel, the total resistance, R, is given by $1/R = 1/r_1 + 1/r_2 + 1/r_3 \ldots$, where r_1, r_2, and r_3 are the resistances of the individual elements. For capacitors in parallel, the total capacitance, C, is given by $C = c_1 + c_2 + c_3 \ldots$.

parallelepiped (parallelopiped) A solid with six faces, all of which are parallelograms.

parallel evolution The development of related organisms along similar evolutionary paths due to strong selection pressures acting on all of them in the same way. It is debatable if the phenomenon really exists: many argue that all evolution is ultimately *convergent or divergent (see adaptive radiation).

parallelogram of forces See parallelogram of vectors.

parallelogram of vectors A method of determining the *resultant of two *vector quantities. The two vector quantities are represented by two adjacent sides of a parallelogram and the resultant is then the diagonal through their point of intersection. The magnitude and direction of the resultant is found by scale drawing or by trigonometry. The method is used for such vectors as forces (*parallelogram of forces*) and velocities (*parallelogram of velocities*). See illustration.

parallelogram of velocities See parallelogram of vectors.

paramagnetism See magnetism.

parametric equation An equation of a curve expressed in the form of the parameters that locate points on the curve. The parametric equations of a straight line are $x = a + bt$, $y = c + dt$. For a circle, they are $x = a\cos\theta$, $y = a\sin\theta$.

paraphyletic See monophyletic.

parasitic capture The absorption of a neutron by a nuclide that does not

A is the velocity of the boat with respect to the water; B is the velocity of the water with respect to the bank

parallelogram of velocities. R is the resultant velocity of the boat with respect to the bank

Parallelogram of vectors

result in either fission or a useful artificial element.

parasitism An association in which one organism (the *parasite*) lives on (*ectoparasitism*) or in (*endoparasitism*) the body of another (the *host*), from which it obtains its nutrients. Some parasites inflict comparatively little damage on their host, but many cause characteristic diseases (these are, however, never immediately fatal, as killing the host would destroy the parasite's source of food). Parasites are usually highly specialized for their way of life, which may involve one host or several (if the *life cycle requires it). They typically produce vast numbers of eggs, very few of which survive to find their way to another suitable host. *Obligate parasites* can only survive and reproduce as parasites; *facultative parasites* can also live as *saprophytes. The parasites of man include fleas and lice (which are ectoparasites), various bacteria, viruses, protozoans, and fungi (endoparasites causing characteristic diseases), and tapeworms (e.g. *Taenia solium*, which lives in the gut).

parasympathetic nervous system Part of the *autonomic nervous system. Its nerve endings release acetylcholine as a *neurotransmitter and its actions tend to antagonize those of the *sympathetic nervous system. For example, the parasympathetic nervous system increases salivary gland secretion, decreases heart rate, promotes digestion (by increasing *peristalsis),

and dilates blood vessels, while the sympathetic nervous system has opposite effects.

parathyroid glands Two pairs of *endocrine glands situated behind, or embedded within, the thyroid gland in higher vertebrates. They produce *parathyroid hormone* (*parathormone*), which controls the amount of calcium in the blood. A low level of calcium stimulates hormone secretion, which increases transfer of calcium from the bones to the blood (*compare* calcitonin).

paraxial ray A ray of light that falls on a reflecting or refracting surface close to and almost parallel to the axis. It is for such rays that simple lens theory can be developed, by means of making small angle approximations.

parenchyma 1. A plant tissue consisting of roughly spherical relatively undifferentiated cells, frequently with air spaces between them. The cortex and pith are composed of parenchyma cells (*see* ground tissues). **2.** Loose *connective tissue formed of large cells. Its function is to pack the spaces between organs in some simple acoelomate animals, such as flatworms (Platyhelminthes).

parent *See* daughter.

parity Symbol P. The property of a *wave function that determines its behaviour when all its spatial coordinates are reversed in direction, i.e. when x,y,z are replaced by $-x,-y,-z$.

If a wave function ψ satisfies the equation $\psi(x,y,z) = \psi(-x,-y,-z)$ it is said to have even parity, if it satisfies $\psi(x,y,z) = -\psi(-x,-y,-z)$ it has odd parity. In general, $\psi(x,y,z) = P\psi(-x,-y,-z)$, where P is a quantum number called parity that can have the value $+1$ or -1. The principle of *conservation of parity* (or *space-reflection symmetry*) would hold if all physical laws could be stated in a coordinate system independent of left- or right-handedness. If parity was conserved there would therefore be no fundamental way of distinguishing between left and right. In electromagnetic and strong interactions, parity is, in fact, conserved. In 1956, however, it was shown that parity is not conserved in weak interactions. In the beta decay of cobalt–60, for example, the electrons from the decay are emitted preferentially in a direction opposite to that of the cobalt spin. This experiment provides a fundamental distinction between left and right.

parsec A unit of length used to express astronomical distance. The distance at which the mean radius of the earth's orbit subtends an angle of one second of arc. One parsec is equal to 3.0857×10^{16} metres or 3.2616 light years.

parthenocarpy The formation of fruit without prior fertilization of the flower by pollen. The resulting fruits are seedless and therefore do not contribute to the reproduction of the plant; examples are bananas and pineapples. Plant *growth substances may have a role in this phenomenon, which can be induced by auxins in the commercial production of tomatoes and other fruits.

parthenogenesis The development of an organism from an unfertilized egg. This occurs sporadically in many plants (e.g. dandelions and hawkweeds) and in a few animals, but in some species it is the main and sometimes only method of repro-

duction. For example, in some species of aphid, males are absent or very rare. The eggs formed by the females contain the full (diploid) number of chromosomes and are genetically identical. Variation is consequently very limited in species that reproduce parthenogenetically.

partial A simple component of a complex tone. When a musical instrument produces a note, say, middle C, it will produce a complex tone in which the fundamental frequency is mixed with a number of partials. Some of these partials, for example, if the note is produced by bowing a taut string, will be *harmonics, i.e. integral multiples of the fundamental. If the string is struck, however, some of the partials can be inexact multiples of the fundamental. Partials are not therefore identical with harmonics.

partial derivative The infinitesimal change in a function consisting of two or more variables when one of the variable changes and the others remain constant. If $z = f(x,y)$, $\partial z/\partial x$ is the partial derivative of z with respect to x, while y remains unchanged. A *partial differential equation*, such as *Laplace's equation, is an equation containing partial derivatives of a function.

partial eclipse *See* eclipse.

partial pressure *See* Dalton's law.

particle 1. (in physics) One of the fundamental components of matter. *See* elementary particles. **2.** (in mechanics) A hypothetical body that has mass but no physical extension. As it is regarded as having no volume, a particle is incapable of rotation and therefore can only have translational motion. Thus a real body may often, for translational purposes, be regarded as a particle located at the body's centre of mass and having a mass equal to that of the whole body.

particle physics The study of *elementary particles.

particulate inheritance The transmission from parent to offspring of separate units that determine characteristics. Gregor Mendel observed that *recessive characteristics, absent in the offspring of a cross in which only one parent possessed them, reappeared repeatedly in the progeny of subsequent crosses. This led him to formulate his theory of inherited 'factors' (now called *alleles) that retain their identity through succeeding generations (*see* Mendel's laws). *Compare* blending inheritance.

partition If a substance is in contact with two different phases then, in general, it will have a different affinity for each phase. Part of the substance will be absorbed or dissolved by one and part by the other, the relative amounts depending on the relative affinities. The substance is said to be *partitioned* between the two phases. For example, if two immiscible liquids are taken and a third compound is shaken up with them, then an equilibrium is reached in which the concentration in one solvent differs from that in the other. The ratio of the concentrations is the *partition coefficient* of the system. The *partition law* states that this ratio is a constant for given liquids.

partition coefficient *See* partition.

parton A pointlike, almost free, particle postulated as a component of nucleons. The parton model enabled the results of very high-energy experiments on nucleons to be understood. *See* quantum chromodynamics.

parturition The act of giving birth to young at the end of the *gestation period. Hormones probably cause the process to start but the mechanism is not fully understood.

pascal The *SI unit of pressure equal to one newton per square metre.

Pascal's law In a confined fluid, externally applied pressure is transmitted uniformly in all directions. In a static fluid, force is transmitted at the speed of sound throughout the fluid and acts at right angles to any surface in or bounding the fluid. This principle is made use of in the hydraulic jack, the pneumatic tyre, and similar devices. The law was discovered in 1647 by Blaise Pascal (1623–62).

Paschen series *See* hydrogen spectrum.

passive Describing a solid that has reacted with another substance to form a protective layer, so that further reaction stops. The solid is said to have been 'rendered passive'. For example, aluminium reacts spontaneously with oxygen in air to form a thin layer of *aluminium oxide, which prevents further oxidation. Similarly, pure iron forms a protective oxide layer with concentrated nitric acid and is not dissolved further.

passive device 1. An electronic component, such as a capacitor or resistor, that is incapable of amplification. **2.** An artificial *satellite that reflects an incoming signal without amplification. **3.** A solar-power device that makes use of an existing structure to collect and utilize solar energy without the use of pumps, fans, etc. **4.** A radar device that provides information for navigation, guidance, surveillance, etc., by receiving the microwave radiation emitted by a warm body or reflected by a body from some other source. Such a passive device emits no microwave energy itself and therefore does not disclose its position. *Compare* active device.

pasteurization The treatment of milk to destroy disease-causing bacteria, such as those of tuberculosis, typhoid, and brucellosis. Milk is heated to 65°C for 30 minutes or to 72°C for 15 minutes followed by rapid cooling to below 10°C. The method was devised by the French microbiologist Louis Pasteur (1822–95).

patella (kneecap) A small rounded movable bone that is situated in a tendon in front of the knee joint in most mammals (including humans). The function of the patella is to protect the knee.

pathogen Any disease-causing microorganism. Pathogens include viruses, rickettsiae, and many bacteria, fungi, and protozoans. *See* infection.

patristic Denoting similarity between organisms resulting from common ancestry. *Compare* homoplastic.

Pauli exclusion principle The quantum-mechanical principle, applying to fermions but not to bosons, that no two identical particles in a system, such as electrons in an atom or quarks in a hadron, can possess an identical set of quantum numbers. It was first formulated by Wolfgang Pauli (1900–58) in 1925. The origin of the Pauli exclusion principle lies in the *spin–statistics theorem of relativistic quantum field theory.

***p*-block elements** The block of elements in the periodic table consisting of the main groups III (B to Tl), IV (C to Pb), V (N to Bi), VI (O to Po), VII (F to At) and 0 (He to Rn). The outer electronic configurations of these elements all have the form ns^2np^x where $x = 1$ to 6. Members at the top and on the right of the *p*-block are nonmetals (C, N, P, O, F, S, Cl, Br, I, At). Those on the left and at the bottom are metals (Al, Ga, In, Tl, Sn, Pb, Sb, Bi, Po). Between the two, from the top left to bottom right, lie an ill-defined group of metalloid elements (B, Si, Ge, As, Te).

PCB *See* polychlorinated biphenyl.

p.d. (potential difference) *See* electric potential.

PDGF *See* growth factor.

peacock ore *See* bornite.

pearl ash *See* potassium carbonate.

pearlite *See* steel.

peat A mass of dark-brown or black fibrous plant debris produced by the partial disintegration of vegetation in wet places. It may accumulate in depressions. When subjected to burial and hence pressure and heat it may be converted to *coal. Peat is used to improve soil and as a fuel, especially in Ireland and Sweden.

peck order *See* dominant.

pectic substances A group of polysaccharides made up primarily of sugar acids. They are important constituents of plant cell walls and the *middle lamella between adjacent cell walls. Normally they are present in an insoluble form, but in ripening fruits and in tissues affected by certain diseases they change into a soluble form, which is evidenced by softening of the tissues.

pectin A type of *pectic substance. It is used in making jam as it forms a gel with sucrose.

pectoral fins *See* fins.

pectoral girdle (shoulder girdle) The bony or cartilaginous structure in vertebrates to which the anterior limbs (pectoral fins, forelegs, or arms) are attached. In mammals it consists of two dorsal *scapulae (shoulder blades) attached to the backbone and two ventral *clavicles (collar bones) attached to the sternum (breastbone).

pedicel The stalk attaching a flower to the main floral axis (*see* peduncle). Some flowers, described as *sessile*, do not have a pedicel and arise directly from the peduncle.

pedology The science of the study of soils, including their origin and characteristics and their utilization.

peduncle The main stalk of a plant that bears the flowers, which may be solitary or grouped in an *inflorescence. *Compare* pedicel.

pegmatite Very coarse-grained igneous rock. Granite pegmatite, the commonest type, consists chiefly of alkali feldspar and quartz; accessory minerals, such as mica, tourmaline, topaz, beryl, fluorite, cassiterite, and garnet, may also be present. Many pegmatites are thus economically important as sources of these minerals. The individual crystals may be extremely large; for example, mica and quartz crystals over 3 m in length have been found.

pelagic Describing organisms that swim or drift in a sea or a lake, as distinct from those that live on the bottom (*see* benthos). Pelagic organisms are divided into *plankton and *nekton.

pellicle The thin outer covering, composed of protein, that protects and maintains the shape of certain unicellular organisms, e.g. *Euglena*. It is transparent and in ciliated organisms, e.g. *Paramecium*, contains small pores through which the cilia emerge.

Peltier effect The change in temperature produced at a junction between two dissimilar metals or semiconductors when an electric current passes through the junction. The direction of the current determines whether the temperature rises or falls. The first metals to be investigated were bismuth and copper; if the current flows from bismuth to copper the temperature rises. If the current is reversed the temperature falls. The effect was discovered in 1834 by J. C. A. Peltier (1785–1845) and has been used recently for small-scale refrigeration. *Compare* Seebeck effect.

pelvic fins *See* fins.

pelvic girdle (pelvis; hip girdle) The bony or cartilaginous structure in vertebrates to which the posterior limbs (pelvic fins or legs) are attached. The pelvic girdle articulates dorsally with the backbone; it is made up of two halves, each produced by the fusion of the *ilium, *ischium, and *pubis.

pelvis 1. *See* pelvic girdle. **2.** The lower part of the abdomen in the region of the pelvic girdle. **3.** A conical chamber in the *kidney into which urine drains from the kidney tubules before passing to the *ureter.

pendulum Any rigid body that swings about a fixed point. The *ideal simple pendulum* consists of a bob of small mass oscillating back and forth through a small angle at the end of a string or wire of negligible mass. Such a device has a period $2\pi\sqrt{(l/g)}$, where l is the length of the string or wire and g is the *acceleration of free fall. This type of pendulum moves with *simple harmonic motion.

The *compound pendulum* consists of a rigid body swinging about a point within it. Its period is given by
$$T = 2\pi\sqrt{[(h^2 + k^2)/hg]},$$
where k is the radius of gyration about an axis through the centre of mass and h is the distance from the pivot to the centre of mass. *See also* Kater's pendulum.

penicillin An *antibiotic derived from the mould *Penicillium notatum*; specifically it is known as *penicillin G* and belongs to a class of similar substances called penicillins. They are all active against a wide variety of bacteria, producing their effects by disrupting synthesis of the bacterial cell wall, and are used to treat a variety of infections caused by these bacteria.

penis The male reproductive organ of mammals (and also of some birds and reptiles) used to introduce sperm into the female reproductive tract to ensure internal fertilization. It contains a duct (the *urethra) through which the sperms pass. The penis becomes erect during precopulatory activity, either by filling with blood or haemolymph or by the action of muscles, and can be inserted into the vagina (or cloaca). In mammals the urine also leaves the body through the penis.

A basic pentadactyl forelimb, as exemplified by the human arm

flipper of whale   foreleg of pig   wing of bat

The modified pentadactyl forelimb of various vertebrates

pentadactyl limb A limb with five digits, characteristic of tetrapod vertebrates (amphibians, reptiles, birds, and mammals). It evolved from the paired fins of primitive fish as an adaptation to locomotion on land and is not found in modern fish. The limb has three parts (see illustration): the upper arm or thigh containing one long bone, the forearm or shank containing two long bones, and the hand or foot, which contains a number of small bones. This basic design is modified in many species, according to the function of the limb, particularly by the loss of fusion of the terminal bones.

pentahydrate A crystalline hydrate containing five molecules of water per molecule of compound.

pentane A straight-chain alkane hydrocarbon, C_5H_{12}; r.d. 0.63; m.p. $-129.7°C$; b.p. 36.1°C. It is obtained by distillation of petroleum.

pentanoic acid (valeric acid) A colourless liquid *carboxylic acid, $CH_3(CH_2)_3COOH$; r.d. 0.9; m.p.

amino acid 1 amino acid 2 dipeptide water

——— peptide bond

Formation of a peptide bond

−34°C; b.p. 185°C. It is used in the perfume industry.

pentavalent (quinquevalent) Having a valency of five.

pentlandite A mineral consisting of a mixed iron–nickel sulphide, $(Fe,Ni)_9S_8$, crystallizing in the cubic system; the chief ore of nickel. It is yellowish-bronze in colour with a metallic lustre. The chief occurrence of the mineral is at Sudbury in Ontario, Canada.

pentode A *thermionic valve with a suppressor grid between the anode and the screen grid of a tetrode. Its purpose is to suppress the loss of electrons from the anode as a result of secondary emission. The suppressor grid is maintained at a negative potential relative to the anode and to the screen grid.

pentose A sugar that has five carbon atoms per molecule. See monosaccharide.

pentyl group *or* **radical** The organic group $CH_3CH_2CH_2CH_2CH_2^-$, derived from pentane.

penumbra See shadow.

pepo See berry.

pepsin An enzyme, secreted by cells lining the interior of the vertebrate stomach, that catalyses the breakdown of proteins.

peptidase See protease.

peptide Any of a group of organic compounds comprising two or more amino acids linked by *peptide bonds*. These bonds are formed by the reaction between adjacent carboxyl $(-COOH)$ and amino $(-NH_2)$ groups with the elimination of water

(see illustration). *Dipeptides* contain two amino acids, *tripeptides* three, and so on. *Polypeptides contain more than ten and usually 100–300. Naturally occurring *oligopeptides* (of less than ten amino acids) include the tripeptide glutathione and the pituitary hormones vasopressin and oxytocin, which are octapeptides. Peptides also result from protein breakdown, e.g. during digestion.

per- Prefix indicating that a chemical compound contains an excess of an element, e.g. a peroxide.

perchlorate See chlorates.

perchloric acid See chloric(VII) acid.

perdisulphuric acid See peroxosulphuric(VI) acid.

perennation The survival of biennial or perennial plants from one year to the next by vegetative means. In biennials and herbaceous perennials the aerial parts of the plant die down and the plants survive by means of underground storage roots (e.g. carrot), *rhizomes (e.g. couch grass, Solomon's seal), *tubers (e.g. dahlia), *bulbs (e.g. daffodil, snowdrop), or *corms (e.g. crocus, gladiolus). These *perennating organs* are also frequently responsible for *vegetative propagation. Woody perennials survive the winter by reducing their metabolic activity (e.g. by leaf loss in deciduous trees and shrubs).

perennial A plant that lives for a number of years. Woody perennials (trees and shrubs) have a permanent aerial form, which continues to grow year after year. Herbaceous (i.e. nonwoody) perennials have aerial shoots that die down each autumn and are replaced in spring by new shoots from an underground structure (*see* perennation). Lupin and rhubarb are examples of herbaceous perennials. *Compare* annual; biennial; ephemeral.

perfect gas *See* ideal gas; gas.

perfect pitch *See* absolute pitch.

perfect solution *See* Raoult's law.

perianth The part of a flower situated outside the stamens and carpels. In dicotyledons it consists of two distinct whorls, the outer of sepals (*see* calyx) and the inner of petals (*see* corolla). In monocotyledons the two whorls are similar and often brightly coloured. In wind-pollinated flowers both whorls may be reduced or absent. In many horticultural varieties the number of perianth parts is multiplied, but the resulting 'double' flowers are often sterile.

pericardial cavity The cavity in vertebrates that contains the heart and is bounded by a membrane (the *pericardium*). It is part of the *coelom.

pericarp (fruit wall) The part of a fruit that develops from the ovary wall of a flower. The type of fruit that develops depends on whether the pericarp becomes dry and hard or soft and fleshy. The pericarp can be made up of three layers. The outer skin (*epicarp* or *exocarp*) may be tough and hard; the middle layer (*mesocarp*) may be succulent as in peach, hard as in almond, or fibrous as in coconut; and the inner layer (*endocarp*) may be hard and stony as in many *drupes, membranous as in citrus fruits, or indistinguishable from the mesocarp, as in many *berries.

pericycle A plant tissue comprising the outermost layer of the root vascular tissue, lying immediately beneath the *endodermis. Lateral roots originate from the pericycle.

pericynthion The point in the orbit around the moon of a satellite launched from the earth that is nearest to the moon. For a satellite launched from the moon the equivalent point is the *perilune*. *Compare* apocynthion.

periderm *See* cork cambium.

perigee *See* apogee.

perihelion The point in the orbit of a planet, comet, or artificial satellite in solar orbit at which it is nearest to the sun. The earth is at perihelion on about 3 January. *Compare* aphelion.

perilymph The fluid of the *inner ear that fills the space between the bony labyrinth and the membranous labyrinth. *Compare* endolymph.

period *See* periodic table.

periodic acid *See* iodic(VII) acid.

periodic law The principle that the physical and chemical properties of elements are a periodic function of their proton number. The concept was first proposed in 1869 by the Russian chemist Dimitri Mendeleev (1834–1907), using relative atomic mass rather than proton number, as a culmination of efforts to rationalize chemical properties by J. W. Döbereiner (1817), J. A. R. Newlands (1863), and Lothar Meyer (1864). One of the major successes of the periodic law was its ability to predict chemical and physical properties of undiscovered elements and unknown compounds that were later confirmed experimentally. *See* periodic table.

periodic motion Any motion of a system that is continuously and identically repeated. The time T that it takes to complete one cycle of an oscillation or wave motion is called the *period*, which is the reciprocal of the *frequency. *See* pendulum; simple harmonic motion.

periodic table A table of elements arranged in order of increasing proton number to show the similarities of chemical elements with related electronic configurations. (The original form was proposed by Dimitri

Mendeleev in 1869 using relative atomic masses.) In the modern *short form* (see Appendix) the *lanthanoids and *actinoids are not shown. The elements fall into vertical columns, known as *groups*. Going down a group, the atoms of the elements all have the same outer shell structure, but an increasing number of inner shells. Traditionally, the alkali metals are shown on the left of the table and the groups are numbered IA to VIIA, IB to VIIB, and 0 (for the noble gases). It is now more common to classify all the elements in the middle of the table as *transition elements and to regard the nontransition elements as *main-group* elements, numbered from I to VII, with the noble gases in group 0. Horizontal rows in the table are *periods*. The first three are called *short periods*; the next four (which include transition elements) are *long periods*. Within a period, the atoms of all the elements have the same number of shells, but with a steadily increasing number of electrons in the outer shell. The periodic table can also be divided into four *blocks* depending on the type of shell being filled: the *s-block, the *p-block, the *d-block, and the *f-block. There are certain general features of chemical behaviour shown in the periodic table. In moving down a group, there is an increase in metallic character because of the increased size of the atom. In going across a period, there is a change from metallic (electropositive) behaviour to nonmetallic (electronegative) because of the increasing number of electrons in the outer shell. Consequently, metallic elements tend to be those on the left and towards the bottom of the table; nonmetallic elements are towards the top and the right.

There is also a significant difference between the elements of the second short period (lithium to fluorine) and the other elements in their respective groups. This is because the atoms in the second period are smaller and their valence electrons are shielded by a small $1s^2$ inner shell. Atoms in the other periods have inner s- and p-electrons shielding the outer electrons from the nucleus. Moreover, those in the second period only have s- and p-orbitals available for bonding. Heavier atoms can also promote electrons to vacant d-orbitals in their outer shell and use these for bonding. *See also* diagonal relationship; inert-pair effect.

periodontal membrane The membrane of connective tissue that surrounds the root of a *tooth and anchors it to its socket in the jawbone. Fibres of the periodontal membrane pass into the *cement covering the root, which provides a firm attachment.

peripheral device Any device, such as an input or output device, connected to the central processing unit of a *computer. Backing store is also usually regarded as a peripheral.

peripheral nervous system All parts of the nervous system excluding the *central nervous system. It consists of all the *cranial and *spinal nerves and their branches, which link the *receptors and *effectors with the central nervous system. *See also* autonomic nervous system.

periscope An optical device that enables an observer to see over or around opaque objects. The simplest type consists of a long tube with mirrors at each end set at 45° to the direction to be viewed. A better type uses internally reflecting prisms instead of plane mirrors. Periscopes are used in tanks (to enable the observer to see over obstacles without being shot at) and in submarines (when the vessel is submerged). Such periscopes are usually quite complicated instruments and include telescopes.

Perissodactyla An order of mammals having hoofed feet with an odd number of toes. They are all herbivores and include the tapirs, rhinoceros, and horse. The teeth are large and specialized for grinding. Cellulose digestion occurs in the caecum and large intestine. Fossils of the Eocene epoch, 60 million years ago, show that these animals were at that time already distinct from the cloven-hoofed *Artiodactyla.

peristalsis Waves of involuntary muscular contraction and relaxation that pass along the alimentary canal, forcing food contents along. It is brought about by contraction of the circular muscles of the gut wall in sequence.

peristome 1. A ring of toothlike structures around the opening of a moss *capsule. The teeth tend to bend and twist in dry weather, so opening the mouth of the capsule and allowing the spores to escape. In wet weather they close over the opening of the capsule. **2.** The area around the mouth in many invertebrates. It sometimes assists in food collecting. Examples are the spirally ciliated groove around the mouth of some ciliate protozoans and the first segment of the earthworm.

peritoneum The thin layer of tissue (*see* serous membrane) that lines the abdominal cavity of vertebrates and covers the abdominal organs. *See also* mesentery.

Permalloys A group of alloys of high magnetic permeability consisting of iron and nickel (usually 40–80%) often with small amounts of other elements (e.g. 3–5% molybdenum, copper, chromium, or tungsten). They are used in thin foils in electronic transformers, for magnetic shielding, and in computer memories.

permanent gas A gas, such as oxygen or nitrogen, that was formerly thought to be impossible to liquefy. A permanent gas is now regarded as one that cannot be liquefied by pressure alone at normal temperatures (i.e. a gas that has a critical temperature below room temperature).

permanent hardness *See* hardness of water.

permanent magnet *See* magnet.

permanent teeth The second and final set of teeth that mammals produce after shedding the *deciduous teeth. An adult human normally has 32 permanent teeth, consisting of incisors, canines, molars, and premolars (*see* dental formula). These usually appear between the ages of approximately 6 and 21 years. *See also* diphyodont.

permanganate *See* manganate(VII).

permeability (magnetic permeability) Symbol μ. The ratio of the magnetic flux density, B, in a substance to the external field strength, H; i.e. $\mu = B/H$. The permeability of free space, μ_0, is also called the *magnetic constant* and has the value $4\pi \times 10^{-7}$ H m^{-1} in *SI units. The relative permeability of a substance, μ_r, is given by μ/μ_0 and is therefore dimensionless. *See* magnetism.

Permian The last geological period in the Palaeozoic era. It extended from the end of the Carboniferous period, about 280 million years ago, to the beginning of the Mesozoic era, about 225 million years ago. It was named by the British geologist Roderick Murchison (1792–1871) in 1841 after the Perm province in Russia. In some areas continental conditions prevailed, which continued into the following period, the Triassic. These conditions resulted in the deposition of the New Red Sandstone. During the period a number of animal groups became extinct, including the trilobites, tabulate and rugose corals, and blastoids. Amphibians and reptiles continued to be the dominant land animals and gymnosperms replaced pteridophytes as the dominant plants.

permittivity Symbol ε. The ratio of the *electric displacement in a medium to the intensity of the electric field producing it. It is important for electrical insulators used as *dielectrics.

If two charges Q_1 and Q_2 are separated by a distance r in a vacuum, the force F between the charges is given by:

$$F = Q_1Q_2/r^24\pi\varepsilon_0$$

In this statement of *Coulomb's law using *SI units, ε_0 is called the absolute permittivity of free space, which is now known as the *electric constant*. It has the value 8.854×10^{-12} F m^{-1}. If the medium between the charges is anything other than a vacuum the equation becomes:

$$F = Q_1Q_2/r^24\pi\varepsilon$$

and the force between the charges is reduced. ε is the *absolute permittivity* of the new medium. The *relative permittivity* (ε_r) of a medium, formerly called the *dielectric constant*, is given by $\varepsilon_r = \varepsilon/\varepsilon_0$.

permonosulphuric(VI) acid *See* peroxosulphuric(VI) acid.

permutations and combinations A combination is any subset of a particular set of objects, regardless of the order of selection. If the set consists of n objects, r objects can be selected giving $n!/r!(n - r)!$ different combinations. This can be written $_nC_r$.

A permutation is an ordered subset (i.e. attention is paid to the order of selection or arrangement) of a particular set of objects. If the set consists of n objects, r such objects can be selected to give $n!/(n - r)!$ permutations. This is written $_nP_r$.

Permutit Trade name for a *zeolite used for water softening.

peroxides A group of inorganic compounds that contain the O_2^{2-} ion. They are notionally derived from hydrogen peroxide, H_2O_2, but these ions do not exist in aqueous solution due to extremely rapid hydrolysis to OH^-.

peroxodisulphuric acid *See* peroxosulphuric(VI) acid.

peroxomonosulphuric(VI) acid *See* peroxosulphuric(VI) acid.

peroxosulphuric(VI) acid The term commonly refers to *peroxomonosulphuric(VI) acid*, H_2SO_5, which is also called *permonosulphuric(VI) acid* and *Caro's acid*. It is a crystalline compound made by the action of hydrogen peroxide on concentrated sulphuric acid. It decomposes in water and the crystals decompose, with melting, above 45°C. The compound *peroxodisulphuric acid*, $H_2S_2O_8$, also exists (formerly called *perdisulphuric acid*). It is made by the high-current electrolysis of sulphate solutions. It decomposes at 65°C (with melting) and is hydrolysed in water to give the mono acid and sulphuric acid. Both peroxo acids are very powerful oxidizing agents. *See also* sulphuric acid (for structural formulas).

perpetual motion 1. Perpetual motion of the first kind. Motion in which a mechanism, once started, would continue indefinitely to perform useful work without being supplied with energy from an outside source. Such a device would contravene the first law of *thermodynamics and is therefore not feasible. Many historical attempts, exercising great ingenuity, were constructed before the concept of energy and its conservation were understood. Some attempts have been made, since the first law of thermodynamics became generally accepted, by inventors seeking to establish loopholes in the laws of nature. 2. Perpetual motion of the second kind. Motion in which a mechanism extracts heat from a source and converts all of it into some other form of energy. An example of such a mechanism would be a ship that utilized the internal energy of the oceans for propulsion. Such a device does not contravene the first law of thermodynamics but it does contravene the second law. In the case of the ship, the sea would have to be at a higher temperature than the ship to establish a useful flow of heat. This could not occur without an external energy source. **3.** Perpetual motion of the third kind. A form of motion that continues indefinitely but without doing any useful work. An example is the random molecular motion in a substance. This type postulates the complete elimination of friction. A mechanism consisting of frictionless bearings maintained in a vacuum could turn indefinitely, once started, without contravening the first or second laws of thermodynamics, provided it did no external work. Experience indicates that on the macroscopic scale such a condition cannot be achieved. On the microscopic scale, however, a superconducting ring of wire will apparently sustain a perpetual current flow without the application of an external force. This could be considered a form of perpetual motion of the third kind, if the energy required to cool the wire to superconducting temperatures is ignored.

Perspex Trade name for a form of *polymethylmethacrylate.

perturbation A departure by a celestial body from the trajectory or orbit it would follow if it moved only under the influence of a single central force. According to *Kepler's law, for example, a single planet orbiting the sun would move in an elliptical orbit. In fact, planets are perturbed from elliptical orbits by the gravitational forces exerted on them by other planets. Similarly, the moon's orbit round the earth is perturbed by the gravitational effect of the sun and the trajectories of comets are perturbed when they pass close to planets.

perturbation theory A method used in calculations in both classical physics (e.g. planetary orbits) and quan-

tum mechanics (e.g. atomic structure), in which the system is divided into a part that is exactly calculable and a small term, which prevents the whole system from being exactly calculable. The technique of perturbation theory enables the effects of the small term to be calculated by an infinite series (which in general is an asymptotic series). Each term in the series is a 'correction term' to the solutions of the exactly calculable system. In classical physics, perturbation theory can be used for calculating planetary orbits. In quantum mechanics, it can be used to calculate the energy levels in molecules. In the many-body problem in quantum mechanics and in relativistic quantum field theory, the terms in perturbation theory may be represented pictorially by Feynman diagrams (*see* quantum electrodynamics).

perversion *See* lateral inversion.

peta- Symbol P. A prefix used in the metric system to denote one thousand million million times. For example, 10^{15} metres = 1 petametre (Pm).

petal One of the parts of the flower that make up the *corolla. Petals of insect-pollinated plants are usually brightly coloured and often scented. Those of wind-pollinated plants are usually reduced or absent. Petals are considered to be modified leaves but their structure is simpler. Epidermal hairs may be present and the cuticle is often covered by lines or dots known as *honey guides*, which direct insects to the *nectar.

petiole The stalk that attaches a *leaf blade to the stem. Leaves without petioles are described as *sessile*.

Petri dish A shallow circular flat-bottomed dish made of glass or plastic and having a fitting lid. It is used in laboratories chiefly for culturing bacteria and other microorganisms. It was invented by the German bacteriologist J. R. Petri (1852–1921).

petrification See fossil.

petrochemicals Organic chemicals obtained from petroleum or natural gas.

petroleum A naturally occurring oil that consists chiefly of hydrocarbons with some other elements, such as sulphur, oxygen, and nitrogen. In its unrefined form petroleum is known as *crude oil* (sometimes *rock oil*). Petroleum is believed to have been formed from the remains of living organisms that were deposited, together with rock particles and biochemical and chemical precipitates, in shallow depressions, chiefly in marine conditions. Under burial and compaction the organic matter went through a series of processes before being transformed into petroleum, which migrated from the source rock to become trapped in large underground reservoirs beneath a layer of impermeable rock. The petroleum often floats above a layer of water and is held under pressure beneath a layer of *natural gas.

Petroleum reservoirs are discovered through geological exploration: commercially important oil reserves are detected by exploratory narrow-bore drilling. The major known reserves of petroleum are in Saudi Arabia, USSR, China, Kuwait, Iran, Iraq, Mexico, USA, United Arab Emirates, Libya, Venezuela, and beneath the North Sea. The oil is actually obtained by the sinking of an oil well. Before it can be used it is separated by fractional distillation in oil refineries. The main fractions obtained are:

(1) *Refinery gas* A mixture of methane, ethane, butane, and propane used as a fuel and for making other organic chemicals.

(2) *Gasoline* A mixture of hydrocarbons containing 5 to 8 carbon atoms, boiling in the range 40–180°C. It is used for motor fuels and for making other chemicals.

(3) *Kerosine* (or *paraffin oil*) A mixture of hydrocarbons having 11 or 12 carbon atoms, boiling in the range 160–250°C. Kerosine is a fuel for jet aircraft and for oil-fired domestic heating. It is also cracked to produce smaller hydrocarbons for use in motor fuels.

(4) *Diesel oil* (or *gas oil*) A mixture of hydrocarbons having 13 to 25 carbon atoms, boiling in the range 220–350°C. It is a fuel for diesel engines. The residue is a mixture of higher hydrocarbons. The liquid components are obtained by vacuum distillation and used in lubricating oils. The solid components (*paraffin wax*) are obtained by solvent extraction. The final residue is a black tar containing free carbon (*asphalt* or *bitumen*).

petroleum ether A colourless volatile flammable mixture of hydrocarbons (not an ether), mainly pentane and hexane. It boils in the range 30–70°C and is used as a solvent.

pewter An alloy of lead and tin. It usually contains 63% tin; pewter tankards and food containers should have less than 35% of lead so that the lead remains in solid solution with the tin in the presence of weak acids in the food and drink. Copper is sometimes added to increase ductility and antimony is added if a hard alloy is required.

pH See pH scale.

PHA See phytohaemagglutinin.

Phaeophyta (brown algae) A division of *algae in which the green chlorophyll pigments are usually masked by the brown pigment fucoxanthin. Brown algae are usually marine (being abundant in cold water) and many species, such as the wracks (*Fucus*), inhabit intertidal zones. They vary in size from small branched filaments to ribbon-like bodies (known as kelps) many metres long.

phage See bacteriophage.

phagocyte A cell that is able to engulf and break down foreign particles, cell debris, and disease-producing microorganisms (*see* phagocytosis). Many protozoans and certain mammalian cells (e.g. *macrophages and *monocytes) are phagocytes. Phagocytes are important elements in the natural defence mechanism of most animals.

phagocytosis The process by which foreign particles invading the body or minute food particles are engulfed and broken down by certain animal cells (known as *phagocytes). The cell membrane of the phagocyte invaginates to capture the particle and then closes around it to form a sac or *vacuole. The vacuole coalesces with a *lysosome, which contains enzymes that break down the particle. *Compare* pinocytosis.

phalanges The bones that make up the *digits of the hand or foot in vertebrates. They articulate with the *metacarpals of the hand or with the *metatarsals of the foot. In the basic *pentadactyl limb there are two phalanges for the first digit (the thumb or big toe in humans) and three for each of the others.

Phanerozoic Geological time since the end of the Precambrian, represented by rock strata containing clearly recognizable fossils. It comprises the *Palaeozoic, *Mesozoic, and *Cenozoic eras and has extended for about 570 million years from the beginning of the Cambrian period. Fossils are extremely rare in Precambrian rocks.

pharmacology The study of the properties of drugs and their effects on living organisms. Clinical pharmacology is concerned with the effects of drugs in treating disease.

pharynx 1. The cavity in vertebrates between the mouth and the *oesophagus and windpipe (*trachea), which serves for the passage of both food and respiratory gases. The presence of food in the pharynx stimulates swallowing (*see* deglutition). In fish and aquatic amphibians the pharynx is perforated by *gill slits. **2.** The corresponding region in invertebrates.

phase 1. A homogeneous part of a heterogeneous system that is separated from other parts by a distinguishable boundary. A mixture of ice and water is a two-phase system. A solution of salt in water is a single-phase system. **2.** A description of the stage that a periodic motion has reached, usually by comparison with another such motion of the same frequency. Two varying quantities are said to be *in phase* if their maximum and minimum values occur at the same instants; otherwise, there is said to be a *phase difference*. See also phase angle. **3.** One of the circuits in an electrical system or device in which there are two or more alternating currents that are not in phase with each other. In a three-phase system the displacement between the currents is one third of a period. **4.** *See* phases of the moon.

phase angle The difference in *phase between two sinusoidally varying quantities. The displacement x_1 of one quantity at time t is given by $x_1 = a\sin\omega t$, where ω is the angular frequency and a is the amplitude. The displacement x_2 of a similar wave that reaches the end of its period T, a fraction β of the period before the first is said to lead the first quantity by a time βT. The value of x_2 is then given by $x_2 = a\sin(\omega t + \phi)$. ϕ is called the phase angle and it is equal to $2\pi\beta$.

phase-contrast microscope A type of *microscope that is widely used for examining such specimens as biological cells and tissues. It makes visible the changes in phase that occur when nonuniformly transparent specimens are illuminated. In passing through an object the light is slowed down and becomes out of phase with the original light. With transparent specimens having some structure *diffraction occurs, causing a larger phase change in light outside the central maximum of the pattern. The phase-contrast microscope provides a means of combining this light with that of the central maximum by means of an annular diaphragm and a *phase-contrast plate*, which produces a matching phase change in the light of the central maximum only. This gives greater contrast to the final image, due to constructive interference between the two sets of light waves. This is *bright contrast*; in *dark contrast* a different phase-contrast plate is used to make the same structure appear dark, by destructive interference of the same waves.

phase diagram A graph showing the relationship between solid, liquid, and gaseous *phases over a range of conditions (e.g. temperature and pressure). *See* steel (illustration).

phase modulation *See* modulation.

phase rule For any system at equilibrium, the relationship $P + F = C + 2$ holds, where P is the number of distinct phases, C the number of components, and F the number of degrees of freedom of the system. The relationship, derived by Josiah Willard Gibbs in 1876, is often called the *Gibbs phase rule*.

phases of the moon The shapes of the illuminated surface of the moon as seen from the earth. The shape changes as a result of the relative positions of the earth, sun, and moon. *New moon* occurs when the nearside is totally unilluminated by the sun. As the moon moves eastwards in its orbit the sunrise *terminator crosses the nearside from east to west producing a *crescent moon*. The moon is half illuminated at *first quarter*. When it is more than half-phase but less than full phase it is said to be a *gibbous moon*. When the moon is at *opposition the nearside is fully

illuminated producing a *full moon*. The sunset terminator then follows to produce a waning gibbous moon, *last quarter*, a waning crescent moon, and eventually the next new moon.

phase space For a system with n degrees of freedom, the $2n$-dimensional space with coordinates $(q_1, q_2, \ldots, q_n, p_1, p_2, \ldots, p_n)$, where the qs describe the degrees of freedom of the system and the ps are the corresponding momenta. Each point represents a state of the system. In a gas of N point particles, each particle has three positional coordinates and three corresponding momentum coordinates, so that the phase space has $6N$-dimensions. If the particles have internal degrees of freedom, such as the vibrations and rotations of molecules, then these must be included in the phase space, which is consequently of higher dimension than that for point particles. As the system changes with time the representative points trace out a curve in phase space known as a *trajectory*. *See also* attractor; configuration space; statistical mechanics.

phase speed (phase velocity) Symbol V_p. The speed of propagation of a pure sine wave, $V_p = \lambda f$, where λ is the wavelength and f is the frequency. The value of the phase speed depends on the nature of the medium through which it is travelling and may also depend on the mode of propagation. For electromagnetic waves travelling through space the phase speed c is given by $c^2 = 1/\varepsilon_0\mu_0$, where ε_0 and μ_0 are the *electric constant and the *magnetic constant respectively.

phasor A rotating *vector that represents a sinusoidally varying quantity. Its length represents the amplitude of the quantity and it is imagined to rotate with angular velocity equal to the angular frequency of the quantity, so that the instantaneous value of the quantity is represented by its projection upon a fixed axis. The concept is convenient for representing the *phase angle between two quantities; it is shown on a diagram as the angle between their phasors.

phellem *See* cork.

phelloderm *See* cork cambium.

phellogen *See* cork cambium.

phenetic Describing a system of *classification of organisms based on similarities and differences in as many observable characteristics as possible. A phenetic system does not aim to reflect evolutionary descent, although it may well do so. *Compare* phylogenetic.

phenol (carbolic acid) A white crystalline solid, C_6H_5OH; r.d. 1.1; m.p. 42°C; b.p. 182°C. It is made by the *cumene process or by the *Raschig process and is used to make a variety of other organic chemicals. *See also* phenols.

phenolphthalein A dye used as an acid-base *indicator. It is colourless below pH 8 and red above pH 9.6. It is used in titrations involving weak acids and strong bases. It is also used as a laxative.

phenols Organic compounds that contain a hydroxyl group (–OH) bound directly to a carbon atom in a benzene ring. Unlike normal alcohols, phenols are acidic because of the influence of the aromatic ring. Thus, phenol itself (C_6H_5OH) ionizes in water:

$$C_6H_5OH \rightarrow C_6H_5O^- + H^+$$

Phenols are made by fusing a sulphonic acid salt with sodium hydroxide to form the sodium salt of the phenol. The free phenol is liberated by adding sulphuric acid.

phenotype The observable characteristics of an organism. These are determined by its genes (*genotype*), the dominance relationships between the *alleles, and by the interaction of the genes with the environment.

phenylalanine *See* amino acid.

phenylamine (aniline; aminobenzene) A colourless oily liquid aromatic *amine, $C_6H_5NH_2$, with an 'earthy' smell; r.d. 1.027; m.p. −6.2°C; b.p. 185°C. The compound turns brown on exposure to sunlight. It is basic, forming the *phenylammonium* (or *anilinium*) *ion*, $C_6H_5NH_3^+$, with strong acids. It is manufactured by the reduction of nitrobenzene or by the addition of ammonia to chlorobenzene using a copper(II) salt catalyst at 200°C and 55 atm. The compound is used extensively in the rubber industry and in the manufacture of drugs and dyes.

phenylammonium ion The ion $C_6H_5NH_3^+$, derived from *phenylamine.

phenylethene (styrene) A liquid hydrocarbon, $C_6H_5CH:CH_2$; r.d. 0.9; m.p. −31°C; b.p. 145°C. It can be made by dehydrogenating ethylbenzene and is used in making polystyrene.

phenyl group The organic group C_6H_5-, present in benzene.

phenylhydrazones *See* hydrazones.

phenylketonuria A genetic disorder in which there is disordered metabolism of the amino acid phenylalanine, leading to severe mental retardation of affected children. The disease is caused by the absence or deficiency of the enzyme phenylalanine hydroxylase, which results in the accumulation of phenylalanine in all body fluids. There are also high levels of the ketone phenylpyruvate in the urine, hence the name of the disease. The disease occurs in individuals who are homozygous for the defective recessive allele; both parents of such individuals are thus heterozygous carriers of the allele. The advent of *gene probes has greatly aided accurate diagnosis, of both phenylketonurics and carriers.

phenylmethanol (benzyl alcohol) A liquid aromatic alcohol, $C_6H_5CH_2OH$; r.d. 1.04; m.p. $-15.3°C$; b.p. $205.4°C$. It is used mainly as a solvent.

3-phenylpropenoic acid *See* cinnamic acid.

pheromone (ectohormone) A chemical substance emitted by an organism into the environment as a specific signal to another organism, usually of the same species. Pheromones play an important role in the social behaviour of certain animals, especially insects and mammals. They are used to attract mates, to mark trails, and to promote social cohesion and coordination in colonies. Pheromones are usually highly volatile organic acids or alcohols and can be effective at minute concentrations.

Phillips process A process for making high-density polyethene by polymerizing ethene at high pressure (30 atmospheres) and 150°C. The catalyst is chromium(III) oxide supported on silica and alumina.

phloem (bast) A tissue that conducts food materials in vascular plants from regions where they are produced (notably the leaves) to regions, such as growing points, where they are needed. It consists of hollow tubes (sieve tubes) that run parallel to the long axis of the plant organ and are formed from elongated cells (*sieve elements) joined end to end. The end walls of these cells are broken down to a greater or lesser extent to allow passage of materials. In young plants and in newly formed tissues of mature plants the phloem is formed by the activity of the *apical meristem. In most plants secondary phloem is later differentiated by the vascular *cambium and this replaces the earlier formed phloem in older regions. *See also* companion cell. *Compare* xylem.

phlogiston theory A former theory of combustion in which all flammable objects were supposed to contain a substance called *phlogiston*, which was released when the object burned. The existence of this hypothetical substance was proposed in 1669 by Johann Becher, who called it 'combustible earth' (*terra pinguis*: literally 'fat earth'). For example, according to Becher, the conversion of wood to ashes by burning was explained on the assumption that the original wood consisted of ash and *terra pinguis*, which was released on burning. In the early 18th century Georg Stahl renamed the substance *phlogiston* (from the Greek for 'burned') and extended the theory to include the calcination (and corrosion) of metals. Thus, metals were thought to be composed of *calx* (a powdery residue) and phlogiston; when a metal was heated, phlogiston was set free and the calx remained. The process could be reversed by heating the metal over charcoal (a substance believed to be rich in phlogiston, because combustion almost totally consumed it). The calx would absorb the phlogiston released by the burning charcoal and become metallic again.

The theory was finally demolished by Antoine Lavoisier, who showed by careful experiments with reactions in closed containers that there was no *absolute* gain in mass – the gain in mass of the substance was matched by a corresponding loss in mass of the air used in combustion. After experiments with Priestley's dephlogisticated air, Lavoisier realized that this gas, which he named oxygen, was taken up to form a calx (now called an oxide). The role of oxygen in the new theory was almost exactly the opposite of phlogiston's role in the old. In combustion and corrosion phlogiston was released; in the modern theory, oxygen is taken up to form an oxide.

phloroglucinol A red dye (usually acidified with hydrochloric acid) that stains lignin in plant cells red.

phon A unit of loudness of sound that measures the intensity of a sound relative to a reference tone of defined intensity and frequency. The reference tone usually used has a frequency of 1 kilohertz and a root-mean-square sound pressure of 2×10^{-5} pascal. The observer listens with both ears to the reference tone and the sound to be measured alternately. The reference tone is then increased until the observer judges it to be of equal intensity to the sound to be measured. If the intensity of the reference tone has been increased by n *decibels to achieve this, the sound being measured is said to have an intensity of n phons. The decibel and phon scales are not identical as the phon scale is subjective and relies on the sensitivity of the ear to detect changes of intensity with frequency.

phonochemistry The use of high-frequency sound (ultrasound, with frequency greater than about 20 kHz) to induce or accelerate certain types of chemical reaction.

phonon A quantum of *crystal-lattice vibrational energy having a magnitude hf, where h is the *Planck constant and f is the frequency of the vibration. Phonons are analogous to the quanta of light, i.e. *photons. The concept of phonons is useful in the treatment of the thermal conductivity of nonmetallic solids and, through consideration of electron–phonon interactions, the temperature dependence of the electrical conductivity of metals.

phosgene *See* carbonyl chloride.

phosphagen A compound found in animal tissues that provides a reserve of chemical energy in the form of high-energy phosphate bonds. The most common phosphagens are *creatine phosphate, occurring in vertebrate muscle and nerves, and arginine phosphate, found in most invertebrates. During tissue activity (e.g. in muscle contraction) phospha-

gens give up their phosphate groups, thereby generating *ATP from ADP. The phosphagens are then reformed when ATP is available.

phosphates Salts based formally on phosphorus(V) oxoacids and in particular salts of *phosphoric(V) acid, H_3PO_4. A large number of polymeric phosphates also exist, containing $P-O-P$ bridges. These are formed by heating the free acid and its salts under a variety of conditions; as well as linear polyphosphates, cyclic polyphosphates and cross-linked polyphosphates or ultraphosphates are known.

phosphatide *See* phospholipid.

phosphide A binary compound of phosphorus with a more electropositive element. Phosphides show a wide range of properties. Alkali and alkaline earth metals form ionic phosphides, such as Na_3P and Ca_3P_2, which are readily hydrolysed by water. The other transition-metal phosphides are inert metallic-looking solids with high melting points and electrical conductivities.

phosphine A colourless highly toxic gas, PH_3; m.p. $-133°C$; b.p. $-87.7°C$; slightly soluble in water. Phosphine may be prepared by reacting water or dilute acids with calcium phosphide or by reaction between yellow phosphorus and concentrated alkali. Solutions of phosphine are neutral but phosphine does react with some acids to give phosphonium salts containing PH_4^+ ions, analogous to the ammonium ions. Phosphine prepared in the laboratory is usually contaminated with diphosphine and is spontaneously flammable but the pure compound is not so. Phosphine can function as a ligand in binding to transition-metal ions. Dilute gas mixtures of very pure phosphine and the rare gases are used for doping semiconductors.

phosphinic acid (hypophosphorus acid) A white crystallline solid,

H_3PO_2; r.d. 1.493; m.p. 26.5°C; decomposes above 130°C. It is soluble in water, ethanol, and ethoxyethane. Salts of phosphinic acid may be prepared by boiling white phosphorus with the hydroxides of group I or group II metals. The free acid is made by the oxidation of phosphine with iodine. It is a weak monobasic acid in which it is the −O−H group that is ionized to give the ion $H_2PO_2^-$. The acid and its salts are readily oxidized to the orthophosphate and consequently are good reducing agents.

phosphite See phosphonic acid.

phospholipid (phosphatide) One of a group of lipids having both a phosphate group and one or more fatty acids. *Glycerophospholipids* (*phosphoglycerides*) are based on *glycerol; the three hydroxyl groups are esterified with two fatty acids and a phosphate group, which may itself be bound to one of a variety of simple organic groups (e.g. in *lecithin it is choline). *Sphingophospholipids* are based on the alcohol sphingosine and contain only one fatty acid linked to an amino group. With their hydrophilic polar phosphate groups and long hydrophobic hydrocarbon 'tails', phospholipids readily form membrane-like structures in water. They are a major component of cell membranes.

phosphonate See phosphonic acid.

phosphonic acid (phosphorous acid; orthophosphorous acid) A colourless to pale-yellow deliquescent crystalline solid, H_3PO_3; r.d. 1.65; m.p. 73.6°C; decomposes at 200°C; very soluble in water and soluble in alcohol. Phosphonic acid may be crystallized from the solution obtained by adding ice-cold water to phosphorus(III) oxide or phosphorus trichloride. The structure of this material is unusual in that it contains one direct P−H bond and is more correctly written $(HO)_2HPO$. The acid is dibasic, giving rise to the ions

$H_2PO_3^-$ and HPO_3^{2-} (*phosphonates*; formerly *phosphites*), and has moderate reducing properties. On heating it gives phosphine and phosphoric(V) acid.

phosphonium ion The ion PH_4^+, or the corresponding organic derivatives of the type R_3PH^+, RPH_3^+. The phosphonium ion PH_4^+ is formally analogous to the ammonium ion NH_4^+ but PH_3 has a much lower proton affinity than NH_3 and reaction of PH_3 with acids is necessary for the production of phosphonium salts.

phosphor A substance that is capable of *luminescence (including phosphorescence). Phosphors that release their energy after a short delay of between 10^{-10} and 10^{-4} second are sometimes called *scintillators*.

phosphor bronze An alloy of copper containing 4% to 10% of tin and 0.05% to 1% of phosphorus as a deoxidizing agent. It is used particularly for marine purposes and where it is exposed to heavy wear, as in gear wheels. See also bronze.

phosphorescence See luminescence.

phosphoric(V) acid (orthophosphoric acid) A white rhombic solid, H_3PO_4; r.d. 1.834; m.p. 42.35°C; loses water at 213°C; very soluble in water and soluble in ethanol. Phosphoric(V) acid is very deliquescent and is generally supplied as a concentrated aqueous solution. It is the most commercially important derivative of phosphorus, accounting for over 90% of the phosphate rock mined. It is manufactured by two methods; the *wet process*, in which the product contains some of the impurities originally present in the rock and applications are largely in the fertilizer industry, and the *thermal process*, which produces a much purer product suitable for the foodstuffs and detergent industries. In the wet process the phosphate rock, $Ca_3(PO_4)_2$, is treated with sulphuric acid and the calcium sulphate removed either as gypsum or the hemihydrate. In the thermal process, molten phosphorus is sprayed and burned in a mixture of air and steam. Phosphoric(V) acid is a weak tribasic acid, which is best visualized as $(HO)_3PO$. Its full systematic name is *tetraoxo-phosphoric(V) acid*. It gives rise to three series of salts containing *phosphate(V)* ions based on the anions $[(HO)_2PO_2]^-$, $[(HO)PO_3]^{2-}$, and PO_4^{3-}. These salts are acidic, neutral, and alkaline in character respectively and phosphate ions often feature in buffer systems. There is also a wide range of higher acids and acid anions in which there is some P–O–P chain formation. The simplest of these is *pyrophosphoric acid* (technically *heptaoxodiphosphoric(V) acid*), $H_4P_2O_7$, produced by heating phosphoric(V) acid (solid) and phosphorus(III) chloride oxide. *Metaphosphoric acid* is a glassy polymeric solid $(HPO_2)_x$.

phosphorous acid *See* phosphonic acid.

phosphorus Symbol P. A nonmetallic element belonging to *group V of the periodic table; a.n. 15; r.a.m. 30.9738; r.d. 1.82 (white), 2.20 (red); m.p. 44.1°C (α-white); b.p. 280°C (α-white). It occurs in various phosphate rocks, from which it is extracted by heating with carbon (coke) and silicon(IV) oxide in an electric furnace (1500°C). Calcium silicate and carbon monoxide are also produced. Phosphorus has a number of allotropic forms. The α-white form consists of P_4 tetrahedra (there is also a β-white form stable below −77°C). If α-white phosphorus is dissolved in lead and heated at 500°C a violet form is obtained. Red phosphorus, which is a combination of violet and white phosphorus, is obtained by heating α-white phosphorus at 250°C with air excluded. There is also a black allotrope, which has a graphite-like structure, made by heating white phosphorus at 300°C with a mercury catalyst. The element is highly reactive. It forms metal *phosphides and covalently bonded phosphorus(III) and phosphorus(V) compounds. Phosphorus is an *essential element for living organisms. It is an important constituent of tissues (especially bones and teeth) and of cells, being required for the formation of nucleic acids and energy-carrying molecules (e.g. ATP) and also involved in various metabolic reactions. The element was discovered by Brandt in 1669.

phosphorus(III) bromide (phosphorus tribromide) A colourless fuming liquid, PBr_3; r.d. 2.85; m.p. −40°C; b.p. 173°C. It is prepared by passing bromine vapour over phosphorus but avoiding an excess, which would lead to the phosphorus(V) bromide. Like the other phosphorus(III) halides, PBr_3 is pyramidal in the gas phase. In the liquid phase the P–Br bonds are labile; for example, PBr_3 will react with PCl_3 to give a mixture of products in which the halogen atoms have been redistributed. Phosphorus(III) bromide is rapidly hydrolysed by water to give phosphonic acid and hydrogen bromide. It reacts readily with many organic hydroxyl groups and is used as a reagent for introducing bromine atoms into organic molecules.

phosphorus(V) bromide (phosphorus pentabromide) A yellow readily sublimable solid, PBr_5, which decomposes below 100°C and is soluble in benzene and carbon tetrachloride (tetrachloromethane). It may be prepared by the reaction of phosphorus(III) bromide with bromine or the direct reaction of phosphorus with excess bromine. It is very readily hydrolysed to give hydrogen bromide and phosphoric(V) acid. An interesting feature of this material is that in the solid state it has the structure $[PBr_4]^+Br^-$. It is used in organic chemistry as a brominating agent.

phosphorus(III) chloride (phosphorus trichloride) A colourless fuming liquid, PCl_3; r.d. 1.57; m.p. $-112°C$; b.p. 75.5°C. It is soluble in ether and in carbon tetrachloride but reacts with water and with ethanol. It may be prepared by passing chlorine over excess phosphorus (excess chlorine contaminates the product with phosphorus(V) chloride). The molecule is pyramidal in the gas phase and possesses weak electron-pair donor properties. It is hydrolysed violently by water to phosphonic acid and hydrogen chloride. Phosphorus(III) chloride is an important starting point for the synthesis of a variety of inorganic and organic derivatives of phosphorus.

phosphorus(V) chloride (phosphorus pentachloride) A yellow-white rhombic solid, PCl_5, which fumes in air; r.d. 3.6; m.p. 148°C (under pressure); sublimes at 160–162°C. It is decomposed by water to give hydrogen chloride and phosphoric(V) acid. It is soluble in organic solvents. The compound may be prepared by the reaction of chlorine with phosphorus(III) chloride. Phosphorus(V) chloride is structurally interesting in that in the gas phase it has the expected trigonal bipyramidal form but in the solid phase it consists of the ions $[PCl_4]^+[PCl_6]^-$. The same ions are detected when phosphorus(V) chloride is dissolved in polar solvents. It is used in organic chemisty as a chlorinating agent.

phosphorus(III) chloride oxide (phosphorus oxychloride; phosphoryl chloride) A colourless fuming liquid, $POCl_3$; r.d. 1.67; m.p. 2°C; b.p. 105.3°C. It may be prepared by the reaction of phosphorus(III) chloride with oxygen or by the reaction of phosphorus(V) oxide with phosphorus(V) chloride. Its reactions are very similar to those of phosphorus(III) chloride. Hydrolysis with water gives phosphoric(V) acid. Phosphorus(III) chloride oxide has a distorted tetrahedral shape and can

act as a donor towards metal ions, thus giving rise to a series of complexes.

phosphorus(III) oxide (phosphorus trioxide) A white or colourless waxy solid, P_4O_6; r.d. 2.13; m.p. 23.8°C; b.p. 173.8°C. It is soluble in ether, chloroform, and benzene but reacts with cold water to give phosphonic acid, H_3PO_3, and with hot water to give phosphine and phosphoric(V) acid. The compound is formed when phosphorus is burned in an oxygen-deficient atmosphere (about 50% yield). As it is difficult to separate from white phosphorus by distillation, the mixture is irradiated with ultra-violet radiation to convert excess white phosphorus into the red form, after which the oxide can be separated by dissolution in organic solvents. Although called a trioxide for historical reasons, phosphorus(III) oxide consists of P_4O_6 molecules of tetrahedral symmetry in which each phosphorus atom is linked to the three others by an oxygen bridge. The chemistry is very complex. Above 210°C it decomposes into red phosphorus and polymeric oxides. It reacts with chlorine and bromine to give oxo-halides and with alkalis to give phosphonates (*see* phosphonic acid).

phosphorus(V) oxide (phosphorus pentoxide; phosphoric anhydride) A white powdery and extremely deliquescent solid, P_4O_{10}; r.d. 2.39; m.p. 580°C under pressure; sublimes at 360°C. It reacts violently with water to give phosphoric(V) acid. It is prepared by burning elemental phosphorus in a plentiful supply of oxygen, then purified by sublimation. The hexagonal crystalline form consists of P_4O_{10} molecular units; these have the phosphorus atoms arranged tetrahedrally, each P atom linked to three others by oxygen bridges and having in addition one terminal oxygen atom. The compound is used as a drying agent and as a dehydrating agent; for example, amides are converted into nitrites and sulphuric acid is converted to sulphur trioxide.

phosphorus oxychloride *See* phosphorus(III) chloride oxide.

phosphorus pentabromide *See* phosphorus(V) bromide.

phosphorus pentachloride *See* phosphorus(V) chloride.

phosphorus tribromide *See* phosphorus(III) bromide.

phosphorus trichloride *See* phosphorus(III) chloride.

phosphorus trioxide *See* phosphorus(III) oxide.

phosphoryl chloride *See* phosphorus(III) chloride oxide.

phot A unit of illuminance equal to 10^4 lux or one lumen per square centimetre.

photic zone The upper layer of a sea or a lake, in which there is sufficient light for photosynthesis. The limit of the photic zone varies from less than a metre to more than 200 metres, depending on the turbidity of the water.

photino *See* supersymmetry.

photocathode A *cathode that emits electrons when light falls upon it, as a result of the *photoelectric effect. *See* photoelectric cell.

photocell *See* photoelectric cell.

photochemical reaction A chemical reaction caused by light or ultraviolet radiation. The incident photons are absorbed by reactant molecules to give excited molecules or free radicals, which undergo further reaction.

photochemical smog A noxious smog produced by the reaction of nitrogen oxides with hydrocarbons in the presence of ultraviolet light from the sun. The reaction is very complex and one of the products is ozone.

photochemistry The branch of chemistry concerned with *photochemical reactions.

photochromism A change of colour occurring in certain substances when exposed to light. Photochromic materials are used in sunglasses that darken in bright sunlight.

photoconductive effect *See* photoelectric effect.

photodiode A *semiconductor diode used to detect the presence of light or to measure its intensity. It usually consists of a $p-n$ junction device in a container that focuses any light in the environment close to the junction. The device is usually biased in reverse so that in the dark the current is small; when it is illuminated the current is proportional to the amount of light falling on it. *See* photoelectric effect.

photodisintegration The decay of a nuclide as a result of the absorption of a gamma-ray photon.

photoelasticity An effect in which certain materials exhibit double refraction when subjected to stress. It is used in a technique for detecting strains in transparent materials (e.g. Perspex, celluloid, and glass). When polarized white light is passed through a stressed sample, the birefringence causes coloured patterns to be seen on the viewing screen of a suitable *polariscope. If monochromatic polarized light is used, a complex pattern of light and dark fringes is produced.

photoelectric cell (photocell) Any of several devices that produce an electric signal in response to exposure to electromagnetic radiation. The original photocells utilized photoemission from a photosensitive cathode (*photocathode*). The electrons emitted are attracted to an anode. A positive potential on the anode enables a current to flow through an external circuit, the current being proportional to the intensity of the illumination on the cathode. The electrodes are enclosed in an evacuated glass tube (*see also* photomultiplier).

More modern light-sensitive devices utilize the photoconductive effect and the photovoltaic effect (*see* photoelectric effect; photodiode; phototransistor; solar cell).

photoelectric effect The liberation of electrons (*see* photoelectron) from a substance exposed to electromagnetic radiation. The number of electrons emitted depends on the intensity of the radiation. The kinetic energy of the electrons emitted depends on the frequency of the radiation. The effect is a quantum process in which the radiation is regarded as a stream of *photons, each having an energy hf, where h is the Planck constant and f is the frequency of the radiation. A photon can only eject an electron if the photon energy exceeds the *work function, ϕ, of the solid, i.e. if $hf_0 = \phi$ an electron will be ejected; f_0 is the minimum frequency (or *threshold frequency*) at which ejection will occur. For many solids the photoelectric effect occurs at ultraviolet frequencies or above, but for some materials (having low work functions) it occurs with light. The maximum kinetic energy, E_m, of the photoelectron is given by *Einstein's equation: $E_m = hf - \phi$ (*see also* photoionization).

Apart from the liberation of electrons from atoms, other phenomena are also referred to as photoelectric effects. These are the *photoconductive effect* and the *photovoltaic effect*. In the photoconductive effect, an increase in the electrical conductivity of a semiconductor is caused by radiation as a result of the excitation of additional free charge carriers by the incident photons. *Photoconductive cells*, using such photosensitive materials as cadmium sulphide, are widely used as radiation detectors and light

switches (e.g. to switch on street lighting).

In the photovoltaic effect, an e.m.f. is produced between two layers of different materials as a result of irradiation. The effect is made use of in *photovoltaic cells*, most of which consist of $p-n$ *semiconductor junctions (*see also* photodiode; phototransistor). When photons are absorbed near a $p-n$ junction new free charge carriers are produced (as in photoconductivity); however, in the photovoltaic effect the electric field in the junction region causes the new charge carriers to move, creating a flow of current in an external circuit without the need for a battery. *See also* photoelectric cell.

photoelectron An electron emitted from a substance by irradiation as a result of the *photoelectric effect or *photoionization.

photoelectron spectroscopy A technique for determining the *ionization potentials of molecules. The sample is a gas or vapour irradiated with a narrow beam of ultraviolet radiation (usually from a helium source at 58.4 nm, 21.21 eV photon energy). The photoelectrons produced in accordance with *Einstein's equation are passed through a slit into a vacuum region, where they are deflected by magnetic or electrostatic fields to give an energy spectrum. The photoelectron spectrum obtained has peaks corresponding to the ionization potentials of the molecule (and hence the orbital energies). The technique also gives information on the vibrational energy levels of the ions formed. *ESCA* (electron spectroscopy for chemical analysis) is a similar analytical technique in which a beam of X-rays is used. In this case, the electrons ejected are from the inner shells of the atoms. Peaks in the electron spectrum for a particular element show characteristic chemical shifts, which depend on the presence of other atoms in the molecule.

photoemission The process in which electrons are emitted by a substance as a result of irradiation. *See* photoelectric effect; photoionization.

photofission A *nuclear fission that is caused by a gamma-ray photon.

photographic density A measure of the opacity of a photographic emulsion (negative or transparency). *See* densitometer.

photography The process of forming a permanent record of an image on specially treated film or paper. In normal black-and-white photography a camera is used to expose a film or plate to a focused image of the scene for a specified time. The film or plate is coated with an emulsion containing silver salts and the exposure to light causes the silver salts to break down into silver atoms; where the light is bright dark areas of silver are formed on the film after development (by a mild reducing agent) and fixing. The negative so formed is printed, either by a contact process or by projection. In either case light passing through the negative film falls on a sheet of paper also coated with emulsion. Where the negative is dark, less light passes through and the resulting positive is light in this area, corresponding with a light area in the original scene. As photographic emulsions are sensitive to ultraviolet and X-rays, they are widely used in studies involving these forms of electromagnetic radiation. *See also* colour photography.

photoionization The *ionization of an atom or molecule as a result of irradiation by electromagnetic radiation. For a photoionization to occur the incident photon of the radiation must have an energy in excess of the *ionization potential of the species being irradiated. The ejected photoelectron will have an energy, E, given by $E = hf - I$, where h is the Planck constant, f is the frequency of the incident radiation, and I is the ionization potential of the irradiated species.

photolithography A technique used in the manufacture of semiconductor components, integrated circuits, etc. It depends on the principle of masking selected areas of a surface and exposing the unmasked areas to such processes as the introduction of impurities, deposition of thin films, removal of material by etching, etc. The technique has been developed for use on tiny structures (typically measured in micrometres), which can only be examined by means of an electron microscope.

photoluminescence *See* luminescence.

photolysis A chemical reaction produced by exposure to light or ultraviolet radiation. Photolytic reactions often involve free radicals, the first step being homolytic fission of a chemical bond. *See* flash photolysis.

photometer An instrument used to measure *luminous intensity, illumination, and other photometric quantities. The older types rely on visual techniques to compare a source of light with a standard source. More modern photometers use *photoelectric cells, of the photoconductive, photoemissive, or photovoltaic typess. The photovoltaic types do not require an external power source and are therefore very convenient to use but are relatively insensitive. The photoemissive type usually incorporates a *photomultiplier, especially for use in astronomy and with other weak sources. Photoconductive units require only low-voltage supplies, which makes them convenient for commercial illumination meters and photographers' exposure meters.

photometric brightness *See* luminance.

photometry The study of visual radiation, especially the calculations and measurements of *luminous intensity, *luminous flux, etc. In some cases photometric calculations and measurements extend into the near infrared and the near ultraviolet.

In photometry, two types of measurement are used: those that measure *luminous* quantities rely on the use of the human eye (for example, to compare the illuminance of two surfaces); those called *radiant* quantities rely on the use of photoelectric devices to measure electromagnetic energy. *See also* photometer.

photomicrography The use of photography to obtain a permanent record (a *photomicrograph*) of the image of an object as viewed through a microscope.

photomultiplier A sensitive type of *photoelectric cell in which electrons emitted from a photocathode are accelerated to a second electrode where several electrons are liberated by each original photoelectron, as a result of *secondary emission. The whole process is repeated as many times as necessary to produce a useful electric current by secondary emission from the last electrode. A photomultiplier is thus a photocathode with the output amplified by an electron multiplier. The initial photocurrent can be amplified by a factor of 10^8. Photomultipliers are thus useful when it is necessary to detect low intensities of light, as in stellar photometry, star and planet tracking in guidance systems, and more mundanely in process control.

photon A particle with zero rest mass consisting of a *quantum of electromagnetic radiation. The photon may also be regarded as a unit of energy equal to hf, where h is the *Planck constant and f is the frequency of the radiation in hertz. Photons travel at the speed of light. They are required to explain the photoelectric effect and other phenomena that require light to have particle character.

photoneutron A neutron emitted by an atomic nucleus undergoing a *photonuclear reaction.

photonuclear reaction A *nuclear reaction that is initiated by a (gamma-ray) photon.

photoperiodism The response of an organism to changes in day length (*photoperiod*). Many plant responses are controlled by day length, the most notable being flowering in many species (*see also* day-neutral plant; long-day plant; short-day plant). In plants the pigment *phytochrome is responsible for the regulation of photoperiodic responses. Activities in animals that are determined by photoperiod include breeding, migration, and other seasonal events.

photopic vision The type of vision that occurs when the cones in the eye are the principal receptors, i.e. when the level of illumination is high. Colours can be identified with photopic vision. *Compare* scotopic vision.

photoreceptor A sensory cell or group of cells that reacts to the presence of light. It usually contains a pigment that undergoes a chemical change when light is absorbed, thus stimulating a nerve. *See* eye.

photosensitive substance 1. Any substance that when exposed to electromagnetic radiation produces a photoconductive, photoelectric, or photovoltaic effect. **2.** Any substance, such as the emulsion of a photographic film, in which electromagnetic radiation produces a chemical change.

photosphere The visible surface of the *sun or other star and the source of its continuous spectrum. It is a gaseous layer several hundreds of kilometres thick with an average temperature of 5780 K. Where the photosphere merges with the *chromosphere the temperature is 4000 K.

photosynthesis The chemical process by which green plants synthesize organic compounds from carbon dioxide and water in the presence of sunlight. It occurs in the *chloroplasts (most of which are in the leaves) and there are two principal series of reactions. In the *light reactions*, which require the presence of light, energy from sunlight is absorbed by *photosynthetic pigments (chiefly the green pigment *chlorophyll) and converted into chemical energy. In the ensuing *dark reactions*, which can take place either in light or darkness, this chemical energy is used in the production of simple organic compounds from carbon dioxide and water. Further chemical reactions convert these compounds into chemicals useful to the plant. Photosynthesis can be summarized by the equation:

$$CO_2 + 2H_2O \rightarrow [CH_2O] + H_2O + O_2$$

Since virtually all other forms of life are directly or indirectly dependent on plants for food, photosynthesis is the basis for all life on earth. Furthermore virtually all the atmospheric oxygen has originated from oxygen released during photosynthesis.

photosynthetic pigments The plant pigments responsible for the capture of light energy during the light reactions of *photosynthesis. The green pigment *chlorophyll is the principal light receptor, absorbing blue and red light. However the *carotenoids and various other pigments also absorb light energy and pass this on to the chlorophyll molecules.

phototaxis The movement of a cell (e.g. a gamete) or a unicellular organism in response to light. For example, certain algae (e.g. *Chlamydomonas*) can perceive light by means of a sensitive eyespot and move to regions of higher light concentration to enhance photosynthesis. *See* taxis.

phototransistor A junction *transistor that is photosensitive. When radiation falls on the emitter-base junction, new free charge carriers are created in the base region and the collector current is increased. Phototransistors are similar to *photodiodes except that the primary photoelectric current is amplified internally and it is therefore more sensitive to light than the photodiode. Some types can be used as switching or bistable devices, a small intensity of radiation switching them from a low to high current state.

phototropism (heliotropism) The growth of plant organs in response to light. Aerial shoots usually grow towards light, while some aerial roots grow away from light. The phototropic response is thought to be controlled by *auxin, a plant growth substance. *See* tropism.

photovoltaic effect *See* photoelectric effect.

pH scale A logarithmic scale for expressing the acidity or alkalinity of a solution. To a first approximation, the pH of a solution can be defined as $-\log_{10}c$, where c is the concentration of hydrogen ions in moles per cubic decimetre. A neutral solution at 25°C has a hydrogen-ion concentration of 10^{-7} mol dm^{-3}, so the pH is 7. A pH below 7 indicates an acid solution; one above 7 indicates an alkaline solution. More accurately, the pH depends not on the concentration of hydrogen ions but on their *activity, which cannot be measured experimentally. For practical purposes, the pH scale is defined by using a hydrogen electrode in the solution of interest as one half of a cell, with a reference electrode (e.g. a calomel electrode) as the other half cell. The pH is then given by $(E - E_R)F/2.303RT$, where E is the e.m.f. of the cell and E_R the standard electrode potential of the reference electrode, and F the Faraday constant. In practice, a glass electrode is more convenient than a hydrogen electrode.

pH stands for 'potential of hydrogen'. The scale was introduced by S. P. Sørensen in 1909.

phthalic acid A colourless crystalline dicarboxylic acid, $C_6H_4(COOH)_2$; r.d. 1.6; m.p. 207°C. The two –COOH groups are substituted on adjacent carbon atoms of the ring, the technical name being *benzene-1,2-dicarboxylic acid*. The acid is made from *phthalic anhydride* (benzene-1,2-dicarboxylic anhydride, $C_8H_4O_3$), which is made by the catalytic oxidation of naphthalene. The anhydride is used in making plasticizers and polyester resins.

phthalic anhydride See phthalic acid.

phycomycetes Primitive *fungi, many of which are found in water (e.g. the water moulds, which may be parasitic on fish) or in damp areas. Many are unicellular but those that form mycelia generally have hyphae lacking cross walls, which distinguishes them from the *ascomycetes and *basidiomycetes.

phyletic See phylogenetic.

phyllotaxis (phyllotaxy) The arrangement of leaves on a plant stem. The leaves may be inserted in whorls or pairs at each node or singly up the stem. When arranged in pairs the two leaves arise on opposite sides of the stem and are usually at right angles to the leaf pairs above and below them. Single leaves may be inserted alternately or in a spiral pattern up the stem. Phyllotaxis generally results in the minimum of shading of leaves by those above them.

phylogenetic (phyletic) Describing a system of *classification of organisms that aims to show their evolutionary history. *Compare* phenetic.

phylogeny The evolutionary history of an organism or group of related organisms. *Compare* ontogeny.

phylum A category used in the *classification of animals that consists of one or several similar or closely related classes. Examples of phyla are the Protozoa, Porifera, Nematoda, Arthropoda, Mollusca, and Chordata. Phyla are grouped into the kingdom Animalia. In plant classification, the *division is usually used instead of the phylum.

physical chemistry The branch of chemistry concerned with the effect of chemical structure on physical properties. It includes chemical thermodynamics and electrochemistry.

physics The study of the laws that determine the structure of the universe with reference to the matter and energy of which it consists. It is concerned not with chemical changes that occur but with the forces that exist between objects and the interrelationship between matter and energy. Traditionally, the study was divided into separate fields: heat, light, sound, electricity and magnetism, and mechanics. Since the turn of the century, however, quantum mechanics and relativistic physics have become increasingly important; the growth of modern physics has been accompanied by the studies of atomic physics, nuclear physics, and particle physics. The physics of astronomical bodies and their interactions is known as *astrophysics*, the physics of the earth is known as *geophysics*, and the study of the physical aspects of biology is called *biophysics*.

physiological saline A liquid medium in which animal tissues may be kept alive for a few hours during experiments without pathological changes or distortion of the cells taking place. Such fluids are salt solutions that are isotonic with and have the same pH as the body fluids of the animal. A well-known example is *Ringer's solution*, formulated by the British physiologist S. Ringer (1835–1910), which is a mixture of sodium chloride, calcium chloride, sodium bicarbonate, and potassium chloride solutions.

physiological specialization The occurrence within a species of several forms that are identical in appearance but differ in physiology: these are termed *physiological races*. For example, many pathogenic fungi develop new physiological races in response to the strong selection pressure exerted when disease-resistant crop varieties are sown over large areas.

physiology The branch of biology concerned with the vital functions of plants and animals, such as nutrition, respiration, reproduction, and excretion.

physisorption *See* adsorption.

phyto- Prefix denoting plants. For example, phytopathology is the study of plant diseases.

phytochrome A protein-based plant pigment present in small quantities in many plant organs. It exists in two interconvertible forms: a physiologically active form, which forms when the plant is illuminated with red light or normal daylight; and an inactive form, formed when the plant is exposed to far-red light or darkness. The active form regulates many plant processes, such as seed germination and the initiation of flowering.

phytogeography *See* plant geography.

phytohaemagglutinin (PHA) Any of various plant-derived compounds that induce changes in lymphocytes normally associated with antigen challenge. These changes include cell enlargement, increased RNA and DNA synthesis, and, finally, cell division. Response to PHAs is used to test for competence of cell-mediated *immunity, for example in patients suffering chronic virus infections.

phytohormone *See* growth substance.

phytoplankton The plant component of *plankton, consisting chiefly of microscopic algae, such as diatoms and dinoflagellates. Near the surface of the sea there may be many millions of such plants per cubic metre. Members of the phytoplankton are of great importance as they form the basis of food for all other forms of aquatic life, being the primary *producers. *Compare* zooplankton.

pi Symbol π. The ratio of the circumference of any circle to its diameter. It is a *transcendental number with the value $3.141\,592\ldots$.

pia mater The innermost of the three membranes (*meninges) that surround the brain and spinal cord of vertebrates. The pia mater lies immediately adjacent to the central nervous system, and the *choroid plexus, which secretes cerebrospinal fluid, is an extension of it.

pi bond *See* orbital.

pico- Symbol p. A prefix used in the metric system to denote 10^{-12}. For example, 10^{-12} farad $= 1$ picofarad (pF).

picornavirus One of a group of small RNA-containing viruses (*pico* = small; hence pico-RNA-virus) commonly present in the alimentary and respiratory tracts of vertebrates. They cause mild infections of these tracts but the group also includes the polioviruses, which attack the central nervous system causing poliomyelitis; and the causal agent of foot and mouth disease in cattle, sheep, and pigs.

picrate A salt or ester of picric acid.

picric acid (2,4,6-trinitrophenol) A yellow highly explosive nitro compound, $C_6H_2(NO_2)_3$; r.d. 1.8; m.p. 122°C.

pie chart A diagram in which percentages are shown as sectors of a circle. If x percent of the prey of a carnivore comprises species X, y percent species Y, and z percent species Z, a pie chart would show three sectors having central angles $3.6x°$, $3.6y°$, and $3.6z°$.

pi electron An electron in a pi orbital. *See* orbital.

piezoelectric effect The generation of a potential difference across the opposite faces of certain nonconducting crystals (*piezoelectric crystals*) as a result of the application of mechanical stress between these faces. The *electric polarization produced is proportional to the stress and the direction of the polarization reverses if the stress changes from compression to tension. The *reverse piezoelectric effect* is the opposite phenomenon: if the opposite faces of a piezoelectric crystal are subjected to a potential difference, the crystal changes its shape. Rochelle salt and quartz are the most frequently used piezoelectric materials. While Rochelle salt produces the greater polarization for a given stress, quartz is more widely used as its crystals have greater strength and are stable at temperatures in excess of $100°C$.

If a quartz plate is subjected to an alternating electric field, the reverse piezoelectric effect causes it to expand and contract at the field frequency. If this field frequency is made to coincide with the natural elastic frequency of the crystal, the plate resonates; the direct piezoelectric effect then augments the applied electric field. This is the basis of the *crystal oscillator and the *quartz clock. *See also* crystal microphone; crystal pickup.

pig iron The impure form of iron produced by a blast furnace, which is cast into pigs (blocks) for converting at a later date into cast iron, steel, etc. The composition depends on the ores used, the smelting procedure, and the use to which the pigs will later be put.

pileus The umbrella-shaped cap of certain fungi, such as mushrooms. Spores are produced from *gills or pores on the lower surface.

piliferous layer The part of the root epidermis that bears *root hairs. It extends over a region about 4 10 mm behind the root tip. Beyond this the piliferous layer is sloughed off to reveal the hypodermis.

Piltdown man Fossil remains, purported to have been found by Charles Dawson at Piltdown, Sussex, in 1912, that were named *Eoanthropus dawsoni* and described as a representative of the true ancestors of modern man. The skull resembled that of a man but the jaw was apelike. In 1953 dating techniques showed the specimen to be a fraud.

pi-meson *See* pion.

pinch effect A magnetic attraction between parallel conductors carrying currents flowing in the same direction. The force was noticed in early induction furnaces. Since the late 1940s it has been widely studied as a means of confining the hot plasma in a *thermonuclear reactor. In an experimental toroidal thermonuclear reactor a large electric current is induced in the plasma by electromagnetic induction; this current both heats the plasma and draws it away from the walls of the tube as a result of the pinch effect.

pineal eye *See* median eye.

pineal gland An outgrowth of the *forebrain. In man its functions are obscure, but in other vertebrates it acts as an endocrine gland, secreting the hormone *melatonin.

pinna (auricle) The visible part of the *outer ear, present in some mammals. It is made of cartilage and its function is to channel sound waves into the external auditory meatus. In some species the pinna is movable and aids in detecting the direction from which a sound originates.

pinocytosis The process by which a living cell engulfs a minute droplet of liquid. It involves a technique similar to *phagocytosis.

pion (pi-meson) An *elementary particle classified as a *meson. It exists in three forms: neutral, positively charged, and negatively charged. The charged pions decay into muons and neutrinos; the neutral pion decays into two gamma-ray photons.

pi orbital *See* orbital.

pipette A graduated tube used for transferring measured volumes of liquid.

Pirani gauge An instrument used to measure low pressures ($1-10^{-4}$ torr; $100-0.01$ Pa). It consists of an electrically heated filament, which is exposed to the gas whose pressure is to be measured. The extent to which heat is conducted away from the filament depends on the gas pressure, which thus controls its equilibrium temperature. Since the resistance of the filament is dependent on its temperature, the pressure is related to the resistance of the filament. The filament is arranged to be part of a *Wheatstone bridge circuit and the pressure is read from a microammeter calibrated in pressure units. As the effect depends on the thermal conductivity of the gas, the calibration has to be made each time the pressure of a different gas is measured.

pirssonite A mineral consisting of a hydrated mixed carbonate of sodium and calcium, $Na_2CO_3.CaCO_3.2H_2O$.

Pisces The fishes: a group of cold-blooded aquatic vertebrates that have *gills and *fins (a few species can move on land and breathe air). The whole body is covered with a tough, usually scaly, skin (*see* scales), which extends over the eye and contains pigments and sometimes slime glands. The circulatory system is a single circuit with blood passing through two sets of capillaries, one at the gills and the other in the body tissues. Three *semicircular canals are present in the inner ear. The sense of smell is particularly well developed and pressure waves are detected by the *lateral line system*.

Fossils of fish date back to the Ordovician period, 500–440 million years ago. Modern fish are divided into the *Elasmobranchii (cartilaginous fishes) and *Osteichthyes (bony fishes).

pistil The female part of a flower, consisting either of a single *carpel (*simple pistil*) or a group of carpels (*compound pistil*).

pitch 1. (in chemistry) A black or dark-brown residue resulting from the distillation of coal tar, wood tar, or petroleum (bitumen). The term is also sometimes used for the naturally occurring petroleum residue (asphalt). Pitch is used as a binding agent (e.g. in road tars), for waterproofing (e.g. in roofing felts), and as a fuel. **2.** (in physics) The property of a sound that characterizes its highness or lowness to an observer. It is related to, but not identical with, frequency. Below about 1000 Hz the pitch is slightly higher than the frequency and above 1000 the position is reversed. The loudness of a sound also affects the pitch. Up to 1000 Hz an increase in loudness causes a decrease in pitch. From about 1000 to 3000 Hz the pitch is independent of loudness, while above 3000 Hz an increase in loudness seems to cause a raising of pitch. Pitch is usually measured in mels; a note of 1000 Hz frequency with a loudness of 40 decibels above the absolute threshold of hearing has a pitch of 1000 mels. **3.** (in mechanics) *See* screw.

pitchblende *See* uraninite.

pith 1. (*or* **medulla**) The cylinder of *parenchyma tissue found in the centre of plant stems to the inside of the vascular tissue. It is light in weight and has been put to various commercial uses, notably the manufacture of pith helmets. **2.** (not in scientific usage) The white tissue below the rind of many citrus fruits. **3.** To destroy the central nervous system of an animal, especially a laboratory animal such as a frog, by severing its spinal cord.

Pithecanthropus See *Homo.*

Pitot tube A device for measuring the speed of a fluid. It consists of two tubes, one with an opening facing the moving fluid and the other with an opening at 90° to the direction of the flow. The two tubes are connected to the opposite sides of a manometer so that the difference between the dynamic pressure in the first tube and the static pressure in the second tube can be measured. The speed v of the flow of an incompressible fluid is then given by: $v^2 = 2(P_2 - P_1)/\rho$, where P_2 is the dynamic pressure, P_1 is the static pressure, and ρ is the density of the fluid. The device has a wide variety of applications. It was devised by Henri Pitot (1695–1771).

pituitary gland (pituitary body; hypophysis) A pea-sized endocrine gland attached by a thin stalk to the *hypothalamus at the base of the brain. It consists of two lobes: the anterior and the posterior. The *anterior pituitary* secretes such hormones as *growth hormone, the *gonadotrophins, *prolactin, thyroid stimulating hormone (see thyroid gland) and *ACTH. Because these hormones regulate the growth and activity of several other endocrine glands, the anterior pituitary is often referred to as the *master endocrine gland.* Activity of the anterior pituitary itself is regulated by specific releasing hormones produced by the hypothalamus (*see also* neuroendocrine system). The *posterior pituitary* secretes the hormones *oxytocin and *vasopressin.

pK value A measure of the strength of an acid on a logarithmic scale. The pK value is given by $\log_{10}(1/K_a)$, where K_a is the acid dissociation constant. pK values are often used to compare the strengths of different acids.

placenta 1. The organ in mammals and other viviparous animals by means of which the embryo is attached to the wall of the uterus. It is composed of embryonic and maternal tissues: extensions of the *chorion and *allantois grow into the uterine wall so that materials (e.g. oxygen, nutrients) can pass between the blood of the embryo and its mother (there is, however, no direct connection between the maternal and embryonic blood). The placenta is eventually expelled as part of the *afterbirth. **2.** A ridge of tissue on the ovary wall of flowering plants to which the ovules are attached. The arrangement of ovules on the placenta (*placentation*) is variable, depending on the number of carpels and whether they are free (*see* apocarpy) or fused (*see* syncarpy).

Placentalia See Eutheria.

placoid scale (denticle) See scales.

plagiotropism The tendency for a *tropism (growth response of a plant) to be orientated at an angle to the line of action of the stimulus concerned. For example, the growth of lateral branches and lateral roots is at an oblique angle to the stimulus of gravity (*plagiogeotropism*). *Compare* orthotropism.

planarians See Turbellaria.

Planck constant Symbol h. The fundamental constant equal to the ratio of the energy E of a quantum of energy to its frequency v: $E = hv$. It has the value $6.626\,176 \times 10^{-34}$ J s. It is named after Max Planck (1858–1947). In quantum-mechanical calculations (especially particle physics) the *rationalized Planck constant* $\hbar = h/2\pi = 1.054\,589 \times 10^{-34}$ J s is frequently used.

Planck length The length scale at which a classical description of gravity ceases to be valid, and *quantum mechanics must be taken into account. It is given by $L_P =$

Planck mass

$\sqrt{(G\hbar/c^3)}$, where G is the gravitational constant, $\hbar$ is the rationalized Planck constant, and c is the speed of light. The value of the Planck length is of order 10^{-35} m (twenty orders of magnitude smaller than the size of a proton, 10^{-15} m).

Planck mass The mass of a particle whose Compton wavelength is equal to the *Planck length. It is given by $m_P = \sqrt{(\hbar c/G)}$, where $\hbar$ is the rationalized Planck constant, c is the speed of light, and G is the gravitational constant. The description of an elementary particle of this mass, or particles interacting with energies per particle equivalent to it (through $E = mc^2$), requires a *quantum theory of gravity. Since the Planck mass is of order 10^{-8} kg (equivalent energy 10^{19} GeV), and, for example, the proton mass is of order 10^{-27} kg and the highest energies attainable in present-day particle accelerators are of order 10^3 GeV, quantum-gravitational effects do not arise in laboratory particle physics. However, energies equivalent to the Planck mass did occur in the early universe according to *big-bang theory, and a quantum theory of gravity is important for discussing conditions there (*see* Planck time).

Planck's radiation law A law giving the distribution of energy radiated by a *black body. It introduced into physics the novel concept of energy as a quantity that is radiated by a body in small discrete packets rather than as a continuous emission. These small packets became known as quanta and the law formed the basis of *quantum theory. The *Planck formula* gives the energy radiated per unit time at frequency ν per unit frequency interval per unit solid angle into an infinitesimal cone from an element of the black-body surface that is of unit area in projection perpendicular to the cone's axis. The expression for this *monochromatic specific intensity* I_ν is:

$$I_\nu = 2hc^{-2}\nu^3/[\exp(h\nu/kT - 1],$$

where h is the *Planck constant, c is the *speed of light, k is the *Boltzmann constant, and T is the thermodynamic temperature of the black body. I_ν has units of watts per square metre per steradian per hertz (W m^{-2} sr^{-1} Hz^{-1}). The monochromatic specific intensity can also be expressed in terms of the energy radiated at wavelength λ per unit wavelength interval; it is then written as I_λ, and the Planck formula is:

$$I_\lambda = 2hc^2\lambda^{-5}/[\exp(hc/\lambda kT) - 1].$$

There are two important limiting cases of the Planck formula. For low frequencies $\nu << kT/h$ (equivalently, long wavelengths $\lambda >> hc/kT$) the *Rayleigh–Jeans formula* is valid:

$$I_\nu = 2c^{-2}\nu^2 kT,$$

or

$$I_\lambda = 2c\lambda^{-4}kT.$$

Note that these expressions do not involve the Planck constant. They can be derived classically and do not apply at high frequencies, i.e. high energies, when the quantum nature of *photons must be taken into account. The second limiting case is the *Wien formula*, which applies at high frequencies $\nu >> kT/h$ (equivalently, short wavelengths $\lambda << hc/kT$):

$$I_\nu = 2hc^{-2}\nu^3 \exp(-h\nu/kT),$$

or

$$I_\lambda = 2hc^2\lambda^{-5} \exp(-hc/\lambda kT).$$

See also Wien's displacement law.

Planck time The time taken for a photon (travelling at the speed of light c) to move through a distance equal to the *Planck length. It is given by $t_P = \sqrt{(G\hbar/c^5)}$, where G is the gravitational constant and $\hbar$ is the rationalized Planck constant. The value of the Planck time is of order 10^{-43} s. In the *big-bang cosmology, up until a time t_P after the initial instant, it is necessary to use a *quantum theory of gravity to describe the evolution of the universe.

Planck units A system of *units, used principally in discussions of *quantum theories of gravity, in which length, mass, and time are expressed as multiples of the *Planck length, mass, and time respectively. This is equivalent to setting the gravitational constant, the speed of light, and the reduced Planck constant all equal to unity. All quantities that ordinarily have dimensions involving length, mass, and time become dimensionless in Planck units. Since, in the subject area where Planck units are used, it is normal to employ *Gaussian or *Heaviside–Lorentz units for electromagnetic quantities, these then become dimensionless. *See also* geometrized units; natural units.

plane A flat surface defined by the condition that any two points in the plane are joined by a straight line that lies entirely in the surface.

plane-polarized light *See* polarization of light.

planet A body that revolves around a central astronomical body, especially one of the nine bodies (Mercury; Venus; earth; Mars; Jupiter; Saturn; Uranus; Neptune; Pluto) that revolve in elliptical orbits around the sun. *See* solar system.

planimeter An instrument used to measure the area of a closed curve. The outline of the curve is followed by a pointer on the instrument and the area is given on a graduated disc.

plankton Minute *pelagic organisms that drift or float passively with the current in a sea or lake. Plankton includes many microscopic animals and plants, such as algae, protozoans, various animal larvae, and some worms. It forms an important food source for many other members of the aquatic community and is divided into *zooplankton and *phytoplankton. *Compare* nekton.

plano-concave lens *See* concave; lens.

plano-convex lens *See* convex; lens.

plant Any living organism of the kingdom Plantae, characterized by an ability to manufacture carbohydrates by *photosynthesis, in which simple inorganic substances are built up into organic compounds. The radiant energy needed for this process is absorbed by *chlorophyll, a complex pigment not found in animals. Most plants, except for certain microscopic algae, are immobile as there is no necessity to search for food. Other characteristics of plants include the possession of *cell walls (usually composed of *cellulose) and slow response to external stimuli. For a classification of the plant kingdom, see Appendix.

plant geography (phytogeography) The study of the distribution of world vegetation, with particular emphasis on the influence of the environmental factors that determine this distribution.

plantigrade Describing the gait of many mammals, including man, in which the whole lower surface of the foot is on the ground. *Compare* digitigrade; unguligrade.

planula The ciliated free-swimming larva of many coelenterates, consisting of a solid mass of cells. It eventually settles on a suitable surface and develops into a *polyp.

plaque A thin layer of organic material covering all or part of the exposed surface of a tooth. It contains dissolved food (mostly sugar) and bacteria. The bacteria in plaque metabolize the sugar and produce acid, which eats into the surface of the enamel of the tooth and eventually causes tooth decay (caries).

plasma 1. (in physics) A highly ionized gas in which the number of free electrons is approximately equal to the number of positive ions. Sometimes described as the fourth state of matter, plasmas occur in interstellar space, in the atmospheres of stars (including the sun), in discharge tubes, and in experimental thermonuclear reactors.

Because the particles in a plasma are charged, its behaviour differs in some respects from that of a gas. Plasmas can be created in the laboratory by heating a low-pressure gas until the mean kinetic energy of the gas particles is comparable to the *ionization potential of the gas atoms or molecules. At very high temperatures, from about 50 000 K upwards, collisions between gas particles cause cascading ionization of the gas. However, in some cases, such as a fluorescent lamp, the temperature remains quite low as the plasma particles are continually colliding with the walls of the container, causing cooling and recombination. In such cases ionization is only partial and requires a large energy input. In *thermonuclear reactors an enormous plasma temperature is maintained by confining the plasma away from the container walls using electromagnetic fields (*see* pinch effect). The study of plasmas is known as *plasma physics*. **2.** (in physiology) *See* blood plasma.

plasma cells Antibody-producing cells found in the epithelium of the lungs and gut and also in bone-forming tissue. They develop in the lymph nodes, spleen, and bone marrow when antigens stimulate lymphocytes to form the precursor cells that give rise to them.

plasmagel The specialized outer gel-like *cytoplasm of living cells (such as *Amoeba*) that move by extruding part of the cell (known as a *pseudopodium) in the direction of motion. A reversible conversion of plasmagel to the more fluid *plasmasol is believed to be responsible for the continuous flow forward of cytoplasm necessary for forming a pseudopodium.

plasmalemma *See* cell membrane.

plasma membrane *See* cell membrane.

plasmasol The specialized inner sol-like *cytoplasm of living cells that move by producing *pseudopodia. *Compare* plasmagel.

plasmid A structure in cells consisting of DNA that can exist and replicate independently of the chromosomes. In bacterial cells plasmids provide genetic instructions for certain cell activities (e.g. resistance to antibiotic drugs). They can be transferred from cell to cell in a bacterial colony. Bacterial plasmids are used to produce recombinant DNA for *gene cloning. Similar blocks of DNA that appear to move from one chromosome to another have been identified in the cells of a number of other species, including man. In organisms (other than man) that have been studied, it appears that they interfere with gene activity (*see* transposon). In man this has not been established, although they are almost identical to *retroviruses, which are known to affect gene action.

plasmolysis The loss of water from a plant cell to the extent that the protoplasm shrinks away from the cell wall. Continual plasmolysis results in wilting. *Compare* turgor.

plasmon A *collective excitation for quantized oscillations of the electrons in a metal.

plaster of Paris The hemihydrate of *calcium sulphate, $2CaSO_4.H_2O$, prepared by heating the mineral gypsum. When ground to a fine powder and mixed with water, plaster of Paris sets hard, forming interlocking crystals of gypsum. The setting results in an increase in volume and so the plaster fits tightly into a mould. It is used in pottery making, as a cast for setting broken bones, and as a constituent of the plaster used in the building industry.

plasticity The property of solids that causes them to change permanently in size or shape as a result of the application of a stress in excess of a certain value, called the *yield point*.

plasticizer A substance added to a synthetic resin to make it flexible. *See* plastics.

plastics Materials that can be shaped by applying heat or pressure. Most plastics are made from polymeric synthetic *resins, although a few are based on natural substances (e.g. cellulose derivatives or shellac). They fall into two main classes. *Thermoplastic materials* can be repeatedly softened by heating and hardened again on cooling. *Thermosetting materials* are initially soft, but change irreversibly to a hard rigid form on heating. Plastics contain the synthetic resin mixed with such additives as pigments, plasticizers (to improve flexibility), antioxidants and other stabilizers, and fillers.

plastid An *organelle within a plant cell, often occurring in large numbers. Apart from the nucleus, plastids are the largest solid inclusions in a plant cell. For convenience they are classified into those containing pigments (*chromoplasts) and those that are colourless (*leucoplasts), although changes from one to the other frequently occur. Plastids develop from *proplastids*, colourless bodies found in meristematic and immature cells; they also arise by division of existing plastids. *See also* chloroplast.

plastron *See* carapace.

platelet (thrombocyte) A small disc of red tissue in the blood. Platelets are formed as fragments of larger cells (*megakaryocytes*) found in red bone marrow; they have no nucleus. They play an important role in *blood clotting and release serotonin, which causes blood vessels to constrict, so reducing bleeding. There are about 250 000 per cubic millimetre of blood.

platelet-derived growth factor *See* growth factor.

plate tectonics The theory that the surface of the earth is made of lithospheric plates, which have moved throughout geological time resulting in the present-day positions of the continents. The theory explains the locations of mountain building as well as earthquakes and volcanoes. The rigid lithospheric plates consist of continental and oceanic crust together with the upper mantle, which lie above the weaker plastic asthenosphere. These plates move relative to each other across the earth. Six major plates (Eurasian, American, African, Pacific, Indian, and Antarctic) are recognized, together with a number of smaller ones. The plate margins coincide with zones of seismic and volcanic activity.

A *constructive* (or *divergent*) plate margin occurs when two plates move away from each other. It is marked by a mid-oceanic ridge where basaltic material wells up from the mantle to form new oceanic crust, in a process known as *sea-floor spreading*. The production of new crust at constructive plate margins is compensated for by the destruction of material along a *destructive* (or *convergent*) plate margin. Along these margins, which are also known as *subduction zones* and marked by an oceanic trench, one plate (usually oceanic) is forced to plunge down beneath the other (which may be continental or oceanic). The crust becomes partially melted and rises to form a chain of volcanoes in the upper plate parallel to the trench. When two continental plates collide the compression results in the formation of mountain chains. A third type of plate margin – the *transform plate margin* – occurs where two plates are slipping past each other.

platinum Symbol Pt. A silvery white

—→ direction of plate movement

·········· transform faults

———— destructive plate margins (subduction zones)

———— constructive plate margins

③ spreading rates (centimetres per year)

The lithospheric plates

+ major earthquake foci   × minor earthquake foci

Cross-section showing a constructive plate margin, where two plates are drawing apart along a mid-ocean ridge, and a destructive plate margin, where the oceanic plate is being subducted below the continental plate. Volcanic activity and earthquakes are associated with these margins

Plate tectonics

metallic *transition element (*see also* platinum metals); a.n. 78; r.a.m. 195.09; r.d. 21.37; m.p. 1772°C; b.p. ~3800°C. It occurs in some nickel and copper ores and is also found native in some deposits. The main source is the anode sludge obtained in copper–nickel refining. The element is used in jewellery, laboratory apparatus (e.g. thermocouples, electrodes, etc.), electrical contacts, and in certain alloys (e.g. with iridium or rhodium). It is also a hydrogenation catalyst. The element does not oxidize nor dissolve in hydrochloric acid. Most of its compounds are platinum(II) or platinum(IV) complexes.

platinum black Black finely divided platinum metal produced by vacuum evaporation and used as an absorbent and a catalyst.

platinum metals The three members of the second and third transition series immediately proceeding silver and gold: ruthenium (Ru), rhodium (Rh), and palladium (Pd); and osmium (Os), iridium (Ir), and platinum (Pt). These elements, together with iron, cobalt, and nickel, were formerly classed as group VIII of the periodic table. The platinum-group metals are relatively hard and resistant to corrosion and are used in jewellery and in some industrial applications (e.g. electrical contacts). They have certain chemical similarities that justify classifying them together. All are resistant to chemical attack. In solution they form a vast range of complex ions. They also form coordination compounds with carbon monoxide and other pi-bonding ligands. A number of complexes can be made in which a hydrogen atom is linked directly to the metal. The metals and their organic compounds have considerable catalytic activity. *See also* transition elements.

Platyhelminthes A phylum of invertebrates comprising the flatworms, characterized by a flattened unsegmented body. The simple nervous system shows some concentration of cells at the head end. The mouth leads to a simple branched gut without an anus. Flatworms are hermaphrodite but self-fertilization is unusual. Many species are parasitic. The phylum contains the classes *Turbellaria (planarians), *Trematoda (flukes), and *Cestoda (tapeworms).

pleiomorphism The existence of distinctly different forms during the life cycle of an individual, e.g. the caterpillar, pupa, and winged adult of a butterfly.

Pleistocene The earlier epoch of the *Quaternary period (*compare* Holocene). It extended from the end of the Pliocene, about 2 million years ago, to the beginning of the Holocene, about 10 000 years ago. The Pleistocene is often known as the *Ice Age* as it was characterized by a series of glacials, in which ice margins advanced towards the equator, separated by interglacials when the ice retreated. *See also* ice age.

pleochroic Denoting a crystal that appears to be of different colours, depending on the direction from which it is viewed. It is caused by polarization of light as it passes through an anisotropic medium.

pleura The double membrane that lines the thoracic cavity and covers the exterior surface of the lungs. It is a *serous membrane forming a closed sac, with a small space (the *pleural cavity*) between the two layers.

plexus A compact branching network of nerves or blood vessels, such as the *brachial plexus* – a network of spinal nerves that supply branches to the forelimbs in vertebrates. *See also* choroid plexus.

Pliocene The fifth and final epoch of the *Tertiary period. Preceded by the Miocene and followed by the Pleistocene, it began about 7 million years ago and lasted for about 5 million years. Mammals similar to modern forms existed during the epoch and the australopithecines (*see Australopithecus*), early forerunners of man, appeared.

plumbago *See* carbon.

plumbane (lead(IV) hydride) An extremely unstable gas, PbH_4, said to be formed by the action of acids on magnesium–lead alloys. It was first reported in 1924, although doubts have since been expressed about the existence of the compound. It demonstrates the declining stability of the hydrides in group IV. More stable organic derivatives are known; e.g. trimethyl plumbane, $(CH_3)_3PbH$.

plumbate A compound formed by reaction of lead oxides (or hydroxides) with alkali. The oxides of lead are amphoteric (weakly acidic) and react to give plumbate ions. With the lead(IV) oxide, reaction with molten alkali gives the plumbate(IV) ion

$$PbO_2 + 2OH^- \rightarrow PbO_3^{2-} + H_2O$$

In fact, various ions are present in which the lead is bound to hydroxide groups, the principal one being the hexahydroxoplumbate(IV) ion $Pb(OH)_6^{2-}$. This is the negative ion present in crystalline 'trihydrates' of the type $K_2PbO_3.3H_2O$. Lead(II) oxide gives the trihydroxoplumbate(II) ion in alkaline solutions

$$PbO(s) + OH^-(aq) + H_2O(l) \rightarrow$$
$$Pb(OH)_3^{2-}(aq)$$

Plumbate(IV) compounds were formerly referred to as *orthoplumbates* (PbO_4^{4-}) or *metaplumbates* (PbO_3^{2-}). Plumbate(II) compounds were called *plumbites*.

plumbic compounds Compounds of lead in its higher (+4) oxidation state; e.g. plumbic oxide is lead(IV) oxide, PbO_2.

plumbite *See* plumbate.

plumbous compounds Compounds of lead in its lower (+2) oxidation state; e.g. plumbous oxide is lead(II) oxide, PbO.

plumule 1. (in zoology) A *down feather. 2. (in botany) The part of a plant embryo that develops into the shoot system. It consists of the stem apex and first leaves. In seedlings showing *epigeal germination the plumule grows above the soil surface together with the cotyledons; in seeds showing *hypogeal germination, the plumule alone emerges. *Compare* radicle.

Pluto The outermost planet in the *solar system and the one whose orbit has the highest eccentricity

(0.25) and inclination (17.2°). Its mean distance from the sun is 5.9×10^9 km but at perihelion, owing to the high eccentricity of the orbit, it is closer to the sun than Neptune at about 4480×10^6 km. Its *sidereal period is 248.4 years. Pluto was discovered in 1930 by Clyde Tombaugh although its presence had been previously predicted. In 1978 Pluto was discovered to have a close satellite, provisionally named Charon, and estimates of its diameter were reduced from about 5.9×10^3 km to about 3.5×10^3 km. The diameter is still not well established. Pluto and Charon appear to be composed mostly of ice with a mean temperature of about 50 K.

plutonium Symbol Pu. A dense silvery radioactive metallic transuranic element belonging to the *actinoids; a.n. 94; mass number of most stable isotope 244 (half-life 7.6×10^7 years); r.d. 19.84; m.p. 641°C; b.p. 3232°C. Thirteen isotopes are known, by far the most important being plutonium–239 (half-life 2.44×10^4 years), which undergoes *nuclear fission with slow neutrons and is therefore a vital power source for *nuclear weapons and some *nuclear reactors. About 20 tonnes of plutonium are produced annually by the world's nuclear reactors, a detailed inventory of every gram of which is kept in order to prevent its military misuse. The element was first produced by Seaborg, McMillan, Kennedy, and Wahl in 1940.

pod *See* legume.

podzol (podsol) A type of soil that forms under cool humid conditions where the vegetation is coniferous or heath. It is characteristic of the taiga zones of Eurasia and North America. The soil has an ashen-coloured upper layer (the A horizon) overlying a layer (the B horizon) containing iron or humus that has been leached down from the A horizon and redeposited.

thin litter layer — coniferous trees — A_0 horizon

partly decomposed litter — A_1

dark-coloured humus

ash-grey sandy layer — A_2

iron pan (not always present)

B

reddish-brown layer

weathering parent material (e.g. sandstone) — C

Podzol soil profile

poikilothermy (cold-bloodedness) The passive variation in the internal body temperature of an animal, which depends on the temperature of the environment. All animals except birds and mammals are cold-blooded. Although unable to maintain a constant body temperature, they can respond to compensate for very low or very high temperatures. For example, the tissue composition (especially cell osmotic pressure) can change to regulate the blood flow to peripheral tissues (and thus increase heat loss or heat absorption), and the animals can actively seek sun or shade. Seasonal changes in metabolism are usually under hormonal control. In particularly hot climates, cold-blooded animals may undergo *aestivation to escape the heat. *Compare* homoiothermy.

point-contact transistor *See* transistor.

point defect *See* defect.

point discharge *See* corona.

poise A *c.g.s. unit of viscosity equal to the tangential force in dynes per square centimetre required to maintain a difference in velocity of one centimetre per second between two parallel planes of a fluid separated by one centimetre. 1 poise is equal to 10^{-1} N s m^{-2}.

Poiseuille's equation An equation relating the volume flow rate, V, of a fluid through a cylindrical tube to the pressure difference, p, between the ends of the tube: $V = \pi p r^4 / 8 l \eta$, where r is the radius and l the length of the tube; η is the viscosity of the fluid. It applies if the Reynolds number is less than 2000 and was first stated by Jean Louis Poiseuille (1799 –1869).

poison 1. (in biology) Any substance that is injurious to the health of a living organism. **2.** (in chemistry) A substance that prevents the activity of a catalyst. **3.** (in physics) A substance that absorbs neutrons in a nuclear reactor and therefore slows down the reaction. It may be added intentionally for this purpose or may be formed as a fission product and need to be periodically removed.

Poisson's ratio The ratio of the lateral strain to the longitudinal strain in a stretched rod. If the original diameter of the rod is d and the contraction of the diameter under stress is Δd, the lateral strain $\Delta d / d = s_d$; if the original length is l and the extension under stress Δl, the longitudinal strain is $\Delta l / l = s_l$. Poisson's ratio is then s_d / s_l. For steels the value is between 0.28 and 0.30 and for aluminium alloys it is about 0.33. It was first introduced by Simeon Poisson (1781–1840).

polar body *See* oogenesis.

polar compound A compound that is either ionic (e.g. sodium chloride) or that has molecules with a large permanent dipole moment (e.g. water).

polar coordinates A system used in analytical geometry to locate a point P, with reference to two or three

two-dimensional coordinates

cylindrical polar coordinates

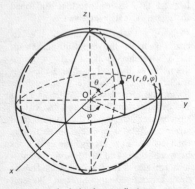

spherical polar coordinates

Polar coordinates

axes. The distance of P from the origin is r, and the angle between the x-axis and the *radius vector OP* is θ, thus in two-dimensional polar coordinates the coordinates of P are (r,θ). If the *Cartesian coordinates of P are (x,y) then, $x = r\cos\theta$ and $y = r\sin\theta$.

In three dimensions the point P may be regarded as lying on the surface of a cylinder, giving *cylindrical polar coordinates*, or on the surface of a sphere, giving *spherical polar coordinates*. In the former the coordinates of P would be (r,θ,z); in the latter they would be (r,θ,ϕ) (see illustration).

polariscope (polarimeter) A device used to study optically active substances (*see* optical activity). The simplest type of instrument consists of a light source, collimator, polarizer, and analyser. The specimen is placed between polarizer and analyser, so that any rotation of the plane of polarization of the light can be assessed by turning the analyser. *See also* photoelasticity.

polarization 1. The process of confining the vibrations of the vector constituting a transverse wave to one direction. In unpolarized radiation the vector oscillates in all directions perpendicular to the direction of propagation. *See* polarization of light. **2.** The formation of products of the chemical reaction in a *voltaic cell in the vicinity of the electrodes resulting in increased resistance to current flow and, frequently, to a reduction in the e.m.f. of the cell. *See also* depolarization. **3.** The partial separation of electric charges in an insulator subjected to an electric field. **4.** The separation of charge in a polar *chemical bond.

polarization of light The process of confining the vibrations of the electric vector of light waves to one direction. In unpolarized light the electric field vibrates in all directions perpendicular to the direction of propagation. After reflection or transmission through certain substances (*see* Polaroid) the electric field is confined to one direction and the radiation is said to be *plane-polarized light*. The plane of plane-polarized light can be rotated when it passes through certain substances (*see* optical activity).

In *circularly polarized light*, the tip of the electric vector describes a circular helix about the direction of propagation with a frequency equal to the frequency of the light. The magnitude of the vector remains constant. In *elliptically polarized light*, the vector also rotates about the direction of propagation but the amplitude changes; a projection of the vector on a plane at right angles to the direction of propagation describes an ellipse. Circularly and elliptically polarized light are produced using a *retardation plate.

polarizing angle *See* Brewster's law.

polar molecule A molecule that has a dipole moment; i.e. one in which there is some separation of charge in the *chemical bonds, so that one part of the molecule has a positive charge and the other a negative charge.

polarography An analytical technique having an electrochemical basis. A dropping-mercury electrode is used as the cathode along with a large nonpolarizable anode, and a dilute solution of the sample. The dropping-mercury electrode consists of a narrow tube through which mercury is slowly passed into the solution so as to form small drops at the end of the tube, which fall away. In this way the cathode can have a low surface area and be kept clean. A variable potential is applied to the cell and a plot of current against potential (a *polarogram*) made. As each chemical species is reduced at the cathode (in order of their electrode potentials) a step-wise increase in current is obtained. The height of each step is proportional to the concentration of the component. The technique is useful for detecting trace amounts of metals and for the investigation of solvated complexes.

Polaroid A doubly refracting material that plane-polarizes unpolarized light passed through it. It consists of a plastic sheet strained in a manner that makes it birefringent by aligning its molecules. Sunglasses incorporating a Polaroid material absorb light that is vibrating horizontally – produced by reflection from horizontal surfaces – and thus reduce glare.

polaron A coupled electron–ion system that arises when an electron is introduced into the conduction band of a perfect ionic crystal and induces lattice polarization around itself.

polar solvent *See* solvent.

pole 1. *See* magnetic poles; magnetic monopole. **2.** The *optical centre of a curved mirror.

pollen The mass of grains containing the male gametes of seed plants, which are produced in large numbers in the *pollen sacs. The pollen grains of insect-pollinated plants may be spiny or pitted and are usually larger than those of wind-pollinated plants, which are usually smooth and light. The pollen grain represents the male *gametophyte generation. The wall of the mature pollen grain consists of the tough outer wall (*exine*) and the more delicate narrower *intine*. The latter gives rise to the *pollen tube. *See also* pollination.

pollen analysis *See* palynology.

pollen sac The structure in seed plants in which pollen is produced. In angiosperms there are usually four pollen sacs in each *anther. In gymnosperms variable numbers of pollen sacs are borne on the microsporophylls that make up the male *cone.

pollen tube An outgrowth of a pollen grain, which transports the male gametes to the ovule. It will only grow if the pollen grain is compatible with the female tissue. In angiosperms, the pollen grain is deposited on the stigma and the pollen tube grows down through the style and into the ovule. In some gymnosperms, e.g. *Pinus* (pines), the pollen tube penetrates the *nucellus but does not develop further until the following year, when the female part of the plant is mature. *See also* embryo sac; fertilization.

pollex The innermost digit on the forelimb of a tetrapod vertebrate. It contains two phalanges (*see* pentadactyl limb) and in man and higher primates it is the thumb, which is opposable (i.e. capable of facing and touching the other digits) and gives the hand greater manipulating ability. In some mammals a pollex is absent. *Compare* hallux.

pollination The transfer of pollen from an anther (the male reproductive organ) to a stigma (the receptive part of the female reproductive organ), either of the same flower (*self-pollination*) or of a different flower of the same species (*cross-pollination*). Cross-pollination involves the action of a pollinating agent to effect transfer of the pollen (*see* anemophily; entomophily; hydrophily). *See also* fertilization; incompatibility.

pollution An undesirable change in the physical, chemical, or biological characteristics of the natural environment, brought about by man's activities. It may be harmful to human or nonhuman life. Pollution may affect the soil, rivers, seas, or the atmosphere. There are two main classes of pollutants: those that are *biodegradable* (e.g. sewage), i.e. can be rendered harmless by natural processes and need therefore cause no permanent harm if adequately dispersed or treated; and those that are *nonbiodegradable* (e.g. *heavy metals (such as lead) in industrial effluents and *DDT and other chlorinated hydrocarbons used as pesticides), which eventually accumulate in the environment and may be concentrated in food chains. Other forms of pollution in the environment include noise (e.g. from jet aircraft, traffic, and industrial processes) and thermal pollution (e.g.

the release of excessive waste heat into lakes or rivers causing harm to wildlife). Recent pollution problems include the disposal of radioactive waste; *acid rain*, formed by reaction of atmospheric moisture with sulphur dioxide from industrial emissions and nitrogen oxides from car exhausts; *photochemical smog; increasing levels of human waste; high levels of carbon dioxide and other greenhouse gases in the atmosphere (*see* greenhouse effect); damage to the *ozone layer by nitrogen oxides, *chlorofluorocarbons (CFCs), and *halons; and pollution of inland waters by agricultural fertilizers and sewage effluent, causing eutrophication (*see* eutrophic). Attempts to contain or prevent pollution include strict regulations concerning factory emissions, the use of smokeless fuels, the banning of certain pesticides, the increasing use of lead-free petrol, restrictions on the use of chlorofluorocarbons, and the introduction, in some countries, of catalytic converters to cut pollutants in car exhausts.

polonium Symbol Po. A rare radioactive metallic element of group VIA of the periodic table; a.n. 84; r.a.m. 210; r.d. 9.32; m.p. 254°C; b.p. 962°C. The element occurs in uranium ores to an extent of about 100 micrograms per 1000 kilograms. It has over 30 isotopes, more than any other element. The longest-lived isotope is polonium−209 (half-life 103 years). Polonium has attracted attention as a possible heat source for spacecraft as the energy released as it decays is 1.4×10^5 J kg^{-1} s^{-1}. It was discovered by Marie Curie in 1898 in a sample of pitchblende.

poly- Prefix indicating a polymer, e.g. polyethene. Sometimes brackets are used in polymer names to indicate the repeated unit, e.g. poly(ethene).

polyamide A type of condensation polymer produced by the interaction of an amino group of one molecule and a carboxylic acid group of another molecule to give a protein-like structure. The polyamide chains are linked together by hydrogen bonding.

polyatomic molecule A molecule formed from several atoms (e.g. pyridine, C_5H_5N, or dinitrogen tetroxide, N_2O_4), as distinguished from diatomic and monatomic molecules.

Polychaeta A class of annelid worms in which each body segment has a pair of flattened fleshy lobes (*parapodia*) bearing numerous bristles (*chaetae). All polychaetes are aquatic and most of them are marine. They include the fanworms (*Sabella*), which construct tubes of sand, etc., in which they live; and the lugworms (*Arenicola*) and ragworms (*Nereis*), which burrow in sand or mud.

polychlorinated biphenyl (PCB) Any of a number of derivatives of biphenyl ($C_6H_5C_6H_5$) in which some of the hydrogen atoms on the benzene rings have been replaced by chlorine atoms. Polychlorinated biphenyls are used in the manufacture of certain polymers, in particular, plastics used as electrical insulators. They are highly toxic and carcinogenic substances and their increasing use has caused concern because they have been shown to accumulate in the food chain.

polychloroethene (PVC; polyvinyl chloride) A tough white solid material, which softens with the application of a plasticizer, manufactured from chloroethene by heating in an inert solvent using benzoyl peroxide as an initiator, or by the free-radical mechanism initiated by heating chloroethene under water with potassium persulphate or hydrogen peroxide. The polymer is used in a variety of ways, being easy to colour and resistant to fire, chemicals, and weather.

polychromatic radiation Electromagnetic radiation that consists of a mix-ture of different wavelengths. This need not refer only to visible radiation. *Compare* monochromatic radiation.

polycyclic Denoting a compound that has two or more rings in its molecules. Polycyclic compounds may contain single rings (as in phenylbenzene, $C_6H_5.C_6H_5$) or fused rings (as in naphthalene, $C_{10}H_8$).

polydioxoboric(III) acid *See* boric acid.

polyembryony 1. The formation of more than one embryo in a plant seed. Often one embryo develops from the fertilized egg cell, while the others have formed asexually from other tissues in the ovule. **2.** The formation of more than one embryo from a single animal zygote. *Identical twins are produced in this way.

polyene An unsaturated hydrocarbon that contains two or more double carbon–carbon bonds in its molecule.

polyester A condensation polymer formed by the interaction of polyhydric alcohols and polybasic acids. Linear polyesters are saturated thermoplastics and linked by dipole–dipole attraction as the carbonyl groups are polarized. They are extensively used as fibres (e.g. *Terylene*). Unsaturated polyesters readily copolymerize to give thermosetting products. They are used in the manufacture of glass-fibre products.

polyethene (polyethylene; polythene) A flexible waxy translucent polyalkene thermoplastic made in a variety of ways producing a polymer of varying characteristics. In the ICI process, ethene containing a trace of oxygen is subjected to a pressure in excess of 1500 atmospheres and a temperature of 200°C. Low-density polyethene (r.d. 0.92) has a formula weight between 50 000 and 300 000, softening at a temperature around 110°C, while the high-density polyethene (r.d. 0.945–0.96) has a formula weight up to 3 000 000, softening around 130°C. The low-density polymer is less crystalline, being more atactic. Polyethene is used as an insulator; it is acid resistant and is easily moulded and blown. *See* Phillips process; Ziegler process.

polyethylene *See* polyethene.

polygene Any of a group of genes influencing a quantitative characteristic, e.g. height in man. *See* multifactorial inheritance.

polygon A plane figure with a number of sides. In a *regular polygon* all the sides and internal angles are equal. For such a polygon with n sides, the interior angle is $(180 - 360/n)$ degrees and the sum of the interior angles is $(180n - 360)$ degrees.

polygon of forces A polygon in which the sides represent, in magnitude and direction, all forces acting on a rigid body. The side required to close the polygon represents the resultant of a system of forces.

polyhedron A solid bounded by polygonal faces. In a *regular polyhedron* all the faces are congruent regular polygons. The cube is one of five possible regular polyhedrons. The others are the *tetrahedron* (four triangular faces), the *octahedron* (eight triangular faces), the *dodecahedron* (twelve pentagonal faces), and the *icosahedron* (twenty triangular faces).

polyhydric alcohol An *alcohol that has several hydroxyl groups per molecule.

polymer A substance having large molecules consisting of repeated units (the monomers). There are a number of natural polymers, such as polysaccharides. Synthetic polymers are extensively used in *plastics. Polymers do not have a definite formula since they consist of chains of different lengths. The various types of polymer are shown in the illustration.

polymerase Any enzyme that catalyses the elongation of a polymeric molecule. *RNA polymerases* catalyse the synthesis of RNA using an existing DNA or RNA strand as a template. *DNA polymerases* catalyse the elongation of a new DNA strand during DNA replication, using an existing DNA strand as template. *RNA-directed DNA polymerase* is more usually known as *reverse transcriptase.

polymerization A chemical reaction in which molecules join together to form a polymer. If the reaction is an addition reaction, the process is *addition polymerization*; condensation reactions cause *condensation polymerization*, in which a small molecule is eliminated during the reaction. Polymers consisting of a single monomer are *homopolymers*; those formed from two different monomers are *copolymers*.

polymethanal A solid polymer of methanal, formed by evaporation of an aqueous solution of methanal.

polymethylmethacrylate A clear thermoplastic acrylic material made by polymerizing methyl methacrylate. The technical name is *poly(methyl 2-methylpropenoate)*. It is used in such materials as *Perspex*.

polymorphism 1. (in biology) The existence of three or more distinctly different forms within a plant or animal species. An example is the *caste system of social insects, in which there are the workers, drones, and queens. This is an *environmental polymorphism*, i.e. the differences are caused by environmental rather than genetic factors, in this case by the larvae receiving different types of food. There are also *heritable* or *genetic polymorphisms*. An example is the occurrence of sickle-cell anaemia, a genetic disease that principally affects Black populations of central Africa and is characterized by an abnormal form of the blood pigment haemoglobin and sickle-shaped red

Addition polymerization of ethene to form polyethene a homopolymer

1,6-diaminoethane   hexanedioic acid

Condensation polymerization to form nylon: a heteropolymer

alternating   A — B — A — B — A   B — A — B — ......

random   A — A — B — A — B — B — B — A — ......

block   A — A — B — B — B — B — A — A —

graft   A — A — A — A — A — A — A — A — A — ......
                    | | |
                    B   B   B
                    | | |
                    B   B   B

Types of copolymer depending on the arrangement of the monomers A and B

isotactic   syndiotactic

Types of stereospecific polymer

Polymers

blood cells. Three different types of individual occur in such populations: those who have two genes (*AA*) for normal haemoglobin and therefore do not suffer from the disease; those with one normal and one abnormal gene (*AS*), who suffer from a relatively mild form; and those with two abnormal genes (*SS*), who suffer a chronic and eventually fatal form of anaemia. Normally such a harmful gene would have been eliminated from the population by the process of natural selection, but it is maintained in this case because people with the mild form of the disease are resistant to a severe form of malaria endemic in central Africa.

See also restriction fragment length polymorphism. **2.** (in chemistry) The existence of chemical substances in two (*dimorphism*) or more physical forms. *See* allotropy.

polynomial A mathematical expression containing three or more terms. It has the general form $a_0x^n + a_1x^{n-1} + a_2x^{n-2} + \ldots + a_n$, where a_0, a_1, etc., are constants and n is the highest power of the variable, called the *degree* of the polynomial.

polyp The sedentary stage in the life cycle of the *Coelenterata, consisting of a cylindrical body fixed at one end to a firm base and having a mouth surrounded by a ring of tentacles at the other. Some polyps (e.g. *Hydra*) are single; others (e.g. the corals and *Obelia*) form colonies. Polyps typically reproduce asexually by budding to form either new polyps or *medusae. The latter reproduce sexually giving rise to new polyps. Sea anemones are solitary polyps that reproduce sexually to form new polyps.

polypeptide A *peptide comprising ten or more amino acids. Polypeptides that constitute proteins usually contain 100–300 amino acids. Shorter ones include certain antibiotics, e.g. gramicidin, and some hormones, e.g. *ACTH, which has 39 amino acids.

The properties of a polypeptide are determined by the type and sequence of its constituent amino acids.

polyphyletic Denoting any group of organisms the members of which have originated from several different ancestors. An example is the group including all insectivorous animals. Polyphyletic groups are not natural groups and do not have any place in natural classifications. *Compare* monophyletic.

polyphyodont Describing a type of dentition in which the teeth are continuously shed and replaced during the lifetime of the animal. Sharks and frogs have a polyphyodont dentition. *Compare* diphyodont; monophyodont.

polyploid Describing a nucleus that contains more than two sets of chromosomes (*see* diploid) or a cell or organism containing such nuclei. For example, *triploid* plants have three sets of chromosomes and *tetraploid* plants have four. Polyploidy is far more common in plants than in animals; many crops, in particular, are polyploid (bread wheat, for example, is hexaploid, i.e. $6n$).

polypropene (**polypropylene**) An isotactic polymer existing in both low and high formula-weight forms. The lower-formula-weight polymer is made by passing propene at moderate pressure over a heated phosphoric acid catalyst spread on an inert material at 200°C. The reaction yields the trimer and tetramer. The higher-formula-weight polymer is produced by passing propene into an inert solvent, heptane, which contains a trialkyl aluminium and a titanium compound. The product is a mixture of isotactic and atactic polypropene, the former being the major constituent. Polypropene is used as a thermoplastic moulding material.

polypropylene *See* polypropene.

polysaccharide Any of a group of carbohydrates comprising long chains of monosaccharide (simple sugar) molecules. *Homopolysaccharides* consist of only one type of monosaccharide; *heteropolysaccharides* contain two or more different types. Polysaccharides may have molecular weights of up to several million and are often highly branched. Some important examples are starch, glycogen, and cellulose.

polyspermy The entry of several sperms into the egg during fertilization although only one sperm nucleus actually fuses with the egg nucleus. Polyspermy occurs in animals with yolky eggs (e.g. birds).

polystyrene A clear glasslike material manufactured by free-radical polymerization of phenylethene (styrene) using benzoyl peroxide as an initiator. It is used as both a thermal and electrical insulator and for packing and decorative purposes.

polysulphides *See* sulphides.

polytetrafluoroethene (PTFE) A thermosetting plastic with a high softening point (327°C) prepared by the polymerization of tetrafluoroethene under pressure (45–50 atmospheres). The reaction requires an initiator, ammonium peroxosulphate. The polymer has a low coefficient of friction and its 'antistick' properties are probably due to its helical structure with the fluorine atoms on the surface of an inner ring of carbon atoms. It is used for coating cooking utensils and nonlubricated bearings.

polythene *See* polyethene.

polythionate A salt of a polythionic acid.

polythionic acids Oxo acids of sulphur with the general formula $HO.SO_2.S_n.SO_2.OH$, where $n = 0–4$. *See also* sulphuric acid.

polyurethane A polymer containing the urethane group $-NH.CO.O-$, prepared by reacting di-isocyanates with appropriate diols or triols. A wide range of polyurethanes can be made, and they are used in adhesives, durable paints and varnishes, plastics, and rubbers. Addition of water to the polyurethane plastics turns them into foams.

polyvinyl chloride *See* polychloroethene.

polyyne An unsaturated hydrocarbon that contains two or more triple carbon–carbon bonds in its molecule.

Polyzoa (Bryozoa) *See* Ectoprocta.

pome A type of fruit characteristic of apples and pears. The flesh of the fruit develops from the *receptacle of the flower, which completely encloses the fused carpels. After fertilization the carpels form the 'core' of the fruit, which contains the seeds. *See also* pseudocarp.

pons (pons Varolii) A thick tract of nerve fibres in the brain that links the medulla oblongata to the midbrain. Its function is to relay impulses between different parts of the brain. The pons is named after its discoverer, the Italian anatomist C. Varoli (1543–75).

population (in ecology) **1.** A group of individuals of the same species within a *community. The nature of a population is determined by such factors as density, *sex ratio, birth and death rates, emigration, and immigration. **2.** The total number of individuals of a given species or other class of organisms in a defined area, e.g. the population of rodents in Britain.

population dynamics The study of the fluctuations that occur in the numbers of individuals in animal and plant populations and the factors controlling these fluctuations. An important distinction is maintained between those factors that are dependent on population density and have a stabilizing effect (e.g. food supply) and those that are independent of population density (e.g. catastrophes, such as flooding).

population genetics The study of the distribution of inherited *variation among a group of organisms of the same species. The potential for change depends on the sum total of alleles that are available to the organisms (the *gene pool*), and estimates of changes in allele frequency in a population give an indication of its response to a changing condition.

population inversion See laser.

population type A method of classifying stars as either population I or population II bodies, devised in 1944 by Wilhelm Baade (1893–1960). Population I stars are the young metal-rich highly luminous stars found in the spiral arms of galaxies. Population II stars are older metal-deficient stars that occur in the centres of galaxies.

Porifera The phylum of marine and freshwater invertebrates that comprise the sponges, which live permanently attached to rocks or other surfaces. The body of a sponge is hollow and consists basically of an aggregation of cells between which there is little nervous coordination. The body is supported by an internal skeleton of spicules of chalk, silica, or fibrous protein (bath sponges have protein skeletons). Flagellated cells (*choanocytes*) cause water to flow in through openings in the body wall and out through openings at the top; food particles are filtered from the water by the choanocytes.

porphyrin Any of a group of related organic compounds characterized by the possession of a cyclic group of four linked nitrogen-containing rings (a *pyrrole* nucleus). Porphyrins differ in the nature of their side-chain groups. They include the *chlorophylls and the *haem groups of haemoglobin, myoglobin, and the cytochromes.

portal vein (portal circulation; portal system) Any vein that collects blood from one network of capillaries and transports it directly to a second capillary network in another region of the body, without returning to the heart. See hepatic portal system.

positive charge See charge.

positive feedback See feedback.

positron The antiparticle of the *electron. See also annihilation; elementary particles; pair production.

postcaval vein See vena cava.

posterior 1. Designating the part of an animal that is to the rear, i.e. that follows when the animal is moving. In man and bipedal animals (e.g. kangaroos) the posterior surface is equivalent to the *dorsal surface. **2.** Designating the side of a flower or axillary bud that faces towards the flower stalk or main stem, respectively. *Compare* anterior.

potash Any of a number of potassium compounds, such as the carbonate or the hydroxide.

potash alum See aluminium potassium sulphate; alums.

potash mica See muscovite.

potassium Symbol K. A soft silvery metallic element belonging to group I of the periodic table (*see* alkali metals); a.n. 19; r.a.m. 39.098; r.d. 0.87; m.p. 63.7°C; b.p. 774°C. The element occurs in seawater and in a number of minerals, such as sylvite (KCl), carnallite ($KCl.MgCl_2.6H_2O$), and kainite ($MgSO_4.KCl.3H_2O$). It is obtained by electrolysis. The metal has few uses but potassium salts are used for a wide range of applications. Potassium is an *essential element for living organisms. The potassium ion, K^+, is the most abundant cation in plant tissues, being absorbed through the roots and being used in such processes as protein synthesis. In animals the passage of potassium and sodium ions across the nerve-cell membrane is responsible for the changes of electrical potential that accompany the transmission of impulses. Chemically, it is highly reactive, resembling sodium in its behaviour and compounds. It also forms an orange-coloured superoxide, KO_2, which contains the O_2^- ion Potassium was discovered by Sir Humphry Davy in 1807.

potassium–argon dating A *dating technique for certain rocks that depends on the decay of the radioisotope potassium-40 to argon-40, a process with a half-life of about 1.27 $\times 10^{10}$ years. It assumes that all the argon-40 formed in the potassium-bearing mineral accumulates within it and that all the argon present is formed by the decay of potassium-40. The mass of argon-40 and potassium-40 in the sample is estimated and the sample is then dated from the equation,

$$^{40}Ar = 0.1102\ ^{40}K(e^{\lambda t} - 1),$$

where λ is the decay constant and t is the time in years since the mineral cooled to about 300°C, when the ^{40}Ar became trapped in the crystal lattice. The method is effective for micas, feldspar, and some other minerals.

potassium bicarbonate See potassium hydrogencarbonate.

potassium bichromate See potassium dichromate.

potassium bromide A white or colourless crystalline solid, KBr, slightly hygroscopic and soluble in water and very slightly soluble in ethanol; cubic; r.d. 2.73; m.p. 734°C; b.p. 1435°C. Potassium bromide may be prepared by the action of bromine on hot potassium hydroxide solution or by the action of iron(III) bromide or hydrogen bromide on potassium carbonate solution. It is used widely in the photographic industry and is also used as a sedative. Because of its range of transparency to infrared radiation, KBr is used both as a matrix for solid samples and as a prism material in infrared spectroscopy.

potassium carbonate (pearl ash; potash) A translucent (granular) or white (powder) deliquescent solid known in the anhydrous and hydrated forms. K_2CO_3 (monoclinic; r.d. 2.4; m.p. 891°C) decomposes without boiling. $2K_2CO_3.3H_2O$ (monoclinic; r.d. 2.04) dehydrates to $K_2CO_3.H_2O$ above 100°C and to K_2CO_3 above 130°C. It is prepared by the Engel–Precht process in which potassium chloride and magnesium oxide react with carbon dioxide to give the compound *Engel's salt*, $MgCO_3.KHCO_3.4H_2O$. This is decomposed in solution to give the hydrogencarbonate, which can then be calcined to K_2CO_3. Potassium carbonate is soluble in water (insoluble in alcohol) with significant hydrolysis to produce basic solutions. Industrial uses include glasses and glazes, the manufacture of soft soaps, and in dyeing and wool finishing. It is used in the laboratory as a drying agent.

potassium chlorate A colourless crystalline compound, $KClO_3$, which is soluble in water and moderately soluble in ethanol; monoclinic; r.d. 2.33; m.p. 360°C; decomposes above 400°C giving off oxygen. The industrial route to potassium chlorate involves the fractional crystallization of a solution of potassium chloride and sodium chlorate but it may also be prepared by electrolysis of hot concentrated solutions of potassium chloride. It is a powerful oxidizing agent finding applications in weedkillers and disinfectants and, because of its ability to produce oxygen, it is used in explosives, pyrotechnics, and matches.

potassium chloride A white crystalline solid, KCl, which is soluble in water and very slightly soluble in ethanol; cubic; r.d. 1.98; m.p. 772°C; sublimes at 1500°C. Potassium chloride occurs naturally as the mineral *sylvite* (KCl) and as *carnallite* (KCl.MgCl$_2$.6H$_2$O); it is produced industrially by fractional crystallization of these deposits or of solutions from lake brines. It has the interesting property of being more soluble than sodium chloride in hot water but less soluble in cold. It is used as a fertilizer, in photography, and as a source of other potassium salts, such as the chlorate and the hydroxide. It has low toxicity.

potassium chromate A bright yellow crystalline solid, K$_2$CrO$_4$, soluble in water and insoluble in alcohol; rhombic; r.d. 2.73; m.p. 971°C; decomposes without boiling. It is produced industrially by roasting powdered chromite ore with potash and limestone and leaching the resulting cinder with hot potassium sulphate solution. Potassium chromate is used in leather finishing, as a textile mordant, and in enamels and pigments. In the laboratory it is used as an analytical reagent and as an indicator. Like other chromium(III) compounds it is toxic when ingested or inhaled.

potassium chromium sulphate (chrome alum) A violet or ruby-red crystalline solid, K$_2$SO$_4$.Cr$_2$(SO$_4$)$_3$.24H$_2$O, that is soluble in water and insoluble in ethanol; cubic or octahedral; r.d. 1.826; m.p. 89°C; loses 10H$_2$O at 100°C, 12H$_2$O at 400°C. Six water molecules surround each of the chromium(III) ions and the remaining ones are hydrogen bonded to the sulphate ions. Like all alums, the compound may be prepared by mixing equimolar quantities of the constituent sulphates. *See* alums.

potassium cyanide (cyanide) A white crystalline or granular deliquescent solid, KCN, soluble in water and in ethanol and having a faint characteristic odour of almonds (due to hydrolysis forming hydrogen cyanide at the surface); cubic; r.d. 1.52; m.p. 634°C. It is prepared industrially by the absorption of hydrogen cyanide in

potassium hydroxide. The compound is used in the extraction of silver and gold, in some metal-finishing processes and electroplating, as an insecticide and fumigant (source of HCN), and in the preparation of cyanogen derivatives. In the laboratory it is used in analysis, as a reducing agent, and as a stabilizing *ligand for low oxidation states. The salt itself is highly toxic and aqueous solutions of potassium cyanide are strongly hydrolysed to give rise to the slow release of equally toxic hydrogen cyanide gas.

potassium dichromate (potassium bichromate) An orange-red crystalline solid, K$_2$Cr$_2$O$_7$, soluble in water and insoluble in alcohol; monoclinic or triclinic; r.d. 2.67; monoclinic changes to triclinic at 241.6°C; m.p. 396°C; decomposes above 500°C. It is prepared by acidification of crude potassium chromate solution (the addition of a base to solutions of potassium dichromate reverses this process). The compound is used industrially as an oxidizing agent in the chemical industry and in dyestuffs manufacture, in electroplating, pyrotechnics, glass manufacture, glues, tanning, photography and lithography, and in ceramic products. Laboratory uses include application as an analytical reagent and as an oxidizng agent. Potassium dichromate is toxic and considered a fire risk on account of its oxidizing properties.

potassium dioxide *See* potassium superoxide.

potassium hydride A white or greyish white crystalline solid, KH; r.d. 1.43–1.47. It is prepared by passing hydrogen over heated potassium and marketed as a light grey powder dispersed in oil. The solid decomposes on heating and in contact with moisture and is an excellent reducing agent. Potassium hydride is a fire hazard because it produces hydrogen on reaction with water.

potassium hydrogencarbonate (potassium bicarbonate) A white crystalline solid, $KHCO_3$, soluble in water and insoluble in ethanol; r.d. 2.17; decomposes about 120°C. It occurs naturally as *calcinite* and is prepared by passing carbon dioxide into saturated potassium carbonate solution. It is used in baking, soft-drinks manufacture, and in CO_2 fire extinguishers. Because of its buffering capacity, it is added to some detergents and also used as a laboratory reagent.

potassium hydrogentartrate (cream of tartar) A white crystalline acid salt, $HOOC(CHOH)_2COOK$. It is obtained from deposits on wine vats (argol) and used in baking powders.

potassium hydroxide (caustic potash; lye) A white deliquescent solid, KOH, often sold as pellets, flakes, or sticks, soluble in water and in ethanol and very slightly soluble in ether; rhombic; r.d. 2.02; m.p. 360.4°C; b.p. 1320°C. It is prepared industrially by the electrolysis of concentrated potassium chloride solution but it can also be made by heating potassium carbonate or sulphate with slaked lime, $Ca(OH)_2$. It closely resembles sodium hydroxide but is more soluble and is therefore preferred as an absorber for carbon dioxide and sulphur dioxide. It is also used in the manufacture of soft soap, other potassium salts, and in Ni–Fe and alkaline storage cells. Potassium hydroxide is extremely corrosive to body tissues and especially damaging to the eyes.

potassium iodate A white crystalline solid, KIO_3, soluble in water and insoluble in ethanol; monoclinic; r.d. 3.9; m.p. 561°C. It may be prepared by the reaction of iodine with hot concentrated potassium hydroxide or by careful electrolysis of potassium iodide solution. It is an oxidizing agent and is used as an analytical reagent. Some potassium iodate is used as a food additive.

potassium iodide A white crystalline solid, KI, with a strong bitter taste, soluble in water, ethanol, and acetone; cubic; r.d. 3.12; m.p. 686°C; b.p. 1330°C. It may be prepared by the reaction of iodine with hot potassium hydroxide solution followed by separation from the iodate (which is also formed) by fractional crystallization. In solution it has the interesting property of dissolving iodine to form the triiodide ion I_3^-, which is brown. Potassium iodide is widely used as an analytical reagent, in photography, and also as an additive to table salt to prevent goitre and other disorders due to iodine deficiency.

potassium manganate(VII) (potassium permanganate) A compound, $KMnO_4$, forming purple crystals with a metallic sheen, soluble in water (intense purple solution), acetone, and methanol, but decomposed by ethanol; r.d. 2.70; decomposition begins slightly above 100°C and is complete at 240°C. The compound is prepared by fusing manganese(IV) oxide with potassium hydroxide to form the manganate and electrolysing the manganate solution using iron electrodes at about 60°C. An alternative route employs production of sodium manganate by a similar fusion process, oxidation with chlorine and sulphuric acid, then treatment with potassium chloride to crystallize the required product.

Potassium manganate(VII) is widely used as an oxidizing agent and as a disinfectant in a variety of applications, and as an analytical reagent.

potassium monoxide A grey crystalline solid, K_2O; cubic; r.d. 2.32; decomposition occurs at 350°C. It may be prepared by the oxidation of potassium metal with potassium nitrate. It reacts with ethanol to form potassium ethoxide (KOC_2H_5), and with liquid ammonia to form potassium hydroxide and potassamide (KNH_2).

potassium nitrate (saltpetre) A colourless rhombohedral or trigonal solid, KNO_3, soluble in water, insoluble in alcohol; r.d. 2.109; transition to trigonal form at 129°C; m.p. 334°C; decomposes at 400°C. It occurs naturally as *nitre* and may be prepared by the reaction of sodium nitrate with potassium chloride followed by fractional crystallization. It is a powerful oxidizing agent (releases oxygen on heating) and is used in gunpowder and fertilizers.

potassium nitrite A white or slightly yellow deliquescent solid, KNO_2, soluble in water and insoluble in ethanol; r.d. 1.91; m.p. 300°C; may explode at 600°C. Potassium nitrite reacts with cold dilute mineral acids to give nitrous acid and is also able to behave as a reducing agent (if oxidized to the nitrate) or as an oxidizing agent (if reduced to nitrogen). It is used in organic synthesis because of its part in diazotization, and in detecting the presence of the amino groups in organic compounds.

potassium permanganate *See* potassium manganate(VII).

potassium sulphate A white crystalline powder, K_2SO_4, soluble in water and insoluble in ethanol; rhombic or hexagonal; r.d. 2.66; m.p. 1072°C. It occurs naturally as *schönite* (Strassfurt deposits) and in lake brines, from which it is separated by fractional crystallization. It has also been produced by the Hargreaves process, which involves the oxidation of potassium chloride with sulphuric acid. In the laboratory it may be obtained by the reaction of either potassium hydroxide or potassium carbonate with sulphuric acid. Potassium sulphate is used in cements, in glass manufacture, as a food additive, and as a fertilizer (source of K^+) for chloride-sensitive plants, such as tobacco and citrus.

potassium sulphide A yellow-red or brown-red deliquescent solid, K_2S, which is soluble in water and in ethanol but insoluble in diethyl ether; cubic; r.d. 1.80; m.p. 840°C. It is made industrially by reducing potassium sulphate with carbon at high temperatures in the absence of air. In the laboratory it may be prepared by the reaction of hydrogen sulphide with potassium hydroxide. The pentahydrate is obtained on crystallization. Solutions are strongly alkaline due to hydrolysis. It is used as an analytical reagent and as a depilatory. Potassium sulphide is generally regarded as a hazardous chemical with a fire risk; dusts of K_2S have been known to explode.

potassium sulphite A white crystalline solid, K_2SO_3, soluble in water and very sparingly soluble in ethanol; r.d. 1.51; decomposes on heating. It is a reducing agent and is used as such in photography and in the food and brewing industries, where it prevents oxidation.

potassium superoxide (potassium dioxide) A yellow paramagnetic solid, KO_2, produced by burning potassium in an excess of oxygen; it is very soluble (by reaction) in water, soluble in ethanol, and slightly soluble in diethyl ether; m.p. 490°C. When treated with cold water or dilute mineral acids, hydrogen peroxide is obtained. The compound is a powerful oxidizing agent and on strong heating releases oxygen with the formation of the monoxide, K_2O.

potential barrier A region containing a maximum of potential that prevents a particle on one side of it from passing to the other side. According to classical theory a particle must possess energy in excess of the height of the potential barrier to pass it. However, in quantum theory there is a finite probability that a particle with less energy will pass through the barrier (*see* tunnel effect). A potential barrier surrounds the atomic nucleus and is important in nuclear physics; a similar but much lower barrier exists at the interface between *semiconductors and metals and between differently doped semiconductors. These barriers are important in the design of electronic devices.

potential difference *See* electric potential.

potential divider *See* voltage divider.

potential energy *See* energy.

Potentiometer

potentiometer 1. *See* voltage divider. **2.** An instrument for measuring, comparing, or dividing small potential differences. A typical example of its use is the measurement of the e.m.f. (E_1) of a cell by comparing it with the e.m.f. (E_2) of a standard cell. In this case a circuit is set up as illustrated, in which AB is a wire of uniform resistance and S is a sliding contact onto this wire. An accumulator X maintains a steady current through the wave. To measure the e.m.f. of a cell C, it is connected up as shown in the diagram and the sliding contact moved until the e.m.f. of C exactly balances the p.d. from the accumulator, as indicated by a zero reading on the galvanometer G. If the length AS is then l_1, the value of E_1 is given by $E_1/E_2 = l_1/l_2$, where l_2 is the length AS when the standard cell is used as the cell C.

potentiometric titration A titration in which the end point is found by measuring the potential on an electrode immersed in the reaction mixture.

potometer An apparatus used to measure the rate of water loss from a shoot (*see* transpiration) under natural or artificial conditions.

pound The unit of mass in the *f.p.s. system of units defined as 0.453 592 37 kilogram. Before 1963 it was defined in terms of a platinum cylinder called the Imperial Standard Pound.

poundal The unit of force in the *f.p.s. system of units equal to the force required to impart to a mass of one pound an acceleration of one foot per second per second.

powder metallurgy A process in which powdered metals or alloys are pressed into a variety of shapes at high temperatures. The process started with the pressing of powdered tungsten into incandescent lamp filaments in the first decade of this century and is now widely used for making self-lubricating bearings and cemented tungsten carbide cutting tools.

The powders are produced by atomization of molten metals, chemical decomposition of a compound of the metal, or crushing and grinding of the metal or alloy. The parts are pressed into moulds at pressures ranging from 140×10^6 Pa to 830×10^6 Pa after which they are heated in a controlled atmosphere to bond the particles together (*see* sintering).

power 1. (in physics) Symbol P. The rate at which work is done or energy is transferred. In SI units it is measured in watts (joules per second). *See also* horsepower. **2.** (in mathematics) The number of times a quantity is multiplied; e.g. x^5 is the fifth power of x. A *power series* is one in which the power of the variable increases with each term, e.g. $a_0 + a_1x + a_2x^2 + a_3x^3 + \ldots + a_nx^n$.

power factor *See* electric power.

power reactor A *nuclear reactor designed to produce electrical power.

poxvirus One of a group of DNA-containing viruses, often enclosed in an outer membrane, that typically produce skin lesions in vertebrates. They include those viruses causing smallpox (variola), cowpox (vaccinia), and myxomatosis. Some poxviruses produce tumours.

praseodymium Symbol Pr. A soft silvery metallic element belonging to the *lanthanoids; a.n. 59; r.a.m. 140.91; r.d. 6.769 (20°C); m.p. 934°C; b.p. 3512°C. It occurs in bastnasite and monazite, from which it is recovered by an ion-exchange process. The only naturally occurring isotope is praseodymium–141, which is not radioactive; however, fourteen radioisotopes have been produced. It is used in mischmetal, a rare-earth alloy containing 5% praseodymium, for use in lighter flints. Another rare-earth mixture containing 30% praseodymium is used as a catalyst in cracking crude oil. The element was discovered by C. A. von Welsbach in 1885.

preamplifier An *amplifier in a radio, record player, etc., providing a first stage of amplification. It is usually located close to the signal source (i.e. the aerial or pick-up) and the signal is then transmitted by cable to the main amplifier. Preamplification at this early stage improves the signal-to-noise ratio of the whole system.

Precambrian The division of geological time from the formation of the earth, believed to be about 4600 million years ago, to the beginning of the Cambrian period, some 570 million years ago. The Precambrian thus represents most of geological time. Fossils are rare, although *stromatolites occur, and subsequent metamorphism of Precambrian rocks makes correlation of rocks and events extremely difficult. The Precambrian is divided by some authorities into the Archaean and Proterozoic eons. The largest areas of exposed Precambrian rocks are the shield areas, such as the Canadian (Laurentian) Shield and the Baltic Shield.

precaval vein See vena cava.

precessional motion A form of motion that occurs when a torque is applied to a rotating body in such a way that it tends to change the direction of its axis of rotation. It arises because the resultant of the angular velocity of rotation and the increment of angular velocity produced by the torque is an angular velocity about a new direction; this commonly changes the axis of the applied torque and leads to sustained rotation of the original axis of rotation.

A spinning top, the axis of which is not exactly vertical, has a torque acting on it as a result of gravity. Instead of falling over, the top precesses about a vertical line through the pivot. The earth also experiences a torque and undergoes a slow precession, primarily as a result of the gravitational attraction of the sun and the moon on its equatorial bulge (*see* precession of the equinoxes).

precession of the equinoxes The slow westward motion of the *equinoxes about the ecliptic as a result of the earth's *precessional motion. The equinoxes move round the ecliptic with a period of 25 800 years.

precipitate A suspension of small solid particles produced in a liquid by chemical reaction.

precipitation 1. All liquid and solid forms of water that are deposited from the atmosphere; it includes rain, drizzle, snow, hail, dew, and hoar frost. **2.** The formation of a precipitate.

precipitin Any *antibody that combines with its specific soluble *antigen to form a precipitate. The term is sometimes applied to the precipitate itself. *See also* agglutination.

precursor A compound that leads to another compound in a series of chemical reactions.

premolar A broad ridged tooth in mammals that is situated behind the *canine teeth (when present) and in front of the *molars. Premolars are adapted for grinding and chewing food and are present in both the deciduous and permanent dentitions.

preons Hypothetical entities that are postulated as being 'building blocks' of quarks and leptons. There is no experimental evidence for preons but the idea has considerable theoretical appeal. It is expected that evidence for preons would only be obtained at much higher energies than are available from present accelerators.

presbyopia A loss of accommodation that normally develops in human eyes over the age of 45–50 years. Vision of distant objects remains unchanged but accommodation of the eye to near objects is reduced as a result of loss of elasticity in the lens of the eye. The defect is corrected by reading glasses using weak converging lenses.

pressure The force acting normally on unit area of a surface or the ratio of force to area. It is measured in *pascals in SI units. *Absolute pressure* is pressure measured on a gauge that reads zero at zero pressure rather than at atmospheric pressure. *Gauge pressure* is measured on a gauge that reads zero at atmospheric pressure.

pressure gauge Any device used to measure pressure. Three basic types are in use: the liquid-column gauge (e.g. the mercury *barometer and the *manometer), the expanding-element gauge (e.g. the *Bourdon gauge and the aneroid *barometer), and the electrical transducer. In the last category the *strain gauge is an example. Capacitor pressure gauges also come into this category. In these devices, the pressure to be measured displaces

one plate of a capacitor and thus alters its capacitance.

pressurized-water reactor *See* nuclear reactor.

presumptive Describing embryonic tissue that is not yet *determined but which will eventually develop into a certain kind of tissue by virtue of its position in the embryo.

Prévost's theory of exchanges A body emits and absorbs radiant energy at equal rates when it is in equilibrium with its surroundings. Its temperature then remains constant. If the body is not at the same temperature as its surroundings there is a net flow of energy between the surroundings and the body because of unequal emission and absorption. The theory was proposed by Pierre Prévost (1751 –1839) in 1791.

prickle A hard sharp protective outgrowth, many of which may cover the surface of a plant. It contains cortical and vascular tissue and is not regarded as an epidermal outgrowth. *Compare* spine; thorn.

primary alcohol *See* alcohols.

primary amine *See* amines.

primary cell A *voltaic cell in which the chemical reaction producing the e.m.f. is not satisfactorily reversible and the cell cannot therefore be recharged by the application of a charging current. *See* Daniell cell; Leclanché cell; Weston cell; mercury cell. *Compare* secondary cell.

primary colour Any one of a set of three coloured lights that can be mixed together to give the sensation of white light as well as approximating all the other colours of the spectrum. An infinite number of such sets exists, the condition being that none of the individual colours of a set should be able to be matched by mixing the other two; however, unless the colours are both intense and very different the range that they can match well will be limited. The set of primary colours most frequently used is red, green, and blue. *See also* colour.

primary growth The increase in size of shoots and roots of plants that results from the activity of the *apical (tip) meristems and subsequent expansion of the cells produced. The tissues thus produced are called *primary tissues* and the resultant plant parts constitute the *primary plant body*. *Compare* secondary growth.

primary winding The winding on the input side of a *transformer or *induction coil. *Compare* secondary winding.

Primates An order of mammals that includes the monkeys, apes, and man. Primates evolved from arboreal insectivores 130 million years ago. They are characterized by thumbs and big toes that are opposable (i.e. capable of facing and touching the other digits), which permits manual dexterity, and forward-facing eyes allowing *binocular vision. The brain, particularly the cerebrum, is relatively large and well-developed, accounting for the intelligence and quick reactions of these mammals. The young are usually produced singly and undergo a long period of growth and development to the adult form.

primordium A group of cells that represents the initial stages in development of a plant organ. Root and shoot primordia are present in a young plant embryo while leaf primordia (or *leaf buttresses*) are seen as small bulges just below the shoot apex.

principal axis *See* optical axis.

principal focus A point through which rays close to and parallel to the axis of a lens or spherical mirror pass, or appear to pass, after refraction or reflection. A mirror has one principal focus, a lens has a principal focus on both sides.

principal plane The plane that is perpendicular to the optical axis of a

lens and that passes through the optical centre. A thick lens has two principal planes, each passing through a *principal point.

principal point Either of two points on the principal axis of a thick lens from which simply related distances can be measured, as from the optical centre of a thin lens.

principle of superposition The resultant displacement at any point in a region through which two waves of the same type pass is the algebraic sum of the displacements that the two would separately produce at that point. Both waves leave the region of superposition unaltered.

printed circuit An electronic circuit consisting of a conducting material deposited (printed) onto the surface of an insulating sheet. These devices are now common in all types of electronic equipment, facilitating batch production and eliminating the unreliability of the hand-soldered joint.

prion *See* slow virus.

prism 1. (in mathematics) A polyhedron with two parallel congruent polygons as bases and parallelograms for all other faces. A *triangular prism* has triangular bases. 2. (in optics) A block of glass or other transparent material, usually having triangular bases. Prisms have several uses in optical systems: they can be used to deviate a ray, to disperse white light into the visible spectrum, or to erect an inverted image (*see* binoculars). Prisms of other materials are used for different kinds of radiation. *See also* Nicol prism; Wollaston prism.

probability The likelihood of a particular event occurring. If there are n equally likely outcomes of some experiment, and a ways in which event E could occur, then the probability of event E is a/n. For instance, if a die is thrown there are 6 possible outcomes and 3 ways in which an odd number may occur.

The probability of throwing an odd number is 3/6 = 1/2.

Proboscidea The order of mammals that comprises the elephants. They are herbivorous, with a muscular trunk (*proboscis) used for drinking, bathing, and collecting food. The tusks are continuously growing upper incisors and the enormous ridged molar teeth are produced in sequence to replace worn teeth throughout life. The order, which evolved in the Eocene epoch, was formerly much larger and more widespread than it is today and included the extinct mammoths. There are only two species of modern elephants: the African and Indian species.

proboscis 1. The trunk of an elephant: a muscular and very flexible elongation of the nose, which has a finger-like extremity and is capable of picking up and moving objects, taking in water, collecting food, etc. **2.** The elongated mouthparts of certain invertebrates, such as the two-winged flies (Diptera).

procambium A plant tissue formed by the *apical meristems of shoots and roots. It consists of cells elongated parallel to the long axis of the plant. The procambium subsequently gives rise to the primary *vascular tissue.

procaryote See prokaryote.

Proconsul A genus of extinct apes known from fossil remains, 50 million years old, found mostly in Kenya and assigned to the species *P. africanus*. A find made in 1948 was the first discovery of a skull of Miocene ape. It is believed that *Proconsul* was similar and possibly closely related to the genus *Dryopithecus*.

producer An organism considered as a source of energy for those above it in a *food chain (i.e. at the next *trophic level). Green plants, which convert energy from sunlight into chemical energy, are *primary* producers; herbivores are *secondary producers*, as they utilize energy from green plants and supply energy for carnivores. *Compare* consumer.

producer gas (air gas) A mixture of carbon monoxide and nitrogen made by passing air over very hot carbon. Usually some steam is added to the air and the mixture contains hydrogen. The gas is used as a fuel in some industrial processes.

product See chemical reaction.

productivity (in ecology) The rate at which an organism, population, or community assimilates energy (*gross productivity*) or makes energy potentially available (as body tissue) to an animal that feeds on it (*net productivity*). The difference between these two rates is due to the rate at which energy is lost through excretion and respiration. Thus *gross primary productivity* is the rate at which plants assimilate light energy, and *net primary productivity* is the rate at which energy is incorporated as plant tissue. In terrestrial plants, much of the net productivity is not actually available to *consumers, e.g. tree roots are not eaten by herbivores. *See* producer.

profundal Occurring in or designating the deep-water zone of an inland lake. Light intensity, oxygen concentration, and (during summer and autumn) temperature are markedly lower than in the surface layer. *Compare* littoral; sublittoral.

progesterone A hormone, produced primarily by the *corpus luteum of the ovary but also by the placenta, that prepares the inner lining of the uterus for implantation of a fertilized egg cell. If implantation fails, the corpus luteum degenerates and progesterone production ceases accordingly. If implantation occurs, the corpus luteum continues to secrete progesterone, under the influence of *luteinizing hormone and *prolactin, for several months of pregnancy, by which time the placenta has taken over this function. During pregnancy, progesterone maintains the constitution of the uterus and prevents further release of eggs from the ovary. Small amounts of progesterone are produced by the testes. *See also* progestogen.

progestogen One of a group of naturally occurring or synthetic hormones that maintain the normal course of pregnancy. The best known is *progesterone. In high doses progestogens inhibit secretion of *luteinizing hormone, thereby preventing ovulation, and alter the consistency of mucus in the vagina so that conception tends not to occur. They are therefore used as major constituents of *oral contraceptives.

program *See* computer.

progressive wave *See* wave.

projectile Any body that is thrown or projected. If the projectile is discharged on the surface of the earth at an angle θ to the horizontal it will describe a parabolic flight path (if $\theta < 90°$ and the initial velocity < the *escape velocity). Neglecting air resistance, the maximum height of this flight path will be $(v^2\sin^2\theta)/2g$, where v is the velocity of discharge and g is the acceleration of free fall. The horizontal distance covered will be $(v^2\sin2\theta)/g$ and the time of the flight will be $(2v\sin\theta)/g$.

projector An optical device for throwing a large image of a two-dimensional object onto a screen. In an *episcope*, light is reflected from the surface of an opaque two-dimensional object (such as a diagram or photographic print) and an enlarged image is thrown onto a distant screen by means of a system of mirrors and lenses. The *diascope* passes light through the two-dimensional object (such as a photographic transparency, slide, or film) and uses a converging projection lens to form an enlarged image on a distant screen. An *epidiascope* is a device that can be used as

both episcope and diascope. An *overhead projector* is a form of diascope that throws its image on a wall or screen behind and above the operator. In a *motion-picture projector* (or *ciné projector*) the film, consisting of a long sequence of transparent pictures, is driven by a motor past the light source in such a way that each picture comes to rest for a brief period in front of the light source. The illusion of motion is created as each image on the screen is replaced by the next; during the picture change the light is interrupted.

prokaryote (procaryote) An organism in which the genetic material is not enclosed in a cell nucleus. Bacteria are prokaryotes. It is believed that eukaryotic cells (*see* eukaryote) probably evolved as symbiotic associations of prokaryotes. *See* cell.

prolactin (lactogenic hormone; luteotrophic hormone; luteotrophin) A hormone produced by the anterior pituitary gland. In mammals it stimulates the mammary glands to secrete milk (*see* lactation) and the corpus luteum of the ovary to secrete the hormone *progesterone. Secretion of prolactin is increased by suckling. In birds prolactin stimulates secretion of crop milk by the crop glands.

proline *See* amino acid.

promethium Symbol Pm. A soft silvery metallic element belonging to the *lanthanoids; a.n. 61; r.a.m. 145; r.d. 7.26 (20°C); m.p. 1042°C; b.p. 3000°C. The only naturally occurring isotope, promethium–147, has a half-life of only 2.52 years. Eighteen other radioisotopes have been produced, but they have very short half-lives. The only known source of the element is nuclear-waste material. Promethium–147 is of interest as a beta-decay power source but the promethium–146 and –148, which emit penetrating gamma radiation, must first be removed. It was discovered by

J. A. Marinsky, L. E. Glendenin, and C. D. Coryell in 1947.

promoter 1. (in chemistry) A substance added to a catalyst to increase its activity. **2.** (in genetics) *See* operon.

prompt neutrons The neutrons emitted during a nuclear fission process within less than a microsecond of fission. *Compare* delayed neutrons.

pronation Rotation of the lower forearm so that the hand faces backwards or downwards with the radius and ulna crossed. *Compare* supination.

proof A measure of the amount of alcohol (ethanol) in drinks. *Proof spirit* contains 49.28% ethanol by weight (about 57% by volume). Degrees of proof express the percentage of proof spirit present, so 70° proof spirit contains 0.7 × 57% alcohol.

propanal (propionaldehyde) A colourless liquid *aldehyde, C_2H_5CHO; m.p. $-185.2°C$; b.p. $48.8°C$.

propane A colourless gaseous hydrocarbon, C_3H_8; m.p. $-190°C$; b.p. $-42°C$. It is the third member of the *alkane series and is obtained from petroleum. Its main use is as bottled gas for fuel.

propanedioic acid (malonic acid) A white crystalline dicarboxylic acid, $HOOCCH_2COOH$; m.p. $132°C$. It decomposes above its melting point to ethanoic acid. Propanedioic acid is used in the synthesis of other dicarboxylic acids.

propanoic acid (propionic acid) A colourless liquid *carboxylic acid, CH_3CH_2COOH; r.d. 0.99; m.p. $-20.8°C$; b.p. $141°C$. It is used to make calcium propanate – an additive in bread.

propanol Either of two *alcohols with the formula C_3H_7OH. Propan-1-ol is $CH_3CH_2CH_2OH$ and propan-2-ol is $CH_3CH(OH)CH_3$. Both are colourless volatile liquids. Propan-2-ol is used in making propanone (acetone).

propanone (acetone) A colourless flammable volatile compound, CH_3COCH_3; r.d. 0.79; m.p. $-95.4°C$; b.p. $56.2°C$. The simplest *ketone, propanone is miscible with water. It is made by oxidation of propan-2-ol (*see* propanol) or is obtained as a by-product in the manufacture of phenol from cumene; it is used as a solvent and as a raw material for making plastics.

propenal (acrolein) A colourless pungent liquid unsaturated aldehyde, CH_2:$CHCHO$; r.d. 0.84; m.p. $-87°C$; b.p. $53°C$. It is made from propene and can be polymerized to give acrylate resins.

propene (propylene) A colourless gaseous hydrocarbon, CH_3CH:CH_2; m.p. $-81°C$; b.p. $48.8°C$. It is an *alkene obtained from petroleum by cracking alkanes. Its main use is in the manufacture of polypropene.

propenoate (acrylate) A salt or ester of *propenoic acid.

propenoic acid (acrylic acid) An unsaturated liquid *carboxylic acid, CH_2:$CHCOOH$; m.p. $7°C$; b.p. $141°C$. It readily polymerizes and it is used in the manufacture of *acrylic resins.

propenonitrile (acrylonitrile; vinyl cyanide) A colourless liquid, H_2C:$CHCN$; r.d. 0.81; m.p. $-83.5°C$. It is an unsaturated nitrile, made from propene and used to make acrylic resins.

proper motion The apparent angular motion of a star on the *celestial sphere. This is motion in a direction that is perpendicular to the line of sight.

prophase The first stage of cell division, during which chromosomes contract and divide along their length (except for the centromeres) into chromatids. In *mitosis, the chromosomes remain separate from each other. In the first division of *meiosis, homologous chromosomes become paired (*see* pairing). By the end of first prophase the two chromosomes begin to move apart.

propionaldehyde *See* propanal.

proplastid *See* plastid.

proportional counter A type of detector for *ionizing radiation in which the size of the output pulse is proportional to the number of ions formed in the initial ionizing event. It operates in a voltage region, called the *proportional region*, intermediate between that of an *ionization chamber and a *Geiger counter, avalanche ionization being limited to the immediate vicinity of the primary ionization rather than the entire length of the central wire electrode.

proportional limit *See* elasticity.

proprioceptor Any *receptor that is sensitive to movement, pressure, or stretching within the body. Proprioceptors occurring in muscles, tendons, and ligaments are important for the coordination of muscular activity and the maintenance of balance and posture.

prop root Any of the modified roots that arise from the stem of certain plants and provide extra support. Such stems are usually tall and slender and the prop roots develop at successively higher levels as the stem elongates, as in the maize plant. *Buttress roots*, which develop at the base of the trunks of many tropical trees, are similar but tend to have a more flattened appearance. *Stilt roots* are stouter than prop roots. Those formed at the base of the mangrove tree provide firm anchorage in the soft mud of the swamps.

propylene *See* propene.

propyl group The organic group $CH_3CH_2CH_2-$.

prosencephalon *See* forebrain.

prostaglandin Any of a group of organic compounds derived from *essential fatty acids and causing a range of physiological effects in animals. Prostaglandins have been detected in most body tissues. They act at very low concentrations to cause the contraction of smooth muscle; natural and synthetic prostaglandins are used to induce abortion or labour in humans and domestic animals. Two prostaglandin derivatives have antagonistic effects on blood circulation: *thromboxane A_2* causes blood clotting while *prostacyclin* causes blood vessels to dilate. Prostaglandins are also involved in inflammation, being released from affected tissues. *See also* aspirin.

prostate gland A gland in male mammals that surrounds and opens into the urethra where it leaves the bladder. During ejaculation it secretes a fluid into the semen that activates the sperms and prevents them from sticking together.

prosthetic group A tightly bound nonpeptide inorganic or organic component of a protein. Prosthetic groups may be lipids, carbohydrates, metal ions, phosphate groups, etc. Some *coenzymes are more correctly regarded as prosthetic groups.

protactinium Symbol Pa. A radioactive metallic element belonging to the *actinoids; a.n. 91; r.a.m. 231.036; r.d. 15.37 (calculated); m.p. $<1600°C$ (estimated). The most stable isotope, protactinium–231, has a half-life of 3.43×10^4 years; at least ten other radioisotopes are known. Protactinium–231 occurs in all uranium ores as it is derived from uranium–235. Protactinium has no practical applications; it was discovered by Lise Meitner and Otto Hahn in 1917.

protamine Any of a group of proteins of relatively low molecular weight found in association with the chromosomal *DNA of vertebrate male germ cells. They contain a single polypeptide chain comprising about 67% arginine. Protamines are thought to protect and support the chromosomes.

protandry 1. The condition in which the male reproductive organs (stamens) of a flower mature before the female ones (carpels), thereby ensuring that self-fertilization does not occur. Examples of protandrous flowers are ivy and rosebay willowherb. *Compare* protogyny; homogamy. *See also* dichogamy. **2.** The condition in some hermaphrodite or colonial invertebrates in which the male gonads or individuals are sexually mature before the female ones. *Compare* protogyny.

protease (peptidase; proteinase; proteolytic enzyme) Any enzyme that catalyses the splitting of proteins into smaller *peptide fractions and amino acids, a process known as *proteolysis*. Examples are *pepsin and *trypsin. Several proteases, acting sequentially, are normally required for the complete digestion of a protein to its constituent amino acids.

protein Any of a large group of organic compounds found in all living organisms. Proteins comprise carbon, hydrogen, oxygen, and nitrogen and most also contain sulphur; molecular weights range from 6000 to several million. Protein molecules consist of one or several long chains (*polypeptides) of *amino acids linked in a characteristic sequence. This sequence is called the *primary structure* of the protein. These polypeptides may undergo coiling or pleating, the nature and extent of which is described as the *secondary structure*. The three-dimensional shape of the coiled or pleated polypeptides is called the *tertiary structure*. *Quaternary structure* specifies the structural relationship of the component polypeptides.

Proteins may be broadly classified into globular proteins and fibrous proteins. Globular proteins have compact rounded molecules and are usually water-soluble. Of prime importance are the *enzymes, proteins that catalyse biochemical reactions. Other globular proteins include the *antibodies, which combine with foreign substances in the body; the carrier proteins, such as *haemoglobin; the storage proteins (e.g. *casein in milk and *albumin in egg white), and certain hormones (e.g. *insulin). Fibrous proteins are generally insoluble in water and consist of long coiled strands or flat sheets, which confer strength and elasticity. In this category are *keratin and *collagen. Actin and myosin are the principal fibrous proteins of muscle, the interaction of which brings about muscle contraction. *Blood clotting involves the fibrous protein called fibrin.

When heated over 50°C or subjected to strong acids or alkalis, proteins lose their specific tertiary structure and may form insoluble coagulates (e.g. egg white). This usually inactivates their biological properties.

protein sequencing The process of determining the amino-acid sequence of a protein or its component polypeptides. The technique most commonly used is Edman degradation (devised by Pehr Edman), in which the terminal amino-acid residues are removed sequentially and identified chromatographically. Each step is automated and the whole process can now be performed by a single machine – the sequenator. Large polypeptides must be cleaved into smaller peptides before sequencing.

The results of this chemical sequencing can often be compared with the amino-acid sequence deduced by *gene sequencing. The gene coding for the protein under investigation may be found by screening a *gene library, for example by *Western blotting. However, the base sequence of the gene gives only the amino-acid sequence of the nascent protein, i.e. before post-translational modification. The sequence of the functional protein can only be found by chemical analysis.

protein synthesis The process by which living cells manufacture proteins from their constituent amino acids, in accordance with the genetic information carried in the DNA of the chromosomes. This information is encoded in messenger *RNA, which is transcribed from DNA in the nucleus of the cell (*see* genetic code; transcription): the sequence of amino acids in a particular protein is determined by the sequence of nucleotides in messenger RNA. At the ribosomes the information carried by messenger RNA is translated into the sequence of amino acids of the protein in the process of *translation.

proteolysis The enzymic splitting of proteins. *See* protease.

proteolytic enzyme *See* protease.

prothallus A small flattened multicellular structure that represents the independent *gametophyte generation of pteridophytes. In some pteridophytes a single prothallus bears both male and female sex organs. In others there are separate male and female prothalli.

prothrombin (Factor II) One of the blood *clotting factors. It is the precursor of the enzyme thrombin, which catalyses the formation of the fibrin matrix of the blood clot. Prothrombin synthesis occurs in the liver and is dependent on adequate supplies of vitamin K. *See also* blood clotting.

Protista A kingdom into which all organisms of simple biological organization can be classified. It was originally proposed by E. Haeckel in 1866 to include the algae, bacteria, fungi, and Protozoa. Some authorities now restrict membership of the kingdom to unicellular organisms, while others have renamed it Protistobionta, a

kingdom that includes the Protozoa, certain algae (e.g. dinoflagellates), and certain fungi (e.g. slime moulds).

protogyny 1. The condition in which the female reproductive organs (carpels) of a flower mature before the male ones (stamens), thereby ensuring that self-fertilization does not occur. Examples of protogynous flowers are plantain and figwort. *Compare* protandry; homogamy. *See also* dichogamy. **2.** The condition in some hermaphrodite or colonial invertebrates in which the female gonads or individuals are sexually mature before the male ones. *Compare* protandry.

proton An *elementary particle that is stable, bears a positive charge equal in magnitude to that of the *electron, and has a mass of $1.672\,614 \times 10^{-27}$ kg, which is 1836.12 times that of the electron. The proton occurs in all atomic nuclei (the hydrogen nucleus consists of a single proton).

proton decay A process of the type

$$p \rightarrow e^+ + \pi^0$$

where a proton decays into a positron and a pion, predicted to occur in *grand unified theories (GUTs) because baryon number is no longer conserved. The lifetime depends on the theory used and is typically 10^{35} years, but a combination of GUTs and *supersymmetry gives a lifetime of about 10^{45} years. Considerable experimental effort has been spent in looking for proton decay, so far with no success.

protonic acid An *acid that forms positive hydrogen ions (or, strictly, oxonium ions) in aqueous solution. The term is used to distinguish 'traditional' acids from Lewis acids or from Lowry–Brønsted acids in non-aqueous solvents.

proton number *See* atomic number.

protoplasm The granular material comprising the living contents of a *cell, i.e. all the substances in a cell except large vacuoles and material recently ingested or to be excreted. It consists of a *nucleus embedded in a jelly-like *cytoplasm.

protoplast (energid) The living unit (protoplasm) of a cell, consisting of the nucleus and cytoplasm bounded by the cell membrane. Protoplasts of bacterial and plant cells can be prepared by removing the cell wall; they are used to study the processes involved in cell metabolism and reproduction.

protostar *See* stellar evolution.

Prototheria A subclass of mammals – the monotremes – that lay large yolky eggs. It contains only the duckbilled platypus and the spiny anteater. After hatching, the young feed on milk from simple mammary glands inside a maternal abdominal pouch. In the anteater the eggs are also incubated in this pouch, while the platypus builds an underground nest. Adult monotremes have no true teeth. Their skeleton resembles that of a reptile, and although they are warm-blooded the body temperature is somewhat variable. They are believed to have originated at least 150 million years ago.

Protozoa A phylum comprising unicellular or acellular, usually microscopic, organisms regarded either as simple animals or members of the kingdom *Protista. They are very widely distributed in marine, freshwater, and moist terrestrial habitats; most protozoans are saprophytes, but some are parasites, including the agents causing malaria (*Plasmodium*) and sleeping sickness (*Trypanosoma*), and a few contain chlorophyll and carry out photosynthesis, like plants. Protozoan cells may be flexible or rigid, with an outer *pellicle or protective *test*. In some (such as *Paramecium* and *Trypanosoma*) *cilia or *flagella are present for locomotion; others (such as *Amoeba*) have

*pseudopodia for movement and food capture. *Contractile vacuoles occur in freshwater protozoans. Reproduction is usually asexual, by binary *fission, but some protozoans undergo a form of sexual reproduction (*see* conjugation).

proventriculus 1. The anterior part of the stomach of a bird, where digestive enzymes are secreted. Food passes from the proventriculus to the *gizzard. **2.** *See* gastric mill.

provirus The intermediate stage in the infection of a host cell by a *retrovirus in which the single RNA strand of the virus is converted into double-stranded DNA. This synthesis is performed by the enzyme *reverse transcriptase, which also digests the original viral RNA. The proviral DNA is then integrated into the host cell DNA and subsequently transcribed to form new RNA viruses. This integration also introduces retroviral oncogenes into the host, with the attendant risk of host-cell transformation to a cancer cell.

proximal Denoting the part of an organ that is nearest to the organ's point of attachment. For example, the knuckles are at the proximal end of the fingers. *Compare* distal.

prussic acid *See* hydrogen cyanide.

pseudoaromatic (antiaromatic) A compound that has a ring of atoms containing alternating double and single bonds, yet does not have the characteristic properties of *aromatic compounds. Such compounds do not obey the Hückel rule. Cyclooctatetraene (C_8H_8), for instance, has a ring of eight carbon atoms with conjugated double bonds, but the ring is not planar and the compound acts like an alkene, undergoing addition reactions.

pseudocarp (false fruit) A fruit that incorporates, in addition to the ovary wall, other parts of the flower, such as the *receptacle. For example, the fleshy part of the strawberry is formed from the receptacle and the 'pips' on the surface are the true fruits. *See also* composite fruit; pome; sorosis; syconus.

pseudohalogens A group of compounds, including cyanogen $(CN)_2$ and thiocyanogen $(SCN)_2$, that have some resemblance to the halogens. Thus, they form hydrogen acids (HCN and HSCN) and ionic salts containing such ions as CN^- and SCN^-.

pseudo order An order of a chemical reaction that appears to be less than the true order because of the experimental conditions used. Pseudo orders occur when one reactant is present in large excess. For example, a reaction of substance A undergoing hydrolysis may appear to be proportional only to [A] because the amount of water present is so large.

pseudoparenchyma A plant tissue that superficially resembles parenchyma but is made up of an interwoven mass of hyphae (in fungi) or filaments (in algae). Examples of pseudoparenchymatous structures are the fruiting bodies (mushrooms, toadstools, etc.) of certain fungi and the thalli of certain red and brown algae.

pseudopodium A temporary outgrowth of the cell of some protozoans (e.g. *Amoeba*), which serves as a feeding and locomotory organ. Pseudopodia may be blunt or threadlike, form a branching network, or be stiffened with an internal supporting rod. Phagocytic white blood cells also form pseudopodia to engulf invading bacteria.

pseudopregnancy A state resembling pregnancy that may occur in some mammals (e.g. rabbits and rodents) in which many of the phenomena of pregnancy are present but there is no fetus developing in the uterus. It is caused by an extended dioestrus (*see* oestrous cycle) in the absence of fertilization.

psi particle (J particle) A *meson discovered in 1974, which led to the extension of the quark model and the hypothesis that a fourth quark existed with the property of charm (*see* elementary particles). The psi particle is believed to consist of a charmed quark and its antiquark.

psychrometer *See* hygrometer.

Pteridophyta A division of the plant kingdom that includes ferns, horsetails, and clubmosses. Pteridophytes possess true stems, leaves, and roots, and have an organized vascular system. They have a wide (mainly terrestrial) distribution with the maximum development in the tropics. Many genera are now extinct, having formed prominent plant groups in the Carboniferous period. There is a marked alternation of generations between gamete-bearing forms (gametophytes) and spore-bearing forms (sporophytes). Unlike the bryophytes, the generations are often independent and the sporophyte is the dominant generation. Pteridophytes differ from gymnosperms and angiosperms in not producing seeds. The division includes the classes *Lycopsida, *Pteropsida, and *Sphenopsida. In some classification schemes these classes are regarded as subdivisions of the *Tracheophyta.

Pteridospermales (seed ferns) An extinct order of the *Gymnospermae that flourished in the Carboniferous period. They possessed characteristics of both the ferns (*see* Filicinae) and the seed plants (*see* Spermatophyta) in reproducing by means of seeds and yet retaining fernlike leaves. Their internal anatomy combined both fern and seed-plant characteristics.

pterodactyls *See* Pterosauria.

Pteropsida A subdivision of the *Tracheophyta containing the ferns (*see* Filicinae) and seed plants (*see* Gymnospermae; Angiospermae), or a class of the *Pteridophyta containing only the ferns.

Pulleys

Pterosauria An extinct order of flying reptiles – the pterodactyls – that lived in the Jurassic and Cretaceous periods (180–70 million years ago). Pterodactyls had beaked jaws and an elongated fourth finger that supported a membranous wing. They had long jointed tails, no feathers, and could probably only fly by soaring.

PTFE *See* polytetrafluoroethene.

Ptolemaic astronomy The system of astronomy originally proposed by Apollonius of Perga in the third century BC and completed by Claudius Ptolemaeus of Alexandria (100–178 AD). It assumed that the earth was at the centre of the universe and that each known planet, the moon, and the sun moved round it in a circular orbit, called the deferent. In addition to this motion the orbiting bodies also described epicycles, small circles about points on the deferent. The system gave moderately good predictions, but was completely replaced by the heliocentric astronomy of Copernicus in the 16th century. The

Ptolemaic system was published by Ptolemaeus in the work known by its Arabic name, the *Almagest*.

ptyalin An enzyme that digests carbohydrates (*see* amylase). It is present in mammalian *saliva and is responsible for the initial stages of starch digestion.

p-type conductivity *See* semiconductor; transistor.

pubis One of the three bones that make up each half of the *pelvic girdle. It is the most anterior of the three pelvic bones. In mammals and many reptiles the pubes are united at a slightly movable joint, the *pubic symphysis. See also* ilium; ischium.

pulley A simple machine consisting of a wheel with a flat, crowned, or grooved rim to take a belt, rope, or chain with which a load can be raised.

In fig (a), assuming the system is frictionless, the force P in any part of the rope is constant, therefore $2P = L$, where L is the load. In general, $nP = L$, where n is the number of supporting ropes. In fig (b), the number of supporting ropes is 4. The mechanical advantage of a pulley system is the ratio of the load, L, to the effort applied to the free end of the rope, P, i.e. mechanical advantage = $L/P = L(L/n)^{-1} = n$. Thus in fig (b) the mechanical advantage is 4. A combination of ropes and pulleys as in fig (b) is called a *block and tackle*.

pulmonary artery The artery that conveys deoxygenated blood from the right ventricle of the heart to the lungs, where it receives oxygen.

pulmonary vein The vein that conveys oxygenated blood from the lungs to the left atrium of the heart.

pulp cavity The central region of a tooth, which is connected by a narrow channel at the tip of the root with the surrounding tissues. The pulp cavity contains the *pulp* – connective tissue in which blood vessels and nerve fibres are embedded, and it is lined with *odontoblasts, which produce the *dentine.

pulsar A celestial source of radiation emitted in brief (0.03 second to 4 seconds) regular pulses. First discovered in 1968, pulsars are believed to be rotating *neutron stars. The strong magnetic field of the neutron star concentrates charged particles in two regions and the radiation is emitted in two directional beams. The pulsing effect occurs as the beams rotate. Most pulsars are radio sources (emit electromagnetic radiation of radio frequencies) but a few that emit light or X-rays have been detected. Over 300 pulsars are now known in our Galaxy.

pulsatance *See* angular frequency.

pulse 1. (in physics) **a.** A brief variation in a quantity, usually for a finite time, especially in a quantity that is normally constant. **b.** A series of such variations having a regular waveform in which the variable quantity rises sharply from a base value to a maximum value and then falls back to the base value in a relatively short time. **2.** (in physiology) A series of waves of dilation that pass along the arteries, caused by pressure of blood pumped from the heart through contractions of the left ventricle. In man it can be felt easily where arteries pass close to the skin surface, e.g. at the wrist.

pulse jet A type of ramjet (*see* jet propulsion) in which a louvred valve at the front of the projectile is blown open by the ram effect of the moving projectile and remains open until pressure has built up in the combustion chamber. Fuel is then admitted and the mixture exploded by spark ignition. This closes the louvred valve and produces thrust at the open rear end of the projectile. The German flying bombs of World War II were powered by pulse jets.

pulse modulation *See* modulation.

pulvinus A group of cells at the base of a leaf or leaflet in certain plants that, by rapidly losing water, brings about changes in the position of the leaves. In the sensitive plant (*Mimosa pudica*), the pulvinus is responsible for the folding of the leaves that occurs at nightfall or when the plant is touched or injured.

pumice A porous volcanic rock that is light and full of cavities due to expanding gases that were liberated from solution in the lava while it solidified. Pumice is often light enough to float on water. It is usually acid (siliceous) in composition, and is used as an abrasive and for polishing.

pump A device that imparts energy to a fluid in order to move it from one place or level to another or to raise its pressure (*compare* vacuum pump). *Centrifugal pumps* and *turbines have rotating impellers, which increase the velocity of the fluid, part of the energy so acquired by the fluid then being converted to pressure energy. Displacement pumps act directly on the fluid, forcing it to flow against a pressure. They include piston, plunger, gear, screw, and cam pumps. *See also* electromagnetic pump.

punctuated equilibrium A theory proposing that plant and animal species usually arise very quickly in terms of geological time (in less than 100 000 years) and seldom through a process of gradual change. It thus questions the traditional Darwinian theory of evolution, citing as evidence the discontinuities observed in the fossil records of certain animal groups (e.g. the ammonites).

pupa The third stage of development in the life cycle of some insects. During the pupal stage locomotion and feeding cease and *metamorphosis from the larva to the adult form takes place. There are three types of pupa. The commonest is the *exarate* or free pupa, in which the wings and other appendages are visible and movable. In the *obtect* type the wings are stuck to the body and immovable, as in the *chrysalis* of a butterfly or moth; and in the *co-arctate* type an exarate pupa develops within a hard barrel-shaped *puparium*, as in the housefly and other Diptera.

pupil See iris.

pure line A population of plants or animals all having a particular feature that has been retained unchanged through many generations. The organisms are *homozygous and are said to 'breed true' for the feature concerned.

Pyramid of numbers for a woodland food chain

Purine

purine An organic nitrogenous base (see formula), sparingly soluble in water, that gives rise to a group of biologically important derivatives, notably *adenine and *guanine, which occur in *nucleotides and nucleic acids (DNA and RNA).

PVC See polychloroethene.

pyramid A solid having a polygonal base with *n* sides, each side forming the base of a triangle. The *n* triangles so formed have a common vertex. The *axis* of the pyramid is a line joining the vertex to the centre of symmetry of the base. If the axis is perpendicular to the base the solid is a *right pyramid*. A *square pyramid* has a square base and a *triangular pyramid* has a triangular base (*see* tetrahedron). The volume of a pyramid is one third of the base area multiplied by the height.

pyramid of numbers The numbers of animals found in an area at ascending *trophic levels of a *food chain (see illustration). Because only a small proportion of the energy taken in by an organism is converted to tissue and is thus available to consumers at the next trophic level, the number of organisms that can be supported at each level is generally much less than the number at the level that supplies its food (i.e. the level below). A *pyramid of biomass*, representing the *biomass of animals at ascending trophic levels, shows a smaller decrease at each trophic level since the animals at the higher levels are generally larger (although fewer) than those at the lower levels.

pyranose A *sugar having a six-membered ring containing five carbon atoms and one oxygen atom.

pyrenocarp See drupe.

pyrenoid A spherical protein body found in the *chloroplasts of many algae. Pyrenoids are associated with the storage of starch: layers of starch are often found around them.

pyridine A colourless liquid with a strong unpleasant smell, C_5H_5N (see formula); r.d. 0.9; m.p. $-42°C$; b.p. $115°C$. Pyridine is an aromatic heterocyclic compound present in coal tar. It is used in making other organic chemicals.

Pyridine

pyridoxine *See* vitamin B complex.

pyrimidine An organic nitrogenous base (see formula), sparingly soluble in water, that gives rise to a group of biologically important derivatives, notably *uracil, *thymine, and *cytosine, which occur in *nucleotides and nucleic acids (DNA and RNA).

$$
\begin{array}{c}
\text{H} \\
\text{C} \\
N_3 \overset{4}{\diagup} \diagdown {}_5\text{CH} \\
| \qquad || \\
HC_2 \diagdown {}_{1} \diagup {}^{6}\text{CH} \\
N
\end{array}
$$

Pyrimidine

pyrite (iron pyrites) A mineral form of iron(II) sulphide, FeS_2. Superficially it resembles gold in appearance, hence it is also known as *fool's gold*, but it is harder and more brittle than gold (which may be cut with a knife). Pyrite crystallizes in the cubic system, is brass yellow in colour, has a metallic lustre, and a hardness of 6–6.5 on the Mohs' scale. It is the most common and widespread of the sulphide minerals and is used as a source of sulphur for the production of sulphuric acid. Sources include the Rio Tinto mines in Spain.

pyro- Prefix denoting an oxo acid that could be obtained from a lower acid by dehydration of two molecules. For example, pyrosulphuric acid is $H_2S_2O_7$ (i.e. $2H_2SO_4$ minus H_2O).

pyroboric acid *See* boric acid.

pyroelectricity The property of certain crystals, such as tourmaline, of acquiring opposite electrical charges on opposite faces when heated. In tourmaline a rise in temperature of 1 K at room temperature produces a polarization of some 10^{-5} C m^{-2}.

pyrolysis Chemical decomposition occurring as a result of high temperature.

pyrometric cones *See* Seger cones.

pyrometry The measurement of high temperatures from the amount of radiation emitted, using a *pyrometer*. Modern *narrow-band* or *spectral* pyrometers use infrared-sensitive *photoelectric cells behind filters that exclude visible light. In the *optical pyrometer* (or disappearing filament pyrometer) the image of the incandescent source is focused in the plane of a tungsten filament that is heated electrically. A variable resistor is used to adjust the current through the filament until it blends into the image of the source, when viewed through a red filter and an eyepiece. The temperature is then read from a calibrated ammeter or a calibrated dial on the variable resistor. In the *total-radiation pyrometer* radiation emitted by the source is focused by a concave mirror onto a blackened foil to which a thermopile is attached. From the e.m.f. produced by the thermopile the temperature of the source can be calculated.

pyrophoric Igniting spontaneously in air. *Pyrophoric alloys* are alloys that give sparks when struck. *See* misch metal.

pyrophosphoric acid *See* phosphoric(V) acid.

pyrosilicate *See* silicate.

pyrosulphuric acid *See* disulphuric(VI) acid.

pyroxenes A group of ferromagnesian rock-forming silicate minerals. They are common in basic igneous rocks but may also be developed by metamorphic processes in gneisses, schists, and marbles. Pyroxenes have a complex crystal chemistry; they are composed of continuous chains of silicon and oxygen atoms linked by a variety of other elements. They are related to the *amphiboles, from which they differ in cleavage angles. The general formula is:

$$X_{1-p}Y_{1+p}Z_2O_6,$$

where X = Ca,Na;
Y = $Mg,Fe^{2+},Mn,Li,Al,Fe^{3+},Ti$;
and Z = Si,Al.
Orthorhombic pyroxenes (*orthopyroxenes*), $(Mg,Fe)_2Si_2O_6$, vary in composition between the end-members enstatite ($Mg_2Si_2O_6$) and orthoferrosilite ($Fe_2Si_2O_6$). Monoclinic pyroxenes (*clinopyroxenes*), the larger group, include:

diopside, $CaMgSi_2O_6$;
hedenbergite, $CaFe^{2+}Si_2O_6$;
johannsenite, $CaMnSi_2O_6$;
augite, $(Ca,Mg,Fe,Ti,Al)_2(Si,Al)_2O_6$;
aegirine, $NaFe^{3+}Si_2O_6$;
jadeite (*see* jade);
pigeonite $(Mg,Fe^{2+},Ca)(Mg,Fe^{2+})Si_2O_6$.

pyruvic acid (2-oxopropanoic acid) A colourless liquid organic acid, $CH_3COCOOH$. It is an important intermediate compound in metabolism, being produced during *glycolysis and converted to acetyl coenzyme A, required for the *Krebs cycle.

Q

QCD *See* quantum chromodynamics.

QED *See* quantum electrodynamics.

QFD Quantum flavourdynamics. *See* electroweak theory.

QSG *See* quasars.

QSO *See* quasars.

QSS *See* quasars.

quadrat An ecological sampling unit consisting of a small square area of ground within which all species of interest are noted or measurements taken. Quadrats may be spaced over a larger area to form an overall view when a total survey would be impracticable, or they may be used to sample along a *transect.

quadrate A paired bone in the upper jaw of bony fishes, amphibians, reptiles, and birds that articulates with the lower jawbone. It is absent in mammals, being reduced to a small bone (the incus) in the middle ear (*see* ear ossicles).

quadratic equation An equation of the second degree having the form $ax^2 + bx + c = 0$. Its roots are:
$$x = [-b \pm \sqrt{(b^2 - 4ac)}]/2a.$$

quadrature The position of the moon or an outer planet when the line joining it to the earth makes a right angle with a line joining the earth to the sun.

quadrivalent Having a valency of four.

qualitative analysis *See* analysis.

qualitative variation *See* discontinuous variation.

quality of sound (timbre) The quality a musical note has as a result of the presence of *harmonics. A pure note consists only of the *fundamental; however, a note from a musical instrument will have several harmonics present, depending on the type of instrument and the way in which it is played. For example, a plucked string (as in a guitar) produces a series of harmonics of diminishing intensity, whereas a struck string (as in a piano) produces a series of harmonics of more nearly equal intensity.

quantitative analysis *See* analysis.

quantitative variation *See* continuous variation.

quantum (*pl.* **quanta**) The minimum amount by which certain properties, such as energy or angular momentum, of a system can change. Such properties do not, therefore, vary continuously, but in integral multiples of the relevant quantum, and are described as *quantized*. This concept forms the basis of the *quantum theory. In waves and fields the quantum can be regarded as an excitation, giving a particle-like interpretation to the wave or field. Thus, the quantum of the electromagnetic field is the *photon and the *graviton is the quantum of the gravitational field. *See* quantum mechanics.

quantum chromodynamics (QCD) A *gauge theory that describes the strong interaction in terms of quarks and antiquarks and the exchange of massless gluons between them (*see also* elementary particles). Quantum chromodynamics is similar to quantum electrodynamics (QED), with colour being analogous to electric charge and the gluon being the analogue of the photon. The gauge group of QCD is non-Abelian and the theory is much more complicated than quantum electrodynamics; the gauge symmetry in QCD is not a *broken symmetry.

QCD has the important property of *asymptotic freedom – that at very high energies (and, hence, short distances) the interactions between quarks tend to zero as the distance between them tends to zero. Because of asymptotic freedom, perturbation theory may be used to calculate the high energy aspects of strong interactions, such as those described by the *parton model.

quantum electrodynamics (QED) The study of the properties of electromagnetic radiation and the way in which it interacts with charged matter in terms of *quantum mechanics. The collision of a moving electron with a proton, in this theory, can be visualized by a space–time diagram

An example of a Feynman diagram for electron-electron scattering

(*Feynman diagram*) in which photons are exchanged (see illustration). *Perturbation-theory calculations using Feynman diagrams enable an agreement between theory and experiment to a greater accuracy than one part in 10^9 to be obtained. Because of this, QED is the most accurate theory known in physical science. Although many of the effects calculated in QED are very small (about 4×10^{-6} eV), such as *energy level splitting in the spectra of *atoms, they are of great significance for demonstrating the physical reality of fluctuations and polarization in the vacuum state. QED is a *gauge theory for which the gauge *group is Abelian.

quantum field theory A quantum-mechanical theory applied to systems that have an infinite number of degrees of freedom. In quantum field theories, particles are represented by fields that have quantized normal modes of oscillation. For instance, *quantum electrodynamics is a quantum field theory in which the photon is emitted or absorbed by particles; the photon is the quantum of the electromagnetic field. *Relativistic quantum field theories* are used to describe fundamental interactions between elementary particles. They predict the existence of *antiparticles and also show the connection between spin and statistics that leads to the Pauli exclusion principle (see spin–statistics theorem). In spite of their success, it is not clear whether a quantum field theory can give a completely unified description of all interactions (including the gravitational interaction).

quantum flavourdynamics (QFD) *See* electroweak theory.

quantum Hall effect A quantum mechanical version of the *Hall effect found at very low temperatures, in which the Hall coefficient R_H is proportional to h/e^2, where h is the Planck constant and e is the charge of the electron. Thus, the Hall coefficient is quantized. There are two types of quantum Hall effect. The *integer quantum Hall effect* has R_H given as an integer with great precision. It can be used for precision measurements of constants such as e and h. In the *fractional quantum Hall effect*, R_H has fractional values.

The integer quantum Hall effect can be understood in terms of noninteracting electrons, whereas the fractional effect is thought to result from many-electron interactions in two-dimensional systems, and be an example of anyons (*see* quantum statistics).

quantum jump A change in a system (e.g. an atom or molecule) from one quantum state to another.

quantum mechanics A system of mechanics that was developed from *quantum theory and is used to explain the properties of atoms and molecules. Using the energy *quantum as a starting point it incorporates Heisenberg's *uncertainty principle and the *de Broglie wavelength to establish the wave–particle duality (*see* complementarity) on which *Schrödinger's equation is based. This form of quantum mechanics is called *wave mechanics*. An alternative but equivalent formalism, *matrix mechanics*, is based on mathematical operators.

quantum number *See* atom; spin.

quantum state The state of a quantized system as described by its quantum numbers. For instance, the state of a hydrogen *atom is described by the four quantum numbers n, l, m, m_s. In the ground state they have values 1, 0, 0, and ½ respectively.

quantum statistics A statistical description of a system of particles that obeys the rules of *quantum mechanics rather than classical mechanics. In quantum statistics, energy states are considered to be quantized. *Bose–Einstein statistics* apply if any number of particles can

occupy a given quantum state. Such particles are called *bosons*. Bosons have an angular momentum $nh/2\pi$, where n is zero or an integer and h is the Planck constant. For identical bosons the *wave function is always symmetric. If only one particle may occupy each quantum state, *Fermi–Dirac statistics* apply and the particles are called *fermions*. Fermions have a total angular momentum $(n + \frac{1}{2})h$ and any wave function that involves identical fermions is always antisymmetric.

The relation between the spin and statistics of particles is given by the *spin–statistics theorem.

In two-space dimensions, it is possible that there are particles (or *quasiparticles) that have statistics intermediate between bosons and fermions. These particles are known as *anyons*; for identical anyons the wave function is not symmetric (a phase sign of +1) or antisymmetric (a phase sign of −1), but interpolates continuously between +1 and −1. Anyons may be involved in the fractional *quantum Hall effect and have been suggested as a mechanism for high-temperature *superconductivity.

quantum theory The theory devised by Max Planck (1858–1947) in 1900 to account for the emission of the black-body radiation from hot bodies. According to this theory energy is emitted in quanta (*see* quantum), each of which has an energy equal to $h\nu$, where h is the *Planck constant and ν is the frequency of the radiation. This theory led to the modern theory of the interaction between matter and radiation known as *quantum mechanics, which generalizes and replaces classical mechanics and Maxwell's electromagnetic theory. In *nonrelativistic quantum theory* particles are assumed to be neither created nor destroyed, to move slowly relative to the speed of light, and to have a mass that does not change with velocity. These assumptions apply to atomic and molecular phenomena and to some aspects of nuclear physics. *Relativistic quantum theory* applies to particles that travel at or near the speed of light.

quark *See* elementary particles.

quark confinement The hypothesis that free quarks can never be seen in isolation. It is a result of *quantum chromodynamics, in which the property of *asymptotic freedom means that the interactions between quarks get weaker as the distance between them gets smaller, and tends to zero as the distance between them tends to zero. Conversely, the attractive interactions between quarks get stronger as the distance between them gets greater, and the quark-confinement hypothesis is that the quarks cannot escape from one another. It is possible that at very high temperatures, such as those in the *early universe, quarks may become free. The temperature at which this occurs is called the *deconfinement temperature*.

quart A unit of capacity equal to one quarter of a *gallon.

quarter-wave plate *See* retardation plate.

quartz The most abundant and common mineral, consisting of crystalline silica (silicon dioxide, SiO_2), crystallizing in the trigonal system. It has a hardness of 7 on the Mohs' scale. Well-formed crystals of quartz are six-sided prisms terminating in six-sided pyramids. Quartz is ordinarily colourless and transparent, in which form it is known as *rock crystal*. Coloured varieties, a number of which are used as gemstones, include *amethyst, citrine quartz (yellow), rose quartz (pink), milk quartz (white), smoky quartz (grey-brown), *chalcedony, *agate, and *jasper. Quartz occurs in many rocks, especially igneous rocks such as granite and quartzite (of which it is the chief constituent), metamorphic rocks such

as gneisses and schists, and sedimentary rocks such as sandstone and limestone. The mineral has the property of being piezoelectric and hence is used to make oscillators for clocks (*see* quartz clock), radios, and radar instruments. It is also used in optical instruments and in glass, glaze, and abrasives.

quartz clock A clock based on a *piezoelectric crystal of quartz. Each quartz crystal has a natural frequency of vibration, which depends on its size and shape. If such a crystal is introduced into an oscillating electronic circuit that resonates at a frequency very close to that of the natural frequency of the crystal, the whole circuit (including the crystal) will oscillate at the crystal's natural frequency and the frequency will remain constant over considerable periods (a good crystal will maintain oscillation for a year with an accumulated error of less than 0.1 second). In a quartz clock or watch the alternating current from the oscillating circuit containing such a crystal is amplified and the frequency subdivided until it is suitable to drive a synchronous motor, which in turn drives a gear train to operate hands. Alternatively it is used to activate a digital display.

quasars A class of astronomical objects that appear on optical photographs as starlike but have large *redshifts quite unlike those of stars. They were first observed in 1961 when it was found that strong radio emission was emanating from many of these starlike bodies. Over 600 such objects are now known and their redshifts can be as high as 4. The redshifts are characteristic of the *expansion of the universe. This *cosmological redshift* is the explanation of the high observed redshifts of quasars favoured by most astronomers. (A few, however, maintain that the redshift could be a local Doppler effect, characteristic of movement relative to the earth and sun of nearby objects in the Galaxy, or a gravitational effect.) If the redshifts are cosmological, quasars are the most distant objects in the universe, some being up to 10^{10} light-years away. The exact nature of quasars is unknown but it is believed that they are the nuclei of galaxies in which there is violent activity. The luminosity of the nucleus is so much greater than that of the rest of the galaxy that the source appears pointlike. It has been proposed that the power source in a quasar is a supermassive *black hole accreting material from the stars and gas in the surrounding galaxy.

The name *quasar* is a contraction of quasistellar object (*QSO*) or quasistellar galaxy (*QSG*). Quasars that are also radio sources are sometimes called quasistellar radio sources (*QSS*).

quasiparticle A long-lived single-particle *excitation in the quantum theory of many-body systems, in which the excitations of the individual particles are modified by their interactions with the surrounding medium.

Quaternary The second period of the Cenozoic era, which began about 2 million years ago, following the *Tertiary period, and includes the present. It is subdivided into two epochs – the *Pleistocene and *Holocene. The beginning of the Quaternary is usually based on the onset of a worldwide cooling. During the period four principal glacial phases occurred in Europe and North America, in which ice advanced towards the equator, separated by interglacials during which conditions became warmer and the ice sheets and glaciers retreated. The last glacial ended about 10 000 years ago. Man became the dominant terrestrial species during the Quaternary. Among the fauna adapted to the cold conditions were the mammoth and the woolly rhinoceros.

quaternary ammonium compounds *See* amine salts.

quenching 1. (in metallurgy) The rapid cooling of a metal by immersing it in a bath of liquid in order to improve its properties. Steels are quenched to make them harder but some nonferrous metals are quenched for other reasons (copper, for example, is made softer by quenching). **2.** (in physics) The process of inhibiting a continuous discharge in a *Geiger counter so that the incidence of further ionizing radiation can cause a new discharge. This is achieved by introducing a quenching vapour, such as methane mixed with argon or with neon, into the tube.

quicklime *See* calcium oxide.

quinhydrone electrode A *half-cell consisting of a platinum electrode in an equimolar solution of quinone (cyclohexadiene-1,4-dione) and hydroquinone (benzene-1,4-diol). It depends on the oxidation–reduction reaction
$$C_6H_4(OH)_2 \rightleftharpoons C_6H_4O_2 + 2H^+ + 2e$$
quinol *See* benzene-1,4-diol.

quinone 1. *See* cyclohexadiene-1,4-dione. **2.** Any similar compound containing $C=O$ groups in an unsaturated ring.

R

race 1. (in biology) A category used in the *classification of organisms that consists of a group of individuals within a species that are geographically, ecologically, physiologically, or chromosomally distinct from other members of the species. The term is frequently used in the same sense as *subspecies. *Physiological races*, for example, are identical in appearance but differ in function. They include strains of fungi adapted to infect different varieties of the same crop spe-

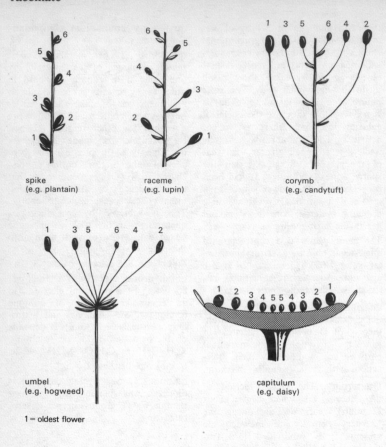

spike
(e.g. plantain)

raceme
(e.g. lupin)

corymb
(e.g. candytuft)

umbel
(e.g. hogweed)

capitulum
(e.g. daisy)

1 = oldest flower

Types of racemose inflorescence

cies. **2.** (in anthropology) A distinct human type possessing several characteristics that are genetically inherited. The major races of man are Mongoloid, Caucasian, Negroid, and Australoid.

racemate *See* racemic mixture.

raceme A type of *racemose inflorescence in which the main flower stalk is elongated and bears stalked flowers. An example is the lupin. *See also* panicle.

racemic mixture (racemate) A mix-

ture of equal quantities of the ($+$)- or *d*- and ($-$)- or *l*- forms of an optically active compound. Racemic mixtures are denoted by the prefix ($\pm$)- or *dl*- (e.g. ($\pm$)-lactic acid). A racemic mixture shows no *optical activity.

racemization A chemical reaction in which an optically active compound is converted into a *racemic mixture.

racemose inflorescence (indefinite inflorescence) A type of flowering shoot (*see* inflorescence) in which the growing region at the tip of the flower stalk continues to produce new flower buds during growth. As a result, the youngest flowers are at the top and the oldest flowers are at the base of the stalk. In a flattened inflorescence, the youngest flowers are in the centre and the oldest flowers are on the outside. Types of racemose inflorescence include the *capitulum, *catkin, *corymb, *raceme, *spadix, *spike, and *umbel (see illustration). *Compare* cymose inflorescence.

rachis (rhachis) 1. The main axis of a compound leaf or an inflorescence. **2.** The shaft of a *feather. **3.** The backbone.

rad *See* radiation units.

radar (radio detection and ranging) A method of detecting the presence, position, and direction of motion of distant objects (such as ships and aircraft) by means of their ability to reflect a beam of electromagnetic radiation of centimetric wavelengths. It is also used for navigation and guidance. It consists of a transmitter producing radio-frequency radiation, often pulsed, which is fed to a movable aerial from which it is transmitted as a beam. If the beam is interrupted by a solid object, a part of the energy of the radiation is reflected back to the aerial. Signals received by the aerial are passed to the receiver, where they are amplified and detected. An echo from a reflection of a solid object is indicated by a sudden rise in the detector output. The time taken for a pulse to reach the object and be reflected back (t) enables the distance away (d) of the target to be calculated from the equation $d = ct/2$, where c is the speed of light. In some systems the speed of the object can be measured using the *Doppler effect. The output of the detector is usually displayed on a cathode-ray tube in a variety of different formats (see illustration).

radial symmetry The arrangement of parts in an organ or organism such that cutting through the centre of the structure in any direction produces two halves that are mirror images of each other. The stems and roots of plants usually show radial symmetry, while all animals belonging to the Coelenterata (e.g. jellyfish) and Echinodermata (e.g. starfish) are radially symmetrical – and typically sessile – in their adult form. The term *actinomorphy* is used to describe radial symmetry in flowers (e.g. a buttercup flower). *Compare* bilateral symmetry.

radial velocity *See* line-of-sight velocity.

radian *See* circular measure.

radiance Symbol L_e. The radiant intensity per unit transverse area, in a given direction, of a source of radiation. It is measured in $W\,sr^{-1}\,m^{-2}$.

radiant energy Energy transmitted as electromagnetic radiation.

radiant exitance *See* exitance.

radiant flux Symbol Φ_e. The total power emitted, received, or passing in the form of electromagnetic radiation. It is measured in watts.

radiant intensity Symbol I_e. The *radiant flux per unit solid angle

plan position indicator (PPI); range is measured radially from the centre

Types of cathode ray tube radar displays

emitted by a point source. It is measured in watts per steradian.

radiation 1. Energy travelling in the form of electromagnetic waves or photons. 2. A stream of particles, especially alpha-or beta-particles from a radioactive source or neutrons from a nuclear reactor.

radiation belts *See* Van Allen belts.

radiation damage Harmful changes that occur to inanimate materials and living organisms as a result of exposure to energetic electrons, nucleons, fission fragments, or high-energy electromagnetic radiation. In inanimate materials the damage may be caused by electronic excitation, ionization, transmutation, or displacement of atoms. In organisms, these mechanisms can cause changes to cells that alter their genetic structure, interfere with their division, or kill them. In humans, these changes can lead to *radiation sickness, radiation burns* (from large doses of radiation), or to long-term damage of several kinds, the most serious of which result in various forms of cancer (especially leukaemia).

radiation pressure The pressure exerted on a surface by electromagnetic radiation. As radiation carries momentum as well as energy it exerts a force when it meets a surface, i.e. the *photons transfer momentum when they strike the surface. The pressure is usually negligible on large bodies, for exa .ple, the pressure of radiation from tne sun on the surface of the earth is of the order of 10^{-5} pascal, but on small bodies it can have a considerable effect, driving them away from the radiation source. Radiation pressure is also important in the interiors of stars of very high mass.

radiation temperature The surface temperature of a celestial body as calculated by *Stefan's law, assuming that the body behaves as a *black body. The radiation temperature is usually measured over a narrow portion of the electromagnetic spectrum, such as the visible range (which gives the *optical temperature*).

radiation units Units of measurement used to express the *activity of a radionuclide and the *dose of ionizing radiation. The units *curie, roentgen, rad*, and *rem* are not coherent with SI units but their temporary use with SI units has been approved while the derived SI units *becquerel, gray*, and *sievert* become familiar.

The becquerel (Bq), the SI unit of activity, is the activity of a radionuclide decaying at a rate, on average, of one spontaneous nuclear transition per second. Thus 1 Bq = 1 s^{-1}. The former unit, the curie (Ci), is equal to 3.7×10^{10} Bq. The curie was originally chosen to approximate the activity of 1 gram of radium-226.

The gray (Gy), the SI unit of absorbed dose, is the absorbed dose when the energy per unit mass imparted to matter by ionizing radiation is 1 joule per kilogram. The former unit, the rad (rd), is equal to 10^{-2} Gy.

The sievert (Sv), the SI unit of dose equivalent, is the dose equivalent when the absorbed dose of ionizing radiation multiplied by the stipulated dimensionless factors is 1 J kg^{-1}. As different types of radiation cause different effects in biological tissue a weighted absorbed dose, called the *dose equivalent*, is used in which the absorbed dose is modified by multiplying it by dimensionless factors stipulated by the International Commission on Radiological Protection. The former unit of dose equivalent, the rem (originally an acronym for *r*oentgen *e*quivalent *m*an), is equal to 10^{-2} Sv.

In SI units, exposure to ionizing radiation is expressed in coulombs per kilogram, the quantity of X- or gamma-radiation that produces ion pairs carrying 1 coulomb of charge of either sign in 1 kilogram of pure dry air. The former unit, the roentgen (R), is equal to 2.58×10^{-4} C kg^{-1}.

radiative collision A collision between charged particles in which part of the kinetic energy is radiated in the form of photons.

radical 1. (in chemistry) A group of atoms, either in a compound or existing alone. *See* free radical; functional group. **2.** (in mathematics) A root of a number or a quantity. The symbol $\sqrt{}$ is called the *radical sign*.

radicle The part of a plant embryo that develops into the root system. The tip of the radicle is protected by a root cap and points towards the micropyle. On germination it breaks through the testa and grows down into the soil. *Compare* plumule.

radio A means of transmitting information in which the transmission medium consists of electromagnetic radiation. Information is transmitted by means of the *modulation of a *carrier wave in a transmitter; the modulated carrier wave is fed to a transmitting aerial from which it is broadcast through the atmosphere or through space. A receiving aerial forms part of a *resonant circuit, which can be tuned to the frequency of the carrier wave, enabling the receiver that it feeds selectively to amplify and then to demodulate the transmitted signal. A replica of the original information is thus produced by the receiver. *See also* radio transmission.

radioactive age The age of an archaeological or geological specimen as determined by a process that depends on a radioactive decay. *See* carbon dating; fission-track dating; potassium–argon dating; rubidium–strontium dating; uranium–lead dating.

radioactive series A series of radioactive nuclides in which each member of the series is formed by the decay of the nuclide before it. The series ends with a stable nuclide. Three radioactive series occur naturally, those headed by thorium–232 (*thorium series*), uranium–235 (*actinium series*), and uranium–238 (*uranium series*). All three series end with an isotope of lead. The neptunium series starts with the artificial isotope plutonium–241, which decays to neptunium–237, and ends with bismuth–209.

radioactive tracing *See* labelling.

radioactive waste (nuclear waste) Any waste material that contains *radionuclides. These wastes are produced in the mining of radioactive ores, the normal running of nuclear power stations and other reactors, the manufacture of nuclear weapons, and in research laboratories. Because high-level radioactive wastes can be extremely dangerous to all living matter and because they may contain radionuclides having half-lives of many thousands of years, their disposal has to be controlled with great stringency.

The first stage in disposal involves extensive reprocessing to retrieve usable material and reduce the activity of the waste. Subsequent final disposal, usually of solid material in strong metal canisters, involves burial of the material. The two currently favoured categories of sites for burial are deep (600 metres) mined cavities in stable geological formations on land and in red clay some 50–100 metres below the seabed away from tectonic plate margins and where the currents at the bottom of the sea are weak.

radioactivity The spontaneous disintegration of certain atomic nuclei accompanied by the emission of alpha-particles (helium nuclei), beta-particles (electrons), or gamma radiation (short-wave electromagnetic waves).

Natural radioactivity is the result of the spontaneous disintegration of naturally occurring radioisotopes. Many radioisotopes can be arranged in three *radioactive series. The rate of disintegration is uninfluenced by chemical changes or any normal changes in their environment. However, radioactivity can be induced in many nuclides by bombarding them with neutrons or other particles. *See also* half-life; ionizing radiation; radiation units.

radio astronomy The study of the radio-frequency radiation emitted by celestial bodies. This branch of astronomy began in 1932 when a US engineer, Karl Jansky (1905–40), first detected radio waves from outside the earth's atmosphere. *See* radio source; radio telescope; radio window.

radiobiology The branch of biology concerned with the effects of radioactive substances on living organisms and the use of radioactive tracers to study metabolic processes (*see* labelling).

radiocarbon dating *See* carbon dating.

radiochemistry The branch of chemistry concerned with radioactive compounds and with ionization. It includes the study of compounds of radioactive elements and the preparation and use of compounds containing radioactive atoms. *See* labelling; radiolysis.

radio frequencies The range of frequencies, between about 3 kilohertz and 300 gigahertz, over which electromagnetic radiation is used in radio transmission. It is subdivided into eight equal bands, known as very low frequency, low frequency, medium frequency, high frequency, very high frequency, and extremely high frequency.

radio galaxies A *radio source outside the Galaxy that has been identified with an optically visible galaxy.

These radio galaxies are distinguished from normal galaxies by having a radio power output some 10^6 times greater (i.e. up to 10^{38} watts rather than 10^{32} W). The source of the radio-frequency energy is associated with violent activity, involving the ejection of relativistic jets of particles from the nucleus of the galaxy. It has been suggested that the sources are powered by supermassive *black holes in the nucleus.

radiography The process or technique of producing images of an opaque object on photographic film or on a fluorescent screen by means of radiation (either particles or electromagnetic waves of short wavelength, such as X-rays and gamma-rays). The photograph produced is called a *radiograph*. The process is widely used in diagnostic *radiology, using X-rays, and in flaw detection in industrial products, using high-energy X-rays, gamma-radiations, neutron beams, and (more recently) beams of charged particles. *See also* autoradiography.

radio interferometry *See* radio telescope.

radioisotope (radioactive isotope) An isotope of an element that is radioactive. *See* labelling.

radiolocation The location of distant objects by means of *radar.

radiology The study and use of X-rays, radioactive materials, and other ionizing radiations for medical purposes, especially for diagnosis (*diagnostic radiology*) and the treatment of cancer and allied diseases (*therapeutic radiology* or *radiotherapy*).

radiolysis The use of ionizing radiation to produce chemical reactions. The radiation used includes alpha particles, electrons, neutrons, X-rays, and gamma rays from radioactive materials or from accelerators. Energy transfer produces ions and excited species, which undergo further reaction. A particular feature of radiolysis is the formation of short-lived solvated electrons in water and other polar solvents.

radiometric dating (radioactive dating) *See* dating techniques; radioactive age.

radionuclide (radioactive nuclide) A *nuclide that is radioactive.

radio-opaque Describing a medium that is opaque to X-rays and gamma rays. Examples are barium salts, used in diagnostic radiology of the digestive tract. *See also* opacity.

radiosonde A meteorological instrument that measures temperature, pressure, humidity, and winds in the upper atmosphere. It consists of a package of instruments and a radio transmitter attached to a balloon. The data is relayed back to earth by the transmitter. The position of the balloon can be found by radar and from its changes in position the wind velocities can be calculated.

radio source An astronomical object that has been observed with a *radio telescope to emit radio-frequency electromagnetic radiation. Radio sources within the Galaxy include Jupiter, the sun, pulsars, and background radiation arising from *synchrotron radiation. Sources outside the Galaxy include spiral galaxies, *radio galaxies, and *quasars. Radio sources were formerly known as *radio stars*.

radio star *See* radio source.

radio telescope An instrument for detecting and measuring electromagnetic radiation of radio frequencies that have passed through the *radio window in the earth's atmosphere and reached the surface of the earth. There are a great diversity of *radio sources within the universe and radio telescopes are required to detect both continuous emissions and specific spectral lines. They therefore require the highest possible angular resolution so that the details of radio sources

can be studied and they should be able to pick up weak signals. The simplest radio telescope consists of a paraboloidal steerable-dish aerial together with ancillary amplifiers. The paraboloidal reflecting dish surface reflects the incoming signal to the principal focus of the reflector. At this point the radio-frequency signals are amplified up to 1000 times and converted to a lower, intermediate, frequency before transmission by cable to the control building. Here the intermediate frequency is amplified again and passed to the detector and display unit. As the radio waves arriving from the surface of the reflector at the focus must be in phase, the surface of the dish must be very accurately constructed; for example, a 100-metre-diameter dish must be accurate to the nearest millimetre, when receiving radiation of 1 cm wavelength. To overcome the problem of constructing large dishes to such a high accuracy, the technique of *radio interferometry* has been developed. In this technique an array of small aerials connected by cable is used to simulate a large dish aerial. In earth-rotation *aperture synthesis*, the individual positions and displacements of an array of only a few such small aerials can be made to simulate an enormous dish aerial as the earth rotates. All but the smallest steerable dishes are constructed from metal mesh so that wind can pass through them. A few very large fixed dishes have been built into the earth's surface.

radiotherapy *See* radiology.

radio transmission The transmission of radio waves from a transmitting aerial to a receiving aerial. The radiation may take several paths (see illustration). The sum of the line-of-sight ground wave, the reflected ground wave, and the surface wave is called the *ground wave* (or *tropospheric wave*). *Sky waves* (or *ionospheric*

communication satellite

ionosphere

sky wave

line-of-sight ground wave

transmitting aerial

receiving aerial

surface wave

reflected ground wave

Radio transmission

waves) are reflected by the *ionosphere and enable long-distance transmissions to be made. The ionization of atoms and molecules in the ionosphere is caused largely by solar ultraviolet and X-radiation and therefore conditions differ between night and day. Ionization in the lower E-region of the ionosphere falls off at night in the absence of sunlight and ions and electrons tend to recombine. However, in the less dense (higher) F-region there are fewer collisions between ions and electrons and therefore fewer recombinations at night. The F-region is therefore a more effective reflector at night.

The UHF and VHF waves used in television broadcasting penetrate the ionosphere with little reflection. Therefore TV broadcasts can only be made over long distances by means of artificial satellites. *See also* radio.

radiotransparent Transparent to radiation, especially to X-rays and gamma-rays.

radio window A region of the electromagnetic spectrum in the radio-frequency band within which radio waves can be transmitted through the earth's atmosphere without significant reflection or attenuation by constituents of the atmosphere. It extends from about 10 megahertz to 100 gigahertz and enables radiation in this range from celestial radio sources to be picked up by *radio telescopes on the earth's surface. Below 100 MHz incoming radio waves are reflected by the *ionosphere and those above 100 GHz are increasingly affected by molecular absorption.

radium Symbol Ra. A radioactive metallic element belonging to *group II of the periodic table; a.n. 88; r.a.m. 226.0254; r.d. ~5; m.p. 700°C; b.p. 1140°C. It occurs in uranium ores (e.g. pitchblende). The most stable isotope is radium–226 (half-life 1602 years), which decays to radon. It is used as a radioactive source in research and, to some extent, in radiotherapy. The element was isolated from pitchblende in 1898 by Marie and Pierre Curie.

radius 1. (in anatomy) The smaller of the two bones in the lower section of the forelimb of a tetrapod vertebrate (*compare* ulna). The radius articulates with some carpal bones and the ulna at the wrist and with the *humerus at the elbow. This sophisticated articulation of the radius enables man (and some other animals) to twist the forearm (*see* pronation; supination). **2.** (in mathematics) *See* circle.

radius of curvature *See* centre of curvature.

radius of gyration Symbol k. The square root of the ratio of the *moment of inertia of a rigid body about an axis to the body's mass, i.e. $k^2 = I/m$, where I is the body's moment of inertia and m is its mass. If a rigid body has a moment of inertia I about an axis and mass m it behaves as if all its mass is rotating at a distance k from the axis.

radius vector *See* polar coordinates.

radon Symbol Rn. A colourless radioactive gaseous element belonging to group 0 of the periodic table (the *noble gases); a.n. 86; r.a.m. 222; d. 9.73 g dm^{-3}; m.p. −71°C. b.p. −61.8°C. At least 20 isotopes are known, the most stable being radon−222 (half-life 3.8 days). It is formed by decay of radium−226 and undergoes alpha decay. It is used in radiotherapy. Radon occurs naturally, particularly in areas underlain by granite, where it is thought to be a health hazard. As a noble gas, radon is practically inert, although a few compounds, e.g. radon fluoride, can be made. It was first isolated by Ramsey and Gray in 1908.

radula A tonguelike organ of molluscs, consisting of a horny strip whose surface is studded with rows of horny teeth for rasping food. In some species it is modified for scraping or boring.

raffinate A liquid purified by solvent extraction.

rainbow An optical phenomenon that appears as an arc of the colours of the spectrum across the sky when falling water droplets are illuminated by sunlight from behind the observer. The colours are produced by the refraction and internal reflection of

Primary rainbow (one internal reflection)

Secondary rainbow (two internal reflections)

the sunlight by the water drops. Two bows may be visible, the inner ring being known as the primary bow and the outer, in which the colours are reversed, as the secondary bow (see illustration).

rainfall *See* precipitation.

r.a.m. *See* relative atomic mass.

Raman effect A type of inelastic *scattering of light and ultraviolet radiation discovered in 1928 by Chandrasekara Raman (1880–1970). If a beam of monochromatic light is passed through a transparent substance some of the radiation will be scattered. Although most of the scattered radiation will be the same as the incident frequency, some will have frequencies above (anti-Stokes radiation) and below (Stokes radiation) that of the incident beam. This effect is known as *Raman scattering* and is due to inelastic collisions between photons and molecules leading to changes in the vibrational and rotational energy levels of the molecules. An increase in frequency represents a loss in molecular energy; a decrease

Ramapithecus

Rankine cycle

in frequency represents a gain in molecular energy.

This effect is used in *Raman spectroscopy* for investigating the vibrational and rotational energy levels of molecules. Because the scattering intensity is low, a laser source is used.

Ramapithecus A genus of extinct primates that lived between 8 and 14 million years ago. Fossil remains of ramapithecines have been found in India and Pakistan, the Near East, and East Africa. Jaw fragments suggest that they chewed from side to side and had fairly short muzzles, both of which are humanoid features. They were possibly the earliest representatives of the hominid line after this split from the pongids (apes) during the Miocene. *See also Dryopithecus*; *Australopithecus*.

ramjet *See* jet propulsion.

Ramsden eyepiece An eyepiece for optical instruments consisting of two identical plano-convex lenses with their convex faces pointing towards each other. They are separated by a distance of two thirds of the focal length of either lens.

Raney nickel A spongy form of nickel made by the action of sodium hydroxide on a nickel–aluminium alloy. The sodium hydroxide dissolves the aluminium leaving a highly active form of nickel with a large surface area. The material is a black pyro-phoric powder saturated with hydrogen. It is an extremely efficient catalyst, especially for hydrogenation reactions at room temperature. It was discovered in 1927 by the American chemist M. Raney.

rank (category) (in biology) The position or status of a *taxon in a *classification hierarchy. Examples of ranks are the family, genus, and species.

Rankine cycle A cycle of operations in a heat engine. The Rankine cycle more closely approximates to the cycle of a real steam engine that does the *Carnot cycle. It therefore predicts a lower ideal thermal efficiency than the Carnot cycle. In the Rankine cycle (see illustration), heat is added at constant pressure p_1, at which water is converted in a boiler to superheated steam; the steam expands at constant entropy to a pressure p_2 in the cylinder; heat is rejected at constant pressure p_2 in a condenser; the water so formed is compressed at constant entropy to pressure p_1 by a feed pump. The cycle was devised by W. J. M. Rankine (1820–70).

Rankine temperature scale An absolute temperature scale based on the Fahrenheit scale. Absolute zero on this scale, $0°R$, is equivalent to $-459.67°F$ and the melting point of ice ($32°F$) is therefore $(459.67 + 32 = 491.67)°R$. The scale was devised by W. J. M. Rankine (1820–70).

ranksite A mineral consisting of a mixed sodium carbonate, sodium sulphate, and potassium chloride, $2Na_2CO_3.9Na_2SO_4.KCl$.

Raoult's law The partial vapour pressure of a solvent is proportional to its mole fraction. If p is the vapour pressure of the solvent (with a substance dissolved in it) and X the mole fraction of solvent (number of moles of solvent divided by total number of moles) then $p = p_0X$, where p_0 is the vapour pressure of the pure solvent. A solution that obeys Raoult's law is said to be an *ideal solution*. In general the law holds only for dilute solutions, although some mixtures of liquids obey it over a whole range of concentrations. Such solutions are *perfect solutions* and occur when the intermolecular forces between molecules of the pure substances are similar to the forces between molecules of one and molecules of the other. Deviations in Raoult's law for mixtures of liquids cause the formation of *azeotropes. The law was discovered by the French chemist François Raoult (1830 –1901).

rare-earth elements *See* lanthanoids.

rarefaction A reduction in the pressure of a fluid and therefore of its density.

rare gases *See* noble gases.

Raschig process An industrial process for making chlorobenzene (and phenol) by a gas-phase reaction between benzene vapour, hydrogen chloride, and oxygen (air) at 230°C

$$2C_6H_6 + 2HCl + O_2 \rightarrow 2H_2O + 2C_6H_5Cl$$

The catalyst is copper(II) chloride. The chlorobenzene is mainly used for making phenol by the reaction

$$C_6H_5Cl + H_2O \rightarrow HCl + C_6H_5OH$$

This reaction proceeds at 430°C with a silicon catalyst.

Raschig synthesis *See* hydrazine.

raster The pattern of scanning lines on the screen of the cathode-ray tube in a television receiver or other device that provides a visual display.

rate constant (velocity constant) Symbol k. The constant in an expression for the rate of a chemical reaction in terms of concentrations (or activities). For instance, in a simple unimolecular reaction

$$A \rightarrow B$$

the rate is proportional to the concentration of A, i.e.

$$\text{rate} = k[A]$$

where k is the rate constant, which depends on the temperature. The equation is the *rate equation* of the reaction, and its form depends on the reaction mechanism.

rate-determining step The slowest step in a chemical reaction that involves a number of steps. In such reactions, there is often a single step that is appreciably slower than the other steps, and the rate of this determines the overall rate of the reaction.

rationalized units A system of units in which the defining equations have been made to conform to the geometry of the system in a logical way. Thus equations that involve circular symmetry contain the factor 2π, while those involving spherical symmetry contain the factor 4π. *SI and *Heaviside–Lorentz units are rationalized; *Gaussian units are unrationalized.

rational number Any number that can be expressed as the ratio of two integers. For example, $0.3333\ldots$ is rational because it can be written as $1/3$. $\sqrt{2}$, however, is *irrational.

Ratitae A group comprising the flightless birds, including the ostrich, kiwi, and emu. They have long legs, heavy bones, small wings, a flat breastbone, and curly feathers. These birds are thought to have descended from a variety of flying birds and are not representatives of a single homologous group.

ray 1. (in optics) A narrow beam of radiation or an idealized representation of such a beam on a *ray diagram*, which can be used to indicate the positions of the object and image in a system of lenses or mirrors. **2.** (in botany) *See* medullary ray.

Rayleigh criterion *See* resolving power.

Rayleigh–Jeans formula *See* Planck's radiation law.

Rayleigh scattering *See* scattering of electromagnetic radiation.

rayon A textile made from cellulose. There are two types, both made from wood pulp. In the viscose process, the pulp is dissolved in carbon disulphide and sodium hydroxide to give a thick brown liquid containing cellulose xanthate. The liquid is then forced through fine nozzles into acid, where the xanthate is decomposed and a cellulose filament is produced. The product is *viscose rayon*. In the acetate process cellulose acetate is made and dissolved in a solvent. The solution is forced through nozzles into air, where the solvent quickly evaporates leaving a filament of *acetate rayon*.

reactance Symbol X. A property of a circuit containing inductance or capacitance that together with any resistance makes up its *impedance. The impedance Z is given by $Z^2 = R^2 + X^2$, where R is the resistance. For a pure capacitance C, the reactance is given by $X_C = 1/2\pi fC$, where f is the frequency of the *alternating current; for a pure inductance L, $X_L = 2\pi fL$. If the resistance, inductance, and capacitance are in series the impedance $Z = \sqrt{[R^2 + (X_L - X_C)^2]}$. Reactance is measured in ohms.

reactant *See* chemical reaction.

reaction 1. A force that is equal in magnitude but opposite in direction to some other force, in accordance with Newton's third law of motion. If a body A exerts a force on body B, then B exerts an equal and opposite force on A. Thus, every force could be described as 'a reaction', and the term is better avoided, although it is still used in such terms as 'reaction propulsion'. **2.** *See* chemical reaction.

reaction propulsion *See* jet propulsion.

reaction time (latent period) The period of time between the detection of a stimulus at a sensory receptor and the performance of the appropriate response by the effector organ. This delay is caused by the time taken for the impulse to travel across the synapses of adjacent neurones. The reaction time for a *reflex response, involving only a single linking synapse, is very short.

reactive dye *See* dyes.

reactor 1. *See* nuclear reactor. **2.** Any device, such as an inductor or capacitor, that introduces *reactance into an electrical circuit.

reagent A substance reacting with another substance. Laboratory reagents are compounds, such as sulphuric acid, hydrochloric acid, sodium hydroxide, etc., used in chemical analysis or experiments.

realgar A red mineral form of arsenic(II) sulphide, As_2S_2.

real gas A gas that does not have the properties assigned to an *ideal gas. Its molecules have a finite size and there are forces between them (*see* equation of state).

real image *See* image.

real-is-positive convention (real-positive convention) A convention used in optical formulae relating to lenses and mirrors. In this convention, distances from optical components to real objects, images, and foci are taken as positive, whereas distances to virtual points are taken as negative.

rearrangement A type of chemical reaction in which the atoms in a molecule rearrange to form a new molecule.

Réaumur temperature scale A temperature scale in which the melting point of ice is taken as 0°R and the boiling point of water as 80°R. It was devised by René Antoine Réaumur (1683–1757).

Recent *See* Holocene.

receptacle 1. (thalamus or torus) The tip of a flower stalk, which bears the petals, sepals, stamens, and carpels. The way the receptacle develops determines the position of the flower parts. It can be dilated and dome-shaped, saucer-shaped, or hollow and enclosing the gynaecium. In some plants it may become part of the fruit (*see* pseudocarp). **2.** A swollen part of the thallus of some algae, e.g. *Fucus*, that bears the conceptacles in which the sex organs are situated.

receptor (in physiology) A cell or group of cells specialized to detect a particular stimulus and to initiate the transmission of impulses via the sensory nerves. The eyes, ears, nose, skin, and other sense organs all contain specific receptors responding to external stimuli; other receptors are sensitive to changes within the body. *See also* baroreceptor; chemoreceptor; mechanoreceptor; proprioceptor.

recessive The *allele that does not function when two different alleles are present in the cells of an organism. The aspect of a characteristic controlled by a recessive allele only appears when two such alleles are present, i.e. in the *double recessive condition*. *Compare* dominant.

recipient An individual who receives tissues or organs of the body from another (the *donor).

reciprocal cross A *cross reversing the roles of males and females to confirm the results obtained from an earlier cross. For example, if the pollen (male) from tall plants is transferred to the stigmas (female) of dwarf plants in one cross, the reciprocal cross would use the pollen of dwarf plants to pollinate the stigmas of tall plants.

reciprocal proportions *See* chemical combination.

recombinant DNA DNA that contains genes from different sources that have been combined by the techni-

ques of *genetic engineering rather than by breeding experiments. Genetic engineering is therefore also known as *recombinant DNA technology*.

recombination The rearrangement of genes that occurs when reproductive cells (gametes) are formed (*see* crossing over). Recombination results in offspring that have a combination of characteristics different from that of their parents. Recombination can also be induced artificially by *genetic engineering techniques.

recombination process The process in which a neutral atom or molecule is formed by the combination of a positive ion and a negative ion or electron; i.e. a process of the type:
$$A^+ + B^- \rightarrow AB$$
or
$$A^+ + e^- \rightarrow A$$
In recombination, the neutral species formed is usually in an excited state, from which it can decay with emission of light or other electromagnetic radiation.

recrystallization A process of repeated crystallization in order to purify a substance or to obtain more regular crystals of a purified substance.

rectification 1. (in physics) The process of obtaining a direct current from an alternating electrical supply. *See* rectifier. **2.** (in chemistry) The process of purifying a liquid by *distillation. *See* fractional distillation.

rectified spirit A constant-boiling mixture of *ethanol (95.6°) and water; it is obtained by distillation.

rectifier An electrical device that allows more current to flow in one direction than the other, thus enabling alternating e.m.f.s to drive only direct current. The device most commonly used for rectification is the semiconductor *diode. In *half-wave rectification*, achieved with one diode, a pulsating current is produced. In

Half-wave rectification

Full-wave rectification

full-wave rectification two diodes are used, one pair conducting during the first half cycle and the other conducting during the second half (see illustration). The full-wave rectified signal can be smoothed using a capacitor or an inductor.

rectum The portion of the *alimentary canal between the *colon and the *anus. Its main function is the storage of *faeces prior to elimination.

red algae *See* Rhodophyta.

red blood cell *See* erythrocyte.

red giant A *giant star thought to be in the later stages of *stellar evolution. It has a surface temperature in the range 2000–3000 K and a diameter 10–100 times that of the sun. *See also* Hertzsprung–Russell diagram.

red lead *See* dilead(II) lead(IV) oxide.

redox *See* oxidation–reduction.

redshift 1. (Doppler redshift) A displacement of the lines in the spectra of certain galaxies towards the red end of the visible spectrum (i.e. towards longer wavelengths). It is usually interpreted as a *Doppler effect resulting from the recession of the galaxies along the line of sight

(*see* expansion of the universe). The redshift is usually expressed as $\Delta\lambda/\lambda$, where $\Delta\lambda$ is the shift in wavelength of radiation of wavelength λ. For relatively low velocities of recession this is equivalent to v/c, where v is the relative velocity of recession and c is the speed of light. If very high velocities of recession are involved, a relativistic version of v/c is required (*see* relativistic velocity). The redshift of spectral lines occurs in all regions of the electromagnetic spectrum; ultraviolet can be shifted into the visible region and visible radiation can be shifted into the infrared region. **2. (gravitational** *or* **Einstein redshift)** A similar displacement of spectral lines towards the red caused not by a Doppler effect but by a high gravitational field. This type of redshift was predicted by Einstein and some astronomers believe that this is the cause of the large redshifts of *quasars, which can be as high as 3.35.

reducing agent (reductant) A substance that brings about reduction in other substances. It achieves this by being itself oxidized. Reducing agents contain atoms with low oxidation numbers; that is the atoms have gained electrons. In reducing other substances, these atoms lose electrons.

reductant *See* reducing agent.

reduction *See* oxidation–reduction.

reduction division *See* meiosis.

refinery gas *See* petroleum.

refining The process of purifying substances or extracting substances from mixtures.

reflectance The ratio of the radiant flux reflected by a surface to that falling on it. This quantity is also known as the *radiant reflectance*. The radiant reflectance measured for a specified wavelength of the incident radiant flux is called the *spectral reflectance*.

reflecting telescope (reflector) *See* telescope.

reflection The return of all or part of a beam of particles or waves when it encounters the boundary between two media. The *laws of reflection* state: (1) that the incident ray, the reflected ray, and the normal to the reflecting interface at the point of incidence are all in the same plane; (2) that the angle of incidence equals the angle of reflection. *See also* mirror; reflectance; total internal reflection.

reflex An automatic and innate response to a particular stimulus, such as withdrawal of the hand from a painful stimulus (such as fire). A reflex response is extremely rapid. This is because it is mediated by a simple nervous circuit called a *reflex arc*, which at its simplest involves only a receptor linked to a sensory neurone, which synapses with a motor neurone (supplying the effector) in the spinal cord or brain. *See also* conditioning.

refluxing A laboratory technique in which a liquid is boiled in a container attached to a condenser (*reflux condenser*), so that the liquid continuously flows back into the container. It is used for carrying out reactions over long periods in organic synthesis.

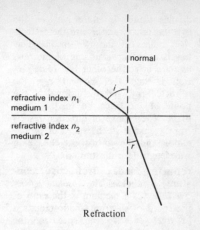

Refraction

reforming The conversion of straight-chain alkanes into branched-chain alkanes by *cracking or by catalytic reaction. It is used in petroleum refining to produce hydrocarbons suitable for use in gasoline. Benzene is also manufactured from alkane hydrocarbons by catalytic reforming. *Steam reforming* is a process used to convert methane (from natural gas) into a mixture of carbon monoxide and hydrogen, which is used to synthesize organic chemicals. The reaction

$$CH_4 + H_2O \rightarrow CO + 3H_2$$

occurs at about 900°C using a nickel catalyst.

refracting telescope (refractor) *See* telescope.

refraction The change of direction suffered by wavefront as it passes obliquely from one medium to another in which its speed of propagation is altered. The phenomenon occurs with all types of waves, but is most familiar with light waves. In optics the direction is changed in accordance with *Snell's law*, i.e. $n_1\sin i = n_2\sin r$, where i and r are the angles made by the incident beam of radiation and the refracted beam to the normal (an imaginary line perpen-

refractive index

dicular to the interface between the two media); n_1 and n_2 are the *refractive indices of the two media. This law is also known as one of the *laws of refraction*. The other law of refraction is that the incident ray, the refracted ray, and the normal at the point of incidence lie in the same plane. The change of direction results from a change in the speed of propagation and the consequent change in wavelength (see illustration).

refractive index (refractive constant) Symbol n. The *absolute refractive index* of a medium is the ratio of the speed of electromagnetic radiation in free space to the speed of the radiation in that medium. As the refractive index varies with wavelength, the wavelength should be specified. It is usually given for yellow light (sodium D-lines; wavelength 589.3 nm). The *relative refractive index* is the ratio of the speed of light in one medium to that in an adjacent medium. *See also* refraction.

refractivity A measure of the extent to which a medium will deviate a ray of light entering its surface. In some contexts it is equal to $(n - 1)$, where n is the *refractive index.

refractometer Any of various instruments for measuring the *refractive index of a substance or medium. An example is the *Pulfrich refractometer*, which is a glass block with a polished top, with a small cell on top of the block for liquid samples. A telescope rotating on a vertical circular scale is used to find the angle (α) between the top of the block and the direction in which the limiting ray (from incident light parallel to the top face) leaves the side of the block. If the refractive index of the block (n_g) is known, that of the liquid can be calculated using $n = \sqrt{(n_g^2 - \sin^2\alpha)}$.

refractory 1. Having a high melting point. Metal oxides, carbides, and silicides tend to be refractory, and are

Refrigerator: vapour–compression cycle

extensively used for lining furnaces. **2.** A refractory material.

refrigerant *See* refrigeration.

refrigeration The process of cooling a substance and of maintaining it at a temperature below that of its surroundings. Most commercial refrigerators use a cycle of operations equivalent to those of a *heat pump (see illustrations). In the *vapour-compression cycle* a volatile liquid (the *refrigerant*) passes to an evaporator where it boils, extracting heat from the refrigerator chamber as it does so. The vapour is then compressed by a compressor, where its temperature and pressure are raised. The compressed vapour then passes to a condenser, where the refrigerant discharges its heat to the surroundings. The liquid so formed then passes to a storage vessel and finally to an expansion valve, through which the high-pressure liquid expands to a low-pressure before entering the evaporator again. The refrigerant used is ammonia, sulphur dioxide, or a haloalkane.

In the *vapour-absorption cycle*, there are no moving parts and energy is supplied as heat, either an electric heater or a gas burner. The refriger-

heat
extracted

condenser · ammonia vapour

hydrogen gas · separator

water

ammonia
vapour
and water

refrigerator

generator

ammonia vapour
and hydrogen

absorber · ammonia
water

heat source

Refrigerator: vapour–absorption cycle

ant is usually ammonia, which is liberated from a water solution and moved through the evaporator by a stream of hydrogen gas under pressure. Heat is applied to the generator, raising the ammonia and water vapour to the separator; the ammonia separates from the water and passes to the condenser, where it cools and liquefies, giving off its latent heat to the surroundings. The liquid ammonia is then mixed with hydrogen gas, which carries it through the evaporator and helps in the process of evaporation. The hydrogen and ammonia vapour then enter the absorber, where the water returned from the separator dissolves the ammonia before returning to the generator.

Small-scale refrigeration is sometimes achieved by means of the *Seebeck effect and the *Peltier effect at junctions of *n*- and *p*-type semiconductors.

regeneration The growth of new tissues or organs to replace those lost or damaged by injury. Many plants can regenerate a complete plant from a shoot segment or a single leaf, this

being the basis of many horticultural propagation methods (*see* cutting). The capacity for regeneration in animals is less marked. Some planarians and sponges can regenerate whole organisms from small pieces, and crustaceans (e.g. crabs), echinoderms (e.g. brittlestars), and some reptiles and amphibians can grow new limbs or tails (*see* autotomy), but in mammals regeneration is largely restricted to wound healing.

regma A dry fruit that is characteristic of the geranium family. It is similar to the *carcerulus but breaks up into one-seeded parts, each of which splits open to release a seed.

Regnault's method A technique for measuring gas density by evacuating and weighing a glass bulb of known volume, admitting gas at known pressure, and reweighing. The determination must be carried out at constant known temperature and the result corrected to standard temperature and pressure. The method is named after the French chemist Henri Victor Regnault (1810–78).

reinforcement (in animal behaviour) Increasing (or decreasing) the frequency of a particular behaviour through *conditioning, by arranging for some biologically important event (the *reinforcer*) always to follow another event. In instrumental conditioning an *appetitive reinforcer*, or *reward* (e.g. food), given after a response made by the animal, increases that response; an *aversive reinforcer*, or *punishment* (e.g. an electric shock) decreases the response.

relative aperture *See* aperture.

relative atomic mass (atomic weight; r.a.m.) Symbol A_r. The ratio of the average mass per atom of the naturally occurring form of an element to 1/12 of the mass of a carbon–12 atom.

relative density (r.d.) The ratio of the *density of a substance to the density of some reference substance. For liquids or solids it is the ratio of the density (usually at 20°C) to the density of water (at its maximum density). This quantity was formerly called *specific gravity*. Sometimes relative densities of gases are used; for example, relative to dry air, both gases being at s.t.p.

relative humidity *See* humidity.

relative molecular mass (molecular weight) Symbol M_r. The ratio of the average mass per molecule of the naturally occurring form of an element or compound to 1/12 of the mass of a carbon–12 atom. It is equal to the sum of the relative atomic masses of all the atoms that comprise a molecule.

relative permeability *See* permeability.

relative permittivity *See* permittivity.

relativistic mass The mass of a moving body as measured by an observer in the same frame of reference as the body. According to the special theory of *relativity the mass m of a body moving at speed v is given by $m = m_0/\sqrt{(1 - v^2/c^2)}$, where m_0 is its *rest mass and c is the speed of light. The relativistic mass therefore only differs significantly from the rest mass if its speed is a substantial fraction of the speed of light. If $v = c/2$, for example, the relativistic mass is 15% greater than the rest mass.

relativistic mechanics An extension of Newtonian mechanics that takes into account the theory of *relativity.

relativistic quantum theory *See* quantum theory; quantum field theory.

relativistic speed A speed that is sufficiently large to make the mass of a body significantly greater than its *rest mass. It is usually expressed as a proportion of the speed of light. *See* relativity; relativistic mass.

relativity One of several theories concerning motion, designed to account for departures from *Newtonian mechanics that occur with very high-speed relative motion. The theory implied is now usually one of two proposed by Albert Einstein (1879–1955). The *special theory* was proposed in 1905 and referred to inertial (nonaccelerated) frames of reference. It assumes that physical laws are identical in all frames of reference and that the speed of light *in vacuo*, c, is constant throughout the universe and is independent of the speed of the observer. The theory develops a system of mathematics to reconcile these apparently conflicting statements. A conclusion of the theory is that the mass of a body, m, increases with its velocity, v, according to the relationship $m = m_0/\sqrt{(1 - v^2/c^2)}$, where m_0 is the *rest mass of the body. Einstein also concluded that if a body loses energy L, its mass diminishes by L/c^2. Einstein generalized this conclusion to the far-reaching postulate that the mass of a body is a measure of its energy content, according to the equation $m = E/c^2$ (or $E = mc^2$ in its more familiar form).

In the *general theory* of 1915, Einstein extended his earlier work to include accelerated systems, which led to his analysis of gravitation. He interpreted the universe in terms of a four-dimensional space–time continuum in which the presence of mass curves space in such a way that the gravitational field is created. The small differences between Newton's interpretation of gravitation and Einstein's interpretation have provided a means of testing the two theories. For example, the motion of the planet Mercury, thought to be anomalous in terms of Newtonian mechanics, can be explained by relativity. Moreover, Einstein's prediction that light rays passing close to the sun would be bent by its gravitational field has also been confirmed, at least qualitatively, by experiments performed during solar eclipses. The general theory is one of several current theories of gravitation; most of these have a good deal in common with it.

relaxation oscillator *See* oscillator.

relay An electrical or electronic device in which a variation in the current in one circuit controls the current in a second circuit. These devices are used in an enormous number of different applications in which electrical control is required. The simplest is the electromechanical relay in which the first circuit energizes an electromagnet, which operates a switch in a second circuit. The *thyratron gas-filled relay found many uses in the past but has now been largely replaced by the *thyristor solid-state relay.

releaser *See* sign stimulus.

reluctance Symbol *R*. The ratio of the magnetomotive force to the total magnetic flux in a magnetic circuit or component. It is measured in henries.

reluctivity The reciprocal of magnetic *permeability.

rem *See* radiation units.

remanence (retentivity) The magnetic flux density remaining in a ferromagnetic substance when the saturating field is reduced to zero. *See* hysteresis.

remote sensing The gathering and recording of information concerning the earth's surface by techniques that do not involve actual contact with the object or area under study. These techniques include photography (e.g. aerial photography), multispectral imagery, infrared imagery, and radar. Remote sensing is generally carried out from aircraft and, increasingly, satellites. The techniques are used, for example, in cartography (map making).

renewable energy sources Sources of energy that do not use up the earth's finite mineral resources. Nonrenewable sources are *fossil fuels and fission fuels (*see* nuclear fission). Various renewable energy sources are being used or investigated. *See* geothermal energy; hydroelectric power; nuclear fusion; solar energy; tides; wind power; wave power.

renin A proteolytic enzyme (*see* protease) that is involved in the formation of the hormone *angiotensin, which raises blood pressure. Renin is secreted into the blood by cells of the kidney glomeruli under the control of the sympathetic nervous system; its release also occurs in response to a fall in blood-sodium levels and to falling blood pressure. Renin catalyses cleavage of circulating α_2-globulin to produce angiotensin I, precursor of the active hormone.

rennin An enzyme secreted by cells lining the stomach in mammals that is responsible for clotting milk. It acts on a soluble milk protein (*caseinogen*), which it converts to the insoluble form *casein. This ensures that milk remains in the stomach long enough to be acted on by protein-digesting enzymes.

renormalization A procedure used in relativistic *quantum field theory to deal with the fact that in *perturbation theory calculations give rise to infinities beyond the first term. Renormalization was first used in *quantum electrodynamics, where the infinities were removed by taking the observed mass and charge of the electron as 'renormalized' parameters rather than the 'bare' mass and charge.

Theories for which finite results for all perturbation-theory calculations exist, by taking a finite number of parameters from experiment and using renormalization, are called *renormalizable*. Only certain types of quantum field theories are renormalizable. Theories that need an infinite number of parameters are said to be *nonrenormalizable* and are regarded as unacceptable as complete consistent physical theories. The *gauge theories that describe the strong, weak, and electromagnetic interactions are renormalizable. The quantum theory of gravitational interactions is a nonrenormalizable theory, which perhaps indicates that gravity needs to be unified with other fundamental interactions before one can have a consistent quantum theory of gravity.

replacing bone *See* cartilage bone.

reproduction The production of new individuals more or less similar in form to the parent organisms. This may be achieved by a number of means (*see* sexual reproduction; asexual reproduction) and serves to perpetuate or increase a species.

Reptilia The class that contains the first entirely terrestrial vertebrates, which can live in dry terrestrial habitats as their skin is covered by a layer of horny scales, preventing water loss. They breathe atmospheric oxygen by means of lungs assisted by respiratory movements principally involving the ribs (there is no diaphragm). Reptiles are cold-blooded (*see* poikilothermy) but behavioural patterns make it possible for them to maintain a fairly even body temperature throughout the day. Fertilization is internal and the majority of reptiles lay eggs on land. These eggs have a porous shell to provide protection from desiccation and allow gas exchange. In some reptiles the eggs are retained within the body of the mother until the young are ready to hatch, thereby greatly reducing juvenile mortality (*see* ovoviviparity). The class includes the modern crocodiles, lizards and snakes (*see* Squamata), and tortoises and turtles, as well as many extinct forms, such as the *dinosaurs and *Pterosauria.

resin A synthetic or naturally occurring *polymer. Synthetic resins are used in making *plastics. Natural resins are acidic chemicals secreted by many trees (especially conifers) into ducts or canals. They are found either as brittle glassy substances or dissolved in essential oils. Their functions are probably similar to those of gums and mucilages, i.e. protective.

resistance Symbol R. The ratio of the potential difference across an electrical component to the current passing through it. It is thus a measure of the component's opposition to the flow of electric charge. In general, the resistance of a metallic conductor increases with temperature, whereas the resistance of a *semiconductor decreases with temperature.

resistance thermometer (resistance pyrometer) A *thermometer that relies on the increase of electrical resistance of a metal wire with rising temperature, according to:
$$R = R_0(1 + aT + bT^2),$$
where R is the resistance of the wire at the unknown temperature T and R_0 is the resistance of the wire at a reference temperature, usually $0°C$; a and b are constants characteristic of the metal of the wire. The metal most frequently used is platinum and the platinum resistance coil is usually incorporated into one arm of a *Wheatstone bridge. The effect of the temperature change on the leads carrying current to the platinum coil is compensated by including a pair of dummy leads within the casing carrying the coil. *See also* thermistor.

resistivity Symbol ρ. A measure of a material's ability to oppose the flow of an electric current. The resistivity of a material is given by RA/l, where R is the resistance of a uniform specimen of the material, having a length l and a cross-sectional area A. It is usually given at $0°C$ or $20°C$ and is measured in ohm metres. It was formerly known as *specific resistance*.

resistor A component in an electrical or electronic circuit that is present because of its electrical resistance. For electronic purposes many resistors are either wire-wound or consist of carbon particles in a ceramic binder. The ceramic coating carries a number or colour code indicating the value of the resistance. Some resistors can be varied manually by means of a sliding contact; others are markedly dependent on temperature or illumination.

resolution 1. (in chemistry) The process of separating a racemic mixture into its optically active constituents. In some cases the crystals of the two forms have a different appearance, and the separation can be done by hand. In general, however, physical methods (distillation, crystallization, etc.) cannot be used because the optical isomers have identical physical properties. The most common technique is to react the mixture with a compound that is itself optically active, and then separate the two. For instance, a racemic mixture of *l*-A and *d*-A reacted with *l*-B, gives two compounds AB that are not optical isomers but diastereoisomers and can be separated and reconverted into the pure *l*-A and *d*-A. Biological techniques using bacteria that convert one form but not the other can also be used. 2. (in mathematics) The separation of a vector quantity into two components, which are usually at right angles to each other. Thus, a force F acting on a body in a vertical plane at an angle θ to the horizontal can be resolved into a horizontal component $F\cos\theta$ and a vertical component $F\sin\theta$, both in the same plane as the original force. 3. (in optics) *See* resolving power.

resolving power A measure of the ability of an optical instrument to form separable images of close objects or to separate close wavelengths of radiation. The *chromatic resolving power* for any spectroscopic instrument is equal to $\lambda/\delta\lambda$,

where $\delta\lambda$ is the difference in wavelength of two equally strong spectral lines that can barely be separated by the instrument and λ is the average wavelength of these two lines. For a telescope forming images of stars the *angular resolving power* is the smallest angular separation of the images; the *linear resolving power* is the linear separation of the images in the focal plane. In a telescope forming images of two stars, as a result of diffraction by the lens aperture each image consists of a bright central blob surrounded by light and dark rings. According to the *Rayleigh criterion* for resolution, the central ring of one image should fall on the first dark ring of the other. The angular resolving power in radians is then $1.22\lambda/d$, where d is the diameter of the objective lens in centimetres and λ is the wavelength of the light (usually taken as 560 nanometres). For microscopes, the resolving power is usually taken as the minimum distance between two points that can be separated. In both cases, the smaller the resolving power, the better the resolution; to avoid this apparent paradox the resolving power is now sometimes taken as the reciprocals of the quantities stated above.

resonance 1. (in physics) An oscillation of a system at its natural frequency of vibration, as determined by the physical parameters of the system. It has the characteristic that large amplitude vibrations will ultimately result from low-power driving of the system. Resonance can occur in atoms and molecules, mechanical systems, and electrical circuits (*see* resonant circuit; resonant cavity). 2. (in particle physics) A very short-lived *elementary particle that can be regarded as an excited state of a more stable particle. Resonances decay by the strong interaction (*see* fundamental interactions) in 10^{-24} second. 3. (in chemistry) The representation of the structure of a molecule by two or more conventional formulae. For example, the formula of methanal can be represented by a covalent structure $H_2C=O$, in which there is a double bond in the carbonyl group. It is known that in such compounds the oxygen has some negative charge and the carbon some positive charge. The true bonding in the molecule is somewhere between $H_2C=O$ and the ionic compound $H_2C^+O^-$. It is said to be a *resonance hybrid* of the two, indicated by

$$H_2C=O \leftrightarrow H_2C^+O^-$$

The two possible structures are called *canonical forms*, and they need not contribute equally to the actual form. Note that the double-headed arrow does not imply that the two forms are in equilibrium.

resonant cavity (cavity resonator) A closed space within a conductor in which an electromagnetic field can be made to oscillate at frequencies above those at which a *resonant circuit will operate. The resonant frequency of the oscillation will depend on the dimensions and the shape of the cavity. The device is used to produce microwaves (*see* klystron; magnetron).

resonant circuit A reactive circuit (*see* reactance) so arranged that it is capable of *resonance. In a *series resonant circuit* a resistor, inductor, and capacitor are arranged in series. Resonance occurs when the *impedance (Z) is a minimum and the current amplitude therefore a maximum. In a *parallel resonant circuit* the inductance and capacitance are in parallel and resonance (with minimal current amplitude) occurs at maximum impedance. The frequency at which resonance occurs is called the *resonant frequency*. In a series resonant circuit

$$Z = R + i[\omega L - 1/\omega C],$$

where $\omega = 2\pi f$ and f is the frequency, R is the resistance, L is the inductance, and C is the capacitance.

At resonance, Z is a minimum and $\omega L = 1/\omega C$, i.e. the circuit behaves as if it is purely resistive. In the parallel circuit, resonance occurs when $R^2 + \omega^2 L^2 = L/C$, which in most cases also approximates to $\omega L = 1/\omega C$. Resonant circuits are widely used in *radio to select one carrier-wave frequency in preference to others.

respiration The metabolic process in animals and plants in which organic substances are broken down to simpler products with the release of energy, which is incorporated into special energy-carrying molecules (*see* ATP) and subsequently used for other metabolic processes. In most plants and animals respiration requires oxygen, and carbon dioxide is an end product. The exchange of oxygen and carbon dioxide between the body tissues and the environment is called *external respiration*. In many animals the exchange of gases takes place at *respiratory organs (e.g. *lungs in air-breathing vertebrates) and is assisted by *respiratory movements (e.g. breathing). In plants oxygen enters through pores on the plant surface and diffuses through the tissues via intercellular spaces or dissolved in tissue fluids.

Respiration at the cellular level is known as *internal* (or *tissue*) *respiration* and can be divided into two stages. In the first, *glycolysis, glucose is broken down to pyruvate. This does not require oxygen and is a form of *anaerobic respiration. In the second stage, the *Krebs cycle, pyruvate is broken down by a cyclic series of reactions to carbon dioxide and water. This is the main energy-yielding stage and requires oxygen. The processes of glycolysis and the Krebs cycle are common to all plants and animals that respire aerobically (*see* aerobic respiration).

respiratory chain *See* electron transport chain.

respiratory movement The muscular movement that enables the passage of air to and from the lungs or other *respiratory organs of an animal. The mechanism of the movement varies with the species. In insects abdominal muscles relax and contract rhythmically to encourage the flow of air through the *tracheae. In amphibians air is drawn into the lungs by a pumping action of the muscles in the floor of the mouth. *Breathing* in mammals involves the muscle of the *diaphragm and the *intercostal muscles between the ribs. Contraction of these muscles lowers the diaphragm and raises the ribs, so that the lungs expand and air is drawn in. Relaxation has the opposite effect and forces air out during exhalation.

respiratory organ Any animal organ across which exchange of carbon dioxide and oxygen takes place. The surface membranes of such organs are always moist, thin, and well supplied with blood. Examples are the *lungs of air-breathing vertebrates, the *gills of fish, and the *tracheae of insects.

respiratory pigment A coloured compound that is capable of reversibly binding with oxygen at high oxygen concentrations and releasing it at low oxygen concentrations. Such pigments are present in the blood, transporting oxygen within the circulatory system from the *respiratory organs to the tissues of the body. In vertebrates the respiratory pigment is *haemoglobin, contained in the erythrocytes (red blood cells). *See also* haemocyanin.

respiratory quotient (RQ) The ratio of the volume of carbon dioxide produced by an organism during respiration to the volume of oxygen consumed. The RQ is usually about 0.8.

response The physiological, muscular, or behavioural activity that can be elicited by a *stimulus.

rest energy The *rest mass of a body expressed in energy terms according to the relationship $E_0 = m_0c^2$, where m_0 is the rest mass of the body and c is the speed of light.

resting potential The difference in electrical potential that exists across the membrane of a nerve cell that is not in the process of transmitting a nerve impulse. The resting potential is maintained by means of the *sodium pump. *Compare* action potential.

restitution coefficient Symbol e. A measure of the elasticity of colliding bodies. For two spheres moving in the same straight line,

$$c = (v_2 - v_1)/(u_1 - u_2),$$

where u_1 and u_2 are the velocities of bodies 1 and 2 before collision ($u_1 > u_2$) and v_1 and v_2 are the velocities of 1 and 2 after impact ($v_2 > v_1$). If the collision is perfectly elastic $e = 1$ and the kinetic energy is conserved; for an inelastic collision $e < 1$.

rest mass The mass of a body at rest when measured by an observer who is at rest in the same frame of reference. *Compare* relativistic mass.

restriction enzyme (restriction endonuclease) A type of enzyme that can cleave molecules of foreign DNA at a particular site. Restriction enzymes are produced by many bacteria and protect the cell by cleaving (and therefore destroying) the DNA of invading viruses. The bacterial cell is protected from attack by its own restriction enzymes by modifying the bases of its DNA during replication. Restriction enzymes are widely used in the techniques of genetic engineering (*see* gene cloning; gene library; gene sequencing; genetic fingerprinting; restriction mapping).

restriction fragment length polymorphism (RFLP) The occurrence of different cleavage sites for *restriction enzymes in the DNA of different individuals of the same species. Cleavage of DNA from different individuals with restriction enzymes thus produces differing sets of restriction fragments. The deletion of existing restriction sites or the creation of new ones is the result of random base changes in the noncoding stretches of DNA (*introns) between genes. RFLPs have provided geneticists with a powerful set of genetic markers for mapping the genome (*see* restriction mapping) and for identifying particular genes (*see* gene tracking).

restriction mapping A technique for determining the sites at which a length of DNA (e.g. from a chromosome) is cleaved by *restriction enzymes. By cleaving the DNA with various such enzymes, both individually and in combination, and analysing the resultant number and size of fragments by electrophoresis, a *restriction map*, indicating the order of restriction sites in the original DNA, can be deduced. Restriction mapping is routinely applied to organisms undergoing genetic investigation. Gene deletions or rearrangements that alter the restriction sites can be detected as changes in the pattern of fragments obtained. This may be used, for instance, to diagnose certain genetic abnormalities in the fetus. The fragments are separated by gel electrophoresis and identified using specific *gene probes, as in the *Southern *blotting technique. The absence of a certain fragment in a fetal DNA digest can be diagnostic of a pathological change in the fetal gene containing the corresponding restriction site.

resultant A *vector quantity that has the same effect as two or more other vector quantities of the same kind. *See* parallelogram of vectors.

retardation (deceleration) The rate of reduction of speed, velocity, or rate of change.

retardation plate A transparent plate of a birefringent material, such as quartz, cut parallel to the optic axis. Light falling on the plate at 90° to the optic axis is split into an ordinary ray and an extraordinary ray (*see* double refraction), which travel through the plate at different speeds. By cutting the plate to different thicknesses a specific phase difference can be introduced between the transmitted rays. In the *half-wave plate* a phase difference of π radians, equivalent to a path difference of half a wavelength, is introduced. In the *quarter-wave plate* the waves are out of step by one quarter of a wavelength.

retina The light-sensitive membrane that lines the interior of the eye. The retina consists of two layers. The inner layer contains nerve cells, blood vessels, and two types of light-sensitive cells (*rods and *cones). The outer layer is pigmented, which prevents the back reflection of light and consequent decrease in visual acuity. Light passing through the lens stimulates individual rods and cones, which generates nerve impulses that are transmitted through the optic nerve to the brain, where the visual image is formed.

retinol *See* vitamin A.

retort 1. A laboratory apparatus consisting of a glass bulb with a long neck. **2.** A vessel used for reaction or distillation in industrial chemical processes.

retrograde motion 1. The apparent motion of a planet from east to west as seen from the earth against the background of the stars. **2.** The clockwise rotation of a planet, as seen from its north pole. *Compare* direct motion.

retrorocket A small rocket motor that produces thrust in the opposite direction to a *rocket's main motor or motors in order to decelerate it.

retrovirus An RNA-containing virus that converts its RNA into DNA by means of the enzyme *reverse transcriptase; this enables it to become integrated into its host's DNA. Retroviruses can cause cancer in animals: they contain oncogenes (cancer-causing genes; *see* oncogenic), which are activated when the virus enters its host cell and starts to replicate. *See also* provirus.

reverberation time The time taken for the energy density of a sound to fall to the threshold of audibility from a value 10^6 times as great; i.e. a fall of 60 decibels. It is an important characteristic of an auditorium. The optimum value is proportional to the linear dimensions of the auditorium.

reverberatory furnace A metallurgical furnace in which the charge to be heated is kept separate from the fuel. It consists of a shallow hearth on which the charge is heated by flames that pass over it and by radiation reflected onto it from a low roof.

reverse osmosis A method of obtaining pure water from water containing a salt, as in *desalination. Pure water and the salt water are separated by a semipermeable membrane and the pressure of the salt water is raised above the osmotic pressure, causing water from the brine to pass through the membrane into the pure water. This process requires a pressure of some 25 atmospheres, which makes it difficult to apply on a large scale.

reverse transcriptase An enzyme, occurring in *retroviruses, that catalyses the formation of double-stranded DNA using the single RNA strand of the viral genome as template. This enables the viral genome to be inserted into the host's DNA and replicated by the host. Reverse transcriptase is thus an RNA-directed DNA *polymerase. The enzyme is used in genetic engineering for producing *complementary DNA from messenger RNA.

reversible process Any process in which the variables that define the state of the system can be made to change in such a way that they pass through the same values in the reverse order when the process is reversed. It is also a condition of a reversible process that any exchanges of energy, work, or matter with the surroundings should be reversed in direction and order when the process is reversed. Any process that does not comply with these conditions when it is reversed is said to be an *irreversible process*. All natural processes are irreversible, although some processes can be made to approach closely to a reversible process.

Reynolds number Symbol *Re*. A dimensionless number used in fluid dynamics to determine the type of flow of a fluid through a pipe, to design prototypes from small-scale models, etc. It is the ratio $v\rho l/\eta$, where v is the flow velocity, ρ is the fluid density, l is a characteristic linear dimension, such as the diameter of a pipe, and η is the fluid viscosity. In a smooth straight uniform pipe, laminar flow usually occurs if $Re < 2000$ and turbulent flow is established if $Re > 3000$.

RFLP *See* restriction fragment length polymorphism.

R_F **value** (in chromatography) The distance travelled by the solvent front divided by the distance travelled by a given component. For a given system at a known temperature, it is a characteristic of the component and can be used to identify components.

rhachis *See* rachis.

rhe A unit of fluidity equal to the reciprocal of the *poise.

rhenium Symbol Re. A silvery-white metallic *transition element; a.n. 75; r.a.m. 186.2; r.d. 21.0; m.p. 3180°C; b.p. 5620°C. The element is obtained as a by-product in refining molybdenum, and is used in certain alloys (e.g. rhenium–molybdenum alloys are superconducting). The element forms a number of complexes with oxidation states in the range 1–7.

rheology The study of the deformation and flow of matter.

rheopexy The process by which certain thixotropic substances set more rapidly when they are stirred, shaken, or tapped. Gypsum in water is such a *rheopectic substance*.

rheostat A variable *resistor, the value of which can be changed without interrupting the current flow. In the common wire-wound rheostat, a sliding contact moves along the length of the coil of wire.

rhesus factor (Rh factor) An *antigen whose presence or absence on the surface of red blood cells forms the basis of the rhesus *blood group system. (The factor was first recognized in rhesus monkeys.) Most people possess the Rh factor, i.e. they are rhesus positive (Rh+). People who lack the factor are Rh−. If Rh+ blood is given to an Rh− patient, the latter develops anti-Rh antibodies. Subsequent transfusion of Rh+ blood results in *agglutination, with serious consequences. Similarly, an Rh− pregnant woman carrying an Rh+ fetus may develop anti-Rh antibodies in her blood; these will react with the blood of a subsequent Rh+ fetus, causing anaemia in the newborn baby.

rhizoid One of a group of delicate and often colourless hairlike outgrowths found in certain algae and the gametophyte generation of bryophytes and some pteridophytes. They anchor the plant to the substrate and absorb water and mineral salts.

rhizome A horizontal underground stem. It enables the plant to survive from one growing season to the next and in some species it also serves to propagate the plant vegetatively. It may be thin and wiry, as in couch grass, or fleshy and swollen, as in *Iris*. Compact upright underground stems, as in rhubarb, strawberry, and primrose, are often called *rootstocks*.

rhodium Symbol Rh. A silvery-white metallic *transition element; a.n. 45; r.a.m. 102.9; r.d. 12.4; m.p. 1966°C; b.p. 3727°C. It occurs with platinum and is used in certain platinum alloys (e.g. for thermocouples) and in plating jewellery and optical reflectors. Chemically, it is not attacked by acids (dissolves only slowly in aqua regia) and reacts with nonmetals (e.g. oxygen and chlorine) at red heat. Its main oxidation state is +3 although it also forms complexes in the +4 state. The element was discovered in 1803 by W. H. Wollaston.

Rhodophyta (red algae) A division of *algae that are often pink or red in colour due to the presence of the pigments phycocyanin and phycoerythrin. Members of the Rhodophyta may be unicellular or multicellular; the latter form branched flattened thalli or filaments. They are commonly found along the coasts of tropical areas.

rhodopsin (visual purple) The light-sensitive pigment found in the *rods of the vertebrate retina. Light falling on the rod causes a chemical change in rhodopsin, which initiates the transmission of a nerve impulse to the brain. The great sensitivity of rhodopsin allows vision in dim light (night vision).

rhombencephalon *See* hindbrain.

rhombus A parallelogram in which all the sides are of equal length.

rhumb line (loxodrome) (in navigation) A line of constant compass direction that cuts across all lines of longitude at the same angle. The rhumb line is not the shortest distance between two points unless the two points are on the same meridian or on the equator. On the Mercator map projection a rhumb line is represented by a straight line.

rhyolite An igneous rock, the volcanic equivalent of *granite. It is usually glassy or cryptocrystalline and consists of quartz, feldspars, and mica or amphibole. It may contain larger crystals (phenocrysts) set in a much finer-grained matrix.

rhytidome *See* bark.

rib One of a series of slender curved bones that form a cage to enclose, support, and protect the heart and lungs (*see* thorax). Ribs occur in pairs, articulating with the *thoracic vertebrae of the spinal column at the back and (in reptiles, birds, and mammals) with the *sternum (breastbone) in front. Movements of the rib cage, controlled by *intercostal muscles* between the ribs, are important in breathing (*see* respiratory movement).

riboflavin *See* vitamin B complex.

ribonuclease *See* RNAase.

ribonucleic acid *See* RNA.

ribose A *monosaccharide, $C_5H_{10}O_5$, rarely occurring free in nature but important as a component of *RNA (ribonucleic acid). Its derivative *deoxyribose*, $C_5H_{10}O_4$, is equally important as a constituent of *DNA (deoxyribonucleic acid), which carries the genetic code in chromosomes.

ribosomal RNA *See* ribosome; RNA.

ribosome A small spherical body within a living cell that is the site of *protein synthesis. Ribosomes consist of a type of RNA (called *ribosomal RNA*) and protein. Usually there are many ribosomes in a cell, either attached to the *endoplasmic reticulum or free in the cytoplasm. During protein synthesis they are associated with messenger RNA in the process of *translation.

Richardson (Richardson–Dushman) equation *See* thermionic emission.

Richter scale A logarithmic scale devised in 1935 by C. F. Richter (1900–85) to compare the magnitude of earthquakes. The scale ranges from 0 to 10 and the Richter scale value is related to the logarithm of the amplitude of the ground motion divided by the period of the dominant wave, subject to certain corrections. On this scale a value of 2 can just be felt as a tremor and damage to buildings occurs for values in excess of 6. The largest shock recorded had a magnitude of 8.9.

rickettsia A very small rod-shaped nonmotile microorganism that resembles a bacterium in its cellular structure and method of asexual reproduction but, like a virus, is totally parasitic, being unable to reproduce outside the cells of its host. Rickettsiae can infect such arthropods as ticks, fleas, lice, and mites, through which they can be transmitted to vertebrates, including man. The group includes the causal agents of trench fever, Rocky Mountain spotted fever, and forms of typhus.

rigidity modulus *See* elastic modulus.

ring A closed chain of atoms in a molecule. In compounds, such as naphthalene, in which two rings share a common side, the rings are *fused rings*. *Ring closures* are chemical reactions in which one part of a chain reacts with another to form a ring, as in the formation of *lactams and *lactones.

r-isomer *See* absolute configuration.

Ringer's solution *See* physiological saline.

ritualization An evolutionary process in which the form or context of an action is altered because it comes to play a role in social communication. For example, many *courtship and greeting ceremonies in animals include ritual food presentation (though the quantities of food may be negligible), derived from the action of feeding the young.

RMS value *See* root-mean-square value.

RNA

Detail of molecular structure of sugar–phosphate backbone. Each ribose unit is attached to a phosphate group and a base, forming a nucleotide

Single-stranded structure of RNA

The four bases of RNA

Molecular structure of RNA

RNA (ribonucleic acid) A complex organic compound (a nucleic acid) in living cells that is concerned with *protein synthesis. In some viruses, RNA is also the hereditary material. Most RNA is synthesized in the nucleus and then distributed to various parts of the cytoplasm. An RNA molecule consists of a long chain of *nucleotides in which the sugar is *ribose and the bases are adenine, cytosine, guanine, and uracil (see illustration; *compare* DNA). *Messen-ger RNA* (*mRNA*) is responsible for carrying the *genetic code transcribed from DNA to specialized sites within the cell (known as *ribosomes), where the information is translated into protein composition (*see* transcription; translation). *Ribosomal RNA* (*rRNA*) is present in ribosomes; it is single-stranded but helical regions are formed by *base pairing within the strand. *Transfer RNA* (*tRNA, soluble RNA, sRNA*) is involved in the assembly of amino acids in a protein

602

chain being synthesized at a ribosome. Each tRNA is specific for an amino acid and bears a triplet of bases complementary with a triplet on mRNA (*see* codon).

RNAase (RNase; ribonuclease) Any enzyme that catalyses the cleavage of nucleotides in RNA. Each RNAase has a specificity for a different cleavage site. For example, RNAase A is a digestive enzyme secreted by the pancreas that hydrolyses phosphodiester bonds in the nucleotide chain. Other RNAases are active at the cellular level, for instance in modifying transfer RNA and ribosomal RNA after transcription.

RNA polymerase *See* polymerase.

Rochelle salt Potassium sodium tartrate tetrahydrate, $KNaC_4H_4O_6.4H_2O$. A colourless crystalline salt used for its piezoelectric properties.

Rochon prism An optical device consisting of two quartz prisms; the first, cut parallel to the optic axis, receives the light; the second, with the optic axis at right angles, transmits the ordinary ray without deviation but the extraordinary ray is deflected and can be absorbed by a screen. The device can be used to produce plane-polarized light and it can also be used with ultraviolet radiation.

rock An aggregate of mineral particles that makes up part of the earth's crust. It may be consolidated or unconsolidated (e.g. sand, gravel, mud, shells, coral, and clay).

rock crystal *See* quartz.

rocket A space vehicle or projectile that is forced through space or the atmosphere by *jet propulsion and that carries its own propellants and oxidizers. It is therefore independent of the earth's atmosphere for lift, thrust, or oxygen and is the only known vehicle for travel outside the earth's atmosphere. Rocket motors (or rocket engines) are currently driven by solid or liquid chemical propellants, which burn in an oxidizer carried within the rocket. Typical liquid bipropellant combinations include hydrazine fuel with dinitrogen tetroxide oxidizer and kerosine fuel with nitric acid oxidizer. Experimental rocket motors have also been tested using ionized gases and plasmas to provide thrust (*see also* ion engine). The measure of a rocket motor's performance is its *specific impulse.

rock salt *See* halite.

rod A type of light-sensitive receptor cell present in the retinas of vertebrates. Rods contain the pigment *rhodopsin and are essential for vision in dim light. They are not evenly distributed on the retina, being absent in the *fovea and occupying all of the retinal margin. *Compare* cone.

Rodentia An order of mammals characterized by a single pair of long curved incisors in each jaw. These teeth are specialized for gnawing: they continue growing throughout life and have enamel only on the front so that they wear to a chisel-shaped cutting edge. Rodents often breed throughout the year and produce large numbers of quickly maturing young. The order includes the squirrels, beavers, rats, mice, and porcupines.

roentgen The former unit of dose equivalent (*see* radiation units). It is named after the discoverer of X-rays, W. K. Roentgen (1845–1923).

rolling friction *Friction between a rolling wheel and the plane surface on which it is rotating. As a result of any small distortions of the two surfaces, there is a frictional force with a component, F_1, that opposes the motion. If N is the normal force, $F_r = N\mu_r$, where μ_r is called the *coefficient of rolling friction*.

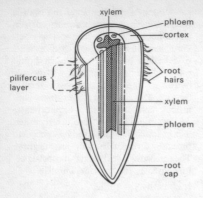

xylem
phloem
cortex
piliferous layer
root hairs
xylem
phloem
root cap

Section through the tip of a plant root

root 1. (in botany) The part of a vascular plant that grows beneath the soil surface in response to gravity and water. It anchors the plant in the soil and absorbs water and mineral salts. Unlike the stem, it never produces leaves, buds, or flowers and never contains chlorophyll. The *radicle (embryonic root) may give rise either to a *tap root system* with a single main *tap root* from which lateral roots develop, or a *fibrous root system*, with many roots of equal size. The *apical meristem at the root tip gives rise to a protective sheath, the *root cap, and to the primary tissues of the root. The vascular tissues usually form a central core (see illustration). This distinguishes roots from stems, in which the vascular tissue often forms a ring. A short distance behind the root tip *root hairs* develop from the epidermis and greatly increase the surface area for absorption of water and minerals. Beyond this, lateral roots develop.

Roots may be modified in various ways. Some are swollen with food to survive the winter, as in the carrot. Certain plants, such as orchids, have absorptive aerial roots; others, such as ivy, have short clasping roots for climbing. The roots of leguminous plants, such as beans and peas, contain nodules due to the presence of symbiotic bacteria that fix free nitrogen from the soil (*see* nitrogen fixation). Other modifications include *prop roots, stilt roots, and buttress roots, which support the plant.

2. (in dentistry) The portion of a *tooth that is not covered with enamel and is embedded in a socket in the jawbone. Incisors, canines, and premolars have single roots; molars normally have several roots.

3. (in anatomy) The point of origin of a nerve in the central nervous system. There are two roots for every *spinal nerve, a dorsal and a ventral root.

4. (in mathematics) **a.** One of the equal factors of a number or quantity, e.g. the cube root of 8 is 2. In general, the nth root of a number or quantity a is a number or quantity x that satisfies the equation $x^n = a$. **b.** The value or values of an independent variable in an equation that satisfies that equation.

root cap (calyptra) A cone-shaped structure that covers the root tip and develops as a result of cell division by a meristem at the root apex (*see* calyptrogen). It protects the root tip as it grows between the soil particles. The cells are constantly worn away by friction and are replaced by the meristem.

root-mean-square value (RMS value) 1. (in statistics) A typical value of a number (n) of values of a quantity $(x_1, x_2, x_3 \ldots)$ equal to the square root of the sum of the squares of the values divided by n, i.e.

$$\text{RMS value} = [\sqrt{(x_1^2 + x_2^2 + x_3^2 \ldots)/n}]$$

2. (in physics) A typical value of a continuously varying quantity, such as an alternating electric current, obtained similarly from many samples taken at regular time intervals during a cycle. Theoretically this can be shown to be the *effective value*, i.e. the value of the equivalent direct current that would produce the same power dissipation in a given resistor. For a sinusoidal current this is equal to $I_m/\sqrt{2}$, where I_m is the maximum value of the current.

rootstock *See* rhizome.

Rose's metal An alloy of low melting point (about 100°C) consisting of 50% bismuth, 25–28% lead, and 22–25% tin.

rot (in mathematics) *See* curl.

rotary converter A device for converting direct current to alternating current or one d.c. voltage to another. It consists of an electric motor coupled to a generator.

rotational motion The laws relating to the rotation of a body about an axis are analogous to those describing linear motion. The *angular displacement* (θ) of a body is the angle in radians through which a point or line has been rotated in a specified sense about a specified axis. The *angular velocity* (ω) is the rate of change of angular displacement with time, i.e. $\omega = d\theta/dt$, and the *angular acceleration* (α) is the rate of change of angular velocity, i.e. $\alpha = d\omega/dt = d^2\theta/dt^2$.
The equations of linear motion have analogous rotational equivalents, e.g.:
$$\omega_2 = \omega_1 + \alpha t$$
$$\theta = \omega_1 t + \alpha t^2/2$$
$$\omega_2^2 = \omega_1^2 + 2\theta\alpha$$
The counterpart of Newton's second law of motion is $T = I\alpha$, where T is the *torque causing the angular acceleration and I is the *moment of inertia of the rotating body.

rotor The rotating part of an electric motor, electric generator, turbine, etc. *Compare* stator.

roundworms *See* Nematoda.

R–S convention *See* absolute configuration.

rubber A polymeric substance obtained from the sap of the tree *Hevea brasiliensis*. Crude natural rubber is obtained by coagulating and drying the sap (latex), and is then modified by *vulcanization and com-

pounding with fillers. It is a polymer of *isoprene containing the unit $-CH_2C(CH_3):CHCH_2-$. Various synthetic rubbers can also be made. *See* neoprene; nitrile rubber; silicones.

rubidium Symbol Rb. A soft silvery-white metallic element belonging to *group I of the periodic table; a.n. 37; r.a.m. 85.47; r.d. 1.53; m.p. 38.4°C; b.p. 688°C. It is found in a number of minerals (e.g. lepidolite) and in certain brines. The metal is obtained by electrolysis of molten rubidium chloride. The naturally occurring isotope rubidium–87 is radioactive (*see* rubidium–strontium dating). The metal is highly reactive, with properties similar to those of other group I elements, igniting spontaneously in air. It was discovered spectroscopically by R. W. Bunsen and G. R. Kirchhoff in 1861.

rubidium–strontium dating A method used to date geological specimens based on the decay of the radioisotope rubidium–87 into the stable isotope strontium–87. Natural rubidium contains 27.85% of rubidium–87, which has a half-life of 4.7 × 10^{10} years. The ratio $^{87}Rb/^{87}Sr$ in a specimen gives an estimate of its age (up to several thousand million years).

ruby The transparent red variety of the mineral *corundum, the colour being due to the presence of traces of chromium. It is a valuable gemstone, more precious than diamonds. The finest rubies are obtained from Mogok in Burma, where they occur in metamorphic limestones; Sri Lanka and Thailand are the only other important sources. Rubies have been produced synthetically by the Verneuil flame-fusion process. Industrial rubies are used in lasers, watches, and other precision instruments.

Ruminantia A suborder of hooved mammals (*see* Artiodactyla) comprising the sheep, cattle, goats, deer, and antelopes. They are characterized by a four-chambered stomach, the first part of which is the *rumen*, which contains microorganisms that break down cellulose. Food is temporarily stored in the rumen, then regurgitated and chewed to a pulp before being swallowed and passed to the rest of the stomach where digestion is continued.

runner A stem that grows horizontally along the soil surface and gives rise to new plants either from axillary or terminal buds. Runners are seen in the creeping buttercup and the strawberry. *Offsets*, e.g. those of the houseleek, are short runners.

rusting Corrosion of iron (or steel) to form a hydrated iron(III) oxide $Fe_2O_3.xH_2O$. Rusting occurs only in the presence of both water and oxygen. It is an electrochemical process in which different parts of the iron surface act as electrodes in a cell reaction. At the anode, iron atoms dissolve as Fe^{2+} ions:
$$Fe(s) \rightarrow Fe^{2+}(aq) + 2e$$
At the cathode, hydroxide ions are formed:
$$O_2(aq) + 2H_2O(l) + 4e \rightarrow 4OH^-(aq)$$
The $Fe(OH)_2$ in solution is oxidized to Fe_2O_3. Rusting is accelerated by impurities in the iron and by the presence of acids or other electrolytes in the water.

rusts A group of parasitic *basidiomycete fungi. Many of these species attack the leaves and stems of cereal crops: characteristic rust-coloured streaks of spores appear on infected plants. The life cycles of some rusts may be complex; many form a number of different types of spore and some require two different host plants. *Compare* smuts.

ruthenium Symbol Ru. A hard white metallic *transition element; a.n. 44; r.a.m. 101.07; r.d. 12.41; m.p. 2310°C; b.p. 3900°C. It is found associated with platinum and is used as a catalyst and in certain platinum alloys. Chemically, it dissolves in fused alkalis but is not attacked by acids. It reacts with oxygen and halogens at high temperatures. It also forms complexes with a range of oxidation states. The element was isolated by K. K. Klaus in 1844.

rutherfordium *See* transactinide elements.

rutile A mineral form of titanium(IV) oxide, TiO_2.

Rydberg constant Symbol R. A constant that occurs in the formulae for atomic spectra and is related to the binding energy between an electron and an atomic nucleus. It is connected to other constants by the relationship $R = \mu_0^2 me^4 c^3/8h^3$, where μ_0 is the magnetic constant (*see* permeability), m and e are the mass and charge of an electron, c is the speed of light, and h is the *Planck constant. It has the value 1.097×10^7 m^{-1}.

Rydberg spectrum An absorption spectrum of a gas in the ultraviolet region, consisting of a series of lines that become closer together towards shorter wavelengths, merging into a continuous absorption region. The absorption lines correspond to electronic transitions to successively higher energy levels. The onset of the continuum corresponds to photoionization of the atom or molecule, and can thus be used to determine the ionization potential.

S

saccharide *See* sugar.

saccharose *See* sucrose.

sacculus (saccule) A chamber of the *inner ear from which the *cochlea arises in reptiles, birds, and mammals. It bears patches of sensory epithelium concerned with balance (*see* macula).

Sachse reaction A reaction of methane at high temperature to produce ethyne:

$$2CH_4 \rightarrow C_2H_2 + 3H_2$$

The reaction occurs at about 1500°C, the high temperature being obtained by burning part of the methane in air.

sacral vertebrae The vertebrae that lie between the lumbar and the caudal vertebrae in the *vertebral column. The function of the sacral vertebrae is to articulate securely with the *pelvic girdle, and they are usually fused to form a single bone (the *sacrum*) to provide a firm support. The number of sacral vertebrae varies from animal to animal. Amphibians have a single sacral vertebra, reptiles have two, and mammals have three or more.

sacrificial protection (cathodic protection) The protection of iron or steel against corrosion (*see* rusting) by using a more reactive metal. A common form is galvanizing (*see* galvanized iron), in which the iron surface is coated with a layer of zinc. Even if the zinc layer is scratched, the iron does not rust because zinc ions are formed in solution in preference to iron ions. Pieces of magnesium alloy are similarly used in protecting pipelines, etc.

safranin A stain used in optical microscopy that colours lignified tissues, cutinized tissues, and nuclei red and chloroplasts pink. It is used mainly for plant tissues, in conjunction with a green or blue counterstain.

sal ammoniac *See* ammonium chloride.

salicylic acid (1-hydroxybenzoic acid) A naturally occurring carboxylic acid, HOC_6H_4COOH, found in certain plants. It is used in making

aspirin and in the foodstuffs and dyestuffs industries.

saline Describing a chemical compound that is a salt, or a solution containing a salt. *See also* physiological saline.

salinometer An instrument for measuring the salinity of a solution. There are two main types: one is a type of *hydrometer to measure density; the other is an apparatus for measuring the electrical conductivity of the solution.

saliva A watery fluid secreted by the *salivary glands in the mouth. Production of saliva is stimulated by the presence of food in the mouth and also by the smell or thought of food. Saliva contains mucin, which lubricates food and eases its passage into the oesophagus, and in some animals the enzyme ptyalin (*see* amylase), which begins the digestion of starch. The saliva of insects is rich in digestive enzymes, and that of bloodsucking animals contains an anticoagulant.

salivary glands Glands in many terrestrial animals that secrete *saliva into the mouth. In man there are three pairs: the sublingual, submandibular, and the submaxillary glands. The salivary gland cells of some insect larvae produce giant (*polytene*) chromosomes, which are widely used in the study of genetics and protein synthesis.

Salmonella A genus of rod-shaped Gram-negative bacteria that inhabit the intestine and cause disease (*salmonellosis*) in humans and animals. They are aerobic or facultatively anaerobic, and most are motile. Salmonellae can exist for long periods outside their host, and may be found, for example, in sewage and surface water. Humans may become infected by consuming contaminated water or food, especially animal products, such as eggs, meat, and milk, or vegetables that have been fertilized with contaminated manure. The bacteria can also be transmitted from human or animal carriers by unhygienic food preparation. Various species of *Salmonella* cause gastroenteritis and septicaemia; typhoid fever and paratyphoid fever are caused by *S. typhi* and *S. paratyphi*, respectively.

sal soda Anhydrous *sodium carbonate, Na_2CO_3.

salt A compound formed by reaction of an acid with a base, in which the hydrogen of the acid has been replaced by metal or other positive ions. Typically, salts are crystalline ionic compounds such as Na^+Cl^- and $NH_4^+NO_3^-$. Covalent metal compounds, such as $TiCl_4$, are also often regarded as salts.

salt bridge An electrical connection made between two half cells. It usually consists of a glass U-tube filled with agar jelly containing a salt, such as potassium chloride. A strip of filter paper soaked in the salt solution can also be used.

salt cake Industrial *sodium sulphate.

saltpetre *See* nitre.

samara A dry single-seeded indehiscent fruit in which the fruit wall hardens and extends to form a long membranous winglike structure that aids dispersal. Examples are ash and elm fruits. The sycamore fruit is a double samara and technically a *schizocarp. *See also* achene.

samarium Symbol Sm. A soft silvery metallic element belonging to the *lanthanoids; a.n. 62; r.a.m. 150.35; r.d. 7.52 (20°C); m.p. 1073°C; b.p. 1791°C. It occurs in monazite and bastnatite. There are seven naturally occurring isotopes, all of which are stable except samarium–147, which is weakly radioactive (half-life 2.5 × 10^{11} years). The metal is used in special alloys for making nuclear-reactor parts as it is a neutron absorber. Samarium oxide (Sm_2O_3) is used in small quantities in special optical glasses. The largest use of the element is in the ferromagnetic alloy $SmCo_5$, which produces permanent magnets five times stronger than any other material. The element was discovered by François Lecoq de Boisbaudran in 1879.

sampling The selection of small groups of entities to represent a large number of entities in *statistics. In *random sampling* each individual of a population has an equal chance of being selected as part of the sample. In *stratified random sampling*, the population is divided into strata, each of which is randomly sampled and the samples from the different strata are pooled. In *systematic sampling*, individuals are chosen at fixed intervals; for example, every tenth animal in a population. In *sampling with replacement*, each individual chosen is replaced before the next selection is made.

sand Particles of rock with diameters in the range 0.06–2.00 mm. Most sands are composed chiefly of particles of quartz, which are derived from the weathering of quartz-bearing rocks.

Sandmeyer reaction A reaction of diazonium salts used to prepare chloro- or bromo-substituted aromatic compounds. The method is to diazotize an aromatic amide at low temperature and add an equimolar solution of the halogen acid and copper(I) halide. A complex of the diazonium salt and copper halide forms, which decomposes when the temperature is raised. The copper halide acts as a catalyst in the reaction of the halide ions from the acid, for example

$$C_6H_5N_2^+(aq) + Cl^-(aq) +$$
$$CuCl(aq) \rightarrow$$
$$C_6H_5Cl(l) + N_2(g) + CuCl(aq)$$

The reaction was discovered in 1884 by the German chemist Traugott Sandmeyer (1854–1922). *See also* Gattermann reaction.

sandwich compound A transition-metal complex in which a metal atom or ion is 'sandwiched' between two rings of atoms. *Ferrocene was the first such compound to be prepared. *See also* metallocene.

saponification The reaction of esters with alkalis to give alcohols and salts of carboxylic acids:

RCOOR′ + OH⁻ → RCOO⁻ +
R′OH

See esterification; soap.

sapphire Any of the gem varieties of *corundum except ruby, especially the blue variety, but other colours of sapphire include yellow, brown, green, pink, orange, and purple. Sapphires are obtained from igneous and metamorphic rocks and from alluvial deposits. The chief sources are Sri Lanka, Kashmir, Burma, Thailand, East Africa, the USA, and Australia. Sapphires are used as gemstones and record-player styluses. They are synthesized by the Verneuil flame-fusion process.

saprophyte Any plant or microorganism that feeds by absorbing dead organic matter. Most saprophytes are bacteria and fungi; the term *saprobe* is frequently used to indicate saprophytic fungi. Saprophytes are important in *food chains as they bring about decay and release nutrients for plant growth. *Compare* parasitism.

sapwood (alburnum) The outer wood of a tree trunk or branch. It consists of living *xylem cells, which both conduct water and provide structural support. *Compare* heartwood.

satellite 1. (*or* **natural satellite**) A relatively small natural body that orbits a planet. For example, the earth's only natural satellite is the moon. **2.** (*or* **artificial satellite**) A man-made spacecraft that orbits the earth, moon, sun, or a planet. Artificial satellites are used for a variety of purposes. *Communication satellites* are used for relaying telephone, radio, and television signals round the curved surface of the earth (*see* synchronous orbit). They are of two types: *passive satellites* reflect signals from one point on the earth's surface to another; *active satellites* are able to amplify and retransmit the signals that they pick up. *Astronomical satellites* are equipped to gather and transmit to earth astronomical information from space, including conditions in the earth's atmosphere, which is of great value in weather forecasting.

satellite DNA A portion of the *DNA in plant and animal cells that consists of repeating sequences of *nucleotides. Some 30% of human DNA consists of such satellite regions. They do not appear to form part of any *genes but their function is unclear. The term is sometimes applied to DNA other than that found in the cell nucleus, in particular mitochondrial DNA, which differs in its density.

saturated 1. (of a solution) Containing the maximum equilibrium amount of solute at a given temperature. In a saturated solution the dissolved substance is in equilibrium with undissolved substance; i.e. the rate at which solute particles leave the solution is exactly balanced by the rate at which they dissolve. A solution containing less than the equilibrium amount is said to be *unsaturated*. One containing more than the equilibrium amount is *supersaturated*. Supersaturated solutions can be made by slowly cooling a saturated solution. Such solutions are metastable; if a small crystal seed is added the excess solute crystallizes out of solution. **2.** (of a vapour) *See* vapour pressure. **3.** (of a ferromagnetic material) Unable to be magnetized more strongly as all the domains are orientated in the direction of the field. **4.** (of a compound) Consisting of molecules that have only single bonds (i.e. no double

saturation

or triple bonds). Saturated compounds can undergo substitution reactions but not addition reactions. *Compare* unsaturated.

saturation 1. *See* colour. **2.** *See* supersaturation.

Saturn A planet having its orbit between those of Jupiter and Uranus. Its mean distance from the sun is 1.43×10^9 km and its equatorial diameter 120 800 km; it has a sidereal period of 29.46 years. Although it is the second largest planet its mean density is lower than any other, its relative density being 0.7. It has at least 22 satellites, of which the largest, Titan, is the only planetary satellite to have a dense atmosphere. Saturn is believed to consist of a dense rocky core, surrounded by hydrogen compressed to such an extent that it behaves like a metal; this layer merges with an atmosphere of hydrogen. The planet is best known for the system of rings that surrounds it in an equatorial plane; the rings have an overall diameter of about 273 000 km. They are believed to consist of millions of particles of ice (possibly with a rock core) with diameters between <1 mm and 10 m. Saturn was given a preliminary examination by the satellite Pioneer II in 1979 and more detailed studies by Voyager I in 1980 and Voyager II in 1981.

sawtooth waveform A waveform in which the variable increases uniformly with time for a fixed period, drops

Sawtooth waveform

sharply to its initial value, and then repeats the sequence periodically. The illustration shows the ideal waveform and the waveform generated by practical electrical circuits. Sawtooth generators are frequently used to provide a time base for electronic circuits, as in the *cathode-ray oscilloscope.

s-block elements The elements of the first two main groups of the *periodic table; i.e. groups IA (Li, Na, K, Rb, Cs, Fr) and IIA (Be, Mg, Ca, Sr, Ba, Ra). The outer electronic configurations of these elements all have inert-gas structures plus outer ns^1 (IA) or ns^2 (IIA) electrons. The term thus excludes elements with incomplete inner *d*-levels (transition metals) or with incomplete inner *f*-levels (lanthanoids and actinoids) even though these often have outer ns^2 or occasionally ns^1 configurations. Typically, the *s*-block elements are reactive metals forming stable ionic compounds containing M^+ or M^{2+} ions. *See* alkali metals; alkaline-earth metals.

scalar product (dot product) The product of two vectors U and V, with components U_1, U_2, U_3 and V_1, V_2, V_3, respectively, given by:
$$U.V = U_1V_1 + U_2V_2 + U_3V_3.$$
It can also be written as $UV\cos\theta$, where U and V are the lengths of U and V, respectively, and θ is the angle between them. *Compare* vector product.

scalar quantity A quantity in which direction is either not applicable (as in temperature) or not specified (as in speed). *Compare* vector quantity.

scalene Denoting a triangle having three unequal sides.

scaler (scaling circuit) An electronic counting circuit that provides an output when it has been activated by a prescribed number of input pulses. A *decade scaler* produces an output pulse when it has received ten or a multiple of ten input pulses; a *binary* *scaler* produces its output after two input pulses.

scales The small bony or horny plates forming the body covering of fish and reptiles. The wings of some insects, notably the Lepidoptera (butterflies and moths), are covered with tiny scales that are modified cuticular hairs.

In fish there are three types of scales. *Placoid scales* (*denticles*), characteristic of cartilaginous fish, are small and toothlike, with a projecting spine and a flattened base embedded in the skin. They are made of *dentine, have a pulp cavity, and the spine is covered with a layer of enamel. Teeth are probably modified placoid scales. *Cosmoid scales*, characteristic of lungfish and coelacanths, have an outer layer of hard *cosmin* (similar to dentine) covered by modified enamel (*ganoine*) and inner layers of bone. The scale grows by adding to the inner layer only. In modern lungfish the scales are reduced to large bony plates. *Ganoid scales* are characteristic of primitive ray-finned fishes, such as sturgeons. They are similar to cosmoid scales but have a much thicker layer of ganoine and grow by the addition of material all round. The scales of modern teleost fish are reduced to thin bony plates.

In reptiles there are two types of scales: horny epidermal *corneoscutes* sometimes fused with underlying bony dermal *osteoscutes*.

scandium Symbol Sc. A rare soft silvery metallic element belonging to group IIIB of the periodic table; a.n. 21; r.a.m. 44.956; r.d. 2.985 (alpha form), 3.19 (beta form); m.p. 1540°C; b.p. 2850°C. Scandium often occurs in *lanthanoid ores, from which it can be separated on account of the greater solubility of its thiocyanate in ether. The only natural isotope, which is not radioactive, is scandium−45, and there are nine radioactive isotopes, all with relatively short half-

lives. Because of the metal's high reactivity and high cost no substantial uses have been found for either the metal or its compounds. Predicted in 1869 by Mendeleev, and then called *ekaboron*, the oxide (called *scandia*) was isolated by Nilson in 1879.

scanning The process of repeatedly crossing a surface or volume with a beam, aerial, or moving detector in order to bring about some change to the surface or volume, to measure some activity, or to detect some object. The fluorescent screen of a television picture tube is scanned by an electron beam in order to produce the picture; an area of the sky may be scanned by the movable dish aerial of a radio telescope in order to detect celestial bodies, etc.

scanning electron microscope *See* electron microscope.

scanning tunnelling microscope (STM) A type of microscope in which a fine conducting probe is held close to the surface of a sample. Electrons tunnel between the sample and the probe, producing an electrical signal. The probe is slowly moved across the surface and raised and lowered so as to keep the signal constant. A profile of the surface is produced, and a computer-generated contour map of the surface is produced. The technique is capable of resolving individual atoms, but works better with conducting materials. *See also* atomic force microscope.

scapula (shoulder blade) The largest of the bones that make up each half of the *pectoral (shoulder) girdle. It is a flat triangular bone, providing anchorage for the muscles of the forelimb and an articulation for the *humerus at the *glenoid cavity. It is joined to the *clavicle (collar bone) in front.

scattering of electromagnetic radiation The process in which electromagnetic radiation is deflected by particles in the matter through which it passes. In *elastic scattering* the photons of the radiation are reflected; i.e. they bounce off the atoms and molecules without any change of energy. In this type of scattering, known as *Rayleigh scattering* (after Lord Rayleigh; 1842–1919), there is a change of phase but no frequency change. In *inelastic scattering* and *superelastic scattering*, there is interchange of energy between the photons and the particles. Consequently, the scattered photons have a different wavelength as well as a different phase. Examples include the *Raman effect and the *Compton effect. *See also* Tyndall effect.

scheelite A mineral form of calcium tungstate, $CaWO_4$, used as an ore of tungsten. It occurs in contact metamorphosed deposits and vein deposits as colourless or white tetragonal crystals.

Schiff's base A compound formed by a condensation reaction between an aromatic amine and an aldehyde or ketone, for example

$$RNH_2 + R'CHO \rightarrow RN{:}CHR' + H_2O$$

The compounds are often crystalline and are used in organic chemistry for characterizing aromatic amines (by preparing the Schiff's base and measuring the melting point). They are named after the German chemist Hugo Schiff (1834–1915).

Schiff's reagent A reagent used for testing for aldehydes and ketones; it consists of a solution of fuchsin dye that has been decolorized by sulphur dioxide. Aliphatic aldehydes restore the pink immediately, whereas aromatic ketones have no effect on the reagent. Aromatic aldehydes and aliphatic ketones restore the colour slowly. It is named after the German chemist Hugo Schiff (1834–1915).

schist A group of coarse-grained metamorphic rocks characterized by the presence of platy minerals (e.g. micas, chlorite, talc, hornblende, and graphite) that show parallel alignment at right angles to the direction of stress. The original bedding of the rock is absent. Schists readily split into layers along schistosity planes, parallel to the alignment of the minerals.

schizocarp A dry indehiscent fruit formed from carpels that develop into separate one-seeded fragments called *mericarps*, which may be dehiscent, as in the *regma, or indehiscent, as in the *cremocarp and *carcerulus.

schizogeny The localized separation of plant cells to form a cavity (surrounded by the intact cells) in which secretions accumulate. Examples are the resin canals of some conifers and the oil ducts of caraway and aniseed fruits. *Compare* lysigeny.

schlieren photography A technique that enables density differences in a moving fluid to be photographed. In the turbulent flow of a fluid, for example, short-lived localized differences in density create differences of refractive index, which show up on photographs taken by short flashes of light as streaks (German: Schliere). Schlieren photography is used in wind-tunnel studies to show the density gradients created by turbulence and the shock waves around a stationary model.

Schmidt camera *See* telescope.

schönite A mineral form of potassium sulphate, K_2SO_4.

Schottky defect *See* defect.

Schrödinger equation An equation used in wave mechanics (*see* quantum mechanics) for the wave function of a particle. The time-independent Schrödinger equation is:

$$\nabla^2\psi + 8\pi^2m(E - U)\psi/h^2 = 0$$

where ψ is the wave function, ∇^2 the Laplace operator (*see* Laplace equation), h the Planck constant, m the particle's mass, E its total energy, and U its potential energy.

Schwann cell A cell that forms the *myelin sheath of a nerve fibre. Each cell is responsible for a single length of the fibre (called an *internode*); adjacent internodes are separated by small gaps (*nodes of Ranvier*). During its development the cell wraps itself around the fibre, so the sheath consists of concentric layers of Schwann cell membrane. These cells are named after the German anatomist T. Schwann (1810–82).

Schwarzschild radius A critical radius of a body of given mass that must be exceeded if light is to escape from that body. It is equal to $2GM/c^2$, where G is the gravitational constant, c is the speed of light, and M is the mass of the body. If the body collapses to such an extent that its radius is less than the Schwarzschild radius the escape velocity becomes equal to the speed of light and the object becomes a *black hole. The Schwarzschild radius is then the radius of the hole's event horizon.

scientific notation *See* standard form.

scintillation counter A type of particle or radiation counter that makes use of the flash of light (scintillation) emitted by an excited atom falling back to its ground state after having been excited by a passing photon or particle. The scintillating medium is usually either solid or liquid and is used in connection with a *photomultiplier, which produces a pulse of current for each scintillation. The pulses are counted with a *scaler. In certain cases, a pulse-height analyser can be used to give an energy spectrum of the incident radiation.

scintillator *See* phosphor.

scion *See* graft.

sclera *See* sclerotic.

sclerometer A device for measuring the hardness of a material by determining the pressure on a standard point that is required to scratch it

or by determining the height to which a standard ball will rebound from it when dropped from a fixed height. The rebound type is sometimes called a *scleroscope*.

scleroprotein Any of a group of proteins found in the exoskeletons of some invertebrates, notably insects. Scleroproteins are formed by conversion of the relatively soft elastic larval protein by a natural tanning process (*sclerotization*) involving orthoquinones. These are secreted and form cross linkages between polypeptides of the proteins, producing a hard rigid covering.

sclerotic (sclera) The tough external layer of the vertebrate eye. At the front of the eye, the sclera is modified to form the *cornea.

scorpions *See* Arachnida.

scotopic vision The type of vision that occurs when the *rods in the eye are the principal receptors, i.e. when the level of illumination is low. With scotopic vision colours cannot be identified. *Compare* photopic vision.

SCP *See* single-cell protein.

screen grid A wire grid in a tetrode or pentode *thermionic valve, placed between the anode and the *control grid to reduce the grid–anode capacitance. *See also* suppressor grid.

screw A simple *machine effectively consisting of an inclined plane wrapped around a cylinder. The mechanical advantage of a screw is $2\pi r/p$, where r is the radius of the thread and p is the *pitch*, i.e. the distance between adjacent threads of the screw measured parallel to its axis.

scrotum The sac of skin and tissue that contains and supports the *testes in most mammals. It is situated outside the body cavity and allows the testes to develop at the optimum temperature, which is slightly lower than body temperature.

S-drop *See* strange matter.

search coil A small coil in which a current can be induced to detect and measure a magnetic field. It is used in conjunction with a *fluxmeter.

Searle's bar An apparatus for determining the thermal conductivity of a bar of material. The bar is lagged and one end is heated while the other end is cooled, by steam and cold water respectively. At two points *d* apart along the length of the bar the temperature is measured using a thermometer or thermocouple. The conductivity can then be calculated from the measured temperature gradient.

seaweeds Large multicellular *algae living in the sea or in the intertidal zone. They are commonly species of the *Chlorophyta, *Phaeophyta, and *Rhodophyta.

sebaceous gland A small gland occurring in mammalian *skin. Its duct opens into a hair follicle, through which it discharges *sebum onto the skin surface.

sebum The substance secreted by *sebaceous glands onto the surface of the *skin. It is a fatty mildly antiseptic material that protects, lubricates, and waterproofs the skin and hair and helps prevent desiccation.

secant 1. See trigonometric functions. **2.** A line that cuts a circle or other curve.

sech See hyperbolic functions.

second 1. Symbol s. The SI unit of time equal to the duration of 9 192 631 770 periods of the radiation corresponding to the transition between two hyperfine levels of the ground state of the caesium–133 atom. **2.** Symbol ″. A unit of angle equal to 1/3600 of a degree or 1/60 of a minute.

secondary alcohol See alcohols.

secondary amine See amines.

secondary cell A *voltaic cell in which the chemical reaction produc-ing the e.m.f. is reversible and the cell can therefore be charged by passing a current through it. See accumulator. Compare primary cell.

secondary colour Any colour that can be obtained by mixing two *primary colours. For example, if beams of red light and green light are made to overlap, the secondary colour, yellow, will be formed. Secondary colours of light are sometimes referred to as the pigmentary primary colours. For example, if transparent yellow and magenta pigments are overlapped in white light, red will be observed. In this case the red is a pigmentary secondary although it is a primary colour of light.

secondary emission The emission of electrons from a surface as a result of the impact of other charged particles, especially as a result of bombardment with (primary) electrons. As the number of secondary electrons can exceed the number of primary electrons, the process is important in *photomultipliers. See also Auger effect.

secondary growth (secondary thickening) The increase in thickness of plant shoots and roots through the activities of the vascular *cambium and *cork cambium. It is seen in most dicotyledons and gymnosperms but not in monocotyledons. The tissues produced by secondary growth are called *secondary tissues* and the resultant plant or plant part is the *secondary plant body*. Compare primary growth.

secondary sexual characteristics External features of a sexually mature animal that, although not directly involved in copulation, are significant in reproductive behaviour. The development of such features is controlled by sex hormones (androgens or oestrogens); they may be seasonal (e.g. the antlers of male deer or the body colour of male sticklebacks) or permanent (e.g. breasts in women or facial hair in men).

secondary thickening *See* secondary growth.

secondary winding The winding on the output side of a *transformer or *induction coil. *Compare* primary winding.

second-order reaction *See* order.

secretin A hormone produced by the anterior part of the small intestine (the *duodenum and *jejunum) in response to the presence of hydrochloric acid from the stomach. It causes the pancreas to secrete alkaline pancreatic juice and stimulates bile production in the liver. Secretin, whose function was first demonstrated in 1902, was the first substance to be described as a hormone.

secretion The manufacture and discharge of specific substances into the external medium by cells in living organisms. (The substance secreted is also called the secretion.) Secretory cells are often specialized and organized in groups to form *glands. The substances produced may be released directly into the blood (*endocrine secretion*; *see* endocrine gland) or through a duct (*exocrine secretion*; *see* exocrine gland). Secretions can be classified according to the manner of their discharge. *Merocrine (eccrine) secretion* occurs without the secretory cells sustaining any permanent change; in *apocrine secretion* the cells release a secretory vesicle incorporating part of the secretory cell membrane; and *holocrine secretion* involves the disruption of the entire cell to release its accumulated secretory vesicles.

secular magnetic variation *See* geomagnetism.

sedimentary rocks A group of rocks formed as a result of the accumulation and consolidation of sediments. Sedimentary rocks are one of the three major rock groups forming the earth's crust (*see also* igneous rocks; metamorphic rocks). Most sedimentary rocks are formed from pre-existing rocks that have been broken down through mechanical processes into small particles, which have then been transported and redeposited; these form the clastic sedimentary rocks. The clastic rocks are subdivided according to the size of the dominant constituent particles: arenaceous being composed largely of sand-grade particles (e.g. sandstone, grit), argillaceous of silt- or clay-grade (e.g. mud, mudstone, clay, shale), and rudaceous of gravel-grade or larger particles (e.g. conglomerate). Chemical sedimentary rocks form the second large division of sedimentary rocks and originate as chemical precipitates at the site of deposition. These include the evaporites and sedimentary iron ores, and the organic sedimentary rocks, which are derived largely from the remains of plants and animals and include coal and many limestones.

sedimentation The settling of the solid particles through a liquid either to produce a concentrated slurry from a dilute suspension or to clarify a liquid containing solid particles. Usually this relies on the force of gravity, but if the particles are too small or the difference in density between the solid and liquid phases is too small, a *centrifuge may be used. In the simplest case the rate of sedimentation is determined by *Stokes's law, but in practice the predicted rate is rarely reached. Measurement of the rate of sedimentation in an *ultracentrifuge can be used to estimate the size of macromolecules.

Seebeck effect (thermoelectric effect) The generation of an e.m.f. in a circuit containing two different metals or semiconductors, when the junctions between the two are maintained at different temperatures. The magnitude of the e.m.f. depends on the nature of the metals and the difference in temperature. The Seebeck effect is the basis of the *thermocouple. It was named after T. J. Seebeck (1770–1831), who actually found that a magnetic field surrounded a circuit consisting of two metal conductors only if the junctions between the metals were maintained at different temperatures. He wrongly assumed that the conductors were magnetized directly by the temperature difference. *Compare* Peltier effect.

seed 1. (in botany) The structure in angiosperms and gymnosperms that develops from the ovule after fertilization. Occasionally seeds may develop without fertilization taking place (*see* apomixis). The seed contains the *embryo and nutritive tissue, either as *endosperm or food stored in the *cotyledons. Angiosperm seeds are contained within a *fruit that develops from the ovary wall. Gymnosperm seeds lack an enclosing fruit and are thus termed *naked*. The seed is covered by a protective layer, the *testa. During development of the testa the seed dries out and enters a resting phase (dormancy) until conditions are suitable for germination.

Annual plants survive the winter or dry season as seeds. The evolution of the seed habit enabled plants to colonize the land, since seed plants do not depend on water for fertilization (unlike the lower plants).
2. (in chemistry) A crystal used to induce other crystals to form from a gas, liquid, or solution.

seed coat *See* testa.

seed ferns *See* Pteridospermales.

seed leaf *See* cotyledon.

seed plants *See* Spermatophyta.

Seger cones (pyrometric cones) A series of cones used to indicate the temperature inside a furnace or kiln. The cones are made from different mixtures of clay, limestone, feldspars, etc., and each one softens at a different temperature. The drooping of the vertex is an indication that the known

softening temperature has been reached and thus the furnace temperature can be estimated.

segmentation 1. *See* metameric segmentation. **2.** *See* cleavage.

segregation The separation of pairs of *alleles during the formation of reproductive cells so that they contain one allele only of each pair. Segregation is the result of the separation of *homologous chromosomes during *meiosis. *See also* Mendel's laws.

seismograph An instrument that records ground oscillations, e.g. those caused by earthquakes, volcanic activity, and explosions. Most modern seismographs are based on the inertia of a delicately suspended mass and depend on the measurement of the displacement between the mass and a point fixed to the earth. Others measure the relative displacement between two points on earth. The record made by a seismograph is known as a *seismogram*.

seismology The branch of geology concerned with the study of earthquakes.

Selachii The major subclass of the Elasmobranchii (cartilaginous fishes), containing the sharks, rays, skates, and similar but extinct forms. Their sharp teeth develop from the toothlike placoid *scales (denticles) and are rapidly replaced as they wear out.

selection pressure The extent to which organisms possessing a particular characteristic are either eliminated or favoured by environmental demands. It indicates the degree of intensity of *natural selection.

selectron *See* supersymmetry.

selenides Binary compounds of selenium with other more electropositive elements. Selenides of nonmetals are covalent (e.g. H_2Se). Most metal selenides can be prepared by direct combination of the elements. Some are well-defined ionic compounds (containing Se^{2-}), while others are nonstoichiometric interstitial compounds (e.g. Pd_4Se, $PdSe_2$).

selenium Symbol Se. A metalloid element belonging to group VI of the periodic table; a.n. 34; r.a.m. 78.96; r.d. 4.79 (grey); m.p. 217°C (grey); b.p. 684.9°C. There are a number of allotropic forms, including grey, red, and black selenium. It occurs in sulphide ores of other metals and is obtained as a by-product (e.g. from the anode sludge in electrolytic refining). The element is a semiconductor; the grey allotrope is light-sensitive and is used in photocells, xerography, and similar applications. Chemically, it resembles sulphur, and forms compounds with selenium in the $+2$, $+4$, and $+6$ oxidation states. Selenium was discovered in 1817 by J. J. Berzelius.

selenium cell Either of two types of *photoelectric cell; one type relies on the photoconductive effect, the other on the photovoltaic effect (see photoelectric effect). In the photoconductive selenium cell an external e.m.f. must be applied; as the selenium changes its resistance on exposure to light, the current produced is a measure of the light energy falling on the selenium. In the photovoltaic selenium cell, the e.m.f. is generated within the cell. In this type of cell, a thin film of vitreous or metallic selenium is applied to a metal surface, a transparent film of another metal, usually gold or platinum, being placed over the selenium. Both types of cell are used as light meters in photography.

selenology The branch of astronomy concerned with the scientific study of the *moon.

self-exciting generator A type of electrical generator in which the magnets are excited by current from the generator output.

self inductance See inductance.

self-sterility The condition found in many hermaphrodite organisms in which male and female reproductive cells produced by the same individual will not fuse to form a zygote, or if they do, the zygote is unable to develop into an embryo. In plants this is usually termed self-incompatibility (see incompatibility).

Seliwanoff's test A biochemical test to identify the presence of ketonic sugars, such as fructose, in solution. It was devised by the Russian chemist F. F. Seliwanoff. A few drops of the reagent, consisting of resorcinol crystals dissolved in equal amounts of water and hydrochloric acid, are heated with the test solution and the formation of a red precipitate indicates a positive result.

semen A fluid containing sperm and nutrients that is produced by a male mammal during copulation and is introduced into the body of the female. Spermatozoa are produced by the *testes and the nutrient liquids by the *prostate gland and *seminal vesicles.

semicarbazones Organic compounds containing the unsaturated group $=C:N.NH.CO.NH_2$. They are formed when aldehydes or ketones react with a semicarbazide ($H_2N.NH.CO.NH_2$). Semicarbazones are crystalline compounds with relatively high melting points. They are used to identify aldehydes and ketones in quantitative analysis: the semicarbazone derivative is made and identified by its melting point. Semicarbazones are also used in separating ketones from reaction mixtures: the derivative is crystallized out and hydrolysed to give the ketone.

semicircular canals The sense organ in vertebrates that is concerned with the maintenance of physical equilibrium (sense of balance). It occurs in the *inner ear and consists of three looped canals set at right angles to each other and attached to the *utriculus. The canals contain a fluid (endolymph) that flows in response to movements of the head and body. A swelling (*ampulla*) at one attachment point of each canal contains sensory cells that respond to movement of the endolymph in any of the three planes. These sensory cells initiate nervous impulses to the brain.

semiconductor A crystalline solid with an electrical conductivity (typically 10^5–10^{-7} siemens per metre) intermediate between that of a conductor (up to 10^9 S m^{-1}) and an insulator (as low as 10^{-15} S m^{-1}). Semiconducting properties are a feature of *metalloid elements, such as silicon and germanium. As the atoms in a crystalline solid are close together, the orbitals of their electrons overlap and their individual *energy levels are spread out into *energy bands. Conduction occurs in semiconductors as the result of a net movement, under the influence of an electric field, of electrons in the conduction band and empty states, called *holes*, in the valence band. A hole behaves as if it was an electron with a positive charge. Electrons and holes are known as the *charge carriers* in a semiconductor. The type of charge carrier that predominates in a particular region or material is called the *majority carrier* and that with the lower concentration is the *minority carrier*. An *intrinsic semiconductor* is one in which the concentration of charge carriers is a characteristic of the material itself; electrons jump to the conduction band from the valence band as a result of thermal excitation, each electron that makes the jump leaving behind a hole in the valence band. Therefore, in an intrinsic semiconductor the charge carriers are equally divided between electrons and holes. In *extrinsic semiconductors* the type of conduction that predominates depends on the number and valence of the impurity atoms present. Germanium and silicon atoms have a valence of four. If impurity atoms with a valence of five, such as arsenic, antimony, or phosphorus, are added to the lattice, there will be an extra electron per atom available for conduction, i.e. one that is not required to pair with the four valence electrons of the germanium or silicon. Thus extrinsic semiconductors doped with atoms of valence five give rise to crystals with electrons as majority carriers, the so-called *n-type conductors*. Similarly, if the impurity atoms have a valence of three, such as boron, aluminium, indium, or gallium, one hole per atom is created by an unsatisfied bond. The majority carriers are therefore holes, i.e. *p-type conductors*.

Semiconductor devices have virtually replaced thermionic devices, because they are several orders of magnitude smaller, cheaper in energy consumption, and more reliable. The basic structure for electronic semiconductor devices is the *semiconductor diode* (*see also* transistor). This consists of a silicon crystal doped in such a way that half is *p*-type and half is *n*-type. At the junction between the two halves there is a depletion layer in which electrons from the *n*-type have filled holes from the *p*-type. This sets up a potential barrier, which tends to keep the remaining electrons in the *n*-region and the remaining holes in the *p*-region. However, if the *p*-region is biased with a positive potential, the height of the barrier is reduced; the diode is said to be forward biased, because the majority holes in the *p*-region can then flow to the *n*-region and majority electrons in the *n*-region flow to the *p*-region. When forward biased there is a good current flow across the barrier. On the other hand if the *p*-region is negatively biased, the height of the potential barrier is increased and there is only a small leakage current of minority electrons from the *p*-region able to flow to the *n*-region. Thus the *p–n* junction acts as an efficient rectifier, for which purpose it is widely used.

semiconductor diode *See* diode; semiconductor.

semimetal *See* metalloid.

seminal receptacle *See* spermatheca.

seminal vesicle 1. A pouch or sac in many male invertebrates and lower vertebrates that is used for storing sperm. **2.** One of a pair of glands in male mammals that secrete a liquid component of semen into the vas deferens.

seminiferous tubules *See* testis.

semipermeable membrane A membrane that is permeable to molecules of the solvent but not the solute in *osmosis. Semipermeable membranes can be made by supporting a film of material (e.g. cellulose) on a wire gauze or porous pot.

semipolar bond *See* chemical bond.

Semtex A nitrogen-based stable odourless plastic *explosive.

senescence The changes that occur in an organism (or a part of an organism) between maturity and death, i.e. ageing. Characteristically there is a deterioration in functioning as the cells become less efficient in maintaining and replacing vital cell components. In animals this results in a decline in physical ability and, in man, there is also often a reduction in mental ability. Not all the parts of the body necessarily become senescent at the same time or age at the same rate. For example, in deciduous trees the shedding of senescent leaves in the autumn is a normal physiological process.

sense organ A part of the body of an animal that contains or consists of a concentration of *receptors that are sensitive to specific stimuli (e.g. sound, light, pressure, heat). Stimulation of these receptors initiates the transmission of nervous impulses to the brain, where sensory information

is analysed and interpreted. Examples of sense organs are the *eye, *ear, *nose, and *taste bud.

sensitization 1. (of a cell) The alteration of the integrity of a cell membrane resulting from the reaction of specific *antibodies with *antigens on the surface of the cell. In the presence of *complement, the cell ruptures. **2.** (of an individual) Initial exposure to a specific antigen such that re-exposure to the same antigen causes a severe immune response (*see* anaphylaxis).

sepal One of the parts of a flower making up the *calyx. Sepals are considered to be modified leaves with a simpler structure. They are usually green and often hairy but in some plants, e.g. monk's hood, they may be brightly coloured.

septivalent (heptavalent) Having a valency of seven.

septum Any dividing wall in a plant or animal. Examples are the septa that separate the chambers of the heart.

sequestration The process of forming coordination complexes of an ion in solution. Sequestration often involves the formation of chelate complexes, and is used to prevent the chemical effect of an ion without removing it from the solution (e.g. the sequestration of Ca^{2+} ions in water softening). It is also a way of supplying ions in a protected form, e.g. sequestered iron solutions for plants in regions having alkaline soil.

series A sequence of terms each of which can be written in a form that is an algebraic function of its position in the series. For example, the *exponential series $1 + x + x^2/2! + x^3/3!$ has an nth term $x^n/n!$. The sum of all the terms from $n = 0$ to $n = \infty$ is written:

$$\sum_{n=0}^{\infty} x^n/n!$$

This series has an infinite number of terms and is therefore called an *infi-

nite series. A *finite series* has a fixed number of terms. *See also* asymptotic series; convergent series; divergent series.

series circuit A circuit in which the circuit elements are arranged in sequence so that the same current flows through each of them in turn. For resistances in series, the total resistance is the sum of the individual resistances. For capacitors in series, the total capacitance, C, is given by $1/C = 1/C_1 + 1/C_2 + 1/C_3 \ldots$.

series-wound machine *See* shunt.

serine *See* amino acid.

serology The laboratory study of blood serum and its constituents, particularly *antibodies and *complement, which play a part in the *immune response.

serotonin (5-hydroxytryptamine) A compound, derived from the amino acid tryptophan, that affects the diameter of blood vessels and also functions as a *neurotransmitter. In the brain it is thought to influence mood: drugs, such as LSD, that alter serotonin levels have hallucinogenic effects.

serous membrane A tissue consisting of a layer of *mesothelium attached to a surface by a thin layer of connective tissue. Serous membrane lines body cavities that do not open to the exterior; the *peritoneum and the *pleura are examples.

serpentine Any of a group of hydrous magnesium silicate minerals with the general composition $Mg_3Si_2O_5(OH)_4$. Serpentine is monoclinic and occurs in two main forms: *chrysotile*, which is fibrous and the chief source of *asbestos; and *antigorite*, which occurs as platy masses. It is generally green or white with a mottled appearance, sometimes resembling a snakeskin – hence the name. It is formed through the metamorphic alteration of ultrabasic rocks rich in olivine, pyroxene, and amphi-

bole. *Serpentinite* is a rock consisting mainly of serpentine; it is used as an ornamental stone.

serum *See* blood serum.

sesqui- Prefix indicating a ratio of 2:3 in a chemical compound. For example, a sesquioxide has the formula M_2O_3.

sessile 1. Describing animals that live permanently attached to a surface, i.e. sedentary animals. Many marine animals, e.g. sea anemones and limpets, are sessile. **2.** Describing any organ that does not possess a stalk where one might be expected. For example, the leaves of the oak (*Quercus robur*) are attached directly to the twigs.

seta 1. A bristle or hair in many invertebrates. Setae are produced by the epidermis and consist either of a hollow projection of cuticle containing all or part of an epidermal cell (as in insects) or are composed of chitin (as in the *chaetae of annelid worms). **2.** *See* sporogonium.

sets Collections of objects or elements that have at least one characteristic in common. For example, the set X may consist of all the elements x_1, x_2, x_3, etc. This is written $\{x_1, x_2, x_3, \ldots\} = X$. A specific element in a set is characterized by $x_1 \in X$, meaning x_1 is a member of set X. A *subset* of set X, say M, would be written $M \subset X$, i.e. M is contained in X. If x_3 is a member of both subsets M and N, then $x_3 \in (M \cap N)$, i.e. x_3 belongs to the *intersection* of M and N. $M \cup N$ means the *union* of M and N. For example, if M consists of $\{1, 4, 5, 8\}$ and N consists of $\{2, 3, 4, 5\}$ then $M \cap N = \{4, 5\}$ and $M \cup N = \{1, 2, 3, 4, 5, 8\}$. In the diagram, the rectangle represents the universal set E, circles represent sets or subsets. These diagrams are called *Venn diagrams*, after John Venn (1834–1923), who invented them.

$x_1 \in X$

$M \subset X$

$x_3 \in (M \cap N)$

$M \cup N$

Sets

sex chromosome A chromosome that operates in the sex-determining mechanism of a species. In most animals there are two kinds of sex chromosomes: the X *chromosome*, similar in size to the other chromosomes, and the smaller Y *chromosome*. In man and many animals a combination of two X chromosomes gives a female while one X and one Y chromosome gives a male. Sex chromosomes carry genes governing the development of sex organs and secondary sexual characteristics. They also carry other genes unrelated to sex (*see* sex linkage). *See also* sex determination.

sex determination The method by which the distinction between males and females is established in a species. It is usually under genetic control. Equal numbers of males and females are produced when sex is determined by *sex chromosomes or by a contrasting pair of alleles. In some species (e.g. bees) females develop from fertilized eggs and males from unfertilized eggs. This does not produce equal numbers.

sex hormones Steroid hormones that control sexual development. The most important are the *androgens and *oestrogens.

sexivalent (hexavalent) Having a valency of six.

sex linkage The tendency for certain inherited characteristics to occur far more frequently in one sex than the other. For example, red–green colour blindness and *haemophilia affect men more often than women. This is because the genes governing normal colour vision and blood clotting occur on the X *sex chromosome. Women have two X chromosomes. If one carries an abnormal allele it is likely that its effects will be masked by a normal allele on the other X chromosome. However, men only have one X chromosome and any abnormal alleles therefore will not be masked. *See also* carrier.

sex ratio The ratio of the number of females to the number of males in a *population. Because the mortality rates in the two sexes may be different, the sex ratios in different age classes may differ.

sextant An instrument used in navigation to measure the altitude of a celestial body. Originally it had an arc of 60° (one sixth of a circle, hence its name) but modern instruments have various angles. The sextant uses two mirrors: the horizon

rays from
celestial body

index mirror

horizon
glass

index arm

telescope

θ

rays
from
horizon

β

micrometer
drum and vernier

A sextant

glass, in which only the lower half is silvered, and the index mirror, which can be rotated about an axis perpendicular to the plane of the instrument (see illustration). An arm attached to the index glass sweeps round the calibrated arc, from which angles are read. The instrument is aimed at the horizon and the index mirror rotated until the celestial object can also be seen through the telescope. After careful adjustment to make the image of the celestial body just touch the horizon, the angle is read off the graduated scale.

sexual reproduction A form of reproduction that involves the fusion of two reproductive cells (*gametes) in the process of *fertilization. Normally, especially in animals, it requires two parents, one male and the other female. However, most plants bear both male and female reproductive organs and self-fertilization may occur, as it does in hermaphrodite animals. Gametes are formed by *meiosis, a special kind of cell division in the parent reproductive organs that both reassorts the genetic material and halves the chromosome number. Meiosis thus ensures genetic variability in the gametes and therefore in the offspring resulting from their subsequent

fusion. Sexual reproduction, unlike *asexual reproduction, therefore generates variability within a species. However, it depends on there being reliable means of bringing together male and female gametes, and many elaborate mechanisms have evolved to ensure this.

sexual selection The means by which it is assumed that certain *secondary sexual characteristics, particularly of male animals, have evolved. Females presumably choose to mate with the male that gives the best courtship display and therefore has the brightest coloration, etc.: these features would be inherited by its male offspring and would thus tend to become exaggerated down the generations.

shadow An area of darkness formed on a surface when an object intercepts the light falling on the surface from a source. In the case of a point source the shadow has a sharply defined outline. If the source has an appreciable size the shadow has two distinct regions; one of full-shadow, called the *umbra*, the other of half-shadow, called the *penumbra* (see illustration).

point source

umbra

large source

penumbra

Shadows

shearing force A force that acts parallel to a plane rather than perpendicularly, as with a tensile or compressive force. A *shear stress* requires a combination of four forces acting over (most simply) four sides

Shearing force

of a plane and produces two equal and opposite couples. It is measured as the ratio of one shearing force to the area over which it acts, $F/(ab)$ in the diagram. The shear strain is the angular deformation, θ, in circular measure. The *shear modulus* is the ratio of the shear stress to the shear strain (*see also* elastic modulus).

shear modulus *See* elastic modulus; shearing force.

shell *See* atom.

shell model 1. *See* atom. 2. A model of the atomic nucleus in which nucleons are assumed to move under the influence of a central field in shells that are analogous to atomic electron shells. The model provides a good explanation of the stability of nuclei that have *magic numbers.

sherardizing The process of coating iron or steel with a zinc corrosion-resistant layer by heating the iron or steel in contact with zinc dust to a temperature slightly below the melting point of zinc. At a temperature of about 371°C the two metals amalgamate to form internal layers of zinc–iron alloys and an external layer of pure zinc. The process was invented by Sherard Cowper-Coles (d. 1935).

shielding 1. A barrier surrounding a region to exclude it from the influence of an energy field. For example, to protect a region from an electric field an earthed barrier is required; to protect it from a magnetic field a shield of high magnetic permeability is needed. 2. A barrier used to sur-

round a source of harmful or unwanted radiations. For example, the core of a *nuclear reactor is surrounded by a cement or lead shield to absorb neutrons and other dangerous radiation.

SHM *See* simple harmonic motion.

shock wave A very narrow region of high pressure and temperature formed in a fluid when the fluid flows supersonically over a stationary object or a projectile flying supersonically passes through a stationary fluid. A shock wave may also be generated by violent disturbances in a fluid, such as a lightning stroke or a bomb blast.

shoot The aerial part of a vascular plant. It develops from the *plumule and consists of a stem supporting leaves, buds, and flowers.

short-day plant A plant in which flowering can be induced or enhanced by short days, usually of less than 12 hours of daylight. Examples are strawberry and chrysanthemum. *See* photoperiodism. *Compare* day-neutral plant; long-day plant.

short period *See* periodic table.

short-sightedness *See* myopia.

shoulder girdle *See* pectoral girdle.

shower *See* cosmic radiation.

shunt An electrical resistor or other element connected in parallel with some other circuit or device, to take part of the current passing through it. For example, a shunt is used across the terminals of a galvanometer to increase the current that can pass through the system. A *shunt-wound* electric generator or motor is one in which the field winding is in parallel with the armature circuit. In a *series-wound* electrical machine the field coils and the armature circuit are in series.

sial The rocks that form the earth's continental crust. These are granite rock types rich in *si*lica (SiO_2) and *al*uminium (Al), hence the name. *Compare* sima.

siblings Individuals that have both parents in common.

sickle-cell anaemia *See* polymorphism.

sideband The band of frequencies above or below the frequency of the carrier wave in a telecommunications system within which the frequency components of the wave produced by *modulation fall. For example, if a carrier wave of frequency f is modulated by a signal of frequency x, the *upper sideband* will have a frequency $f + x$ and the *lower sideband* a frequency $f - x$.

side chain *See* chain.

side reaction A chemical reaction that occurs at the same time as a main reaction but to a lesser extent, thus leading to other products mixed with the main products.

sidereal day *See* day.

sidereal period The time taken for a planet or satellite to complete one revolution of its orbit measured with reference to the background of the stars. *See also* day; synodic period; year.

siderite A brown or grey-green mineral form of iron(II) carbonate, $FeCO_3$, often with magnesium and manganese substituting for the iron. It occurs in sedimentary deposits or in hydrothermal veins and is an important iron ore. It is found in England, Greenland, Spain, N Africa, and the USA.

siemens Symbol S. The SI unit of electrical conductance equal to the conductance of a circuit or element that has a resistance of 1 ohm. $1 S = 10^{-1} \Omega$. The unit was formerly called the mho or reciprocal ohm. It is named after Ernst Werner von Siemens (1816–92).

sieve element A type of plant cell occurring within the *phloem. Sieve elements combine to form a series of tubes (*sieve tubes*) connecting the leaves, shoots, and roots in a fine network. Food materials are transported from one element to another via perforations termed *sieve areas* or *sieve plates*. This is an active (energy-requiring) process but the mechanism is uncertain. Sieve elements contain little cytoplasm and no nucleus. It is believed that their metabolic activities may be controlled by *companion cells in angiosperms and by albuminous cells in gymnosperms.

sievert The SI unit of dose equivalent (*see* radiation units).

sieve tube A tube within the *phloem tissue of a plant, composed of joined *sieve elements.

sigma bond *See* orbital.

sigma electron An electron in a sigma orbital. *See* orbital.

signal The variable parameter that contains information and by which information is transmitted in an electronic system or circuit. The signal is created by the *signal generator*, often a voltage source in which the amplitude, frequency, and waveform can be varied.

sign convention A set of rules determined by convention for giving plus or minus signs to distances in the formulae involving lenses and mirrors. The *real-is-positive is the convention now usually adopted. The *New Cartesian convention* is now not widely used. In this convention distances to the right of the pole are treated as positive, those to the left as negative. This system has the advantage of conforming to the sign convention used with Cartesian coordinates in mathematics and is therefore preferred by some for the more complicated calculations.

significant figures The number of digits used in a number to specify its accuracy. The number 6.532 is a value taken to be accurate to four significant figures. The number 7.3 × 10^3 is accurate only to three significant figures. Similarly 0.0732 is also only accurate to three significant figures. In these cases the zeros only indicate the order of magnitude of the number, whereas 7.065 is accurate to four significant figures as the zero in this case is significant in expressing the value of the number.

sign stimulus (releaser) The essential feature of a stimulus, which is necessary to elicit a response. For example, a red belly (characteristic of courting male sticklebacks) is the sign stimulus necessary to provoke an attack from a rival male; even a very crude model fish is attacked if it has a red undersurface.

silane (silicane) 1. A colourless gas, SiH_4, which is insoluble in water; d. 1.44 g dm^{-3}; r.d. 0.68 (liquid); m.p. $-185°C$; b.p. $-112°C$. Silane is produced by reduction of silicon with lithium tetrahydridoaluminate(III). It is also formed by the reaction of magnesium silicide (Mg_2Si) with acids, although other silicon hydrides are also produced at the same time. Silane itself is stable in the absence of air but is spontaneously flammable, even at low temperatures. It is used for controlled silicon doping of semiconductors in the electronics industry. It is also a reducing agent and has been used for the removal of corrosion in inaccessible plants (e.g. pipes in nuclear reactors). **2.** (*or* **silicon hydride**) Any of a class of compounds of silicon and hydrogen. They have the general formula Si_nH_{2n+2}. The first three in the series are silane itself (SiH_4), *disilane* (Si_2H_6), and *trisilane* (Si_3H_8). The compounds are analogous to the alkanes but are much less stable and only the lower members of the series can be prepared in any quantity (up to Si_6H_{14}). No silicon hydrides containing double or triple bonds exist (i.e.

there are no analogues of the alkenes and alkynes).

silica *See* silicon(IV) oxide.

silica gel A rigid gel made by coagulating a sol of sodium silicate and heating to drive off water. It is used as a support for catalysts and also as a drying agent because it readily absorbs moisture from the air. The gel itself is colourless but, when used in desiccators, etc., a blue cobalt salt is added. As moisture is taken up, the salt turns pink, indicating that the gel needs to be regenerated (by heating).

silicane *See* silane.

silicate Any of a group of substances containing negative ions composed of silicon and oxygen. The silicates are a very extensive group and natural silicates form the major component of most rocks (*see* silicate minerals). The basic structural unit is the tetrahedral SiO_4 group. This may occur as a simple discrete SiO_4^{4-} anion as in the *orthosilicates*, e.g. *phenacite* (Be_2SiO_4) and *willemite* (Zn_2SiO_4). Many larger silicate species are also found (see illustration). These are composed of SiO_4 tetrahedra linked by sharing oxygen atoms as in the *pyrosilicates*, $Si_2O_7^{6-}$, e.g. $Sc_2Si_2O_7$. The linking can extend to such forms as benitoite, $BaTiSi_3O_9$, or alternatively infinite chain anions, which are single strand (*pyroxenes*) or double strand (*amphiboles*). Spodumene, LiAl-$(SiO_3)_2$, is a pyroxene and the asbestos minerals are amphiboles. Infinite two-dimensional sheets are also possible, as in the various *micas (see illustration), and the linking can extend to full three-dimensional framework structures, often with substituted trivalent atoms in the lattice. The *zeolites are examples of this.

silicate minerals A group of rock-forming minerals that make up the bulk of the earth's outer crust (about 90%) and constitute one-third of all minerals. All silicate minerals are based on a fundamental structural

silicide

SiO_4^{4-} as in Be_2SiO_4 (phenacite)

$Si_2O_5^{2-}$ as in $Sc_2Si_2O_7$ (thortveitite)

$Si_3O_9^{6-}$ as in $BaTiSi_3O_9$ (bentonite)

$Si_6O_{18}^{12-}$ as in $Be_3Al_2Si_6O_{18}$ (beryl)

Structure of some discrete silicon ions

single chain : pyroxenes

double chain : amphiboles

sheet : micas

Structure of some polymeric silicate ions

unit – the SiO_4 tetrahedron (see silicate). They consist of a metal (e.g. calcium, magnesium, aluminium) combined with silicon and oxygen. The silicate minerals are classified on a structural basis according to how the tetrahedra are linked together. The six groups are: nesosilicates (e.g. olivine and *garnet); sorosilicates (e.g. hemimorphite); cyclosilicates (e.g. axinite, *beryl, and *tourmaline); inosilicates (e.g. *amphiboles and *pyroxenes); phyllosilicates (e.g. *micas, *clay minerals, and *talc); and tektosilicates (e.g. *feldspars and *feldspathoids). Many silicate minerals are of economic importance.

silicide A compound of silicon with a more electropositive element. The silicides are structurally similar to the interstitial carbides but the range encountered is more diverse. They react with mineral acids to form a range of *silanes.

silicon Symbol Si. A metalloid element belonging to *group IV of the periodic table; a.n. 14; r.a.m. 28.086; r.d. 2.33; m.p. 1410°C; b.p. 2355°C. Silicon is the second most abundant element in the earth's crust

(25.7% by weight) occurring in various forms of silicon(IV) oxide (e.g. *quartz) and in *silicate minerals. The element is extracted by reducing the oxide with carbon in an electric furnace and is used extensively for its semiconductor properties. It has a diamond-like crystal structure; an amorphous form also exists. Chemically, silicon is less reactive than carbon. The element combines with oxygen at red heat and is also dissolved by molten alkali. There is a large number of organosilicon compounds (e.g. *siloxanes) although silicon does not form the range of silicon–hydrogen compounds and derivatives that carbon does (*see* silanes). The element was identified by Lavoisier in 1787 and first isolated in 1823 by Berzelius.

silicon carbide (carborundum) A black solid compound, SiC, insoluble in water and soluble in molten alkali; r.d. 3.217; m.p. *c.* 2700°C. Silicon carbide is made by heating silicon(IV) oxide with carbon in an electric furnace (depending on the grade required sand and coke may be used). It is extremely hard and is widely used as an abrasive. The solid exists in both zinc blende and wurtzite structures.

silicon chip A single crystal of a semiconducting silicon material, typically having millimetre dimensions, fabricated in such a way that it can perform a large number of independent electronic functions (*see* integrated circuit).

silicon dioxide *See* silicon(IV) oxide.

silicones Polymeric compounds containing chains of silicon atoms alternating with oxygen atoms, with the silicon atoms linked to organic groups. A variety of silicone materials exist, including oils, waxes, and rubbers. They tend to be more resistant to temperature and chemical attack than their carbon analogues.

silicon hydride *See* silane.

silicon(IV) oxide (silicon dioxide; silica) A colourless or white vitreous solid, SiO_2, insoluble in water and soluble (by reaction) in hydrofluoric acid and in strong alkali; m.p. 1610°C; b.p. 2230°C. The following forms occur naturally: *cristobalite* (cubic or tetragonal crystals; r.d. 2.32); *tridymite* (rhombic; r.d. 2.26); *quartz* (hexagonal; r.d. 2.63–2.66); *lechatelierite* (r.d. 2.19). Quartz has two modifications: α-quartz below 575°C and β-quartz above 575°C; above 870°C β-quartz is slowly transformed to tridymite and above 1470°C this is slowly converted to cristobalite. Various forms of silicon(IV) oxide occur widely in the earth's crust; yellow sand for example is quartz with iron(III) oxide impurities and flint is essentially amorphous silica. The gemstones amethyst, opal, and rock crystal are also forms of quartz.

Silica is an important commercial material in the form of *silica brick*, a highly refractive furnace lining, which is also resistant to abrasion and to corrosion. Silicon(IV) oxide is also the basis of both clear and opaque silica glass, which is used on account of its transparency to ultraviolet radiation and its resistance to both thermal and mechanical shock. A certain proportion of silicon(IV) oxide is also used in ordinary glass and in some glazes and enamels. It also finds many applications as a drying agent in the form of *silica gel.

silicula A type of *capsule formed from a bicarpellary ovary. It is longitudinally flattened and divided lengthwise into two cavities (*loculi*). It is broader than a *siliqua. Examples include the fruits of *Alyssum* and candytuft.

siliqua A type of *capsule formed from a bicarpellary ovary. It resembles a *silicula but is longer than it is broad; an example is the fruit of the wallflower. *See also* lomentum.

siloxanes A group of compounds containing silicon atoms bound to oxygen atoms, with organic groups linked to the silicon atoms, e.g. $R_3SiOSiR_3$, where R is an organic group. *Silicones are polymers of siloxanes.

Silurian A geological period of the Palaeozoic era following the Ordovician period and extending until the beginning of the Devonian period. It began about 440 million years ago and lasted for about 45 million years. The Silurian was named by Roderick Murchison (1792–1871) after an ancient British tribe that inhabited South Wales, where he observed rocks of this period. The majority of Silurian life was marine but during the later part of the period primitive plants began to make their appearance on land. Trilobites and graptolites became less common, brachiopods were numerous and varied, crinoids became common for the first time, and corals also increased. The only known vertebrates during the Silurian were primitive fish; the first jawed fish appeared later in the period. The Caledonian orogeny (mountain-building period) reached its peak towards the end of the Silurian.

silver Symbol Ag. A white lustrous soft metallic *transition element; a.n. 47; r.a.m. 107.87; r.d. 10.5; m.p. 961.93°C; b.p. 2212°C. It occurs as the element and as the minerals argentite (Ag_2S) and horn silver (AgCl). It is also present in ores of lead and copper, and is extracted as a by-product of smelting and refining these metals. The element is used in jewellery, tableware, etc., and silver compounds are used in photography. Chemically, silver is less reactive than copper. A dark silver sulphide forms when silver tarnishes in air because of the presence of sulphur compounds. Silver(I) ionic salts exist (e.g. $AgNO_3$, AgCl) and there are a number of silver(II) complexes.

silver(I) bromide A yellowish solid compound, AgBr; r.d. 6.5; m.p. 432°C. It can be precipitated from silver(I) nitrate solution by adding a solution containing bromide ions. It dissolves in concentrated ammonia solutions (but, unlike the chloride, does not dissolve in dilute ammonia). The compound is used in photographic emulsions.

silver(I) chloride A white solid compound, AgCl; r.d. 5.6; m.p. 455°C; b.p. 1550°C. It can be precipitated from silver(I) nitrate solution by adding a solution of chloride ions. It dissolves in ammonia solution (due to formation of the complex ion $[Ag(NH_3)_2]^+$). The compound is used in photographic emulsions.

silver(I) iodide A yellow solid compound, AgI; r.d. 6.01; m.p. 556°C; b.p. 1506°C. It can be precipitated from silver(I) nitrate solutions by adding a solution of iodide ions. Unlike the chloride and bromide, it does not dissolve in ammonia solutions.

silver-mirror test *See* Tollen's reagent.

silver(I) nitrate A colourless solid, $AgNO_3$; r.d. 4.3; m.p. 212°C. It is an important silver salt because it is water-soluble. It is used in photography. In the laboratory, it is used as a test for chloride, bromide, and iodide ions and in volumetric analysis of chlorides using an *absorption indicator.

silver(I) oxide A brown slightly water-soluble amorphous powder, Ag_2O; r.d. 7.14. It can be made by adding sodium hydroxide solution to silver(I) nitrate solution. Silver(I) oxide is strongly basic and is also an oxidizing agent. It is used in certain reactions in preparative organic chemistry; for example, moist silver(I) oxide converts haloalkanes into alcohols; dry silver oxide converts haloalkanes into ethers. The compound decomposes to the elements at

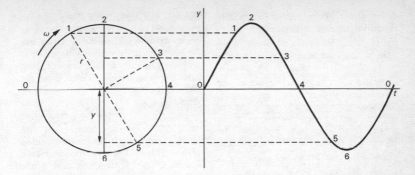

Simple harmonic motion

160°C and can be reduced by hydrogen to silver. With ozone it gives the oxide AgO (which is diamagnetic and probably $Ag^IAg^{III}O_2$).

sima The rocks that form the earth's oceanic crust and underlie the upper crust. These are basaltic rock types rich in *si*lica (SiO_2) and *ma*gnesium (Mg), hence the name. The sima is denser and more plastic than the *sial that forms the continental crust.

simple harmonic motion (SHM) A form of periodic motion in which a point or body oscillates along a line about a central point in such a way that it ranges an equal distance on either side of the central point and that its acceleration towards the central point is always proportional to its distance from it. One way of visualizing SHM is to imagine a point rotating around a circle of radius r at a constant angular velocity ω. If the distance from the centre of the circle to the projection of this point on a vertical diameter is y at time t, this projection of the point will move about the centre of the circle with simple harmonic motion. A graph of y against t will be a sine wave, whose equation is $y = r\sin\omega t$ (see diagram). *See also* pendulum.

simultaneous equations A set of equations that jointly specify the values of the variables they contain. If the number of variables is equal to the number of equations, each variable has a unique value, i.e. the equations can be solved.

sine *See* trigonometric functions.

sine wave (sinusoidal wave) Any waveform that has an equation in which one variable is proportional to the sine of the other. Such a waveform can be generated by an oscillator that executes *simple harmonic motion.

single bond *See* chemical bond.

single-cell protein (SCP) Protein produced by microorganisms, such as bacteria, yeasts, and unicellular algae, that is extracted for use as a component of human and animal foods.

sinh *See* hyperbolic functions.

sintered glass Porous glass made by sintering powdered glass, used for filtration of precipitates in gravimetric analysis.

sintering The process of heating and compacting a powdered material at a temperature below its melting point in order to weld the particles together into a single rigid shape. Materials commonly sintered include metals and alloys, glass, and ceramic oxides. Sintered magnetic materials, cooled in a magnetic field, make especially retentive permanent magnets.

sinus A saclike cavity or organ in an animal, e.g. the *sinus venosus* in the heart of lower vertebrates.

sinusoid A tiny blood vessel or blood-filled space in an organ. Sinusoids replace capillaries in certain organs, notably the liver; they allow more direct contact between the blood and the tissue it is supplying.

sinusoidal oscillator *See* oscillator.

sinusoidal wave *See* sine wave.

siphon An inverted U-tube with one limb longer than the other. Liquid will be transferred from a reservoir at the base of the shorter limb to the end of the longer limb, provided that the U-tube is filled with liquid (see illustration). The device is useful for emptying an inaccessible container, such as a car's petrol tank. The pressure (p_1) on the liquid at the base of the short limb (length h_1) is $p - h_1k$, where p is the atmospheric pressure and k is a constant equal to the product of the density of the liquid and the acceleration of free fall. The pressure (p_2) on the liquid at the base of the long limb (length h_2) is $p - h_2k$. Thus for fluid flow to occur

Siphon

through the tube, from short limb to long limb, $p_1 > p_2$, and for this to occur $h_2 > h_1$. Thus if the limbs are of equal length no flow will occur; it will only occur if the limb dipping into the reservoir is shorter than the delivering limb.

Siphonaptera An order of wingless insects comprising the fleas. The body of a flea is laterally compressed and bears numerous backward-directed spines. Fleas live as blood-sucking ectoparasites of mammals and birds, having mouthparts adapted to piercing their host, injecting saliva to prevent clotting, and sucking up the blood. The long bristly legs can transmit energy stored in the elastic body wall to leap relatively long distances (over 300 mm horizontally). Apart from causing irritation, fleas can transmit disease organisms, most notably bubonic plague bacteria, which can be carried from rats to humans by the rat flea (*Xenopsylla cheopsis*). The whitish wormlike legless larvae feed on organic matter. After two moults the larva spins a cocoon and undergoes metamorphosis into the adult.

Siphunculata (Anoplura) An order of wingless insects comprising the sucking lice: blood-sucking ectoparasites of mammals, with piercing and sucking mouthparts forming a snoutlike proboscis. They constitute an irritating pest to humans and domestic animals and can transmit diseases, including typhoid. The human louse (*Pediculus humanus*) exists in two forms: the head louse (*P. humanus capitis*) and the body louse (*P. humanus corporis*).

s-isomer *See* absolute configuration.

SI units Système International d'Unités: the international system of units now recommended for all scientific purposes. A coherent and rationalized system of units derived from the *m.k.s. units, SI units have now replaced *c.g.s. units and *Imperial

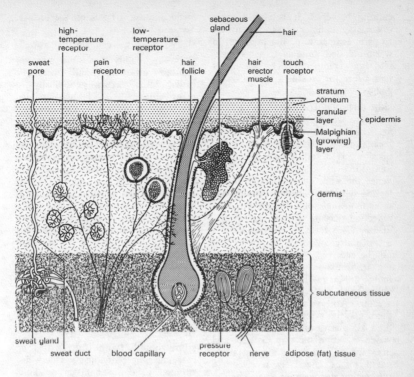

Structure of mammalian skin

units. The system has seven *base units* and two *supplementary units* (see Appendix), all other units being derived from these nine units. There are 18 derived units with special names. Each unit has an agreed symbol (a capital letter or an initial capital letter if it is named after a scientist, otherwise the symbol consists of one or two lower-case letters). Decimal multiples of the units are indicated by a set of prefixes; whenever possible a prefix representing 10 raised to a power that is a multiple of three should be used.

skeletal muscle *See* voluntary muscle.

skeleton The structure in an animal that provides mechanical support for the body, protection for internal organs, and a framework for anchoring the muscles. The skeleton may be external (*see* exoskeleton) or internal (*see* endoskeleton). Both types require *joints to allow locomotion. The skeleton of higher vertebrates consists of a system of *bones.

skin The outer layer of the body of a vertebrate (see illustration). It is composed of two layers, the *epidermis and *dermis, with a complex nervous and blood supply. The skin may bear a variety of specialized structures, including *hair, *scales, and *feathers. This skin has an important role in protecting the body from mechanical injury, water loss, and the entry of harmful agents (e.g. disease-causing bacteria). It is also a sense organ, containing receptors sensitive to pain, temperature, and pressure. In warm-blooded animals it helps regulate body temperature by means of hair, fur, or feathers and *sweat glands.

skip distance The minimum distance from the transmitter of a radio wave at which reception is possible by means of a sky wave (*see* radio transmission). If a radio wave strikes the ionosphere at a small angle of incidence the wave passes through it and is not reflected. There is therefore a minimum angle of incidence at which reflection occurs for a given frequency. This leads to a region around a transmitter in which sky waves cannot be received. As the frequency of the transmission increases the minimum angle of incidence at which ionospheric reflection occurs becomes greater. Above about 4 megahertz there may be a region of several hundred kilometres around a transmitter, which is within the skip distance and in which ground waves are too attenuated to be effectively received. In this region no reception is possible.

skull The skeleton of the head. In mammals it consists of a *cranium enclosing the brain and the bones of the face and jaw. All the joints between the individual bones of the skull are immovable (*see* suture) except for the joint between the mandible (lower jaw) and the rest of the skull. There is a large opening (*foramen magnum*) at the base of the skull through which the spinal cord passes from the brain.

sky wave *See* radio transmission.

slag Material produced during the *smelting or refining of metals by reaction of the flux with impurities (e.g. calcium silicate formed by reaction of calcium oxide flux with silicon dioxide impurities). The liquid slag can be separated from the liquid

metal because it floats on the surface. *See also* basic slag.

slaked lime *See* calcium hydroxide.

sleep A readily reversible state of reduced awareness and metabolic activity that occurs periodically in many animals. Usually accompanied by physical relaxation, the onset of sleep in humans and other mammals is marked by a change in the electrical activity of the brain, which is recorded by an *electroencephalogram as waves of low frequency and high amplitude (*slow-wave sleep*). This is interspersed by short bouts of high-frequency low-amplitude waves (similar to wave patterns produced when awake) associated with restlessness, dreaming, and rapid eye movement (REM); this is called *REM* (or *paradoxical*) *sleep*. Several regions of the brain are involved in sleep, especially the reticular formation of the *brainstem.

sleep movements *See* nyctinasty.

slepton *See* supersymmetry.

slime moulds (slime fungi) A group of small simple organisms widely distributed in damp habitats on land. They exist either as free cells or as multinucleate aggregates of cells (*plasmodia*). Cell walls are generally absent. Slime moulds show amoeboid movement and feed by ingesting small particles of food. They reproduce by means of spores. Since these organisms exhibit characteristics of both simple plants and simple animals their taxonomic position is uncertain. They are usually classified as a division, Myxomycota, either of fungi or of protists (*see* Protista).

slow neutron A neutron with a kinetic energy of less than 10^2 eV (10^{-17} joule). *See also* fast neutron; thermalization.

slow virus (prion) Any of a group of quasi-viral or subviral agents that have some characteristics of viruses but show much greater resistance to certain physical and chemical agents (e.g. radiation). Their structure remains uncertain, but they may be unusual proteins rather than nucleic acid-containing particles. Slow viruses are responsible for the diseases scrapie in sheep and bovine spongiform encephalopathy (BSE) in cattle.

slug 1. (in physics) An f.p.s. unit of mass equal to the mass that will acquire an acceleration of $1 \, \text{ft sec}^{-2}$ when acted on by a force of one pound-force. **2.** (in zoology) See Gastropoda.

slurry A paste consisting of a suspension of a solid in a liquid.

small intestine The portion of the *alimentary canal between the stomach and the large intestine. It is subdivided into the *duodenum, *jejunum, and *ileum. It plays an essential role in the final digestion and absorption of food.

smectic See liquid crystal.

smelting The process of separating a metal from its ore by heating the ore to a high temperature in a suitable furnace in the presence of a reducing agent, such as carbon, and a fluxing agent, such as limestone. Iron ore is smelted in this way so that the metal melts and, being denser than the molten *slag, sinks below the slag, enabling it to be removed from the furnace separately.

smoke A fine suspension of solid particles in a gas.

smooth muscle See involuntary muscle.

smuts A group of parasitic *basidiomycete fungi. Many of these species attack the ears of cereal crops, replacing the grain by a mass of dark spores. Compare rusts.

S_N1 reaction See nucleophilic substitution.

S_N2 reaction See nucleophilic substitution.

snakes See Squamata.

Snell's law See refraction.

SNG Substitute (or synthetic) natural gas; a mixture of gaseous hydrocarbons produced from coal, petroleum, etc., and suitable for use as a fuel. Before the discovery of natural gas *coal gas was widely used as a domestic and industrial fuel. This gave way to natural gas in the early part of this century in the US and other countries where natural gas was plentiful. The replacement of coal gas occurred somewhat later in the UK and other parts of Europe. More recently, interest has developed in ways of manufacturing hydrocarbon gas fuels. The main sources are coal and the naphtha fraction of petroleum. In the case of coal three methods have been used: (1) pyrolysis – i.e. more efficient forms of destructive distillation, often with further hydrogenation of the hydrocarbon products; (2) heating the coal with hydrogen and catalysts to give hydrocarbons – a process known as *hydroliquefaction* (see also Bergius process); (3) producing carbon monoxide and hydrogen and obtaining hydrocarbons by the *Fischer–Tropsch process. SNG from naptha is made by steam *reforming.

soap A substance made by boiling animal fats with sodium hydroxide. The reaction involves the hydrolysis of *glyceride esters of fatty acids to glycerol and sodium salts of the acids present (mainly the stearate, oleate, and palmitate), giving a soft semisolid with *detergent action. Potassium hydroxide gives a more liquid product (*soft soap*). By extension, other metal salts of long-chain fatty acids are also called soaps. See also saponification.

soda Any of a number of sodium compounds, such as caustic soda (NaOH) or, especially, washing soda ($Na_2CO_3.10H_2O$).

soda ash Anhydrous *sodium carbonate, Na_2CO_3.

soda lime A mixed hydroxide of sodium and calcium made by slaking lime with caustic soda solution (to give $NaOH + Ca(OH)_2$) and recovering greyish white granules by evaporation. The material is produced largely for industrial adsorption of carbon dioxide and water, but also finds some applications in pollution and effluent control. It is also used as a laboratory drying agent.

sodamide See sodium amide.

sodium Symbol Na. A soft silvery reactive element belonging to group I of the periodic table (see alkali metals); a.n. 11; r.a.m. 22.9898; r.d. 0.97; m.p. 97.5°C; b.p. 892°C. Sodium occurs as the chloride in sea water and in the mineral halite. It is extracted by electrolysis in a *Downs cell. The metal is used as a reducing agent in certain reactions and liquid sodium is also a coolant in nuclear reactors. Chemically, it is highly reactive, oxidizing in air and reacting violently with water (it is kept under oil). It dissolves in liquid ammonia to form blue solutions containing solvated electrons. Sodium is a major *essential element required by animals. It is important in maintaining the *acid–base balance and in controlling the volume of extracellular fluid and functions in the transmission of nerve impulses (see sodium pump). The element was first isolated by Humphry Davy in 1807.

sodium acetate See sodium ethanoate.

sodium aluminate A white solid, $NaAlO_2$ or $Na_2Al_2O_4$, which is insoluble in ethanol and soluble in water giving strongly alkaline solutions; m.p. 1800°C. It is manufactured by heating bauxite with sodium carbonate and extracting the residue with water, or it may be prepared in the laboratory by adding excess aluminium to hot concentrated sodium hydroxide. In solution the ion $Al(OH)_4^-$ predominates. Sodium aluminate is used as a mordant, in the production of zeolites, in effluent treatment, in glass manufacture, and in cleansing compounds.

sodium amide (sodamide) A white crystalline powder, $NaNH_2$, which decomposes in water and in warm ethanol, and has an odour of ammonia; m.p. 210°C; b.p. 400°C. It is produced by passing dry ammonia over metallic sodium at 350°C. It reacts with red-hot carbon to give sodium cyanide and with nitrogen(I) oxide to give sodium azide.

sodium azide A white or colourless crystalline solid, NaN_3, soluble in water and slightly soluble in alcohol; hexagonal; r.d. 1.846; decomposes on heating. It is made by the action of nitrogen(I) oxide on hot sodamide ($NaNH_2$) and is used as an organic reagent and in the manufacture of detonators.

sodium benzenecarboxylate (sodium benzoate) An either colourless crystalline or white amorphous powder, C_6H_5COONa, soluble in water and slightly soluble in ethanol. It is made by the reaction of sodium hydroxide with benzoic acid and is used in the dyestuffs industry and as a food preservative. It was formerly used as an antiseptic.

sodium benzoate See sodium benzenecarboxylate.

sodium bicarbonate See sodium hydrogencarbonate.

sodium bisulphate See sodium hydrogensulphate.

sodium bisulphite See sodium hydrogensulphite.

sodium bromide A white crystalline solid, NaBr, known chiefly as the dihydrate (monoclinic; r.d. 2.17), and as the anhydrous salt (cubic; r.d. 3.20; m.p. 747°C; b.p. 1390°C). The dihydrate loses water at about 52°C and is very slightly soluble in alcohol. Sodium bromide is prepared by the reaction of bromine on hot sodium hydroxide solution or of hydrogen bromide on sodium carbonate solution. It is a sedative and is also used in photographic processing and in analytical chemistry.

sodium carbonate Anhydrous sodium carbonate (*soda ash*, *sal soda*) is a white powder, which cakes and aggregates on exposure to air due to the formation of hydrates. The monohydrate, $Na_2CO_3.H_2O$, is a white crystalline material, which is soluble in water and insoluble in alcohol; r.d. 1.55; loses water at 109°C; m.p. 851°C.

The decahydrate, $Na_2CO_3.10H_2O$ (*washing soda*), is a translucent efflorescent crystalline solid; r.d. 1.44; loses water at 32–34°C to give the monohydrate; m.p. 851°C. Sodium carbonate may be manufactured by the *Solvay process or by suitable crystallization procedures from any one of a number of natural deposits, such as:

trona ($Na_2CO_3.NaHCO_3.2H_2O$),
natron ($Na_2CO_3.10H_2O$),
ranksite ($2Na_2CO_3.9Na_2SO_4.KCl$),
pirsonnite ($Na_2CO_3.CaCO_3.2H_2O$),
gaylussite ($Na_2CO_3.CaCO_3.5H_2O$).

The method of extraction is very sensitive to the relative energy costs and transport costs in the region involved. Sodium carbonate is used in photography, in cleaning, in pH control of water, in textile treatment, glasses and glazes, and as a food additive and volumetric reagent. *See also* sodium sesquicarbonate.

sodium chlorate(V) A white crystalline solid, $NaClO_3$; cubic; r.d. 2.49; m.p. 250°C. It decomposes above its melting point to give oxygen and sodium chloride. The compound is soluble in water and in ethanol and is prepared by the reaction of chlorine on hot concentrated sodium hydroxide. Sodium chlorate is a powerful oxidizing agent and is used in the manufacture of matches and soft explosives, in calico printing, and as a garden weedkiller.

sodium chloride (common salt) A colourless crystalline solid, NaCl, soluble in water and very slightly soluble in ethanol; cubic; r.d. 2.17; m.p. 801°C; b.p. 1413°C. It occurs as the mineral *halite (rock salt) and in natural brines and sea water. It has the interesting property of a solubility in water that changes very little with temperature. It is used industrially as the starting point for a range of sodium-based products (e.g. Solvay process for Na_2CO_3, Castner–Kellner process for NaOH), and is known universally as a preservative and seasoner of foods. Sodium chloride has a key role in biological systems in maintaining electrolyte balances.

sodium cyanide A white or colourless crystalline solid, NaCN, deliquescent, soluble in water and in liquid ammonia, and slightly soluble in ethanol; cubic; m.p. 564°C; b.p. 1496°C. Sodium cyanide is now made by absorbing hydrogen cyanide in sodium hydroxide or sodium carbonate solution. The compound is extremely poisonous because it reacts with the iron in haemoglobin in the blood, so preventing oxygen reaching the tissues of the body. It is used in the extraction of precious metals and in electroplating industries. Aqueous solutions are alkaline due to salt hydrolysis.

sodium dichromate A red crystalline solid, $Na_2Cr_2O_7.2H_2O$, soluble in water and insoluble in ethanol. It is usually known as the dihydrate (r.d. 2.52), which starts to lose water above 100°C; the compound decomposes above 400°C. It is made by melting chrome iron ore with lime and soda ash and acidification of the chromate thus formed. Sodium dichromate is cheaper than the corresponding potassium compound but has the disadvantage of being hygroscopic. It is used as a mordant in dyeing, as an oxidizing agent in organic chemistry, and in analytical chemistry.

sodium dihydrogenorthophosphate *See* sodium dihydrogenphosphate(V).

sodium dihydrogenphosphate(V) (sodium dihydrogenorthophosphate) A colourless crystalline solid, NaH_2PO_4, which is soluble in water and insoluble in alcohol, known as the monohydrate (r.d. 2.04) and the dihydrate (r.d. 1.91). The dihydrate loses one water molecule at 60°C and the second molecule of water at 100°C, followed by decomposition at 204°C. The compound may be prepared by treating sodium carbonate with an equimolar quantity of phosphoric acid or by neutralizing phosphoric acid with sodium hydroxide. It is used in the preparation of sodium phosphate (Na_3PO_4), in baking powders, as a food additive, and as a constituent of buffering systems. Both sodium dihydrogenphosphate and trisodium phosphate enriched in ^{32}P have been used to study phosphate participation in metabolic processes.

sodium dioxide *See* sodium superoxide.

sodium ethanoate (sodium acetate) A colourless crystalline compound, CH_3COONa, which is known as the anhydrous salt (r.d. 1.52; m.p. 324°C) or the trihydrate (r.d. 1.45; loses water at 58°C). Both forms are soluble in water and in ethoxyethane, and slightly soluble in ethanol. The compound may be prepared by the reaction of ethanoic acid (acetic acid) with sodium carbonate or with sodium hydroxide. Because it is a salt of a strong base and a weak acid, sodium ethanoate is used in buffers for pH control in many laboratory applications, in foodstuffs, and in electroplating. It is also used in dyeing, soaps, pharmaceuticals, and in photography.

sodium fluoride A crystalline compound, NaF, soluble in water and very slightly soluble in ethanol; cubic; r.d. 2.56; m.p. 993°C; b.p. 1695°C. It occurs naturally as villiaumite and may be prepared by the reaction of sodium hydroxide or of sodium carbonate with hydrogen fluoride. The reaction of sodium fluoride with concentrated sulphuric acid may be used as a source of hydrogen fluoride. The compound is used in ceramic enamels and as a preservative agent for fermentation. It is highly toxic but in very dilute solution (less than 1 part per million) it is used in the fluoridation of water for the prevention of tooth decay on account of its ability to replace OH groups with F atoms in the material of dental enamel.

sodium formate *See* sodium methanoate.

sodium hexafluoraluminate A colourless monoclinic solid, Na_3AlF_6, very slightly soluble in water; r.d. 2.9; m.p. 1000°C. It changes to a cubic form at 580°C. The compound occurs naturally as the mineral *cryolite but a considerable amount is manufactured by the reaction of aluminium fluoride wth alumina and sodium hydroxide or directly with sodium aluminate. Its most important use is in the manufacture of aluminium in the *Hall–Heroult cell. It is also used in the manufacture of enamels, opaque glasses, and ceramic glazes.

sodium hydride A white crystalline solid, NaH; cubic; r.d. 0.92; decomposes above 300°C (slow); completely decomposed at 800°C. Sodium hydride is prepared by the reaction of pure dry hydrogen with sodium at 350°C. Electrolysis of sodium hydride in molten LiCl/KCl leads to the evolution of hydrogen; this is taken as evidence for the ionic nature of NaH and the presence of the hydride ion (H^-). It reacts violently with water to give sodium hydroxide and hydrogen, with halogens to give the halide and appropriate hydrogen halide, and ignites spontaneously with oxygen at

230°C. It is a powerful reducing agent with several laboratory applications.

sodium hydrogencarbonate (bicarbonate of soda; sodium bicarbonate) A white crystalline solid, $NaHCO_3$, soluble in water and slightly soluble in ethanol; monoclinic; r.d. 2.159; loses carbon dioxide above 270°C. It is manufactured in the *Solvay process and may be prepared in the laboratory by passing carbon dioxide through sodium carbonate or sodium hydroxide solution. Sodium hydrogencarbonate reacts with acids to give carbon dioxide and, as it does not have strongly corrosive or strongly basic properties itself, it is employed in bulk for the treatment of acid spillage and in medicinal applications as an antacid. Sodium hydrogencarbonate is also used in baking powders (and is known as *baking soda*), dry-powder fire extinguishers, and in the textiles, tanning, paper, and ceramics industries. The hydrogencarbonate ion has an important biological role as an intermediate between atmospheric CO_2/H_2CO_3 and the carbonate ion CO_3^{2-}. For water-living organisms this is the most important and in some cases the only source of carbon.

sodium hydrogensulphate (sodium bisulphate) A colourless solid, $NaHSO_4$, known in anhydrous and monohydrate forms. The anhydrous solid is triclinic (r.d. 2.435; m.p. >315°C). The monohydrate is monoclinic and deliquescent (r.d. 2.103; m.p. 59°C). Both forms are soluble in water and slightly soluble in alcohol. Sodium hydrogensulphate was originally made by the reaction between sodium nitrate and sulphuric acid, hence its old name of *nitre cake*. It may be manufactured by the reaction of sodium hydroxide with sulphuric acid, or by heating equimolar proportions of sodium chloride and concentrated sulphuric acid. Solutions of sodium hydrogensulphate are acidic. On heating the compound decomposes (via $Na_2S_2O_7$) to give sulphur trioxide. It is used in paper making, glass making, and textile finishing.

sodium hydrogensulphite (sodium bisulphite) A white solid, $NaHSO_3$, which is very soluble in water (yellow in solution) and slightly soluble in ethanol; monoclinic; r.d. 1.48. It decomposes on heating to give sodium sulphate, sulphur dioxide, and sulphur. It is formed by saturating a solution of sodium carbonate with sulphur dioxide. The compound is used in the brewing industry and in the sterilization of wine casks. It is a general antiseptic and bleaching agent. *See also* aldehydes.

sodium hydroxide (caustic soda) A white transluscent deliquescent solid, $NaOH$, soluble in water and ethanol but insoluble in ether; r.d. 2.13; m.p. 318°C; b.p. 1390°C. Hydrates containing 7, 5, 4, 3.5, 3, 2, and 1 molecule of water are known.

Sodium hydroxide was formerly made by the treatment of sodium carbonate with lime but its main source today is from the electrolysis of brine using mercury cells or any of a variety of diaphragm cells. The principal product demanded from these cells is chlorine (for use in plastics) and sodium hydroxide is almost reduced to the status of a by-product. It is strongly alkaline and finds many applications in the chemical industry, particularly in the production of soaps and paper. It is also used to adsorb acidic gases, such as carbon dioxide and sulphur dioxide, and is used in the treatment of effluent for the removal of heavy metals (as hydroxides) and of acidity. Sodium hydroxide solutions are extremely corrosive to body tissue and are particularly hazardous to the eyes.

sodium iodide A white crystalline solid, NaI, very soluble in water and soluble in both ethanol and ethanoic acid. It is known in both the anhydrous form (cubic; r.d. 3.67; m.p. 661°C; b.p. 1304°C) and as the dihydrate (monoclinic; r.d. 2.45). It is prepared by the reaction of hydrogen iodide with sodium carbonate or sodium hydroxide in solution. Like potassium iodide, sodium iodide in aqueous solution dissolves iodine to form a brown solution containing the I_3^- ion. It finds applications in photography and is also used in medicine as an expectorant and in the administration of radioactive iodine for studies of thyroid function and for treatment of diseases of the thyroid.

sodium methanoate (sodium formate) A colourless deliquescent solid, HCOONa, soluble in water and slightly soluble in ethanol; monoclinic; r.d. 1.92; m.p. 253°C; decomposes on further heating. The monohydrate is also known. The compound may be produced by the reaction of carbon monoxide with solid sodium hydroxide at 200°C and 10 atmospheres pressure; in the laboratory it can be conveniently prepared by the reaction of methanoic acid and sodium hydroxide. Its uses are in the production of oxalic acid (ethanedioic acid) and methanoic acid and in the laboratory it is a convenient source of carbon monoxide.

sodium monoxide A whitish-grey deliquescent solid, Na_2O; r.d. 2.27; sublimes at 1275°C. It is manufactured by oxidation of the metal in a limited supply of oxygen and purified by sublimation. Reaction with water produces sodium hydroxide. Its commercial applications are similar to those of sodium hydroxide.

sodium nitrate (Chile saltpetre) A white solid, $NaNO_3$, soluble in water and in ethanol; trigonal; r.d. 2.261; m.p. 306°C; decomposes at 380°C. A rhombohedral form is also known. It is obtained from deposits of caliche or may be prepared by the reaction of nitric acid with sodium hydroxide

or sodium carbonate. It was previously used for the manufacture of nitric acid by heating with concentrated sulphuric acid. Its main use is in nitrate fertilizers.

sodium nitrite A yellow hygroscopic crystalline compound, $NaNO_2$, soluble in water, slightly soluble in ether and in ethanol; rhombohedral; r.d. 2.17; m.p. 271°C; decomposes above 320°C. It is formed by the thermal decomposition of sodium nitrate and is used in the preparation of nitrous acid (reaction with cold dilute hydrochloric acid). Sodium nitrite is used in organic *diazotization and as a corrosion inhibitor.

sodium orthophosphate *See* trisodium phosphate(V).

sodium peroxide A whitish solid (yellow when hot), Na_2O_2, soluble in ice-water and decomposed in warm water or alcohol; r.d. 2.80; decomposes at 460°C. A crystalline octahydrate (hexagonal) is obtained by crystallization from ice-water. The compound is formed by the combustion of sodium metal in excess oxygen. At normal temperatures it reacts with water to give sodium hydroxide and hydrogen peroxide. It is a powerful oxidizing agent reacting with iodine vapour to give the iodate and periodate, with carbon at 300°C to give the carbonate, and with nitrogen(II) oxide to give the nitrate. It is used as a bleaching agent in wool and yarn processing, in the refining of oils and fats, and in the production of wood pulp.

sodium pump A mechanism by which sodium ions are transported out of a neurone across the cell membrane. The process requires energy in the form of ATP, being a form of *active transport. It maintains the different concentration of sodium ions on either side of the neurone membrane necessary for establishing the *resting potential of the neurone.

sodium sesquicarbonate A white crystalline hydrated double salt, $Na_2CO_3.NaHCO_3.2H_2O$, soluble in water but less alkaline than sodium carbonate; r.d. 2.12; decomposes on heating. It may be prepared by crystallizing equimolar quantities of the constituent materials; it also occurs naturally as *trona* and in Searles Lake brines. It is widely used as a detergent and soap builder and, because of its mild alkaline properties, as a water-softening agent and bath-salt base. *See also* sodium carbonate.

sodium sulphate A white crystalline compound, Na_2SO_4, usually known as the anhydrous compound (orthorhombic; r.d. 2.67; m.p. 888°C) or the decahydrate (monoclinic; r.d. 1.46; which loses water at 100°C). The decahydrate is known as *Glauber's salt*. A metastable heptahydrate $(Na_2SO_4.7H_2O)$ also exists. All forms are soluble in water, dissolving to give a neutral solution. The compound occurs naturally as

  mirabilite ($Na_2SO_4.10H_2O$),
  thenardite (Na_2SO_4), and
  glauberite ($Na_2SO_4.CaSO_4$).

Sodium sulphate may be produced industrially by the reaction of magnesium sulphate with sodium chloride in solution followed by crystallization, or by the reaction of concentrated sulphuric acid with solid sodium chloride. The latter method was used in the *Leblanc process for the production of alkali and has given the name *salt cake* to impure industrial sodium sulphate. Sodium sulphate is used in the manufacture of glass and soft glazes and in dyeing to promote an even finish. It also finds medicinal application as a purgative and in commercial aperient salts.

sodium sulphide A yellow-red solid, Na_2S, formed by the reduction of sodium sulphate with carbon (coke) at elevated temperatures. It is a corrosive and readily oxidized material of variable composition and usually contains polysulphides of the type Na_2S_2, Na_2S_3, and Na_2S_4, which cause the variety of colours. It is known in an anhydrous form (r.d. 1.85; m.p. 1180°C) and as a nonahydrate, $Na_2S.9H_2O$ (r.d. 1.43; decomposes at 920°C). Other hydrates of sodium sulphide have been reported. The compound is deliquescent, soluble in water with extensive hydrolysis, and slightly soluble in alcohol. It is used in wood pulping, dyestuffs manufacture, and metallurgy on account of its reducing properties. It has also been used for the production of sodium thiosulphate (for the photographic industry) and as a depilatory agent in leather preparation. It is a strong skin irritant.

sodium sulphite A white solid, Na_2SO_3, existing in an anhydrous form (r.d. 2.63) and as a heptahydrate (r.d. 1.59). Sodium sulphite is soluble in water and because it is readily oxidized it is widely used as a convenient reducing agent. It is prepared by reacting sulphur dioxide with either sodium carbonate or sodium hydroxide. Dilute mineral acids reverse this process and release sulphur dioxide. Sodium sulphite is used as a bleaching agent in textiles and in paper manufacture. Its use as an antioxidant in some canned foodstuffs gives rise to a slightly sulphurous smell immediately on opening, but its use is prohibited in meats or foods that contain vitamin B_1. Sodium sulphite solutions are occasionally used as biological preservatives.

sodium superoxide (sodium dioxide) A whitish-yellow solid, NaO_2, formed by the reaction of sodium peroxide with excess oxygen at elevated temperatures and pressures. It reacts with water to form hydrogen peroxide and oxygen.

sodium thiosulphate (hypo) A colourless efflorescent solid, $Na_2S_2O_3$, soluble in water but insoluble in ethanol, commonly encountered as the pentahydrate (monoclinic; r.d. 1.73; m.p. 42°C), which loses water at 100°C to give the anhydrous form (r.d. 1.66). It is prepared by the reaction of sulphur dioxide with a suspension of sulphur in boiling sodium hydroxide solution. Aqueous solutions of sodium thiosulphate are readily oxidized in the presence of air to sodium tetrathionate and sodium sulphate. The reaction with dilute acids gives sulphur and sulphur dioxide. It is used in the photographic industry and in analytical chemistry.

sodium-vapour lamp A form of *electric lighting that gives a yellow light as a result of the luminous discharge obtained by the passage of a stream of electrons between tungsten electrodes in a tube containing sodium vapour. To facilitate starting, the tube also contains some neon; for this reason, until the lamp is warm the neon emits a characteristic pink glow. As the sodium vaporizes, the yellow light predominates. Sodium-vapour lamps are widely used as street lights because of their high luminous efficiency and because the yellow light is less absorbed than white light by fog and mist. Low-pressure sodium lamps emit a characteristic yellow light; in high-pressure lamps the atoms are sufficiently close to each other to interact and broaden the spectral lines into the orange and green regions.

soft iron A form of iron that contains little carbon, has high relative permeability, is easily magnetized and demagnetized, and has a small hysteresis loss. Soft iron and other *soft ferromagnetic materials*, such as silicon steel, are used in making parts exposed to rapid changes of magnetic flux, such as the cores of electromagnets, motors, generators, and transformers.

By comparison, *hard ferromagnetic materials*, such as cobalt steel and various alloys of nickel, aluminium, and cobalt, have low relative permeability, are difficult to magnetize, and have a high hysteresis loss. They are used in making permanent magnets.

soft radiation Ionizing radiation of low penetrating power, usually used with reference to X-rays of long wavelength. *Compare* hard radiation.

soft soap *See* soap.

software *See* computer.

soft water *See* hardness of water.

softwood *See* wood.

soil The layer of unconsolidated particles derived from weathered rock, organic material (*humus*), water, and air that forms the upper surface over much of the earth and supports plant growth. The formation of soil depends on the parent material (i.e. the original material from which the soil is derived), the climate and topography of the area, the organisms present in the soil, and the time over which the soil has been developing. Soils are often classified in terms of their structure and texture. The structure of a soil is the way in which the individual soil particles are bound together to form aggregates or peds. The structure types include platy, blocky, granular, and crumbs. The texture of a soil denotes the proportion of the various particle sizes that it contains. The four main texture classes are sand, silt, clay, and loam, of which loams are generally the best agricultural soils as they contain a mixture of all particle sizes. A number of distinct horizontal layers can often be distinguished in a vertical section (profile) of soil – these are known as *soil horizons*. Four basic horizons are common to most soils: an uppermost A horizon containing the organic matter; an underlying B horizon, which contains little organic material and is strongly leached; a C horizon consisting of weathered rock; and a D horizon comprising the bedrock. *See also* brown earth; chernozem; podzol.

sol A *colloid in which small solid particles are dispersed in a liquid continuous phase.

solar cell An electric cell that uses the sun's radiation to produce usable electric current. Most solar cells consist of a single-crystal silicon $p-n$ junction. When photons of light energy from the sun fall on or near the *semiconductor junction the electron-hole pairs created are forced by the electric field at the junction to separate so that the holes pass to the p-region and the electrons pass to the n-region. This displacement of free charge creates an electric current when a load is connected across the terminals of the device (see illustration). Individual silicon solar cells cannot be made with a surface area much in excess of 4000 mm^2 and the maximum power delivered by such a cell is approximately 0.6 W at about 0.5 V in full sun. The efficiency of such devices is about 15%. For practical use, therefore, solar cells have to be assembled in arrays. Panels of solar cells have been the exclusive source of power for satellites and space capsules. Their use on earth has been largely limited by their high cost, a reduction in the cost by a factor of 10 being required to make

Silicon p-n junction

them competitive with other energy sources at present.

solar constant The rate at which solar energy is received per unit area at the outer limit of the earth's atmosphere at the mean distance between the earth and the sun. The value is 1.353 kW m^{-2}.

solar day *See* day.

solar energy The electromagnetic energy radiated from the sun. The tiny proportion (about 5×10^{-10} of the total) that falls on the earth is indicated by the *solar constant. The total quantity of solar energy falling on the earth in one year is about 4×10^{18} J, whereas the total annual energy consumption of the earth's inhabitants is only some 3×10^{14} J. The sun, therefore, could provide all the energy needed. The direct ways of making use of solar energy can be divided into thermal methods (*see* solar heating) and nonthermal methods (*see* solar cell).

solar heating A form of domestic or industrial heating that relies on the direct use of solar energy. The basic form of *solar heater* is a thermal device in which a fluid is heated by the sun's rays in a collector (see illustrations) and pumped or allowed to flow round a circuit that provides some form of heat storage and some form of auxiliary heat source for use when the sun is not shining. More complicated systems are combined heating-and-cooling devices, providing heat in the winter and air-conditioning in the summer. The simplest form of solar collector is the flat-plate collector, in which a blackened receiving surface is covered by one or more glass plates that acts like a greenhouse (*see* greenhouse effect) and traps the maximum amount of solar energy. Tubes attached to the receiving surface carry air, water, or some other fluid to which the absorbed heat is transferred. The whole panel is insulated at the back and can thus

Typical solar heating system

Flat plate solar collector

form part of the roof of a building. More sophisticated collectors focus the sun's rays using reflectors. *See also* solar cell.

solar parallax The angle subtended by the earth's equatorial radius at the centre of the sun at the mean distance between the earth and the sun (i.e. at 1 astronomical unit). It has the value 8.794 148 arc seconds.

solar prominence A cloud of gas that forms temporarily in the upper *chromosphere or inner *corona of the sun. It has a lower temperature but higher density than its surroundings and is observed as a bright projection.

solar system The sun, the nine major planets (Mercury, Venus, Earth, Mars, Jupiter, Saturn, Uranus, Neptune, and Pluto) and their natural satellites, the asteroids, the comets, and meteoroids. Over 99% of the mass of the system is concentrated in the sun. The solar system as a whole moves in an approximately circular orbit about the centre of the Galaxy, taking about 2.2×10^8 years to complete its orbit.

solar wind A continuous outward flow of charged particles, mostly protons and electrons, from the sun's *corona into interplanetary space. The particles are controlled by the sun's magnetic field and are able to escape from the sun's gravitational field because of their high thermal energy. The average velocity of the particles in the vicinity of the earth is about 450 km s^{-1} and their density at this range is about 8×10^6 protons per cubic metre.

solar year *See* year.

solder An alloy used to join metal surfaces. A *soft solder* melts at a temperature in the range 200–300°C and consists of a tin–lead alloy. The tin content varies between 80% for the lower end of the melting range and 31% for the higher end. *Hard solders* contain substantial quantities of silver in the alloy. *Brazing solders* are usually alloys of copper and zinc, which melt at over 800°C.

solenoid A coil of wire wound on a cylindrical former in which the length of the former is greater than its diameter. When a current is passed through the coil a magnetic field is produced inside the coil parallel to its axis. This field can be made to operate a plunger inside the former so that the solenoid can be used to operate a circuit breaker, valve, or other electromechanical device.

solid A state of matter in which there is a three-dimensional regularity of structure, resulting from the proximity of the component atoms, ions, or molecules and the strength of the forces between them. True solids are crystalline (*see also* amorphous). If a crystalline solid is heated, the kinetic energy of the components increases. At a specific temperature, called the *melting point*, the forces between the components become unable to contain them within the crystal structure. At this temperature, the lattice breaks down and the solid becomes a liquid.

solid angle Symbol Ω. The three-dimensional 'angle' formed by the vertex of a cone. When this vertex is the centre of a sphere of radius r and the base of the cone cuts out an area s on the surface of the sphere, the solid angle in *steradians is defined as s/r^2.

solid solution A crystalline material that is a mixture of two or more components, with ions, atoms, or molecules of one component replacing some of the ions, atoms, or molecules of the other component in its normal crystal lattice. Solid solutions are found in certain alloys. For example, gold and copper form solid solutions in which some of the copper atoms in the lattice are replaced by gold atoms. In general, the gold atoms are distributed at random, and a range of gold–copper compositions is possible. At a certain composition, the gold and copper atoms can each form regular individual lattices (referred to as *superlattices*). Mixed crystals of double salts (such as alums) are also examples of solid solutions. Compounds can form solid solutions if they are isomorphous (*see* isomorphism).

solid-state detector *See* junction detector.

solid-state physics The study of the physical properties of solids, with special emphasis on the electrical properties of semiconducting materials in relation to their electronic structure. *Solid-state devices* are electronic components consisting entirely of solids (e.g. semiconductors, transistors, etc.) without heating elements, as in thermionic valves.

Recently the term *condensed-matter physics* has been introduced to include the study of crystalline solids, amorphous solids, and liquids.

soliton A stable particle-like solitary wave state that is a solution of certain equations for propagation. Solitons are thought to occur in many areas of physics and applied mathematics, such as plasmas, fluid mechanics, lasers, optics, solid-state physics, and elementary-particle physics.

solstice 1. Either of the two points on the *ecliptic midway between the *equinoxes, at which the sun is at its greatest angular distance north (*summer solstice*) or south (*winter solstice*) of the celestial equator. **2.** The time at which the sun reaches either of these points. The summer solstice occurs on June 21 and the winter solstice on December 21 in the northern hemisphere; the dates are reversed in the southern hemisphere.

solubility The quantity of solute that dissolves in a given quantity of solvent to form a saturated solution. Solubility is measured in kilograms per metre cubed, moles per kilogram of solvent, etc. The solubility of a substance in a given solvent depends on the temperature. Generally, for a solid in a liquid, solubility increases with temperature; for a gas, solubility decreases. *See also* concentration.

solubility product Symbol K_s. The product of the concentrations of ions in a saturated solution. For instance, if a compound A_xB_y is in equilibrium with its solution

$$A_xB_y(s) \leftrightharpoons xA^+(aq) + yB^-(aq)$$

the equilibrium constant is

$$K_c = [A^+]^x[B^-]^y/[A_xB_y]$$

Since the concentration of the undissolved solid can be put equal to 1, the solubility product is given by

$$K_s = [A^+]^x[B^-]^y$$

The expression is only true for sparingly soluble salts. If the product of ionic concentrations in a solution exceeds the solubility product, then precipitation occurs.

solute The substance dissolved in a solvent in forming a *solution.

solution A homogeneous mixture of a liquid (the *solvent) with a gas or solid (the *solute*). In a solution, the molecules of the solute are discrete and mixed with the molecules of solvent. There is usually some interaction between the solvent and solute molecules (*see* solvation). Two liquids that can mix on the molecular level are said to be *miscible*. In this case, the solvent is the major component and the solute the minor component. *See also* solid solution.

solvation The interaction of ions of a solute with the molecules of a solvent. For instance, when sodium chloride is dissolved in water the sodium ions attract polar water molecules, with the negative oxygen atoms pointing towards the positive Na^+ ion. Solvation of transition-metal ions can also occur by formation of coordinate bonds, as in the hexaquocopper(II) ion $[Cu(H_2O)_6]^{2+}$. Solvation is the process that causes ionic solids to dissolve, because the energy released compensates for the energy necessary to break down the crystal lattice. It occurs only with polar solvents. Solvation in which the solvent is water is called *hydration*.

Solvay process (ammonia–soda process) An industrial method of making sodium carbonate from calcium carbonate and sodium chloride. The calcium carbonate is first heated to give calcium oxide and carbon dioxide, which is bubbled into a solution of sodium chloride in ammonia. Sodium hydrogencarbonate is precipitated:

$$H_2O + CO_2(g) + NaCl(aq) + NH_3(aq) \rightarrow$$
$$NaHCO_3(s) + NH_4Cl(aq)$$

The sodium hydrogencarbonate is heated to give sodium carbonate and carbon dioxide. The ammonium chloride is heated with calcium oxide (from the first stage) to regenerate the ammonia. The process was patented in 1861 by the Belgian chemist Ernest Solvay (1838–1922).

solvent A liquid that dissolves another substance or substances to form a *solution. *Polar solvents* are compounds such as water and liquid ammonia, which have dipole moments and consequently high dielectric constants. These solvents are capable of dissolving ionic compounds or covalent compounds that ionize (*see* solvation). *Nonpolar solvents* are compounds such as ethoxyethane and benzene, which do not have permanent dipole moments. These do not dissolve ionic compounds but will dissolve nonpolar covalent compounds. Solvents can be further categorized according to their proton-donating and accepting properties. *Amphiprotic solvents* self-ionize and can therefore act both as proton donators and acceptors. A typical example is water:

$$2H_2O \leftrightharpoons H_3O^+ + OH^-$$

Aprotic solvents neither accept nor donate protons; tetrachloromethane (carbon tetrachloride) is an example.

solvent extraction The process of separating one constituent from a mixture by dissolving it in a solvent in which it is soluble but in which the other constituents of the mixture are not. The process is usually carried out in the liquid phase, in which case it is also known as *liquid–liquid extraction*. In liquid–liquid extraction, the solution containing the desired constituent must be immiscible with the rest of the mixture. The process is widely used in extracting oil from oil-bearing materials.

solvolysis A reaction between a compound and its solvent. *See* hydrolysis.

somatic 1. Relating to all the cells of an animal or plant other than the reproductive cells. Thus a somatic *mutation is one that is not heritable.

2. Relating to organs and tissues of the body other than the gut and its associated structures. The term is applied especially to voluntary muscles, the sense organs, and the nervous system.

somatotrophin *See* growth hormone.

sonar *See* echo.

sonic boom A strong *shock wave generated by an aircraft when it is flying in the earth's atmosphere at supersonic speeds. This shock wave is radiated from the aircraft and where it intercepts the surface of the earth a loud booming sound is heard. The loudness depends on the speed and altitude of the aircraft and is lower in level flight than when the aircraft is undertaking a manoeuvre. The maximum increase of pressure in the shock wave during a transoceanic flight of a commercial supersonic transport (SST) is 120 Pa, equivalent to 136 decibels.

sonometer A device consisting essentially of a hollow sounding box with two bridges attached to its top. The string, fixed to the box at one end, is stretched between the two bridges so that the free end can run over a pulley and support a measured load. When the string is plucked the frequency of the note can be matched with that of another sound source, such as a tuning fork. It can be used to verify that the frequency (f) of a stretched string is given by $f = (1/2l)\sqrt{(T/m)}$, where l is the length of the string, m is its mass per unit length, and T is its tension.

Originally called the *monochord*, the sonometer was widely used as a tuning aid, but is now used only in teaching laboratories.

sorosis A type of *composite fruit formed from an entire inflorescence spike. Mulberry and pineapple fruits are examples.

sorption *Absorption of a gas by a solid.

sorption pump A type of vacuum pump in which gas is removed from a system by absorption on a solid (e.g. activated charcoal or a zeolite) at low temperature.

sorus 1. Any of the spore-producing structures on the undersurface of a fern frond, visible as rows of small brown dots. **2.** A reproductive area on the thallus of some algae, e.g. *Laminaria*. **3.** Any of various spore-producing structures in certain fungi.

sound A vibration in an elastic medium at a frequency and intensity that is capable of being heard by the human ear. The frequency of sounds lie in the range 20–20 000 Hz, but the ability to hear sounds in the upper part of the frequency range declines with age (*see also* pitch). Vibrations that have a lower frequency than sound are called *infrasounds* and those with a higher frequency are called *ultrasounds*.

Sound is propagated through an elastic fluid as a longitudinal *sound wave*, in which a region of high pressure travels through the fluid at the *speed of sound in that medium. At a frequency of about 10 kilohertz the maximum excess pressure of a sound wave in air lies between 10^{-4} Pa and 10^3 Pa. Sound travels through solids as either longitudinal or transverse waves.

source The electrode in a field-effect *transistor from which electrons or holes enter the interelectrode space.

Southern blotting A chromatographic technique for isolating and identifying specific fragments of DNA, such as the fragments formed as a result of DNA cleavage by *restriction enzymes. The mixture of fragments is subjected to electrophoresis through an agarose gel, followed by denaturation to form single-stranded fragments. These are transferred, or 'blotted', onto a nitrocellulose filter where they are immobilized in their relative positions. Specific *gene probes labelled with a radioisotope are then added. These hybridize with any complementary fragments on the filter, which are subsequently revealed by autoradiography. The technique was devised by E. M. Southern. A similar technique for detecting RNA fragments is called *Northern blotting*, by analogy. *See also* Western blotting.

space 1. A property of the universe that enables physical phenomena to be extended into three mutually perpendicular directions. In Newtonian physics, space, time, and matter are treated as quite separate entities. In Einsteinian physics, space and time are combined into a four-dimensional continuum (*see* space–time) and in the general theory of *relativity matter is regarded as having an effect on space, causing it to curve. **2.** (*or* **outer space**) The part of the universe that lies outside the earth's atmosphere.

space probe An unmanned spacecraft that investigates features within the solar system. A *planetary probe* examines the conditions on or in the vicinity of one or more planets and a *lunar probe* is designed to obtain information about the moon. Probes are propelled by rocket motors and once out of the earth's gravitational field use their propulsion systems for course changes. Many are powered by panels of *solar cells, for both internal operation and radio communications.

space-reflection symmetry *See* parity.

space–time (space–time continuum) A geometry that includes the three dimensions and a *fourth dimension* of time. In Newtonian physics, space and time are considered as separate entities and whether or not events are simultaneous is a matter that is regarded as obvious to any competent observer. In Einstein's concept of the physical universe, based on a system of geometry devised by

H. Minkowski (1864–1909), space and time are regarded as entwined, so that two observers in relative motion could disagree regarding the simultaneity of distant events. In Minkowski's geometry, an event is identified by a *world point* in a four-dimensional continuum.

spadix A flowering shoot (a type of *spike) with a large fleshy floral axis bearing small, usually unisexual, flowers. It is protected by a large petal-like bract, the *spathe*, and is characteristic of plants of the family Araceae (e.g. calla lily).

spallation A type of nuclear reaction in which the interacting nuclei disintegrate into a large number of protons, neutrons, and other light particles, rather than exchanging nucleons between them.

spark *See* electric spark.

spark chamber A device for detecting charged particles. It consists of a chamber, filled with helium and neon at atmospheric pressure, in which a stack of 20 to 100 plates are placed; the plates are connected alternately to the positive and negative terminals of a source of high potential (10 000 V or more). An incoming particle creates ion pairs in its track, causing the gas to become conducting and sparks to jump between the plates. The light from the sparks is focused to obtain stereoscopic photographs of the particles' tracks. It can also function as a counter (called a *spark counter*) when connected to suitable counting circuits. Some versions use crossed sets of parallel wires rather than plates; simple patterns may have a single wire near a plate, in the open atmosphere.

spark counter *See* spark chamber.

spathe *See* spadix.

special creation The belief, in accordance with the Book of Genesis, that every species was individually created by God in the form in which it exists today and is not capable of undergoing any change. It was the generally accepted explanation of the origin of life until the advent of *Darwinism. The idea has recently enjoyed a revival, especially among members of the fundamentalist movement in the USA, partly because there still remain problems that cannot be explained entirely by Darwinian theory. However, special creation is contradicted by fossil evidence and genetic studies, and the pseudoscientific arguments of *creation science* cannot stand up to logical examination.

specialization 1. Increasing *adaptation of an organism to a particular environment. **2.** *See* physiological specialization.

special theory of relativity *See* relativity.

speciation The development of one or more species from an existing species. It occurs when *sympatric or *allopatric populations diverge so much from the parent population that interbreeding can no longer occur between them.

species 1. (in biology) A category used in the *classification of organisms that consists of a group of similar individuals that can usually breed among themselves and produce fertile offspring. Similar or related species are grouped into a genus. Within a species groups of individuals may become reproductively isolated because of geographical or behavioural factors. Such populations may, because of different selection pressures, develop different characteristics from the main population and so form a distinct *subspecies. *See also* binomial nomenclature. **2.** (in chemistry) A chemical entity, such as a particular atom, ion, or molecule.

specific 1. (in physics) **a.** Denoting that an extensive physical quantity so described is expressed per unit mass. For example, the *specific latent heat*

of a body is its latent heat per unit mass. When the extensive physical quantity is denoted by a capital letter (e.g. *L* for latent heat), the specific quantity is denoted by the corresponding lower-case letter (e.g. *l* for specific latent heat). **b.** In some older physical quantities the adjective 'specific' was added for other reasons (e.g. specific gravity, specific resistance). These names are now no longer used. **2.** (in biology) Relating to a species.

specific activity *See* activity.

specific charge The ratio of the charge of an *elementary particle or other charged body to its mass.

specific gravity *See* relative density; specific.

specific heat capacity *See* heat capacity.

specific humidity *See* humidity.

specific impulse A measure of the thrust available from a rocket propellant. It is the ratio of the thrust produced to the fuel consumption.

specific intensity *See* Planck's radiation law.

specific latent heat *See* latent heat.

specific resistance *See* resistivity; specific.

specific surface The surface area of a particular substance per unit mass, expressed in $m^2 \ kg^{-1}$. It provides a measure of the surface area available for a process, such as adsorption, for a given mass of a powder or porous substance.

specific volume The volume of a substance per unit mass. The reciprocal of density, it has the units $m^3 \ kg^{-1}$.

spectral class (spectral type) A form of classification used for stars, based on their spectra. The *Harvard classification*, introduced in 1890 and modified in the 1920s, is based on the seven star types known as O, B, A, F, G, K, M:

O hottest blue stars; ionized helium lines dominant

B hot blue stars; neutral helium lines dominant, no ionized helium

A blue blue-white stars; hydrogen lines dominant

F white stars; metallic lines strengthen, hydrogen lines weaken

G yellow stars; ionized calcium lines dominant

K orange-red stars; neutral metallic lines dominant, some molecular bands

M coolest red stars; molecular bands dominant

spectrograph *See* spectroscope.

spectrometer Any of various instruments for producing a spectrum and measuring the wavelengths, energies, etc., involved. A simple type, for visible radiation, is a spectroscope equipped with a calibrated scale allowing wavelengths to be read off or calculated. In the X-ray to infrared region of the electromagnetic spectrum, the spectrum is produced by dispersing the radiation with a prism or diffraction grating (or crystal, in the case of hard X-rays). Some form of photoelectric detector is used, and the spectrum can be obtained as a graphical plot, which shows how the intensity of the radiation varies with wavelength. Such instruments are also called *spectrophotometers*. Spectrometers also exist for investigating the gamma-ray region and the microwave and radio-wave regions of the spectrum (*see* electron-spin resonance; nuclear magnetic resonance). Instruments for obtaining spectra of particle beams are also called spectrometers (*see* spectrum; photoelectron spectroscopy).

spectrophotometer *See* spectrometer.

Spectroscope

spectroscope An optical instrument that produces a *spectrum for visual observation. The first such instrument was made by R. W. Bunsen; in its simplest form it consists of a hollow tube with a slit at one end by which the light enters and a collimating lens at the other end to produce a parallel beam, a prism to disperse the light, and a telescope for viewing the spectrum (see illustration). In the *spectrograph*, the spectroscope is provided with a camera to record the spectrum. For a broad range of spectroscopic work, from the ultraviolet to the infrared, a diffraction grating is used instead of a prism. *See also* spectrometer.

spectroscopic binary *See* binary stars.

spectroscopy The study of methods of producing and analysing *spectra using *spectroscopes, *spectrometers, spectrographs, and spectrophotometers. The interpretations of the spectra so produced can be used for chemical analysis, examining atomic and molecular energy levels and molecular structures, and for determining the composition and motions of celestial bodies (*see* redshift).

spectrum (*pl.* **spectra**) **1.** A distribution of entities or properties arrayed in order of increasing or decreasing magnitude. For example, a beam of ions passed through a mass spectrograph, in which they are deflected according to their charge-to-mass ratios, will have a range of masses called a *mass spectrum*. A *sound spectrum* is the distribution of energy over a range of frequencies of a particular source. **2.** A range of electromagnetic energies arrayed in order of increasing or decreasing wavelength or frequency (*see* electromagnetic spectrum). The *emission spectrum* of a body or substance is the characteristic range of radiations it emits when it is heated, bombarded by electron or

648

ions, or absorbs photons. The *absorption spectrum* of a substance is produced by examining, through the substance and through a spectroscope, a continuous spectrum of radiation. The energies removed from the continuous spectrum by the absorbing medium show up as black lines or bands. With a substance capable of emitting a spectrum, these are in exactly the same positions in the spectrum as some of the lines and bands in the emission spectrum.

Emission and absorption spectra may show a *continuous spectrum*, a *line spectrum*, or a *band spectrum*. A continuous spectrum contains an unbroken sequence of frequencies over a relatively wide range; it is produced by incandescent solids, liquids, and compressed gases. Line spectra are discontinuous lines produced by excited atoms and ions as they fall back to a lower energy level. Band spectra (closely grouped bands of lines) are characteristic of molecular gases or chemical compounds. *See also* action spectrum; spectroscopy.

speculum An alloy of copper and tin formerly used in reflecting telescopes to make the main mirror as it could be cast, ground, and polished to make a highly reflective surface. It has now been largely replaced by silvered glass for this purpose.

speed The ratio of a distance covered by a body to the time taken. Speed is a *scalar quantity, i.e. no direction is given. Velocity is a *vector quantity, i.e. both the rate of travel and the direction are specified.

speed of light Symbol c. The speed at which electromagnetic radiation travels. The speed of light in a vacuum is $2.997\,924\,58 \times 10^8$ m s^{-1}. When light passes through any material medium its speed is reduced (*see* refractive index). The speed of light in a vacuum is the highest speed attainable in the universe (*see* relativity; Cerenkov radiation). It is a uni-

versal constant and is independent of the speed of the observer. Since October 1983 it has formed the basis of the definition of the *metre.

speed of sound Symbol c or c_s. The speed at which sound waves are propagated through a material medium. In air at 20°C sound travels at 344 m s^{-1}, in water at 20°C it travels at 1461 m s^{-1}, and in steel at 20°C at 5000 m s^{-1}. The speed of sound in a medium depends on the medium's modulus of elasticity (E) and its density (ρ) according to the relationship $c = \sqrt{(E/\rho)}$. For longitudinal waves in a narrow solid specimen, E is the Young modulus; for a liquid E is the bulk modulus (*see* elastic modulus); and for a gas $E = \gamma p$, where γ is the ratio of the principal specific *heat capacities and p is the pressure of the gas. For an ideal gas the relationship takes the form $c = \sqrt{(\gamma r T)}$, where r is the gas constant per unit mass and T is the thermodynamic temperature. This equation shows how the speed of sound in a gas is related to its temperature. This relationship can be written $c = c_0\sqrt{(1 + t/273)}$, where c_0 is the speed of sound in a particular gas at 0°C and t is the temperature in °C.

sperm *See* spermatozoon.

spermatheca (seminal receptacle) A sac or receptacle in some female or hermaphrodite animals (e.g. earthworms) in which sperm from the mate is stored until the eggs are ready to be fertilized.

spermatogenesis The series of cell divisions in the testis that results in the production of spermatozoa. Within the seminiferous tubules of the testis germ cells grow and divide by mitosis to produce spermatogonia. These divide by mitosis to produce *spermatocytes*, which divide by meiosis to produce *spermatids*. The spermatids, which thus have half the number of chromosomes of the original germ cells, then develop into spermatozoa.

Spermatophyta A division of the plant kingdom containing plants that reproduce by means of *seeds. These species make up the bulk of the present-day flora – trees, shrubs, grasses, and herbaceous plants. The division is subdivided into the *Angiospermae and the *Gymnospermae. In some classifications seed plants are included within the *Pteropsida.

spermatozoid *See* antherozoid.

spermatozoon (sperm) The mature mobile reproductive cell (*see* gamete) of male animals, which is produced by the testis (*see* spermatogenesis). It consists of a head section containing a *haploid nucleus and an *acrosome*, which allows the sperm to penetrate the egg at fertilization; a middle section containing *mitochondria to provide the energy for movement; and a tail section, which lashes to drive the sperm forward.

sphalerite (zinc blende) A mineral form of zinc sulphide, ZnS, crystallizing in the cubic system; the principal ore of zinc. It is usually yellow-brown to brownish-black in colour and occurs, often with galena, in metasomatic deposits and also in hydrothermal veins and replacement deposits. Sphalerite is mined on every continent, the chief sources including the USA, Canada, Mexico, the USSR, Australia, Peru, and Poland.

Sphenopsida A class of the *Pteridophyta or a subdivision of the *Tracheophyta, the only living members of which are the horsetails (*Equisetum*). Horsetails have a perennial creeping rhizome supporting erect jointed stems bearing whorls of thin leaves. Spores are produced by terminal conelike structures. The group has a fossil record extending back to the Palaeozoic with its greatest development in the Carboniferous period, when giant tree forms were the dominant vegetation with the *Lycopsida.

sphere The figure generated when a circle is rotated about a diameter.

The volume of a sphere is $4\pi r^3/3$ and its surface area is $4\pi r^2$, where r is its radius. In Cartesian coordinates the equation of a sphere centred at the origin is $x^2 + y^2 + z^2 = r^2$.

spherical aberration *See* aberration.

spherical mirror *See* mirror.

spherical polar coordinates *See* polar coordinates.

Spherometer

spherometer An instrument for measuring the curvature of a surface. The usual instrument for this purpose consists of a tripod, the pointed legs of which rest on the spherical surface at the corners of an equilateral triangle. In the centre of this triangle is a fourth point, the height of which is adjusted by means of a micrometer screw (see illustration). If the distance between each leg and the axis through the micrometer screw is l, and the height of the micrometer point above (or below) a flat surface is x, the radius (r) of the sphere is given by $r = (l^2 + x^2)/2x$.

sphincter A specialized muscle encircling an opening or orifice. Contraction of the sphincter tends to close the orifice. Examples are the anal sphincter (round the opening of the anus) and the pyloric sphincter (at the lower opening of the stomach).

spiders *See* Arachnida.

spiegel (spiegeleisen) A form of *pig iron containing 15–30% of manganese and 4–5% of carbon. It is added to steel in a Bessemer converter as a deoxidizing agent and to raise the manganese content of steel.

spike A type of *racemose inflorescence in which stalkless flowers arise from an undivided floral axis, as in plantain and *Orchis*. In the family Gramineae (sedges and grasses) the flowers are grouped in clusters called *spikelets*, which may be arranged to form a compound spike (as in wheat).

spin (intrinsic angular momentum) Symbol *s*. The part of the total angular momentum of a particle, atom, nucleus, etc., that can continue to exist even when the particle is apparently at rest, i.e. when its translational motion is zero and therefore its orbital angular momentum is zero. A molecule, atom, or nucleus in a specified energy level, or a particular elementary particle, has a particular spin, just as it has a particular charge or mass. According to *quantum theory, this is quantized and is restricted to multiples of $h/2\pi$, where h is the *Planck constant. Spin is characterized by a quantum number s. For example, for an electron $s = \pm\frac{1}{2}$, implying a spin of $+ h/4\pi$ when it is spinning in one direction and $-h/4\pi$ when it is spinning in the other. Because of their spin, particles also have their own intrinsic *magnetic moments and in a magnetic field the spin of the particles lines up at an angle to the direction of the field, precessing around this direction. *See also* electron spin resonance; nuclear magnetic resonance.

spinal column *See* vertebral column.

spinal cord The part of the vertebrate central nervous system that is posterior to the brain and enclosed within the *vertebral column. It consists of a hollow core of *grey matter (H-shaped in cross section) surrounded by an outer layer of white matter; the central cavity contains *cerebrospinal fluid. Paired *spinal nerves arise from the spinal cord.

spinal nerves Pairs of nerves that arise from the *spinal cord (*compare* cranial nerves). In man there are 31 pairs (one from each of the vertebrae). Each nerve contains both motor and sensory fibres (i.e. they are mixed nerves), and the spinal nerves form an important part of the *peripheral nervous system.

spindle A structure formed from fine fibres of protein in the cytoplasm of cells at the beginning of the *metaphase stage of cell division (*see* meiosis; mitosis). Two poles become established at diametrically opposite points close to the nucleus. Fibres radiating from each pole towards the opposite one diverge to their greatest extent midway between. This widest part of the spindle is the *equator*. Chromosomes are attached by their centromeres to the spindle fibres at the equator.

spindle attachment *See* centromere.

spine 1. *See* vertebral column. **2.** A hard pointed protective structure on a plant that is formed through modification of a leaf, part of a leaf, or a stipule. The edge of the holly leaf is drawn out into spines, but in cacti the whole leaf is modified as a spine. *Compare* prickle; thorn.

spinel A group of oxide minerals with the general formula $F^{2+}R_2^{3+}O_4$, where F^{2+} = Mg, Fe, Zn, Mn, or Ni and R^{3+} = Al, Fe, or Cr, crystallizing in the cubic system. The spinels are divided into three series: spinel $(MgAl_2O_4)$, *magnetite, and *chromite. They occur in high-temperature igneous or metamorphic rocks.

spin glass An alloy of a small amount of a magnetic metal (0.1–10%) with a nonmagnetic metal, in which the atoms of the magnetic element are randomly distributed through the crystal lattice of the nonmagnetic element. Examples are AuFe and CuMn. Theories of the magnetic and other properties of spin glasses are complicated by the random distribution of the magnetic atoms.

spinneret A small tubular appendage from which silk is produced in spiders and some insects. Spiders have four to six spinnerets on the hind part of the abdomen, into which numerous silk glands open. The silk is secreted as a fluid and hardens on contact with the air. Various types of silk are produced depending on its use (e.g. for webs, egg cocoons, etc.). The spinnerets that produce the cocoons of insects are not homologous with those of spiders. For example, the spinneret of the silkworm is in the pharynx and the silk is produced by modified salivary glands.

spin–statistics theorem A fundamental theorem of relativistic *quantum field theory that states that half-integer *spins can only be quantized consistently if they obey Fermi–Dirac statistics and even-integer spins can only be quantized consistently if they obey Bose–Einstein statistics (*see* quantum statistics). This theorem enables one to understand the result of quantum statistics that wave functions for bosons are symmetric and wave functions for fermions are antisymmetric. It also provides the foundation for the *Pauli exclusion principle. It was first proved by Pauli in 1940.

spin wave (magnon) A *collective excitation associated with magnetic systems. Spin waves occur in both ferromagnetic and antiferromagnetic systems (*see* magnetism).

spiracle 1. A small paired opening that occurs on each side of the head in cartilaginous fish. It is the reduced first *gill slit, its small size resulting from adaptations of the skeleton for the firm attachment of the jaws. In modern teleosts (bony fish) the spira-cle is closed up. In tetrapods the first gill slit develops into the middle ear cavity. **2.** Any of the external openings of the *tracheae along the side of the body of an insect.

spiral galaxy *See* galaxy.

spirillum Any rigid spiral-shaped bacterium. Generally, spirilla are Gram-negative (*see* Gram's stain), aerobic, and highly motile, bearing one or more tufts of flagella. They occur in soil and water, feeding on organic matter.

spirits of salt Hydrogen chloride, so-called because it can be made by adding sulphuric acid to common salt (sodium chloride).

spirochaete Any nonrigid corkscrew-shaped bacterium that moves by means of muscular flexions of the cell. Most spirochaetes are Gram-negative (*see* Gram's stain), anaerobic, and feed on dead organic matter. They are very common in sewage-polluted waters. Some, however, can cause disease in man; *Treponema*, the agent of syphilis, is an example.

spleen A vertebrate organ, lying behind the stomach, that is basically a collection of *lymphoid tissue. Its functions include producing lymphocytes and destroying foreign particles. It acts as a reservoir for erythrocytes and can regulate the number in circulation. It is also the site for the breakdown of worn-out erythrocytes and it stores the iron they contain.

sponges *See* Porifera.

spongy mesophyll *See* mesophyll.

spontaneous combustion Combustion in which a substance produces sufficient heat within itself, usually by a slow oxidation process, for ignition to take place without the need for an external high-temperature energy source.

spontaneous generation The discredited belief that living organisms can somehow be produced by nonliving matter. For example, it was once thought that microorganisms arose by the process of decay and even that vermin spontaneously developed from household rubbish. Controlled experiments using sterilized media by Pasteur and others finally disproved these notions. *Compare* biogenesis. *See also* biopoiesis.

sporangium A reproductive structure in plants that produces asexual spores. *See* sporophyll.

spore A reproductive cell that can develop into an individual without first fusing with another reproductive cell (*compare* gamete). Spores are produced by plants, fungi, bacteria, and some Protozoa. A spore may develop into an organism resembling the parent or into another stage in the life cycle, either immediately or after a period of dormancy. In plants showing *alternation of generations, spores are formed by the *sporophyte generation and give rise to the *gametophyte generation. In ferns, the rows of brown reproductive structures on the undersurface of the fronds are spore-producing bodies.

spore mother cell (sporocyte) A diploid cell that gives rise to four haploid spores by meiosis.

sporocyte *See* spore mother cell.

sporogonium The *sporophyte generation in bryophytes. It is made up of an absorptive *foot*, a stalk (*seta*), and a spore-producing *capsule*. It may be completely or partially dependent on the *gametophyte.

sporophore (fructification) The aerial spore-producing part of certain fungi; for example, the stalk and cap of a mushroom.

sporophyll A leaf that bears *sporangia* (spore-producing structures). In ferns the sporophylls are the normal foliage leaves, but in higher plants the sporophylls are modified and arise in specialized structures such as the strobilus (cone) of certain pteridophytes and gymnosperms and the flower of

angiosperms. Angiosperms, gymnosperms, and many pteridophytes produce spores of two different sizes (small *microspores* and large *megaspores*). The sporophylls bearing these are called *microsporophylls* and *megasporophylls* respectively.

sporophyte The generation in the life cycle of a plant that produces spores. The sporophyte is *diploid but its spores are *haploid. It is either completely or partially dependent on the *gametophyte generation in bryophytes but is the dominant plant in the life cycle of pteridophytes and seed plants. *See also* alternation of generations.

spring balance A simple form of *balance in which a force is measured by the extension it produces in a helical spring. The extension, which is read off a scale, is directly proportional to the force, provided that the spring is not overstretched. The device is often used to measure the weight of a body approximately.

sputtering The process by which some of the atoms of an electrode (usually a cathode) are ejected as a result of bombardment by heavy positive ions. Although the process is generally unwanted, it can be used to produce a clean surface or to deposit a uniform film of a metal on an object in an evacuated enclosure.

Squamata An order of reptiles comprising the lizards and snakes. They appeared at the end of the Triassic period, about 170 million years ago, and have invaded a wide variety of habitats. Most lizards have four legs and a long tail, eardrums, and movable eyelids. Snakes are limbless reptiles that lack eardrums; the eyes are covered by transparent immovable eyelids and the articulation of the jaws is very loose, enabling a wide gape to facilitate swallowing prey whole.

square-planar Describing a coordination compound in which four ligands

square wave

positioned at the corners of a square coordinate to a metal ion at the centre of the square. *See* complex.

square wave A train of rectangular voltage pulses that alternate between two fixed values for equal lengths of time. The time of transition between each fixed value is negligible compared to the duration of the fixed value. See diagram.

Square wave

squark *See* supersymmetry.

stabilization energy The amount by which the energy of a delocalized chemical structure is less than the theoretical energy of a structure with localized bonds. It is obtained by subtracting the experimental heat of formation of the compound (in $kJ\ mol^{-1}$) from that calculated on the basis of a classical structure with localized bonds.

stabilizer 1. A substance used to inhibit a chemical reaction, i.e. a negative catalyst. **2.** A substance used to prevent a colloid from coagulating.

stable equilibrium *See* equilibrium.

staggered conformation *See* conformation.

staining A technique in which cells or thin sections of biological tissue that are normally transparent are immersed in one or more coloured dyes (*stains*) to make them more clearly visible through a microscope. Staining heightens the contrast between the various cell or tissue components. Stains are usually organic salts with a positive and negative ion. If the colour comes from the negative ion (organic anion), the stain is described as *acidic*, e.g. *eosin. If the colour comes from the positive ion (organic cation), the stain is described as *basic*, e.g. *haematoxylin. *Neutral stains* have a coloured cation and a coloured anion; an example is *Leishman's stain. Cell constituents are described as being *acidophilic* if they are stained with acidic dyes, *basophilic* if receptive to basic dyes, and *neutrophilic* if receptive to neutral dyes. *Vital stains* are used to colour the constituents of living cells without harming them (*see* vital staining); *nonvital stains* are used for dead tissue.

Counterstaining involves the use of two or more stains in succession, each of which colours different cell or tissue constituents. *Temporary staining* is used for immediate microscopical observation of material, but the colour soon fades and the tissue is subsequently damaged. *Permanent staining* does not distort the cells and is used for tissue that is to be preserved for a considerable period of time.

Electron stains, used in the preparation of material for electron microscopy, are described as *electron-dense* as they interfere with the transmission of electrons. Examples are lead citrate, phosphotungstic acid (PTA), and uranyl acetate (UA).

stainless steel A form of *steel containing at least 11–12% of chromium, a low percentage of carbon, and often some other elements, notably nickel and molybdenum. Stainless steel does not rust or stain and therefore has a wide variety of uses in industrial, chemical, and domestic environments. A particularly successful alloy is the steel known as 18–8, which contains 18% Cr, 8% Ni, and 0.08% C.

stalactites and stalagmites Accretions of calcium carbonate in limestone caves. Stalactites are tapering cones or pendants that hang down from the roofs of caves; stalagmites are upward projections from the cave floor and tend to be broader at their bases than stalactites. Both are formed from drips of water containing calcium carbonate in solution and may take thousands of years to grow.

stumen One of the male reproductive parts of a flower. It consists of an upper fertile part (the *anther) on a thin sterile stalk (the *filament*).

staminode A sterile stamen.

standard cell A *voltaic cell, such as a *Clark cell, or *Weston cell, used as a standard of e.m.f.

standard deviation A measure of the dispersion of data in statistics. For a set of values a_1, a_2, a_3, ... a_n, the mean m is given by $(a_1 + a_2 + \ldots + a_n)/n$. The *deviation* of each value is absolute value of the difference from the mean $|m - a_1|$, etc. The standard deviation is the square root of the mean of the squares of these values, i.e.
$$\sqrt{[(|m - a_1|^2 + \ldots |m - a_n|^2)/n]}$$
When the data is continuous the sum is replaced by an integral.

standard electrode An electrode (a half cell) used in measuring electrode potential. *See* hydrogen half cell.

standard electrode potential *See* electrode potential.

standard form (scientific notation) A way of writing a number, especially a large or small number, in which only one integer appears before the decimal point, the value being adjusted by multiplying by the appropriate power of 10. For example, 236,214 would be written in the standard form as $2.362\ 14 \times 10^5$; likewise 0.006821047 would be written $6.821\ 047 \times 10^{-3}$. Note that in the standard form, commas are not used, the digits are grouped into threes and a space is left between groups.

standard model The combination of *quantum chromodynamics to describe strong interactions, *electroweak theory to give a unified description of the electromagnetic interaction and the weak interaction, and the general theory of *relativity to describe classical gravitational interactions. Although the standard model, in principle, gives a complete description of all known phenomena, it is regarded by most physicists as being incomplete since it has many arbitrary features.

standard solution A solution of known concentration for use in volumetric analysis.

standard state A state of a system used as a reference value in thermodynamic measurements. Standard states involve a reference value of pressure (usually one atmosphere, 101.325 kPa) or concentration (usually 1 M). Thermodynamic functions are designated as 'standard' when they refer to changes in which reactants and products are all in their standard and their normal physical state. For example, the standard molar enthalpy of formation of water at 298 K is the enthalpy change for the reaction
$$H_2(g) + \tfrac{1}{2}O_2(g) \rightarrow H_2O(l)$$
$$\Delta H_{298}^{\ominus} = -285.83 \text{ kJ mol}^{-1}.$$
Note the superscript $\ominus$ is used to denote standard state and the temperature should be indicated.

standard temperature and pressure *See* s.t.p.

standing crop The total amount of living material in a specified population, expressed as *biomass or its equivalent in terms of energy. The standing crop may vary at different times of the year; for example, in a population of deciduous trees between summer and winter.

standing wave *See* stationary wave.

stannane *See* tin(IV) hydride.

stannate A compound formed by reaction of tin oxides (or hydroxides) with alkali. Tin oxides are amphoteric (weakly acidic) and react to give stannate ions. Tin(IV) oxide with molten alkali gives the stannate(IV) ion:

stannic compounds

$$SnO_2 + 2OH^- \rightarrow SnO_3^{2-} + H_2O$$

In fact, there are various ions present in which the tin is bound to hydroxide groups, the commonest one being the hexahydroxostannate(IV) ion, $Sn(OH)_6^{2-}$. This is the negative ion present in crystalline 'trihydrates' of the type $K_2Sn_2O_3.3H_2O$.

Tin(II) oxide gives the trihydroxostannate(II) ion in alkaline solutions

$$SnO(s) + OH^-(aq) + H_2O(l) \rightarrow Sn(OH)_3^-(aq)$$

Stannate(IV) compounds were formerly referred to as *orthostannates* (SnO_4^{4-}) or *metastannates* (SnO_3^{2-}). Stannate(II) compounds were called *stannites*.

stannic compounds Compounds of tin in its higher ($+4$) oxidation state; e.g. stannic chloride is tin(IV) chloride.

stannite *See* stannate.

stannous compounds Compounds of tin in its lower ($+2$) oxidation state; e.g. stannous chloride is tin(II) chloride.

stapes (stirrup) The third of the three *ear ossicles of the mammalian middle ear.

Staphylococcus A genus of spherical nonmotile Gram-positive bacteria that occur widely as saprophytes or parasites. The cells occur in grapelike clusters. Many species inhabit the skin and mucous membranes, and some cause disease in humans and animals. *S. aureus* infection can lead to boils and abscesses in humans; this species also produces *toxins that irritate the gastrointestinal tract and result in staphylococcal food poisoning.

star A self-luminous celestial body, such as the *sun, that generates nuclear energy within its core. Stars are not distributed uniformly throughout the universe, but are collected together in *galaxies. The age and lifetime of a star are related to its mass (*see* stellar evolution; Hertzsprung–Russell diagram).

starch A *polysaccharide consisting of various proportions of two glucose polymers, *amylose and *amylopectin. It occurs widely in plants, especially in roots, tubers, seeds, and fruits, as a carbohydrate energy store. Starch is therefore a major energy source for animals. When digested it ultimately yields glucose. Starch granules are insoluble in cold water but disrupt if heated to form a gelatinous solution. This gives an intense blue colour with iodine solutions and starch is used as an *indicator in certain titrations.

star cluster A group of stars that are sufficiently close to each other for them to be physically associated. Stars belonging to the cluster are formed together from the same cloud of interstellar gas and have approximately the same age and initial chemical composition. Because of this, and since the stars in a given cluster are at roughly the same distance from earth, observations of star clusters are of great importance in studies of stellar evolution.

There are two types of star cluster. *Open* (or *galactic*) *clusters* are fairly loose systems of between a few hundred and a few thousand members. The stars in open clusters are quite young by astronomical standards (some as young as a few million years) and have relatively high abundances of heavy elements. *Globular clusters* are approximately spherical collections of between ten thousand and a million stars. These are very old (of order 10^{10} years) and have low heavy-element abundances.

stat- A prefix attached to the name of a practical electrical unit to provide a name for a unit in the electrostatic system of units, e.g. statcoulomb, statvolt. *Compare* ab-. In modern practice both absolute and electrostatic units have been replaced by *SI units.

state of matter One of the three physical states in which matter can exist, i.e. *solid, *liquid, or *gas. *Plasma is sometimes regarded as the fourth state of matter.

static electricity The effects produced by electric charges at rest, including the forces between charged bodies (*see* Coulomb's law) and the field they produce (*see* electric field).

statics The branch of mechanics concerned with bodies that are acted upon by balanced forces and couples so that they remain at rest or in unaccelerated motion. *Compare* dynamics.

stationary orbit *See* synchronous orbit.

stationary phase *See* chromatography.

stationary state A state of a system when it has an energy level permitted by *quantum mechanics. Transitions from one stationary state to another can occur by the emission or absorption of an appropriate quanta of energy (e.g. in the form of photons).

stationary wave (standing wave) A form of *wave in which the profile of the wave does not move through the medium but remains stationary. This is in contrast to a *travelling* (or *progressive*) *wave*, in which the profile moves through the medium at the speed of the wave. A stationary wave results when a travelling wave is reflected back along its own path. In a stationary wave there are points at which the displacement is zero; these are called *nodes*. Points of maximum displacement are called *antinodes*. The distance between a node and its neighbouring antinode is one quarter of a wavelength. In a stationary wave all the points along the wave have different amplitudes and the points between successive nodes are in phase; in a travelling wave every point vibrates with the same amplitude and the phase of vibration changes for different points along its path.

statistical mechanics The branch of physics in which statistical methods are applied to the microscopic constituents of a system in order to predict its macroscopic properties. The earliest application of this method was Boltzmann's attempt to explain the thermodynamic properties of gases on the basis of the statistical properties of large assemblies of molecules.

In classical statistical mechanics, each particle is regarded as occupying a point in *phase space, i.e. to have an exact position and momentum at any particular instant. The probability that this point will occupy any small volume of the phase space is taken to be proportional to the volume. The Maxwell–Boltzmann law gives the most probable distribution of the particles in phase space.

With the advent of quantum theory, the exactness of these premises was disturbed (by the Heisenberg uncertainty principle). In the *quantum statistics that evolved as a result, the phase space is divided into cells, each having a volume h^f, where h is the Planck constant and f is the number of degrees of freedom of the particles. This new concept led to Bose–Einstein statistics, and for particles obeying the Pauli exclusion principle, to Fermi–Dirac statistics.

statistics The branch of mathematics concerned with the inferences that can be drawn from numerical data on the basis of probability. A *statistical inference* is a conclusion drawn about a population as a result of an analysis of a representative sample. *See* sampling.

statocyst (otocyst) A balancing organ found in many invertebrates. It consists of a fluid-filled sac lined with sensory hairs and contains granules of calcium carbonate, sand, etc. (*statoliths* or *otoliths*). As the animal moves the statoliths stimulate different hairs, giving a sense of the position of the body or part of it. The *semicircular canals in the ears of vertebrates act on the same principle and have a similar function.

stator The stationary electromagnetic structure of an electric motor or electric generator. *Compare* rotor.

steady-state theory The cosmological theory that the universe has always existed in a steady state, that it had no beginning, will have no end, and has a constant mean density. To compensate for the observed *expansion of the universe this theory postulates that matter is created throughout the universe at a rate of about 10^{-10} nucleon per metre cubed per year as a property of space. Because it has failed to account for the *microwave background radiation or the evidence of evolution in the universe it has lost favour to the *big-bang theory. It was first proposed by Hermann Bondi (1919–), Thomas Gold (1920–), and Fred Hoyle (1915–) in 1948.

steam distillation A method of distilling liquids that are immiscible with water by bubbling steam through them. It depends on the fact that the vapour pressure (and hence the boiling point) of a mixture of two immiscible liquids is lower than the vapour pressure of either pure liquid.

steam engine A *heat engine in which the thermal energy of steam is converted into mechanical energy. It consists of a cylinder fitted with a piston and valve gear to enable the high-pressure steam to be admitted to the cylinder when the piston is near the top of its stroke. The steam forces the piston to the bottom of its stroke and is then exhausted from the cylinder usually into a condenser. The reciprocating motion of the piston is converted to rotary motion of the flywheel by means of a connecting rod, crosshead, and crank. The steam

engine reached its zenith at the end of the 19th century, since when it has been replaced by the steam turbine and the internal-combustion engine. *See also* Rankine cycle.

steam point The temperature at which the maximum vapour pressure of water is equal to the standard atmospheric pressure (101 325 Pa). On the Celsius scale it has the value 100°C.

stearate (octadecanoate) A salt or ester of stearic acid.

stearic acid (octadecanoic acid) A solid saturated *fatty acid, $CH_3(CH_2)_{16}COOH$; r.d. 0.94; m.p. 69.6°C; b.p. 376°C (with decomposition). It occurs widely (as *glycerides) in animal and vegetable fats.

steel Any of a number of alloys consisting predominantly of iron with varying proportions of carbon (up to 1.7%) and, in some cases, small quantities of other elements (*alloy steels*), such as manganese, silicon, chromium, molybdenum, and nickel. Steels containing over 11–12% of chromium are known as *stainless steels.

Carbon steels exist in three stable crystalline phases: *ferrite* has a body-

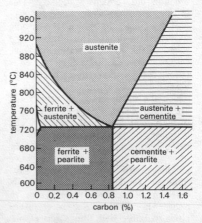

Phase diagram for steel

centred cubic crystal, *austenite* has a face-centred cubic crystal, and *cementite* has an orthorhombic crystal. *Pearlite* is a mixture of ferrite and cementite arranged in parallel plates. The phase diagram shows how the phases form at different temperatures and compositions.

Steels are manufactured by the *basic-oxygen process (L–D process), which has largely replaced the *Bessemer process and the *open-hearth process, or in electrical furnaces.

Stefan's law (Stefan–Boltzmann law) The total energy radiated per unit surface area of a *black body in unit time is proportional to the fourth power of its thermodynamic temperature. The constant of proportionality, the *Stefan constant* (or *Stefan–Boltzmann constant*) has the value $5.6697 \times 10^{-8} \, J\,s^{-1}\,m^{-2}\,K^{-4}$. The law was discovered by Joseph Stefan (1853–93) and theoretically derived by Ludwig Boltzmann (1844–1906).

stele The vascular tissue (i.e. *xylem and *phloem) of seed plants and pteridophytes, together with the endodermis and pericycle (when present). The arrangement of stelar tissues is very variable. In roots the stele often forms a solid core, which better enables the root to withstand tension and compression. In stems it is often a hollow cylinder separating the cortex and pith. This arrangement makes the stem more resistant to bending stresses. Monocotyledons and dicotyledons can usually be distinguished by the pattern of their stelar tissue. In monocotyledons the vascular bundles are scattered throughout the stem whereas in dicotyledons (and gymnosperms) they are arranged in a circle around the pith.

stellar evolution The changes that occur to a *star during its lifetime, from birth to final extinction. A star is believed to form from a condensation of interstellar matter, which collects either by chance or for unexplained reasons, and grows by attracting other matter towards itself as a result of its gravitational field. This initial cloud of cold contracting matter, called a *protostar*, builds up an internal pressure as a result of its gravitational contraction. The pressure raises the temperature until it reaches $5-10 \times 10^6$ K, at which temperature the thermonuclear conversion of hydrogen to helium begins. In our *sun, a typical star, hydrogen is converted at a rate of some $10^{11} \, kg\,s^{-1}$ with the evolution of some 6×10^{25} $J\,s^{-1}$ of energy. It is estimated that the sun contains sufficient hydrogen to burn at this rate for 10^{10} years and that it still has half its life to live as a main-sequence star (*see* Hertzsprung –Russell diagram). Eventually, however, this period of stability comes to an end, because the thermonuclear energy generated in the interior is no longer sufficient to counterbalance the gravitational contraction. The core, which is now mostly helium, collapses until a sufficiently high temperature is reached in a shell of unburnt hydrogen round the core to start a new phase of thermonuclear reaction. This burning of the shell causes the star's outer envelope to expand and cool, the temperature drop changes the colour from white to red and the star becomes a *red giant* or a *supergiant* if the original star was very large. The core now contracts, reaching a temperature of 10^8 K, and the helium in the core acts as the thermonuclear energy source. This reaction produces carbon, but a star of low mass relatively soon runs out of helium and the core collapses into a *white dwarf, while the outer regions drift away into space, possibly forming a *planetary nebula*. Larger stars (several times larger than the sun) have sufficient helium for the process to continue so that heavier elements, up to iron, are formed. But iron is the heaviest element that can be formed with the production of energy and

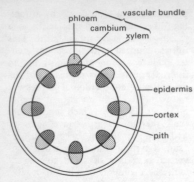

Transverse section through a
herbaceous stem

when the helium has all been consumed there is a catastrophic collapse
of the core, resulting in a *supernova
explosion, blowing the outer layers
away. The current theory suggests
that thereafter the collapsed core
becomes a *neutron star or a *black
hole.

stem The part of a plant that usually
grows vertically upwards towards the
light and supports the leaves, buds,
and reproductive structures (see illustration). The leaves develop at the
*nodes and side or branch stems
develop from buds at the nodes. The
stems of certain species are modified
as bulbs, corms, rhizomes, and tubers.
Some species have twining stems;
others have horizontal stems, such as
*runners. Another modification is the
*cladode. Erect stems may be cylin-
drical or angular; they may be
covered with hairs, prickles, or spines
and many exhibit secondary growth
and become woody (*see* growth ring).
In addition to its supportive function,
the stem contains *vascular tissue
that conducts food, water, and mineral salts between the roots and
leaves. It may also contain
chloroplasts and carry out photosynthesis.

step A single stage in a chemical
reaction. For example, the addition of

hydrogen chloride to ethene involves
three steps:
$$HCl \rightarrow H^+ + Cl^-$$
$$H^+ + C_2H_4 \rightarrow CH_3CH_2^+$$
$$CH_3CH_2^+ + Cl^- \rightarrow CH_3CH_2Cl$$

steradian Symbol sr. The supplementary *SI unit of solid angle equal to
the solid angle that encloses a surface
on a sphere equal to the square of
the radius of the sphere.

stere A unit of volume equal to 1 m^3.
It is not now used for scientific purposes.

stereochemistry The branch of
chemistry concerned with the structure of molecules and the way the
arrangement of atoms and groups
affects the chemical properties.

stereoisomerism *See* isomerism.

stereoregular Describing a *polymer
that has a regular pattern of side
groups along its chain.

stereospecific Describing chemical
reactions that give products with a
particular arrangement of atoms in
space. An example of a stereospecific
reaction is the *Ziegler process for
making polyethene.

steric effect An effect in which the
rate or path of a chemical reaction
depends on the size or arrangement
of groups in a molecule.

steric hindrance An effect in which
a chemical reaction is slowed down
or prevented because large groups on
a reactant molecule hinder the
approach of another reactant molecule.

sternum (breastbone) 1. A shield-
shaped or rod-shaped bone in terrestrial vertebrates, on the ventral side
of the thorax, that articulates with the
*clavicle (collar bone) of the pectoral
girdle and with most of the ribs. It is
absent in fish, and in birds it bears a
*keel. **2.** The ventral portion of each
segment of the exoskeleton of arthropods.

steroid nucleus

HO

cholesterol (a sterol)

OH

O

testosterone (an androgen)

Steroid structure

steroid Any of a group of lipids derived from a saturated compound called cyclopentanoperhydrophenanthrene, which has a nucleus of four rings (see formulae). Some of the most important steroid derivatives are the steroid alcohols, or *sterols. Other steroids include the *bile acids, which aid digestion of fats in the intestine; the sex hormones (*androgens and *oestrogens); and the *corticosteroid hormones, produced by the adrenal cortex. *Vitamin D is also based on the steroid structure.

sterol Any of a group of *steroid-based alcohols having a hydrocarbon side-chain of 8–10 carbon atoms.

Sterols exist either as free sterols or as esters of fatty acids. Animal sterols (*zoosterols*) include *cholesterol and lanosterol. The major plant sterol (*phytosterol*) is beta-sitosterol, while fungal sterols (*mycosterols*) include *ergosterol.

stigma 1. The glandular sticky surface at the tip of a carpel of a flower, which receives the pollen. In insect-pollinated plants the stigmas are held within the flower, whereas in wind-pollinated species they hang outside it. **2.** *See* eyespot.

stilt root *See* prop root.

stimulated emission *See* laser.

stimulus Any change in the external or internal environment of an organism that provokes a physiological or behavioural response in the organism. In an animal specific *receptors are sensitive to stimuli

stipe 1. The stalk that forms the lower portion of the fruiting body of certain fungi, such as mushrooms, and supports the umbrella-shaped cap. **2.** The stalk between the holdfast and blade (*lamina*) of certain brown algae, notably kelps.

stipule An outgrowth from the petiole or leaf base of certain plants. Those of the garden pea are leaflike photosynthetic organs. The stipules of the lime tree are scalelike and protect the winter buds, whereas those of the false acacia (*Robinia*) are modified as spines.

STM *See* scanning tunnelling microscope.

stochastic process Any process in which there is a random variable.

stock *See* graft.

stoichiometric Describing chemical reactions in which the reactants combine in simple whole-number ratios.

stoichiometric coefficient *See* chemical equation.

stoichiometric compound A compound in which atoms are combined in exact whole-number ratios. *Compare* nonstoichiometric compound.

stoichiometric mixture A mixture of substances that can react to give products with no excess reactant.

stoichiometric sum *See* chemical equation.

stoichiometry The relative proportions in which elements form compounds or in which substances react.

stokes Symbol St. A c.g.s. unit of kinematic viscosity equal to the ratio of the viscosity of a fluid in poises to its density in grams per cubic centimetre. 1 stokes = 10^{-4} m^2 s^{-1}. It is named after Sir George Stokes (1819–1903).

Stokes' law A law that predicts the frictional force F on a spherical ball moving through a viscous medium. According to this law $F = 6\pi r \eta v$, where r is the radius of the ball, v is its velocity, and η is the viscosity of the medium. The sphere accelerates until it reaches a steady terminal speed. For a falling ball, F is equal to the gravitational force on the sphere, less any upthrust. The law was discovered by Sir George Stokes (1819–1903).

stolon A long aerial side stem that gives rise to a new daughter plant when the bud at its apex touches the soil. Plants that multiply in this way include blackberry and currant bushes. Gardeners often pin down stolons to the soil to aid the propagation of such plants. This process is termed *layering*.

stoma (*pl.* **stomata**) A pore, large numbers of which are present in the epidermis of leaves (especially on the undersurface) and young shoots. Stomata function in gas exchange between the plant and the atmosphere. Each stoma is bordered by two semicircular *guard cells* (specialized epidermal cells), whose movements (due to changes in water content) control the size of the aperture. The term stoma is also used to mean both the pore and its associated guard cells.

stomach The portion of the vertebrate *alimentary canal between the oesophagus and the small intestine. It is a muscular organ, capable of dramatic changes in size and shape, in which ingested food is stored and undergoes preliminary digestion. Cells lining the stomach produce *gastric juice, which is thoroughly mixed with the food by muscular contractions of the stomach. The resultant acidic partly digested food mass (*chyme) is discharged into the *duodenum through the pyloric *sphincter for final digestion and absorption. Some herbivorous animals (ruminants) have multichambered stomachs from which food is regurgitated, rechewed, and swallowed again.

stomium A region of thin-walled cells in certain spore-producing structures that ruptures to release the spores. For example, in the sporangium of the fern *Dryopteris* the stomium ruptures when the annulus dries out.

stop A circular aperture that limits the effective size of a lens in an optical system. It may be adjustable, as the iris diaphragm in a camera, or have a fixed diameter, as the disc used in some telescopes.

stopping power A measure of the ability of matter to reduce the kinetic energy of a particle passing through it. The *linear stopping power*, $-dE/dx$, is energy loss of a particle per unit distance. The *mass stopping power*, $(1/\rho)dE/dx$, is the linear stopping power divided by the density (ρ) of the substance. The *atomic stopping power*, $(1/n)dE/dx = (A/\rho N)dE/dx$, is the energy loss per atom per unit area perpendicular to the particle's motion, i.e. n is the number of atoms in unit volume of the substance, N is the Avogadro number, and A is the relative atomic mass of the substance.

The relative stopping power is the ratio of the stopping power of a substance to that of a standard substance, usually aluminium, oxygen, or air.

storage ring A large evacuated toroidal ring forming a part of some particle accelerators. The rings are designed like *synchrotrons, except that they do not accelerate the particles circling within them but supply just sufficient energy to make up for losses (mainly *synchrotron radiation). The storage rings are usually built tangentially to the associated accelerator so that particles can be transferred accurately between them. At *CERN in Geneva, two interlaced storage rings are used, containing protons rotating in opposite directions. At the intersections very high collision energies (up to 1700 GeV) can be achieved.

s.t.p. Standard temperature and pressure, formerly known as *N.T.P.* (normal temperature and pressure). The standard conditions used as a basis for calculations involving quantities that vary with temperature and pressure. These conditions are used when comparing the properties of gases. They are 273.15 K (or 0°C) and 101 325 Pa (or 760.0 mmHg).

straight chain See chain.

strain A measure of the extent to which a body is deformed when it is subjected to a *stress. The *linear strain* or *tensile strain* is the ratio of the change in length to the original length. The *bulk strain* or *volume strain* is the ratio of the change in volume to the original volume. The *shear strain* is the angular distortion in radians of a body subjected to a *shearing force. See also elasticity; elastic modulus.

strain gauge A device used to measure a small mechanical deformation in a body (see strain). The most widely used devices are metal wires or foils or semiconductor materials, such as a single silicon crystal, which are attached to structural members; when the members are stretched under tensile *stress the resistance of the metal or semiconductor element increases. By making the resistance a component in a *Wheatstone-bridge circuit an estimate of the strain can be made by noting the change in resistance. Other types of strain gauge rely on changes of capacitance or the magnetic induction between two coils, one of which is attached to the stressed member.

strain hardening (work hardening) An increase in the resistance to the further plastic deformation of a body as a result of a rearrangement of its internal structure when it is strained, particularly by repeated stress. See also elasticity.

strange attractor See attractor.

strange matter Matter composed of up, down, and strange quarks (rather than the up and down quarks found in normal nucleons). It has been suggested that strange matter may have been formed in the *early universe, and that pieces of this matter (called *S-drops*) may still exist.

strangeness Symbol *s*. A property of certain elementary particles called hadrons (K-mesons and hyperons) that decay more slowly than would have been expected from the large amount of energy released in the process. These particles were assigned the quantum number *s* to account for this behaviour. For nucleons and other nonstrange particles $s = 0$; for strange particles *s* does not equal zero but has an integral value. In quark theory (see elementary particles) hadrons with the property of strangeness contain a strange quark or its antiquark.

stratigraphy The branch of geology concerned with the origin, composition, sequence, and correlation of rock strata. It forms the basis of historical geology and has also found practical application in mineral exploration, especially that of petroleum.

stratosphere *See* earth's atmosphere.

stratum corneum The layers of dead keratinized cells that form the outermost layers of mammalian *epidermis. The stratum corneum provides a water-resistant barrier between the external environment and the living cells of the *skin.

streamline flow A type of fluid flow in which no *turbulence occurs and the particles of the fluid follow continuous paths, either at constant velocity or at a velocity that alters in a predictable and regular way (*see also* laminar flow).

Streptococcus A genus of spherical Gram-positive bacteria occurring widely in nature, typically as chains or pairs of cells. Many are saprophytic and exist as usually harmless commensals inhabiting the skin, mucous membranes, and intestine of humans and animals. Others are parasites, some of which cause diseases, including scarlet fever (*S. pyogenes*; group A streptococci), endocarditis (*S. viridans*), and pneumonia (*S. pneumoniae*).

stress The force per unit area on a body that tends to cause it to deform (*see* strain). It is a measure of the internal forces in a body between particles of the material of which it consists as they resist separation, compression, or sliding in response to externally applied forces. *Tensile stress* and *compressive stress* are axial forces per unit area applied to a body that tend either to extend it or compress it linearly. *Shear stress* is a tangential force per unit area that tends to shear a body. *See also* elasticity; elastic modulus.

striated muscle *See* voluntary muscle.

stridulation The production of sounds by insects rubbing one part of the body against another. The parts of the body involved vary from species to species. Stridulation is typical of the Orthoptera (grasshoppers, crickets, cicadas), in which the purpose of the sounds is usually to bring the sexes together, although they are also used in territorial behaviour, warning, etc.

string A one-dimensional object used in theories of elementary particles and in cosmology (*cosmic string*). *String theory* replaces the idea of a pointlike elementary particle (used in quantum field theory) by a line or loop (a closed string). States of a particle may be produced by standing waves along this string. The combination of string theory with supersymmetry leads to *superstring theory.

strobilus 1. A type of *composite fruit that is formed from a complete inflorescence. It produces *achenes enclosed in bracts and when mature becomes cone-shaped. The hop fruit is an example. **2.** *See* cone.

stroboscope A device for making a moving body intermittently visible in order to make it appear stationary. It may consist of a lamp flashing at regular intervals or a shutter that enables it to be seen intermittently. *Stroboscopic photography* is the taking of very short-exposure pictures of moving objects using an electronically controlled flash lamp.

stroma Tissue that forms the framework of an organ; for example, the tissue of the ovary that surrounds the reproductive cells.

stromatolite A rocky cushion-like mass formed by the unchecked growth of millions of lime-secreting blue-green algae. Stromatolites are found only in areas where other organisms that would normally keep down the algal numbers cannot survive, such as extremely salty bays. It is thought that such formations were abundant 2000 million years ago, when blue-green algae were the most advanced form of life on earth, and that the white rings of fossilized microorganisms found in rocks of this age are the remains of stromatolites.

strong acid An *acid that is completely dissociated in aqueous solution.

strong interaction See fundamental interactions.

strontia See strontium oxide.

strontianite A mineral form of *strontium carbonate, $SrCO_3$.

strontium Symbol Sr. A soft yellowish metallic element belonging to group II of the periodic table (see alkaline-earth elements); a.n. 38; r.a.m. 87.62; r.d. 2.54; m.p. 800°C; b.p. 1300°C. The element is found in the minerals strontianite ($SrCO_3$) and celestine ($SrSO_4$). It can be obtained by roasting the ore to give the oxide, followed by reduction with aluminium (i.e. the *Goldschmidt process). The element, which is highly reactive, is used in certain alloys and as a vacuum getter. The isotope strontium–90 is present in radioactive fallout (half-life 28 years), and can be metabolized with calcium so that it collects in bone. Strontium was discovered by Klaproth and Hope in 1798 and isolated by Humphry Davy in 1808.

strontium bicarbonate See strontium hydrogencarbonate.

strontium carbonate A white solid, $SrCO_3$; orthorhombic; r.d. 3.7; decomposes at 1770°C. It occurs naturally as the mineral *strontianite* and is prepared industrially by boiling celestine (strontium sulphate) with ammonium carbonate. It can also be prepared by passing carbon dioxide over strontium oxide or hydroxide or by passing the gas through a solution of strontium salt. It is a phosphor, used to coat the glass of cathode-ray screens, and is also used in the refining of sugar, as a slagging agent in certain metal furnaces, and to provide a red flame in fireworks.

strontium chloride A white compound, $SrCl_2$. The anhydrous salt (cubic; r.d. 3.05; m.p. 872°C; b.p. 1250°C) can be prepared by passing chlorine over heated strontium. It is deliquescent and readily forms the hexahydrate, $SrCl_2.6H_2O$ (r.d. 2.67). This can be made by neutralizing hydrochloric acid with strontium carbonate, oxide, or hydroxide. Strontium chloride is used for military flares.

strontium hydrogencarbonate (strontium bicarbonate) A compound, $Sr(HCO_3)_2$, which is stable only in solution. It is formed by the action of carbon dioxide on a suspension of strontium carbonate in water. On heating, this process is reversed.

strontium oxide (strontia) A white compound, SrO; r.d. 4.7; m.p. 2430°C, b.p. 3000°C. It can be prepared by the decomposition of heated strontium carbonate, hydroxide, or nitrate, and is used in the manufacture of other strontium salts, in pigments, soaps and greases, and as a drying agent.

strontium sulphate A white solid, $SrSO_4$; r.d. 3.96; m.p. 1605°C. It can be made by dissolving strontium oxide, hydroxide, or carbonate in sulphuric acid. It is used as a pigment in paints and ceramic glazes and to provide a red colour in fireworks.

structural formula See formula.

structural isomerism See isomerism.

strychnine A colourless poisonous crystalline alkaloid found in certain plants.

style The stalk of a carpel, between the stigma and the ovary. In many plants it is elongated to aid pollination.

styrene See phenylethene.

subarachnoid space The space between the *arachnoid membrane and the *pia mater, two of the membranes (*meninges) that surround the brain and spinal cord. It is filled with *cerebrospinal fluid.

subatomic particle *See* elementary particles.

subclavian artery A paired artery that passes beneath the collar bone (clavicle) and branches to supply blood to the arm. The left subclavian artery arises from the aorta; the right from the innominate artery.

subcritical *See* critical mass; critical reaction; multiplication factor.

subcutaneous tissue The tissue that lies immediately beneath the *dermis (*see* skin). It is made up of loose fibrous *connective tissue, muscle, and fat (*see* adipose tissue), which in some animals (e.g. whales and hibernating mammals) forms an insulating layer or an important food store.

suberin A mixture of waxy substances, similar to *cutin, present in the thickened cell walls of many trees and shrubs, particularly in corky tissues. The deposition of suberin (*suberization*) provides a protective water-impermeable layer.

sublimate A solid formed by sublimation.

sublimation A direct change of state from solid to gas.

sublittoral 1. Designating or occurring in the shallow-water zone of a sea, over the continental shelf and below the low tide mark. **2.** Designating or occurring in the zone of a lake below the littoral zone, to a depth of 6–10 metres.

subshell *See* atom.

subsonic speed A speed that is less than *Mach 1.

subspecies A group of individuals within a *species that breed more freely among themselves than with other members of the species and resemble each other in more characteristics. Reproductive isolation of a subspecies may become so

extreme that a new species is formed (*see* speciation). Subspecies are sometimes given a third Latin name, e.g. the mountain gorilla, *Gorilla gorilla beringei* (*see also* binomial nomenclature).

substantive dye *See* dyes.

substantivity The affinity of a dye for its substrate.

substituent 1. An atom or group that replaces another in a substitution reaction. **2.** An atom or group regarded as having replaced a hydrogen atom in a chemical derivative. For example, dibromobenzene ($C_6H_4Br_2$) is a derivative of benzene with bromine substituents.

substitution reaction (displacement reaction) A reaction in which one atom or molecule is replaced by another atom or molecule. *See* electrophilic substitution; nucleophilic substitution.

substrate 1. The substance that is affected by the action of a catalyst; for example, the substance upon which an *enzyme acts in a biochemical reaction. **2.** The substance on which some other substance is adsorbed or in which it is absorbed. Examples include the material to which a dye is attached, the porous solid absorbing a gas, and the *matrix trapping isolated atoms, radicals, etc. **3.** (in biology) The material on which a sedentary organism (such as a barnacle or a plant) lives or grows. The substrate may provide nutrients for the organism or it may simply act as a support.

subtractive process *See* colour.

succession (in ecology) The sequence of communities that develops in an area from the initial stages of colonization until a stable mature *climax community* is achieved. Many factors, including climate and changes brought about by the colonizing organisms, influence the nature of a succession; for example, after many years shrubs produce soil deep enough to support trees, which then shade out the shrubs.

succinic acid (butanedioic acid) A crystalline solid, $HOOC(CH_2)_2COOH$, that is soluble in water. It occurs in living organisms as an intermediate in metabolism, especially in the *Krebs cycle.

succulent A plant that conserves water by storing it in fleshy leaves or stems. Succulents are found either in dry regions or in areas where there is sufficient water but it is not easily obtained, as in salt marshes. Such plants are often modified to reduce water loss by transpiration. For example, the leaves of cacti are reduced to spines.

succus entericus A mixture of enzymes produced by glands within the small intestine. These enzymes are responsible for most of the digestive processes that occur in the alimentary canal and include *proteases, *lipases, and *amylases.

sucker (turion) A shoot that arises from an underground root or stem and grows at the expense of the parent plant. Suckers can be dug up with a portion of root attached and used to propagate a plant. If, however, a plant is grafted onto a different rootstock, as many roses are, any suckers will be of the wild rootstock, rather than the ornamental scion, and must be removed.

sucrose (cane sugar; beet sugar; saccharose) A sugar comprising one molecule of glucose linked to a fructose molecule. It occurs widely in plants and is particularly abundant in sugar cane and sugar beet (15–20%), from which it is extracted and refined for table sugar. If heated to 200°C, sucrose becomes caramel.

sugar (saccharide) Any of a group of water-soluble *carbohydrates of relatively low molecular weight and typically having a sweet taste. The

simple sugars are called *monosaccharides. More complex sugars comprise between two and ten monosaccharides linked together: *disaccharides contain two, trisaccharides three, and so on. The name is often used to refer specifically to *sucrose (cane or beet sugar).

sugar of lead *See* lead(II) ethanoate.

sulpha drugs *See* sulphonamides.

sulphanes Compounds of hydrogen and sulphur containing chains of sulphur atoms. They have the general formula H_2S_n. The simplest is hydrogen sulphide, H_2S; other members of the series are H_2S_2, H_2S_3, H_2S_4, etc. *See* sulphides.

sulphate A salt or ester of sulphuric(VI) acid. Organic sulphates have the formula R_2SO_4, where R is an organic group. Sulphate salts contain the ion SO_4^{2-}.

sulphides 1. Inorganic compounds of sulphur with more electropositive elements. Compounds of sulphur with nonmetals are covalent compounds, e.g. hydrogen sulphide (H_2S). Metals form ionic sulphides containing the S^{2-} ion; these are salts of hydrogen sulphide. *Polysulphides* can also be produced containing the polymeric ion S_x^{2-}. **2.** (*or* **thio ethers**) Organic compounds that contain the group $-S-$ linked to two hydrocarbon groups. Organic sulphides are named from the linking groups, e.g. dimethyl sulphide (CH_3SCH_3), ethyl methyl sulphide ($C_2H_5SCH_3$). They are analogues of ethers in which the oxygen is replaced by sulphur (hence the alternative name) but are generally more reactive than ethers. Thus they react with halogen compounds to form *sulphonium compounds and can be oxidized to *sulphoxides.

sulphinate (dithionite; hyposulphite) A salt that contains the negative ion $S_2O_4^{2-}$, usually formed by the reduction of sulphites with excess SO_2. Solutions are not very

sulphininic acid

stable and decompose to give thiosulphate and hydrogensulphite ions. The structure is $^-O_2S\text{-}SO_2^-$.

sulphininic acid (dithionous acid; hyposulphurous acid) An unstable acid, $H_2S_2O_4$, known in the form of its salts (sulphinates). *See also* sulphuric acid.

sulphite A salt or ester derived from sulphurous acid. The salts contain the trioxosulphate(IV) ion SO_3^{2-}. The sulphites generally have reducing properties.

sulphonamides Organic compounds containing the group $-SO_2.NH_2$. The sulphonamides are amides of sulphonic acids. Many have antibacterial action and are also known as *sulpha drugs*, including sulphadiazine, $NH_2C_6H_4SO_2NHC_4H_3N_2$, sulphathiazole, $NH_2C_6H_4SO_2NHC_3H_2NS$, and several others. They act by preventing bacteria from reproducing and are used to treat a variety of bacterial infections, especially of the gut and urinary system.

sulphonate A salt or ester of a sulphonic acid.

sulphonation A type of chemical reaction in which a $-SO_3H$ group is substituted on a benzene ring to form a *sulphonic acid. The reaction is carried out by refluxing with concentrated sulphuric(VI) acid for a long period. It can also occur with cold disulphuric(VI) acid ($H_2S_2O_7$). Sulphonation is an example of electrophilic substitution in which the electrophile is a sulphur trioxide molecule, SO_3.

sulphonic acids Organic compounds containing the $-SO_2.OH$ group. Sulphonic acids are formed by reaction of aromatic hydrocarbons with concentrated sulphuric acid. They are strong acids, ionizing completely in solution to form the sulphonate ion, $-SO_2.O^-$.

Examples of organic sulphur compounds

sulphonium compounds Compounds containing the ion R_3S^+ (sulphonium ion), where R is any organic group. Sulphonium compounds can be formed by reaction of organic sulphides with halogen compounds. For example, diethyl sulphide, $C_2H_5SC_2H_5$, reacts with chloromethane, CH_3Cl, to give diethylmethylsulphonium chloride, $(C_2H_5)_2.CH_3.S^+Cl^-$.

sulphoxides Organic compounds containing the group $=S=O$ (*sulphoxide group*) linked to two other groups, e.g. dimethyl sulphoxide, $(CH_3)_2SO$.

sulphur Symbol S. A yellow nonmetallic element belonging to *group VI of the periodic table; a.n. 16; r.a.m. 32.06; r.d. 2.07 (rhombic); m.p. 112.8°C; b.p. 444.674°C. The element occurs in many sulphide and sulphate minerals and native sulphur is also sulphur (VI) dichloride dioxide

sulphuryl group

sulphur (IV) dichloride oxide

thionyl group

Oxychlorides of sulphur

found in Sicily and the USA (obtained by the *Frasch process). Sulphur has various allotropic forms. Below 95.6°C the stable crystal form is rhombic; above this temperature the element transforms into a triclinic form. These crystalline forms both contain cyclic S_8 molecules. At temperatures just above its melting point, molten sulphur is a yellow liquid containing S_8 rings (as in the solid form). At about 160°C, the sulphur atoms form chains and the liquid becomes more viscous and dark brown. If the molten sulphur is cooled quickly from this temperature (e.g. by pouring into cold water) a reddish-brown solid known as *plastic sulphur* is obtained. Above 200°C the viscosity decreases. Sulphur vapour contains a mixture of S_2, S_4, S_6, and S_8 molecules. *Flowers of sulphur* is a yellow powder obtained by subliming the vapour. It is used as a plant fungicide. The element is also used to produce sulphuric acid and other sulphur compounds.

Sulphur is an *essential element in living organisms, occurring in the amino acids cysteine and methionine and therefore in many proteins. It is also a constituent of various cell metabolites, e.g. coenzyme A. Sulphur is absorbed by plants from the soil as the sulphate ion (SO_4^{2-}).

sulphur dichloride See disulphur dichloride.

sulphur dichloride dioxide (sulphuryl chloride) A colourless liquid, SO_2Cl_2; r.d. 1.67; m.p. −54.1°C; b.p. 69°C. It decomposes in water but is soluble in benzene. The compound is formed by the action of chlorine on sulphur dioxide in the presence of an iron(III) chloride catalyst or sunlight. It is used as a chlorinating agent and a source of the related fluoride, SO_2F_2.

sulphur dichloride oxide (thionyl chloride) A colourless fuming liquid, $SOCl_2$; m.p. −99.5°C; b.p. 75.7°C. It hydrolyses rapidly in water but is soluble in benzene. It may be prepared by the direct action of sulphur on chlorine monoxide or, more commonly, by the reaction of phosphorus(V) chloride with sulphur dioxide. It is used as a chlorinating agent in synthetic organic chemistry (replacing −OH groups with Cl).

sulphur dioxide (sulphur(IV) oxide) A colourless liquid or pungent gas, SO_2, formed by sulphur burning in air; r.d. 1.43 (liquid); m.p. −72.7°C; b.p. −10°C. It can be made by heating iron sulphide (pyrites) in air. The compound is a reducing agent and is used in bleaching and as a fumigant and food preservative. Large quantities are also used in the *contact process for manufacturing sulphuric acid. It dissolves in water to give a mixture of sulphuric and sulphurous acids.

sulphuretted hydrogen See hydrogen sulphide.

sulphuric acid (oil of vitriol) A colourless oily liquid, H_2SO_4; r.d. 1.84; m.p. 10.36°C; b.p. 338°C. The pure acid is rarely used; it is commonly available as a 96–98% solution (m.p. 3.0°C). It also forms the hydrates: $H_2SO_4.H_2O$ (m.p. 8.62°C); $H_2SO_4.2H_2O$ (m.p. −38/39°C); $H_2SO_4.6H_2O$ (m.p. −54°C); $H_2SO_4.8H_2O$ (m.p. −62°C). Its full systematic name is *tetraoxosulphuric(VI) acid*.

Structures of some oxo acids of sulphur

Until the 1930s, sulphuric acid was manufactured by the *lead-chamber process, but this has now been replaced by the *contact process (catalytic oxidation of sulphur dioxide). More sulphuric acid is made in the UK than any other chemical product; production levels (UK) are commonly 12 000 to 13 000 tonnes per day. It is extensively used in industry, the main applications being fertilizers (32%), chemicals (16%), paints and pigments (15%), detergents (11%), and fibres (9%).

In concentrated sulphuric acid there is extensive hydrogen bonding and several competing equilibria, to give species such as H_3O^+, HSO_4^-, $H_3SO_4^+$, and $H_2S_2O_7$. Apart from being a powerful protonating agent (it protonates chlorides and nitrates producing hydrogen chloride and

nitric acid), the compound is a moderately strong oxidizing agent. Thus, it will dissolve copper:

$$Cu(s) + H_2SO_4(l) \rightarrow CuO(s) + H_2O(l) + SO_2(g)$$

$$CuO(s) + H_2SO_4(l) \rightarrow CuSO_4(aq) + H_2O(l)$$

It is also a powerful dehydrating agent, capable of removing H_2O from many organic compounds (as in the production of acid *anhydrides). In dilute solution it is a strong dibasic acid forming two series of salts, the sulphates and the hydrogensulphates.

sulphuric(IV) acid *See* sulphurous acid.

sulphur monochloride *See* disulphur dichloride.

sulphurous acid (sulphuric(IV) acid) A weak dibasic acid, H_2SO_3, known in the form of its salts: the sulphites and hydrogensulphites. It is considered to be formed (along with sulphuric acid) when sulphur dioxide is dissolved in water. It is probable, however, that the molecule H_2SO_3 is not present and that the solution contains hydrated SO_2. It is a reducing agent. The systematic name is *trioxosulphuric(IV) acid*. *See also* sulphuric acid.

sulphur(IV) oxide *See* sulphur dioxide.

sulphur(VI) oxide *See* sulphur trioxide.

sulphur trioxide (sulphur(VI) oxide) A colourless fuming solid, SO_3, which has three crystalline modifications. In decreasing order of stability these are: α, r.d. 1.97; m.p. 16.83°C; b.p. 44.8°C; β, m.p. 16.24°C; sublimes at 50°C; r.d. 2.29; γ, m.p. 16.8°C; b.p. 44.8°C. All are polymeric, with linked SO_4 tetrahedra: the γ-form has an icelike structure and is obtained by rapid quenching of the vapour; the β-form has infinite helical chains; and the α-form has infinite chains with some crosslinking of the SO_4 tetrahedra. Even in the vapour, there are polymeric species, and not discrete sulphur trioxide molecules (hence the compound is more correctly called by its systematic name *sulphur(VI) oxide*).

Sulphur trioxide is prepared by the oxidation of sulphur dioxide with oxygen in the presence of a vanadium(V) oxide catalyst. It may be prepared in the laboratory by distilling a mixture of concentrated sulphuric acid and phosphorus(V) oxide. It reacts violently with water to give sulphuric(VI) acid and is an important intermediate in the preparation of sulphuric acid and oleum.

sulphuryl chloride *See* sulphur dichloride dioxide.

sulphuryl group The group $=SO_2$, as in *sulphur dichloride oxide.

sulphydryl group *See* thiols.

summation *See* synergism.

sun The *star at the centre of the *solar system. A typical main-sequence dwarf star (*see* Hertzsprung –Russell diagram; stellar evolution), the sun is some 149 600 000 km from earth. It has a diameter of about 1 392 000 km and a mass of 1.9×10^{30} kg. Hydrogen and helium are the primary constituents (about 75% hydrogen, 25% helium), with less than 1% of heavier elements. In the central core, some 400 000 km in diameter, hydrogen is converted into helium by thermonuclear reactions, which generate vast quantities of energy. This energy is radiated into space and provides the earth with all the light and heat necessary to have created and maintained life (*see* solar constant). The surface of the sun, the *photosphere, forms the boundary between its opaque interior and its transparent atmosphere. It is here that *sunspots occur. Above the photosphere is the *chromosphere and above this the *corona, which extends tenuously into interplanetary space. *See also* solar wind.

sunspot A dark patch in the sun's *photosphere resulting from a localized fall in temperature to about 4000 K. Most spots have a central very dark umbra surrounded by a lighter penumbra. Sunspots tend to occur in clusters and to last about two weeks. The number of sunspots visible fluctuates over an eleven-year cycle – often called the *sunspot cycle*. The cause is thought to be the presence of intense localized magnetic fields, which suppress the convection currents that bring hot gases to the photosphere.

supercluster *See* galaxy cluster.

superconductivity The absence of measurable electrical resistance in certain substances at temperatures close to 0 K. First discovered in 1911 in mercury, superconductivity is now known to occur in some 26 metallic elements and many compounds and alloys. The temperature below which a substance becomes superconducting is called the *transition temperature* (or *critical temperature*). Compounds are now known that show superconductivity at liquid-nitrogen temperatures. The theoretical explanation of the phenomenon was given by J. Bardeen (1908–), L. N. Cooper (1930–), and J. R. Schrieffer (1931–) in 1957 and is known as the *BCS theory*. According to this theory an electron moving through an elastic crystal lattice creates a slight distortion of the lattice as a result of Coulomb forces between the positively charged lattice and the negatively charged electron. If this distortion persists for a finite time it can affect a second passing electron. In 1956 Cooper showed that the effect of this phenomenon is for the current to be carried in superconductors not by individual electrons but by bound pairs of electrons, the *Cooper pairs*. The BCS theory is based on a *wave function in which all the electrons are paired. Because the total momentum of a Cooper pair is unchanged by the interaction between one of its electrons and the lattice, the flow of electrons continues indefinitely.

Superconducting coils in which large currents can circulate indefinitely can be used to create powerful magnetic fields and are used for this purpose in some particle accelerators and in other devices.

Superconductivity can also occur by a slightly more complicated mechanism than BCS theory in *heavy-fermion systems. In 1986, Bednorz and Müller found an apparently completely different type of superconductivity. This is called *high-temperature superconductivity*, since the critical temperature is very much higher than for BCS superconductors; some high-temperature superconductors have critical temperatures greater than 100 K. A typical high-temperature superconductor is $YBa_2Cu_3O_{1-7}$.

At the present time a theory of high-temperature superconductivity has not been established in spite of a great deal of effort, which is still going on. The BCS mechanism, and minor modifications of it, almost certainly do not apply. One model for high-temperature superconductivity uses the concept of anyons, but at present there is no evidence for this mechanism.

supercooling 1. The cooling of a liquid to below its freezing point without a change from the liquid to solid state taking place. In this metastable state the particles of the liquid lose energy but do not fall into place in the lattice of the solid crystal. If the liquid is seeded with a small crystal, crystallization usually takes place and the temperature returns to the freezing point. Crystallization can also be induced by the presence of particles of dust, by mechanical vibration, or by rough surfaces. This is a common occurrence in the atmosphere where water droplets frequently remain unfrozen at temperatures well below 0°C until disturbed, following which they rapidly freeze. The supercooled droplets, for example, rapidly freeze on passing aircraft forming 'icing', which can be a hazard. **2.** The analogous cooling of a vapour to make it supersaturated until a disturbance causes condensation to occur, as in the Wilson *cloud chamber.

supercritical *See* critical mass; critical reaction; multiplication factor.

superficial expansivity *See* expansivity.

superfluidity The property of liquid helium at very low temperatures that enables it to flow without friction. Both helium isotopes possess this property, but ^4He becomes superfluid at 2.172 K, whereas ^3He does not become superfluid until a temperature of 0.00093 K is reached. There is a basic connection between superfluidity and *superconductivity, so that sometimes a superconductor is called a charged superfluid. The temperature at which superfluidity occurs is called the *lambda point*.

supergiant The largest and most luminous type of star. They are formed from the most massive stars and are therefore very rare. They lie above the giants on the *Hertzsprung –Russell diagram. *See also* stellar evolution.

supergravity A *unified-field theory for all the known fundamental interactions that involves *supersymmetry. Supergravity is most naturally formulated as a *Kaluza–Klein theory in eleven dimensions. The theory contains particles of spin 2, spin 3/2, spin 1, spin 1/2, and spin 0. Although supersymmetry means that the infinities in the calculations are less severe than in other attempts to construct a quantum theory of gravity, it is probable that supergravity still contains infinities that cannot be removed by the process of *renormalization. It is thought by many physicists that to obtain a consistent quantum theory of gravity one has to abandon *quantum field theories, since they deal with point objects, and move to theories based on extended objects, such as *superstrings and *supermembranes, and therefore that supergravity is not a complete theory of the fundamental interactions.

superheating The heating of a liquid to above its normal boiling point by increasing the pressure.

superheterodyne receiver A widely used type of radio receiver in which the incoming radio-frequency signal is mixed with an internally generated signal from a local oscillator. The output of the mixer has a carrier frequency equal to the difference between the transmitted carrier frequency and the locally generated frequency, still retains the transmitted modulation, and is called the *intermediate frequency* (IF). The IF signal is amplified before being passed to the audio-frequency amplifier. This system enables the IF signal to be amplified with less distortion, greater gain, better selectivity, and easier elimination of noise than can be achieved by amplifying the radio-frequency signal.

super high frequency (SHF) A radio frequency in the range 3–30 gigahertz.

superior Describing a structure that is positioned above or higher than another structure in the body. For example, in flowering plants the ovary is described as superior when located above the other organs of the flower (*see* hypogyny). *Compare* inferior.

superlattice *See* solid solution.

supermembrane theory A unified theory of the *fundamental interactions involving *supersymmetry, in which the basic entities are two-dimensional extended objects (*supermembranes*). They are thought to have about the same length scale as *superstrings, i.e. 10^{-35} m. At the present time there is no experimental evidence for supermembranes.

supernatant liquid The clear liquid remaining when a precipitate has settled.

supernova An explosive brightening of a star in which the energy radiated by it increases by a factor of 10^{10}. It takes several years to fade and while it lasts dominates the whole galaxy in which it lies. It is estimated that there could be a supernova explosion in the Milky Way every 30 years, although only six have actually been observed in the last 1000 years. A supernova explosion occurs when a star has burnt up all its available nuclear fuel and the core collapses catastrophically (*see* stellar evolution). *Compare* nova.

superoxides A group of inorganic compounds that contain the O_2^- ion. They are formed in significant quantities only for sodium, potassium, rubidium, and caesium. They are very powerful oxidizing agents and react vigorously with water to give oxygen gas and OH^- ions. The superoxide ion has an unpaired electron and is paramagnetic and coloured (orange).

superphosphate A commercial phosphate mixture consisting mainly of monocalcium phosphate. Single-superphosphate is made by treating phosphate rock with sulphuric acid; the product contains 16–20% 'available' P_2O_5:
$$Ca_{10}(PO_4)_6F_2 + 7H_2SO_4 =$$
$$3Ca(H_2PO_4)_2 + 7CaSO_4 + 2HF$$
Triple-superphosphate is made by using phosphoric(V) acid in place of sulphuric acid; the product contains 45–50% 'available' P_2O_5:
$$Ca_{10}(PO_4)_6F_2 + 14H_3PO_4 =$$
$$10Ca(H_2PO_4)_2 + 2HF$$

superplasticity The ability of some metals and alloys to stretch uniformly by several thousand percent at high temperatures, unlike normal alloys, which fail after being stretched 100%

or less. Since 1962, when this property was discovered in an alloy of zinc and aluminium (22%), many alloys and ceramics have been shown to possess this property. For superplasticity to occur, the metal grain must be small and rounded and the alloy must have a slow rate of deformation.

supersaturated solution *See* saturated.

supersaturation 1. The state of the atmosphere in which the relative humidity is over 100%. This occurs in pure air where no condensation nuclei are available. Supersaturation is usually prevented in the atmosphere by the abundance of condensation nuclei (e.g. dust, sea salt, and smoke particles). **2.** The state of any vapour whose pressure exceeds that at which condensation usually occurs (at the prevailing temperature).

supersonic *See* Mach number.

superstring theory A unified theory of the *fundamental interactions involving supersymmetry, in which the basic objects are one-dimensional objects (*superstrings*). Superstrings are thought to have a length scale of about 10^{-35} m and, since very short distances are associated with very high energies, they should have energy scales of about 10^{19} GeV, which is far beyond the energy of any accelerator that can be envisaged.

Strings associated with bosons are only consistent as quantum theories in a 26-dimensional *space–time; those associated with fermions are only consistent as quantum theories in 10-dimensional space–time. It is thought that four macroscopic dimensions arise by a Kaluza–Klein mechanism, with the remaining dimensions being 'curled up' to become very small.

One of the most attractive features of the theory of superstrings is that it leads to spin 2 particles, which are identified as *gravitons. Thus, a superstring theory automatically contains a quantum theory of the gravitational interaction. It is thought that superstrings are free of the infinities that cannot be removed by *renormalization, which plague attempts to construct a quantum field theory incorporating gravity. There is some evidence that superstring theory is free of infinities but not a complete proof yet.

Although there is no direct evidence for superstrings, some features of superstrings are compatible with the experimental facts of *elementary particles, such as the possibility of particles that do not respect *parity, as found in the weak interactions. Another attractive aspect of the theory is that it reveals the existence of particles that do not conserve parity, as in the weak interaction.

supersymmetry A *symmetry that can be applied to elementary particles so as to include both bosons and fermions. In the simplest supersymmetry theories, every boson has a corresponding fermion partner and every fermion has a corresponding boson partner. The boson partners of existing fermions have names formed by adding 's' to the beginning of the name of the fermion, e.g. *selectron*, *squark*, and *slepton*. The fermion partners of existing bosons have names formed by replacing '-on' at the end of the boson's name by '-ino' or by adding '-ino', e.g. *gluino*, *photino*, *wino*, and *zino*.

The infinities that cause problems in relativistic quantum field theories (*see* renormalization) are less severe in supersymmetry theories because infinities of bosons and fermions can cancel one another out.

If supersymmetry is relevant to observed elementary particles then it must be a *broken symmetry, although there is no convincing evidence at present to show at what energy it would be broken. There is, in fact, no experimental evidence for the theory, although it is thought that it may form part of a unified theory of interactions. This would not necessarily be a *unified-field theory; the idea of strings with supersymmetry may be the best approach to unifying the four fundamental interactions (*see* superstring theory).

supination Rotation of the lower forearm so that the hand faces forwards or upwards with the radius and ulna parallel. *Compare* pronation.

supplementary units *See* SI units.

suppressor grid A wire grid in a pentode *thermionic valve placed between the *screen grid and the anode to prevent electrons produced by *secondary emission from the anode from reaching the screen grid.

supramolecular chemistry A recent field of chemical research, the study and synthesis of very large molecules able to be used in complex chemical reactions in a fashion similar to, for example, the actions of the naturally occurring haemoglobin and nucleic acid molecules. Such molecules will have great potential in such areas as medicine, electronics, and optics. The *helicate and *texaphyrin molecules are typical examples of compounds of interest in this field.

suprarenal glands *See* adrenal glands.

surd A quantity that cannot be expressed as a *rational number. It consists of the root of an arithmetic member (e.g. $\sqrt{3}$), which cannot be exactly determined, or the sum or difference of such roots.

surface tension Symbol γ. The property of a liquid that makes it behave as if its surface is enclosed in an elastic skin. The property results from intermolecular forces: a molecule in the interior of a liquid experiences a force of attraction from other molecules equally from all sides, whereas a molecule at the surface is only attracted by molecules below it in the liquid. The surface tension is defined as the force acting over the surface per unit length of surface perpendicular to the force. It is measured in newtons per metre. It can equally be defined as the energy required to increase the surface area by one square metre, i.e. it can be measured in joules per metre squared (which is equivalent to $N\,m^{-1}$).

The surface tension of water is very strong, due to the intermolecular hydrogen bonding, and is responsible for the formation of drops, bubbles, and meniscuses, as well as the rise of water in a capillary tube (*capillarity*), the absorption of liquids by porous substances, and the ability of liquids to wet a surface. Capillarity is very important in plants as it is largely responsible for the transport of water, against gravity, within the plant.

surfactant (surface active agent) A substance, such as a *detergent, added to a liquid to increase its spreading or wetting properties by reducing its *surface tension.

surveying The practice of accurately measuring and recording the relative altitudes, angles, and distances of features on, above, or below the land surface from which maps and plans can be plotted. Surveying is necessary to locate and measure property lines; to lay out buildings, bridges, roads, dams, and other constructions; and to obtain topographic information for mapping and charting. A number of methods are used depending on the degree of precision that is required. The chief methods include triangulation, trilateration, levelling (*see* level), plane tabling, and traversing.

susceptance Symbol B. The reciprocal of the *reactance of a circuit and thus the imaginary part of its *admittance. It is measured in siemens.

susceptibility 1. (*or magnetic susceptibility*) Symbol χ_m. The dimensionless quantity describing the contribution made by a substance when subjected to a magnetic field to the total magnetic flux density present. It is equal to $\mu_r - 1$, where μ_r is the relative *permeability of the material. Diamagnetic materials have a low negative susceptibility, paramagnetic materials have a low positive susceptibility, and ferromagnetic materials have a high positive value. **2.** (*or* **electric susceptibility**) Symbol χ_e. The dimensionless quantity referring to a *dielectric equal to $P/\varepsilon_0 E$, where P is the electric polarization, E is the electric intensity producing it, and ε_0 is the electric constant. The electric susceptibility is also equal to $\varepsilon_r - 1$, where ε_r is the relative *permittivity of the dielectric.

suspension A mixture in which small solid or liquid particles are suspended in a liquid or gas.

suture The line marking the junction of two body structures. Examples are the immovable joints between the bones of the skull and, in plants, the seam along the edge of a pea or bean pod.

swallowing *See* deglutition.

sweat The salty fluid secreted by the *sweat glands onto the surface of the skin. Excess body heat is used to evaporate sweat, thereby resulting in cooling of the skin surface. Small amounts of urea are excreted in sweat.

sweat gland A small gland in mammalian skin that secretes *sweat. The distribution of sweat glands on the body surface varies between species: they occur over most of the body surface in man and higher primates but have a more restricted distribution in other mammals.

swim bladder (air bladder) An air-filled sac lying above the alimentary canal in bony fish that regulates the buoyancy of the animal. Air enters or leaves the bladder either via a pneumatic duct opening into the oesopha-gus or stomach or via capillary blood vessels, so that the specific gravity of the fish always matches the depth at which it is swimming. This makes the fish weightless, so less energy is required for locomotion. In lungfish it also has a respiratory function. The lungs of tetrapods are homologous with the swim bladder, which has developed its hydrostatic function by specialization.

syconus A type of *composite fruit formed from a hollow fleshy inflorescence stalk inside which tiny flowers develop. Small *drupes, the 'pips', are produced by the female flowers. An example is the fig.

sylvite (sylvine) A mineral form of *potassium chloride, KCl.

symbiosis An interaction between individuals of different species. The term symbiosis is usually restricted to interactions in which both species benefit (*see* mutualism), but it may be used for other close associations, such as *commensalism. Many symbioses are obligatory (i.e. the participants cannot survive without the interaction); for example, a lichen is an obligatory symbiotic relationship between an alga and a fungus.

symmetry The set of invariances of a system. Upon application of a symmetry operation on a system, the system is unchanged. Symmetry is studied mathematically using *group theory. Some of the symmetries are directly physical. Examples include reflections and rotation for molecules and translation in crystal lattices. Symmetries can be *discrete* (i.e. have a finite number), such as the set of rotations for an octahedral molecule, or *continuous* (i.e. do not have a finite number), such as the set of rotations for atoms or nuclei. More general and abstract symmetries can occur, as in CPT invariance and in the symmetries associated with *gauge theories. *See also* broken symmetry; supersymmetry.

sympathetic nervous system Part of the *autononomic nervous system. Its nerve endings release noradrenaline or adrenaline as a neurotransmitter and its actions tend to antagonize those of the *parasympathetic nervous system, thus achieving a balance in the organs they serve. For example, the sympathetic nervous system decreases salivary gland secretion, increases heart rate, and constricts blood vessels, while the parasympathetic nervous system has opposite effects.

sympatric Describing groups of similar organisms that, although in close proximity and theoretically capable of interbreeding, do not interbreed because of differences in behaviour, flowering time, etc. *Compare* allopatric.

symphysis A *joint that is only slightly movable; examples are the joints between the vertebrae of the vertebral column and that between the two pubic bones in the pelvic girdle. The bones at a symphysis articulate by means of smooth layers of cartilage and strong fibres.

sympodium The composite primary axis of growth in such plants as lime and horse chestnut. After each season's growth the shoot tip of the main stem stops growing (sometimes terminating in a flower spike); growth is continued by the tip of one or more of the lateral buds. *Compare* monopodium.

synapse The junction between two adjacent neurones (nerve cells), i.e. between the axon ending of one and the dendrites of the next. At a synapse, the membranes of the two cells are in close contact, with only a minute gap between them. A nerve *impulse is transmitted across the synapse by the release from the tip of the *axon of a *neurotransmitter substance, which diffuses across the synaptic space to the *dendrites of the second neurone. This triggers the propagation of the impulse from the dendrite along the length of the next neurone. Most neurones have more than one synapse.

synapsis (in genetics) *See* pairing.

syncarpy The condition in which the female reproductive organs (*carpels) of a flower are joined to each other. It occurs, for example, in the primrose. *Compare* apocarpy.

synchrocyclotron A form of *cyclotron in which the frequency of the accelerating potential is synchronized with the increasing period of revolution of a group of the accelerated particles, resulting from their relativistic increase in mass as they reach *relativistic speeds. The accelerator is used with protons, deuterons, and alpha-particles.

synchronous motor *See* electric motor.

synchronous orbit (geosynchronous orbit) An orbit of the earth made by an artificial *satellite with a period exactly equal to the earth's period of rotation on its axis, i.e. 23 hours 56 minutes 4.1 seconds. If the orbit is inclined to the equatorial plane the satellite will appear from the earth to trace out a figure-of-eight track once every 24 hours. If the orbit lies in the equatorial plane and is circular, the satellite will appear to be stationary. This is called a *stationary orbit* (or *geostationary orbit*) and it occurs at an altitude of 35 900 km. Most communication satellites are in stationary orbits, with three or more spaced round the orbit to give worldwide coverage.

synchronous rotation The rotation of a natural satellite in which the period of rotation is equal to its orbital period. The moon, for example, is in synchronous rotation about the earth.

synchrotron A particle accelerator used to impart energy to electrons and protons in order to carry out experiments in particle physics and in some cases to make use of the *synchrotron radiation produced. The particles are accelerated in closed orbits (often circular) by radio-frequency fields. Magnets are spaced round the orbit to bend the trajectory of the particles and separate focusing magnets are used to keep the particles in a narrow beam. The radio-frequency accelerating cavities are interspersed between the magnets. The motion of the particles is automatically synchronized with the rising magnetic field, as the field strength has to increase as the particle energy increases; the frequency of the accelerating field also has to increase synchronously.

synchrotron radiation (magneto-bremsstrahlung) Electromagnetic radiation that is emitted by charged particles moving at relativistic speeds in circular orbits in a magnetic field. The rate of emission is inversely proportional to the product of the radius of curvature of the orbit and the fourth power of the mass of the particles. For this reason, synchrotron radiation is not a problem in the design of proton *synchrotrons but it is significant in electron synchrotrons. The greater the circumference of a synchrotron, the less important is the loss of energy by synchrotron radiation. In *storage rings, synchrotron radiation is the principal cause of energy loss.

However, since the 1950s it has been realized that synchrotron radiation is itself a very useful tool and many accelerator laboratories have research projects making use of the radiation on a secondary basis to the main high-energy research. The radiation used for these purposes is primarily in the ultraviolet and X-ray frequencies.

Much of the microwave radiation from celestial radio sources outside the Galaxy is believed to originate from electrons moving in curved paths in celestial magnetic fields; it is also called synchrotron radiation as it is analogous to the radiation occurring in a synchrotron.

syndiotactic See polymer.

synergism (summation) 1. The phenomenon in which the combined action of two substances (e.g. drugs or hormones) produces a greater effect than would be expected from adding the individual effects of each substance. **2.** The combined action of one muscle (the *synergist*) with another (the *agonist*) in producing movement. *Compare* antagonism.

syngamy See fertilization.

synodic month (lunation) The interval between new *moons. It is equal to 29 days, 12 hours, and 44 minutes.

synodic period The mean time taken by any object in the solar system to move between successive returns to the same position, relative to the sun as seen from the earth. Since a planet is best observed at opposition the synodic period of a planet, S, is easier to measure than its *sidereal period, P. For inferior planets $1/S = 1/P - 1/E$; for superior planets $1/S = 1/E - 1/P$, where E is the sidereal period of the earth.

synoptic chart See synoptic meteorology.

synoptic meteorology The branch of meteorology concerned with those atmospheric processes connected with the weather. It is based on the analysis of the *synoptic chart*, which is built up from simultaneous observations at weather stations of such elements as wind velocity and direction, air temperature, cloud cover, and air pressure over an area at a particular time. Synoptic meteorology is applied chiefly to weather forecasting.

synovial membrane The membrane that lines the ligament surrounding a freely movable joint (such as that at the hip or elbow). It secretes a fluid (*synovial fluid*) that lubricates the layers of cartilage forming the articulating surfaces of the joint.

synthesis The formation of chemical compounds from more simple compounds.

synthesis gas *See* Haber process.

synthetic Describing a substance that has been made artificially; i.e. one that does not come from a natural source.

syrinx The sound-producing organ of a bird, situated at the lower end of the trachea where it splits into the bronchi. It has a complex structure with a number of vibrating membranes.

systematics The study of the diversity of organisms and their natural relationships. It is sometimes used as a synonym for *taxonomy. The term *biosystematics* describes the experimental study of diversity, especially at the species level. Biosystematic methods include breeding experiments, biochemical work (known as *chemosystematics*), and cytotaxonomy.

Système International d'Unités *See* SI units.

systole The phase of the heart beat during which the ventricles of the heart contract to force blood into the arteries. *Compare* diastole. *See* blood pressure.

syzygy The situation that occurs when the sun, the moon (or a planet), and the earth are in a straight line. This occurs when the moon (or planet) is at *conjunction or *opposition.

T

2,4,5-T 2,4,5-trichlorophenoxyacetic acid: a synthetic *auxin frequently used as a herbicide and defoliant. There are now moves to restrict its use, as it tends to become con-

taminated with the toxic chemical dioxin.

tachometer An instrument for measuring angular speed, especially the number of revolutions made by a rotating shaft in unit time. Various types of instrument are used, including mechanical, electrical, and electronic devices. The widely used electrical-generator tachometer consists of a small generator in which the output voltage is a measure of the rate of rotation of the shaft that drives it.

tachyon A hypothetical particle that has a speed in excess of the *speed of light. According to electromagnetic theory, a charged particle travelling through a medium at a speed in excess of the speed of light in that medium emits *Cerenkov radiation. A charged tachyon would emit Cerenkov radiation even in a vacuum. No such particle has yet been detected. According to the special theory of *relativity, it is impossible to accelerate a particle up to the speed of light because its energy E, given by $E = mc^2/\sqrt{(1 - v^2/c^2)}$, would have to become infinite. The theory, however, does not forbid the existence of particles with $v > c$ (where c is the speed of light). In such cases the expression in the brackets becomes negative and the energy would be imaginary.

tactic movement *See* taxis.

tactic polymer *See* polymer.

talc A white or pale-green mineral form of magnesium silicate, $Mg_3Si_4O_{10}(OH)_2$, crystallizing in the triclinic system. It forms as a secondary mineral by alteration of magnesium-rich olivines, pyroxenes, and amphiboles of ultrabasic rocks. It is soaplike to touch and very soft, having a hardness of 1 on the Mohs' scale. Massive fine-grained talc is known as *soapstone* or *steatite*. Talc in powdered form is used as a lubricant, as a filler in paper, paints, and rubber, and in cosmetics, ceramics, and

French chalk. It occurs chiefly in the USA, USSR, France, and Japan.

tandem generator A type of particle generator, essentially consisting of a *Van de Graaff generator that maintains one electrode at a high positive potential; this electrode is placed between two earthed electrodes. Negative ions are accelerated from earth potential to the positively charged electrode, where surplus electrons are stripped from the ions to produce positive ions, which are accelerated again from the positive electrode back to earth. Thus the ions are accelerated twice over by a single high potential. This tandem arrangement enables energies up to 30 MeV to be achieved.

tangent 1. A line that touches a curve or a plane that touches a surface. **2.** *See* trigonometric functions.

tangent galvanometer A type of galvanometer, now rarely used, in which a small magnetic needle is pivoted horizontally at the centre of a vertical coil that is adjusted to be parallel to the horizontal component of the earth's magnetic field. When a current I is passed through the coil, the needle is deflected so that it makes an angle θ with its equilibrium position parallel to the earth's field. The value of I is given by $I = (2Hr\tan\theta)/n$, where H is the strength of the earth's horizontal component of magnetizing force, r is the radius of the coil, and n is the number of turns in the coil. Although not now used for measuring current, the instrument provides a means of measuring the earth's magnetizing force.

tanh *See* hyperbolic functions.

tannic acid A yellowish complex organic compound present in certain plants. It is used in dyeing as a mordant.

tannin One of a group of complex organic chemicals commonly found in leaves, unripe fruits, and the bark of trees. Their function is uncertain though the unpleasant taste may discourage grazing animals. Some tannins have commercial uses, notably in the production of leather and ink.

tantalum Symbol Ta. A heavy blue-grey metallic *transition element; a.n. 73; r.a.m. 180.948; r.d. 16.63; m.p. 2996°C; b.p. 5427°C. It is found with niobium in the ore columbite-tantalite $(Fe,Mn)(Ta,Nb)_2O_6$. It is extracted by dissolving in hydrofluoric acid, separating the tantalum and niobium fluorides to give K_2TaF_7, and reduction of this with sodium. The element contains the stable isotope tantalum–181 and the long-lived radioactive isotope tantalum–180 (0.012%; half-life $>10^7$ years). There are several other short-lived isotopes. The element is used in certain alloys and in electronic components. Tantalum parts are also used in surgery because of the unreactive nature of the metal (e.g. in pins to join bones). Chemically, the metal forms a passive oxide layer in air. It forms complexes in the +2, +3, +4, and +5 oxidation states. Tantalum was identified in 1802 by Ekeberg and first isolated in 1820 by Berzelius.

tapetum A reflecting layer, containing crystals of guanine, in the *choroid of the eye of many nocturnal vertebrates. It reflects light back onto the retina, thus improving vision and causing the eyes to shine in the dark.

tapeworms *See* Cestoda.

tap root *See* root.

tar Any of various black semisolid mixtures of hydrocarbons and free carbon, produced by destructive distillation of *coal or by *petroleum refining.

tarsal (tarsal bone) One of the bones that form the ankle (*see* tarsus) in terrestrial vertebrates.

tar sand *See* oil sand.

tarsus The ankle (or corresponding part of the hindlimb) in terrestrial vertebrates, consisting of a number of small bones (*tarsals*). The number of tarsal bones varies with the species: humans, for example, have seven.

tartaric acid A white crystalline naturally occurring carboxylic acid, $(CHOH)_2(COOH)_2$; r.d. 1.8; m.p. 170°C. It can be obtained from tartar (potassium hydrogen tartrate) deposits from wine vats, and is used in baking powders and as a foodstuffs additive. The compound is optically active (*see* optical activity). The systematic name is *2,3-dihydroxybutanedioic acid*.

tartrate A salt or ester of *tartaric acid.

taste bud A small sense organ in most vertebrates, specialized for the detection of taste. In terrestrial animals taste buds are concentrated on the upper surface of the *tongue. They are sensitive to four types of taste: sweet, salt, bitter, or sour. The taste bud transmits information about a particular type of taste to the brain via nerve fibres. The four types of taste bud show distinct distribution patterns on the surface of the human tongue.
In fishes, taste buds are distributed over the entire surface of the body and provide information about the surrounding water.

tau particle *See* elementary particles; lepton.

tautomerism A type of *isomerism in which the two isomers (*tautomers*) are in equilibrium. *See* keto–enol tautomerism.

taxis (tactic movement) The movement of a cell (e.g. a gamete) or a microorganism in response to an external stimulus. Certain microorganisms have a light-sensitive region that enables them to move towards or away from high light intensities (positive and negative *phototaxis* respectively). Many bacteria move in response to chemical stimuli (*chemotaxis*); a specific example is *aerotaxis*, in which atmospheric oxygen is the stimulus. Tactic movements are restricted to cells that possess cilia, flagella, or some other means of locomotion. The term is usually not applied to the movements of higher animals. *Compare* kinesis; tropism.

taxon Any named taxonomic group of any *rank in the hierarchical *classification of organisms. Thus the taxa Papilionidae, Lepidoptera, Insecta, and Arthropoda are named examples of a family, order, class, and phylum, respectively.

taxonomy The study of the theory, practice, and rules of *classification of living and extinct organisms. The naming, description, and classification of a given organism draws on evidence from a number of fields. *Classical taxonomy* is based on morphology and anatomy. *Biochemical taxonomy* studies similarities in the structure of certain proteins and nucleic acids. *Cytotaxonomy* compares the size, shape, and number of chromosomes of different organisms. *Numerical taxonomy* uses mathematical procedures to assess similarities and differences and establish taxonomic groups. *See also* systematics.

Taylor series The infinite power series of derivatives into which a function f(x) can be expanded, for a fixed value of the variable $x = a$:

$$f(x) = f(a) + f'(a)(x - a) + f''(a)(x - a)^2/2! + \ldots$$

When $a = 0$, the series formed is known as *Maclaurin's series*:

$$f(x) = f(0) + f'(0)x + f''(0)x^2/2! + \ldots$$

The series was discovered by Brook Taylor (1685–1731) and the special case was named after Colin Maclaurin (1698–1746).

TCA cycle *See* Krebs cycle.

T cell (T lymphocyte) Any of a population of *lymphocytes that are the principal agents of cell-mediated *immunity. T cells are derived from the bone marrow but migrate to the thymus to mature (hence *T* cell). Subpopulations of T cells play different roles in the immune response and can be characterized by their surface antigens (*see* CD). *Helper T cells* recognize foreign antigens provided these are presented by inducer cells (such as macrophages and B lymphocytes) bearing class II *histocompatibility antigens. The helper T cell carries *T-cell receptors*, which recognize the class II antigens on the inducer cell. Interleukin-1 released by inducer cells stimulates helper T cells to release interleukin-2, which in turn triggers the release of other cytokines (*see* interleukin). Consequently there is a proliferation of B lymphocytes and the generation of *effector T cells*, i.e. cytotoxic T cells, delayed hypersensitivity T cells, and suppressor T cells.

Cytotoxic T cells recognize foreign antigen on the surface of virus-infected cells and destroy the cell by releasing cytolytic proteins. *Suppressor T cells* are important in regulating the activity of other lymphocytes and are crucial in maintaining tolerance to self tissues. *Delayed hypersensitivity T cells* mediate delayed hypersensitivity by releasing various *lymphokines.

technetium Symbol Tc. A radioactive metallic *transition element; a.n. 43; m.p. 2171°C; b.p. 4876°C. The element can be detected in certain stars and is present in the fission products of uranium. It was first made by Perrier and Segré by bombarding molybdenum with deuterons to give technetium–97. The most stable isotope is technetium–99 (half-life 2.6×10^6 years); this is used to some extent in labelling for medical diagnosis. There are sixteen known isotopes. Chemically, the metal has properties intermediate between manganese and rhenium.

Teflon Trade name for a form of *polytetrafluoroethene.

tektite A small black, greenish, or yellowish glassy object found in groups on the earth's surface and consisting of a silicaceous material unrelated to the geological formations in which it is found. Tektites are believed to have formed on earth as a result of the impact of meteorites.

telecommunications The study and application of means of transmitting information, either by wires or by electromagnetic radiation.

Teleostei The major subclass of the *Osteichthyes (bony fish), containing about 20 000 species. Teleosts have colonized an extensive variety of habitats and show great diversity of form. The group includes the eel, seahorse, plaice, and salmon. They have been the dominant fish since the Cretaceous period (about 70 million years ago).

telescope An instrument that collects radiation from a distant object in order to produce an image of it or enable the radiation to be analysed. An *optical telescope* uses visible radiation (*see also* radio telescope). Optical *astronomical telescopes* fall into two main classes: *refracting telescopes* and *reflecting telescopes*. The refracting telescopes use a converging lens to collect the light and the resulting image is magnified by the eyepiece, a lens of short focal length. This type of instrument was first constructed in 1608 by Hans Lippershey (1587–1619) in Holland and developed in the following year as an astronomical instrument by Galileo, who used a diverging lens as eyepiece (see illustration (a)). The *Galilean telescope* was later improved by Johannes Kepler (1571–1630), who substituted a converging eyepiece lens, and this form of the instrument is still in use for small astronomical telescopes. For terrestrial telescopes an additional lens is usually inserted in order to provide an upright image (see illustration (b)). However the Galilean

(a) Galilean

(b) Keplerian

(c) Newtonian

(d) Gregorian

(e) Cassegrain

(f) Coudé

(g) Maksutov

(h) Schmidt

Telescopes

design provides an upright image originally, and so is still used for the compact telescopes of opera glasses.

The first reflecting telescope was produced by Newton in 1668. This used a concave mirror to collect and focus the light and a small secondary mirror at an angle of 45° to the main beam to reflect the light into the magnifying eyepiece. The *Gregorian telescope*, designed by James Gregory (1638–75), and the *Cassegrainian telescope*, designed by N. Cassegrain (*fl.* 1670s), use different secondary optical systems (see illustrations (d) and (e)). The *coudé telescope* (French: angled) is sometimes used with larger instruments as it increases their focal lengths (see illustration (f)).

Catadioptic telescopes use both lenses and mirrors. The most widely used astronomical instruments in this class are the *Maksutov telescope* and the *Schmidt camera*. The Maksutov instrument has a thick spheroidal meniscus lens to introduce spherical aberration to the image, which is corrected by the spheroidal principal mirror (see illustration (g)). The Schmidt camera has a correcting plate of complicated aspherical shape close to the centre of curvature of the spheroidal principal mirror. This instrument is used primarily for photographic purposes (see illustration (h)).

television The transmission and reception of moving images by means of radio waves or cable. The scene to be transmitted is focused onto a photoelectric screen in the television *camera. This screen is scanned by an electron beam. The camera produces an electric current, the instantaneous magnitude of which is proportional to the brightness of the portion of the screen being scanned. In Europe the screen is scanned by 625 lines and 25 such frames are produced every second. In the USA 525 lines and 30 frames per second are used. The picture signal so produced is used to modulate a VHF or UHF carrier wave and is transmitted with an independent sound signal, but with colour information (if any) incorporated into the brief gaps between the picture lines. The signals received by the receiving aerial are demodulated in the receiver; the demodulated picture signal controls the electron beam in a cathode-ray tube, on the screen of which the picture is reconstructed. *See also* colour television.

television tube *See* cathode rays.

tellurides Binary compounds of tellurium with other more electropositive elements. Compounds of tellurium with nonmetals are covalent (e.g. H_2Te). Metal tellurides can be made by direct combination of the elements and are ionic (containing Te^{2-}) or nonstoichiometric interstitial compounds (e.g. Pd_4Te, $PdTe_2$).

tellurium Symbol Te. A silvery metalloid element belonging to *group VI of the periodic table; a.n. 52; r.a.m. 127.60; r.d. 6.24 (crystalline); m.p. 451°C; b.p. 1390±3°C. It occurs mainly as *tellurides in ores of gold, silver, copper, and nickel and it is obtained as a by-product in copper refining. There are eight natural isotopes and nine radioactive isotopes. The element is used in semiconductors and small amounts are added to certain steels. Tellurium is also added in small quantities to lead. Its chemistry is similar to that of sulphur. It was discovered by Franz Müller in 1782.

telophase The fourth stage of cell division. In *mitosis the chromatids that separated from each other at *anaphase collect at the poles of the spindle. A nuclear membrane forms around each group, producing two daughter nuclei with the same number and kind of chromosomes as the original cell nucleus. In the first telophase of *meiosis, complete chromosomes from the pairs that separated at first anaphase form the daughter nuclei. The number of chromosomes in these nuclei is therefore half the number in the original one. In the second telophase, daughter nuclei are formed from chromatids (as in mitosis).

temperament The way in which the intervals between notes on keyboard instruments are distributed throughout the scale to ensure that music in all keys sounds in tune. The problem can be illustrated by a piano keyboard. Taking a low C and a high C seven octaves above, the interval should be $2^7 = 128$. However, in passing through the cycle of 12 keys, each using as its fundamental the fifth of its predecessor, the interval between Cs becomes $(3/2)^{12} = 129.75$. The difference between 129.75 and 128 is known as the *comma of Pythagoras*. The *equal-temperament scale*, which has been in use since the time of J. S. Bach, distributes the comma of Pythagoras equally between the 12 intervals of the scale over seven octaves. Thus each fifth becomes $(128)^{1/12} = 1.4983$. All forms of temperament involve a measure of compromise; this system is now regarded as the best.

temperature The property of a body or region of space that determines whether or not there will be a net flow of heat into it or out of it from a neighbouring body or region and in which direction (if any) the heat will flow. If there is no heat flow the bodies or regions are said to be in *thermodynamic equilibrium* and at the same temperature. If there is a flow of heat, the direction of the flow is from the body or region of higher temperature. Broadly, there are two methods of quantifying this property. The empirical method is to take two or more reproducible temperature-dependent events and assign *fixed points* on a scale of values to these events. For example, the Celsius temperature scale uses the freezing point and boiling point of water as the two fixed points, assigns the values 0 and 100 to them, respectively, and divides the scale between them into 100 degrees. This method is serviceable for many practical purposes (*see* temperature scales), but lacking a theoretical basis it is awkward to use in many scientific contexts. In the 19th century, Lord Kelvin proposed a thermodynamic method to specify temperature, based on the measurement of the quantity of heat flowing between bodies at different temperatures. This concept relies on an absolute scale of temperature with an *absolute zero of temperature, at which no body can give up heat. He also used Sadi Carnot's concept of an ideal frictionless perfectly efficient heat engine (*see* Carnot cycle). This Carnot engine takes in a quantity of heat q_1 at a temperature T_1, and exhausts heat q_2 at T_2, so that $T_1/T_2 = q_1/q_2$. If T_2 has a value fixed by definition, a Carnot engine can be run between this fixed temperature and any unknown temperature T_1, enabling T_1 to be calculated by measuring the values of q_1 and q_2. This concept remains the basis for defining *thermodynamic temperature*, quite independently of the nature of the working substance. The unit in which thermodynamic temperature is now expressed is the *kelvin. In practice thermodynamic temperatures cannot be measured directly; they are usually inferred from measurements with a gas thermometer containing a nearly ideal gas. This is possible because another aspect of thermodynamic temperature is its relationship to the *internal energy of a given amount of substance. This can be shown most simply in the case of an ideal monatomic gas, in which the internal

energy per mole (U) is equal to the total kinetic energy of translation of the atoms in one mole of the gas (a monatomic gas has no rotational or vibrational energy). According to *kinetic theory, the thermodynamic temperature of such a gas is given by $T = 2U/3R$, where R is the universal *gas constant.

temperature coefficient A coefficient that determines the rate of change of some physical property with change in temperature. For example, the dependence of the resistance (R) of a material on the Celsius temperature t, is given by $R = R_0 + \alpha t + \beta t^2$, where R_0 is the resistance at $0°C$ and α and β are constants. If β is negligible, then α is the *temperature coefficient of resistance*.

temperature scales A number of empirical scales of *temperature have been in use: the *Celsius scale is widely used for many purposes and in certain countries the *Fahrenheit scale is still used. These scales both rely on the use of *fixed points*, such as the freezing point and the boiling point of water, and the division of the *fundamental interval* between these two points into units of temperature (100 degrees in the case of the Celsius scale and 180 degrees in the Fahrenheit scale).

However, for scientific purposes the scale in use is the *International Practical Temperature Scale* (*IPTS*), which is designed to conform as closely as possible to thermodynamic temperature and is expressed in the unit of thermodynamic temperature, the *kelvin. The 1968 version of the table (known as IPTS-68) had 11 fixed points defined by both Celsius and thermodynamic temperatures. The most recent version (IPTS-90), introduced in 1990, has 16 fixed points with temperatures expressed in kelvins:

Triple point of hydrogen: 13.8033

Boiling point of hydrogen (33 321.3 Pa): 17.035
Boiling point of hydrogen (101 292 Pa): 20.27
Triple point of neon: 24.5561
Triple point of oxygen: 54.3584
Triple point of argon: 83.8058
Triple point of mercury: 234.3156
Triple point of water: 273.16 (0.01°C)
Melting point of gallium: 302.9146
Freezing point of indium: 429.7485
Freezing point of tin: 505.078
Freezing point of zinc: 692.677
Freezing point of aluminium: 933.473
Freezing point of silver: 1234.93
Freezing point of gold: 1337.33
Freezing point of copper: 1357.77
Methods for measuring intermediate temperatures between these fixed points are specified; for example, at low temperatures (0–5 K) they are measured by means of vapour-pressure determinations of ^{3}He and ^{4}He; at high temperatures (above 1234.93 K) a radiation pyrometer is used.

tempering The process of increasing the toughness of an alloy, such as steel, by heating it to a predetermined temperature, maintaining it at this temperature for a predetermined time, and cooling it to room temperature at a predetermined rate. In steel, the purpose of the process is to heat the alloy to a temperature that will enable the excess carbide to precipitate out of the supersaturated solid solution of *martensite and then to cool the saturated solution fast enough to prevent further precipitation or grain growth. For this reason steel is quenched rapidly by dipping into cold water.

temporary hardness *See* hardness of water.

temporary magnetism Magnetism in a body that is present when the body is in a magnetic field but that largely disappears when it is removed from the field.

tendon A thick strand or sheet of tissue that attaches a muscle to a

bone. Tendons consist of *collagen fibres and are therefore inelastic: they ensure that the force exerted by muscular contraction is transmitted to the relevant part of the body to be moved.

tendril A slender branched or unbranched structure found in many climbing plants. It may be a modified stem, leaf, leaflet, or petiole. Tendrils respond to contact with solid objects by twining around them (*see* haptotropism). The cells that touch the object lose water and decrease in volume in comparison to the outer cells, thus causing the tendril to curve.

tensile strength A measure of the resistance that a material offers to tensile *stress. It is defined as the stress, expressed as the force per unit cross-sectional area, required to break it.

tensimeter A form of differential manometer with two sealed bulbs attached to the limbs. It is used to measure the difference in vapour pressure between liquids sealed into the bulbs. If one liquid has a known vapour pressure (often water is used) that of the other can be determined.

tensiometer Any apparatus for measuring *surface tension.

tera- Symbol T. A prefix used in the metric system to denote one million million times. For example, 10^{12} volts = 1 teravolt (TV).

teratogen Any environmental factor that acts on a fetus to cause congenital abnormality. Examples include ionizing radiation (e.g. X-rays), nutritional deficiencies, drugs (e.g. thalidomide), toxic chemicals, and virus infections (e.g. rubella).

terbium Symbol Tb. A silvery metallic element belonging to the *lanthanoids; a.n. 65; r.a.m. 158.92; r.d. 8.23 (20°C); m.p. 1365°C; b.p. 3230°C. It occurs in apatite and xenotime, from which it is obtained by an ion-exchange process. There is only one natural isotope, terbium–159, which is stable. Seventeen artificial isotopes have been identified. It is used as a dopant in semiconducting devices. It was discovered by C. G. Mosander in 1843.

terminal 1. The point at which electrical connection is made to a device or system. **2.** A device at which data is put into a *computer or taken from it.

terminal speed The constant speed finally attained by a body moving through a fluid under gravity when there is a zero resultant force acting on it. *See* Stokes's law.

terminator The boundary, on the surface of the moon or a planet, between the sunlit area and the dark area.

ternary compound A chemical compound containing three different elements.

terpenes A group of unsaturated hydrocarbons present in plants (*see* essential oil). Terpenes consist of isoprene units, $CH_2:C(CH_3)CH:CH_2$. Monoterpenes have two units, $C_{10}H_{16}$, sesquiterpenes three units, $C_{15}H_{24}$, diterpenes four units, $C_{20}H_{32}$, etc.

terrestrial magnetism *See* geomagnetism.

territory A fixed area that an animal or group of animals defends against intrusion from others of its species. Outside the territory (which may contain food sources, hiding places, and nesting sites) others are not threatened. Many mammals indicate their territory boundaries with scent markings, while birds sing territorial songs that repel would-be intruders. Animals in neighbouring territories normally respect each other's boundaries, which reduces overt *aggression. Some animals are territorial only at certain times of the year, usually the breeding season (*see* courtship).

Tertiary The older geological period of the Cenozoic era (*compare* Quaternary). It began about 65 million years ago, following the Cretaceous period, and extended to the beginning of the Quaternary, about 2 million years ago. It is subdivided into the *Palaeocene, *Eocene, *Oligocene, *Miocene, and *Pliocene epochs in ascending order. The Tertiary period was characterized by the rise of the modern mammals and the development of shrubs, grasses, and other flowering plants.

tertiary alcohol *See* alcohols.

tertiary amine *See* amines.

tertiary colour A colour obtained by mixing two *secondary colours.

tervalent (trivalent) Having a valency of three.

Terylene Trade name for a type of *polyester used in synthetic fibres.

tesla Symbol T. The SI unit of magnetic flux density equal to one weber of magnetic flux per square metre, i.e. $1\ T = 1\ Wb\ m^{-2}$. It is named after Nikola Tesla (1870–1943), Croatian-born US electrical engineer.

Tesla coil A device for producing a high-frequency high-voltage current. It consists of a *transformer with a high turns ratio, the primary circuit of which includes a spark gap and a fixed capacitor; the secondary circuit is tuned by means of a variable capacitor to resonate with the primary. It was devised by Nikola Tesla (1870–1943). Tesla coils are commonly used to excite luminous discharges in glass vacuum apparatus, in order to detect leaks.

testa (seed coat) The lignified or fibrous protective covering of a seed that develops from the integuments of the ovule after fertilization. *See also* hilum; micropyle.

test cross *See* back cross.

testicle *See* testis.

testis (testicle) The reproductive organ in male animals in which spermatozoa are produced. In vertebrates

Human male reproductive system

there are two testes; as well as sperm, they produce steroid hormones (*see* androgen). In most animals the testes are within the body cavity but in mammals, although they develop within the body near the kidneys, they come to hang outside the body cavity in a *scrotum (see illustration). Most of the vertebrate testis is made up of a mass of *seminiferous tubules*, in which the sperms develop. It is connected to the outside by means of the *vas deferens.

testosterone The principal male sex hormone. *See* androgen.

tetrachloroethene A colourless non-flammable volatile liquid, $CCl_2{:}CCl_2$; r.d. 1.6; m.p. $-22°C$; b.p. $121°C$. It is used as a solvent.

tetrachloromethane (carbon tetra-chloride) A colourless volatile liquid with a characteristic odour, virtually insoluble in water but miscible with many organic liquids, such as ethanol and benzene; r.d. 1.586; m.p. $-23°C$; b.p. $76.8°C$. It is made by the chlorination of methane (previously by chlorination of carbon disulphide). The compound is a good solvent for waxes, lacquers, and rubbers and the main industrial use is as a solvent, but safer substances (e.g. 1,1,1-trichloroethane) are increasingly being used. Moist carbon tetrachloride is

partly decomposed to phosgene and hydrogen chloride and this provides a further restriction on its use.

tetrad A group of four *haploid cells formed at the end of the second division of *meiosis.

tetraethyl lead *See* lead(IV) tetra-ethyl.

tetragonal *See* crystal system.

tetrahedral angle 1. (in geometry) The solid angle bounded by three faces of a tetrahedron. **2.** (in chemistry) The angle between the bonds in a *tetrahedral compound (approximately 109° for a regular tetrahedron).

tetrahedral compound A compound in which four atoms or groups situated at the corners of a tetrahedron are linked (by covalent or coordinate bonds) to an atom at the centre of the tetrahedron. *See also* complex.

tetrahedron A polyhedron with four triangular faces. In a *regular tetrahedron* all four triangles are congruent equilateral triangles. It constitutes a regular triangular *pyramid.

tetrahydrate A crystalline hydrate containing four molecules of water per molecule of compound.

tetrahydrofuran (THF) A colourless volatile liquid, C_4H_8O; r.d. 0.89; m.p. $-65°C$; b.p. $67°C$. It is made by the acid hydrolysis of polysaccharides in oat husks, and is widely used as a solvent.

tetrahydroxomonoxodiboric(III) acid *See* boric acid.

tetraoxophosphoric(V) acid *See* phosphoric(V) acid.

tetraoxosulphuric(VI) acid *See* sulphuric acid.

tetrapod A vertebrate animal with four limbs. Tetrapods include amphibians, reptiles, birds, and mammals. The skeleton of the limbs of all tetrapods is based on the same five-digit pattern (*see* pentadactyl limb).

tetravalent (quadrivalent) Having a valency of four.

tetrode A *thermionic valve with a *screen grid* placed between the anode and the control grid of a *triode to reduce the capacitance between these two electrodes and so improve the valve's performance as an amplifier or oscillator at high frequencies. The screen grid is maintained at a fixed potential.

texaphyrin A synthetic molecule similar to a porphyrin but containing five central nitrogen atoms rather than four, thus increasing the size of the central 'hole' and enabling larger cations, such as cadmium, to be bound stably. *See also* supramolecular chemistry.

thalamus 1. (in anatomy) Part of the vertebrate *forebrain that lies above the hypothalamus. It relays sensory information to the cerebral cortex and is also concerned with the translation of impulses into conscious sensations. **2.** (in botany) *See* receptacle.

thallium Symbol Tl. A greyish metallic element belonging to *group IIIA of the periodic table; a.n. 81; r.a.m. 204.39; r.d. 11.85 (20°C); m.p. 303.3°C; b.p. 1460°C. It occurs in zinc blende and some iron ores and is recovered in small quantities from lead and zinc concentrates. The naturally occurring isotopes are thallium–203 and thallium–205; eleven radioisotopes have been identified. It has few uses – experimental alloys for special purposes and some minor uses in electronics. The sulphate has been used as a rodenticide. Thallium(I) compounds resemble those of the alkali metals. Thallium(III) compounds are easily reduced to the thallium(I) state and are therefore strong oxidizing agents. The element was discovered by Sir William Crookes in 1861.

Thallophyta A former division of the plant kingdom containing relatively simple plants, i.e. those with no leaves, stems, or roots. It included the *algae, *bacteria, *fungi, and *lichens.

thallus A relatively undifferentiated plant body with no true roots, stems, leaves, or vascular system. It is found in the algae, lichens, and bryophytes and in the gametophyte generation of pteridophytes.

theca *See* capsule.

theodolite An optical surveying instrument for measuring horizontal and vertical angles. It consists of a sighting telescope, with crosshairs in the eyepiece for focusing on the target, which can be rotated in both the horizontal and vertical planes. It is mounted on a tripod and a spirit level is used to indicate when the instrument is horizontal. The angles are read off graduated circles seen through a second eyepiece in the instrument.

therm A practical unit of energy defined as 10^5 British thermal units. 1 therm is equal to 1.055×10^8 joules.

thermal analysis A technique for chemical analysis and the investigation of the products formed by heating a substance. In *differential thermal analysis (DTA)* a sample is heated, usually in an inert atmosphere, and a plot of weight against temperature made. In *differential scanning calorimetry (DSC)* heat is added to or removed from a sample electrically as the temperature is increased, thus allowing the enthalpy changes due to thermal decomposition to be studied.

thermal capacity *See* heat capacity.

thermal conductivity *See* conductivity.

thermal diffusion The diffusion that takes place in a fluid as a result of a temperature gradient. If a column of gas is maintained so that the lower end is cooler than the upper end, the heavier molecules in the gas will tend to remain at the lower-temperature end and the lighter molecules will diffuse to the higher-temperature end. This property has been used in the separation of gaseous isotopes (*see* Clusius column).

thermal equilibrium *See* equilibrium.

thermal expansion *See* expansivity.

thermalization The reduction of the kinetic energy of neutrons in a thermal *nuclear reactor by means of a *moderator; the process of producing thermal neutrons.

thermal neutrons *See* moderator; nuclear reactors; thermalization.

thermal reactor *See* nuclear reactor.

thermionic emission The emission of electrons, usually into a vacuum, from a heated conductor. The emitted current density, J, is given by the *Richardson* (or *Richardson–Dushman*) *equation*, i.e. $J = AT^2\exp(-W/kT)$, where T is the thermodynamic temperature of the emitter, W is its *work function, k is the Boltzmann constant, and A is a constant. Thermionic emission is the basis of the *thermionic valve and the *electron gun in cathode-ray tubes.

thermionics The branch of electronics concerned with the study and design of devices based on the emission of electrons from metal or metal-oxide surfaces as a result of high temperatures. The primary concern of thermionics is the design of *thermionic valves and the electron guns of cathode-ray tubes and other devices.

thermionic valve An electronic valve based on *thermionic emission. In such valves the cathode is either directly heated by passing a current through it or indirectly heated by placing it close to a heated filament. Directly heated cathodes are usually made of tungsten wire, whereas indirectly heated cathodes are usually coated with barium and strontium oxides. Most electronic valves are thermionic vacuum devices, although a few have cold cathodes and some

are gas-filled (*see* thyratron). *See* diode; triode; tetrode; pentode.

thermistor A semiconductor electronic device the resistance of which decreases as its temperature increases. It consists of a bead, rod, or disc of various oxides of manganese, nickel, cobalt, copper, iron, or other metals. Thermistors are used as thermometers, often forming one element in a resistance bridge. They are used for this purpose in such applications as bearings, cylinder heads, and transformer cores. They are also used to compensate for the increased resistance of ordinary resistors when hot, and in vacuum gauges, time-delay switches, and voltage regulators.

thermite A stoichiometric powdered mixture of iron(III) oxide and aluminium for the reaction:
$$2Al + Fe_2O_3 \rightarrow Al_2O_3 + 2Fe$$
The reaction is highly exothermic and the increase in temperature is sufficient to melt the iron produced. It has been used for localized welding of steel objects (e.g. railway lines) in the *Thermit process*. Thermite is also used in incendiary bombs.

thermochemistry The branch of physical chemistry concerned with heats of chemical reaction, heats of formation of chemical compounds, etc.

thermocouple A device consisting of two dissimilar metal wires or semiconducting rods welded together at their ends. A thermoelectric e.m.f. is generated in the device when the ends are maintained at different temperatures, the magnitude of the e.m.f. being related to the temperature difference. This enables a thermocouple to be used as a thermometer over a limited temperature range. One of the two junctions, called the *hot* or *measuring junction*, is exposed to the temperature to be measured. The other, the *cold* or *reference junction*, is maintained at a known reference temperature. The e.m.f. generated is measured by a suitable milli-voltmeter or potentiometer incorporated into the circuit. *See* Seebeck effect; thermopile.

thermodynamics The study of the laws that govern the conversion of energy from one form to another, the direction in which heat will flow, and the availability of energy to do work. It is based on the concept that in an isolated system anywhere in the universe there is a measurable quantity of energy called the *internal energy (U) of the system. This is the total kinetic and potential energy of the atoms and molecules of the system of all kinds that can be transferred directly as heat; it therefore excludes chemical and nuclear energy. The value of U can only be changed if the system ceases to be isolated. In these circumstances U can change by the transfer of mass to or from the system, the transfer of heat (Q) to or from the system, or by the work (W) being done on or by the system. For an adiabatic ($Q = 0$) system of constant mass, $\Delta U = W$. By convention, W is taken to be positive if work is done on the system and negative if work is done by the system. For nonadiabatic systems of constant mass, $\Delta U = Q + W$. This statement, which is equivalent to the law of conservation of energy, is known as the *first law of thermodynamics*.

All natural processes conform to this law, but not all processes conforming to it can occur in nature. Most natural processes are irreversible, i.e. they will only proceed in one direction (*see* reversible process). The direction that a natural process can take is the subject of the *second law of thermodynamics*, which can be stated in a variety of ways. R. Clausius (1822–88) stated the law in two ways: "heat cannot be transferred from one body to a second body at a higher temperature without producing some other effect" and "the entropy of a closed system increases with time". These statements introduce the thermodynamic concepts of *temperature (T) and *entropy (S), both of which are parameters determining the direction in which an irreversible process can go. The temperature of a body or system determines whether heat will flow into it or out of it; its entropy is a measure of the unavailability of its energy to do work. Thus T and S determine the relationship between Q and W in the statement of the first law. This is usually presented by stating the second law in the form $\Delta U = T\Delta S - W$.

The second law is concerned with changes in entropy (ΔS). The *third law of thermodynamics* provides an absolute scale of values for entropy by stating that for changes involving only perfect crystalline solids at *absolute zero, the change of the total entropy is zero. This law enables absolute values to be stated for entropies.

One other law is used in thermodynamics. Because it is fundamental to, and assumed by, the other laws of thermodynamics it is usually known as the *zeroth law of thermodynamics*. This states that if two bodies are each in thermal equilibrium with a third body, then all three bodies are in thermal equilibrium with each other. *See also* enthalpy; free energy.

thermodynamic temperature *See* temperature.

thermoelectricity An electric current generated by temperature difference. There are three interrelated *thermoelectric effects. See* Peltier effect; Seebeck effect; Thomson effect.

thermograph 1. A recording thermometer used in meteorology to obtain a continuous record of temperature changes over a period on a graph. **2.** A record so obtained. **3.** (*or* **thermogram**) A record obtained by the technique of *thermography.

thermography A medical technique that makes use of the infrared radiation from the human skin to detect an area of elevated skin temperature that could be associated with an underlying cancer. The heat radiated from the body varies according to the local blood flow, thus an area of poor circulation produces less radiation. A tumour, on the other hand, has an abnormally increased blood supply and is revealed on the *thermogram* (or *thermograph*) as a 'hot spot'. The technique is used particularly in mammography, the examination of the infrared radiation emitted by human breasts in order to detect breast cancer.

thermoluminescence *Luminescence produced in a solid when its temperature is raised. It arises when free electrons and *holes, trapped in a solid as a result of exposure to ionizing radiation, unite and emit photons of light. The process is made use of in *thermoluminescent dating*, which assumes that the number of electrons and holes trapped in a sample of pottery is related to the length of time that has elapsed since the pottery was fired. By comparing the luminescence produced by heating a piece of pottery of unknown age with the luminescence produced by heating similar materials of known age, a fairly accurate estimate of the age of an object can be made.

thermoluminescent dating *See* thermoluminescence.

thermometer An instrument used for measuring the *temperature of a substance. A number of techniques and forms are used in thermometers depending on such factors as the degree of accuracy required and the range of temperatures to be measured, but they all measure temperature by making use of some property of a substance that varies with temperature. For example, *liquid-in-glass thermometers* depend on the expansion of a liquid, usually mercury or alcohol coloured with dye. These consist of a liquid-filled glass bulb attached to a partially filled capillary tube. In the *bimetallic thermometer* the unequal expansion of two dissimilar metals that have been bonded together into a narrow strip and coiled is used to move a pointer round a dial. The *gas thermometer*, which is more accurate than the liquid-in-glass thermometer, measures the variation in the pressure of a gas kept at constant volume. The *resistance thermometer* is based on the change in resistance of conductors or semiconductors with temperature change. Platinum, nickel, and copper are the metals most commonly used in resistance thermometers. *See also* pyrometry; thermistor; thermocouple.

thermonuclear reaction *See* nuclear fusion; thermonuclear reactor.

thermonuclear reactor (fusion reactor) A reactor in which *nuclear fusion takes place with the controlled release of energy. Although thermonuclear reactors do not yet exist, intense research in many parts of the world is being carried out with a view to achieving such a machine. There are two central problems in the creation of a self-sustaining thermonuclear reactor: heating the reacting nuclides to the enormous *ignition temperature (about 40×10^6 K for a deuterium–tritium reaction) and containing the reacting nuclides for long enough for the fusion energy released to exceed the energy required to achieve the ignition temperature (*see* Lawson criterion). The two methods being explored are *magnetic containment* and *pellet fusion*.

In the closed magnetic-containment device the fusion *plasma is contained in a toroidal-shaped reactor, called a *tokamak*, in which strong magnetic fields guide the charged plasma particles round the toroid without allowing them to contact the container walls. In open-ended magnetic systems the plasma is trapped between magnetic mirrors (strong magnetic fields) at the two ends of a straight containment vessel.

In pellet fusion the objective is to heat and compress a tiny pellet of the nuclear fuels, by means of a laser or an electron beam, so rapidly that fusion is achieved before the pellet flies apart. Results with this type of equipment have been comparable to those achieved by magnetic confinement.

thermonuclear weapon *See* nuclear weapons.

thermopile A device used to detect and measure the intensity of radiant energy. It consists of a number of *thermocouples connected together in series to achieve greater sensitivity. The hot junctions of the thermocouples are blackened and exposed to the radiation to be detected or measured, while the cold junctions are shielded from the radiation. The thermoelectric e.m.f. generated enables the hot junction excess temperature to be calculated and the radiant intensity to be deduced. They are used in various applications, from a safety device that produces an electric signal if a pilot light blows out to an instrument to measure the heat radiation received from the sun.

thermoplastic *See* plastics.

thermosetting *See* plastics.

thermostat A device that controls the heating or cooling of a substance in order to maintain it at a constant temperature. It consists of a temperature-sensing instrument connected to a switching device. When the temperature reaches a predetermined level the sensor switches the heating or cooling source on or off according to a predetermined program. The sensing thermometer is often a *bimetallic strip that triggers a simple electrical switch. Thermostats are used for space-heating controls, in water

heaters and refrigerators, and to maintain the environment of a scientific experiment at a constant temperature.

THF *See* tetrahydrofuran.

thiamin(e) *See* vitamin B complex.

thigmotropism (haptotropism) The growth of an aerial plant organ in response to localized physical contact. For example, when a tendril of sweet pea touches a supporting structure, it curves in the direction of the support and coils around it. *See* tropism.

thin-layer chromatography A technique for the analysis of liquid mixtures using *chromatography. The stationary phase is a thin layer of an absorbing solid (e.g. alumina) prepared by spreading a slurry of the solid on a plate (usually glass) and drying it in an oven. A spot of the mixture to be analysed is placed near one edge and the plate is stood upright in a solvent. The solvent rises through the layer by capillary action carrying the components up the plate at different rates (depending on the extent to which they are absorbed by the solid). After a given time, the plate is dried and the location of spots noted. It is possible to identify constituents of the mixture by the distance moved in a given time. The technique needs careful control of the thickness of the layer and of the temperature. *See also* R_F value.

thiocyanate A salt or ester of cyanic acid.

thiocyanic acid An unstable gas, HSCN.

thio ethers *See* sulphides.

thiol group *See* thiols.

thiols (mercaptans; thio alcohols) Organic compounds that contain the group −SH (called the *thiol group*, *mercapto group*, or *sulphydryl group*). Thiols are analogues of alcohols in which the oxygen atom is replaced by a sulphur atom. They are named according to the parent hydrocarbon;

e.g. ethane thiol (C_2H_5SH). A characteristic property is their strong disagreeable odour. For example the odour of garlic is produced by ethane thiol. Unlike alcohols they are acidic, reacting with alkalis and certain metals to form saltlike compounds. The older name, mercaptan, comes from their ability to react with ('seize') mercury.

thionyl chloride See sulphur dichloride oxide.

thionyl group The group $=SO$, as in *sulphur dichloride oxide.

thiosulphate A salt containing the ion $S_2O_3^{2-}$ formally derived from thiosulphuric acid. Thiosulphates readily decompose in acid solution to give elemental sulphur and hydrogensulphite (HSO_3^-) ions.

thiosulphuric acid An unstable acid, $H_2S_2O_3$, formed by the reaction of sulphur trioxide with hydrogen sulphide. See also sulphuric acid.

thiourea A white crystalline solid, $(NH_2)_2CS$; r.d. 1.4; m.p. 180°C. It is used as a fixer in photography.

thixotropy See Newtonian fluid.

Thomson effect (Kelvin effect) When an electric current flows through a conductor, the ends of which are maintained at different temperatures, heat is evolved at a rate approximately proportional to the product of the current and the temperature gradient. If either the current or the temperature gradient is reversed heat is absorbed rather than being evolved. It is named after Sir William Thomson (later Lord Kelvin; 1824–1907).

Thomson scattering The scattering of electromagnetic radiation by free charged particles, especially electrons, when the photon energy is small compared with the energy equivalent to the *rest mass of the charged particles. The energy lost by the radiation is accounted for by classical theory as a result of the radiation emitted by the charged particles when they are accelerated in the transverse electric field of the radiation. It is named after Sir J. J. Thomson (1856–1940).

thoracic duct The main collecting vessel of the *lymphatic system, running longitudinally in front of the backbone. The thoracic duct drains its lymph into the superior vena cava.

thoracic vertebrae The *vertebrae of the upper back, which articulate with the *ribs. They lie between the *cervical vertebrae and the *lumbar vertebrae and are distinguished by a number of articulating facets for attachment of the ribs. In man there are 12 thoracic vertebrae.

thorax The anterior region of the body trunk of animals. In vertebrates it contains the heart and lungs within the rib cage. It is particularly well-defined in mammals, being separated from the *abdomen by the *diaphragm. In insects the thorax is divided into an anterior *prothorax*, a middle *mesothorax*, and a posterior *metathorax*, each of which bears a pair of legs; the hindmost two segments also both carry a pair of wings. In other arthropods, especially crustaceans and arachnids, the thorax is fused with the head to form a *cephalothorax*.

thoria See thorium.

thorium Symbol Th. A grey radioactive metallic element belonging to the *actinoids; a.n. 90; r.a.m. 232.038; r.d. 11.5–11.9 (17°C); m.p. 1740–1760°C; b.p. 4780–4800°C. It occurs in monazite sand in Brazil, India, and USA. The isotopes of thorium have mass numbers from 223 to 234 inclusive; the most stable isotope, thorium –232, has a half-life of 1.39×10^{10} years. It has an oxidation state of ($+4$) and its chemistry resembles that of the other actinoids. It can be used as a nuclear fuel for breeder reactors as thorium–232 captures slow neutrons to breed uranium–233. Thorium dioxide (*thoria*, ThO_2) is used on gas mantles and in special refractories. The element was discovered by J. J. Berzelius in 1829.

thorium series *See* radioactive series.

thorn A hard side stem with a sharp point at the tip, replacing the growing point. In some plants the development of thorns and subsequent suppression of the growing points may be a response to dry conditions. Examples are the thorns of gorse and hawthorn. *Compare* prickle; spine.

thread cell (nematoblast; cnidoblast) A specialized cell found only in the ectoderm of the *Coelenterata. It contains a *nematocyst*, a fluid-filled sac within which lies a long hollow coiled thread. When a small sensory projection (*cnidocil*) on the surface of the thread cell is touched, e.g. by prey, the thread is shot out and adheres to the prey, coils round it, or injects poison into it. Numerous thread cells on the tentacles of jellyfish produce their sting.

threnardite A mineral form of *sodium sulphate, Na_2SO_4.

threonine *See* amino acid.

threshold 1. (in physics) The minimum value of a parameter or variable that will produce a specified effect. **2.** (in physiology) The minimum intensity of a stimulus that is necessary to initiate a response.

threshold frequency *See* photoelectric effect.

thrombin An enzyme that catalyses the conversion of fibrinogen to fibrin. *See* blood clotting; prothrombin.

thrombocyte *See* platelet.

thrust The propelling force generated by an aircraft engine or rocket. It is usually calculated as the product of the rate of mass discharge and the velocity of the exhaust gases relative to the vehicle.

thulium Symbol Tm. A soft grey metallic element belonging to the *lanthanoids; a.n. 69; r.a.m. 168.934; r.d. 9.321 (20°C); m.p. 1545°C; b.p. 1950°C. It occurs in apatite and xenotime. There is one natural isotope, thulium–169, and seventeen artificial isotopes have been produced. There are no uses for the element, which was discovered by P. T. Cleve in 1879.

Cross-section through a thunderstorm cell

thunderstorm A convective storm accompanied by *lightning and thunder and a variety of weather conditions, especially heavy rain or hail, high winds, and sudden temperature changes. Thunderstorms originate when intense heating causes a parcel of moist air to rise, leading to instability and the development of cumulonimbus cloud – a towering cloud with a characteristic anvil-shaped top (see illustration). The exact mechanisms of thunderstorms are not fully understood. They occur most frequently in the tropics but are also common in the mid-latitudes.

thymidine A nucleoside consisting of one thymine molecule linked to a D-ribose sugar molecule.

thymine A *pyrimidine derivative and one of the major component bases of *nucleotides and the nucleic acid *DNA.

thymus An organ, present only in vertebrates, that is concerned with development of *lymphoid tissue and hence the antibody-producing white blood cells (*see* T cell) and the *immune response. In mammals it is a bilobed organ in the region of the lower neck, above and in front of the heart. It shrinks in size after sexual maturity and is therefore believed to function only during early life.

thyratron A thermionic valve (usually a triode) that functions as a gas-filled relay. A positive pulse fed to a correctly biased thyratron causes a discharge to start and to continue until the anode voltage has been reduced. It has now been replaced by its solid-state counterpart, the silicon-controlled rectifier.

thyristor A silicon-controlled rectifier whose anode–cathode current is controlled by a signal applied to a third electrode (the gate) in much the same way as in a thyratron valve. It consists usually of a four-layer chip comprising three p–n junctions.

thyrocalcitonin *See* calcitonin.

thyroid gland A bilobed endocrine gland in vertebrates, situated in the base of the neck. It secretes two iodine-containing hormones, *thyroxine* and *triiodothyronine*, which control the rate of all metabolic processes in the body and influence physical development. Growth and activity of the thyroid is controlled by a hormone, *thyrotrophin* (or *thyroid-stimulating hormone*), secreted by the anterior *pituitary gland.

thyroxine The principal hormone of the *thyroid gland.

tibia 1. The larger of the two bones of the lower hindlimb of terrestrial vertebrates (*compare* fibula). It articulates with the *femur at the knee and with the *tarsus at the ankle. The tibia is the major load-bearing bone of the lower leg. **2.** The fourth segment of an insect's leg, which is attached to the femur.

tidal energy *See* tides.

tides The regular rise and fall of the water level in the earth's oceans as a result of the gravitational forces between the earth, moon, and sun. The forces involved are complex, but the moon is approximately twice as effective as the sun in causing tides. In illustration (a) the resultant gravitational forces between the moon and various points on the earth (solid lines) are shown as the vector sums of the tide-generating forces (broken lines) and a constant force (dotted lines) that is the same at all points on the earth and is equal to the moon's attraction on the earth's centre. The resultant force when the moon is in zenith (Z in the illustration) is greater than that at nadir (N)

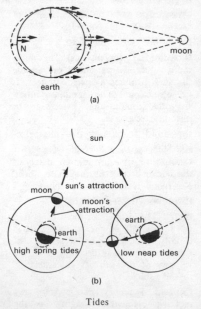

Tides

because Z is closer to the moon than N and the force is inversely proportional to the square of the distance according to *Newton's law of gravitation. Illustration (b) shows how at full and new moon the sun and moon act together to produce the high-range *spring tides*, while at quarter moon the forces are at right angles to each other causing the low-range *neap tides*.

The use of *tidal energy*, estimated at some 4×10^{18} J per annum at known tidal sites, dates back to medieval tidal mills. Modern tidal power stations use specially designed turbines, operated by tidal waters, to drive generators.

timbre *See* quality of sound.

time A dimension that enables two otherwise identical events that occur at the same point in space to be distinguished (*see* space–time). The interval between two such events forms the basis of time measurement. For general purposes, the earth's rotation on its axis provides the units of the clock (*see* day) and the earth's orbit round the sun (*see* year) provides the units of the calendar. For scientific purposes, intervals of time are now defined in terms of the frequency of a specified electromagnetic radiation (*see* second). *See also* time dilation; time reversal.

time dilation (time dilatation) The principle, predicted by Einstein's special theory of *relativity, that intervals of time are not absolute but are relative to the motion of the observers. If two identical clocks are synchronized and placed side by side in an inertial frame of reference they will read the same time for as long as they both remain side by side. However, if one of the clocks has a velocity relative to the other, which remains beside a stationary observer, the travelling clock will show, to that observer, that less time has elapsed than the stationary clock. In general,

the travelling clock goes more slowly by a factor $\sqrt{(1 - v^2/c^2)}$, when measured in a frame of reference travelling at a velocity v relative to another frame of reference; c is the speed of light. The principle has been verified in a number of ways; for example, by comparing the lifetimes of fast muons, which increase with the speed of the particles to an extent predicted by this factor.

time-lapse photography A form of ciné photography used to record a slow process, such as plant growth. A series of single exposures of the object is made on ciné film at predetermined regular intervals. The film produced is then projected at normal speeds and the process appears to be taking place at an extremely high rate.

time reversal Symbol T. The operation of replacing time t by time $-t$. The *symmetry of time reversal is known as T invariance. As with CP violation, T violation occurs in weak interactions of kaon decays. *See also* CP invariance; CPT theorem.

tin Symbol Sn. A silvery malleable metallic element belonging to *group IV of the periodic table; a.n. 50; r.a.m. 118.69; r.d. 7.29; m.p. 231.97°C; b.p. 2270°C. It is found as tin(IV) oxide in ores, such as cassiterite, and is extracted by reduction with carbon. The metal (called *white tin*) has a powdery nonmetallic allotrope *grey tin*, into which it changes below 18°C. The formation of this allotrope is called *tin plague*; it can be reversed by heating to 100°C. The natural element has 21 isotopes (the largest number of any element); five radioactive isotopes are also known. The metal is used as a thin protective coating for steel plate and is a constituent of a number of alloys (e.g. phosphor bronze, gun metal, solder, Babbitt metal, and pewter). Chemically it is reactive. It combines directly with chlorine and oxygen and

tin(II) chloride

displaces hydrogen from dilute acids. It also dissolves in alkalis to form *stannates. There are two series of compounds with tin in the +2 and +4 oxidation states.

tin(II) chloride A white solid, $SnCl_2$, soluble in water and ethanol. It exists in the anhydrous form (rhombic; r.d. 3.95; m.p. 246°C; b.p. 652°C) and as a dihydrate, $SnCl_2.2H_2O$ (monoclinic; r.d. 2.71; m.p. 37.7°C). The compound is made by dissolving metallic tin in hydrochloric acid and is partially hydrolysed in solution.

$$Sn^{2+} + H_2O \rightleftharpoons SnOH^+ + H^+$$

Excess acid must be present to prevent the precipitation of basic salts. In the presence of additional chloride ions the pyramidal ion $[SnCl_3]^-$ is formed; in the gas phase the $SnCl_2$ molecule is bent. It is a reducing agent in acid solutions and oxidizes slowly in air:

$$Sn^{2+} \rightarrow Sn^{4+} + 2e$$

tin(IV) chloride A colourless fuming liquid, $SnCl_4$, hydrolysed in cold water, decomposed by hot water, and soluble in ethers; r.d. 2.226; m.p. -33°C; b.p. 114°C. Tin(IV) chloride is a covalent compound, which may be prepared directly from the elements. It dissolves sulphur, phosphorus, bromine, and iodine, and there is evidence for the presence of species such as $SnCl_2I_2$. In hydrochloric acid and in chloride solutions the coordination is extended from four to six by the formation of the $SnCl_6^{2-}$ ion.

tincture A solution with alcohol as the solvent (e.g. tincture of iodine).

tin(IV) hydride (stannane) A highly reactive and volatile gas (b.p. -53°), SnH_4, which decomposes on moderate heating (150°C). It is prepared by the reduction of tin chlorides using lithium tetrahydridoaluminate(III) and is used in the synthesis of some organotin compounds. The compound has reducing properties.

tin(IV) oxide (tin dioxide) A white solid, SnO_2, insoluble in water; tetra-hedral; r.d. 6.95; m.p. 1127°C; sublimes at about 1850°C. Tin(IV) oxide is trimorphic: the common form, which occurs naturally as the ore *cassiterite, has a rutile lattice but hexagonal and rhombic forms are also known. There are also two so-called dihydrates, $SnO_2.2H_2O$, known as α- and β-stannic acid. These are essentially tin hydroxides. Tin(IV) oxide is amphoteric, dissolving in molten alkalis to form *stannates; in the presence of sulphur, thiostannates are produced.

tin plague *See* tin.

tin(II) sulphide A grey-black cubic or monoclinic solid, SnS, virtually insoluble in water; r.d. 5.22; m.p. 882°C; b.p. 1230°C. It has a layer structure similar to that of black phosphorus. Its heat of formation is low and it can be made by heating the elements together. Above 265°C it slowly decomposes (disproportionates) to tin(IV) sulphide and tin metal. The compound reacts with hydrochloric acid to give tin(II) chloride and hydrogen sulphide.

tin(IV) sulphide (mosaic gold) A bronze or golden yellow crystalline compound, SnS_2, insoluble in water and in ethanol; hexagonal; r.d. 4.5; decomposes at 600°C. It is prepared by the reaction of hydrogen sulphide with a soluble tin(IV) salt or by the action of heat on thiostannic acid, H_2SnS_3. The golden-yellow form used for producing a gilded effect on wood is prepared by heating tin, sulphur, and ammonium chloride.

T invariance *See* time reversal.

tissue A collection of similar cells organized to carry out one or more particular functions. For example, in animals nervous tissue is specialized to perceive and transmit stimuli. An organ, such as a lung or kidney, contains many different types of tissues.

tissue culture The growth of the tissues of living organisms outside the body in a suitable culture medium. Culture (or nutrient) media contain a mixture of nutrients either in solid form (e.g. in *agar) or in liquid form (e.g. in *physiological saline). Tissue culture has proved to be invaluable for gaining information about factors that control the growth and differentiation of cells. Culture of plant tissues has resulted in the regeneration of complete plants, enabling commercial propagation (e.g. of orchids) and – through culture of meristem tissues – the production of virus-free crop plants. *See also* explantation.

Titan *See* Saturn.

titania *See* titanium(IV) oxide.

titanium Symbol Ti. A white metallic *transition element; a.n. 22; r.a.m. 47.9; r.d. 4.507; m.p. $1660\pm10°C$; b.p. 3280°C. The main sources are rutile (TiO_2) and, to a lesser extent, ilmenite ($FeTiO_3$). The element also occurs in numerous other minerals. It is obtained by heating the oxide with carbon and chlorine to give $TiCl_4$, which is reduced by the *Kroll process. The main use is in a large number of strong light corrosion-resistant alloys for aircraft, ships, chemical plant, etc. The element forms a passive oxide coating in air. At higher temperatures it reacts with oxygen, nitrogen, chlorine, and other nonmetals. It dissolves in dilute acids. The main compounds are titanium(IV) salts and complexes; titanium(II) and titanium(III) compounds are also known. The element was first discovered by Gregor in 1789.

titanium dioxide *See* titanium(IV) oxide.

titanium(IV) oxide (titania; titanium dioxide) A white oxide, TiO_2, occurring naturally in various forms, particularly the mineral rutile. It is used as a white pigment and as a filler for plastics, rubber, etc.

titration A method of volumetric analysis in which a volume of one reagent is added to a known volume of another reagent slowly from a burette until an end point is reached (*see* indicator). The volume added before the end point is reached is noted. If one of the solutions has a known concentration, that of the other can be calculated.

titre 1. The number of infectious virus particles present in a suspension. 2. A measure of the amount of *antibody present in a sample of serum, given by the highest dilution of the sample that results in the formation of visible clumps with the appropriate antigen (*see* agglutination).

TNT *See* trinitrotoluene.

toads *See* Amphibia.

tobacco mosaic virus (TMV) A rigid rod-shaped RNA-containing virus that causes distortion and blistering of leaves in a wide range of plants, especially the tobacco plant. It is transmitted by insects when they feed on plant tissue. TMV was the first virus to be discovered.

tocopherol *See* vitamin E.

tokamak *See* thermonuclear reactor.

Tollen's reagent A reagent used in testing for aldehydes. It is made by adding sodium hydroxide to silver nitrate to give silver(I) oxide, which is dissolved in aqueous ammonia (giving the complex ion $[Ag(NH_3)_2]^+$). The sample is warmed with the reagent in a test tube. Aldehydes reduce the complex Ag^+ ion to metallic silver, forming a bright silver mirror on the inside of the tube (hence the name *silver-mirror test*). Ketones give a negative result.

toluene *See* methylbenzene.

tomography The use of X-rays to photograph a selected plane of a human body with other planes eliminated. The *CT* (*computerized tomography*) *scanner* is a ring-shaped X-ray machine that rotates through 180° around the horizontal patient,

Section through an incisor tooth

making numerous X-ray measurements every few degrees. The vast amount of information acquired is built into a three-dimensional image of the tissues under examination by the scanner's own computer. The patient is exposed to a dose of X-rays only some 20% of that used in a normal diagnostic X-ray.

tone (tonus) The state of sustained tension in muscles that is necessary for the maintenance of posture. In a tonic muscle contraction, only a certain proportion of the muscle fibres are contracting at any given time; the rest are relaxed and recovering for subsequent contractions. The fibres involved in tone contract more slowly than the fast fibres used for rapid responses by the same muscle. The proportions of slow and fast fibres depends on the function of the muscle.

tongue A muscular organ of vertebrates that in most species is attached to the floor of the mouth. It plays an important role in manipulating food during chewing and swallowing and in terrestrial species it bears numerous *taste buds on its upper surface. In some advanced vertebrates the tongue is used in the articulation of sounds, particularly in human speech.

tonsil A mass of *lymphoid tissue, several of which are situated at the back of the mouth and throat in higher vertebrates. In humans there are the *palatine tonsils* at the back of the mouth, *lingual tonsils* below the tongue, and *pharyngeal tonsils* (or *adenoids*) in the pharynx. They are concerned with the production of *lymphocytes and therefore with defence against infection.

tonus *See* tone.

tooth Any of the hard structures in vertebrates that are used principally for biting and chewing food but also for attack, grooming, and other functions. In fish and amphibians the teeth occur all over the palate, but in higher vertebrates they are concentrated on the jaws. They evolved in cartilaginous fish as modified placoid *scales, and this is reflected in their structure: a body of bony *dentine with a central *pulp cavity and an outer covering of *enamel on the exposed surface (*crown*). The portion of the tooth embedded in the jawbone is the *root* (see illustration).

In mammals there are four different types of teeth, specialized for different functions (*see* canine; incisor; molar; premolar). Their number varies with the species (*see* dental formula). *See also* deciduous teeth; permanent teeth.

topaz A variably coloured aluminium silicate mineral, $Al_2(SiO_4)(OH\,F)_2$, that forms orthorhombic crystals. It occurs chiefly in acid igneous rocks, such as granites and pegmatites. Topaz is valued as a gemstone because of its transparency, variety of colours (the wine-yellow variety being most highly prized), and great hardness (8 on the Mohs' scale). When heated, yellow or brownish topaz often becomes a rose-pink colour. The main sources of topaz are Brazil, the USSR, and the USA.

topology The branch of geometry concerned with the properties of geometrical objects that are unchanged by continuous deformations, such as twisting or stretching. Mathematical approaches employing topology are of great importance in modern theories of the *fundamental interactions.

tornado A violently rotating column of air, 10–100 m in diameter, usually made visible by a funnel cloud, which may reach the ground surface. Wind speeds of up to 100 metres per second may occur, but the damage to meteorological instruments caused by the passage of tornadoes makes exact measurement difficult. Tornadoes frequently occur in groups and are most common in the central USA and Australia, where they cause considerable destruction.

torque (moment of a force or couple) The product of a force and its perpendicular distance from a point about which it causes rotation or *torsion. The unit of torque is the newton metre, a vector product, unlike the joule, also equal to a newton metre, which is a scalar product. A turbine produces a torque on its central rotating shaft. *See also* couple.

torr A unit of pressure, used in high vacuum technology, defined as 1 mmHg. 1 torr is equal to 133.322 pascals. The unit is named after Evangelista Torricelli (1609–47).

Torricellian vacuum The vacuum formed when a long tube, closed at one end and filled with mercury, is inverted into a mercury reservoir so that the open end of the tube is below the surface of the mercury. The pressure inside the Torricellian vacuum is the vapour pressure of mercury, about 10^{-3} torr.

torsion A twisting deformation produced by a *torque or *couple. A *torsion bar* is a form of spring in which one end of a bar is fixed and a torque is applied to the other end. Torsion bars are used in the suspension systems of motor vehicles.

torsion balance An instrument for measuring very weak forces. It consists of a horizontal rod fixed to the end of a vertical wire or fibre or to the centre of a taut horizontal wire. The forces to be measured are applied to the end or ends of the rod. The turning of the rod may be measured by the displacement of a beam of light reflected from a plane mirror attached to it. The best-known form is that used by Henry Cavendish (1731–1810) and later by Sir Charles Boys (1855–1944) to determine the *gravitational constant; in this form the balance is calibrated by determining the torsional coefficient of the suspension by treating the device as a torsional pendulum.

torus 1. (in mathematics) A solid generated by rotating a circle about an external line in its plane, also called an *anchor ring*. It has the shape of the inner tube of a tyre. If r is the radius of the rotating circle and R the distance between the centre of the circle and the axis of rotation, the volume of the torus is $2\pi^2 Rr^2$ and the surface area is $4\pi^2 Rr$. In Cartesian coordinates, if the z-axis is the axis of rotation, the equation of the torus is $[\sqrt{(x^2 + y^2)} - R]^2 + z^2 = r^2$. **2.** (in botany) *See* receptacle.

total internal reflection The total reflection of a beam of light at the interface of one medium and another medium of lower refractive index, when the angle of incidence to the second medium exceeds a specific *critical angle*.

If a beam of light passing through a medium A (say glass) strikes the boundary to a medium B of lower refractive index (say air) with a small angle of incidence i, part will be refracted, with an angle of refraction r, and part will be reflected (see illustration a). If i is increased it will reach a critical angle c, at which $r = 90°$ (see illustration b). If i is now increased further, no refraction can occur and all the light energy is reflected by the interface (see illustration c). This total internal reflection

Total internal reflection

occurs when c (given by $n\sin c = 1$) is exceeded (n is the refractive index of A relative to B). The critical angle of optical glass is usually about 40° and total internal reflection is made use of by incorporating prisms in some optical instruments instead of mirrors.

totality The period during a total *eclipse of the sun in which the view of the sun's surface from a point on the earth is totally obscured by the moon. The maximum duration of totality is 7.67 minutes, but it is usually less.

total-radiation pyrometer *See* pyrometry.

tourmaline A group of minerals composed of complex cyclosilicates containing boron with the general formula $NaR_3^{2+}Al_6B_3Si_6O_{27}(H,F)_4$, where $R = Fe^{2+}$, Mg, or (Al + Li). The crystals are trigonal, elongated, and variably coloured, the two ends of the crystals often having different colours. Tourmaline is used as a gemstone and because of its double refraction and piezoelectric properties is also used in polarizers and some pressure gauges.

toxin A poison produced by a living organism, especially a bacterium. An *endotoxin* is released only when the bacterial cell dies or disintegrates. An *exotoxin* is secreted by a bacterial cell into the surrounding medium. In the body a toxin acts as an *antigen, producing an *immune response.

trace element *See* essential element.

trace fossil *See* fossil.

trachea 1. The windpipe in air-breathing vertebrates: a tube that conducts air from the throat to the *bronchi. It is strengthened with incomplete rings of cartilage. **2.** An air channel in insects and most other terrestrial arthropods. Tracheae occur as ingrowths of the body wall. They open to the exterior by *spiracles* and branch into finer channels (*tracheoles*) that terminate in the tissues. Pumping movements of the abdominal muscles cause air to be drawn into and out of the tracheae.

tracheid A type of cell occurring within the *xylem of conifers, ferns, and related plants. Tracheids are elongated and their walls are usually extensively thickened by deposits of lignin. Water flows from one tracheid to another through unthickened regions (pits) in the cell walls. *Compare* vessel element.

Tracheophyta A group (division) of the plant kingdom containing all plants possessing organized *vascular tissue. This classification is an alternative to other classifications, which divide vascular plants into the *Pteridophyta and *Spermatophyta. The Tracheophyta comprises the subdivisions Psilopsida (mostly extinct), *Lycopsida, *Sphenopsida, and *Pteropsida.

tracing (radioactive tracing) *See* labelling.

trajectory *See* phase space.

transactinide elements Elements with an atomic number greater than 103, i.e. elements above lawrencium in the *periodic table. So far, elements 104 and 105 have been synthesized and element 106 has been claimed. The names proposed for element 104 are *kurchatovium* (symbol Ku), *rutherfordium* (symbol Rf), and the recommended name *unnilquadium* (symbol Unq), which has proved unpopular and is rarely encountered. Element 105 has been called *hahnium* (symbol Ha) and *nielsbohrium*. All these elements are unstable and have very short half-lives.

transamination A biochemical reaction in amino acid metabolism in which an amine group is transferred from an amino acid to a keto acid to form a new amino acid and keto acid. The coenzyme required for this reaction is pyridoxal phosphate.

transcendental number A number that is not algebraic, such as π or e. A *transcendental function* is also nonalgebraic, such as a^x, sin$x$, or log$x$.

transcriptase *See* reverse transcriptase.

transcription The process in living cells in which the genetic information of *DNA is transferred to a molecule of messenger *RNA (mRNA) as the first step in *protein synthesis (*see also* genetic code). Transcription takes place in the cell nucleus or nuclear region. It involves the action of RNA *polymerase enzymes in assembling the nucleotides necessary to form a complementary strand of mRNA from the DNA template, and (in eukaryote cells) the subsequent removal of the noncoding sequences from this primary transcript (*see* gene splicing) to form a functional mRNA molecule. *See also* reverse transcriptase. *Compare* translation.

transducer A device for converting a nonelectrical signal, such as sound, light, heat, etc., into an electrical signal, or vice versa. Thus microphones and loudspeakers are *electroacoustic transducers*. An *active transducer* is one that can itself introduce a power gain and has its own power source. A *passive transducer* has no power source other than the actuating signal and cannot introduce gain.

transect A straight line across an expanse of ground along which ecological measurements are taken, continuously or at regular intervals. Thus an ecologist wishing to study the numbers and types of organisms at different distances above the low-tide line might sample at five-metre intervals along a number of transects perpendicular to the shore.

trans effect An effect in the substitution of inorganic square-planar complexes, in which certain ligands in the original complex are able to direct the incoming ligand into the trans position. The order of ligands in decreasing trans-directing power is: $CN^- > NO_2 > I^- > Br^- > Cl^- > NH_3 > H_2O$.

transfer RNA *See* RNA.

transformer A device for transferring electrical energy from one alternating-current circuit to another with a change of voltage, current, phase, or impedance. It consists of a primary winding of N_p turns magnetically linked by a ferromagnetic core or by proximity to the secondary winding of N_s turns. The *turns ratio* (N_s/N_p) is approximately equal to V_s/V_p and to I_p/I_s, where V_p and I_p are the voltage and current fed to the primary winding and V_s and I_s are the voltage and current induced in the secondary winding, assuming that there are no power losses in the core. In practice, however, there are *eddy-current and *hysteresis losses in the core, incomplete magnetic linkage between the coils, and heating losses in the coils

Transistors

themselves. By the use of a *laminated core and careful design, transformers with 98% efficiency can be achieved.

transgenic Describing an organism whose genome incorporates and expresses genes from another species. Transgenic individuals are created by genetic engineering, using suitable *vectors to insert the desired foreign gene into the fertilized egg or early embryo of the host. For example, the gene for rat growth hormone can be inserted into fertilized mouse eggs to produce mice with cells that produce rat growth hormone. Transgenic organisms offer considerable commercial potential; for example, transgenic farm animals may be engineered to produce therapeutic proteins in their milk or extra growth hormone for improved meat production. Similarly, plants may have disease resistance conferred upon them by transgenic techniques.

transient A brief disturbance or oscillation in a circuit caused by a sudden rise in the current or e.m.f.

trans-isomer *See* isomerism.

transistor A *semiconductor device capable of amplification in addition to rectification. It is the basic unit in radio, television, and computer circuits, having almost completely replaced the *thermionic valve. The *point-contact transistor*, which is now obsolete, was invented in 1948. It consists of a small germanium crystal with two rectifying point contacts attached to it; a third contact, called the *base*, makes a low-resistance nonrectifying (ohmic) connection with the crystal. Current flowing through the device between the point contacts is modulated by the signal fed to the base. This type of transistor was replaced by the *junction transistor*, which was developed in 1949–50. The *field-effect transistor* (*FET*) was a later invention. *Bipolar transistors*, such as the junction transistor, depend on the flow of both majority and minority carriers, whereas in *unipolar transistors*, such as the FET, the current is carried by majority carriers only.

In the bipolar junction transistor, two *p*-type semiconductor regions are separated by a thin *n*-type region, making a *p*–*n*–*p* structure. Alternatively, an *n*–*p*–*n* structure can also be used. In both cases the thin central region is called the base and one outer region of the sandwich is called the *emitter*, the other the *collector*. The emitter–base junction is forward-biased and the collector–base junction is reverse-biased. In the *p*–*n*–*p* transistor, the forward bias causes holes in the emitter region to flow across the junction into the base; as the base is thin, the majority of holes are swept right across it (helped by the reverse bias), into the collector. The minority of holes that do not flow from the base to the collector combine with electrons in the *n*-type base. This recombination is balanced by a small electron flow in the base circuit. The diagram illustrates the (conventional) current flow using the *common-base* type of connection. If the emitter, base, and collector currents are I_e, I_b, and I_c, respectively, then $I_e = I_b + I_c$ and the current gain is I_c/I_b.

Field-effect transistors are of two kinds, the *junction FET* (*JFET* or *JUGFET*) and the *insulated-gate FET* (*IGFET*; also known as a *MOSFET*, i.e. metal-oxide-semiconductor FET). Both are unipolar devices and in both the current flows through a narrow *channel* between two electrodes (the *gate*) from one region, called the *source*, to another, called the *drain*. The modulating signal is applied to the gate. In the JFET, the channel consists of a semiconductor material of relatively low conductivity sandwiched between two regions of high conductivity of the opposite

polarity. When the junctions between these regions are reverse-biased, *depletion layers form, which narrow the channel. At high bias the depletion layers meet and pinch-off the channel completely. Thus the voltage applied to the two gates controls the thickness of the channel and thus its conductivity. JFETs are made with both *n*-type and *p*-type channels.

In the IGFET, a wafer of semiconductor material has two highly doped regions of opposite polarity diffused into it, to form the source and drain regions. An insulating layer of silicon dioxide is formed on the surface between these regions and a metal conductor is evaporated on to the top of this layer to form the gate. When a positive voltage is applied to the gate, electrons move along the surface of the *p*-type substrate below the gate, producing a thin surface of *n*-type material, which forms the channel between the source and drain. This surface layer is called an *inversion layer*, as it has opposite conductivity to that of the substrate. The number of induced electrons is directly proportional to the gate voltage, thus the conductivity of the channel increases with gate voltage. IGFETs are also made with both *p*-type and *n*-type channels.

transition elements A set of elements in the *periodic table in which filling of electrons in an inner *d*- or *f*-level occurs. With increasing proton number, electrons fill atomic levels up to argon, which has the electron configuration $1s^22s^22p^63s^23p^6$. In this shell, there are 5 *d*-orbitals, which can each contain 2 electrons. However, at this point the subshell of lowest energy is not the 3*d* but the 4*s*. The next two elements, potassium and calcium, have the configurations [Ar]4s^1 and [Ar]4s^2 respectively. For the next element, scandium, the 3*d* level is of lower energy than the 4*p* level, and scandium has the configuration

[Ar]$3d^14s^2$. This filling of the inner d-level continues up to zinc [Ar]$3d^{10}4s^2$, giving the first transition series. There is a further series of this type in the next period of the table: between yttrium ([Kr]$4d5s^2$) and cadmium ([Kr]$4d^{10}5s^2$). This is the second transition series. In the next period of the table the situation is rather more complicated. Lanthanum has the configuration [Xe]$5d^16s^2$. The level of lowest energy then becomes the $4f$ level and the next element, cerium, has the configuration [Xe]$4f^15d^16s^2$. There are 7 of these f-orbitals, each of which can contain 2 electrons, and filling of the f-levels continues up to lutetium ([Xe]$4f^{14}5d^16s^2$). Then the filling of the $5d$ levels continues from hafnium to mercury. The series of 14 elements from cerium to lutetium is a 'series within a series' called an *inner transition series*. This one is the *lanthanoid series. In the next period there is a similar inner transition series, the *actinoid series, from thorium to lawrencium. Then filling of the d-level continues from element 104 onwards.

In fact, the classification of chemical elements is valuable only in so far as it illustrates chemical behaviour, and it is conventional to use the term 'transition elements' in a more restricted sense. The elements in the inner transition series from cerium (58) to lutetium (71) are called the lanthanoids; those in the series from thorium (90) to lawrencium (103) are the actinoids. These two series together make up the f-block in the periodic table. It is also common to include scandium, yttrium, and lanthanum with the lanthanoids (because of chemical similarity) and to include actinium with the actinoids. Of the remaining transition elements, it is usual to speak of three *main transition series*: from titanium to copper; from zirconium to silver; and from hafnium to gold. All these elements have similar chemical properties that

result from the presence of unfilled d-orbitals in the element or (in the case of copper, silver, and gold) in the ions. The elements from 104 to 109 and the undiscovered elements 110 and 111 make up a fourth transition series. The elements zinc, cadmium, and mercury have filled d-orbitals both in the elements and in compounds, and are usually regarded as nontransition elements (*see* group II elements).

The elements of the three main transition series are all typical metals (in the nonchemical sense), i.e. most are strong hard materials that are good conductors of heat and electricity and have high melting and boiling points. Chemically, their behaviour depends on the existence of unfilled d-orbitals. They exhibit variable valency, have coloured compounds, and form *coordination compounds. Many of their compounds are paramagnetic as a result of the presence of unpaired electrons. Many of them are good catalysts. They are less reactive than the s-and p-block metals.

transition point (transition temperature) 1. The temperature at which one crystalline form of a substance changes to another form. **2.** The temperature at which a substance changes phase. **3.** The temperature at which a substance becomes superconducting (*see* superconductivity). **4.** The temperature at which some other change, such as a change of magnetic properties (*see also* Curie temperature), takes place.

transition state (activated complex) The association of atoms of highest energy formed during a chemical reaction. The transition state can be regarded as a short-lived intermediate that breaks down to give the products. For example, in a S_N2 substitution reaction, one atom or group approaches the molecule as the other leaves. The transition state is an intermediate state in which both

(a)

(b)

(c)

The stages of translation in protein synthesis

attacking and leaving groups are partly bound to the molecule, e.g.

B + RA → B---R---A → BR + A

In the theory of reaction rates, the reactants are assumed to be in equi-librium with this activated complex, which decomposes to the products.

transition zone *See* hypocotyl.

translation 1. (in biochemistry) The process in living cells in which the genetic information encoded in mes-senger *RNA (mRNA) in the form of a sequence of base triplets (*codons) is translated into a sequence of amino acids in a polypeptide chain during *protein synthesis (see illustration). Translation takes place on *ribosomes in the cell cytoplasm. The ribosomes move along the mRNA 'reading' each codon in turn. Molecules of transfer RNA (tRNA), each bearing a particu-lar amino acid, are brought to their correct positions along the mRNA molecule: base pairing occurs between the bases of the codons and the com-plementary base triplets of tRNA. In this way amino acids are assembled in the correct sequence to form the polypeptide chain. **2.** (in physics) Motion of a body in which all the points in the body follow parallel paths.

translocation The movement of minerals and chemical compounds within a plant. There are two main processes. The first is the uptake of soluble minerals from the soil and their passage upwards from the roots to various organs by means of the water-conducting vessels (*xylem). The second is the transfer of organic compounds, synthesized by the leaves, both upwards and downwards to vari-ous organs, particularly the growing points. This movement occurs within the *phloem tubes.

translucent Permitting the passage of radiation but not without some scat-tering or diffusion. For example, frosted glass allows light to pass through it but an object cannot be seen clearly through it because the light rays are scattered by it. *Compare* transparent.

transmission coefficient *See* trans-mittance.

transmission electron microscope
See electron microscope.

transmittance (transmission coefficient) The ratio of the energy of some form of radiation transmitted through a surface to the energy falling on it. The reciprocal of the transmittance is the *opacity*.

transmitter 1. The equipment used to generate and broadcast radio-frequency electromagnetic waves for communication purposes. It consists of a carrier-wave generator, a device for modulating the carrier wave in accordance with the information to be broadcast, amplifiers, and an aerial system. **2.** The part of a telephone system that converts sound into electrical signals.

transmutation The transformation of one element into another by bombardment of nuclei with particles. For example, plutonium is obtained by the neutron bombardment of uranium.

transparent Permitting the passage of radiation without significant deviation or absorption. *Compare* translucent. A substance may be transparent to radiation of one wavelength but not to radiation of another wavelength. For example, some forms of glass are transparent to light but not to ultraviolet radiation, while other forms of glass may be transparent to all visible radiation except red light. *See also* radiotransparent.

transpiration The loss of water vapour by plants to the atmosphere. It occurs mainly from the leaves through pores (stomata) whose primary function is gas exchange. The water is replaced by a continuous column of water moving upwards from the roots within the *xylem vessels. *See also* potometer.

transplantation *See* graft.

transport number Symbol *t*. The fraction of the total charge carried by a particular type of ion in the conduction of electricity through electrolytes.

transposon A mobile genetic element that can become integrated at many different sites along a chromosome. The simplest types of transposon are known as *insertion sequences*; these are well known in *E. coli* and other bacteria and comprise just one or two genes. Their presence can cause deletions, inversions, and other major rearrangements of adjacent genes. Certain bacterial *plasmids can also behave as transposons. Complex, or composite, transposons carry other genes, such as antibiotic resistance genes. Comparable elements in eukaryotes are called *transposable elements*. Maize, for example, has several families of such elements (also known as *controlling elements*). These can undergo transposition, causing chromatid breaks and modifying the action of neighbouring genes.

transuranic elements Elements with an atomic number greater than 92, i.e. elements above uranium in the *periodic table. Most of these elements are unstable and have short half-lives. *See also* transactinide elements.

transverse wave *See* wave.

travelling wave *See* wave.

Trematoda A class of parasitic flatworms (*see* Platyhelminthes) comprising the flukes, such as *Fasciola* (liver fluke). Flukes have suckers and hooks to anchor themselves to the host and their body surface is covered by a protective cuticle. The whole life cycle may either occur within one host or require one or more intermediate hosts to transmit the infective eggs or larvae. *Fasciola hepatica*, for example, undergoes larval development in a land snail (the intermediate host) and infects sheep (the primary host) when contaminated grass containing the larvae is swallowed.

triacylglycerol *See* triglyceride.

triangle of vectors A triangle constructed so that each of its sides represents one of three coplanar *vectors acting at a point with no resultant. If the triangle is completed, with the sides representing the vectors in both magnitude and direction, so that there are no gaps between the sides, then the vectors are in equilibrium. If the three vectors are forces, the figure is called a *triangle of forces*; if they are velocities, it is a *triangle of velocities*.

Triassic The earliest period of the Mesozoic era. It began about 225 million years ago, following the Permian, the last period of the Palaeozoic era, and extended until about 190 million years ago when it was succeeded by the Jurassic. It was named, by F. von Alberti in 1834, after the sequence of three divisions of strata that he studied in central Germany – Bunter, Muschelkalk, and Keuper. The Triassic rocks are frequently difficult to distinguish from the underlying Permian strata and the term *New Red Sandstone* is often applied to rocks of the Permo-Triassic. During the period marine animals diversified: molluscs were the dominant invertebrates – ammonites were abundant and bivalves replaced the declining brachiopods. Reptiles were the dominant vertebrates and included turtles, phytosaurs, dinosaurs, and the marine ichthyosaurs.

triatomic molecule A molecule formed from three atoms (e.g. H_2O or CO_2).

triazine *See* azine.

tribe A category used in the *classification of plants and animals that consists of several similar or closely related genera within a family. For example the Bambuseae, Oryzeae, Paniceae, and Aveneae are tribes of grasses.

triboelectricity *Static electricity produced as a result of friction.

tribology The study of friction, lubrication, and lubricants.

triboluminescence *Luminescence caused by friction; for example, some crystalline substances emit light when they are crushed as a result of static electric charges generated by the friction.

tribromomethane (bromoform) A colourless liquid *haloform, $CHBr_3$; r.d. 2.9; m.p. 8°C; b.p. 150°C.

tricarbon dioxide (carbon suboxide) A colourless gas, C_3O_2, with an unpleasant odour; r.d. 1.114 (liquid at 0°C); m.p. –111.3°C; b.p. 7°C. It is the acid anhydride of malonic acid, from which it can be prepared by dehydration using phosphorus(V) oxide. The molecule is linear (O:C:C:C:O).

tricarboxylic acid cycle *See* Krebs cycle.

trichloroethanal (chloral) A liquid aldehyde, CCl_3CHO; r.d. 1.51; m.p. –57.5°C; b.p. 97.8°C. It is made by chlorinating ethanal and used in making DDT. *See also* 2,2,2-trichloroethanediol.

2,2,2-trichloroethanediol (chloral hydrate) A colourless crystalline solid, $CCl_3CH(OH)_2$; r.d. 1.91; m.p. 57°C; b.p. 96.3°C. It is made by the hydrolysis of trichloroethanal and is unusual in having two –OH groups on the same carbon atom. Gem diols of this type are usually unstable; in this case the compound is stabilized by the presence of the three Cl atoms. It is used as a sedative.

trichloromethane (chloroform) A colourless volatile sweet-smelling liquid *haloform, $CHCl_3$; r.d. 1.48; m.p. –63.5°C; b.p. 61°C. It can be made by chlorination of methane (followed by separation of the mixture of products) or by the haloform reaction. It is an effective anaesthetic but can cause liver damage and it has now been replaced by other halogenated hydrocarbons. Chloroform is used as a solvent and raw material for making other compounds.

trichome A hairlike projection from a plant epidermal cell. Examples include root hairs and the stinging hairs of nettle leaves.

triclinic *See* crystal system.

tricuspid valve A valve, consisting of three flaps, situated between the right atrium and the right ventricle of the mammalian heart. When the right ventricle contracts, forcing blood into the pulmonary artery, the tricuspid valve closes the aperture to the atrium, thereby preventing any backflow of blood. The valve reopens to allow blood to flow from the atrium into the ventricle. *Compare* mitral valve.

tridymite A mineral form of *silicon(IV) oxide, SiO_2.

triglyceride (triacylglycerol) An ester of glycerol (propane-1,2,3-triol) in which all three hydroxyl groups are esterified with a fatty acid. Triglycerides are the major constituent of fats and oils and provide a concentrated food energy store in living organisms as well as cooking fats and oils, margarines, soaps, etc. Their physical and chemical properties depend on the nature of their constituent fatty acids. In *simple triglycerides* all three fatty acids are identical; in *mixed triglycerides* two or three different fatty acids are present.

trigonal bipyramid *See illustration at* complex.

trigonometric functions Functions defined in terms of a right-angled triangle (see diagram) and widely used in the solution of many mathematical problems. They are defined as:
tangent of angle A, written $\tan A = a/b$
sine of angle A, written $\sin A = a/c$

Trigonometric functions

cosine of angle A, written $\cos A = b/c$,
where a is the length of the side opposite the angle A, b is the length of the side opposite the angle B, and c is the hypotenuse of the triangle.
The reciprocal functions are:
cotangent of angle A, written $\cot A = 1/\tan A = b/a$
secant of angle A, written $\sec A = 1/\cos A = c/b$
cosecant of angle A, written $\csc A = 1/\sin A = c/a$.

trihydric alcohol *See* triol.

tri-iodomethane (iodoform) A yellow volatile solid sweet-smelling *haloform, CHI_3; r.d. 4.1; m.p. 115°C. It is made by the haloform reaction.

tri-iron tetroxide (ferrosoferric oxide) A black magnetic oxide, Fe_3O_4; r.d. 5.2. It is formed when iron is heated in steam and also occurs naturally as the mineral *magnetite. The oxide dissolves in acids to give a mixture of iron(II) and iron(III) salts.

trilobite An extinct marine arthropod belonging to the class Trilobita (some 4000 species), fossils of which are found in deposits dating from the Precambrian to the Permian period (590–280 million years ago). Trilobites were typically small (1–7 cm long); the oval flattened body comprised a head (covered by a semicircular dorsal shield) and a thorax and abdomen, which were protected by overlapping dorsal plates with a raised central part and flattened lateral portions, presenting a three-lobed appearance. The head bore a pair of antenna-like appendages and a pair of compound eyes; nearly all body segments bore a pair of Y-shaped (biramous) appendages – one branch for locomotion and the other fringed for respiratory exchange. Trilobites were bottom-dwelling scavengers.

trimethylaluminium   (aluminium trimethyl) A colourless liquid, $Al(CH_3)_3$, which ignites in air and reacts with water to give aluminium hydroxide and methane, usually with extreme vigour; r.d. 0.752; m.p. 0°C; b.p. 130°C. Like other aluminium alkyls it may be prepared by reacting a Grignard reagent with aluminium trichloride. Aluminium alkyls are used in the *Ziegler process for the manufacture of high-density polyethene (polythene).

2,4,6-trinitrophenol *See* picric acid.

trinitrotoluene (TNT) A yellow highly explosive crystalline solid, $CH_3C_6H_2(NO_2)_3$; r.d. 1.6; m.p. 81°C. It is made by nitrating toluene (methylbenzene), the systematic name being 1-methyl-2,4,6-trinitrobenzene.

triode A *thermionic valve with three electrodes. Electrons produced by the heated cathode flow to the anode after passing through the negatively biased *control grid. Small voltage fluctuations superimposed on the grid bias cause large fluctuations in the anode current. The triode was thus the first electronic device capable of amplification. Its role has now been taken over by the transistor.

triol (trihydric alcohol) An *alcohol containing three hydroxyl groups per molecule.

trioxoboric(III) acid *See* boric acid.

trioxosulphuric(IV) acid *See* sulphurous acid.

trioxygen *See* ozone.

triple bond *See* chemical bond.

triple point The temperature and pressure at which the vapour, liquid, and solid phases of a substance are in equilibrium. For water the triple point occurs at 273.16 K and 611.2 Pa. This value forms the basis of the definition of the *kelvin and the thermodynamic *temperature scale.

triploblastic Describing an animal having a body composed of three embryonic cell layers: the *ectoderm, *mesoderm, and *endoderm. Most multicellular animals are triploblastic; the coelenterates, which are *diploblastic, are an exception.

trisilane *See* silane.

trisodium   phosphate(V)   (sodium orthophosphate) A colourless crystalline compound, Na_3PO_4, soluble in water and insoluble in ethanol. It is known both as the decahydrate (octagonal; r.d. 2.54) and the dodecahydrate (trigonal; r.d. 1.62) The dodecahydrate loses water at about 76°C and the decahydrate melts at 100°C. Trisodium phosphate may be prepared by boiling sodium carbonate with the stoichiometric amount of phosphoric acid and subsequently adding sodium hydroxide to the disodium salt thus formed. It is useful as an additive for high-pressure boiler feed water (for removal of calcium and magnesium as phosphates), in emulsifiers, as a water-softening agent, and as a component in detergents and cleaning agents. Sodium phosphate labelled with the radioactive isotope ^{32}P is used in the study of the role of phosphate in biological processes and is also used (intravenously) in the treatment of polycythaemia.

tritiated compound *See* labelling.

tritium Symbol T. An isotope of hydrogen with mass number 3; i.e. the nucleus contains 2 neutrons and 1 proton. It is radioactive (half-life 12.3 years), undergoing beta decay to helium–3. Tritium is used in *labelling.

triton The nucleus of a tritium atom.

trivalent (tervalent) Having a valency of three.

trochanter 1. Any of several bony knobs on the femur of vertebrates to which muscles are attached. 2. The second segment of an insect's leg, between the *coxa and the *femur.

trona A mineral form of sodium sesquicarbonate, consisting of a mixed hydrated sodium carbonate and sodium hydrogencarbonate, Na_2CO_3.$NaHCO_3.2H_2O$.

trophic level The position that an organism occupies in a *food chain. For example, green plants (which obtain their energy directly from sunlight) are the primary *producers; herbivores are primary *consumers (and secondary producers). A carnivore that eats only herbivores is a secondary consumer and a tertiary producer. Many animals feed at several different trophic levels.

tropical year *See* year.

tropism The directional growth of a plant organ in response to an external stimulus, such as light, touch, or gravity. Growth towards the stimulus is a *positive tropism*; growth away from the stimulus is a *negative tropism*. *See also* geotropism; hydrotropism; orthotropism; phototropism; plagiotropism; thigmotropism. *Compare* nastic movements; taxis.

troposphere *See* earth's atmosphere.

tropylium ion The positive ion $C_7H_7^+$, having a ring of seven carbon atoms. The ion is symmetrical and has characteristic properties of *aromatic compounds.

trypsin An enzyme that digests proteins (*see* protease). It is secreted in an inactive form (*trypsinogen*) by the pancreas into the duodenum. There, trypsinogen is acted on by an enzyme (*enterokinase*) produced in the duodenum to yield trypsin. The active enzyme plays an important role in the digestion of proteins in the anterior portion of the small intestine.

trypsinogen *See* trypsin.

tryptophan *See* amino acid.

tuber A swollen underground stem or root in certain plants. It enables the plant to survive the winter or dry season and is also a means of propagation. A *stem tuber*, such as the potato, forms at the end of an underground stem. Each tuber represents several nodes and internodes. The following season several new plants develop from the terminal and axillary buds (eyes). *Root tubers*, such as those of the dahlia, are modified food-storing adventitious roots and may also give rise to new plants.

tungsten Symbol W. A white or grey metallic *transition element (formerly called *wolfram*); a.n. 74; r.a.m. 183.85; r.d. 19.3; m.p. 3410°C; b.p. 5660°C. It is found in a number of ores, including the oxides wolframite, $(Fe,Mn)WO_4$, and scheelite, $CaWO_4$. The ore is heated with concentrated sodium hydroxide solution to form a soluble *tungstate*. The oxide WO_3 is precipitated from this by adding acid, and is reduced to the metal using hydrogen. It is used in various alloys, especially high-speed steels (for cutting tools) and in lamp filaments. Tungsten forms a protective oxide in air and can be oxidized at high temperature. It does not dissolve in dilute acids. It forms compounds in which the oxidation state ranges from $+2$ to $+6$. The metal was first isolated by J. J. and F. d'Elhuyer in 1783.

tungsten carbide A black powder, WC, made by heating powdered tungsten metal with lamp black at 1600°C. It is extremely hard (9.5 on Mohs' scale) and is used in dies and cutting tools. A ditungsten carbide, W_2C, also exists.

stationary nozzle · rotor

Impulse turbine

nozzle attached to rotor · rotor

Reaction turbine

Principle of the turbine

tuning fork A metal two-pronged fork that when struck produces an almost pure tone of a predetermined frequency. It is used for tuning musical instruments and in experiments in acoustics.

tunnel diode (Esaki diode) A semiconductor diode, discovered in 1957 by L. Esaki (1925–), based on the *tunnel effect. It consists of a highly doped *p–n* semiconductor junction, which short circuits with negative bias and has negative resistance over part of its range when forward biased. Its fast speed of operation makes it a useful device in many electronic fields.

tunnel effect An effect in which electrons are able to tunnel through a narrow *potential barrier that would constitute a forbidden region if the electrons were treated as classical particles. That there is a finite probability of an electron tunnelling from one classically allowed region to another arises as a consequence of *quantum mechanics. The effect is made use of in the *tunnel diode. Alpha decay (*see* alpha particle) is an example of a tunnelling process.

Turbellaria A class of free-living flatworms (*see* Platyhelminthes) comprising the planarians, which occur in wet soils, fresh water, and marine environments. Their undersurface is covered with cilia, used for gliding over stones and weeds. Planarians can also swim by means of undulations of the body.

turbine A machine in which a fluid is used to produce rotational motion. The most widely used turbines are the *steam turbines* and *water turbines* that provide some 95% of the world's electric power (in the form of *turbogenerators) and the *gas turbines* that power all the world's jet-propelled aircraft. In the *impulse turbine* a high-pressure low-velocity fluid is expanded through stationary nozzles, producing low-pressure high-velocity jets, which are directed onto the blades of a rotor. The rotor blades reduce the speed of the jets and thus convert some of the fluid's kinetic energy into rotational kinetic energy of the rotor shaft. In the *reaction turbine* the discharge nozzles are themselves attached to the rotor. The acceleration of the fluid leaving the nozzles produces a force of reaction on the pipes, causing the rotor to move in the opposite direction to that of the fluid. (See illustrations.) Many turbines work on a combination of the impulse and reaction principles.

turbogenerator A steam turbine driving an electric generator. This is the normal method of generating electricity in power stations. In a conventional power station the steam is raised by burning a fossil fuel (coal, oil, or natural gas); in a nuclear power station the steam is raised by heat transfer from a nuclear reactor.

turbojet *See* jet propulsion.

turbulence A form of fluid flow in which the particles of the fluid move in a disordered manner in irregular paths, resulting in an exchange of momentum from one portion of a fluid to another. Turbulent flow takes over from *laminar flow when high values of the *Reynolds number are reached.

turgor The condition in a plant cell when its *vacuole is distended with water, pushing the protoplasm against the cell wall. In this condition the osmotic pressure forcing water in is balanced by pressure exerted by the cell wall. Turgidity assists in maintaining the rigidity of plants; a decrease in turgidity leads to wilting. *Compare* plasmolysis.

turion 1. A winter bud, covered with scale leaves and mucilage, that is produced by certain aquatic plants, such as frogbit. Turions become detached and remain dormant on the pond or lake bottom during the winter before developing into new plants the following season. **2.** *See* sucker.

Turner's syndrome A genetic disorder of women caused by the absence of the second *sex chromosome (such women are XO, rather than the normal XX). It is characterized by a lack of ovaries and menstrual cycle. Affected women are sterile and lack secondary sexual characteristics, although the external genitalia are present. The syndrome is named after the US endocrinologist H. H. Turner (1892–1970), who first described it.

turns ratio *See* transformer.

turpentine An oily liquid extracted from pine resin. It contains pinene, $C_{10}H_{16}$, and other terpenes and is mainly used as a solvent.

turquoise A mineral consisting of a hydrated phosphate of aluminium and copper, $CuAl_6(PO_4)_4(OH)_8.4H_2O$), that is prized as a semiprecious stone. It crystallizes in the triclinic system and is generally blue in colour, the 'robin's egg' blue variety being the most sought after. It usually occurs in veinlets and as masses and is formed by the action of surface waters on aluminium-rich rocks. The finest specimens are obtained from Iran.

T violation *See* time reversal.

tweeter A small loudspeaker capable of reproducing sounds of relatively high frequency, i.e. 5 kilohertz

upwards. In high-fidelity equipment a tweeter is used in conjunction with a *woofer.

tympanic cavity *See* middle ear.

tympanum (tympanic membrane; eardrum) The membrane that separates the *outer ear from the *middle ear. It vibrates in response to sound waves and transmits these vibrations via the *ear ossicles of the middle ear to the site of hearing (the *cochlea of the *inner ear). In amphibians and some reptiles there is no external ear and the tympanum is exposed at the skin surface.

Tyndall effect The scattering of light as it passes through a medium containing small particles. If a polychromatic beam of light is passed through a medium containing particles with diameters less than one-twentieth of the wavelength of the light, the scattered light appears blue. This accounts for the blue appearance of tobacco smoke. At higher particle diameters, the scattered light remains polychromatic. It is named after John Tyndall (1820–93). *See also* scattering of electromagnetic radiation.

type specimen The specimen used for naming and describing a *species or subspecies. If this is the original specimen collected by the author who named the species it is termed a *holotype*. The type specimen is not necessarily the most characteristic representative of the species. The term *type* is also applied to any taxon selected as being representative of the rank to which it belongs. For example, the genus *Solanum* (potato) is said to be the type genus of the family Solanaceae.

typhoon A tropical cyclone or *hurricane that occurs in the W and N Pacific Ocean and the South China Sea.

tyrosine *See* amino acid.

U

ubiquinone *See* coenzyme Q.

ulna The larger of the two bones in the forearm of vertebrates (*compare* radius). It articulates with the outer carpals at the wrist and with the humerus at the elbow.

ultracentrifuge A high-speed centrifuge used to measure the rate of sedimentation of colloidal particles or to separate macromolecules, such as proteins or nucleic acids, from solutions. Ultracentrifuges are electrically driven and are capable of speeds up to 60 000 rpm.

ultrahigh frequency (UHF) A radio frequency in the range $3 \times 10^9 - 0.3 \times 10^9$ Hz; i.e. having a wavelength in the range 10 cm to 1 m.

ultramicroscope A form of microscope that uses the *Tyndall effect to reveal the presence of particles that cannot be seen with a normal optical microscope. Colloidal particles, smoke particles, etc, are suspended in a liquid or gas in a cell with a black background and illuminated by an intense cone of light that enters the cell from the side and has its apex in the field of view. The particles then produce diffraction-ring systems, appearing as bright specks on the dark background.

ultramicrotome *See* microtome.

ultrasonics The study and use of pressure waves that have a frequency in excess of 20 000 Hz and are therefore inaudible to the human ear. *Ultrasonic generators* make use of the *piezoelectric effect, *ferroelectric materials, or *magnetostriction to act as transducers in converting electrical energy into mechanical energy. Ultrasonics are used in medicine for diagnosis, particularly in conditions such as pregnancy, in which X-rays could have a harmful effect, and for treatment, the vibratory effect being used to break up kidney stones, etc. Ultra-

sonic techniques are also used industrially to test for flaws in metals, to clean surfaces, to test the thickness of parts, and to form colloids.

ultrastructure The submicroscopic, almost molecular, structure of living cells, which is revealed by the use of an electron microscope.

ultraviolet microscope A *microscope that has quartz lenses and slides and uses *ultraviolet radiation as the illumination. The use of shorter wavelengths than the visible range enables the instrument to resolve smaller objects and to provide greater magnification than the normal optical microscope. The final image is either photographed or made visible by means of an *image converter.

ultraviolet radiation (UV) Electromagnetic radiation having wavelengths between that of violet light and long X-rays, i.e. between 400 nanometres and 4 nm. In the range 400–300 nm the radiation is known as the *near ultraviolet*. In the range 300–200 nm it is known as the *far ultraviolet*. Below 200 nm it is known as the *extreme ultraviolet* or the *vacuum ultraviolet*, as absorption by the oxygen in the air makes the use of evacuated apparatus essential. The sun is a strong emitter of UV radiation but only the near UV reaches the surface of the earth as the *ozone layer of the atmosphere absorbs all wavelengths below 290 nm.

Most UV radiation for practical use is produced by various types of *mercury-vapour lamps. Ordinary glass absorbs UV radiation and therefore lenses and prisms for use in the UV are made from quartz.

ultraviolet–visible spectroscopy (UV–visible spectroscopy) A technique for chemical analysis and the determination of structure. It is based on the principle that electronic transitions in molecules occur in the visible and ultraviolet regions of the electromagnetic spectrum, and that a given transition occurs at a characteristic wavelength. The spectrometer has two sources, one of ultraviolet and the other of white visible light, which together cover the whole wavelength range of the instrument. If the whole wavelength range is used, the source is changed over at the appropriate point. The radiation from the source is split into two beams of equal intensity. One beam is passed through a dilute solution of the sample while the other is passed through the pure solvent and is used as a reference against which the first is compared after transmittance. The cells used for the sample and reference solutions are usually made of silica and are matched.

umbel A type of *racemose inflorescence in which stalked flowers arise from the same point on the flower axis, resembling the spokes of an umbrella. An involucre (cluster) of bracts may occur at the point where the stalks emerge. This arrangement is characteristic of the family Umbelliferae (e.g. carrot, hogweed, parsley, parsnip), in which the inflorescence is usually a compound umbel.

umbilical cord The cord that connects the embryo to the *placenta in mammals. It contains a vein and two arteries that carry blood between the embryo and placenta. It is severed after birth to free the newly born animal from the placenta, and shrivels to leave a scar, the navel, on the animal.

umbra See shadow.

uncertainty principle (Heisenberg uncertainty principle; principle of indeterminism) The principle that it is not possible to know with unlimited accuracy both the position and momentum of a particle. This principle, discovered in 1927 by Werner Heisenberg (1901–76), is usually stated in the form: $\Delta x \Delta p_x \geqslant h/4\pi$, where Δx is the uncertainty in the x-coordinate of the particle, Δp_x is the uncertainty in the x-component of the particle's momentum, and h is the *Planck constant. An explanation of the uncertainty is that in order to locate a particle exactly, an observer must be able to bounce off it a photon of radiation; this act of location itself alters the position of the particle in an unpredictable way. To locate the position accurately, photons of short wavelength would have to be used. These would have associated large momenta and cause a large effect on the position. On the other hand, using long-wavelength photons would have less effect on the particle's position, but would be less accurate because of the longer wavelength. The principle has had a profound effect on scientific thought as it appears to upset the classical relationship between cause and effect at the atomic level.

underdamped See damping.

ungulate A herbivorous mammal with hoofed feet (see unguligrade). Ungulates are grouped into two orders: *Artiodactyla and *Perissodactyla.

unguligrade Describing the gait of ungulates (e.g. horses and cows), in which only the tips of the digits (i.e. the hooves) are on the ground and the rest of the foot is off the ground. Compare digitigrade; plantigrade.

uniaxial crystal A double-refracting crystal (see double refraction) having only one *optic axis.

unicellular Describing tissues, organs, or organisms consisting of a single cell. For example, the reproductive organs of some algae and fungi are unicellular. Unicellular organisms, e.g. bacteria, protozoans, and certain algae and fungi, are sometimes placed in a separate kingdom, the *Protista. Compare acellular; multicellular.

unified-field theory A comprehensive theory that would relate the electromagnetic, gravitational, strong, and weak interactions (*see* fundamental interactions) in one set of equations. In its original context the expression referred only to the unification of general *relativity and classical electromagnetic theory. No such theory has yet been found but some progress has been made in the unification of the electromagnetic and weak interactions (*see* electroweak theory).

Einstein attempted to derive *quantum mechanics from unified-field theory, but it is now thought that any unified-field theory has to start with quantum mechanics. Attempts to construct unified-field theories, such as *supergravity and *Kaluza–Klein theory, have run into great difficulties. At the present time it is not clear whether the framework of relativistic *quantum field theory is adequate to give a unified theory for all the known fundamental interactions and elementary particles, or whether one has to go to extended objects, such as superstrings or supermembranes.

Unified-field theories and other fundamental theories, such as *superstring theory and *supermembrane theory, are of great importance in understanding cosmology, particularly the *early universe. In turn cosmology puts constraints on unified-field theories. *See also* grand unified theory.

unimolecular reaction A chemical reaction or step involving only one molecule. An example is the decomposition of dinitrogen tetroxide:

$$N_2O_4 \rightarrow 2NO_2$$

Molecules colliding with other molecules acquire sufficient activation energy to react, and the activated complex only involves the atoms of a single molecule.

union *See* set theory.

unisexual Describing animals or plants with either male or female reproductive organs but not both. Most of the more advanced animals are unisexual but plants are often

*hermaphrodite. Flowers that contain either stamens or carpels but not both are also described as unisexual. *See also* monoecious; dioecious.

unit A specified measure of a physical quantity, such as length, mass, time, etc., specified multiples of which are used to express magnitudes of that physical quantity. For many scientific purposes previous systems of units have now been replaced by *SI units.

unit cell The group of particles (atoms, ions, or molecules) in a crystal that is repeated in three dimensions in the *crystal lattice. *See also* crystal system.

unit magnetic pole *See* magnetic poles.

univalent (monovalent) Having a valency of one.

universal constants *See* fundamental constants.

universal indicator A mixture of acid–base *indicators that changes colour (e.g. red-yellow-orange-green-blue) over a range of pH.

universal motor *See* electric motor.

universe All the matter, energy, and space that exists. *See* cosmology; heat death of the universe.

unnilquadium *See* transactinide elements.

unnilunium *See* mendelevium.

unsaturated 1. (of a compound) Having double or triple bonds in its molecules. Unsaturated compounds can undergo addition reactions as well as substitution. *Compare* saturated. **2.** (of a solution) *See* saturated.

unstable equilibrium *See* equilibrium.

upper atmosphere The upper part of the *earth's atmosphere above about 300 km. This is the part of the atmosphere that cannot be reached by balloons.

upthrust *See* Archimedes' principle.

UPVC Unplasticized PVC: a tough hardwearing form of PVC used for window frames and similar applications.

uracil A *pyrimidine derivative and one of the major component bases of *nucleotides and the nucleic acid *RNA.

uraninite A mineral form of uranium(IV) oxide, containing minute amounts of radium, thorium, polonium, lead, and helium. When uraninite occurs in a massive form with a pitchy lustre it is known as *pitchblende*, the chief ore of uranium. Uraninite occurs in Saxony (East Germany), Romania, Norway, the UK (Cornwall), E Africa (Zaîre), USA, and Canada (Great Bear Lake).

uranium Symbol U. A white radioactive metallic element belonging to the *actinoids; a.n. 92; r.a.m. 238.03; r.d. 19.05 (20°C); m.p. $1132\pm1°C$; b.p. 3818°C. It occurs as *uraninite, from which the metal is extracted by an ion-exchange process. Three isotopes are found in nature: uranium–238 (99.28%), uranium–235 (0.71%), and uranium–234 (0.006%). As uranium–235 undergoes *nuclear fission with slow neutrons it is the fuel used in *nuclear reactors and *nuclear weapons; uranium has therefore assumed enormous technical and political importance since their invention. It was discovered by M. H. Klaproth in 1789.

uranium(VI) fluoride (uranium hexafluoride) A volatile white solid, UF_6; r.d. 5.1; m.p. 64.5°C. It is used in the separation of uranium isotopes by gas diffusion.

uranium hexafluoride See uranium(VI) fluoride.

uranium–lead dating A group of methods of *dating certain rocks that depends on the decay of the radioisotopes uranium–238 to lead–206 (half-life 4.5×10^9 years) or the decay of uranium–235 to lead–207 (half-life 7.1×10^8 years). One form of uranium–lead dating depends on measuring the ratio of the amount of helium trapped in the rock to the amount of uranium present (since the decay $^{238}U \rightarrow ^{206}Pb$ releases eight alpha-particles). Another method of calculating the age of the rocks is to measure the ratio of radiogenic lead (^{206}Pb, ^{207}Pb, and ^{208}Pb) present to nonradiogenic lead (^{204}Pb). These methods give reliable results for ages of the order 10^7–10^9 years.

uranium(IV) oxide A black solid, UO_2; r.d. 10.9; m.p. 3000°C. It occurs naturally as *uraninite and is used in nuclear reactors.

uranium series See radioactive series.

Uranus A planet having its orbit between Saturn and Neptune. Its mean distance from the sun is 2869.6 $\times 10^6$ km and its mean diameter is 51 800 km; it has a *sidereal period of 84.01 years. The equator of Uranus is tilted at 98° with respect to its orbit, so that each pole is almost facing the sun during part of each orbit. The temperature of Uranus is very low, about 35 K, and its atmosphere is thought to contain methane and hydrogen. The planet itself is believed to have an ice mantle some 8000 km thick surrounding a rocky core. The planet has 15 known satellites and a system of about 20 rings, nine of which were discovered in 1977 with the rest being photographed in 1986 by the US Voyager 2 probe.

urea (carbamide) A white crystalline solid, $CO(NH_2)_2$; r.d. 1.3; m.p. 133°C. It is soluble in water but insoluble in certain organic solvents. Urea is the major end product of nitrogen excretion in mammals, being synthesized by the *urea cycle. Urea is synthesized industrially from ammonia and carbon dioxide for use in *urea–formaldehyde resins and pharmaceuticals, as a source of nonprotein nitrogen for ruminant live-stock, and as a nitrogen fertilizer.

urea cycle (ornithine cycle) The series of biochemical reactions that converts ammonia to *urea during the excretion of metabolic nitrogen. Urea formation occurs in mammals and, to a lesser extent, in some other animals. The liver converts ammonia to the much less toxic urea, which is excreted in solution in *urine.

urea–formaldehyde resins Synthetic resins made by copolymerizing urea with formaldehyde (methanal). They are used as adhesives or thermosetting plastics.

ureter The duct in vertebrates that conveys urine from the *kidney to the *bladder.

urethane resins (polyurethanes) Synthetic resins containing the repeating group –NH–CO–O– There are numerous types made by copolymerizing isocyanate esters with polyhydric alcohols. They have a variety of uses in plastics, paints, and solid foams.

urethra The duct in mammals that conveys urine from the *bladder to be discharged to the outside of the body. In males the urethra passes through the penis and is joined by the *vas deferens, it therefore also serves as a channel for sperm.

uric acid The end product of purine breakdown in most mammals, birds, terrestrial reptiles, and insects and also (except in mammals; *see* urea) the major form in which metabolic nitrogen is excreted. Being fairly insoluble, uric acid can be expelled in solid form, which conserves valuable water in arid environments. The accumulation of uric acid in the synovial fluid of joints causes gout.

uridine A nucleoside consisting of one uracil molecule linked to a D-ribose sugar molecule. The derived nucleotide uridine diphosphate (UDP) is important in carbohydrate metabolism.

urine The aqueous fluid formed by the excretory organs of animals for the removal of metabolic waste products. In higher animals, urine is produced by the *kidneys, stored in the *bladder, and excreted through the *urethra or *cloaca. Apart from water, the major constituents of urine are one or more of the end products of nitrogen metabolism – ammonia, urea, uric acid, and creatinine. It may also contain various inorganic ions, the pigments urochrome and urobilin, amino acids, and purines. Precise composition depends on many factors, especially the habitat of a particular species: aquatic animals produce copious volumes; terrestrial animals need to conserve water and produce much less (about 1.0–1.5 litres per day in humans).

uriniferous tubule *See* nephron.

uterus (womb) The organ of female mammals in which the embryo develops. Paired in most mammals but single in humans, it is situated between the bladder and rectum and is connected to the *fallopian tubes and to the *vagina. The lining shows cyclical changes (*see* menstrual cycle; oestrous cycle) associated with egg production and provides a thick spongy layer in which the fertilized egg becomes embedded. The outer wall of the uterus is thick and muscular; by contracting, it forces the fully grown fetus through the vagina to the outside.

utriculus (utricle) A chamber of the *inner ear from which the *semicircular canals arise. It bears patches of sensory epithelium concerned with detecting changes in the direction and speed of movement (*see* macula).

UV *See* ultraviolet radiation.

UV–visible spectroscopy *See* ultraviolet–visible spectroscopy.

V

vacancy *See* defect.

vaccination *See* immunization.

vaccine A liquid preparation of treated disease-producing microorganisms or their products used to stimulate an *immune response in the body and so confer resistance to the disease (*see* immunization). Vaccines are administered orally or by injection (*inoculation*). They take the form of dead viruses or bacteria that can still act as antigens, live but weakened microorganisms (*see* attenuation), specially treated *toxins, or antigenic extracts of the microorganism.

vacuole A space within the cytoplasm of a living *cell that is filled with air, water or other liquid, sap, or food particles. In plant cells there is usually one large vacuole; animal cells usually have several small ones. *See also* contractile vacuole.

vacuum A space in which there is a low pressure of gas, i.e. relatively few atoms or molecules. A *perfect vacuum* would contain no atoms or molecules, but this is unobtainable as all the materials that surround such a space have a finite *vapour pressure. In a *soft* (or *low*) *vacuum* the pressure is reduced to about 10^{-2} pascal, whereas a *hard* (or *high*) vacuum has a pressure of $10^{-2}-10^{-7}$ pascal. Below 10^{-7} pascal is known as an *ultrahigh vacuum*. *See also* vacuum pump.

vacuum distillation Distillation under reduced pressure. The depression in the boiling point of the substance distilled means that the temperature is lower, which may prevent the substance from decomposing.

vacuum pump A pump used to reduce the gas pressure in a container. The normal laboratory rotary oil-seal pump can maintain a pressure of 10^{-1} Pa. For pressures down to 10^{-7} Pa a *diffusion pump is required. *Ion pumps can achieve a pressure of 10^{-9} Pa and a *cryogenic pump combined with a diffusion pump can reach 10^{-13} Pa.

vacuum state The ground state in a relativistic *quantum field theory. A vacuum state does not mean a state of nothing. Because one is dealing with *quantum mechanics, the vacuum state has a *zero-point energy, which gives rise to *vacuum fluctuations*. The existence of vacuum fluctuations has observable consequences in *quantum electrodynamics.

vacuum tube *See* thermionic valve.

vagina The tube leading from the uterus to the outside. Sperm are deposited in the vagina during copulation and the fully developed fetus is born through it. In a number of mammals the vagina may be sealed when the animal is not sexually receptive and only open during oestrus. Its lining produces mucus, which prevents friction and the entry of infective organisms.

vagus nerve The tenth *cranial nerve: a paired nerve that supplies branches to many major internal organs. It carries motor nerve fibres to the heart, lungs, and viscera and sensory fibres from the viscera.

valence *See* valency.

valence band *See* energy bands.

valence electron An electron in one of the outer shells of an atom that takes part in forming chemical bonds.

valency (valence) The combining power of an atom or radical, equal to the number of hydrogen atoms that the atom could combine with or displace in a chemical compound (hydrogen has a valency of 1). It is equal to the ionic charge in ionic compounds; for example, in Na_2S, sodium has a valency of 1 (Na^+) and sulphur a valency of 2 (S^{2-}). In covalent compounds it is equal to the number of bonds formed; in CO_2

oxygen has a valency of 2 and carbon has a valency of 4.

valine *See* amino acid.

valve 1. (in anatomy) Any of various structures for restricting the flow of a fluid through an aperture or along a tube to one direction. Valves in the heart (*see* mitral valve; tricuspid valve), veins, and lymphatic vessels consist of two or three flaps of tissue (*cusps*) fastened to the walls. The cusps are flattened to the walls to allow the normal passage of blood or lymph, but a reverse flow causes them to block the vessel or aperture, so preventing further backflow. 2. (in botany) **a.** Any of the parts that make up a capsule or other dry fruit that sheds its seeds. **b.** One of the two halves of the cell wall of a diatom. 3. (in electronics) *See* thermionic valve.

vanadium Symbol V. A silvery-white metallic *transition element; a.n. 23; r.a.m. 50.94; r.d. 6.1; m.p. 1890°C; b.p. 3380°C. It occurs in a number of complex ores, including vanadinite ($Pb_5Cl(VO_4)_3$) and carnotite ($K_2(ClO_2)_2(VO_4)_2$). The pure metal can be obtained by reducing the oxide with calcium. The element is used in a large number of alloy steels. Chemically, it reacts with nonmetals at high temperatures but is not affected by hydrochloric acid or alkalis. It forms a range of complexes with oxidation states from $+2$ to $+5$. Vanadium was discovered in 1801 by del Rio, who allowed himself to be persuaded that what he had discovered was an impure form of chromium. The element was rediscovered and named by Sefström in 1880.

vanadium(V) oxide (vanadium pentoxide) A crystalline compound, V_2O_5, used extensively as a catalyst in industrial gas-phase oxidation processes.

vanadium pentoxide *See* vanadium(V) oxide.

Van Allen belts

Van Allen belts (radiation belts) Belts that are sources of intense radiation surrounding the earth, consisting of high-energy charged particles trapped in the earth's magnetic field within which they follow roughly helical paths. They were discovered in 1958 by James Van Allen (1914–) as a result of radiation detectors carried by Explorer satellites. The lower belt, extending from 1000 to 5000 km above the equator, contains electrons and protons, while the upper belt, 15 000–25 000 km above the equator, contains mainly electrons (see illustration).

Van de Graaff generator An electrostatic generator used to produce a high voltage, usually in the megavolt range. It consists of a large metal dome-shaped terminal mounted on a hollow insulating support. An endless insulating belt runs through the support from the base to a pulley within the spherical terminal. In the original type, charge is sprayed by point discharge from metal needles, held at a potential of about 10 kV, on to the bottom of the belt. A row of needles near the upper belt pulley removes the charge from the belt and passes it to the outer surface of the spherical terminal. The voltage achieved by the device is proportional to the radius of the spherical terminal. A typical device with a terminal having a radius of 1 m will produce about 1 MV. However, terminals can be made smaller, for a given voltage, by enclosing the apparatus in nitrogen at

van der Waals' equation

a pressure of 10–20 atmospheres (1–2 MPa) to reduce sparking. Generators having a positive-ion source are fitted with an evacuated tube through which the particles can be accelerated for research purposes. Machines having an electron source are used for various medical and industrial purposes. The generator was invented by R. J. Van de Graaff (1901–67).

Modern patterns of the generator have a chainlike belt of alternate links of metal and insulator. The metal links are charged by contact with a metal pulley, and discharge to the dome in the same way. This permits much higher current drain that the point discharge.

van der Waals' equation *See* equation of state.

van der Waals' force An attractive force between atoms or molecules, named after J. D. van der Waals (1837–1923). The force accounts for the term a/V^2 in the van der Waals equation (*see* equation of state). These forces are much weaker than those arising from valence bonds and are inversely proportional to the seventh power of the distance between the atoms or molecules. They are the forces responsible for nonideal behaviour of gases and for the lattice energy of molecular crystals. There are three factors causing such forces: (1) dipole–dipole interaction, i.e. electrostatic attractions between two molecules with permanent dipole moments; (2) dipole-induced dipole interactions, in which the dipole of one molecule polarizes a neighbouring molecule; (3) dispersion forces arising because of small instantaneous dipoles in atoms.

van't Hoff factor Symbol i. A factor appearing in equations for *collative properties, equal to the ratio of the number of actual particles present to the number of undissociated particles. It was first suggested by Jacobus van't Hoff (1852–1911).

van't Hoff's isochore An equation for the variation of equilibrium constant with temperature

$$(d \log_e K)/dT = \Delta H/RT^2$$

where T is the thermodynamic temperature and ΔH the enthalpy of the reaction.

vapour density The density of a gas or vapour relative to hydrogen, oxygen, or air. Taking hydrogen as the reference substance, the vapour density is the ratio of the mass of a particular volume of a gas to the mass of an equal volume of hydrogen under identical conditions of pressure and temperature. Taking the density of hydrogen as 1, this ratio is equal to half the relative molecular mass of the gas.

vapour pressure The pressure exerted by a vapour. All solids and liquids give off vapours, consisting of atoms or molecules of the substances that have evaporated from the condensed forms. These atoms or molecules exert a vapour pressure. If the substance is in an enclosed space, the vapour pressure will reach an equilibrium value that depends only on the nature of the substance and the temperature. This equilibrium value occurs when there is a dynamic equilibrium between the atoms or molecules escaping from the liquid or solid and those that strike the surface of the liquid or solid and return to it. The vapour is then said to be a *saturated vapour* and the pressure it exerts is the *saturated vapour pressure*.

variation 1. (in biology) The differences between individuals of a plant or animal species. Variation may be the result of environmental conditions; for example, water supply and light intensity affect the height and leaf size of a plant. Differences of this kind, acquired during the lifetime of an individual, are not transmitted to succeeding generations since the genes are not affected. Variations due to differences in genetic constitution are inherited (*see* continuous variation; discontinuous variation). Wide genetic variation improves the ability of a species to survive in a changing environment, since the chances that some individuals will tolerate a particular change are increased. Such individuals will survive and transmit the advantageous genes to their offspring. **2.** (in magnetism) *See* geomagnetism.

variety A category used in the *classification of plants and animals below the *species level. A variety consists of a group of individuals that differ distinctly from but can interbreed with other varieties of the same species. The characteristics of a variety are genetically inherited. Examples of varieties include breeds of domestic animals and the *races of man. *See also* cultivar. *Compare* subspecies.

variometer 1. A variable inductor consisting of two coils connected in series and able to move relative to each other. It can be used to measure inductance. **2.** Any of several devices for detecting and measuring changes in the geomagnetic elements (*see* geomagnetism).

varve dating (geochronology) An absolute *dating technique using thin sedimentary layers of clays called *varves*. The varves, which are particularly common in Scandinavia, have alternate light and dark bands corresponding to winter and summer deposition. Most of them are found in the Pleistocene series, where the edges of varve deposits can be correlated with the annual retreat of the ice sheet, although some varve formation is taking place in the present day. By counting varves it is possible to establish an absolute time scale for fossils up to about 20 000 years ago.

vascular bundle (fascicle) A long continuous strand of conducting (vascular) tissue in seed plants and pteridophytes that extends from the roots through the stem and into the leaves.

It consists of *xylem and *phloem, which are separated by a *cambium in plants that undergo secondary thickening. *See* vascular tissue.

vascular cambium *See* cambium.

vascular plants All plants possessing organized *vascular tissue. They may be grouped in a single division, the *Tracheophyta, or separated into two divisions, the *Pteridophyta and *Spermatophyta.

vascular system 1. A specialized network of vessels for the circulation of fluids throughout the body tissues of an animal. All animals, apart from protozoans and other simple invertebrate groups, possess a *blood vascular system*, which enables the passage of respiratory gases, nutrients, excretory products, and other metabolites into and out of the cells. In vertebrates it consists of a muscular *heart, which pumps blood through major blood vessels (*arteries) into increasingly finer branches until in the *capillaries it is in intimate contact with tissues. It then returns to the heart via another network of vessels (the *veins). This circulation also enables a stable *internal environment for tissue function (*see* homeostasis), the transmission of chemical messengers (*hormones) around the body, and a means of defending the body against pathogens and damage via the immune system. See illustration.

A *water vascular system* is characteristic of the *Echinodermata.

2. The system of *vascular tissue in plants.

vascular tissue (vascular system) The tissue that conducts water and nutrients through the plant body in higher plants (ferns and seed plants). It consists of *xylem and *phloem. Since the xylem and phloem tissues are always in close proximity to each other, distinct regions of vascular tissue can be identified (*see* vascular bundle). The possession of vascular

Blood vascular system of a mammal

tissue has enabled the higher plants to attain a considerable size and dominate most terrestrial habitats.

vas deferens One of a pair of ducts carrying sperm from the testis (or *epididymis) to the outside, in mammals through the *urethra.

vas efferens Any of various small ducts carrying sperm. In reptiles, birds, and mammals they convey sperm from the seminiferous tubules of the testis to the *epididymis; in invertebrates they carry sperm from the testis to the vas deferens.

vasoactive intestinal peptide *See* VIP.

vasoconstriction The reduction in the internal diameter of blood vessels, especially arterioles or capillaries. The constriction of arterioles is mediated by the action of nerves on the smooth muscle fibres of the arteriole walls and results in an increase in blood pressure.

vasodilation (vasodilatation) The increase in the internal diameter of blood vessels, especially arterioles or capillaries. The vasodilation of arterioles is mediated by the action of nerves on the smooth muscle fibres of the arteriole walls and results in a decrease in blood pressure.

vasomotor nerves The nerves of the *autonomic nervous system that control the diameter of blood vessels. *Vasoconstrictor nerves* decrease the diameter (*see* vasoconstriction); *vasodilator nerves* increase it (*see* vasodilation).

vasopressin (antidiuretic hormone; ADH) A hormone, secreted by the posterior *pituitary gland, that stimulates absorption of water by the kidneys and thus controls the concentration of body fluids. Vasopressin is produced by specialized nerve cells in the hypothalamus of the brain and is transported to the posterior pituitary in the bloodstream. Deficiency of ADH results in a disorder known as *diabetes insipidus*, in which large volumes of urine are excreted; it is treated by administration of natural or synthetic hormone.

vector 1. (in mathematics) A quantity in which both the magnitude and the direction must be stated (*compare* scalar quantity). Force, velocity, and field strength are examples of vector quantities. Note that distance and speed are scalar quantities, whereas displacement and velocity are vector quantities. Vector quantities must be treated by *vector algebra*, for example, the resultant of two vectors may be found by a *parallelogram of vectors. A (three-dimensional) vector V may be written in terms of components V_1, V_2, and V_3 along the x, y, and z axes (say) as $V_1i + V_2j + V_3k$, where i, j, and k are *unit vectors* (i.e. vectors of unit length) along the x, y, and z axes. *See also* triangle of vectors. **2.** (in medicine) An animal, usually an insect, that passively transmits disease-causing microorganisms from one animal or plant to another or from an animal to man. *Compare* carrier. **3.** (in genetics) (**cloning vector**) A vehicle used in *gene cloning to insert a foreign DNA fragment into the genome of a host cell. For bacterial hosts three different types of vector are used: *bacteriophages, *plasmids, and their hybrid derivatives, *cosmids. The foreign DNA is spliced into the vector using specific *restriction enzymes and *ligases to cleave the vector DNA and join the foreign DNA to the two ends created (*insertional vectors*). In some phage vectors, part of the viral genome is enzymically removed and replaced with the foreign DNA (*replacement vectors*). *Retroviruses can be effective vectors for introducing recombinant DNA into mammalian cells. In plants, derivatives of the tumour-inducing (Ti) plasmid of the crown gall bacterium, *Agrobacterium tumefaciens*, are used as vectors.

vector product (cross product) The product of two *vectors U and V, with components U_1, U_2, U_3 and V_1, V_2, V_3, respectively, given by:
$$U \times V = (U_2V_3 - U_3V_2)i + (U_3V_1 - U_1V_3)j + (U_1V_2 - U_2V_1)k.$$
It is itself a vector, perpendicular to both U and V, and of length $UV\sin\theta$, where U and V are the lengths of U and V, respectively, and θ is the angle between them. *Compare* scalar product.

vegetative propagation (vegetative reproduction) 1. A form of *asexual reproduction in plants whereby new individuals develop from specialized multicellular structures (e.g. *tubers, *bulbs) that become detached from the parent plant. Examples are the production of strawberry plants from *runners and of gladioli from daughter *corms. Artificial methods of vegetative propagation include grafting (*see* graft), *budding, and making *cuttings. **2.** Asexual reproduction in animals, e.g. budding in *Hydra*.

vein 1. A blood vessel that carries blood towards the heart. Most veins carry deoxygenated blood (the *pulmonary vein is an exception). The largest veins are fed by smaller ones, which are formed by the merger of *venules. Veins have thin walls and a relatively large internal diameter. *Valves within the veins ensure that the flow of blood is always towards the heart. *Compare* artery. **2.** A vascular bundle in a leaf (*see* venation). **3.** Any of the tubes of chitin that strengthen an insect's wing.

velamen A whitish spongy sheath of dead empty cells that surrounds the aerial roots of epiphytic plants, such as certain orchids. It absorbs any surface water on the roots.

velocity Symbol *v*. The rate of displacement of a body. It is the *speed of a body in a specified direction. Velocity is thus a *vector quantity, whereas speed is a scalar quantity.

velocity modulation *See* klystron.

velocity ratio (distance ratio) The ratio of the distance moved by the point of application effort in a simple *machine to the distance moved by the point of application load in the same time.

velum *See* annulus.

vena cava Either of the two large veins that carry deoxygenated blood into the right atrium of the heart. The *precaval vein* (*anterior* or *superior vena cava*) receives blood from the head and forelimbs; the *postcaval vein* (*posterior* or *inferior vena cava*) drains blood from the trunk and hindlimbs.

venation 1. The arrangement of veins (vascular bundles) in a leaf. The leaves of dicotyledons have a central main vein (midrib) with side branches that themselves further subdivide to form a network (*net* or *reticulate venation*). The leaves of monocotyledons have parallel veins (*parallel venation*). **2.** The arrangement of the veins in an insect's wing, which is often important in classification.

Venn diagram *See* set theory.

ventral Describing the surface of a plant or animal that is nearest or next to the ground or other support, i.e. the lower surface. In bipedal animals, such as man, it is the forward-directed (*anterior) surface. *Compare* dorsal.

ventricle 1. A chamber of the *heart that receives blood from an *atrium and pumps it into the arterial system. Amphibians and fish have a single ventricle, but mammals, birds, and reptiles have two, pumping deoxygenated blood to the lungs and oxygenated blood to the rest of the body, respectively. **2.** Any of the four linked fluid-filled cavities in the brain of vertebrates. One of these cavities is in the *medulla oblongata, two are in the cerebral hemispheres (*see* cerebrum), and the fourth is in the posterior part of the *forebrain. The ventricles contain cerebrospinal fluid filtered from the blood by the *choroid plexus.

Venturi tube A device for mixing a fine spray of liquid with a gas or measuring a flow rate of a gas. It consists of two tapered sections of pipe joined by a narrow throat. The fluid velocity in the throat is increased and the pressure is therefore reduced. By attaching manometers to the three sections of the tube, the pressure drop can be measured and the flow rate through the throat can be calculated. In a carburettor, the petrol from the float chamber is made into a fine spray by being drawn through a jet into the low pressure in the throat of a Venturi tube, where it mixes with the air being drawn into the engine. The device was invented by G. B. Venturi (1746–1822).

venule A small blood vessel that receives blood from the capillaries and transports it to a vein.

Venus A planet having its orbit between Mercury and the earth. Its mean distance from the sun is 108 × 10^6 km and its mean diameter 12 100 km. It has a *sidereal period of 224.7 days. US Mariner and Pioneer spacecraft and Soviet Venera spacecraft have provided a considerable amount of information about the planet. The atmosphere is 98% carbon dioxide and most of the remainder is nitrogen. There is a dense layer of cloud about 20 km thick covering the whole of the planet. This cloud layer consists largely of concentrated sulphuric acid. Some photographs of the surface have been obtained by a Venera spacecraft that functioned for a short time after landing on Venus, but the dim sunlight revealed only the presence of many rocks rounded by windblown sand. The high surface temperature of 730 K is a result of the *greenhouse effect, not proximity to the sun. The atmospheric pressure on the surface of Venus is about 90 times that on earth. Venus is almost unique among the planets in rotating in the opposite direction to the earth's rotation. It also rotates very slowly with a period of 243 days.

Vernier scale

verdigris A green patina of basic copper salts formed on copper. The composition of verdigris varies depending on the atmospheric conditions, but includes the basic carbonate $CuCO_3.Cu(OH)_2$, the basic sulphate $CuSO_4.Cu(OH)_2.H_2O$, and in some cases the basic chloride $CuCl_2.Cu(OH)_2$.

vermiculite *See* clay minerals.

vermiform appendix *See* appendix.

vernalization The application of a cold treatment to germinating seeds or seedlings to ensure flowering. Many plants, including winter cereals, will not flower unless subjected to a period of chilling early in their development. Winter cereals are therefore sown in the autumn for flowering the following year. However, if germinating seeds are artificially vernalized they can be sown in the spring for flowering the same year.

vernier A short auxiliary scale placed beside the main scale on a measuring instrument to enable subdivisions of the main scale to be read accurately. The vernier scale is usually calibrated so that each of its divisions is 0.9 of the main scale divisions. The zero on the vernier scale is set to the observed measurement on the main scale and by noting which division on the vernier scale is exactly in line with a main scale division, the second decimal place of the measurement is obtained (see illustration). The device was invented by Pierre Vernier (1580 –1637) in about 1630.

vertebra Any of the bones that make up the *vertebral column. In mammals each vertebra typically consists of a main body, or *centrum*, from which arises a *neural arch* through which the spinal cord passes, and *transverse processes* projecting from the side. There are five groups of vertebrae, specialized for various functions and varying in number with the species. In man, for example, there are 7 *cervical vertebrae, 12 *thoracic vertebrae, 5 *lumbar vertebrae, 5 fused *sacral vertebrae, and 5 fused *caudal vertebrae (forming the *coccyx).

vertebral column (backbone; spinal column; spine) A flexible bony column in vertebrates that extends down the long axis of the body and provides the main skeletal support. It also encloses and protects the *spinal cord and provides attachment for the muscles of the back. The vertebral column consists of a series of bones (*see* vertebra) separated by discs of cartilage (*intervertebral discs*). It articulates with the skull by means of the *atlas vertebra, with the ribs at the *thoracic vertebrae, and with the pelvic girdle at the sacrum (*see* sacral vertebrae).

Vertebrata (Craniata) The largest subphylum of the *Chordata, comprising all those animals with backbones (*see* vertebral column). The replacement of the rigid notochord by the more flexible backbone has permitted the vertebrates a greater degree of movement and subsequent improvement in the sense organs and enlargement of the brain. The brain is enclosed in a skeletal case, the cranium. There are seven extant classes: *Agnatha (jawless fishes), *Elasmobranchii (cartilaginous fishes), *Osteichthyes (bony fishes), *Amphibia, *Reptilia, *Aves (birds), and *Mammalia.

very high frequency (VHF) A radio frequency in the range 3×10^8 – 0.3×10^8 Hz, i.e. having a wavelength in the range 1–10 m.

very low frequency (VLF) A radio frequency in the range 3×10^4 – 0.3×10^4 Hz, i.e. having a wavelength in the range 10–100 km.

vesicle A small, usually fluid-filled, membrane-bound sac within the cytoplasm of a living cell. Vesicles form part of the *Golgi apparatus.

vessel 1. (in botany) A tube within the *xylem composed of joined *vessel elements. Vessels facilitate the efficient movement of water from the roots to the shoots and leaves of a plant. **2.** (in zoology) Any of various tubular structures through which sub-stances are transported, especially a blood vessel or a lymphatic vessel.

vessel element A type of cell occurring within the *xylem of flowering plants, many of which, end to end, form water-conducting *vessels*. Vessel elements are frequently very broad and have side walls thickened by deposits of lignin over most of the surface area. However, the end walls are broken down to provide connections with the cells both above and below them. *Compare* tracheid.

vestigial organ Any part of an organism that has diminished in size during its evolution because the function it served decreased in importance or became totally unnecessary. Examples are the human appendix and the wings of the ostrich.

vibrio Any comma-shaped bacterium. Generally, vibrios are Gram-negative (*see* Gram's stain), motile, and aerobic. They are widely distributed in soil and water and while most feed on dead organic matter some are parasitic, e.g. *Vibrio cholerae*, the causal agent of cholera.

vicinal (vic) Designating a molecule in which two atoms or groups are linked to adjacent atoms. For example, 1,2-dichloroethane (CH_2ClCH_2Cl) is a vicinal (or vic) dihalide and can be named *vic*-dichloroethane.

Victor Meyer's method A method of measuring vapour density, devised by Victor Meyer (1848–97). A weighed sample in a small tube is dropped into a heated bulb with a long neck. The sample vaporizes and displaces air, which is collected over water and the volume measured. The vapour density can then be calculated.

villiaumite A mineral form of sodium fluoride, NaF.

villus A microscopic outgrowth from the surface of some tissues and organs, which serves to increase the surface area of the organ. Numerous villi line the interior of the small intestine. Their shape may vary from finger-like (in the *duodenum) to spadelike (in the *ileum). Intestinal villi are specialized for the absorption of soluble food material: each contains blood vessels and a lymph vessel (*see* lacteal).

Chorionic villi occur on the chorion of the mammalian placenta, where they increase the surface area for the exchange of materials between the fetal and maternal blood.

vinyl acetate *See* ethenyl ethanoate.

vinyl chloride *See* chloroethene.

vinyl group The organic group $CH_2:CH-$.

VIP (vasoactive intestinal peptide) A peptide hormone secreted by endocrine cells of the upper part of the small intestine in response to the entry of partially digested food from the stomach. VIP, along with *secretin, stimulates the pancreas to produce a thin watery secretion containing bicarbonate. This raises the pH in the intestine in preparation for secretion of pancreatic enzymes. VIP also occurs as a neuropeptide in central nervous tissue.

virial equation A gas law that attempts to account for the behaviour of real gases, as opposed to an ideal gas. It takes the form

$$pV = RT + Bp + Cp^2 + Dp^3 + \ldots,$$

where B, C, and D are known as *virial coefficients*.

virion *See* virus.

virology The scientific study of *viruses. *See* microbiology.

virtual image *See* image.

virtual state The state of the *virtual particles* that are exchanged between two interacting charged particles. These particles, called *photons, are not in the real state, i.e. directly observable; they are constructs to enable the phenomenon to be

explained in terms of *quantum mechanics.

Virtual work

virtual work The imaginary work done when a system is subjected to infinitesimal hypothetical displacements. According to the *principle of virtual work*, the total work done by all the forces acting on a system in equilibrium is zero. This principle can be used to determine the forces acting on a system in equilibrium. For example, the illustration shows a ladder leaning against a wall, with the bottom of the ladder attached to the wall by a horizontal weightless string. The tension, T, in the string can be calculated by assuming that infinitesimal movement dx and dy take place as shown. Then by applying the principle of virtual work, $Tdx + Wdy = 0$. As dx and dy can be calculated from the geometry, T can be found.

virulence The disease-producing ability of a microorganism. *See also* pathogen.

virus A particle that is too small to be seen with a light microscope or to be trapped by filters but is capable of independent metabolism and reproduction within a living cell. Outside its host cell a virus is completely inert. A mature virus (a *virion*) ranges in size from 20 to 400 nm in diameter. It consists of a core of nucleic acid (DNA or RNA) surrounded by a protein coat (*capsid*). Some bear an outer envelope (*enveloped viruses*). Inside its host cell the virus initiates the synthesis of viral proteins and undergoes replication. The new virions are released when the host cell disintegrates. Viruses are parasites of animals, plants, and some bacteria (*see* bacteriophage). Viral diseases of animals include the common cold, influenza, smallpox, AIDS, herpes, hepatitis, polio, and rabies (*see* adenovirus; arbovirus; herpesvirus; HIV; myxovirus; papovavirus; picornavirus; poxvirus); some viruses are also implicated in the development of cancer (*see* retrovirus). Plant viral diseases include various forms of yellowing and blistering of leaves and stems (*see* tobacco mosaic virus). Antibiotics are ineffective against viral diseases but *vaccines (if available) provide good protection.

viscometer An instrument for measuring the viscosity of a fluid. In the *Ostwald viscometer*, used for liquids, a bulb in a capillary tube is filled with the liquid and the time taken for the meniscus to reach a mark on the capillary, below the bulb, is a measure of the viscosity. The *falling-sphere viscometer*, based on *Stokes' law, enables the speed of fall of a ball falling through a sample of the fluid to be measured. Various other devices are used to measure viscosity.

viscosity A measure of the resistance to flow that a fluid offers when it is subjected to shear stress. For a *Newtonian fluid, the force, F, needed to maintain a velocity gradient, dv/dx, between adjacent planes of a fluid of area A is given by: $F = \eta A(dv/dx)$, where η is a constant, the coefficient of viscosity. In *SI units it has the unit pascal second (in the c.g.s. system it is measured in *poise). Non-Newtonian fluids, such as clays, do not conform to this simple model. *See also* kinematic viscosity.

visible spectrum The *spectrum of electromagnetic radiations to which the human eye is sensitive. *See* colour.

visual binary *See* binary stars.

visual-display unit (VDU) The part of a *computer system or word processor on which text or diagrams are displayed. It consists of a *cathode-ray tube and usually has its own input keyboard attached.

visual purple *See* rhodopsin.

vital staining A technique in which a harmless dye is used to stain living tissue for microscopical observation. The stain may be injected into a living animal and the stained tissue removed and examined (*intravital staining*) or the living tissue may be removed directly and subsequently stained (*supravital staining*). Microscopic organisms, such as Protozoa, may be completely immersed in the dye solution. Vital stains include trypan blue, vital red, and Janus green, the latter being especially suitable for observing mitochondria.

vitamin One of a number of organic compounds required by living organisms in relatively small amounts to maintain normal health. There are some 14 generally recognized major vitamins: the water-soluble *vitamin B complex (containing 9) and *vitamin C and the fat-soluble *vitamin A, *vitamin D, *vitamin E, and *vitamin K. Most B vitamins and vitamin C occur in plants, animals, and microorganisms; they function typically as *coenzymes. Vitamins A, D, E, and K occur only in animals, especially vertebrates, and perform a variety of metabolic roles. Animals are unable to manufacture many vitamins themselves and must have adequate amounts in the diet. Foods may contain vitamin precursors (called *provitamins*) that are chemically changed to the actual vitamin on entering the body. Many vitamins are destroyed by light and heat, e.g. during cooking.

vitamin A (retinol) A fat-soluble vitamin that cannot be synthesized by mammals and other vertebrates and must be provided in the diet. Green plants contain precursors of the vitamin, notably carotenes, that are converted to vitamin A in the intestinal wall and liver. The aldehyde derivative of vitamin A, *retinal*, is a constituent of the visual pigment *rhodopsin. Deficiency affects the eyes, causing night blindness, xerophthalmia, and eventually total blindness. The role of vitamin A in other aspects of metabolism is less clear; it may be involved in controlling *ATP production and the growth of epithelial cells.

vitamin B complex A group of water-soluble vitamins that characteristically serve as components of *coenzymes. Plants and many microorganisms can manufacture B vitamins but dietary sources are essential for most animals. Heat and light tend to destroy B vitamins.

Vitamin B₁ (thiamin(e)) is a precursor of the coenzyme thiamine pyrophosphate, which functions in carbohydrate metabolism. Deficiency leads to beriberi in humans and to polyneuritis in birds. Good sources include brewer's yeast, wheatgerm, beans, peas, and green vegetables.

Vitamin B₂ (riboflavin) occurs in green vegetables, yeast, liver, and milk. It is a constituent of the coenzymes *FAD and FMN, which have an important role in the metabolism of all major nutrients as well as in the oxidative phosphorylation reactions of the *electron transport chain. Deficiency of B₂ causes inflammation of the tongue and lips and mouth sores.

Vitamin B₆ (pyridoxine) is widely distributed in cereal grains, yeast, liver, milk, etc. It is a constituent of a coenzyme (pyridoxal phosphate) involved in amino acid metabolism. Deficiency causes retarded growth, dermatitis, convulsions, and other symptoms.

Vitamin B₁₂ (cyanocobalamin) is manufactured only by microorganisms and natural sources are entirely of animal origin. Liver is especially rich

in it. One form of B₁₂ functions as a coenzyme in a number of reactions, including the oxidation of fatty acids and the synthesis of DNA. It also works in conjunction with *folic acid (another B vitamin) in the synthesis of the amino acid methionine and it is required for normal production of red blood cells. Vitamin B₁₂ can only be absorbed from the gut in the presence of a glycoprotein called intrinsic factor; lack of this factor or deficiency of B₁₂ results in pernicious anaemia.

Other vitamins in the B complex include *nicotinic acid, *pantothenic acid, *biotin, and *lipoic acid. *See also* choline; inositol.

vitamin C (ascorbic acid) A colourless crystalline water-soluble vitamin found especially in citrus fruits and green vegetables. Most organisms synthesize it from glucose but man and other primates and various other species must obtain it from their diet. It is required for the maintenance of healthy connective tissue; deficiency leads to scurvy. Vitamin C is readily destroyed by heat and light.

vitamin D A fat-soluble vitamin occurring in the form of two steroid derivatives: *vitamin D₂ (ergocalciferol, or calciferol)*, found in yeast; and *vitamin D₃ (cholecalciferol)*, which occurs in animals. Vitamin D₂ is formed from a steroid by the action of ultraviolet light and D₃ is produced by the action of sunlight on a cholesterol derivative in the skin. Fish-liver oils are the major dietary source. The active form of vitamin D is manufactured in response to the secretion of *parathyroid hormone, which occurs when blood calcium levels are low. It causes increased uptake of calcium from the gut, which increases the supply of calcium for bone synthesis. Vitamin D deficiency causes rickets in growing animals and osteomalacia in mature animals. Both conditions are characterized by weak deformed bones.

vitamin E (tocopherol) A fat-soluble vitamin consisting of several closely related compounds, deficiency of which leads to a range of disorders in different species, including muscular dystrophy, liver damage, and infertility. Good sources are cereal grains and green vegetables. Vitamin E prevents the oxidation of unsaturated fatty acids in cell membranes, so maintaining their structure.

vitamin K A fat-soluble vitamin consisting of several related compounds that act as coenzymes in the synthesis of several proteins (including prothrombin) necessary for blood clotting. Deficiency of vitamin K, which leads to extensive bleeding, is rare because a form of the vitamin is manufactured by intestinal bacteria. Green vegetables and egg yolk are good sources.

vitelline membrane *See* egg membrane.

vitreous Having a glasslike appearance or structure.

vitreous humour The colourless jelly that fills the space between the lens and the retina of the vertebrate eye.

viviparity 1. (in zoology) A form of reproduction in animals in which the developing embryo obtains its nourishment directly from the mother via a *placenta or by other means. Viviparity occurs in some insects and other arthropods, in certain fishes, amphibians, and reptiles, and in the majority of mammals. *Compare* oviparity; ovoviviparity. **2.** (in botany) **a.** A form of *asexual reproduction in certain plants, such as the onion, in which the flower develops into a budlike structure that forms a new plant when detached from the parent. **b.** The development of young plants on the inflorescence of the parent plant, as seen in certain grasses and the spider plant.

Voltage divider

vocal cords A pair of elastic membranes that project into the *larynx in air-breathing vertebrates. Vocal sounds are produced when expelled air passing through the larynx vibrates the cords. The pitch of the sound produced depends on the tension of the cords, which is controlled by muscles and cartilages in the larynx.

volt Symbol V. The SI unit of electric potential, potential difference, or e.m.f. defined as the difference of potential between two points on a conductor carrying a constant current of one ampere when the power dissipated between the points is one watt. It is named after Alessandro Volta (1745–1827).

voltage Symbol V. An e.m.f. or potential difference expressed in volts.

voltage divider (potential divider; potentiometer) A resistor or a chain of resistors connected in series that can be tapped at one or more points to obtain a known fraction of the total voltage across the whole resistor or chain. In the illustration, V is the total voltage across the divider and v is required voltage, then

$$v/V = R_2/(R_1 + R_2).$$

voltaic cell (galvanic cell) A device that produces an e.m.f. as a result of chemical reactions that take place within it. These reactions occur at the surfaces of two electrodes, each of which dips into an electrolyte. The first voltaic cell, devised by Alessandro Volta (1745–1827), had electrodes of two different metals dipping into brine. *See* primary cell; secondary cell.

voltaic pile An early form of battery, devised by Alessandro Volta, consisting of a number of flat *voltaic cells joined in series. The liquid electrolyte was absorbed into paper or leather discs.

voltameter (coulometer) 1. An electrolytic cell formerly used to measure quantity of electric charge. The increase in mass (m) of the cathode of the cell as a result of the deposition on it of a metal from a solution of its salt enables the charge (Q) to be determined from the relationship $Q = m/z$, where z is the electrochemical equivalent of the metal. **2.** Any other type of electrolytic cell used for measurement.

voltmeter An instrument used to measure voltage. *Moving-coil instruments are widely used for this purpose; generally a galvanometer is used in series with a resistor of high values (sometimes called a *multiplier*). To measure an alternating potential difference a rectifier must be included in the circuit. A moving-iron instrument can be used for either d.c. or a.c. without a rectifier. *Cathode-ray oscilloscopes are also used as voltmeters. The electronic *digital voltmeter* displays the value of the voltage in digits. The input is repeatedly sampled by the voltmeter and the instantaneous values are displayed.

volume Symbol V. The space occupied by a body or mass of fluid.

volumetric analysis A method of quantitative analysis using measurement of volumes. For gases, the main technique is in reacting or absorbing gases in graduated containers over mercury, and measuring the volume

Structure and action of a voluntary muscle

changes. For liquids, it involves *titrations.

voluntary muscle (skeletal, striped, or striated muscle) Muscle that is under the control of the will and is generally attached to the skeleton. An individual muscle consists of bundles of long *muscle fibres*, each containing many nuclei, the whole muscle being covered with a strong connective tissue sheath (*epimysium*) and attached at each end to a bone by inextensible *tendons. Each fibre contains smaller fibres (*myofibrils*) having alternate light and dark bands (*sarcomeres*), which are responsible for the muscle's contractile ability and its typical striped appearance under the microscope. See illustration.

The end of the muscle that is attached to a nonmoving bone is called the *origin* of the muscle; the end attached to a moving bone is the *insertion*. As a muscle contracts it becomes shorter and fatter, moving one bone closer to the other. Since a muscle cannot expand, another muscle (the *extensor*, or *antagonist*) is required to move the bone in the opposite direction and stretch the first muscle (known as the *flexor*, or *agonistic muscle*). See illustration.

vulcanite (ebonite) A hard black insulating material made by the vulcanization of rubber with a high proportion of sulphur (up to 30%).

vulcanization A process for hardening rubber by heating it with sulphur or sulphur compounds.

vulva The external opening of the *vagina, comprising in women the female external genitalia – two pairs of fleshy folds of tissue (the *labia*).

W

Wacker process A process for the manufacture of ethanal by the air oxidation of ethene. A mixture of air and ethene is bubbled through a solution containing palladium(II) chloride and copper(II) chloride. The Pd^{2+} ions form a complex with the ethene in which the ion is bound to the pi electrons in the $C=C$ bond. This decreases the electron density in the bond, making it susceptible to nucleophilic attack by water molecules. The complex formed breaks down to ethanal and palladium metal. The Cu^{2+} ions oxidize the palladium back to Pd^{2+}, being reduced to Cu^+ ions in the process. The air present oxidizes Cu^+ back to Cu^{2+}. Thus the copper(II) and palladium(II) ions effectively act as catalysts in the process, which is now the main source of ethanal and, by further oxidation, ethanoic acid. It can also be applied to other alkenes.

Wallace's line An imaginary line that runs between the Indonesian islands of Bali and Lombok and represents the separation of the Australian and Oriental faunas. It was proposed by the naturalist A. R. Wallace (1823–1913), who had noted that the mammals in Southeast Asia are different from and more advanced than their Australian counterparts. He suggested this was because the Australian continent had split away from Asia before the better adapted placental mammals evolved in Asia. Hence the isolated Australian marsupials and monotremes were able to thrive while those in Asia were driven to extinction by competition from placental mammals. *See also* zoogeography.

wall effect Any effect resulting from the nature or presence of the inside wall of a container on the system it encloses.

warfarin 3-(alpha-acetonylbenzyl)-4-hydroxycoumarin: a synthetic *anticoagulant used both therapeutically in clinical medicine and, in lethal doses, as a rodenticide.

warm-bloodedness *See* homoiothermy.

warning coloration (aposematic coloration) The conspicuous markings of an animal that make it easily recognizable and warn would-be predators that it is a poisonous, foul-tasting, or dangerous species. For example, the yellow-and-black striped abdomen of the wasp warns of its sting. *See also* mimicry.

washing soda *Sodium carbonate decahydrate, $Na_2CO_3.10H_2O$.

water A colourless liquid, H_2O; r.d. 1.000 (4°C); m.p. 0.000°C; b.p. 100.000°C. In the gas phase water consists of single H_2O molecules in which the H–O–H angle is 105°. The structure of liquid water is still controversial; hydrogen bonding of the type $H_2O \ldots H-O-H$ imposes a high degree of structure and current models supported by X-ray scattering studies have short-range ordered regions, which are constantly disintegrating and re-forming. This ordering of the liquid state is sufficient to make the density of water at about 0°C higher than that of the relatively open-structured ice; the maximum density occurs at 3.98°C. This accounts for the well-known phenomenon of ice floating on water and the contraction of water below ice, a fact of enormous biological significance for all aquatic organisms. Ice has nine distinct structural modifications of which ordinary ice, or ice I, has an open structure built of puckered six-membered rings in which each H_2O unit is tetrahedrally surrounded by four other H_2O units.

Because of its angular shape the water molecule has a permanent dipole moment and in addition it is strongly hydrogen bonded and has a high dielectric constant. These properties combine to make water a powerful solvent for both polar and ionic compounds. Species in solution are frequently strongly hydrated and in fact ions frequently written as, for example, Cu^{2+} are essentially $[Cu(H_2O)_6]^{2+}$. Crystalline *hydrates are also common for inorganic substances; polar organic compounds, particularly those with O–H and N–H bonds, also form hydrates.

Pure liquid water is very weakly dissociated into H_3O^+ and OH^- ions by self ionization:

$$H_2O \leftrightharpoons H^+ + OH^-$$

(*see* ionic product) and consequently any species that increases the concentration of the positive species, H_3O^+, is acidic and species increasing the concentration of the negative species, OH^-, are basic (*see* acid). The phenomena of ion transport in water and the division of materials into *hydrophilic* (water loving) and *hydrophobic* (water hating) substances are central features of almost all biological chemistry. A further property of water that is of fundamental importance to the whole planet is its strong absorption in the infrared range of the spectrum and its transparency to visible and near ultraviolet radiation. This allows solar radiation to reach the earth during hours of daylight but restricts rapid heat loss at night. Thus atmospheric water prevents violent diurnal oscillations in the earth's ambient temperature. *See also* greenhouse effect.

water gas A mixture of carbon monoxide and hydrogen produced by passing steam over hot carbon (coke):

$$H_2O(g) + C(s) \rightarrow CO(g) + H_2(g)$$

The reaction is strongly endothermic but the reaction can be used in conjunction with that for *producer gas for making fuel gas. The main use of water gas before World War II was in producing hydrogen for the *Haber process. Here the above reaction was combined with the *water-gas shift reaction* to increase the amount of hydrogen:

$$CO + H_2O \leftrightharpoons CO_2 + H_2$$

Most hydrogen for the Haber process is now made from natural gas by steam *reforming.

water glass A viscous colloidal solution of sodium silicates in water, used to make silica gel and as a size and preservative.

water of crystallization Water present in crystalline compounds in definite proportions. Many crystalline salts form hydrates containing 1, 2, 3, or more molecules of water per molecule of compound, and the water may be held in the crystal in various ways. Thus, the water molecules may simply occupy lattice positions in the crystal, or they may form bonds with the anions or the cations present. In the pentahydrate of copper sulphate ($CuSO_4.5H_2O$), for instance, each copper ion is coordinated to four water molecules through the lone pairs on the oxygen to form the *complex $[Cu(H_2O)_4]^{2+}$. Each sulphate ion has one water molecule held by hydrogen bonding. The difference between the two types of bonding is demonstrated by the fact that the pentahydrate converts to the monohydrate at 100°C and only becomes anhydrous above 250°C. *Water of constitution* is an obsolete term for water combined in a compound (as in a metal hydroxide $M(OH)_2$ regarded as a hydrated oxide $MO.H_2O$).

water potential Symbol Ψ. The tendency of water to migrate across a semipermeable membrane during *osmosis. The water potential of pure water is zero; aqueous solutions of increasing concentration have increasingly negative values. Water tends to move from a less concentrated to a more concentrated solution, i.e. from areas of low negative potential to areas of high negative potential. Because water potential more accurately describes the phenomenon of water uptake in plants, it is replacing the term 'osmotic pressure' in plant physiology.

water softening *See* hardness of water.

Watson–Crick model The double-stranded twisted ladder-like molecular structure of *DNA as determined by the US biochemist James Watson (1928–) and the British biochemist Francis Crick (1916–) at Cambridge, England, in 1953. It is commonly known as the *double helix.*

watt Symbol W. The SI unit of power, defined as a power of one joule per second. In electrical contexts it is equal to the rate of energy transformation by an electric current of one ampere flowing through a conductor the ends of which are maintained at a potential difference of one volt. The unit is named after James Watt (1736–1819).

wattmeter An instrument for measuring the power in watts in an alternating-current electric circuit. In a direct-current circuit, power is usually determined by separate measurements of the voltage and the current. The *electrodynamic wattmeter* consists of two coils, one fixed (current) coil and one movable (potential) coil. The fixed coil carries the load current, and the movable coil carries a current proportional to the voltage applied to the measured circuit. The deflection of the needle attached to the movable coil indicates the power.

wave A periodic disturbance in a medium or in space. In a *travelling wave* (or *progressive wave*) energy is transferred from one place to another by the vibrations (*see also* stationary wave). In a wave passing over the surface of water, for example, the water rises and falls as the wave passes but the particles of water on average do not move forward with the wave. This is called a *transverse wave* because the disturbances are at right angles to the direction of propagation. The water surface moves up and down while the waves travel across the surface of the water. Elec-

Sine wave

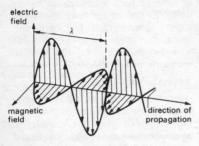

Electromagnetic waves

tromagnetic waves (see diagram) are also of this kind, with electric and magnetic fields varying in a periodic way at right angles to each other and to the direction of propagation. In sound waves, the air is alternately compressed and rarefied by displacements in the direction of propagation. Such waves are called *longitudinal waves*.

The chief characteristics of a wave are its *speed of propagation*, its *frequency*, its *wavelength*, and its *amplitude*. The speed of propagation is the distance covered by the wave in unit time. The frequency is the number of complete disturbances (cycles) in unit time, usually expressed in *hertz. The wavelength is the distance in metres between successive points of equal phase in a wave. The amplitude is the maximum difference of the disturbed quantity from its mean value.

Generally, the amplitude (a) is half the peak-to-peak value. There is a simple relationship between the wavelength (λ) and the frequency (f), i.e. λ

$= c/f$, where c is the speed of propagation. The energy transferred by a progressive *sine wave (see diagram) is proportional to $a^2 f^2$. *See also* simple harmonic motion.

wave equation A partial differential equation of the form:

$$\nabla^2 u = (1/c^2)\partial^2 u/\partial t^2$$

where

$$\nabla^2 = \partial^2/\partial x^2 + \partial^2/\partial y^2 + \partial^2/\partial z^2$$

is the Laplace operator (*see* Laplace equation). It represents the propagation of a wave, where u is the displacement and c the speed of propagation. *See also* Schrödinger's equation.

wave form The shape of a wave or the pattern representing a vibration. It can be illustrated by drawing a graph of the periodically varying quantity against distance for one complete wavelength. *See also* sine wave.

wavefront A line or surface within a two- or three-dimensional medium through which waves are passing, being the locus of all adjacent points at which the disturbances are in phase. At large distances from a small source in a uniform medium, the fronts are small parts of a sphere of very large radius and they can be considered as plane. For example, sunlight reaches the earth with plane wavefronts.

wave function A function $\psi(x,y,z)$ appearing in the *Schrödinger equation in *quantum mechanics. The wave function is a mathematical expression involving the coordinates of a particle in space. If the Schrödinger equation can be solved for a particle in a given system (e.g. an electron in an atom) then, depending on the boundary conditions, the solution is a set of allowed wave functions (*eigenfunctions*) of the particle, each corresponding to an allowed energy level (*eigenvalue*). The physical significance of the wave function is that the square of its absolute value,

$|\psi|^2$, at a point is proportional to the probability of finding the particle in a small element of volume, $dxdydz$, at that point. For an electron in an atom, this gives rise to the idea of atomic and molecular *orbitals.

wave guide A hollow tube through which microwave electromagnetic radiation can be transmitted with relatively little attenuation. They often have a rectangular cross section, but some have a circular cross section. In transverse electric (TE) modes the electric vector of the field has no component in the direction of propagation. In transverse magnetic (TM) modes, the magnetic vector has no such component.

wavelength See wave.

wave mechanics See quantum mechanics.

wavemeter A device for measuring the wavelength of electromagnetic radiation. For frequencies up to about 100 MHz a wavemeter consists of a tuned circuit with a suitable indicator to establish when resonance occurs. Usually the tuned circuit includes a variable capacitor calibrated to read wavelengths and resonance is indicated by a current-detecting instrument. At higher frequencies a cavity-resonator in a waveguide is often used. The cavity resonator is fitted with a piston, the position of which determines the resonant frequency of the cavity.

wave number Symbol k. The number of cycles of a wave in unit length. It is the reciprocal of the wavelength (see wave).

wave–particle duality The concept that waves carrying energy may have a corpuscular aspect and that particles may have a wave aspect; which of the two models is the more appropriate will depend on the properties the model is seeking to explain. For example, waves of electromagnetic radiation need to be visualized as

particles, called *photons, to explain the *photoelectric effect while electrons need to be thought of as de Broglie waves in *electron diffraction. See also complementarity; de Broglie wavelength; light.

wave power The use of wave motion in the sea to generate energy. The technique used is to anchor a series of bobbing floats offshore; the energy of the motion of the floats is used to turn a generator. It has been estimated that there are enough suitable sites to generate over 100 GW of electricity in the UK.

wave theory See light.

wax Any of various solid or semi-solid substances. There are two main types. Mineral waxes are mixtures of hydrocarbons with high molecular weights. Paraffin wax, obtained from *petroleum, is an example. Waxes secreted by plants or animals are mainly esters of fatty acids and usually have a protective function. Examples are the beeswax forming part of a honeycomb and the wax coating on some leaves, fruits, and seed coats, which acts as a protective water-impermeable layer supplementing the functions of the cuticle. The seeds of a few plants contain wax as a food reserve.

W boson (W particle) Either of a pair of elementary particles (W^+ or W^-), classified as intermediate vector bosons, that are believed to transmit the weak interaction (see fundamental interactions) in much the same way as photons transmit the electromagnetic interaction. They are not, however, massless like photons, and are believed to have a rest mass of the order of 10^{-25} kg (806 GeV). W bosons were discovered at CERN in 1983 with the expected mass. See also Z boson.

weak acid An *acid that is only partially dissociated in aqueous solution.

weak interaction *See* fundamental interactions.

weakly interacting massive particle (WIMP) *See* missing mass.

weather The state of atmospheric conditions, including humidity, precipitation (e.g. rain, snow, hail), temperature, cloud cover, visibility, and wind, at any one place and time. A *weather forecast* is a prediction of the weather conditions to be expected at a particular place over a given period. These may be short-range (1–2 days), medium-range (5–7 days), or long-range (1 month or a season). The two chief methods employed in obtaining short-range and medium-range forecasts are *synoptic forecasting* and *numerical forecasting*. Synoptic forecasting involves the simultaneous observation of weather elements at a series of weather stations, the collection of data, and the plotting of the information obtained on weather maps, from which forecasts can be made. Since the early 1960s data collected by satellites, such as pictures of cloud cover and infrared measurements, have been increasingly used in forecasting. Numerical forecasting involves the numerical solution of equations governing the motions and changes of atmospheric conditions. Computers are used to carry out the vast number of calculations. Both synoptic and numerical methods are unsuitable for long-range forecasts and instead statistical and analogue methods are used.

weathering The process of breakdown and alteration of rocks on the earth's surface by mechanical or chemical processes. Mechanical (physical) weathering includes the splitting of rocks through the action of frost and extreme temperature changes. Chemical weathering includes solution (the dissolving of solid materials by water); carbonation (the dissolving of soluble rocks and minerals by a weak carbonic acid formed by the combination of water with atmospheric carbon dioxide); oxidation (the combination of atmospheric oxygen with rock materials); and hydration (the chemical combination of rock materials with water). Organic weathering, which may involve both chemical and mechanical processes, is caused by plants and animals. For example, burrowing animals and plant roots may physically break up rocks; lichens, which can exist on bare rock surfaces, cause decomposition through the removal of nutrients.

weber Symbol Wb. The SI unit of magnetic flux equal to the flux that, linking a circuit of one turn, produces in it an e.m.f. of one volt as it is reduced to zero at a uniform rate in one second. It is named after Wilhelm Weber (1804–91).

weight The force by which a body is attracted to the earth. *See* mass.

weightlessness A condition of a body when it is an infinite distance from any other body. In practice the appearance of weightlessness occurs in space when the gravitational attraction of the earth on a body in space is equal to the centripetal force required by its orbital motion so that the body is effectively in free fall. Weightlessness can also be simulated for short periods in an aircraft flying a parabolic flight path, so that its occupants are again in free fall.

Weinberg–Salam model (WS model) *See* electroweak theory.

Weismannism The theory of the *continuity of the germ plasm* published by August Weismann (1834–1914) in 1886. It proposes that the contents of the reproductive cells (sperms and ova) are passed on unchanged from one generation to the next, unaffected by any changes undergone by the rest of the body. It thus rules out any possibility of the inheritance of acquired characteristics, and has

Western blotting

become fundamental to neo-Darwinian theory.

Western blotting (protein blotting) An *immunoassay for determining very small amounts of a particular protein in tissue samples or cells. The sample is subjected to electrophoresis on SDS-polyacrylamide gel to separate constituent proteins. The resultant protein bands are then 'blotted' onto a polymer sheet. A radiolabelled antibody specific for the target protein is added; this binds to the protein, which can then be detected by autoradiography. A variation of this technique is used to screen bacterial colonies containing cDNA clones in order to isolate those colonies expressing a particular protein. The name is derived by analogy to that of *Southern blotting.

Weston cell (cadmium cell) A type of primary *voltaic cell, which is used as a standard; it produces a constant e.m.f. of 1.0186 volts at 20°C. The cell is usually made in an H-shaped glass vessel with a mercury anode covered with a paste of cadmium sulphate and mercury(I) sulphate in one leg and a cadmium amalgam cathode covered with cadmium sulphate in the other leg. The electrolyte, which connects the two electrodes by means of the bar of the H, is a saturated solution of cadmium sulphate. In some cells sulphuric acid is added to prevent the hydrolysis of mercury sulphate.

wet-and-dry bulb hygrometer *See* hygrometer.

whalebone (baleen) Transverse horny plates hanging down from the upper jaw on each side of the mouth of the toothless whales (*see* Cetacea), forming a sieve. Water, containing plankton on which the whale feeds, enters the open mouth and is then expelled with the mouth slightly closed, so that food is retained on the baleen plates.

whales *See* Cetacea.

Wheatstone bridge

Wheatstone bridge An electrical circuit for measuring the value of a resistance. In the illustration, R_1 is a resistance of unknown value, R_2 is a fixed resistance of known value, R_3 and R_4 are variable resistances with known values. When no current flows between A and B the bridge is said to be balanced, the galvanometer registers no deflection, and $R_1/R_2 = R_3/R_4$. R_1 can therefore be calculated. The Wheatstone bridge is used in various forms. In the *metre bridge*, a wire 1 metre long of uniform resistance is attached to the top of a board alongside a metre rule. A sliding contact is run along the wire, which corresponds to R_3 and R_4, until the galvanometer registers zero. Most practical forms use one or more rotary rheostats to provide the variation. The device was popularized though not invented by Sir Charles Wheatstone (1802–75).

white arsenic *See* arsenic(III) oxide.

white blood cell *See* leucocyte.

white dwarf A compact stellar object that is supported against collapse under self-gravity by the *degeneracy pressure of electrons. White dwarfs are formed as the end products of the evolution of stars of relatively low mass (about that of the sun); high-mass stars may end up as *neutron stars or *black holes (*see* stellar evolution). White dwarfs consist of helium nuclei (and carbon and oxygen nuclei in the more massive cases) and a *degenerate gas of electrons. A typical white-dwarf density is 10^9 kg m^{-3}; white dwarf masses and radii are in the region of 0.7 solar masses and 10^3 km respectively. There is a maximum mass for white dwarfs, above which they are unstable to gravitational collapse – this is known as the *Chandrasekhar limit and is about 1.4 solar masses.

white matter Part of the tissue that makes up the central nervous system of vertebrates. It consists chiefly of nerve fibres enclosed in whitish *myelin sheaths. *Compare* grey matter.

white mica *See* muscovite.

white spirit A liquid mixture of hydrocarbons obtained from petroleum, used as a solvent for paint ('turpentine substitute').

Wiedemann–Franz law The ratio of the thermal conductivity of any pure metal to its electrical conductivity is approximately constant at a given temperature. The law is fairly well obeyed, except at low temperatures.

Wien formula *See* Planck's radiation law.

Wien's displacement law For a *black body, $\lambda_m T$ = constant, where λ_m is the wavelength corresponding to the maximum radiation of energy and T is the thermodynamic temperature of the body. Thus as the temperature rises the maximum of the spectral energy distribution curve is displaced towards the short-wavelength end of the spectrum. The law was stated by Wilhelm Wien (1864–1928).

Wigner energy Energy stored in a crystalline substance as a result of irradiation. This phenomenon is known as the *Wigner effect*. For example, some of the energy lost by neutrons in a *nuclear reactor is stored by the graphite moderator. As a result, the crystal lattice is changed and there is a consequent change in the physical dimensions of the moderator. It is named after E. P. Wigner (1902–).

Wigner nuclides Pairs of isobars with odd nucleon numbers in which the atomic number and the neutron number differ by one. ^{3}H and ^{3}He are examples.

wild type Describing the form of an *allele possessed by most members of a population in their natural environment. Wild-type alleles are usually *dominant.

Williamson's synthesis Either of two methods of producing ethers, both named after the British chemist Alexander Williamson (1824–1904).
1. The dehydration of alcohols using concentrated sulphuric acid. The overall reaction can be written

$$2ROH \rightarrow H_2O + ROR$$

The method is used for making ethoxyethane ($C_2H_5OC_2H_5$) from ethanol by heating at 140°C with excess of alcohol (excess acid at 170°C gives ethene). Although the steps in the reaction are all reversible, the ether is distilled off so the reaction can proceed to completion. This is *Williamson's continuous process*. In general, there are two possible mechanisms for this synthesis. In the first (favoured by primary alcohols), an alkylhydrogen sulphate is formed

$$ROH + H_2SO_4 \rightleftharpoons ROSO_3H + H_2O$$

This reacts with another alcohol molecule to give an oxonium ion

$$ROH + ROSO_3H \rightarrow ROHR^+$$

This loses a proton to give ROR. The second mechanism (favoured by tertiary alcohols) is formation of a carbonium ion

$$ROH + H^+ \rightarrow H_2O + R^+$$

This is attacked by the lone pair on the other alcohol molecule

$$R^+ + ROH \rightarrow ROHR^+$$

and the oxonium ion formed again gives the product by loss of a proton.

Wilson cloud chamber

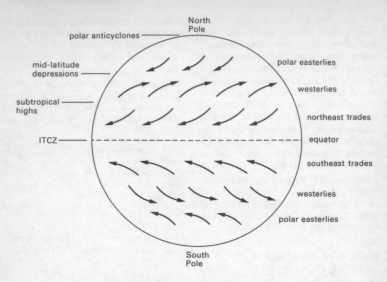

Global surface wind and pressure systems

The method can be used for making symmetric ethers (i.e. having both R groups the same). It can successfully be used for mixed ethers only when one alcohol is primary and the other tertiary (otherwise a mixture of the three possible products results).

2. A method of preparing ethers by reacting a haloalkane with an alkoxide. The reaction, discovered in 1850, is a nucleophilic substitution in which the negative alkoxide ion displaces a halide ion; for example:

$$RI + {}^-OR' \rightarrow ROR' + I^-$$

A mixture of the reagents is refluxed in ethanol. The method is particularly useful for preparing mixed ethers, although a possible side reaction under some conditions is an elimination to give an alcohol and an alkene.

Wilson cloud chamber *See* cloud chamber.

WIMP Weakly interacting massive particle. *See* missing mass.

Wimshurst machine A laboratory electrostatic generator. It consists of two insulating discs to which radial strips of metal foil are attached. After a few strips have been charged individually, the discs are rotated in opposite directions and the charge produced on the strips by induction is collected by metal combs or brushes. It was invented by J. Wimshurst (1836–1903).

wind The motion (usually horizontal) of air relative to the earth's surface. The general circulation of the atmosphere – the large-scale patterns of wind and pressure that persist throughout the year or recur seasonally – results largely from differences in the net radiation received at the earth's surface. However, it is modified by the rotation of the earth, the presence of mountain barriers, the relative distribution of land and sea, and the positions of the ocean currents. Although the resulting wind circulation patterns are complex, they can be simplified into a series of belts (see illustration). Surface heating, which is at its greatest along the equator, creates a belt of low pressure

– the intertropical convergence zone (ITCZ) – towards which the airstreams of the northern and southern hemispheres converge. On either side of the ITCZ lie the trade wind belts: the northeast trade winds in the northern hemisphere and the southeast trade winds of the southern hemisphere. The trade winds are separated from the westerlies – the predominant winds in the mid-latitudes of both hemispheres – by the subtropical high-pressure belt (the horse latitudes), which lies between about 30° and 35° latitude. Near to the poles are the polar easterlies, separated from the westerlies by the subpolar low-pressure troughs.

window 1. A band of electromagnetic wavelengths that is able to pass through a particular medium with little reflection or absorption. For example, there is a *radio window* in the atmosphere allowing radio waves of wavelengths 5 mm to 30 m to pass through. This radio window enables *radio telescopes to be used on the surface of the earth. **2.** A period of time during which an event may occur in order to achieve a desired result. For example a *launch window* is the period during which a space vehicle must be launched to achieve a planned encounter.

wind power The use of winds in the earth's atmosphere to drive machinery, especially to drive an electrical generator. Practical land-based *wind generators (aerogenerators)* are probably capable of providing some 10^{20} J (10^{14} kW h) of energy per year throughout the world and interest in this form of renewable energy is increasing. The power, P, available to drive a wind generator is given by $P = kd^2v^3$, where k is the air density, d is the diameter of the blades, and v is the average wind speed.

wino *See* supersymmetry.

witherite A mineral form of *barium carbonate, $BaCO_3$.

Wöhler's synthesis A synthesis of urea performed by the German chemist Friedrich Wöhler (1800–82) in 1828. He discovered that urea ($CO(NH_2)_2$) was formed when a solution of ammonium isocyanate (NH_4NCO) was evaporated. At the time it was believed that organic substances such as urea could be made only by living organisms, and its production from an inorganic compound was a notable discovery. It is sometimes (erroneously) cited as ending the belief in vitalism.

wolfram *See* tungsten.

wolframite (iron manganese tungsten) A mineral consisting of a mixed iron–manganese tungstate, $(FeMn)WO_4$, crystallizing in the monoclinic system; the principal ore of tungsten. It commonly occurs as blackish or brownish tabular crystal groups. It is found chiefly in quartz veins associated with granitic rocks. China is the major producer of wolframite.

Wollaston prism A type of quartz prism for producing plane-polarized light. It deviates the ordinary and extraordinary rays in opposite directions by approximately the same amount. The Wollaston prism, like the *Rochon prism, can be used with ultraviolet radiation. It is named after the inventor W. H. Wollaston (1766–1828).

womb *See* uterus.

wood The hard structural and water-conducting tissue that is found in many perennial plants and forms the bulk of trees and shrubs. It is composed of secondary *xylem and associated cells, such as fibres. The wood of angiosperms is termed *hardwood*, e.g. oak and mahogany, and that of gymnosperms *softwood*, e.g. pine and fir. New wood is added to the outside of the old wood each growing season by divisions of the vascular cambium (*see* growth ring). Only the outermost new wood (*sapwood) functions in water conduction; the inner wood (*heartwood) provides only structural support.

wood alcohol *See* methanol.

Wood's metal A low-melting (71°C) alloy of bismuth (50%), lead (25%), tin (12.5%), and cadmium (12.5%). It is used for fusible links in automatic sprinkler systems. The melting point can be changed by varying the composition.

Woodward–Hoffmann rules Rules governing the formation of products during certain types of organic concerted reactions. The theory of such reactions was put forward in 1969 by the American chemists Robert Burns Woodward (1917–79) and Roald Hoffmann (1937–), and is concerned with the way that orbitals of the reactants change continuously into orbitals of the products during reaction and with conservation of orbital symmetry during this process. It is sometimes known as *frontier-orbital theory*.

woofer A large loudspeaker designed to reproduce sounds of relatively low frequency, in conjunction with a *tweeter and often a mid-range speaker, in a high-fidelity sound reproducing system.

word A number of *bits, often 32, 48, or 64, processed by a computer as a single unit.

work The work done by a force acting on a body is the product of the force and the distance moved by its point of application in the direction of the force. If a force F acts in such a way that the displacement s is in a direction that makes an angle θ with the direction of the force, the work done is given by: $W = Fs\cos\theta$. Work is the scalar product of the force and displacement vectors. It is measured in joules.

work function A quantity that determines the extent to which *thermionic or *photoelectric emission will occur according to the Richardson equation or Einstein's photoelectric equation (*see* Einstein equation). It is sometimes expressed as a potential difference (symbol ϕ) in volts and sometimes as the energy required to remove an electron (symbol W) in electronvolts or joules. The former has been called the *work function potential* and the latter the *work function energy*.

work hardening An increase in the hardness of metals as a result of working them cold. It causes a permanent distortion of the crystal structure and is particularly apparent with iron, copper, aluminium, etc., whereas with lead and zinc it does not occur as these metals are capable of recrystallizing at room temperature.

W particle *See* W boson.

wrought iron A highly refined form of iron containing 1–3% of slag (mostly iron silicate), which is evenly distributed throughout the material in threads and fibres so that the product has a fibrous structure quite dissimilar to that of crystalline cast iron. Wrought iron rusts less readily than other forms of metallic iron and it welds and works more easily. It is used for chains, hooks, tubes, etc.

WS model Weinberg–Salam model. *See* electroweak theory.

Wurtz reaction A reaction to prepare alkanes by reacting a haloalkane with sodium:

$$2RX + 2Na \rightarrow 2NaX + RR$$

The haloalkane is refluxed with sodium in dry ether. The method is named after the French chemist Charles-Adolphe Wurtz (1817–84). The analogous reaction using a haloalkane and a haloarene, for example:

$$C_6H_5Cl + CH_3Cl + 2Na \rightarrow 2NaCl + C_6H_5CH_3$$

is called the *Fittig reaction* after the German chemist Rudolph Fittig (1835–1910).

X

xanthates Salts or esters containing the group −SCS(OR), where R is an organic group. Cellulose xanthate is an intermediate in the manufacture of *rayon by the viscose process.

xanthophyll A member of a class of oxygen-containing *carotenoid pigments, which provide the characteristic yellow and brown colours of autumn leaves.

X chromosome *See* sex chromosome.

xenon Symbol Xe. A colourless odourless gas belonging to group 0 of the periodic table (*see* noble gases); a.n. 54; r.a.m. 131.30; d. 5.887 g dm^{-3}; m.p. −111.9°C; b.p. −107.1°C. It is present in the atmosphere (0.00087%) from which it is extracted by distillation of liquid air. There are nine natural isotopes with mass numbers 124, 126, 128–132, 134, and 136. Seven radioactive isotopes are also known. The element is used in fluorescent lamps and bubble chambers. Liquid xenon in a supercritical state at high temperatures is used as a solvent for infrared spectroscopy and for chemical reactions. The compound Xe$^+$PtF$_6^-$ was the first noble-gas compound to be synthesized. Several other compounds of xenon are known, including XeF$_2$, XeF$_4$, XeSiF$_6$, XeO$_2$F$_2$, and XeO$_3$. Recently, compounds have been isolated that contain xenon–carbon bonds, such as [C$_6$H$_5$Xe][B(C$_6$H$_5$)$_3$F] (pentafluorophenylxenon fluoroborate), which is stable under normal conditions. The element was discovered in 1898 by Ramsey and Travers.

xeromorphic Describing the structural modifications of certain plants (*xerophytes) that enable them to reduce water loss, particularly from their leaves and stems.

xerophyte A plant that is adapted to live in conditions in which there is either a scarcity of water in the soil, or the atmosphere is dry enough to provoke excessive transpiration, or both. Xerophytes have special structural (*xeromorphic*) and functional modifications, including swollen water-storing stems and specialized leaves that may be hairy, rolled, or reduced to spines or have a thick cuticle to lower the rate of transpiration. Examples of xerophytes are desert cacti and many species growing on sand dunes and exposed moorlands. *Compare* mesophyte; hydrophyte.

X-ray astronomy The study of *X-ray sources by rockets and balloons in the earth's atmosphere and by satellites beyond it. The first nonsolar X-ray source was detected during a rocket flight in 1962, and this observation heralded an entirely new branch of astronomy which developed rapidly with the availability of satellites in the 1970s.

X-ray crystallography The use of *X-ray diffraction to determine the structure of crystals or molecules, such as nucleic acids. The technique involves directing a beam of X-rays at a crystalline sample and recording the diffracted X-rays on a photographic plate. The diffraction pattern consists of a pattern of spots on the plate, and the crystal structure can be worked out from the positions and intensities of the diffraction spots. X-rays are diffracted by the electrons in the molecules and if molecular crystals of a compound are used, the electron density distribution in the molecule can be determined. *See also* neutron diffraction.

X-ray diffraction The diffraction of X-rays by a crystal. The wavelengths of X-rays are comparable in size to the distances between atoms in most crystals, and the repeated pattern of the crystal lattice acts like a diffrac-

X-ray fluorescence

tion grating for X-rays. Thus, a crystal of suitable type can be used to disperse X-rays in a spectrometer. X-ray diffraction is also the basis of X-ray crystallography. *See also* Bragg's law.

X-ray fluorescence The emission of *X-rays from excited atoms produced by the impact of high-energy electrons, other particles, or a primary beam of other X-rays. The wavelengths of the fluorescent X-rays can be measured by an X-ray spectrometer as a means of chemical analysis. X-ray fluorescence is used in such techniques as *electron-probe microanalysis.

X-rays Electromagnetic radiation of shorter wavelength than ultraviolet radiation produced by bombardment of atoms by high-quantum-energy particles. The range of wavelengths is 10^{-11} m to 10^{-9} m. Atoms of all the elements emit a characteristic *X-ray spectrum* when they are bombarded by electrons. The X-ray photons are emitted when the incident electrons knock an inner orbital electron out of an atom. When this happens an outer electron falls into the inner shell to replace it, losing potential energy (ΔE) in doing so. The wavelength λ of the emitted photon will then be given by $\lambda = ch/\Delta E$, where c is the speed of light and h is the Planck constant. *See also* Bremsstrahlung.

X-rays can pass through many forms of matter and they are therefore used medically and industrially to examine internal structures. X-rays are produced for these purposes by an *X-ray tube.

X-ray sources Sources of X-radiation from outside the solar system. Some 100 sources within the Galaxy have been observed as objects that emit most of their energy in the X-ray region of the electromagnetic spectrum and only a relatively small proportion of their energy in the visible spectrum. Many of these X-ray sources appear to be members of a binary system, consisting of one optically visible star and one very compact object; it is thought that the latter is either a *neutron star or (if very massive) a *black hole. Owing to the absorption of X-rays by the earth's atmosphere these X-ray sources are only visible by *X-ray telescopes* carried by space probes and satellites, although some high-energy X-rays can penetrate the upper atmosphere and are detectable by X-ray telescopes mounted on balloons.

X-ray spectrum *See* X-rays.

X-ray tube

X-ray tube A device for generating *X-rays by accelerating electrons to a high energy by an electrostatic field and making them strike a metal target either in a tube containing a low-pressure gas or, as in modern tubes, in a high vacuum. The target is made from a heavy metal, usually tungsten, and is backed by a massive metal anode to conduct the heat away (see illustration showing a liquid-cooled copper anode). The electron beam is produced by heating the cathode by means of a white-hot tungsten filament. A transformer supplies the high voltage, often 100 kV, the tube acting as its own rectifier. On the half-cycles when the target is negative nothing happens. When the target becomes positive, the electrons bombarding it generate X-rays.

xylem A tissue that transports water and dissolved mineral nutrients in vascular plants. In flowering plants it consists of hollow *vessels* that are formed from cells (*vessel elements) joined end to end. The end walls of the vessel elements are perforated to allow the passage of water. In less advanced vascular plants, such as conifers and ferns, the constituent cells of the xylem are called *tracheids. In young plants and at the shoot and root tips of older plants the xylem is formed by the apical meristems. In plants showing secondary growth this xylem is replaced in most of the plant by secondary xylem, formed by the vascular *cambium. The walls of the xylem cells are thickened with lignin, the extent of this thickening being greatest in secondary xylem. Xylem contributes greatly to the mechanical strength of the plant: *wood is mostly made up of secondary xylem. *See also* fibre. *Compare* phloem.

xylenes *See* dimethylbenzenes.

Y

Yagi aerial A directional aerial array widely used for television and *radio telescopes. It consists of one or two dipoles, a parallel reflector, and a series of closely spaced directors (0.15 –0.25 wavelength apart) in front of the dipole. When used for reception this arrangement focuses the incoming signal on the dipole. For transmission, the output of the dipole is reinforced by the directors. It is named after Hidetsuga Yagi (1886–1976).

Yang–Mills theory *See* gauge theory.

yard The former Imperial standard unit of length. In 1963 the yard was redefined as 0.9144 metre exactly.

Y chromosome *See* sex chromosome.

year The measure of time on which the calendar is based. It is the time taken for the earth to complete one orbit of the sun. The *calendar year* consists of an average of 365.25 mean solar days – three successive years of 365 days followed by one (leap) year of 366 days. The *solar year* (or *astronomical year*) is the average interval between two successive returns of the sun to the first point of Aries; it is 365.242 mean solar days. The *sidereal year* is the average period of revolution of the earth with respect to the fixed stars; it is 365.256 mean solar days. The *anomalistic year* is the average interval between successive perihelions; it is 365.259 mean solar days. *See also* ephemeris time.

yeasts A group of unicellular fungi many of which are *ascomycetes. Certain species of the genus *Saccharomyces* are used in the baking and brewing industries.

yield point *See* elasticity.

yolk The food stored in an egg for the use of the embryo. It can consist mainly of protein (*protein yolk*) or of phospholipids and fats (*fatty yolk*). The eggs of oviparous animals (e.g. birds) contain a relatively large yolk.

Young modulus of elasticity *See* elastic modulus.

Young's slits *See* interference.

ytterbium Symbol Yb. A silvery metallic element belonging to the *lanthanoids; a.n. 70; r.a.m. 173.04; r.d. 6.966 (20°C); m.p. 819°C; b.p. 1196°C. It occurs in gadolinite, monazite, and xenotime. There are seven natural isotopes and ten artificial isotopes are known. It is used in certain steels. The element was discovered by J. D. G. Marignac in 1878.

yttrium Symbol Y. A silvery-grey metallic element belonging to group IIIB of the periodic table; a.n. 39; r.a.m. 88.905; r.d. 4.469 (20°C); m.p. 1522°C; b.p. 3338°C. It occurs in uranium ores and in *lanthanoid ores, from which it can be extracted by an ion exchange process. The natural isotope is yttrium–89, and there are 14 known artificial isotopes. The metal is used in superconducting alloys and in alloys for strong permanent magnets (in both cases, with cobalt). The oxide (Y_2O_3) is used in colour-television phosphors, neodymium-doped lasers, and microwave components. Chemically it resembles the lanthanoids, forming ionic compounds containing Y^{3+} ions. The metal is stable in air below 400°C. It was discovered in 1828 by Friedrich Wöhler.

Z

Z boson An electrically neutral elementary particle, Z^0, which – like *W bosons – is thought to mediate the weak interactions in the *electroweak theory. The Z^0 boson was discovered at CERN in 1983 and has a mass of about 90 GeV as had been predicted from theory.

Zeeman effect The splitting of the lines in a spectrum when the source of the spectrum is exposed to a magnetic field. It was discovered in 1896 by P. Zeeman (1865–1943). In the *normal Zeeman effect* a single line is split into three if the field is perpendicular to the light path or two lines if the field is parallel to the light path. This effect can be explained by classical electromagnetic principles in terms of the speeding up and slowing down of orbital electrons in the source as a result of the applied field. The *anomalous Zeeman effect* is a complicated splitting of the lines into several closely spaced lines, so called because it does not agree with classical predictions. This effect is explained by quantum mechanics in terms of electron spin.

Zener diode A type of semiconductor diode, consisting of a *p-n* junction with high doping concentrations on either side of the junction. It acts as a rectifier until the applied reverse voltage reaches a certain value, the *Zener breakdown voltage*, when the device becomes conducting. This effect occurs as a result of electrons being excited directly from the valence band into the conduction band (*see* energy bands). Zener diodes are used in voltage-limiting circuits; they are named after C. M. Zener (1905–).

zenith The point on the *celestial sphere that lies directly above an observer. *Compare* nadir.

zeolite A natural or synthetic hydrated aluminosilicate with an open three-dimensional crystal structure, in which water molecules are held in cavities in the lattice. The water can be driven off by heating and the zeolite can then absorb other molecules of suitable size. Zeolites are used for separating mixtures by selective absorption – for this reason they are often called *molecular sieves*. They are also used in sorption pumps for vacuum systems and certain types (e.g. *Permutit*) are used in ion-exchange (e.g. water-softening).

zero order *See* order.

zero-point energy The energy remaining in a substance at the *absolute zero of temperature (0 K). This is in accordance with quantum theory, in which a particle oscillating with simple harmonic motion does not have a stationary state of zero kinetic energy. Moreover, the *uncertainty principle does not allow such a particle to be at rest at exactly the centrepoint of its oscillations.

zeroth law of thermodynamics *See* thermodynamics.

Ziegler process An industrial process for manufacturing high-density polyethene using catalysts of titanium(IV) chloride (TiCl₄) and aluminium alkyls (e.g. triethylaluminium, Al(C₂H₅)₃). The process was introduced in 1953 by the German chemist Karl Ziegler (1898–1973). It allowed the manufacture of polythene at lower temperatures (about 60°C) and pressures (about 1 atm.) than used in the original process. Moreover, the polyethene produced had more straight-chain molecules, giving the product more rigidity and a higher melting point than the earlier low-density polyethene. The reaction involves the formation of a titanium alkyl in which the titanium can coordinate directly to the pi bond in ethene.

In 1954 the process was developed further by the Italian chemist Giulio Natta (1903–79), who extended the use of Ziegler's catalysts (and similar catalysts) to other alkenes. In particular he showed how to produce stereospecific polymers of propene.

Ziesel reaction A method of determining the number of methoxy (−OCH₃) groups in an organic compound. The compound is heated with excess hydriodic acid, forming an alcohol and iodomethane

$$R-O-CH_3 + HI \rightarrow ROH + CH_3I$$

The iodomethane is distilled off and led into an alcoholic solution of silver nitrate, where it precipitates silver iodide. This is filtered and weighed, and the number of iodine atoms and hence methoxy groups can be calculated. The method was developed by S. Ziesel in 1886.

zinc Symbol Zn. A blue-white metallic element; a.n. 30; r.a.m. 65.38; r.d. 7.1; m.p. 419.57°C; b.p. 907°C. It occurs in sphalerite (or zinc blende, ZnS), which is found associated with the lead sulphide, and in smithsonite (ZnCO₃). Ores are roasted to give the oxide and this is reduced with carbon (coke) at high temperature, the zinc vapour being condensed. Alternatively, the oxide is dissolved in sul-phuric acid and the zinc obtained by electrolysis. There are five stable isotopes (mass numbers 64, 66, 67, 68, and 70) and six radioactive isotopes are known. The metal is used in galvanizing and in a number of alloys (brass, bronze, etc.). Chemically it is a reactive metal, combining with oxygen and other nonmetals and reacting with dilute acids to release hydrogen. It also dissolves in alkalis to give *zincates. Most of its compounds contain the Zn²⁺ ion.

zincate A salt formed in solution by dissolving zinc or zinc oxide in alkali. The formula is often written ZnO₂²⁻ although in aqueous solution the ions present are probably complex ions in which the Zn²⁻ is coordinated to OH⁻ ions. ZnO₂²⁻ ions may exist in molten sodium zincate, but most solid 'zincates' are mixed oxides.

zinc blende A mineral form of *zinc sulphide, ZnS, the principal ore of zinc (see sphalerite). The *zinc-blende structure* is the crystal structure of this compound (and of other compounds). It has zinc atoms surrounded by four sulphur atoms at the corners of a tetrahedron. Each sulphur is similarly surrounded by four zinc atoms. The crystals belong to the cubic system.

zinc chloride A white crystalline compound, ZnCl₂. The anhydrous salt, which is deliquescent, can be made by the action of hydrogen chloride gas on hot zinc; r.d. 2.9; m.p. 290°C; b.p. 732°C. It has a relatively low melting point and sublimes easily, indicating that it is a molecular compound rather than ionic. Various hydrates also exist. Zinc chloride is used as a catalyst, dehydrating agent, and flux for hard solder. It was once known as *butter of zinc.*

zinc group The group of elements in the periodic table consisting of zinc (Zn), cadmium (Cd), and mercury (Hg). See group II elements.

zincite A mineral form of *zinc oxide, ZnO.

zinc oxide A powder, white when cold and yellow when hot, ZnO; r.d. 5.5; m.p. 1975°C. It occurs naturally as a reddish orange ore *zincite*, and can also be made by oxidizing hot zinc in air. It is amphoteric, forming *zincates with bases. It is used as a pigment (*Chinese white*) and a mild antiseptic in zinc ointments. An archaic name is *philosopher's wool*.

zinc sulphate A white crystalline water-soluble compound made by heating zinc sulphide ore in air and dissolving out and recrystallizing the sulphate. The common form is the heptahydrate, $ZnSO_4.7H_2O$; r.d. 1.9. This loses water above 30°C to give the hexahydrate and more water is lost above 100°C to form the monohydrate. The anhydrous salt forms at 450°C and this decomposes above 500°C. The compound, which was formerly called *white vitriol*, is used as a mordant and as a styptic (to check bleeding).

zinc sulphide A yellow-white water-soluble solid, ZnS. It occurs naturally as *sphalerite (*see also* zinc blende) and wurtzite. The compound sublimes at 1180°C. It is used as a pigment and phosphor.

zino *See* supersymmetry.

zirconia *See* zirconium.

zirconium Symbol Zr. A grey-white metallic *transition element; a.n. 40; r.a.m. 91.22; r.d. 6.44; m.p. 1853°C; b.p. 4376°C. It is found in zircon ($ZrSiO_4$; the main source) and in baddeleyite (ZnO_2). Extraction is by chlorination to give $ZrCl_4$ which is purified by solvent extraction and reduced with magnesium (Kroll process). There are five natural isotopes (mass numbers 90, 91, 92, 94, and 96) and six radioactive isotopes are known. The element is used in nuclear reactors (it is an effective neutron absorber) and in certain alloys. The metal forms a passive layer of oxide in air and burns at 500°C. Most of its compounds are complexes of zirconium(IV). *Zirconium(IV) oxide* (*zirconia*) is used as an electrolyte in fuel cells. The element was identified in 1789 by Klaproth and was first isolated by Berzelius in 1824.

zirconium(IV) oxide *See* zirconium.

zodiac A band that passes round the *celestial sphere, extending 9° on either side of the *ecliptic. It includes the apparent paths of the sun, moon, and planets (except Pluto). The band is divided into the twelve *signs of the zodiac*, each 30° wide. These signs indicate the sun's position each month in the year and were named by the ancient Greeks after the *zodiacal constellations* that occupied the signs some 2000 years ago. However, as a result of the *precession of the equinoxes the constellations have since moved eastwards by over 30° and no longer coincide with the signs.

zodiacal light A faint luminous glow in the sky that can be observed on a moonless night on the western horizon after sunset or on the eastern horizon before sunrise. It is caused by the scattering of sunlight by dust particles in interplanetary space.

zone fossil *See* index fossil.

zone refining A technique used to reduce the level of impurities in certain metals, alloys, semiconductors, and other materials. It is based on the observation that the solubility of an impurity may be different in the liquid and solid phases of a material. To take advantage of this observation, a narrow molten zone is moved along the length of a specimen of the material, with the result that the impurities are segregated at one end of the bar and the pure material at the other. In general, if the impurities lower the melting point of the material they are moved in the same direction as the molten zone moves, and vice versa.

zoogeography The study of the geographical distributions of animals. The earth can be divided into several zoogeographical regions separated by natural barriers, such as oceans, deserts, and mountain ranges. The characteristics of the fauna of each region are believed to depend particularly on the process of *continental drift and the stage of evolution reached when the various land masses became isolated. For example Australia, which has been isolated since Cretaceous times, has the most primitive native mammalian fauna, consisting solely of marsupials and monotremes. *See also* Wallace's line.

zoology The scientific study of animals, including their anatomy, physiology, biochemistry, genetics, ecology, evolution, and behaviour.

zooplankton The animal component of *plankton. All major animal phyla are represented in zooplankton, as adults, larvae, or eggs; some are just visible to the naked eye but most cannot be seen without magnification. Near the surface of the sea there may be many thousands of such animals per cubic metre.

zwitterion (ampholyte ion) An ion that has a positive and negative charge on the same group of atoms. Zwitterions can be formed from compounds that contain both acid groups and basic groups in their molecules. For example, aminoethanoic acid (the amino acid glycine) has the formula $H_2N.CH_2.COOH$. However, under neutral conditions, it exists in the different form of the zwitterion $^+H_3N.CH_2.COO^-$, which can be regarded as having been produced by an internal neutralization reaction (transfer of a proton from the carboxyl group to the amino group). Aminoethanoic acid, as a consequence, has some properties characteristic of ionic compounds; e.g. a high melting point and solubility in water. In acid solutions, the positive ion $^+H_3NCH_2COOH$ is formed. In basic solutions, the negative ion $H_2NCH_2COO^-$ predominates. The name comes from the German *zwei*, two.

zygomorphy *See* bilateral symmetry.

zygospore A zygote with a thick resistant wall, formed by some algae and fungi. It results from the fusion of two gametes, neither of which is retained by the parent in any specialized sex organ (such as an oogonium). It enters a resting phase before germination. *Compare* oospore.

zygote A fertilized female *gamete: the product of the fusion of the nucleus of the ovum or ovule with the nucleus of the sperm or pollen grain. *See* fertilization.

zymogen Any inactive enzyme precursor that, following secretion, is chemically altered to the active form of the enzyme. For example, the protein-digesting enzyme *trypsin is secreted by the pancreas as the zymogen trypsinogen. This is changed in the small intestine by the action of another enzyme, enterokinase, to the active form.

Appendix 1 SI units

Table 1.1 Base and supplementary SI units

Physical quantity	Name	Symbol
length	metre	m
mass	kilogram	kg
time	second	s
electric current	ampere	A
thermodynamic temperature	kelvin	K
luminous intensity	candela	cd
amount of substance	mole	mol
*plane angle	radian	rad
*solid angle	steradian	sr

*supplementary units

Table 1.2 Derived SI units with special names

Physical quantity	Name of SI unit	Symbol of SI unit
frequency	hertz	Hz
energy	joule	J
force	newton	N
power	watt	W
pressure	pascal	Pa
electric charge	coulomb	C
electric potential difference	volt	V
electric resistance	ohm	Ω
electric conductance	siemens	S
electric capacitance	farad	F
magnetic flux	weber	Wb
inductance	henry	H
magnetic flux density (magnetic induction)	tesla	T
luminous flux	lumen	lm
illuminance	lux	lx
absorbed dose	gray	Gy
activity	becquerel	Bq
dose equivalent	sievert	Sv

Table 1.3 Decimal multiples and submultiples to be used with SI units

Submultiple	Prefix	Symbol	Multiple	Prefix	Symbol
10^{-1}	deci	d	10	deca	da
10^{-2}	centi	c	10^2	hecto	h
10^{-3}	milli	m	10^3	kilo	k
10^{-6}	micro	μ	10^6	mega	M
10^{-9}	nano	n	10^9	giga	G
10^{-12}	pico	p	10^{12}	tera	T
10^{-15}	femto	f	10^{15}	peta	P
10^{-18}	atto	a	10^{18}	exa	E

Table 1.4 Conversion of units to SI units

From	To	Multiply by
in	m	2.54×10^{-2}
ft	m	0.3048
sq. in	m^2	6.4516×10^{-4}
sq. ft	m^2	9.2903×10^{-2}
cu. in	m^3	1.63871×10^{-5}
cu. ft	m^3	2.83168×10^{-2}
l(itre)	m^3	10^{-3}
gal(lon)	m^3	$4.546\ 09 \times 10^{-3}$
gal(lon)	l(itre)	4.546 09
miles/hr	m s^{-1}	0.477 04
km/hr	m s^{-1}	0.277 78
lb	kg	0.453 592
g cm^{-3}	kg m^{-3}	10^3
lb/in^3	kg m^{-3}	$2.767\ 99 \times 10^4$
dyne	N	10^{-5}
kgf	N	9.806 65
poundal	N	0.138 255
lbf	N	4.448 22
mmHg	Pa	133.322
atmosphere	Pa	$1.013\ 25 \times 10^5$
hp	W	745.7
erg	J	10^{-7}
eV	J	$1.602\ 10 \times 10^{-19}$
kW h	J	3.6×10^6
cal	J	4.1868

Appendix 2 Fundamental constants

Constant	Symbol	Value in SI units
acceleration of free fall	g	$9.806\,65$ m s^{-2}
Avogadro constant	L, N_A	$6.022\,1367(36) \times 10^{23}$ mol^{-1}
Boltzmann constant	$k = R/N_A$	$1.380\,658(12) \times 10^{-23}$ J K^{-1}
electric constant	ε_0	$8.854\,187\,817 \times 10^{-12}$ F m^{-1}
electronic charge	e	$1.602\,177\,33(49) \times 10^{-19}$ C
electronic rest mass	m_e	$9.109\,3897(54) \times 10^{-31}$ kg
Faraday constant	F	$9.648\,5309(29) \times 10^4$ C mol^{-1}
gas constant	R	$8.314\,510(70)$ J K^{-1} mol^{-1}
gravitational constant	G	$6.672\,59(85) \times 10^{-11}$ m^3 kg^{-1} s^{-2}
Loschmidt's constant	N_L	$2.686\,763(23) \times 10^{25}$ m^{-3}
magnetic constant	μ_0	$4\pi \times 10^{-7}$ H m^{-1}
neutron rest mass	m_n	$1.674\,9286(10) \times 10^{-27}$ kg
Planck constant	h	$6.626\,0755(40) \times 10^{-34}$ J s
proton rest mass	m_p	$1.672\,6231(10) \times 10^{-27}$ kg
speed of light	c	$2.997\,924\,58 \times 10^8$ m s^{-1}
Stefan–Boltzmann constant	σ	$5.670\,51(19) \times 10^{-8}$ W m^{-2} K^{-4}

Appendix 3 The solar system

Planet	Equatorial diameter (km)	Mean distance from sun (10^6 km)	Sidereal period
Mercury	4878	57.91	87.969 days
Venus	12 100	108	224.7 days
Earth	12 756	149.60	365.256 days
Mars	6762	227.94	686.980 days
Jupiter	142 700	778	11.86 years
Saturn	120 800	1430	29.46 years
Uranus	51 800	2869.6	84.01 years
Neptune	49 400	4496	164.8 years
Pluto	3500	5900	248.4 years

Appendix 4　Geological time scale

Era	Period	Epoch	millions of years ago
Cenozoic	Quaternary	Holocene	
		Pleistocene	1.8
	Tertiary	Pliocene	
		Miocene	
		Oligocene	
		Eocene	
		Palaeocene	
Mesozoic	Cretaceous		65
	Jurassic		136
	Triassic		190
Palaeozoic	Permian		225
	Carboniferous		280
	Devonian		345
	Silurian		395
	Ordovician		440
	Cambrian		500
Precambrian	Precambrian		570
			4600

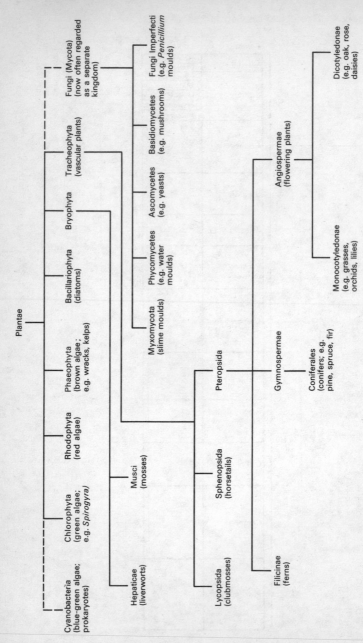

A simplified classification of the plant kingdom (excluding extinct and mostly extinct groups)

A simplified classification of the animal kingdom (down to class level (subclass or major groups) and including only the major phyla)

Appendix 7   The periodic table

IA	IIA	IIIA	IVA	VA	VIA	VIIA	VIII	VIII	VIII	IB	IIB	IIIB	IVB	VB	VIB	VIIB	O
1 H																	2 He
3 Li	4 Be											5 B	6 C	7 N	8 O	9 F	10 Ne
11 Na	12 Mg											13 Al	14 Si	15 P	16 S	17 Cl	18 Ar
19 K	20 Ca	21 Sc	22 Ti	23 V	24 Cr	25 Mn	26 Fe	27 Co	28 Ni	29 Cu	30 Zn	31 Ga	32 Ge	33 As	34 Se	35 Br	36 Kr
37 Rb	38 Sr	39 Y	40 Zr	41 Nb	42 Mo	43 Tc	44 Ru	45 Rh	46 Pd	47 Ag	48 Cd	49 In	50 Sn	51 Sb	52 Te	53 I	54 Xe
55 Cs	56 Ba	57* La	72 Hf	73 Ta	74 W	75 Re	76 Os	77 Ir	78 Pt	79 Au	80 Hg	81 Tl	82 Pb	83 Bi	84 Po	85 At	86 Rn
87 Fr	88 Ra	89† Ac															

s – block   d – block   p – block

Transition elements

*Lanthanides

57 La	58 Ce	59 Pr	60 Nd	61 Pm	62 Sm	63 Eu	64 Gd	65 Tb	66 Dy	67 Ho	68 Er	69 Tm	70 Yb	71 Lu

†Actinides

89 Ac	90 Th	91 Pa	92 U	93 Np	94 Pu	95 Am	96 Cm	97 Bk	98 Cf	99 Es	100 Fm	101 Md	102 No	103 Lr

f – block